Phyllosphere Microbial Plant Pathogens: Detection and Crop Disease Management

This book provides comprehensive knowledge of the methods of detection and identification of phyllosphere microbial pathogens and the management of different kinds of diseases caused by them in various crops. Interactions between pathogens and host plants result in the induction of defense responses expressed via molecular signals, from initiation of infection to systemic progression of pathogen invasion in susceptible plants and contrasting signals in resistant plants, leading to inhibition of pathogen development through activation of preinfectional and postinfectional defense responses. These are critically discussed.

The author describes the intricate and complex competitive activities of the pathogens and host plants in a molecular warfare that the host plant must win to break the link in the pathogens' life cycle, allowing the development of disease management strategies based on the principles of exclusion, eradication, and immunization. Integration of strategies concerning the development of cultivars resistant to pathogens through breeding and biotechnological techniques, application of biotic and abiotic inducers of resistance to pathogens, and use of disease-free seeds and propagules that are complementary to each other along with effective cultural practices are emphasized.

This book presents information gathered through an extensive literature search to help researchers and graduate students in agricultural sciences identify research gaps and successfully complete their research projects.

Phyllosphere Microbial Plant Pathogens: Detection and Crop Disease Management

Volume 1 Nature and Biology

Prof. P. Narayanasamy
Former Professor and Head
Department of Plant Pathology
Tamil Nadu Agricultural University
Coimbatore, INDIA

CRC Press
Taylor & Francis Group
Boca Raton London New York

CRC Press is an imprint of the
Taylor & Francis Group, an **informa** business

Cover image: P. Narayanasamy

First edition published 2025
by CRC Press
2385 NW Executive Center Drive, Suite 320, Boca Raton FL 33431

and by CRC Press
4 Park Square, Milton Park, Abingdon, Oxon, OX14 4RN

CRC Press is an imprint of Taylor & Francis Group, LLC

ISBN: 978-1-032-59918-2 (hbk)
ISBN: 978-1-032-59919-9 (pbk)
ISBN: 978-1-003-45682-7 (ebk)

DOI: 10.1201/9781003456827

Typeset in Times
by Apex CoVantage, LLC

To the memory of my parents for
their love and affection

Contents

Volume 1: Nature and Biology

Preface

Innumerable living and nonliving entities are present in the air and then they are moved by wind short or long distances. The existence of all organisms ranging from simple subcellular forms to highly evolved human beings is possible because of the presence of oxygen in the air. Spores of fungal pathogens and cells of bacterial pathogens capable of infecting above-ground organs of plants are dispersed by wind. On the other hand, the vectors of some of the fungal and bacterial pathogens and entire groups of phytoplasmal and viral pathogens transmit them to aerial organs of plants to initiate infection. Phyllosphere microbial pathogens cause many destructive diseases of different crops such as potato late blight, coffee rust, and rice blast diseases with potential to become epidemic, impacting the social and economic conditions in different countries. Phyllosphere pathogens occur in the form of several strains/races differing in the levels of virulence, whereas crop cultivars/genotypes differ in the levels of resistance/susceptibility, emphasizing the need for early detection and precise identification of microbial pathogens inducing diseases in various crops. Morphological, biochemical, immunological, and genetic characteristics of pathogens form the bases for development of techniques that are useful for detection, identification, differentiation, and quantification of microbial pathogens present in infected plants and planting materials. Investigations on the biology and infection process of the pathogens have highlighted the weak links in the life cycles of pathogens that could be used to disrupt pathogen development and disease spread.

Principles of crop disease management are essentially based on exclusion, eradication, and immunization. The production of disease-free seeds and propagules is essential to prevent introduction of the pathogens in new locations in the same country or other countries. Plant quarantine and certification programs have played an important role in preventing the introduction of exotic pathogens into the country. Breeding for cultivars resistant to diseases is the most preferred strategy for reducing disease incidence and severity, since this approach is cost-effective and environmentally friendly. Biotechnological techniques are useful to transfer resistance genes from diverse sources, which could not be used by conventional breeding methods. The usefulness of biotic and abiotic biocontrol agents to arrest pathogens led to development of commercial preparations. Biocontrol agents may promote plant growth, in addition to pathogen suppression. The development of resistance in pathogens to chemicals; the presence of residues in grains, vegetables, and fruits; and possible environmental pollution strongly suggested the reduction or replacement of chemicals substantially, although chemicals are applied because of the market preference to susceptible cultivars. Integration of compatible disease management strategies to derive additive effect, providing more effective protection to crops, has been shown to be a desirable approach. Efforts to develop integrated disease management (IDM) systems by combining two or more strategies may be useful to protect the environment and provide healthy food materials to consumers.

Information presented in this book representing an extensive literature search is expected to be very useful to researchers, teachers, and graduate students of the departments of plant pathology, microbiology, plant breeding and genetics, agriculture, and horticulture, as well as to quarantine and certification personnel.

P. Narayanasamy

Acknowledgement

The author places on record his sincere respects and salutation to his alma mater, which remains a source of great inspiration, providing direction to fulfill his goal of sharing his experience and thoughts with those who have contributed to the advancement of plant pathology. He expresses his immense appreciation to the staff and students of the Department of Plant Pathology, Tamil Nadu Agricultural University, Coimbatore, India, for their assistance in one way or the other. It would not have been possible without the immeasurable, loving, and unstinting support of his wife, Mrs. N. Rajakumari, for him to gather and present the information useful to the audience. Further, his family members, Mr. N. Kumar Perumal, Mrs. Nirmala Suresh, Mr. T. R. Suresh, and Mr. S. Varun Karthik, provided their kind and affectionate support to strengthen his spirits during the periods that demanded focused attention for preparing this book. Sincere appreciation is expressed to the publisher, CRC Press, for providing the opportunity to share his concepts on various aspects of plant pathology.

P. Narayanasamy

Author Biography

Prof. P. Narayanasamy was awarded the BSc (Ag.) (1958), MSc (Ag.) (1960), and PhD (1963) degrees by the University of Madras, and later with the Rockefeller Foundation Fellowship, he pursued postdoctoral research on rice virus diseases at the International Rice Research Institute, Philippines, during 1966–1967. He served as the virus pathologist at the Indian Agricultural Research Institute, New Delhi, during 1969–1970. He returned to his alma mater, which was upgraded as the Tamil Nadu Agricultural University (TNAU), Coimbatore in 1970. He was appointed associate professor and then promoted to professor and head of the Department of Plant Pathology. He was elected a fellow of the Indian Phytopathological Society, New Delhi. He functioned as the editor of the *Madras Agricultural Journal* published from the Tamil Nadu Agricultural University Campus and as a member of editorial committees of the plant pathology journals published in India. He was invited to participate as the lead speaker and chairman of sessions in the national seminars held in India.

As a researcher in plant pathology, he was the leader of the projects on diseases of rice, legumes, and oilseeds. He organized the National Seminar for the Management of Diseases of Oilseed Crops. His research on antiviral principles yielded practical solutions for the management of virus diseases affecting various crops. He published over 250 research papers in national and international journals. He was the principal investigator of several research projects funded by the Department of Science and Technology, Government of India, New Delhi, and the International Crops Research Institute for Semi-Arid Tropics, Hyderabad, India. He continues to share his experience and knowledge with the staff and graduate students of the Department of Plant Pathology, TNAU, Coimbatore. As a teacher, he taught courses on plant virology, molecular biology, physiopathology, and crop disease management for master's and doctoral programs. Under his guidance, 25 graduate students and 15 research scholars earned master's and doctoral degrees, respectively. With long experience and in-depth knowledge of various aspects of microbial plant pathogens and crop disease management, he has authored 20 books published by leading publishers such as Marcel Dekker, John Wiley & Sons, The Haworth Press, Springer Science, and CRC Press. These publications cover various aspects of plant pathology and are well received by their intended audience.

As a social obligation, he is deeply involved in social welfare activities to help orphaned, old, and infirm people, as well as children through Udavum Karangal (Coimbatore), HelpAge (New Delhi), Global Cancer Concern (New Delhi), Children Rights and You (CRY, Bangalore), and The Hindu Mission Hospital (Chennai) to lessen the sufferings of needy persons. As one interested in literature and spirituality, he has composed poems in Tamil and English and published two collections of poems to inculcate self-confidence in the young and spiritual exploration for older people.

1 Introduction

1.1 INTRODUCTION

Phyllosphere microbial plant pathogens—oomycetes, fungi, bacteria, phytoplasmas, viruses, and viroids—cause diseases in plants at different stages of their growth and affect production potential. Microbial pathogens present in the spermosphere and rhizosphere cause, respectively, seedborne and soilborne diseases that account for serious losses in various crops (Narayanasamy 2017, 2020). Various aspects of phyllosphere microbial pathogens infecting all above-ground (epigeal) plant organs such as stem, leaves, inflorescences, and spikes/earheads are elucidated to develop effective management systems to enhance levels of yield and quality of the produce. History is replete with many examples of plant diseases that have been responsible for major social and economic impacts. The Irish potato famine that occurred in the 19th century led to mass migration of people from Europe to North America and other countries. This situation arose primarily due to potato late blight disease incited by airborne spores of *Phytophthora infestans* infecting foliage and then spreading to tubers in the soil. Wind dissemination is the major mode of pathogen dispersal to locations several thousands of kilometers away, breaching the boundaries of countries, oceans, and mountains. The phyllosphere may present a harsh environment for the survival of pathogens and initiation of infection in susceptible plant hosts. The leaf surface is particularly exposed to conditions such as lack of moisture, ultraviolet irradiation, strong winds, and heat. Phyllosphere pathogens have evolved mechanisms either to avoid or tolerate these stresses, and they reach population levels that are required to initiate infection culminating in development of different levels of disease severity, depending on the level of resistance/susceptibility of the cultivars/genotypes.

1.2 ECONOMIC IMPORTANCE AND DISTRIBUTION OF PHYLLOSPHERE MICROBIAL PLANT PATHOGENS

Sustainable agriculture includes food security while minimizing pressure on the environment. Crop diseases form a group of major and important constraints that place formidable obstacles to agricultural development and global food security. Assessment of crop losses by the Food and Agriculture Organization (FAO), indicated that crop diseases globally reduce crop yields by about 20–40% (Yin and Qiu 2019). Wheat yellow rust (also known as stripe rust) caused by *Puccinia striiformis* f.sp. *tritici* (Pst) occurred in epiphytotic form in Washington and Oregon states of the United States in 1961, and the extent of loss was estimated to be $45 million. Incidence of wheat stripe rust was observed in more than 64 countries and global damages to due to the disease were placed at over 5 million tons of wheat grains equivalent to $979 million per year (Wellings 2011; Beddow et al. 2015). Wheat stripe rust was considered one of the most economically important factors threatening food security worldwide (Chen 2020). *Leptosphaeria maculans* (oilseed rape Phoma stem canker disease) occurred in devastating proportions all over the world, being responsible for total loss (100%) of the crop in some fields, with monetary loss of $1,000 million, and the area cropped to rapeseed nose-diving from 49,000 ha in 1972 to just 2,000 ha in 1974. Fusarium head blight (FHB) complex caused by *Fusarium graminearum* and other *Fusarium* spp. was considered a reemerging disease, and it caused a serious reduction in cereal grain production, in addition to contamination of grains with mycotoxins produced by these pathogens harmful to human beings and animals. Severe epidemics of FHB were recorded in the North America during 1991–1996 and they resulted in unparalleled economic and social impact in the Northern Great Plains region in the United States. Direct and secondary economic losses due to FHB disease complex during 1993–2001 were estimated to be $7.7 billion in nine states of the United States. It was difficult to precisely estimate the quantitative and qualitative losses due to FHB disease in other countries (McMullen et al. 2012).

Citrus canker disease caused by the bacterial pathogen *Xanthomonas axonopodis* pv. *citri* (*Xac*) spread rapidly through infected budwood materials, demanding imposition of plant quarantine regulations and implementation of a massive eradication program involving eradication of 20 million citrus trees at a cost of $20 million (Schumann 1991). Tomato crops in Michigan state, United States, suffered heavily due to *Clavibacter michiganensis* subsp. *michiganensis* (*Cmm*) infection, the estimated loss being $300,000 in one year for individual processing tomato growers (Hausbeck et al. 2000). Operation of the Good Seed and Plant Practices Organization in the Netherland and France emphasized the need for adopting disease management strategies for preventing infection of tomato seeds and avoiding planting infected seedlings in the main field (Sen et al. 2015). Huanglongbing (HLB), earlier known as greening disease, affecting *Citrus* spp. reduced citrus production severely, and its incidence in Asian countries was known for over a century and the disease spread to several countries later. The causal agent '*Candidatus* Liberibacter asiaticus' is spread through infected nursery stock and the Asian citrus psyllid *Diaphorina citri* to short and long distances. HLB disease could cost growers over $4 billion each year and also resulted in the loss of jobs for thousands of workers (Gottwald 2010). Citrus variegated chlorosis (CVC) disease

DOI: 10.1201/9781003456827-1

is caused by xylem-limited bacterium *Xylella fastidiosa* and transmitted by the planthoppers *Homolodisca vitripennis* and *Graphocephala atropunctata*. The disease occurred in epidemic proportions, resulting in elimination of 6 million trees in Sao Paulo state in Brazil (Fundecitrus-Fundo de defesa da citricultura 2011).

Plant viruses transmitted through infected seeds, propagules, and different biological vectors infect different aerial plant organs and tissues. Seedborne viruses can multiply in developing seedlings and move to various plant tissues causing different characteristic symptoms, and also they may be transmitted vertically to subsequent generations of plants via infected seeds. Viruses transmitted via propagules may remain asymptomatic, since they have long latent period in mother plants infected late in growing season. Such infected plants, especially perennial trees, may exhibit a decline of growth and the produce from such plants may be of poor quality. Thus infection by viruses may result in enormous quantitative and qualitative losses which may be difficult to determine precisely in fruit tree crops. As the infection persists for several growing seasons, the yield is reduced progressively, until maintenance of such infected trees in orchards becomes uneconomical. Decreased marketability of the produce contributes to a yield loss of 15% of global food production to plant diseases. Virus diseases account for 47% of this loss (Narayanasamy 2011, Boualem et al. 2016).

Sharka disease caused by *Plum pox virus* (PPV) is one of the most destructive diseases affecting stone fruits in all countries. Incidence of sharka was limited to European countries in the last century, but it spread to Africa, South Africa, North America, and Asia in due course (OEPP/EPPO 2004). Propagative materials prepared from mother plants that remained asymptomatic were responsible for long-distance dissemination of PPV, whereas aphid vectors might be responsible for local spread in some areas. Pollen transmission of PPV was also observed in some locations (Isac et al. 1998). In Europe, cost of sharka disease eradication program was calculated at >10 million Euros (Cambra et al. 2006). Spread of *Citrus tristeza virus* (CTV) occurs primarily through infected plants and budwood. *Toxoptera citricida*, and other aphid species are responsible for secondary spread. Citrus decline due to CTV ruined cultivation in several countries and killed >10 million trees propagated widely on sour orange rootstock in South America, the United States, and Israel (Moreno et al. 2008). Vegetable crops particularly are seriously affected by large number of viruses, leading to significant losses in quantity and quality in all countries. Infection by about 40 different viruses on potatoes has been observed worldwide, resulting in yield reduction up to 75%, with *Potato virus X* (PVX) alone accounting for 15–30% of the loss. Other viruses *Potato leaf roll virus* (PLRV), *Potato virus A* (PVA), *Potato virus M* (PVM), *Potato virus S* (PVS), and *Potato virus Y* (PVY) also are responsible for a significant reduction in yield. PVY is more adaptive to conditions prevalent in different countries due to some genetic modification of viral genome, resulting in greater genetic diversity and consequent ability to infect cultivars which showed resistance earlier. Potato tuber necrosis-inducing strain causes greater damage to the crops worldwide, affecting the marketability of tubers (Davie et al. 2017). Attempts have been made to eradicate virus propagation through seed tubers by establishing standards for producing certified seed potato stocks from in vitro virus-free plantlets for distribution to commercial production (Danci et al. 2012).

1.3 DETECTION AND IDENTIFICATION OF MICROBIAL PLANT PATHOGENS

Microbial plant pathogens occur in the form of different strains/races with varying levels of virulence on cultivars/genotypes with different levels of susceptibility/resistance. Early detection and precise identification of isolates of a pathogen infecting a crop is the basic step to study various aspects of host-pathogen interactions, leading to the development of management systems for diseases affecting various crops. Phyllosphere microbial plant pathogens are detected, identified, differentiated, and quantified based on their morphological, physiological, immunological, and genetic characteristics. Various detection techniques are applied to detect and identify fungal pathogens (Chapter 2), bacterial and phytoplasmal pathogens (Chapter 4), and viral and viroid pathogens (Chapter 6). Morphological characteristics alone are insufficient to identify the pathogens precisely. Further, they need a long time and require expert knowledge in taxonomy. Immunoassays have been useful for detection, differentiation, and quantification of pathogen populations. However, availability of specific antisera has limited their wide application. Nucleic acid–based techniques are widely employed, because the genetic constitution of pathogens remains unaltered at all stages of their development. Different polymerase chain reaction (PCR)–based assays are sensitive in detecting low populations of the pathogens present in plants and environmental samples (from spore traps) and provide reliable results rapidly. Real-time quantitative PCR assay and the loop-mediated isothermal amplification (LAMP) technique are more frequently employed for early detection and accurate quantification of microbial pathogens. Multiplex molecular diagnostic techniques are helpful in detecting and differentiating pathogens involved in complexes caused by two or more pathogens.

1.4 CONCEPTS AND IMPLICATION OF INFECTION BY PHYLLOSPHERE MICROBIAL PATHOGENS

Microbial plant pathogens adopt different strategies to infect plants which possess different kinds of preformed and postinfectional mechanisms to counter the invasion of the pathogens. Pathogens have specific virulence genes, whereas the saprophytes present in the phyllosphere lack virulence genes but compete with pathogens for the nutrients available on plant surfaces. Pathogens may be of two types viz., biotrophs requiring presence of living cells/tissues for completion of their life cycle and necrotrophs capable of killing host tissue and deriving nutrition from the dead host cells/tissues. Fungal pathogens causing powdery mildews, downy mildews, and rusts are biotrophs, whereas those inducing leaf spots, blasts, blights, and anthracnose symptoms are necrotrophs. Viruses and viroids form the large group of biotrophs depending on their biological vectors for their dispersal. Among the phytobacterial pathogens, phytoplasmas causing yellows type of diseases has a biotrophic mode of life, and they have not been cultured in any artificial media so far. All other bacterial pathogens are necrotrophic in nature causing leaf spots, blights, and cankers in aerial plant organs/tissues. During interactions between the microbial pathogen and host plant, the pathogen produces various kinds of enzymes and toxins that attack preformed host cellular structures like the cuticle and epidermis of host plants. The surveillance mechanism of resistant plants recognizing the presence of the pathogen-derived molecules activates the host defense system. On the other hand, susceptible plants fail to sense the presence of pathogen-derived molecules and allow initiation of infection. Thus the pathogen-associated molecular patterns (PAMPs) determine the future course of development or failure of disease symptoms characteristic of the pathogen (pathogenesis).

Microbial plant pathogens, on reaching the phyllosphere, have to survive the initial epiphytic phase on leaf surface, where they are likely to be exposed to desiccation, UV irradiation, and nutrient limitation. Epiphytes complete their life cycle on the leaf surface, whereas microbial pathogens have to enter leaf tissue and cross the cuticle and epidermis for initiation of infection. Fungal pathogens may either enter through stomata or directly penetrate the epidermis, producing cuticle- and cell wall-degrading enzymes (CWDEs), whereas bacterial pathogens enter the leaf surface through natural openings or wounds on leaf surface. Viruses require to be introduced directly into the susceptible cells by specific vectors or sometimes through fresh wounds created through leaf contact and by abrasives under experimental conditions. As soon as contact between the pathogen or nonpathogen and host cells/tissues is established, ion fluxes, phosphorylation/dephosphorylation of proteins, and production of signaling molecules such as salicylic acid (SA), jasmonic acid (JA), and ethylene (ET) and reactive oxygen species (ROS) are activated. Regulation of gene expression and induction of defense responses such as cell wall strengthening and accumulation of pathogenesis-related (PR) proteins and phytoalexins soon follow in resistant plants. On the other hand, pathogens produce effector molecules which contribute to virulence, resulting in suppression of plant defense. The virulent pathogen overcoming the suppressive activities of the host plant may spread to other plant tissues or organs in compatible (susceptible) interaction (Jones and Dangl 2006; Garcia-Brugger et al. 2006).

Plant cells deploy extracellular and intracellular receptors to detect the pathogen and its invasion pattern and subsequently mount immune responses. Pathogens, in turn, secrete effector proteins to deregulate host responses. Effectors may exert their activity either in the host's extracellular environment (apoplastic effectors) or inside host cells (cytoplasmic effectors) (Giraldo and Valent 2013). Effectors may be translocated subsequently into the host cytoplasm. Pathogen genomes may encode several effectors with different functions (Ghibriel et al. 2016). Resistance of a cultivar/genotype may be due to its ability to recognize the effector protein molecules of pathogen origin by plant receptor proteins. In incompatible interaction, the host plant arrests/prevents pathogen activity commencing from entry into the host cells or inhibits pathogen proliferation and invasion to other tissues/organs. Successful defense is based on the early recognition of avirulent/virulent strains which produce elicitors of resistance or effectors contributing to virulence, respectively, and rapid activation of defense-related enzymes and phenolic compounds. In addition, a localized programmed cell death (PCD) resulting in formation of visible necrotic spots may effectively arrest the spread of the biotrophic pathogens (Narayanasamy 2008). The molecular battle between pathogens and susceptible plants may lead to development of disease symptoms. On the other hand, failure of infection at different stages of pathogenesis in resistant plants may represent effective suppression of pathogen development by enhanced host defense response. Interactions between pathogen and host plant are expressed as different levels of severity of symptoms that are graded based on a scale to categorize them as susceptible, moderately susceptible, moderately resistant, highly resistant, and immune. Cultivars and genotypes of host plant species differ in their levels of resistance, whereas pathogen strains/races show variations in their virulence. Sets of differentials are used to determine the infection type or virulence levels and identify the strain/race of the pathogen(s) infecting different crops. On the other hand, crop varieties/genotypes may show vertical resistance or horizontal resistance to microbial pathogens. Resistance of a cultivar may remain higher over a period of several years, or the growing season is termed durable resistance, even when disease pressure is high.

1.5 PATHOGEN BIOLOGY AS A FOUNDATION FOR DISEASE MANAGEMENT

Investigations on the biology of microbial plant pathogens have provided an insight into biological characteristics such as disease cycles of pathogens, host range, modes of dispersal, survival in the absence of primary host plants (crops), ability to produce a plethora of enzymes, toxins, effectors that aid in the establishment of infection and disease progression in susceptible plants, and their inability to develop in resistant plants that possess effective counteracting immune responses. The molecular warfare operating between host plant and pathogen reveals the race between them to gain a foothold by the pathogen in the susceptible plant and the range of weapons used by the host plant by activating the expression of resistance genes involved in the production of defense-related PR-proteins, phytoalexins, and reactive oxygen species at different stages after entry of the pathogen into the plant. Genetic diversity in crop plant species and the pathogen indicates the possible choice(s) to select the appropriate cultivar that has built-in resistance to target pathogen(s). Knowledge on the weak links in the life cycle of pathogens may be useful to identify suitable tactics to keep the fungal pathogen (Chapter 3), bacterial and phytoplasmal pathogens (Chapter 5), and viral and viroid pathogens (Chapter 7) at bay. Assessment of variability in modes of dispersal and responses to environmental conditions such as heat, UV irradiation, and chemicals forms the basis for utilization of these factors to eliminate pathogens from seeds and propagules. Influence of cultural practices on disease incidence and spread has been determined to eradicate all sources of infection and to adopt suitable practices such as application of optimum levels of nutrition and irrigation methods to reduce adverse of effects of diseases caused by phyllosphere pathogens. Movement of microbial pathogens from seeds/propagules to developing seedlings and dissemination to short and long distances necessitated production of disease-free seeds and asexually propagated plant materials by applying tissue culture methods as well as molecular methods to eliminate infected plant materials (Chapter 8).

The presence of strains/races of microbial pathogens with differing virulence levels in different locations and seasons indicated the requirement of identification of cultivars that show resistance to many strains present in the location concerned. The development of cultivars resistant to diseases through conventional breeding methods and biotechnological techniques has been an effective strategy to restrict incidence and spread of major diseases of different crops, as this strategy is cost-effective and environmentally friendly (Chapter 9). Among the nonchemical methods, usefulness of application of biotic and abiotic biocontrol agents has been assessed for large-scale application under field conditions, and some commercial formulations are effective against certain diseases. However, the limitation of the biotic biocontrol agents (BCAs) such as lack of adaptation of the BCAs to different environments and short shelf life restrict their wide application. Inducing resistance in cultivars to pathogens using various resistance activators from different organic and inorganic sources may be an effective strategy for reducing crop losses to some extent (Chapter 10). Fungal pathogens produce, at different stages in their life cycle, asexual spores, sexual spores, and other structures like sclerotia and chlamydospores which differ in their sensitivity to chemicals that are applied to suppress development of diseases. Based on the concentration required for pathogen inhibition, the chemicals are selected for application under field/greenhouse conditions. Since the effectiveness of chemicals for control of some diseases was greater, their use was considered indispensable, in the absence of other alternatives. However, development of resistance in microbial pathogens to systemic chemicals with single-site activity, when applied repeatedly, became common. In vitro assessments showed that insensitivity to chemicals may be due to mutation in the pathogen gene, and it was persistent for several generations of the pathogen. Growing awareness of the possible pollution of environment and chemical residues left in grains, vegetables, and fruits outweighing the benefits of chemicals applied led to reduction or avoidance of chemical use. Chemical resistance management practices have been suggested to overcome the problem of development of resistance in pathogens to chemicals (Chapter 11).

1.6 DEVELOPMENT OF INTEGRATED DISEASE MANAGEMENT SYSTEMS

Investigations on various aspects of phyllosphere microbial plant pathogens are performed with a view to gathering required information that can be used for developing tactics for effective management of various diseases caused by them. Crop disease management systems based on various principles aim to (i) to reduce/prevent introduction of the pathogen/disease (exclusion), (ii) to eliminate/reduce the amount of inoculum present initially in a location/region (eradication), and (iii) to enhance levels of resistance to disease(s) (immunization). Strategies of crop disease management may be grouped as host management practices useful to reduce the rate of increase of pathogen populations and pathogen management practices that lead to suppression of pathogen development directly. They may also be grouped as biological and nonbiological methods or chemical and nonchemical methods. Parameters of differentiation may overlap each other. The disease management decision is taken primarily based on economic consideration (benefit-cost ratio), and hence, it may not be possible to adopt some tactics due to lack cost-effectiveness, albeit they may be effective in suppressing disease development.

Many disease management practices have been shown to be effective against selected disease(s) under a set of environmental conditions. However, usefulness and economic feasibility of several practices have not been convincingly demonstrated. Further, no single disease management strategy is known to be effective under different environmental conditions prevailing in various locations/regions. Hence, selection of suitable practices that are individually effective and their integration into appropriate crop production system becomes essential to derive benefits of their synergistic activity, resulting in a higher level of protection to the crops against target disease(s). Various combinations of cultural practices, use of disease-free seeds/propagules, planting resistant or tolerant cultivars, development of cultivars incorporated with disease resistance gene(s) through conventional breeding methods and/biotechnological techniques, employing biotic and abiotic biocontrol agents, induction of resistance using abiotic inducers, and restriction of chemical application to the minimum level have been evaluated for their efficacy to minimize disease incidence and spread. Efforts to integrate effective tactics have to increase the derived benefits of available strategies and preserve the environment in addition to providing chemical residue-free food materials to consumers (Chapter 12).

This book presents information derived from an extensive literature search for academic upgradation to graduate students and researchers in the departments of plant pathology, microbiology, genetics and plant breeding, biotechnology, agriculture, horticulture and personnel of quarantine and certification programs seeking additional input to make a kick-start of their academic and research projects to reach the targeted goal.

REFERENCES

Beddow JM, Pardey PG, Chai Y, Hurley TM, Kriticos DJ et al. (2015) Research investment implications of shifts in the global geography of wheat stripe rust. Nat Plants 1: 15132. https://doi.org/10.1038/nplank.2015.132

Boualem A, Dogimont C, Bendahmane A (2016) The battle for survival between viruses and their host plants. Curr Opin Virol 17: 32–38.

Cambra M, Capote N, Myrta, Llacer G (2006) *Plum pox virus* and estimated costs associated with sharka disease. OEPP/EPPO Bulletin 36: 202–204.

Chen XM (2020) Pathogens which threaten food security: *Puccinia striiformis*, the wheat stripe rust pathogen. Food Sck 12: 239–251.

Danci O, Oana D, Luiza B, Anca O, Daniela P et al. (2012) Production of virus-free potato plantlets. J Hortic Forestry Biotechnol 16: 232–238.

Davie K, Holmes R, Pickup J, Lacomme C (2017) Dynamics of PVY strains in field grown potato: impact of strain competition and ability to overcome host resistance mechanisms. Virus Res 241: 95–104.

Fundecitrus-Fundo de defesa da citricultura (2011) www.fundeicitrus.comb.br

Garcia-Brugger A, Lamotte O, Vandella E, Bourque S, Lecourieux D et al. (2006) Early signaling events induced by elicitors of plant defenses. Molec Plant-Microbe Interact 19: 711–724.

Ghibriel HAY, Thomma BPHJ, Seidle MF (2016) The age of effectors: genome-based discovery and applications. Phytopathology 106: 1206–1212.

Giraldo MC, Valent B (2013) Filamentous plant pathogens effectors in action. Nature Rev Microbiol 11: 800–814.

Gottwald TR (2010) Current epidemiological understanding of citrus huanglongbing. Annu Rev Phytopathol 48: 119–139.

Hausbeck MK, Bell J, Medina-Mora C, Podolsky R, Fulbright DW (2000) Effect of bactericides on population size and spread of *Clavibacter michiganensis* subsp. *michiganensis* on tomatoes in the greenhouse and on disease development and crop yield in the field. Phytopathology 90: 38–44.

Isac M, Preda S, Marcu M (1998) Aphid species, vector of *Plum pox virus*. Acta Virol 42: 233–234.

Jones JD, Dangl JL (2006) The plant immune system. Nature 444: 323–329.

McMullen M, Bergstrom G, De Wolf F, Dill-Macky R, Heshman D et al. (2012) A unified effort to fight an enemy of wheat and barley: fusarium head blight. Plant Dis 96: 1712–1728.

Moreno P, Ambros S, Albaich-Marti M, Guerri J, Pena L (2008) *Citrus tristeza virus*: a pathogen that changed the course of citrus industry. Molec Plant Pathol 9: 251–268.

Narayanasamy P (2008) *Molecular Biology in Plant Pathogenesis and Disease Management*. Springer Science, Heidelberg, Germany.

Narayanasamy P (2011) *Microbial Plant Pathogens—Detection and Disease Diagnosis*. Springer Science, Heidelberg, Germany.

Narayanasamy P (2017) *Microbial Plant Pathogens—Detection and Management in Seeds and Propagules*. Wiley-Blackwell, Chichester, UK.

Narayanasamy P (2020) *Soilborne Microbial Plant Pathogens—Detection and Disease Management*. CRC Press, Florida, USA.

OEPP/EPPO (2004) EPPO standards: diagnostic protocols for regulated pests: *plum pox virus*. OEPP/EPPO Bulletin 34: 247–256.

Schumann GL (1991) *Plant Diseases—Their Biology and Social Impact*. The American Phytopathological Society, St. Paul, MN, USA.

Sen Y, van der Wolf, Visser RGR, van Heusden S (2015) Bacterial canker of tomato: current knowledge of detection, management of resistance and interactions. Plant Dis 99: 4–13.

Wellings CR (2011) Global status of stripe rust: a review of historical and current threats. Euphytica 179: 129–141. https://doi.org/10.1007/s10681-011-0360-y

Yin K, Qiu J-L (2019) Genome editing for plant disease resistance applications and perspectives. Phil Trans Royal Soc B 374: 20180322. https://doi.org/10.1098/rstb.2018.0322

2 Detection and Identification of Phyllosphere Fungal Pathogens

2.1 INTRODUCTION

Among the microbial plant pathogens, oomycetes, fungi (except obligate fungi), and bacteria are capable of existing in a free state in the environment, whereas phytoplasmas and viruses require live plants for propagation and vector(s) for their transmission. Viroids, the smallest of the microbial pathogens, are generally transmitted through seeds and propagules. All plant species, including crops, are exposed to phyllosphere plant pathogens under natural conditions at all stages of development. The susceptible plants, when infected by fungal pathogens, may exhibit distinct symptoms. Phyllosphere fungal plant pathogens are present also in seed, air, soil, and water in a free state or in infected plant debris/residues left on the soil. Spores/mycelium of the pathogens are carried to short or long distances by wind, and new infections are initiated in susceptible plants in different locations. It is essential to detect infections by fungal pathogens preferably before symptom expression, and their identity has to be established precisely and rapidly, as they exist in the form of strains with varying virulence. Based on the morphological, biochemical, immunological, and genetic characteristics of the fungal pathogens, various techniques have been developed for detection, identification, differentiation, and quantification in plant phyllosphere and seeds/propagules. Comparative sensitivity, reliability, and rapidity of the various techniques are indicated for selecting the technique to suit the individual need.

The plant phyllosphere includes all aerial plant organs such as stem, leaves, inflorescences, and earheads which are exposed to microbial plant pathogens at different stages of plant development. Phyllosphere fungal pathogens commonly induce leaf spots, blights, anthracnose, rusts, downy mildews, powdery mildews, head blight, and dieback diseases on various crops causing a different magnitude of crop losses depending on the levels of virulence of the pathogen and levels of susceptibility/resistance of the cultivars and the environmental conditions prevailing in the geographical location concerned. Early detection and precise identification of the pathogen(s) involved in the disease is the basic step required for development of effective disease management systems. Detection, identification, differentiation and quantification of air-dispersed fungal plant pathogens are essential to (i) apply preventive and curative measures; (ii) determine the effectiveness of different disease management tactics, when applied alone or in combination; (iii) implement seed certification and quarantine programs; (iv) differentiate emerging new strains with greater virulence or resistance to chemicals; (v) identify new sources of resistance to disease(s) that can be used in breeding programs; (vi) differentiate components of disease complexes induced by two or more pathogens; and (vii) have an insight into the complex interactions between microbial pathogens and crop cultivars with different levels of susceptibility/resistance. Various aspects of fungal pathogens dispersed via seeds and propagules, soil and water that may infect aerial plant organs, and the diseases caused by them have been elaborately discussed in earlier publications (Narayanasamy 2011, 2017, 2020).

2.2 DETECTION OF FUNGAL PATHOGENS IN PLANTS

2.2.1 Biological Methods

Fungal pathogens are the largest among microbial pathogens with a well-organized thallus (body) composed of mycelium and spores formed during asexual and sexual phases of their life cycles. Morphological characteristics are useful to identify the fungal pathogens up to the generic level and in some cases up to the species level. In the case of obligate pathogens, in situ examination of infected plant tissues may be performed by preparing prefixed temporary or semipermanent mounts.

2.2.1.1 Histological Methods

Fungal hyphae and sporulating structures present in infected plant tissues may be clearly viewed by employing different stains. Early recognition of infection of wheat leaves by *Pyrenophora tritici-repentis* (tan spot) was achieved by clearing the leaf pieces and staining with the dyes. Infection by *P. tritici-repentis* could be detected by using bright field microscopy, even in the early stage of infection. Aniline blue fluorescence was more sensitive, revealing the presence of the pathogen in the infected epidermal cells rapidly (Dushnicky et al. 1998). *Pernospora tabacina*, causing tobacco downy mildew, could be detected by treating leaf samples with 80% methyl alcohol and placing in NaClO solution followed by staining with lactophenol (Caiazzo et al. 2006). Oospores of *Peronospora effusa*, causing spinach downy mildew, were detected in the seed-wash, using a light microscope. Shrinking of the cytoplasm characteristic of plasmolysis occurred at 15 min after suspending oospores in NaCl solution (Figure 2.1). In the trypan blue staining test, the oospores were stained with 0.4% trypan blue. Viable oospores were not stained with trypan blue (Kunjeti et al. 2016b).

DOI: 10.1201/9781003456827-2

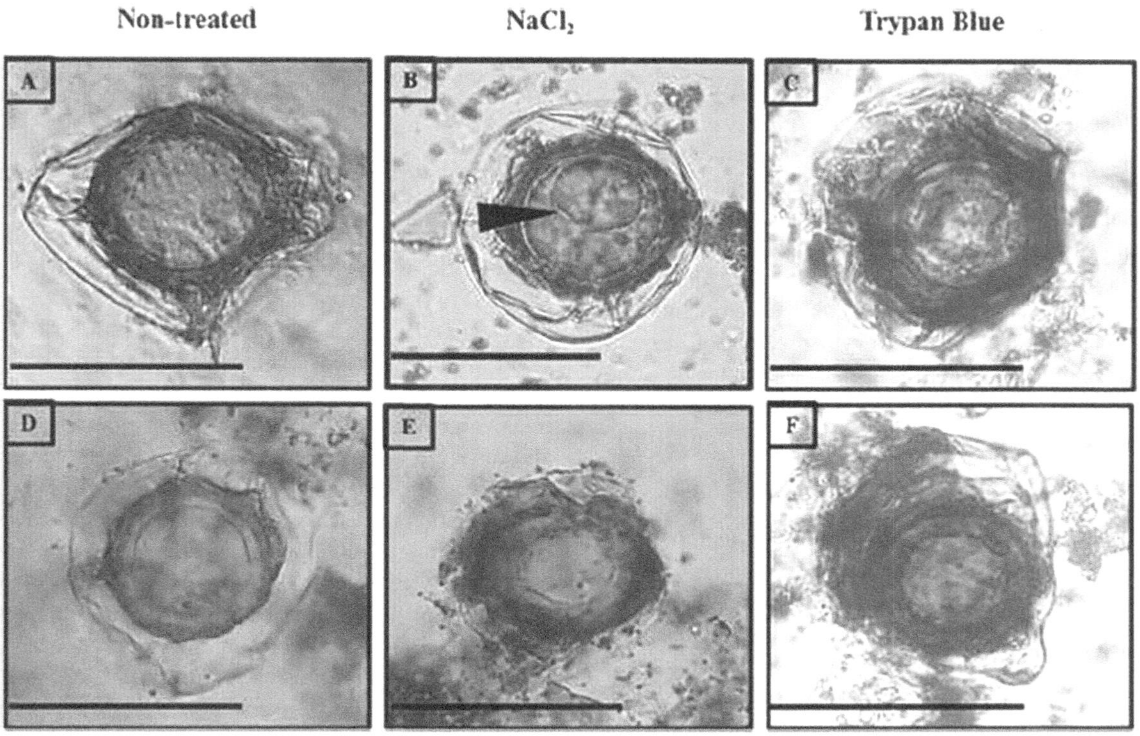

FIGURE 2.1 Effects of treatment with 4 M NaCl on oospores of *Peronospora effusa*. Plasmolysis of oospores occurred at 45 min after exposure to NaCl; viable spores were stained with trypan blue (B), while nonviable spores excluded the stain (C); (A and D) untreated oospores and (E) nonplasmolyzed oospore.

Source: (Courtesy of Kunjeti et al. 2016b and with kind permission of the American Phytopathological Society, MN, USA)

2.2.1.2 Isolation-Based Methods

2.2.1.2.1 Cultural Characteristics

Phyllosphere fungal pathogens may infect different aerial organs of susceptible plants, causing leaf spots, blight, anthracnose, downy mildew, powdery mildew, rust, blast, and head blight diseases in various crops. The causal agent can be isolated on appropriate cell-free media that may differentially support the development of target pathogen(s). Standard methods for isolation of fungal pathogens have been described in earlier publications (Narayanasamy 2002, 2011). Some of the fungal pathogens may need specific semi-selective or selective media that differentially support their development and sporulation. *Alternaria brassicola* infecting cruciferous crops was preferentially encouraged by CW medium. This medium suppressed the development of other saprophytic fungi present, while permitting the target pathogen to proliferate well from the infected seed (Wu and Chen 1999). *Alternaria dauci* and *A. radicina* could be successfully isolated from carrot seeds, using a semi-selective medium containing carrot leaf extract, despite the presence of other fungi associated with carrot seeds (Strandberg 2002). The effectiveness of a semi-selective medium for isolation of *Colletotrichum gloeosporioides* (yam anthracnose disease) was demonstrated by Ekefan et al. (2000). Likewise, Botrytis selective medium (BSM) was useful for isolation of *B. cinerea* from plant tissues. BSM medium significantly allowed formation of greater number of *B. cinerea* colony-forming units (CFUs) on individual petri plates (Edwards and Seddon 2001). The morphological characteristics of fungus-like and fungal pathogens obtained from pure cultures at asexual and sexual stages are determined and used for their identification and classification into kingdom, phylum, class, order, family, genus, and species. About 10,000 of 100,000 species have been identified as plant pathogens (Agrios 2005). Morphological characters of *Phomopsis longicolla* conidia were employed to identify the pathogen infecting soybean, and conidia formed abundantly on medium prepared from soybean seeds and mature stems after exposure to low temperature (–10° C) (Vidic et al. 2013). *Colletotrichum gloeosporioides* causing brown spot on leaves of the tea-oil tree (*Camellia oleifera*) in China was isolated and identified based on morphological characteristics and confirmed by similarity in an internal transcribed spacer sequence (ITS) with that of *C. gloeosporioides* (Xu et al. 2017). *Alternaria alternata* was shown to be the causal agent of pear dead flower buds disease in the Netherlands based on morphological characters (Winneker et al. 2019).

2.2.1.3 Biochemical Methods

2.2.1.3.1 Activity of β-Glucuronase

A diagnostic method for early diagnosis of strawberry anthracnose caused by *Colletotrichum theobromicola* was developed, based on analysis of the metabolome by the gas chromatography–mass spectrometry technique. The metabolic profiles of control and inoculated strawberry plants exhibited, respectively, 189 and 202 total ion chromatogram peaks. Biomarkers of *C. theobromicola* were identified to indicate infection of strawberry plants. These potential biomarkers, citric acid, D-xylose, erythrose, galactose, gallic acid and malic acid, methyl-α-galactopyranoside, phosphate, and shikimic acid, might be employed for early diagnosis of strawberry anthracnose disease (Dai et al. 2019).

2.2.1.3.2 Isozyme Analysis

Isozyme analysis is a powerful biochemical technique that is applied to detect, differentiate, and identify morphologically similar fungal pathogens. Isozyme analysis has been applied for identification of *Puccinia graminis* f.sp. *tritici* (Burdon et al. 1986), *Pyricularia oryzae* (Leung and Williams 1986), *Uromyces appendiculatus* (Lu and Groth 1987), and *Erysiphe graminis* f.sp. *tritici* (Koch and Kohler 1990). Isozyme patterns of *P. sorghi, P. sacchari*, and *P. sacchari-philippinensis* are useful for their identification. *Leptosphaeria maculans* isolates fell into two groups, ET1 and ET2, reflecting fast or slow movement of glucose phosphate isomerase (GPI) in the starch gel. Highly virulent strains had fast-moving GPI (ET1), whereas the GPI of weakly virulent isolates could move only slowly (ET2) (Sippell and Hall 1995). Isozymes of nicotinamide adenine dinucleotide dehydrogenase (NADH) and diaphorase (DIA) of *Colletotrichum gloeosporioides* yielded a large number of electrophoretic phenotypes that clustered as three major groups (I, II, and III) (Kaufmann and Weidemann 1996). Cellulose acetate electrophoresis (CAE) was applied to resolve allozymes genotypes of *Phytophthora infestans* (late blight diseases of potato and tomato) at two loci, GPI and peptidase (Pep). The CAE system could be employed to predict mating-types and metalaxyl sensitivities of the field isolates of *P. infestans* (Goodwin et al. 1995). The fatty acid methylesterases (FAMEs) of fungal pathogens are estimated using a gel chromatography technique. Fatty acid profiling is a relatively rapid, efficient, and cost-effective method for identifying isolates of *Phytophthora infestans*. All isolates of *P. infestans* representing US-1, US-6, US-7, US-8, US-11, US-14, and US-17 genotypes produced eight primary fatty acids. Profiles of individual isolates of *P. infestans* maintained consistency and reproducibility (Larkin and Groves 2003).

2.2.1.4 Physical Methods

An automatic procedure was developed to detect and determine disease severity of coffee rust caused by *Hemileia vastatrix* and passion fruit leaf spot induced by *Xanthomonas axonopodis*. The algorithm was formulated based on the categories of four categories of symptoms (yellow, light brown, dark brown, and black). Image analysis was performed using diseased coffee and passion fruit leaves. This method provided accurate estimates of disease severity (Barbedo 2014). Early detection of *Parastagonospora nodorum* and monitoring progression of infection in Italian durum wheat cultivars was achieved by employing hyperspectral imaging analyses which showed that this technique could be employed for differentiation of healthy and infected tissues, based on differences in hyperspectral values (Iori et al. 2015). Software packages like ASSESS 2.0, 'Leaf Doctor', and Scion Image software in custom-made modules are used for digital image analysis (www.plant-image analysis.org). In ASSESS 2.0, parameters for healthy and diseased areas may be adjusted by the user in a well-organized graphical user interface. ASSESS 2.0 was useful for evaluation of disease severity on single leaves and well-arranged images. Changes in reflectance due to plant pathogens and disease symptoms induced and impairments in the leaf structure and chemical composition of tissue during pathogenesis could be assessed. Formation of chlorotic and necrotic lesions and fungal structures produced by powdery mildew hyphae and conidia, and rust urediniospores could be detected by reflectance from infected leaves. Unique disease patterns could be used as the basis of differentiation of foliar pathogens inducing leaf spots, scab, and blight diseases (Mahlein 2016).

Hyperstructural imaging technique at the leaf scale was developed to detect and distinguish wheat brown rust and leaf rust disease under controlled conditions. The pure spectra of rust spores of *Puccinia triticina* and *P. striiformis*, inducing brown rust and yellow rust, respectively, on wheat leaves were employed as a spectral fingerprint for detection of respective diseases. By employing reference spectra on spore-scale observations, hyperspectral imaging could be applied for detection of crop diseases without pixel-wise labeling at the leaf scale (Bohnenkamp et al. 2019). Digital infrared thermography was developed for presymptomatic detection of infection by *Pseudoperonospora cubensis*, causing downy mildew disease in cucumber. The maximum temperature difference (MTD) values of infected leaves were greater than those for noninfected leaves. However, the MTD values decreased rapidly as the symptoms became visible. Differences between spot temperatures of the symptomatic areas in healthy and infected leaves were significant at 4 days after inoculation. Differentiation of healthy and infected leaves would be possible, based on thermal infrared imaging and Fourier-transform infrared spectra (Wen et al. 2019).

Diagnosis of the disease based on visual symptoms alone was difficult due to occurrence of other foliar diseases and abiotic stress. Automated detection and classification of sugar beet Cercospora leaf spot (CLS) and other foliar diseases was developed. An algorithm based on an image database captured with smartphone cameras was developed for identification of sugar beet foliar diseases. This procedure combines image acquisition and segmentation on the smartphone and advanced image data processing on a server. Diseased and nondiseased leaves may be separated. Plants with foliar diseases and healthy plants can also be differentiated. Accuracy of classification/differentiation of CLS, Ramularia leaf spot, Phoma leaf spot, beet rust, and bacterial blight was high (Hallan et al. 2018).

2.2.2 Immunoassays

Immunodiagnosis has been useful for the rapid detection and identification of fungal pathogens in symptomatic and asymptomatic plants/seeds, facilitating the elimination of infected plants. Principles of immunological reactions and their applications are discussed in earlier publications (Narayanasamy 2005, 2011). Polyclonal antibodies (PAbs) and monoclonal antibodies (MAbs) are used in various immunoassays. MAbs are more specific and provide more reliable results than PAbs, although they are expensive and difficult to prepare. Phage-displayed recombinant antibodies involving the use of the M13 phage infecting *Escherichia coli* can be prepared more rapidly and at lower cost compared to MAbs. The recombinant antibodies are equally specific as MAbs in identifying microbial plant pathogens.

2.2.2.1 Enzyme-Linked Immunosorbent Assay

Enzyme-linked immunosorbent assay (ELISA) and its variants have been more frequently employed among the immunoassays for detection, quantification, and differentiation of fungal pathogens and their strains in plant tissues with varying specificity and sensitivity. Double antibody sandwich (DAS)-ELISA, the standard format, is highly strain-specific and requires the use of a different antibody conjugate for each strain of the pathogen. In contrast, the indirect ELISA method can be performed, using a common conjugate for different pathogen species, and the antirabbit globulin conjugate or antigoat globulin available commonly may be employed. Other variants, direct antigen-coating (DAC)-ELISA, and protein A (extracted from cells of *Streptococcus aureus*) coating (PAC)-ELISA have also been employed for detection of fungal pathogens. Monoclonal or polycloanal antibodies (MAbs or PAbs) have been used in different formats of ELISA tests for the detection of *Botrytis cinerea* in grapevine, strawberry, and tomato (Bossi and Dewey 1992); *Leptosphaeria maculans* in canola (Stace-Smith et al. 1993); *Cladosporiun fulvum* in tomato (Karpovich-Tate et al. 1998); and *Colletotrichum acutatum* in strawberry (Krátká et al. 2002). Phage-displayed antibody (ScFv) was employed for the detection of *Alternaria alternata* (Carzaniga et al. 2002). MAb 57D3 could specifically bind with a 16-kDa protein that was produced by *Pyricularia oryze*, causing rice blast disease. *P. oryzae* races occurring predominantly in the United States could be detected by ELISA tests. MAb57D3 reacted with intercellular and cell wall antigens of *P. oryzae*, and also with extracts of blast lesions on rice leaves (Nannapeneni et al. 2000).

Stagonospora nodorum (leaf and glume blotch disease) could be detected by using pathogen-specific MAbs in plate-trapped antigen (PTA)-ELISA format (Zelinger et al. 2004). MAbs raised using partially purified extracts of *Plasmopara halstedii* (sunflower downy mildew) differentiated three pathogen antigens with molecular weight (MW) of 68, 140, and 192 kDa. By employing specific MAbs in ELISA tests, infection of sunflower seeds with *P. halstedii* could be detected (Bouterige et al. 2000). PAbs detected homologous *Erysiphe necator* (grapevine powdery mildew) conidial antigens in the PTA-ELISA format. By using isolate-specific MAb, grapevine cultivars could be categorized according to the disease severity induced by the powdery mildew pathogen (Markovic et al. 2002). An MAb reactive specifically to a high MW antigen from *M. fijiensis* mycelial single ascospore was employed as a capturing reagent and the PAbs as secondary antibody in the triple antibody system (TAS)-ELISA format (Otero et al. 2007). Comparative efficacies of plate-trapped antigen (PTA)-ELISA and subtractive inhibition ELISA formats in detecting *Phytophthora infestans* using specific MAbs were assessed. Both formats were specific in detecting *P. infestans* (Skottrup et al. 2007b). A library of MAbs was generated against urediniospores of *Puccinia striiformis*, causing wheat stripe rust. The subtractive inhibition ELISA was more sensitive than the competitive ELISA format (Skottrup et al. 2007a).

2.2.2.2 Immunofluorescence Assay

An immunofluorescence (IF) assay may be performed using PAbs and MAbs for detection of fungal pathogens in infected plants. MAbs against whole conidia, their extracellular matrix (ECM), and a putative cutin esterase isolated from the conidia of *Botrytis cinerea* were generated. The IF assay was effective in detecting *B. cinerea* in different plant species (Salinas and Schots 1994) and *Mycosphaerella brassicola* in crucifers (Kennedy et al. 1999). Polyclonal antisera were generated against intact nongerminated (SBR1A) or germinated (SBR2) urediniospores of *Puccinia pachyrhizi* (soybean rust disease). Indirect immunofluorescence spore assay (IFSA) detected *P. pachyrhizi*, using both SBR1A and SBR2 antisera. Both DBR1A and SBR2 bound to *P. pachyrhizi* and *P. meibomiae* were detected using a Leica DM 1RB epifluorescence microscope (Baysal-Gurel et al. 2008). Indirect IF assay in combination with propidium iodide (PI) staining was employed to detect viable urediniospores of *P. pachyrhizi* using fluorescence isothiocyanate (FITC)–labeled secondary antibodies. The IF assay including PI counterstaining was rapid and reliably provided differentiation of live and dead urediniospores of *P. pachyrhizi*, offering a potential application in forecasting soybean rust disease. Dual staining of freshly collected live spores of *P. pachyrhizi* with CFDA and PI resulted in viable spores fluorescing green indicating membrane integrity. In heat-killed dead spores, the nuclei fluoresced red, due to compromised membranes, indicating nonviability (Figure 2.2). Viability of urediniospores was assessed using

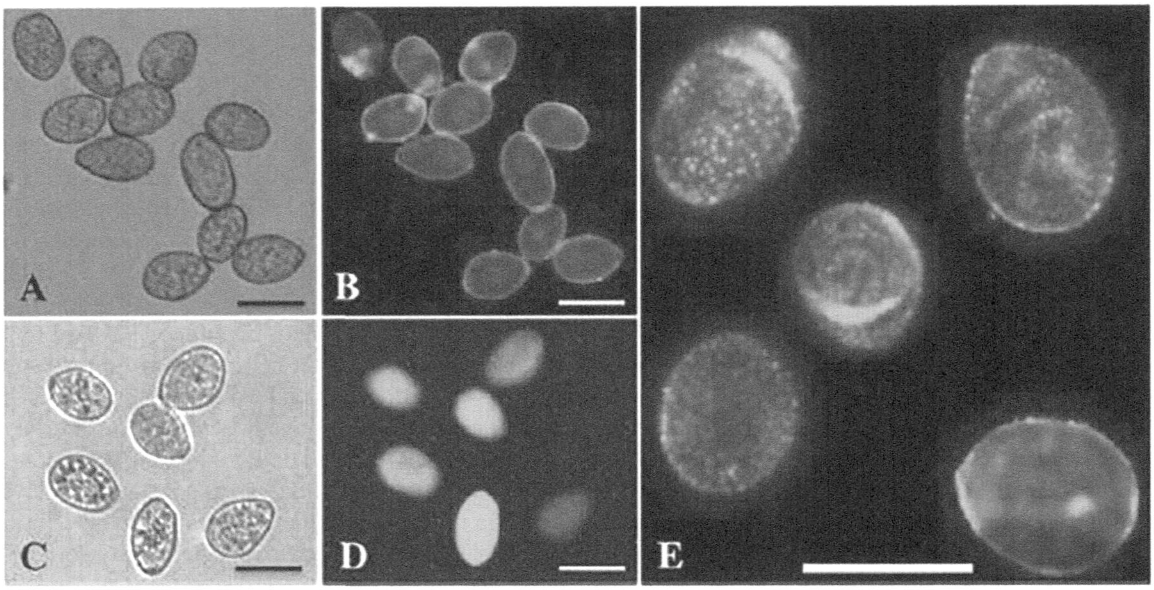

FIGURE 2.2 Assessment of viability of urediniospores of *Phakopsora pachyrhizi* stained with 5-carboxyfluorescein diacetate (CFDA). (A and B) Viable spores emitting green fluorescence; (C and D) dead spores lacking ability to emit green fluorescence.

Source: (Courtesy of Vittal et al. 2012 and with kind permission of the American Phytopathological Society, MN, USA)

FUN1 and Calcofluor White M2R as in the CFDA-PI dual staining method. Dead spores with no metabolic activity had a diffuse, faint fluorescence. Calcofluor White M2R did not stain ungerminated, live and heat-killed dead spores, indicating the absence of cellulose and chitin in the cell walls of urediniospores. However, the germ tubes of germinating spores stained blue (Figure 2.3) (Vittal et al. 2012).

2.2.2.3 Lateral-Flow Immunoassay

Lateral-flow immunoassay (LIFA), also known as immunochromatography assay, is adapted to operate along a single axis to suit the test strip format, and it is useful for onsite analysis. Antibody-based diagnostic tests rely on enzyme reaction for visualization of results. To remove unbound enzyme, washing steps are necessary, thereby making the tests more time-consuming. Enzymes can be replaced by binding antibodies to particulate labels such as latex, colloidal gold, or silica. Using these labels, the target antigen is visualized. In the field competitive immunochromatographic lateral flow devices (LFDs) assay was developed to detect *Alternaria brassicae*, causing dark leaf spot of *Brassica* crops in air sampled above the crop canopy. Visual examination had a threshold of about 50 conidia of dark leaf spot pathogen. The competitive immunochromatographic lateral flow device (CLFD) assay could detect infection earlier. Test sensitivity was increased by about 5-fold by using an optical density (OD) reader, providing quantitative assessment of *A. brassicae* in air samples (Wakeham et al. 2016).

2.2.2.4 Immunolabeling Method

Antibodies generated against fungal pathogens labeled with protein-A complexes are employed to locate the pathogen/antigen in plant tissues after infection/inoculation. Symptomless infection of lettuce by *Botrytis cinerea* was investigated by employing the immunolabeling technique. *Botrytis*-specific monoclonal antibody BC-12-CA4 was used to detect the pathogen in the lettuce plant tissues. As plants developed, invasion of plant tissues progressed from roots to stem and leaves of plants growing in greenhouses. Surface sterilization of seeds reduced the number of infected symptomless plants, indicating the spread of infection from seeds (Sowley et al. 2010).

2.2.3 Nucleic Acid–Based Techniques

Deoxyribonucleic acid (DNA) and/ribonucleic acid (RNA) form the vital constituents of all organisms, and they are the determinants of genetic characteristics and production of several enzymes and biological compounds required for their metabolic activities and survival. The nature/structure of DNA or RNA in the spores/mycelium remains unchanged in all stages of pathogen development. Hence, the nucleic acid (NA)–based techniques may be more reliably employed for detection of fungal pathogens at all stages of their life cycle. In general, the sensitivity of NA-based techniques is greater, and they are suitable for detection of fungal pathogens even at early stages of infection. NA-based techniques are preferred because of their

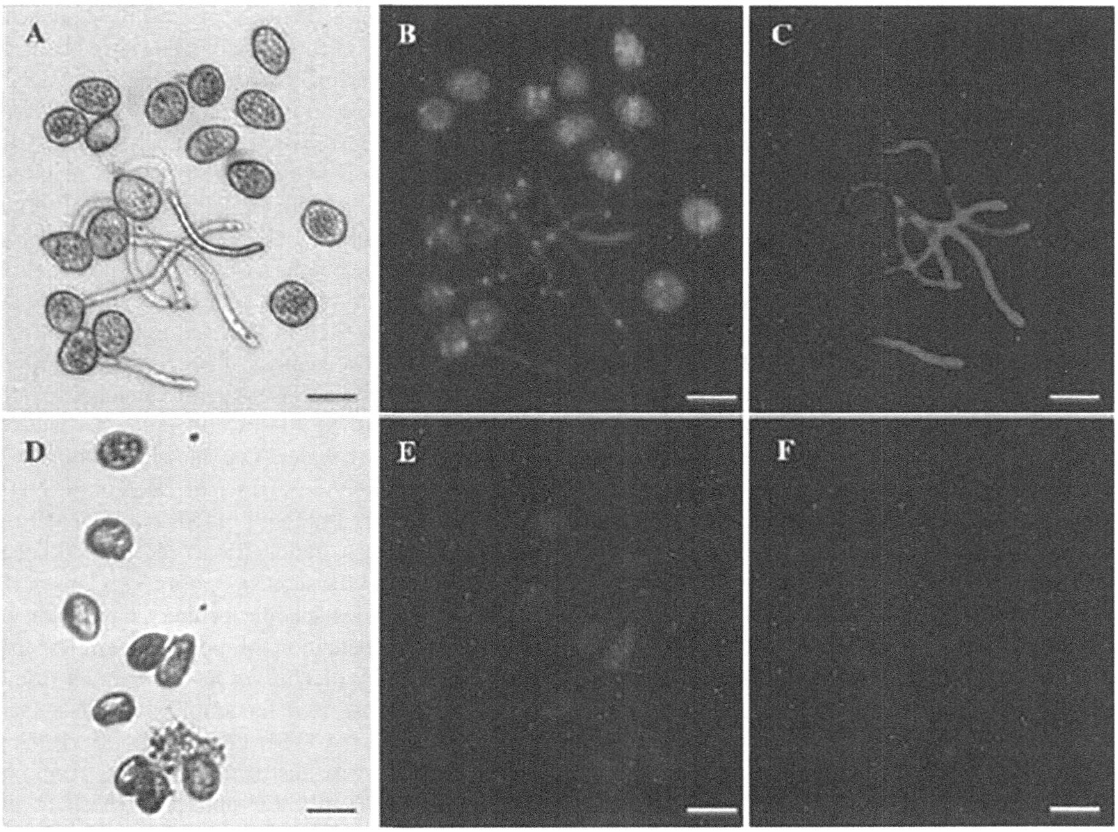

FIGURE 2.3 Assessment of viability of urediniospores of *Phakospsora pachyrhizi* by staining with FUN1 and Calcofluor White M2R. (A, B, and C) Viable spores emitting fluorescence detected using different filter sets under the bright field of the Olympus BX51 microscope. (D, E, and F) Heat-killed dead spores lacked the ability to fluoresce with both stains.

Source: (Courtesy of Vittal et al. 2012 and with kind permission of the American Phytopathological Society, MN, USA)

higher levels of sensitivity, rapidity, reliability, and specificity for detection and identification of fungal pathogens and diagnosis of the diseases induced by them.

2.2.3.1 Hybridization-Based Nucleic Acid Techniques

Mutations may occur in the pathogen nucleic acid (NA) over a period of time, resulting in random changes in NA sequences, and some of the changes may accumulate, resulting in recognizable variations. Characteristic genetic constitution of individual organisms/strains is due to many generations of mutations and recombinations. Closely related organisms share a greater nucleotide (nt) sequence similarity than distantly related organisms that exhibit less nucleotide similarity. Highly specific nucleotide(s) present in a particular strain or isolate of the microbial plant pathogen species, but absent in other strains or isolates of the same species, is identified, and this sequence is employed as the probe to detect/identify the target species/strain/isolate. Detection of fungal plant pathogens by NA hybridization assays is based on the formation of double-stranded (ds) nucleic acid molecules by specific hybridization between single-stranded (ss) target *nt* sequences (denatured DNA or RNA) and complimentary ss-NA probe. Restriction fragment length polymorphism (RFLP) analysis is based on the natural variations in the genomes of different genera, species, or strains of fungal pathogens. Variations (polymorphisms) in fragment sizes may be generated by loss or gain of by other events that influence fragment sizes (such as deletion or insertions) in the DNA sequences.

Field isolates of rice blast pathogen *Magnaporthe oryzae* were studied to determine their genetic relationships. The probe specific for a dispersed repeated sequence (known as MGR) was employed to determine the genetic relationships of isolates of *M. oryzae* from rice and other grasses by RFLP analysis by using the repetitive probe MGR586. Isolates of rice blast pathogen representing four distinct races exhibited multiple bands hybridizing to the probe MGR586. The results indicated the possibility of *M. oryzae* population from nonrice hosts being the sources of inoculum for the rice crops (Han et al. 1995). A pathotype-specific DNA probe effectively identified R-type isolates of *Pseudocercosporella herpotrichoides*. Rye seedlings infected by R-type isolates could be identified by hybridization of the probe to DNA extracted from infected rye plants (Nicholson et al. 1994). RFLP analysis was useful in establishing a correlation between molecular and biological characteristics of the biotypes of *Stagonospora nodorum* from barley and wheat (Ueng et al. 1995). Isolates of *Glomerella cingulata*, *Colletotrichum gloesoporioides*, and *C. acutatum* were grouped based on RFLP analysis as well as using morphological characteristics and vegetative compatibility, using mitochondrial (mt) DNA. Seven mtDNA RFLP haplotypes were identified and differentiated among isolates of *G. cingulata* (Gonzáles et al. 2006). The majority of the isolates of *Alternaria solani* and *A. alternata* pathogenic to potato could not be differentiated based on morphology alone, and they were consistently grouped based on restriction patterns (Tymon et al. 2016). A procedure to identify *P. infestans* at the molecular level in planta using terminal-RFLP (T-RFLP) of the RG57 loci was developed. This technique was sensitive to detect and differentiate *Phytophthora infestans* genotypes directly in planta without isolation of the pathogen and propagation (Wijekoon et al. 2017).

2.2.3.2 Polymerase Chain Reaction–Based Techniques

Polymerase chain reaction (PCR) involves amplification of specific sequences of target DNA exponentially by in vitro DNA synthesis. Amplification is performed rapidly with high specificity and fidelity using oligonucleotide primers and a DNA polymerase in an automated reaction. Designing suitable primers is the primary requirement for performing PCR assay. In general, short sequences (100–1,000 bp) are more effectively amplified and resolved by agarose gel electrophoresis. Specific primers are designed based on the sequences of either amplified or cloned DNA (cDNA) or RNA from target pathogen(s). Three DNA sequences commonly found on fungal pathogens are employed to develop diagnostic techniques based on PCR. The nuclear DNA region, which usually consists of three genes (the large subunit gene, the small subunit gene and 5.8S gene) are separated by internal transcribed spacer (ITS) regions. Sequences of ß-tubulin gene and the mating-type gene of ascomycetes are the other genes utilized for designing primers. The ITS region of fungal DNA is the most widely sequenced DNA region (Narayanasamy 2008, 2011).

It is essential to extract the fungal pathogen DNA devoid of all inhibitors likely to inhibit PCR assay. The DNA from healthy and infected plants tissues and fungal pathogens has to be extracted and purified to remove all inhibitors of DNA polymerase, such as polysaccharides, phenolic compounds, or humic substances. The commercially available Qiagen DNeasy Plant Tissue Kit was employed as per the manufacturer's instructions to purify the DNA (Griffin et al. 2002). Magnetic microparticles specific to DNA molecules were employed apparently to separate inhibitors and the DNA. Biomagnetic separation was effective for handling a large number of samples (Scott et al. 2002). Sequences of ITS regions of DNAs of fungal pathogens have been commonly used to design primers for amplification of DNA of target pathogens. A region in the ITS1 and ITS2 specific to *Phytophthora infestans* was used to construct PCR primers (PINF) which could detect the pathogen in infected field samples of potato and tomato (Tooley et al. 1998). A primer pair (INF FW2 and INF REV) specific to *P. infestans* could be employed to detect the pathogen in leaves, stems and also in tubers of potato prior to appearance of visible symptoms (Hussain et al. 2005). PCR-based assays employing primers based on ITS sequences of *Eutypa lata* (Eutypa dieback disease of grapevine) DNA and sequence characterized amplified regions (SCARs) were applied to detect diseased grapevine wood in water (Leconte et al. 2000). PCR amplification of *Pyrenophora brassicae* DNA of the NH10 isolate using the primers PbITSF/PbITSR could detect the pathogen in symptomless canola tissues (Karolewski et al. 2006).

Differences in the sequences of ITS regions or rDNA and PCR using species-specific primers have been employed effectively for detection of powdery mildew pathogens. Whole-cell DNA of *Oidium neolycopersici* (tomato powdery mildew) was used for amplifying the nuclear rDNA region spanning the ITS1, ITS2, and 5.8S rRNA by PCR. This pathogen was distinctly different from any other species of the family Erysiphaceae (Kiss et al. 2005). Sunflower powdery mildew is induced by *Podosphaera xanthii*, *Golovinomyces cichoraceaurm*, and *Leveilulla taurica*. Based on the sequence differences in the ITS region, species-specific primers S1/S2, G1/G2, and L1/L2 for *P. xanthii*, *G. cichoracearum*, and *L. Taurica*, respectively, were applied for their detection and differentiation (Chen et al. 2008). A pathogen-specific primer pair was employed to detect *Peronospora effusa* in spinach seeds using PCR assay. An amplicon of 67-bp was detected in the extracts of seed lots (Figure 2.4) (Kunjeti et al. (2016).

Based on the ITS sequences of the powdery mildew pathogens infecting sugar beet in Europe and North America, they were identified as *Erysiphe betae* (Francis et al. 2007). The genomic DNA of *Peronospora tabacina* (tobacco downy mildew disease) extracted from herbarium dry leaf of tobacco was used as a template in PCR amplification with specific primers for ITS of *P. tabacina*. Sequence analysis of PCR products confirmed the identity of the pathogen in the herbarium as *P. tabacina* in Europe (Carrieri et al. 2017). *Puccinia striiformis* f.sp. *tritici (Pst)* (wheat stripe rust) was detected from asymptomatic wheat leaves inoculated and naturally infected by the pathogen (Zhao et al. 2007). In another investigation, using primers PST1 and PST2, *Pst* was detected in dormant stage in young wheat plants in the areas where the pathogen overwintered. All isolates of *Pst* produced a distinct band of a fragment of 470-bp amplified by the primer pair (Wang et al. 2008). The primers ITS1-F and ITS4-B showed good specificity to the genus *Puccinia*, amplifying the pathogen-specific product. The primers SR-1 and SR-2 were specific to *Puccinia helianthi*, causing sunflower rust (Guo et al.

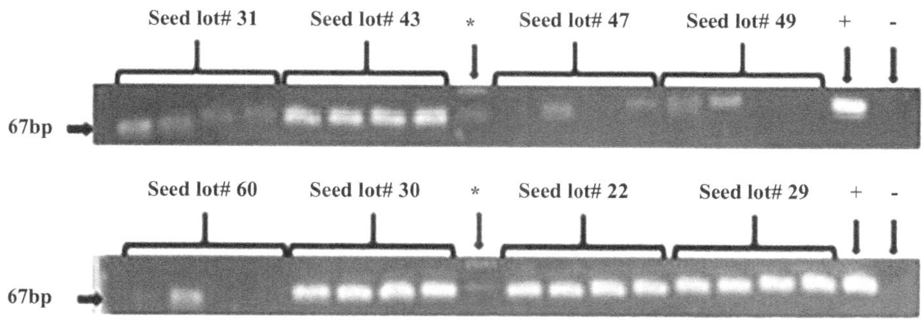

FIGURE 2.4 Detection of *Peronospora effusa* in spinach seeds using PCR assay with a pathogen-specific primer pair. Note the presence of 67-bp amplicon in the extracts of infected seed lots; the molecular ladder is at the far right to identify the pathogen-specific amplicon.

Source: (Courtesy of Kunjeti et al. 2016 and with kind permission of the American Phytopathological Society, MN, USA)

2016). *Colletotrichum truncatum, C. gloeosporioides*, and *C. scovillei* causing chili anthracnose were detected and differentiated using primers designed based on sequences of the nuclear ribosomal DNA region. Amplicons of 66-, 112-, and 168-bp were specific to *C. oeosporioides, C. truncatum*, and *C. scovillet*, respectively. Primers selected could amplify and identify DNA at very low concentrations. Primer sets efficiently amplified DNA directly from infected seeds and fruit tissues (Abdelrazig et al. 2020).

Primers have been designed based on the sequences of genes that control specific metabolic activity of the target pathogen. Two primer sets were designed based on ITS sequences of rDNA of *Rynchosporium secalis* and from the amplified fragment length polymorphism (AFLP) marker set RS1/RS9, which detected *R. secalis* in infected barley leaves and seeds, whereas RS8/RS9 efficiently detected *P. teres* in infected barley leaves (Gubis et al. 2004). PCR assay using primers designed based on the sequences of the *AMT* gene involved in production of the AM-toxin (AMT) by *Alternaria alternata* (Alternaria blotch) detected specifically AM toxin-producing isolates of the *A. alternata* pathotype (Johnson et al. 2000). Specific PCR primers were designed to target the Smp450–2 gene in the gibberellins biosynthetic gene cluster of *Spaceloma manihoticola* (cassava super-elongation disease). The presence of the pathogen was detected by the PCR assay in leaves and cankers of infected plants (Alleyne et al. 2015). Wheat blast occurring in South American countries is caused by *Magnaporthe oryzae Triticum* pathotype. A genomics-based approach to identify molecular markers to *M. oryzae Triticum* pathotype was applied. Primers based on the sequences of MoT3 consistently amplified a product of 361 bp from *M. oryzae Triticum* pathotype isolates (Pieck et al. 2017).

Three sets of primers designed from β-tubulin, squalene monooxygenase, and ketopantoate reductases from the *Puccinia striiformis* f.sp. *tritici (Pst)* genome were employed in the PCR assays. A fragment of 1518 bp unique to *Pst* pathotypes was amplified by the primer set *Pst* KetoF130/ Pst KetoR1. The PCR/QPCR assay using the marker might be suitable for accurate quantification of pathogen DNA, reflecting the population of stripe rust pathogen (Aggarwal et al. 2017a). Detection of *Alternaria alternata* (sunflower leaf blight) was possible by PCR assay employing partial ITS regions, translation elongation factor, polymerase second subunit, glyceraldehyde-3-phophate dehydrogenase, and *Alternaria* allergen gene regions (Kgatle et al. 2018). Specificity of primer pair APN1 was tested against virulent strains of *Chondrostereum purpureum* (apple silverleaf disease) from infected apple trees. DNA from inoculated nursery plants were assayed. APN1 primers could discriminate the isolates of the pathogen from artificially inoculated plants as well as naturally infected plants. The PCR format employed was rapid and minimally destructive in terms of diagnostic toll to detect *C. purpureum* (Grinbergs et al. 2020). DNA of *Peronosclerospora maydis* derived from maize seeds was genotyped with multiplexing using maize microsatellite (bnlg1189) and downy mildew pathogen microsatellite (DM38) primer. Alleles in the size of 153-bp could be observed after amplification of infected samples with DM38 (Lukman et al. 2016). Primers based on the *mat1–1* and *mat1–2* genes of *Mycosphaerella fijiensis* (banana Sigatoka leaf spot) were employed for PCR amplification (Conde-Ferráez et al. 2010). Primers derived from *Cladosporium fulvum* avirulence (*Avr*) genes were employed in the multiplex PCR assay to detect the pathogen. All *Avr* genes could be amplified in single or multiplex PCR using DNA isolated from in vitro grown monospore cultures (Medina et al. 2015). In order to detect *Puccinia striiformis* f.sp. *tritici (Pst)* early, inter-simple sequence repeat (ISSR) procedure was applied to develop a sequence-characterized amplified region (SCAR) marker. One fragment unique to *Pst* was identified by ISSR and then sequenced. The SCAR primers were able to amplify the fragment unique to *Pst* from all isolates tested. The pathogen could be detected in asymptomatic leaves of wheat plants inoculated with *P. striiformis* f.sp *tritici* under greenhouse conditions (Wang et al. 2017a). The shotgun cloning procedure was performed for detection of *Kabatiella zeae* (eye spot disease).

K. zeae-specific sequences formed the basis of designing primers which were highly specific for *K. zeae* (Tillessen et al. 2018).

2.2.3.2.1 Arbitrarily Primed (AP)–PCR Assay

The arbitrarily primed (AP)-PCR technique was employed for the detection of *Colletotrichum acutatum* (almond anthracnose disease). The AP-PCR procedure was applied to detect the presence of *C. acutatum* in naturally diseased blueberry bushes (*Vaccnium* spp.). The primers CAG and GACAC amplified *C. acutatum*–specific sequences in samples from diseased and symptomless bushes indicating the potential of the blueberry bushes to function as sources of infection for almond trees. The isolates of *C. acutatum* from blueberry were genetically identical to those from almond trees (Yoshida et al. 2007). The AP-PCR assay was successfully applied to detect *Colletotrichum gloeosporioides* in lychee plants and fruits. A specific genotype identified by the AP-PCR assay was associated with the production of teleomorph in culture (Anderson et al. 2013).

2.2.3.2.2 Flow Strip–Based Polymerase Recombinase Amplification Assay

A lateral flow strip–based recombinase polymerase amplification (LF-RPA) assay was applied for rapid, equipment-free detection of *Phytophthora infestans* infection of potato and tomato in the field. This assay targeted the Ras-related protein (*Ypt 1*) gene, and it could be performed over a wide range of temperatures (25–45° C). All 12 *P. infestans* isolates were detected by LF-RPA assay, and no cross-reaction occurred with related oomycetes or fungal pathogens tested. By combining a simplified polyethylene glycol-NaOH procedure for DNA extraction from plant samples, the entire LF-RPA assay could be completed within 30 min without the requirement of any special equipment. *P. infestans* was detected in naturally infected field samples from different areas in China (Lu et al. 2020).

2.2.3.2.3 Nested Polymerase Chain Reaction

In order to enhance the sensitivity of the conventional PCR assay, a nested PCR format was developed, employing two pairs of primers. The nested PCR format was developed using specific primers for the detection of grapevine trunk pathogens (GFTPs), Botryosphaeriaceae species, *Eutypa lata, Cadophora luteo-olivacea, Phaeoacremonium* spp., and *Phaeomoniella chlamydospora* in infected plants and also pruning tools indicating the possibility of contaminated pruning shears transferring the inoculum through pruning wounds (Agusti-Brisach et al. 2015). The nested PCR format could detect *Colletotrichum acutatum* in inoculated leaves and also in field-collected asymptomatic strawberry leaves, providing reliable results under field conditions (Perez-Hernandez et al. 2008). *Puccinia striiformis* f.sp. *tritici (Pst)* was detected in infected wheat plants by the nested PCR format involving the use of primer pair Psta/Psts designed based on the genome-specific sequence of *Pst*. The sensitivity of detection of *Pst* by the nested PCR assay was enhanced by 1,000-fold. The pathogen could be detected prior to symptom appearance (Xiaojie et al. 2009). A nested PCR assay with a higher level of sensitivity was developed for detection of *Blumeria graminis* f.sp. *tritici (Bgt)* (wheat powdery mildew) by employing the DNA primer sets including external and internal primer pairs. Latent infection of wheat seedlings could be detected by the nested PCR. The nested PCR format was about 10,000-fold more sensitive than the conventional PCR format. The nested PCR assay developed in this investigation was useful to efficiently estimate the initial inoculum potential of wheat powdery mildew epidemics in the fields (Zeng et al. 2010).

The nested PCR assay was developed based on one pair of genus-specific primers derived from the ribosomal DNA ITS conservative sequence of the genus *Valsa* and one pair of species-specific primers from apple Valsa canker pathogen *Valsa mali* var. *mali (Vmm)*-specific sequence. The nested PCR assay detected a distinct band of 348-bp in all *Vmm* isolates (Zang et al. 2012). Primers CGF and CGR designed based on rDNA ITS sequences were employed in the conventional PCR assay to obtain a product of ~390-bp from the DNA extracted from *Colletotrichum gloeosporioides* (anthracnose disease of guava) isolates. The nested PCR format was developed by combining specific and universal ITS primers (ITS1/ITS4). The pathogen could be detected in infected plant tissues in the early stage of infection (Yao et al. 2018).

2.2.3.2.4 Reverse Transcription (RT)–PCR Assay

Reverse transcription (RT)-PCR assay is an elegant diagnostic technique which detects targeted mRNA in plant tissues. RT-PCR detected the presence of *Mycosphaerella graminicola* at least 4 days before symptom expression in inoculated wheat plants. The ability of RT-PCR to detect pathogen presence prior to symptom appearance was very useful to determine extent of infection, since *M. graminicola* required a latent period of at least 4 weeks for exhibiting visible symptoms of infection under natural conditions in Germany (Verreet et al. 2000). In a later investigation, primers were designed from the β-tubulin gene sequences for detection of *M. graminicola*, because they were the most conserved genes with high copy numbers (Guo et al. 2005). *Oidium neolycopersici* (tomato powdery mildew) was detected by RT-PCR assay. The transcripts expressed in single conidium were amplified by the RT-PCR format. *O. lycopersici* or *Erysiphe trifolii* pathogenic and nonpathogenic, respectively, to tomato were differentiated and identified by RT-PCR assay (Matsuda et al. 2005).

2.2.3.2.5 Kinetic-PCR Assay

As DNA replication is intimately linked to cell division, kinetic-PCR (kPCR) assay could be applied to quantify pathogen DNA at different stages of pathogenesis. This technique offered an advantage over conventional PCR by its ability to differentiate two rust pathogen species present in mixed infections. The k-PCR procedure was effective in detecting and quantifying DNA of *Melampsora* spp. (poplar rust disease). *Melampsora medusa* f.sp. *delloidae* and *M. larici-populina* were distinguished by employing species-specific primers (Boyle et al. 2005).

2.2.3.2.6 Repetitive Element (Rep)–Based PCR Assay

Rice genotypes with the *Pi-ta* gene can effectively resist infection by strains of *Magnaporthe oryzae*, causing blast disease. The genetic identity of *M. oryzae* isolates from rice cultivars with and without *Pi-ta* gene was analyzed, using an international differential system, repetitive element (Rep)-PCR, and sequence analysis of PCR products of *AVR-Pita1*. These isolates belonged to races IA1, IB1, IB17, IC1, and IC17 of *M. oryzae*. Rep-PCR further differentiated those isolates into 15 distinct groups. A predominant group within each race was identified using the Rep-PCR format. Sequence analysis of 90 PCR products revealed ten *AVR-Pita1* haplotypes, four of which were new (Xing et al. 2013).

2.2.3.2.7 Multiplex PCR Assays

The multiplex PCR assay provides the advantage of acceleration and identification process of fungal pathogens, in addition to the nonrequirement of their isolation in pure culture. *Pseudoperonospora humuli* and *Sphaerotheca humuli* cause, respectively, downy mildew and powdery mildew diseases in hop. Specific PCR assays were developed to detect *P. humuli* and *S. humuli* in naturally infected hop plants. A specific PCR assay with combinations of primers P1 + P2 and S1 + S2 amplified specific ITS sequences from *P. humuli* and *S. humuli*. A multiplex method was developed for detecting two or more fungal pathogens in addition to *P. humuli* and *S. humuli*. It was possible to detect and differentiate *S. humuli, P. humuli, V. albo-atrum*, and *F. sambucinum* infecting hop by amplifying PCR combinations of primers rapidly and precisely (Patzak 2003). Sunflower powdery mildew pathogens, *Podosphaera xanthii* and *Golovinomyces cichoracearum*, were detected and differentiated by applying a multiplex PCR assay rapidly and precisely in field-infected sunflower plants (Chen et al. 2008). A single-step multiplex PCR assay was applied to detect wheat stem rust (*Puccinia graminis* f.sp. *tritici*), wheat leaf rust (*P. triticina*), and wheat powdery mildew (*Blumeria graminis* f.sp. *tritici*) pathogens simultaneously. Three specific bands of 395-bp, 151-bp, and 464-bp were formed following amplification with a three reactions system. The procedure developed in this investigation was useful for simultaneous detection of three important pathogens at early stages of infection of wheat crops (Chen et al. 2015).

Duplex PCR assay was developed to detect *Colletotrichum lindemuthianum* infection in common bean. A single amplicon of 442-bp was amplified by the primer pair CY1/CY2 only from *C. lindemuthianum* and *C. orbiculare*. This product was cloned, sequenced and employed for designing a species-specific primer pair CD1/CD2, which could differentiate *C. lindemuthianum* from *C. orbiculare* by forming distinct 638-bp product. A duplex method employing both pairs of primers CY1/CY2 and CD1/CD2, was applied to detect *C. lindemuthianum* infection in the common bean (Wang et al. 2008). Real-time multiplex PCR assay based on double-stranded DNA-binding dye coupled with dissociation curve analysis was designed, using a region of the cytochrome c oxidase subunit 1 (*cox1*) gene to differentiate four species of *Colletotrichum*. The duplex real-time PCR assays were rapid and effective for detection and differentiation of most of the *Colletotrichum* spp. infecting soybean (Yang and Hartman 2015). DNA templates were produced by automated grinding and robotic paramagnetic particle capture. DNA was extracted from freshly collected symptomatic seeds. The duplex PCR format was applied for the simultaneous detection of DNA templates from *Colletotrichum truncatum* and *Fusarium* sp. infecting green and red bell pepper. Duplex PCR using ITS219-F/ITS419-R and Fa/Ra primer sets simultaneously detected and identified *C. truncatum* and *Fusarium* sp. in symptomatic seeds of infected red bell pepper fruit. The optimized PCR was sensitive and rapid in detecting dual infections of seeds of green and red bell pepper seeds very efficiently (Villafana et al. 2019).

2.2.3.2.8 Combination of PCR Assay with Other Techniques

The PCR assay may be applied in combination with other nucleic acid–based assays to enhance the sensitivity and specificity of detection of fungal pathogens. The magnetic capture-hybridization (MCH)-PCR format was developed to reduce the adverse effects of PCR inhibitors coextracted with DNA from lignified apple tissues for detection of *Nectria galligena*, causing Nectria canker disease in apple and pear. Sequences of ITS1 region of the rRNA repeats were used as a target. The trapping agent used to coat the magnetic beads was a ss-DNA oligonucleotide biotin-labeled on the 5′ terminal and designed to be complementary to part of the rRNA gene ITS1 region of *N. galligena*. PCR amplification was performed after the magnetic capture hybridization using species-specific primers ch1 and ch2. The sensitivity of the MCH-PCR assay was greater than that of the conventional PCR assay (Langrell and Barbara 2001). *Fusarium graminearum* (*Fg*) (Fusarium head blight disease) is toxigenic, producing trichothecene mycotoxins in grains, necessitating monitoring disease incidence and accumulation of toxin in grains. PCR-RFLP procedure was developed based on DNA sequences of the translational elongation factor 1 α (*TEF1α*) gene. Partial amplification of the *TEF1α* gene with primers mh1/mh2 generated a 459-bp PCR fragment. Digestion of PCR product with *Hpa*II generated a unique restriction pattern that could be used to rapidly and precisely differentiate *Fg* from other *Fusarium* spp. (Hafez et al. 2020).

2.2.3.2.9 Quantitative PCR Assay

In addition to qualitative detection of fungal pathogens by PCR formats, quantitative assessment of the population present in a sample is essential. *Phakopsora pachyrhizi* and *P. meibomiae* (soybean rust disease) were detected and identified by quantitative (Q)-PCR assay, using specific primer sequences and a fluorigenic 5′-exonuclease linear hydrolysis probe. The DNA of single urediniospores could be detected, indicating the extremely high level of sensitivity of the assay. To overcome the effects of inhibitors of PCR present in urediniospores, three synthetic DNA molecules of arbitrary sequences were constructed as multiplexed internal controls (ICs) to eliminate false-negative results (Haudenshield and Hartman 2011). The qPCR format was applied for detection and quantification of *Ramularia beticola* (sugar beet Ramularia leaf spot disease). Specific primers derived from the ITS 2 region of the pathogen DNA were employed to identify the DNA from spores trapped on tape from Burkard spore traps placed under artificially inoculated and natural disease incidence conditions. *R. beticola* DNA of air samples from fields with natural infections was detected at significant and increasing levels from development of the first symptoms, indicating spore production within the crop played a major role in development of the disease. Under inoculated conditions, *R. beticola* infection could be detected in asymptomatic leaves at least 10 days before symptom appearance (Wieczorek et al. 2014). Quantification of an airborne *Bremia lactucae* (lettuce downy mildew disease) population near the lettuce crop was undertaken by qPCR-based assay using a target sequence in mitochondrial (mt) DNA for specific detection of the pathogen. Primers developed for the TaqMan assays were species-specific and amplified templates only from *B. lactucae*. DNA from a single sporangium could be detected (Kunjeti et al. 2016a). For the detection of *Peronospora effusa* (spinach downy mildew), conventional PCR assay was applied to detect the pathogen DNA, employing a pathogen-specific primer pair AS1 and Per1. The assay detected *P. effusa* in a subset of 17 seed lots containing different levels of pathogen populations (Figure 2.4). The TaqMan assay detected *P. effusa* in seed lots, using the DNA template derived from three seed lots (Kunjeti et al. 2016b).

Two TaqMan probe-based qPCR assays were optimized for detection of *Diaporthe caulivora* and *D. longicolla* (soybean stem canker disease). Two primer sets, DPC-3 specific for *D. caulivora* and PL-3 specific for *D. longicolla*, could detect the pathogens efficiently (Kontz et al. 2016). A qPCR assay was applied to detect, differentiate, and quantify pycnidia formed by *Guignardia citricarpa* and *G. mangiferae* (citrus black spot) present in partially decomposed leaves (Hu et al. 2014). *Pyrenophora tritici-repentis (Ptr)* (barley tan or yellow spot) infected barley by producing Ptr ToxB toxin. The qPCR assay was used to quantify pathogen biomass in planta at different stages of disease development along with cytological analysis. The results indicated the process of infection of barley by *Ptr* was similar to that in wheat (Aboukhaddour and Strelkov 2016). Infections of nectarine (*Prunus persica* var. *nucipersica*) flowers and fruits by *Monilinia* spp. remained latent for long time. The qPCR assay was developed for early detection and quantification of brown rot infection in the blossoms and fruits of nectarine trees and for differentiating *Monilinia* spp. associated with the disease. The qPCR format detected greater amounts of *M. fructicola* DNA than that of other *Monilinia* spp. DNA in latently infected flowers. The qPCR format was sensitive, reliable, and rapid for detecting latent brown rot infections in nectarines (Garcia-Benitez et al. 2017). Species of Botryosphaeriaceae cause Botryosphaeria dieback of grapevine. Pathogen DNA was extracted from Burkard volumetric spore sampler placed in naturally infected and artificially inoculated grapevine orchard. Multispecies primers amplified DNA of ten Botryosphaeriaceae species (Billones-Baaijens et al. 2018). PCR and qPCR assays were developed to detect and quantify *Peronospora somniferi* and *P. meconopsidis* causing downy mildew of the poppy. Primers targeting the *coxI* and *coxII* gene regions were designed for detection, differentiation, and quantification of *P. somniferi* and *P. meconopsidis* in poppy seed. Both PCR and qPCR formats provided reliable results rapidly and could be employed for routine tests of seed lots (Thangavel et al. 2018). A specific primer pair based on sequences of a part of the ITS region of peach powdery mildew pathogen *Podosphaera pannosa* ribosomal DNA was employed in the qPCR assay. It was possible to quantify conidia placed on plastic tapes commonly used in volumetric air samplers. Quantification of conidia contents by microscopic and molecular techniques yielded comparable results. Molecular assay was applied to detect latent fungal inoculum in different plant parts of peach trees. The presence of *P. pannosa* was detected more frequently on the bark of affected twigs and to a lesser extent in foliar buds (Marimon et al. 2020).

2.2.3.3 Real-Time PCR Assay

Real-time PCR assay quantifies diagnostic amplicons on time with a fluorescence detection system. An additional dual-labeled fluorigenic probe is employed for hybridizing to target DNA within the region defined by the two fluorescent dyes, a reporter (6-carboxy-fluorescein, FSM) at the 5′ end and a quencher (6-carboxy-tetramethylrhodamine, TAMARA) at the 3′ end, which is blocked by a phosphate group in order to prevent PCR-derived elongation. Both fluorescent dyes are excited with an argon-ion-laser at 488 nm. Based on the measurable light emission of the reporter dye, synthesis of amplicons could be monitored continuously during thermocycling, and no post-PCR handling is required for quantification of PCR amplicons. The Ct value is the number of cycles at which a statistically significant increase in the reporter fluorescence can be detected for the first time. The Ct values are then used to calculate the starting copy number of the target for each sample. The principal advantage of real-time PCR is that amplification products can be monitored, as they are accumulated in the long-linear phase of amplification. In addition, the rapidity and greater accuracy of real-time PCR are the other advantages over conventional PCR assay (Böhm et al. 1999; Schaad et al. 2003; Vandemark and Barker 2003). The real-time PCR assay may be either amplicon sequence-nonspecific (employing SYBR Green) or sequence-specific (TaqMan Molecular Beacon or Scorpion PCR) (Mumford et al. 2006). Real-time PCR performed via specific or nonspecific methods is more sensitive, specific, and rapid compared with conventional PCR assay, in addition to a reduction in the risk of false positives and

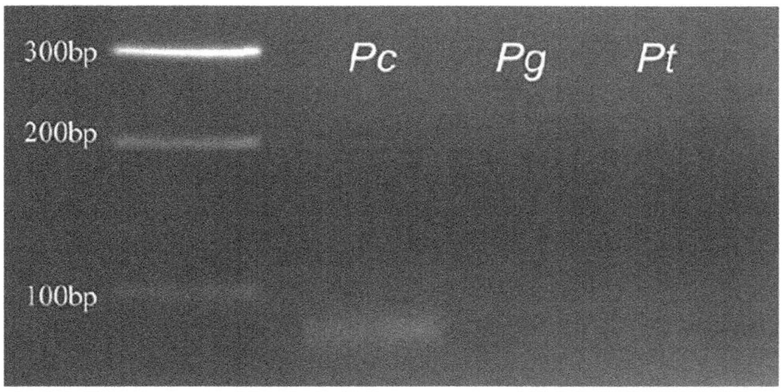

FIGURE 2.5 Real-time PCR format detecting pathogen-specific band amplified only from the DNA of *Puccinia coronata*, but not from *P. graminis* and *P. triticina* extracted from the respective urediniospores. Usefulness of primer specificity for visualization of reactions containing DNAs of all three rust pathogen is indicated by the detection technique.

Source: (Courtesy of Jackson et al. 2006 and with kind permission of the American Phytopathological Society, MN, USA)

absence of the need for postamplification steps. Real-time PCR techniques provide quantitative and multiplex analysis to obtain estimates of microbial plant pathogens (Schena et al. 2006). Real-time PCR assay was effective in detecting and quantifying *Phytophthora infestans*. Stem samples of field-infected plants contained a much greater starting copy number of the pathogen target DNA/mg of host tissue (wet weight) than samples from artificially inoculated tubers under in vitro conditions (Böhm et al. 1999). Real-time PCR assay was developed to detect *P. infestans* in leaves and tubers of potato, employing the primer pair (PinfTQF/PinTQR) and a fluorigenic probe (PinfTQPR). The assay detected *P. infestans* specifically as well as closely related *P. mirabilis, P. phaseoli*, and *P. ipomoeae*, but not other potato pathogens *P. erythroseptica* and *P. nicotianae*. The assay efficiently detected *P. infestans* in the asymptomatic leaves and tubers of potato (Lees et al. 2012).

Three different real-time PCR assay viz., ß-tubulin TaqMan probe, cytochrome *b* SYNR Green I, and cytochrome *b* LNA probe assays were highly specific in detecting *Rhynchosporium secalis* (winter barley leaf blotch disease). The cytochrome *b* LNA was the most sensitive in detecting *R. secalis* DNA in the presence of barley leaf DNA (Fountaine et al. 2007). Two forms of *Pyrenophora teres* f. *teres* (PTT) and *P. teres* f. *maculata* (PTM) cause net blotch and spot type symptoms, respectively, in barley. TaqMan MGB (Minor Groove Binder) primers and probes were designed on the *RacB* gene sequences that exhibited high specificity for the two forms of *P. teres*. The real-time PCR assay could be used to monitor the dynamics of PTT and PTM forms during barley crop season (Leisova et al. 2006). In another investigation, the primer pair PtrMulti F/R that targeted the multilocus indel DNA region was evaluated for species specificity and sensitivity. The PtrMulti SYBR qPCR assay was effective in identifying and quantifying *Ptr*. At early growth stages, qPCR assay detected *P. tritici-repentis* DNA on the younger leaves in the absence of visible tan spot lesions (See et al. 2016). A real-time quantitative PCR assay was applied to detect and quantify fungal DNA (FDNA) in oat cultivars infected by *Puccinia coronata* f.sp. *avenae* (oat crown rust disease). A real-time PCR assay with pathogen-specific TaqMan primers/probe set was applied to detect *P. coronata* template DNA. A single band of about 75-bp was amplified in the reaction containing the *P. coronata* DNA template, while no band was detectable in reactions containing DNA templates from both *P. graminis* f.sp. *avenae* and *P. triticina*. After development of *P. cornata* after inoculation, fungal DNA (FDNA)/mg of oat tissue was estimated. Amounts of FDNAs determined by the assay correlated with levels of resistance of oat cultivars. A negative relationship between FDNA accumulation and levels of resistance of oat genotypes to *P. coronata* was observed (Figure 2.6) (Jackson et al. 2006).

Rust pathogens exhibit sequence similarities and differences in the ITS1 region of their DNAs, based on which duplex real-time PCR assay was developed for detection and differentiation of the pathogens infecting cereals. The variable ITS1 region of the nuclear rDNA gene was selected

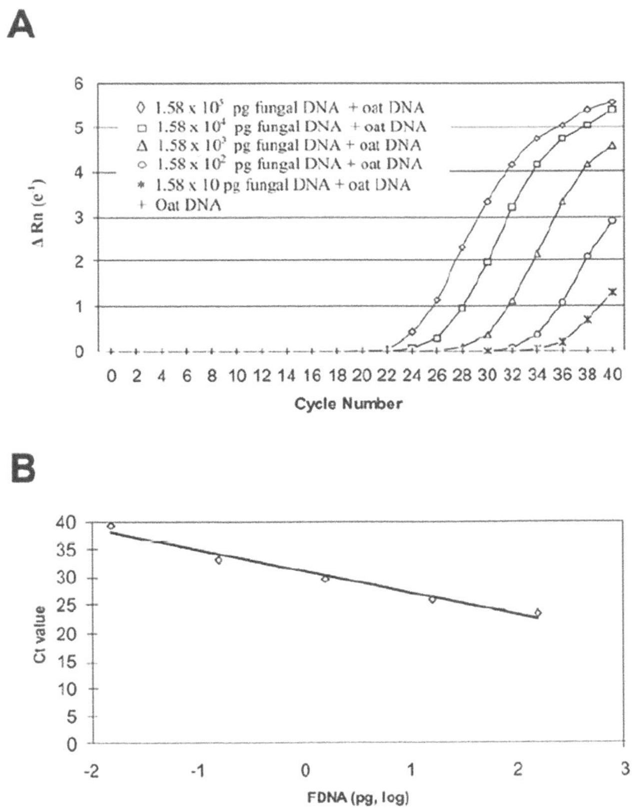

FIGURE 2.6 Standard curve developed using different concentrations of FDNA of *Puccinia coronata* useful for correlation with levels of resistance to oat genotypes to *P. coronata*. (A) Amplification profiles from 10-fold dilutions of *P. coronata* DNA extracted from urediniospores. (B) Standard curve for conversion of cycle threshold (Ct) into FDNA.

Source: (Courtesy of Jackson et al. 2006 and with kind permission of the American Phytopathological Society, MN, USA)

to differentiate rust pathogens, *Puccinia graminis, P. striiformis, P. triticina*, and *P. recondita*. The sequence alignment showed the ITS1 to have a higher level of variation than ITS2 and, therefore, the ITS1 region was employed for primer and probe design. The primer pair flanking the ITS1 (ITS1rustF10d and ITS1rustR3c) amplified all the rust species tested. Design of species-specific *Puccinia*-specific TaqMan 6-carboxy-fluorescein (FAM) probes was based on selective sites identified to maximize sequence differences between the four *Puccinia* spp. (Figure 2.7). The species-specific ITS1 primer/probe combination was very effective in detecting *Puccinia* spp. in multiple collections (Barnes and Szabo 2007). Latent infection of wheat by stripe rust pathogen *Puccinia striiformis* f.sp. *tritici (Pst)* was observed commonly in China. A real-time PCR assay was developed to investigate field distribution of latent infection and the corresponding disease incidence. The molecular disease index (MDX), based on pathogen DNA contents, was calculated. Disease indices (DX) were assessed at 1–3 weeks after leaf sampling in the corresponding experimental fields. Using the real-time PCR format, the disease foci for latent infections were accurately detected in two provinces of China (Yan et al. 2012).

Presence of *Colletotrichum acutatum* (ripe rot) and *Greeneria uvicola* (bitter rot) was detected by the assay in artificially inoculated and naturally infected grape inflorescences and mature berries. Disease progression occurred only on mature berries under field conditions, the infection remaining latent in other tissues. Real-time PCR assay was highly specific, rapid, and sensitive in detecting and differentiating the pathogen species (Samuelian et al. 2011). The real-time PCR format was effective in detecting *C. lagenarium* infection in squash, watermelon, cucumber, and muskmelon. The assay amplified different ITS sequences from *C. lagenarium*, using the primer set co-m-3371F1/R1 differentially enabling discrimination of *C. lagenarium* from other pathogens (Kuan et al. 2011). A PCR assay based on the SCAR marker was developed for detection of *Colletotrichum capsici* in infected pepper fruits and seeds. SCAR primers specifically amplified a 250-bp fragment from the genomic DNA of seven isolates of *C. capsici* and also from DNA extracted from infected fruits. A real-time PCR assay using SYBR Green detected *C. capsici* in infected fruits and seeds of pepper (Srinivasan et al. 2014).

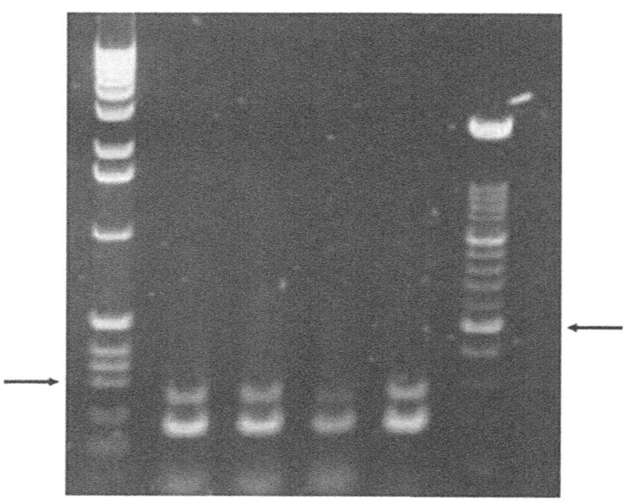

FIGURE 2.7 Duplex real-time PCR amplification of rDNA of *Puccinia graminis* (*Pg*), *P. recondita* (*Pr*), *P. striiformis* (*Ps*), and *P. triticina* (*Pt*), using a primer pair from ITS1 sequences. Lane 1: 1-kb molecular marker, Lane 2: *Pg*, Lane 3: *Pr*, Lane 4: *Ps* and Lane 5: *Pt*.

Source: (Courtesy of Barnes and Szabo 2007 and with kind permission of the American Phytopathological Society, MN, USA)

Real-time PCR assays were developed for specific detection, differentiation, and quantification of *Colletotrichum acutatum, C. gloeosporioides,* and *C. fragariae* (strawberry anthracnose or crown rot). Primers and probes were designed based on the sequences of the most divergent area of ITS1, ITS2, and 5.8S rRNA gene regions. The real-time PCR format provided rapid, sensitive, and reliable detection, allowing high-throughput and inexpensive screening of plant samples directly without the need for pathogen isolation and culturing on artificial media (Garrido et al. 2009). The extent of infection of the common bean (*Phaseolus vulgaris*) by *C. lindemuthianum* was assessed by real-time PCR assay. One primer set and a probe were selected and based on the aligned sequence matrix for the SYBR Green real-time PCR and TaqMan MGB (minor groove binder) formats. The primer was highly specific for *C. lindemuthianum* (Chen et al. 2013). The comparative efficacy of the PCR primers designed based on sequences of β-tubulin gene, ITS region, and a microsatellite region of *Cladosporium fulvum* (tomato leaf mold) was assessed. The PCR primer set CfF1/CfR1 from satellite DNA was 100-fold more sensitive than the primer pair CfF3/CfR3 from ß-tubulin sequences. The most sensitive primer pair CfF1/CfR1 was employed in real-time PCR assay to detect *C. fulvum* in tomato leaves (Yan et al. 2008). Chrysanthemum crops are affected by white rust caused by *Puccinia horiana* and brown rust induced by *P. chrysanthemi*. A multiplex real-time PCR format was developed, using species-specific hydrolysis probe labeled with different fluorescent dyes, based on the rDNA ITS1 region. *P. horiana* and *P. chrysanthemi* were precisely detected in all samples. The multiplex assay could discriminate between *P. horiana* and *P. chrysanthemi* (Demers et al. 2015).

Many genes potentially involved in the virulence of *Fusarium graminearum (Fg)* (FHB) have been identified. The SYBR-based real-time PCR assay was developed to quantify the total genomic DNA in plant samples, as well as the total genomic DNA of *Fg* that contributed from a strain containing a common selectable marker used to create deletion mutants. This method allowed correlation of amounts of disease recorded in wheat field trials with mutants of

F. graminearum under investigation (Dyer et al. 2006). *F. graminearum* (*Fg*), could infect also noncereal crops such as potato and sugar beet. Isolates from these crops induced typical FHB symptoms in susceptible wheat cultivar and also produced a wide range of trichothecene toxins. Most of the potato and sugar beet isolates induced equal or even greater FHB severity, FgDNA content, and higher accumulation of mycotoxins. The real-time PCR format was developed based on intergenic sequences specific to trichodiene synthase gene (*Tri5*) from *F. graminearum*. All isolates contained the *Tri5* gene and were virulent to wheat. Significant correlation was observed between FgDNA and FHB severity or deoxynivalenol (DON) production. *F. graminearum* causing dry rot and sugar beet decay could be potential sources of FHB epidemics (Burlakoti et al. 2007). The real-time PCR format detected *Alternaria brassicola* in naturally contaminated radish seeds, and it could be used for quantification of natural infection levels in cabbage and radish seeds (Guillemete et al. 2004). The real-time PCR-based TaqMan probe assay was developed based on sequences of the 156-bp ITS region of *Alternaria helianthi* (sunflower leaf blight) DNA to detect the pathogen in infected plant tissues and seed of the sunflower. The probe and primers were specific to *A. helianthi* as indicated by the absence of cross-reactivity with closely related *Alternaria* spp. (Chavhan et al. 2015). The qPCR assay was applied to detect and quantify *Diaporthe helianthi* and *D. gulyae* in Phomopsis stem canker affected sunflower plants. The qPCR format differentiated *D. helianthi* and *D. gulyae* with high efficiency. The DSI of the sunflower accessions correlated with the amount of pathogen DNA determined by qPCR assays (Elverson et al. 2020).

Real-time PCR assay using SYBR Green dye was applied to quantify the biomass of *Leveillula taurica* (pepper powdery mildew) in infected leaf tissues. Relative amounts of pathogen DNA in infected pepper tissues was quantified by determining both fungal and plant DNA from the CaActin gene in the same sample. A good correlation was observed between real-time PCR assessments and DSI values (Zheng et al. 2013). A real-time PCR assay was developed to quantify latent infection levels of wheat powdery mildew pathogen *Blumeria graminis* f.sp. *tritici (Bgt)* in artificially inoculated and naturally infected wheat leaves in fields. The real-time PCR format was useful to rapidly and precisely quantify latent infection levels of wheat leaves (Zheng et al. 2013). In another investigation, real-time PCR assay was employed to quantify airborne inoculum of *B. graminis* f.sp. *tritici* (*Bgt*) collected using Burkard 7-day recording spore traps. The real-time PCR assay could be applied to rapidly and reliably estimate the airborne pathogen inoculum over wheat fields (Cao et al. 2016). The extent of latent infection of *Bgt* in volunteer wheat plants during the oversummering cycle was assessed using the real-time PCR format. Latent infections in volunteer plants were more at lower altitude, whereas volunteer plants at higher elevation showed powdery mildew symptoms. Volunteer wheat played a key role in the oversummering cycle of *Bgt*. The pathogen could oversummer by infecting generations of volunteer wheat in summer, thereby functioning as primary sources of infection for autumn-sown wheat crops (Liu et al. 2019). *Magnaporthe oryzae* (rice blast disease) was detected in rice plants and seeds using the SYBR Green I real-time PCR format. The PCR assay was effective in quantifying *M. oryzae* DNA in leaves of field-infected rice plants using rice actin as a housekeeping gene (Sun et al. 2015). The importance of diagnostic tests to prevent introduction of *M. oryzae* was realized by the importing countries. Genomic analysis revealed that the *Triticum* lineage of *M. oryzae* was associated with wheat blast disease. A real-time PCR assay was developed to target a polymorphism specific to the *Triticum* lineage. The C17 quantitative PCR assay detected isolates belonging to the *Triticum* lineage with high sensitivity (Thierry et al. 2020).

A real-time PCR assay was developed using 5′ fluorogenic exonuclease (TaqMan) chemistry to detect and quantify *Peronospora tabacina* DNA in infected tobacco tissues. The PtabBM TaqMan assay was effective in detecting *P. tabacina* in field-infected plants, artificially inoculated plant leaves and roots, and systemically infected tobacco seedlings (Blanco-Meneses and Ristaino 2011). Conventional and SYBR Green I real-time PCR formats were developed based on SCAR markers for detection and quantification of *Peronosclerospora sorghi* (maize downy mildew). The real-time PCR format detected the pathogen in naturally infected plants (Sireesha and Velazhahan 2018). The conventional PCR format detected *Peronospora effusa* (spinach downy mildew) infection in asymptomatic leaves at 7 days prior to symptom appearance. Quantitative PCR with TaqMan methodology detected *P. effusa* in asymptomatic spinach leaves with greater level of sensitivity than the real-time PCR assay (Subbarao et al. 2015). The real-time PCR assay could detect *Peronospora belbahrii*, a causal agent of basil downy mildew, in all *Ocimum* spp. genotypes (Wyenandt et al. 2018). q-PCR was employed to assess the population levels of *Pseudoperonospora cubensis* (cucurbit downy mildew, CDM) and *P. humuli* (hop downy mildew) in spore trap collections. The qPCR assay could detect single sporangium of each species on spore trap samples collected in the fields, and it also allowed detection of *P. cubensis* and *P. humuli* in samples containing sporangia of both pathogens. *P. cubensis* sporangia were detected 5–10 days before CDM symptoms could be observed in cucumber fields (Bello et al. 2021).

qPCR was developed, based on dual-labeled hybridization probes, allowing simultaneous detection of apple and *Venturia inaequalis* DNA within a single reaction. The fungal biomass measured using qPCR assay significantly correlated with the levels of host resistance. Disease severity grades determined by visual scoring were correlated with results of the qPCR assay measured (Gusberti et al. 2012). A reliable Eva Green–based qPCR assay was developed for relative quantification of *Botrytis cinerea* in *Pelargonium* x *hortorum* and plant DNA in infected plant tissues. Primers designed based on RPB2 (DNA-dependent RNA polymerase subunit II) gene sequences directed amplification of a specific 149-bp product of *B. cinerea* without any cross-reaction with closely related fungal pathogens. The presence of *B. cinerea* could be detected in both symptomatic and asymptomatic plants (Moretti et al. 2015). Conventional and one-tube multiplex PCR formats were developed for detection and quantification of *Elsinoe fawcettii*, *E. australis*, and *Pseudocercospora angolensis* (citrus scab and spot diseases). Primers and hybridization probes were designed to target the single-copy protein-coding gene MS204 for *E. fawcettii* and *E. australis* and the translation elongation factor (Tef-1α) gene for *P. angolensis*. The assay proved to be cost-effective and capable of high-throughput detection of three targets directly in citrus samples (Ahmed et al. 2019).

qPCR assays were employed to detect and differentiate *Diaporthe helianthi* and *D. gulyae* (sunflower Phomopsis stem canker disease) from other *Diaporthe* spp. The assays detected the pathogens in plant samples from commercial fields with high efficiency. Further, the qPCR assay was effective in assessing the DSI of sunflower accessions, based on the amount of pathogen DNA. The qPCR assays identified PI664232 and PI561918 as less susceptible to *D. helianthi* and *D. gulyae,* respectively. The assays were useful for detection, differentiation of the pathogens, and to screen the sunflower accessions for their levels of susceptibility to Phomopsis stem canker disease (Elverson et al. 2020). Transmission of *Colletotrichum truncatum, Corynespora cassiicola,* and *Sclerotinia sclerotiorum* occurred through infected soybean seeds. A multiplex quantitative real-time PCR format was developed for simultaneous detection of all three pathogens in soybean seeds. Species-specific primers and probes for *C. truncatum* and *C. cassiicola* were designed based on *GAPDH* and *TEF1* genes, respectively, to be combined with a qPCR detection protocol. The multiplex qPCR efficiently detected these pathogens along with host internal control. The pathogens were detected in artificially inoculated and naturally infected soybean seeds (Ciampi-Guillardi et al. 2020).

2.2.3.4 Loop-Mediated Isothermal Amplification Assay

The loop-mediated isothermal amplification (LAMP) assay involves use of a strand-displacement DNA polymerase and four to six primers targeting a specific DNA region with predetermined secondary structures. The principles of LAMP reaction are described in an earlier publication (Narayanasamy 2017). The LAMP procedure may be performed inexpensively, using a simple water bath or heating block. It works at isothermal temperature for the synthesis of large amounts of Bst DNA polymerase, which has strand displacement 5′– 3′ polymerase activity, but lacks 5′ → 3′ exonuclease activity. Two sets of inner primers, (forward inner primer, FIP) and backward inner primer, BIP) and outer primer (F3 and B3) are employed to target a specific genome region. Isothermal nucleic acid amplification provides a significant advantage over PCR-based methods, because they can be applied in a single-step process at a constant temperature. Further, nonrequirement of thermal cycling allows for on-site diagnostics to be carried out using small portable instruments. Integration of LAMP with lateral flow devices (LFDs) are more efficient, less time-consuming, and have potential for in-field application. LFDs detect the labels incorporated into the amplification products. LAMP products can be detected by agarose gel electrophoresis in real time using intercalation of turbidity or color changes.

Detection of latent infection and quantification of urediniospores of *Puccinia graminis* f.sp. *striiformis (Pst)* (stripe rust) in wheat leaves could be accomplished by the LAMP assay. The marker could detect latent infection in wheat leaves at 48 h postinoculation. The LAMP protocol was efficient and sensitive and rapid in providing results without ambiguity (Aggarwal et al. 2017b). Several *Phytophthora* sp.–specific assays have been developed using different methods to detect LAMP products. Colorimetric LAMP assay for detection of *P. infestans* was developed using hydroxynaphthol blue in a closed tube reaction (Hansen et al. 2016b). Real-time LAMP and recombinase polymerase amplification (RPA) were developed targeting the ITS2 region of rDNA of *P. infestans*. Both LAMP and RPA assays showed specificity to *P. infestans*. LAMP assay effectively detected *P. infestans* in asymptomatic potato leaves at 24 h postinoculation. Comparative efficacies of LAMP, conventional PCR, real-time LAMP, and droplet digital PCR assays were assessed for detection of late blight disease in potato and tomato crops. The LAMP assay was specific for *P. infestans* on potato and tomato and did not amplify other *Phytophthora spp.* infecting potato and tomato (Amour et al. 2017). The LAMP assay was adapted using SYBR Green and a mobile reader (mReader) for use in the field. The adapted assay was more sensitive than visualization with the human eye (Ristaino et al. 2020).

Rapid diagnosis of soybean anthracnose disease caused by *Colletotrichum truncatum* was achieved by applying the LAMP assay. Species specific-primers derived from the sequences of *Rpb1* gene that encodes large subunit of RNA polymerase were employed for amplification of pathogen DNA. *C. truncatum* could be detected in field samples also, showing the usefulness of the assay for rapid diagnosis of soybean anthracnose disease (Tian et al. 2017). In another study, the LAMP assay was developed for the detection of *Colletotrichum gloeosporioides* (soybean anthracnose disease). A glutamine synthase gene sequence Gs-Cg-LAMP was employed for designing primers for use in the assay. The color of the amplicon changed from orange to yellow-green following addition of SYBR Green I to the reaction products. The LAMP assay detected *C. gloeosporioides* in field-infected soybean plants and soybean seeds obtained from a farmers' market. The assay was found to be simple, sensitive, and reliable for obtaining results rapidly (Wang et al. 2017b). Primer sets for conventional PCR and LAMP techniques were designed based on a specific genomic segment amplified with strawberry anthracnose pathogen *Colletotrichum nymphaeae* only.

Genus- and species-specific PCR primers discriminated the genus *Colletotrichum* and *C. nymphaeae* from other fungi in assays of diseased detached leaves. The LAMP assay yielded more precise results than the conventional PCR assay. Asymptomatic strawberry plants carrying the pathogen were identified by both PCR and LAMP assays (Karimi et al. 2020).

The LAMP assay was designed to rapidly detect airborne inoculum of grapevine powdery mildew pathogen, *Erysiphe necator* (Thiessen et al. 2016). The LAMP assay was based on a specific region of the *UvG-β1* gene sequences of the rice false smut pathogen *Ustilaginoidea viren*. The LAMP assay detected *U. virens* in spikelets and rice seed samples, indicating the effectiveness of primers based on *Urg-β1* gene sequences (Yang et al. 2018). LAMP-based methods were developed to detect *Monilina laxa* and *M. fructicola*. LAMP assays could be applied for monitoring brown rot pathogens in the field and in storage (Ortega et al. 2019). Cytochrome *b* DNA sequences of sugar beet rust pathogen *Uromyces betae* were used to construct primers for the LAMP assay. The assay could detect rapidly and specifically all strains of *U. betae* occurring in the UK and Denmark. The LAMP method was applied to quantify urediniospores in air-spore tape samples (Kaczmarek et al. 2019).

Primers for the LAMP assay were designed to target the PoT2 and MoT3 loci specific for *Magnaporthe oryze* and *M. oryzae* Triticum type, respectively. The primers showed high specificity as indicated by the negative reaction with *M. oryzae* isolates of diverse origin and temporal variation. Assays with PoT2 and MoT3 detected and differentiated *M. oryzae* Triticum in wheat seed and spikes. By using a field DNA extraction kit and portable Genie II system, *M. oryzae* and *M. oryzae* Triticum pathotypes were identified under field conditions. Mitochondrial NADH-dehydrogenase (*nad5*) gene, an internal control for plant DNA, was multiplexed with PoT2 and MoT3, and the results were comparable with those of individual assays (Yasuhara-Bell et al. 2018). The LAMP procedure was established for early detection of the asymptomatic infection of rice plants by *M. oryzae*. Primers were derived from AlbI, a hypothetical protein MGG_04322, a nuclear shuttling factor involved in ribosome and melanin biosynthesis. The quantitative LAMP (q-LAMP) detection system was applied to quantify conidia produced by *M. oryzae*. The q-LAMP assay detected and differentiated virulent (wild-type) strains of *M. oryzae* in comparison with that of two mutant strains (Li et al. 2019).

2.2.3.5 Combination of PCR-Based Assays with Other Techniques

PCR-based assays are combined with other techniques to overcome the problem of inhibition of PCR reaction by compounds coextracted with pathogen DNA from plants and other substrates or to improve the efficiency of PCR-based assays. *Didymella bryoniae* and the bacterial pathogen *Acidovorax avenae* pv. *citrulli (Aac)* (bacterial fruit blotch) affecting melon and watermelon were detected simultaneously by applying the MCH multiplex real-time PCR technique. Single-stranded DNA hybridization capture probes targeted DNA of *D. bryoniae* and *Aac* that were covalently attached to magnetic beads and employed for selective concentration of template DNA, which were subsequently amplified by multiple real-time PCR, using pathogen-specific TaqMan PCR assays. The MCH multiplex real-time PCR assay provided the advantage of simultaneous detection of *D. bryoniae* and *A. avenae* ssp. *citrulli* in naturally infected melon seed lots (Ha et al. 2009).

In order to detect and differentiate *Colletotrichum acutatum* and *C. gloeosporioides*, restriction fragment length polymorphism (RFLP) of a 1-kb introns of glutamine synthetase (*GS*) gene was evaluated, using isolates of diverse origin. Unique band patterns of the 1-kb GS introns were obtained for *C. acutatum* and *C. gloesporioides* based on *PstI* enzyme digestion of the amplified PCR products. Both species of *Colletotrichum* could be differentiated by multiplex PCR. CaInt1, CgInt, and ITS4 in one PCR distinguished *C. acutatum* and *C. gloeosporioides*. RFLP analysis of the introns of *Colletotrichum* spp. provided a sound basis for differentiation of the commonly occurring *Colletotrichum* spp. (Liu et al. 2012). Infection of flowers and fruits by *Monilinia laxa* and *M. fructicola* may be initiated in the orchards. The propidium monoazide (PMA)–quantitative PCR method was optimized, validated, and applied to quantify viable cells of *M. fructicola* in artificially inoculated and naturally infected samples. *M. fructicola* live cells were quantified specifically, and the PMA-qPCR method showed the potential for quantifying viable conidia of brown rot pathogen providing vital information for epidemiological investigation (Vilanova et al. 2017). Single-spore isolates of *Botrytis cinerea* were obtained from vineyards in different locations in China, and all were heterothallic. They were assigned to MAT1-1 type or MAT1-2. All isolates belonged to Group II of *Bc-hch* locus (Zhang et al. 2018).

2.2.3.6 Random Amplified Polymorphic DNA Technique

The random amplified polymorphic DNA (RAPD) technique uses arbitrary selection of sequences to distinguish races, strains and pathogenic or nonpathogenic isolates of a pathogen species. Following electrophoretic separation of the product, distinct banding patterns may be recognized in the gels, and some of these patterns may prove to be specific to certain species or varieties or strains. Efaw-1 and Efaw-2 primer sets identified *Elsinoe fawcettti (Ef)* and *E. ausatralis (Ea)* (citrus scab and sweet orange scab) pathotype isolates by using RAPD products unique to each species or pathotype present in lesions on leaves as well as in fruits (Hyun et al. 2007). RAPD analysis employing universal primers, Q17 and Q13, facilitated discrimination of isolates of *Corynespora cassiicola* virulent on sweet pepper (Shimomoto et al. 2011). Fusaria associated with mango malformation disease (MMD) were analyzed by RAPD and other methods to determine the genetic diversity of the pathogen isolates. Most of the isolates were identified as *F. tupiense*. Three exotic MMD pathogenic species were present in southern Spain (Crespo et al. 2016). The DNA SCAR marker was identified in isolates of *Peronosclerospora sorghi* (sorghum downy mildew) using PCR. The RAPD primer OPB15 consistently amplified a 1,000-bp product in PCR from DNA of *P. sorghi* isolates from corn, but not from sorghum isolates. The PCR product was cloned and sequenced. Specific primers were designed based on the sequence of the SCAR maker to identify maize isolates of *P. sorghi*. SCAR primers amplified a 800-bp fragment only from the isolates from corn. The RAPD procedure was useful for detection and identification of *P. sorghi* isolates from corn (Ladhalakshmi et al. 2009). Rapid and reliable detection of *Puccinia striiformis* f.sp. *tritici (Pst)* in latently infected wheat leaves was achieved by employing SYBR Green assay with SCAR markers. A 362-bp amplicon was obtained using the primer pair Bt2a/Bt2b from *Pst* and a 486-bp amplicon from *P. graminis* f.sp. *trici (Pgt)* and *P. graminis* f.sp. *triticina* (leaf rust pathogen). Both SCAR markers could be detected in wheat leaves at 9 h postinoculation (*hpi*). An SYBR Green RT-PCR assay was also applied to detect *P. striiformis* f.sp. *tritici* in symptomatic and asymptomatic leaf samples (Gao et al. 2016).

2.2.3.7 DNA Array Technique

DNA array technology is essentially a reverse blot technique useful for characterization of DNA fragments. Membrane-based assays have a matrix-specific DNA bound to a flexible membrane such as nylon, whereas the other class involves use of higher-density chips on a rigid support such as silicon or gels. Oligonucleotides are immobilized on a solid support and used for detecting the target pathogen. PCR-amplified target DNA is labeled using consensus primers spanning a genomic region harboring specific sequences of the test pathogen. The labeled PCR products are then hybridized to the array (Lievens and Thomma 2005). A chromogenic method for sensitive detection of *Stemphylium solani* and *Phytophthora capsici* comparable with luminescent method was employed, using digoxigenin (DIG)-labeled targets. The ITS region of the nuclear rDNA, including ITS1, 5.8S rDNA, and ITS2, was used as a target gene and PCR amplified as in the DNA microarray procedure. Different quantities of species-specific oligonucleotides on the array, amounts of DIG-labeled ITS amplicon, and hybridization temperatures were tested. Incubation of the hybridized array with anti-DIG-conjugated alkaline phosphate substrates, NBT/BCIP, produced visible target signals between 1 and 3 h, compared with 1 h in chemiluminescent detection (Wong and Smart 2012). Young vine decline (YVD), occurring in all grapevine-growing countries, is a complex disease induced by over 50 different fungi. A DNA microarray-based reverse-blot hybridization containing 102 oligonucleotides complementary to portions of the β-tubulin region was developed for detection of pathogenic fungi associated with YVD. The macroarray detected and identified 34 YVD species. A high incidence of YVD fungi in commercial nursery plants was detected, and this could not be recognized by conventional isolation methods (Úrbez-Torres et al. 2015).

2.3 VARIABILITY IN FUNGAL PATHOGENS

2.3.1 Assessment of Variability in Biological Characteristics

2.3.1.1 Use of Differential Plant Species/Cultivars

Differential interactions between host resistance gene (*R*) and pathogen virulence gene as an important characteristic of gene-for-gene relationships were proposed by Flor (1946, 1956). Several host-pathogen interactions, showing a gene-for-gene relationship such as apple-*Venturia inaequalis*, barley-*Erysiphe graminis* f.sp. *hordei*, coffee-*Hemileia vastatrix*, rice-*Magnaporthe oryzae* and maize-*Puccinia sorghi* are known. Almond anthracnose is induced by *C. acutatum* species complex but *Colletotrichum fiorinae* and *C. godetiae* were also associated with the disease. Based on morphological characteristics, isolates of *Colletotrichum* were grouped. Pathogenicity tests in almond, olive, and apple fruit revealed differences in virulence and some degree of pathogenic specialization among isolates. Olive isolates were identified as *C. godetiae* and *C. nymphaeae* (López-Moral et al. 2017). Pathogenicity and virulence assays of 40 isolates of *Corynespora cassiicola* from different hosts inoculated onto cotton, soybean, tomato, and cucumber showed that isolates from cotton were more aggressive to cotton than those from other hosts. Soybean and tomato were more susceptible to isolates from cotton, which were genetically distinct from isolates from other hosts (Sumabat et al. 2018).

Emergence of new variants of a wheat stem rust pathogen *Puccinia graminis* f.sp. *tritici (Pgt)* with changes in virulence has been observed periodically due to introduction of wheat cultivars incorporated with new resistance genes. In Kenya, six races of *Pgt* were identified including three new races not detected prior to 2013. The new races, TTHST, TTKTK, and TTKTT, belonged to the Ug99 race group (Newcomb et al. 2016). Six races of *Pgt* were identified in 48 isolates collected from German wheat fields, based on infection types induced in differentials, TKTTF, TKKTF, TKPTF, TKKTP, PKPTF, and MMMTF. None of these races belonged to the TTKS (Ug99) race group. German isolates of race TKTTF were phenotypically different from those that induced epidemics in Ethiopia. Races existing in Germany indicated the presence of virulence to resistance genes that are effective against the race TTKSK. The race TKKTP was of special concern because of its combined virulence to stem rust resistance genes *Sr24, SrTmp*, and *Sr1RS*^{Amigo} (Firpo et al. 2017). Wheat stripe rust pathogen *Puccinia striiformis* f.sp. *tritici (Pst)* produced new races capable of overcoming resistance in existing wheat cultivars. Race V26 with virulence to *Yr26* and many other stripe rust resistance genes caused epidemics in China. A highly diverse population might be produced from a single isolate by selfing on a barberry plant, and sexually produced population could be used to genetically characteristic virulence of stripe rust pathogen *P. striiformis* f.sp. *tritici* (Tian et al. 2016). Central European and 50 Australian isolates of *Blumeria graminis* f.sp. *hordei (Bgh)* were tested on 50 barley differential varieties with different specific resistance genes. All central European isolates differed from each other in their virulence combination and belonged to 50 various pathotypes. There was no similarity between the central European and Australian pathotypes (Dreiseitl 2014).

Isolates of *Magnaporthe oryzae* (122) from sub-Saharan African countries were evaluated for virulence diversity and genetic relatedness. Virulence spectrum was assessed by pathotype analysis with a panel of 43 rice genotypes. Cultivar 75–1-127, the Pi9R gene donor, was resistant to all pathogen isolates (100%) tested, followed by four African rice cultivars. By pyramiding Pi9, along with other promising *R* genes into desired rice cultivars that are adapted to East and West African countries, rice blast disease could be contained effectively (Mutiga et al. 2017). The major *R* gene-mediated resistance is effective in recognizing strains of *Magnaporthe oryzae* that contain corresponding avirulence (*AVR*) gene. Twelve *AVR* genes in *M. oryzae* were cloned. Distribution and variation of *AVR-Pii* gene in China were analyzed to determine the haplotype diversity of *AVR-Pii* under field conditions. Field isolates of *M. oryzae* were tested by using gene-specific PCR markers and contained *AVR-Pii*. Three *AVR-Pii* haplotypes were identified. The haplotype H1 was identical with the sequence in GenBank, while H2 and H3 were novel variants. All DNA sequence variations were found to occur in the protein-coding region, resulting in amino acid substitutions (Lu et al. 2019). Variability in the pathogenicity of isolates of *Colletotrichum acutatum* (pepper anthracnose disease) was assessed. Pathogenicity assays showed that the isolate Coll-524 was highly virulent and Coll-153 was moderately virulent to three commercial pepper cultivars. Coll-365 was highly virulent to tomato fruit and mango leaves on which it induced lesions with acervuli and conidia. The most virulent C-524 showed stronger laccase activity and was more resistant to capsaicin, compared to C-153 and Coll-36 (Liao et al. 2012). Variability in the pathogenicity of isolates of *Botrytis cinerea* collected from seven host plant species in various locations was assessed. Significant variability was observed in saprophytic growth rate, necrotrophic growth rate on tomato leaves and stems, and in infection on leaves of whole plants. Correlation between saprophytic growth rate and necrotrophic growth on leaves was significant. Based on normalized necrotrophic growth rate, existence virulence mechanisms independent of growth rate was inferred (Srivastava et al. 2018).

2.3.1.2 Variability in Production of Toxins by Fungal Pathogens

Fungal pathogens are known to produce metabolites toxic to host plants as well as to other organisms. Mycotoxins produced by *Fusarium graminearum* species complex are in involved in production of symptoms of FHB disease in wheat and also they are harmful to humans and animals. Chemotypes of isolates were determined by chemotype-specific PCR assay. *F. asiaticum* isolates may be of nivalenol (NIV) or 15-acetyldeoxynivalenol (15-ADON) type or 3-acetyldeoxynivalenol (3-ADON) type (Puri et al. 2012). Races of *Pyrenophora tritici-repentis (Ptr)*, causing wheat tan spot disease, showed variations in their virulence. The *ToxA* gene was identified in isolates classified as race 1 or race 2, but not in race 3, using PCR assay and Southern blot analysis. None of the isolates of race 1 and 2 had a *ToxB* gene. In contrast, race 3 isolates contained the *ToxB* homologue (Aboukhaddour et al. 2013).

2.3.2 Variability in Biochemical Characteristics

2.3.2.1 Isozyme Analysis

Isozyme analysis is useful for analysis of genetic variability in pathogen populations and as a disease diagnostic tool. Isozyme analysis by electrophoresis was employed to differentiate highly virulent and weakly virulent strains of *Leptosphaeria maculans* (canola blackleg or stem canker disease). Highly virulent isolates had a fast-moving isozyme (electrophoresis type, ET1), whereas weakly virulent isolates had a slow-moving isozyme (ET2) (Sippell and Hall 1995). The electrophoresis technique showed the presence of a maximum number phenotypes in the isozyme of nicotinamide adenine dinucleotide dehydrogenase (NADH) and diaphorase (DIA) of *Colletotrichum gloeosporioides* (anthracnose disease). The ETs of isolates clustered together on the basis of host origin (Kaufmann and Weidemann 1996). Variability in isozymes produced by isolates of sunflower downy mildew pathogen *Plasmopara halstedii* was studied using the cellulase acetate gel electrophoresis (CAE) method. Isolates of *P. halstedii* from sunflower, cocklebur, *Helianthus* x *laetiflorus*, representing ten different virulence phenotypes, were subjected to CAE analysis. Three enzymes, isocitrate dehydrogenase, malate dehydrogenase, and phosphoglucomutase, forming16 isozyme systems provided clear and reproducible banding patterns and revealed certain polymorphisms among isolates (Komjati et al. 2008). The biodiversity and phylogeny of isolates of *F. graminearum* of different geographical origins were analyzed using isozyme variation, ß-tubulin, and IGS sequences. Diversity among Asian isolates was greater than in European isolates. Correlations between genetic clusters of IGS sequences and geographical origin of isolates of *F. graminearum* were observed (Gagkaeva and Yli-Mattila 2004). Variability in protein profiles of the isolates of *F. avenaceum* was assessed. Specific signature protein bands representing each isolate were identified. Protein profiles of each group of isolates correlated with virulence (Vujanovic et al. 2009).

2.3.2.2 Cellular Fatty Acid Analysis

Presence and variations in fatty acids have been used as a basis for differentiation of *Phytophthora* spp. The potential of using whole cell fatty acid methyl ester (FAME) profiles for identification and differentiation of isolates of *P. infestans*, incitant of late blight diseases of potato and tomato, was assessed. FAME profiles could be applied for differentiation of subgroups within a species. When the culture conditions were kept constant, profiles of individual isolates of *P. infestans* were the same and reproducible over time. FAME profiling is a relatively rapid, cost-effective, and effective approach applicable for assessment of variability in biochemical characteristics of *P. infestans* (Larkin and Groves 2003).

2.3.3 Variability in Immunological Characteristics

Pathogenic and nonpathogenic isolates of *Venturia inaequalis* (apple scab disease) could be differentiated by the immunocytochemical technique. Clear differentiation could be observed by examining the appearance of the cell wall. Distribution of fimbrial epitopes labeled with specific antiserum and immunogold complex could be visualized on the cell walls (Svircev et al. 2000). Chicken egg yolk was used for generating antibodies (IgY) against *Colletotrichum falcatum* and *Fusarium subglutinans* infecting sugarcane. Strains of these pathogens were differentiated and also quantified by employing specific IGY (Vöhringer and Sander 2001). The antiserum against exoantigens (ExAgs), a soluble mixture of extracellular products of FHB pathogen *Gibberella zeae* (anamorph: *Fusarium graminearum*) was employed to detect and differentiate *G. zeae* isolates which showed distinct variation in host tissue colonization and presence of DON in infected tissues. A linear relationship was established between ExAgs of *G. zeae* and DON content in wheat plant tissues by applying an indirect ELISA format (Cumagun et al. 2004).

2.3.4 Variability in Genetic Characteristics

Genetic characteristics of fungal pathogens are governed by specific genes or gene clusters. But their expression during pathogen development outside or with plant hosts may be influenced by environmental conditions. The role of enzymes and toxins in the pathogenesis has been investigated in different pathosystems.

2.3.4.1 Restriction Fragment Length Polymorphism Technique

The restriction fragment length polymorphism (RFLP) technique has been applied to assess the extent of natural variations in the genome of different pathogenic species, biotypes, strains, or races affecting various crops. Based on the variations in the virulence, *Pseudocercosporella herpotrichoides* populations were differentiated into R, N, C, or S pathotypes. The DNA fragment of a 6.7-kb pathotype-specific probe for R-type isolates hybridized with R-type isolates, but not with other pathotypes of *P. herpotrichoides* (Nicholson et al. 1994). The dispersed repeated sequence (called MGR) was employed as the probe to construct genotype-specific *Eco*RI RFLP profiles (MGR-DNA fragments) to differentiate field isolates of rice blast pathogen *Magnaporthe grisea* (Levy et al. 1999). RFLP analysis was useful to establish a correlation between molecular and biological characteristics of biotypes of *Stagonospora nodorum* from barley and wheat (Ueng et al. 1995). Variations in strains of *C. fragariae, C. acutatum*, and *C. gloeosporioides* infecting strawberry plants were assessed based on rDNA and DNA restriction patterns. Strains of *C. acutatum* and *C. fragariae* were placed into four and two groups, respectively. In contrast, all strains of *C. gloeosporioides* generated identical patterns with each of the four restriction enzymes used (Mills et al. 1998). *C. gloeosporioides* isolates infecting water yam (*Dioscorea alata*) were grouped into four forms as slow-growing gray (SGG), fast-growing salmon (FGS), fast-growing olive (FGO), and fast-growing gray (FGG), based on cultural characteristics. FGG isolates generated unique ITS-RFLP banding patterns, whereas the isolates of other three forms produced RFLP patterns identical to those of *C. gloeosporioides*. Restriction patterns generated following digestion with endonuclease enzymes *Alu*I, *Hha*I, and *Hae*III could be used as the basis for rapid differentiation of FGG isolates from other forms of *Colletotrichum* from yam (Abang et al. 2002).

2.3.4.2 Polymerase Chain Reaction–Based Techniques

PCR-based techniques are useful for detection and quantification and differentiation of strains, races, or biotypes of fungal pathogens that differ in their virulence and production of toxins and enzymes involved in pathogenesis. Isolates of *Fusarium culmorum*, causing FBH disease of cereals, vary in their ability to produce mycotoxins. The PCR assay, using primer pairs specific to *Tri7* and *Tri13* genes, was applied to determine molecular variability among selected isolates (Tóth et al. 2004). Primers targeting nuclear rDNA repeat sequences were employed for detection and differentiation of *Phytophthora* spp. RFLP and sequence analysis of PCR-amplified rDNA were effective in identifying isolates of *Phytopththora infestans* infecting tomatoes. The pathogen isolates were characterized based on the sequences of PCR-amplified nuclear rDNA (Camele et al. 2005). Citrus black spot (CBS) disease is induced by *Guignardia citricarpa*. Nonpathogenic *G. mangiferae* also could colonize citrus fruit, and it was often misidentified as *G. citricarpa*, creating obstacles for the export of citrus fruits from countries where CBS was prevalent. A primer set designed based on ITS region sequences proved to be highly specific in detecting *G. citricarpa* in different types of black spot lesions (Peres et al. 2007).

Pyrenophora teres occurs in two forms as net form (NF) net blotch induced by *P. teres* f. *teres (Ptt)* and spot form (SF) caused by *P. teres f. maculata (Ptm)* in barley. MAT-specific single nucleotide polymorphism (SNP) primers were designed for PCR. The NF and SF forms were differentiated by distinct PCR products. Both *MAT1-1* and *MAT1-2* SNP primers showed their specificity when used in duplex PCR format (Lu et al. 2010). *Bipolaris maydis* (southern corn leaf blight, SCLB) isolates were assessed for mating type distribution, genetic diversity, and pathogenicity to local sweet corn cultivars. Intersimple sequence repeat (ISSR) markers (13) within and among two mating types MAT-1 and MAT-2 revealed a high level of DNA polymorphism for all combined isolates. The MAT-1–2 population showed more diversity, based on DNA polymorphism, than the MAT-1 population. The pathogenicity assay showed that both MAT-1 and MAT-2 populations were pathogenic to all local sweet corn cultivars tested (Dai et al. 2020).

Banana freckle disease was induced by *Phyllosticta musarum, P. maculata*, and *P. cavendish*. High-resolution melting (HRM) analysis was applied to detect and differentiate the pathogens involved in the disease. HRM analysis assay was rapid in identifying and differentiating three *Phyllosticta* spp., using a single primer pair in a one-step closed-tube system without labeled fluorescent probes (Wong et al. 2013). The genetic structure of *Mycosphaerella musicola* (banana yellow Sigatoka leaf spot) was investigated. The frequency of MAT1-1 or MAT1-2-1 idiomorphs and polymorphism at nine microsatellite loci was determined. Eight genetic groups were detected in *M. musicola*. Both sexual reproduction and gene flow affected genetic variability (Gomes et al. 2018). Genetic diversity of populations of *Pseudocercosporella fijiensis* infecting banana from two regions of Mexico was assessed by employing MAT-specific PCR assay and ten simple sequence repeat (SSR) loci (Manzo-Sanchez et al. 2019).

Phytophthora infestans occurs in different genotypes with varying virulence all over the world. Several regions of pathogen DNA containing single nucleotide polymorphisms (SNPs) were identified within both nuclear genes and flanking sequences of microsatellite loci, following mining of *P. infestans* genome. Allele-specific oligonucleotide (ASO)-PCR assays were developed for detection of SNPs (Gagnon et al. 2016). Mating type is an important characteristic of strains affecting population structure of the pathogens. *Phytophthora infestans* reproduced asexually and sexually between strains of two mating types A1 and A2. PCR markers W16, S1, and PHYB for *P. infestans* mating type determination were identified. Molecular markers as tools for determination of *P. infestans* mating type could be applied (Brylinska et al. 2018). Real-time PCR assay was applied to assess the species composition of *Alternaria* spp. causing early blight/brown spot (EB/BS) in potato. Small spored virulent *A. alternata* and *A. arborescens* were detected predominantly than *A. solani*. As the disease progressed, the proportion of *A. solani* also increased (Vandecasteele et al. 2018). *Diaporthe* spp. causing Phomopsis cane and leaf spot disease in grapevine were identified based on morphological characters and sequences of ITS1 and 2 for PCR assays. *D. ampelina, D. ambigua*, and *D. foeniculina* were identified by PCR assays (Lesuthu et al. 2019). Trichothecene profiles and genetic structure of *Gibberella zeae* isolates were studied. Trichothecene biosynthesis gene (*TRI*)–based PCR assays and ten variable tandem repeat (VNTR) markers were used to determine the genetic diversity and compare trichothecene profiles of an older population of *G. zeae*. Low genetic differentiation and high gene flow was observed between 3-ADON and 15-ADON populations, indicating minor influence of chemotype of *G. zeae* population subdivision (Burlakotti et al. 2011). Quantitative PCR assays based on ITS rDNA locus were developed for rapid detection and differentiation of *Phakopsora pachyrhizi* strains infecting soybean. A large amount of intraspecific and intragenomic variations in ITS was observed, including the presence of polymorphic ITS copies within single leaf samples and within single rust sorus (Rush et al. 2019).

2.3.4.3 Random Amplified Polymorphic DNA Technique

Differentiation of races and strains has been effectively achieved by applying the random amplified polymorphic DNA (RAPD) technique. Further, pathogenic and nonpathogenic isolates of fungal species can also be discriminated using the RAPD approach. A specific fragment of the mitochondrial (mt) cytochrome *b* (*cytb*) gene from rust pathogens infecting crops was sequenced. On the basis of *cytb* and ITS sequences, *Puccinia recondita* f.sp. *tritici, P. graminis* f.sp. *striiformis*, and other rusts infecting cereals were considered to be related (Grasso et al. 2006). Cluster analysis of genetic distances estimated from 12 polymorphic molecular markers (universally primed-PCR and RAPD) revealed that two groups of isolates of *A. solani* and *A. tomatophila* were the causal agents of early blight disease. The majority of *A. solani* isolates contained mating type locus 1 (MAT1) idiomorph MAT1-1, whereas a majority of isolates of *A. tomatophila* had the MAT1-2 idiomorph (Gannibal et al. 2014). *Bipolaris oryzae* strains (36) from infected rice seed collected from different locations in India were subjected to genetic diversity assessments, using molecular marker systems viz., RAPD, universal rice primers (URPs) and ISSR. The results showed differences in strains of different geographical origins, and strains clustered based on these differences. Combined application of three marker systems was recommended for obtaining more sensitive and reliable results for the assessment of genetic diversity of strains of *B. oryzae* (Kandan et al. 2015).

2.3.4.4 Amplified Fragment Length Polymorphism Technique

The AFLP technique is useful to overcome certain limitations of isozyme analysis with RFLP and RAPD methods. The AFLP procedure involves selective amplification of restriction fragments from a digest of total genomic DNA of test organism using PCR assay. The DNA fragments are identified on polyacrylamide gel and designated AFLP markers. High levels of sensitivity and reproducibility of results are the hallmarks of AFLP technique. Molecular fingerprints were prepared to assess the possibility of development of isolates of *Phytophthora infestans* virulent to potato cultivars. The AFLP technique detected 89 bands within the collection of *P. infestans* French isolates. Five

AFLP genotypes were identified based on the genomic fingerprinting patterns of *P. infestans* isolates tested against all recognized genes except *R5* tested. No relationship between AFLP genotypes and races could be established (Pilet et al. 2005). Isolates of *P. infestans* from four solanaceous species, black nightshade, hairy nightshade, petunia, and potato, growing within and around fields of late blight-affected potato in different locations were characterized. Close similarity could be observed between the isolates of petunia, hairy nightshade and black nightshade isolates, and potato isolates. The results indicated the possibility of weeds and crops as potential sources of *P. infestans* inoculum for potato crops (Deahl et al. 2006). Genetic diversity of *Fusarium graminearum* and *F. asiaticum* (wheat FHB or scab) was assessed by subjecting isolates collected from wheat spikes to sequence characterized amplified regions (SCAR) analysis which discriminated both pathogens. The AFLP profiles clearly separated two groups, A and B, that were entirely congruent with both species. Isolates from the same location belonged to different AFLP haplotypes, indicating the presence of a mixture of genetically different isolates. The results indicated that SCAR and AFLP analyses were corroborative of each other (Qu et al. 2008). Genetic diversity of *Gibberella fujikuroi* and *Fusarium sterilihyphosum* (mango malformation disease, MMD) was assessed by employing AFLP and vegetative compatibility analyses. Six AFLP and VCGs could be differentiated, and the isolates formed a new *Fusarium* lineage associated with MMD (Lima et al. 2009).

Pyrenophora teres f. *teres* (PTT-net form) and *P. teres* f. *maculata* (PTM-spot form) infecting barley could be more reliably differentiated using the AFLP technique. Isolates of *P. teres* were grouped into two divergent groups corresponding to the net and spot forms by determining AFLP patterns formed by the pathogen isolates. Two distinct groups representing PTT and PTM forms were identified using diagnostic markers 85- and 134-bp specific for PTT and PTM forms, respectively (Leišova et al. 2005). *Pyriculalria oryzae* isolates (rice blast disease) were analyzed for their genetic diversity, based on 160 polymorphic markers. Isolates of *P. oryzae* were grouped into 12 AFLP genetic groups. Isolates of *P. oryzae* from *japonica* and *indica* cultivars were placed in two different groups, based on DNA sequence similarity and cluster analysis (Thuan et al. 2006). Variability in pathogenic potential of 96 isolates of *Cochlibolus sativus* from the Hill and Palain areas of Nepal and North Dakota in the United States was assessed by inoculating on 12 differential wheat lines. Eight selected AFLP primer combinations were employed to determine DNA polymorphism in these isolates. The isolates were classified into three distinct groups with low, intermediate, and high virulence. The results showed that pathogenic and molecular diversity were high in populations of *C. sativus* from wheat foliar infections causing spot blotch disease of wheat (Gurung et al. 2013). Isolates of *Puccinia striiformis* f.sp. *tritici* from key areas for oversummering were analyzed by the AFLP technique. Forty AFLP genotypes were shared by populations of the two provinces of China. Gene flow between these two regions in autumn and spring subpopulations was also observed (Liang et al. 2013).

Variability of *Venturia inaequalis* (apple scab disease) between and within orchards was investigated, using 212 isolates from two mixed orchards in the UK. The isolates were genotyped with AFLP and simple sequence repeat (SSR) markers. Isolates did not differ significantly in terms of allele frequencies at the screened AFLP and SSR loci. However, groups of isolates from individual cultivars differed significantly within each orchard. Further, significant differences were recorded between groups of isolates from individual trees of the same cultivar in the same orchard. The tree-to-tree pathogen variability indicated possible role for conidia as source of primary inoculum (Xu et al. 2013). The AFLP technique was applied to assess the variability in 112 monopustule isolates of *Hemileia vastatrix* (coffee rust disease) from different locations and coffee genotypes. The pathogen population exhibited a low level of genotype diversity. Evidence for recombination was obtained for two subpopulations from two different locations (Cabral et al. 2016).

2.3.4.5 DNA Sequence Analysis

Combinations of morphological and molecular characteristics such as conidial morphology and ITS rDNA sequence data are used for identification of *Magnaporthe* (*Pyricularia*) spp. The ITS region containing the ITS1 region, 5.8S gene, and ITS2 region were amplified from *Pyricularia* spp. and related species. The results suggested that *Pyricularia* spp. isolated from different hosts were genetically distinct (Bussaban et al. 2005). SSRs or microsatellites are multiallelic, codominant, and highly polymorphic, and they have been used as molecular markers to investigate the genetic structure and reproductive biology of fungal pathogens. This method is based on the anchored PCR format in which microsatellite amplification is carried out with one primer complementary to the flanking sequence and one that is specific to the repeat motif. Markers amplified by anchored PCR are called sequence-tagged microsatellites (STMs). They are generally developed using selectively amplified microsatellite (SAM) analysis and sequence tagged microsatellite profiling (Hayden et al. 2006). Genetic differentiation of isolates of *Puccinia triticina*, (wheat leaf rust) was taken up. Virulence of *P. triticina* isolates was assessed using 20 wheat lines with single genes for leaf rust resistance, and 23 SSR markers were employed for molecular genotyping. After elimination of isolates with identical virulence and SSR genotypes, five different groups of isolates based on SSR genotypes, differing in virulence to leaf rust genes, were identified. All pairs of groups, except for one pair, were significantly differentiated for SSR genotypes. All but two pairs of groups were significantly differentiated for virulence phenotype. The high degree of similarity for SSR genotype of isolates from South America and North America suggested a common European origin of *P. triticina* that was introduced into both continents (Ordoñez et al. 2010).

Genetic differentiation of isolates of *P. triticina* occurring in Europe was taken up to determine their virulence on 20 wheat lines with single leaf rust resistance genes and for molecular genotypes with 23 SSR markers. Correlation between the SSR genotypes and virulence phenotypes was significant. The European population of *P. triticina* had higher levels of genetic differentiation compared to other continental populations (Kolmer et al. 2013). Twenty single uredinial isolates of *P. triticina* isolates were virulent to *Lr26*, whereas none of the isolates was virulent to *Lr18* or *Lr24*. Isolates with identical virulence phenotypes and SSR genotypes were detected. The SSR genotypes of *P. triticina* showed high levels of linkage disequilibrium and heterozygosity, whereas the SSR genotypes exhibited significant correlation with virulence phenotypes indicating clonal reproduction (Kolmer et al. 2015). In another investigation, single uredinial isolates of *P. triticina* from host plants with the same ploidy were genetically similar. Collections from hexaploid species were highly similar to those from common wheat (Gultyaeva et al. 2018). The relationship between virulence and molecular patterns using SSR markers on selected isolates of *Puccinia graminis* f.sp. *tritici (Pgt)* collected during a 52-year period was examined. SSR variability among the isolates within a putative clonal linkage based on pathotype 21-10 provided evidence of clonality over the 52-year period, coupled with single-step acquisition of virulence for resistance genes (Zhang et al. 2017). An Australian *P. graminis* f.sp. *avenae (Pga)* (oat stem rust) isolate was sequenced and the abundance of SSR was determined. The genome-derived SSR markers were effective to determine diversity in *Pga* populations occurring in different countries (Gnocato et al. 2018). Genetic variability of *Mycosphaerella graminicola* (wheat Septoria tritici blotch) isolates was assessed using SSR markers. Single-spore isolates representing 11 populations from spring and winter crops were analyzed for mating type frequency and for genetic variability at 17 microsatellite or SSR loci. Equal distribution of both mating types in the populations from Kansas, Indiana, and North Dakota was indicated. High levels of genetic diversity were recorded within the 11 populations. Sexual recombination occurred frequently in the *M. graminicola* populations, and most populations were genetically differentiated over the major spring and winter wheat-growing regions of the United States (Gurung et al. 2011). SSR marker analysis was applied to assess genetic structure and diversity of isolates of *Pyrenophora teres* f. *teres (Ptt)* and *P. teres f. maculata (Ptm)* (net and spot forms of barley net blotch disease). Thirteen SSR loci were polymorphic within *Ptt* and *Ptm* populations. Isolates clustered in two groups conforming to *Ptt* and *Ptm* forms. (Akhavan et al. 2016).

In order to understand genetic diversity and population stratification at the intraspecific level, isolates of *Colletotrichum gloeosporioides (Cg)* collected from papaya were examined using the inter SSR-PCR technique. Three genetic subpopulations, group 1, 2, and 3, were differentiated in the isolates present in southern and northern parts of Trinidad (Rampersad 2013). Specific primers ITS40Cg effectively identified all the 54 isolates of *Colletotrichum gloeosporioides (Cg)* species complex (citrus wither-tip of twigs). Genotypic analysis using 10 SSR markers showed a moderate diversity level in *Cg* populations and indicated that the pathogen reproduced mainly clonally (Bahri et al. 2019). SSR markers were used to assess the temporal change of *Venturia inaequalis* populations inducing apple scab disease (Passey et al. 2016). SSR analysis of *Phyllosticta citricarpa* (citrus black spot disease) showed the presence of a single multilocus genotype (MLG) in the Floridian population and 11 MLG within the Australian population (Wang et al. 2016). Genetic homogeneity of *Phyllosticta ampelicida* (black rot disease) infecting *Vitis* spp. was investigated. Analysis of nuclear microsatellites applied for strains of *P. ampelicida* confirmed the existence of a high degree of genetic variability in the pathogen population (Rinaldi et al. 2017).

Isolates of *Phytophthora infestans* from British gardens were analyzed using 12 multiple SSR markers to monitor disease incidence in commercial potato crops. Seven new clonal lineages were detected in garden populations, and higher frequencies of unique clonal lineage were present compared with populations from British commercial potato crops. British garden populations on tomato and potato were much more diverse than those on commercial potato crops (Stroud et al. 2016). Genetic diversity of *P. infestans* isolates collected from infected potato and tomato plants was assessed based on using markers such as mating type, *Gpi* allozyme, mtDNA haplotype, sensitivity to metalaxyl-M, and tuber pathogenicity microsatellites. All isolates belonged to the A2 mating type and Ia mtDNA haplotype. Tuber pathogenicity, characterized as severity of tuber late blight, was variable among isolates. Microsatellites (or SSR) showed a shift in the population, characterized by two clusters differentiated over time. The results suggested that US-8 genotype might have been displaced by US-22 from 2008 to 2010 in Michigan (Rojas and Kirk 2016). *Pseudoperonospora cubensis* causes CDM disease in cucurbitaceous plants. Genetic diversity of isolates of the pathogen was high, as indicated by SSR marker analysis. Populations of *P. cubensis* present in Michigan and Ontario were significantly different, and a unique cluster was identified in Michigan population of *P. cubensis* (Naegele et al. 2016). The isolates of grapevine powdery mildew pathogen *Erysiphe necator* collected from vineyards in New York (NY) and Winchester (VA) were genotyped for mating type and by using 11 microsatellite markers. The spatial aggregation and clonal genotype that arise during the asexual phase of the epidemic might contribute to the persistent linkage disequilibrium even though populations undergo sexual reproduction annually (Brewer et al. 2012).

An extremely low level of allelic diversity was discernible in a geographically diverse set of *Puccinia horiana* (chrysanthemum white rust) isolates using complexity reduction of polymorphic sequences (CROPS) technology. Thirty SSRs were translated into molecular markers and employed to genotype 45 isolates originating from North and South America, Asia, and Europe. In most cases, phylogenetic clustering was related to geographic origin, indicating local establishment (De Baker et al. 2013). Genetic diversity of identified and unidentified races of *Hemileia vastatrix* (coffee leaf rust disease) were analyzed based on sequence-related amplified polymorphism (SRAP) markers (Kosaraju et al. 2017). The population genetic structure of *Leptoshphaeria maculans* was investigated by employing ten minisatellite markers. Bayesian clustering analyses of multiloci inferred two distinct clusters in all subpopulations examined. A relatively high degree of gene exchange occurred between different *L. maculans* isolates (Zou et al. 2018). Genetic diversity of isolates of *Moniliophthora roreri* inducing cocoa frosty pod rot (FPR) disease from different locations was analyzed using SSR markers. High levels of diversity might be attributed to high mutation rates in the SSR and gene flow. Cocoa genotypes and environmental variables did not contribute to the genetic differentiation of the groups (Jaimes et al. 2016). SSR markers were used to genotype 10–20 single spores from each of four blossoms/canker pairs per orchard. Individual blossoms and cankers showed the presence of four to five genotypes. A single blossom might be infected by one or more genotypes of *M. fructicola*, and the diversity was retained in the corresponding canker (Dowling et al. 2019).

2.3.4.6 Phylogenetic Analysis

Morphologically based identification of fungal plant pathogens has been often unreliable. Multigene phylogenetics are employed to systematically characterize *Colletotrichum* species relationships and to serve as a basis for species diagnosis. Isolates of *Colletotrichum* species collected from leaves of small cardamom (*Elettaria cardamomum*) were differentiated by multilocus phylogenetic analysis based on the sequences of *ITS, ACT, GAPDH, TUB22, CHS-1, CYLH3, GS*, and *ApMAT* gene regions. Six species including *C. gloeosporioides* were identified as the incitants of anthracnose disease of small cardamom. All the six species were found to be pathogenic (Chethana et al. 2016). Isolates of *Colletotrichum* collected from pepper (chili) causing anthracnose disease were analyzed. Based on partial sequences of six gene regions of *ITS, GAPDH, ACT, CHS-1, TUB2*, and *HIS3* for *C. acutatum* complex and four gene regions of *ITS, TUB2, ApMAT*, and *GS* for *C. gloeosporioides*, phylogenetic analysis delineated four *Colletotrichum* species. Three species *C. queenslandicum, C. simmondsii*, and *C. siamense* were also found to be associated with pepper anthracnose (De Silva et al. 2017). Genetic structure and pathogenic relatedness of *Pseudoperonospora cubensis* infecting cucurbits and *P. humuli* infecting hop were investigated. Phylogenetic analyses of two nuclear and three mitochondrial loci of isolates of *P. cubensis* and *P. humuli* were compared. A consistent separation of the majority of *P. humuli* isolates and *P. cubensis* isolates in nuclear and mitochondrial and ITS phylogenetic analyses was observed. PCR analysis of inoculum obtained from hop infected with *P. cubensis* confirmed that the pathogen contained SNPs in the *cox2* region, which corresponded with all except two *P. cubensis* isolates (CDM-24 and CDM-248). Hop plants had more hypersensitive reaction (HR) lesions. Cucumber plants exhibited typical disease symptoms in plants inoculated with *P. cubensis*. However, sporulation, while abundant, was not as profuse as in many of the positive control cucumber plants in the host-specificity experiments. The results showed that biological characteristics could differentiate *P. cubensis* and *P. humuli* (Mitchell et al. 2011).

The *Fusarium graminearum* species complex (FGSC) causes wheat FHB and maize stalk rot (MSR) diseases. Phylogenetic analyses and PCR trichothecene chemotype assay were applied to establish species identity and chemotype of members of the FGSC from wheat spikes and maize stalks. *F. graminearum* sensu stricto was the major component of the FGSC isolates associated with FHB and MSR diseases. The 15-ADON trichothecene chemotype was predominant among the isolates from wheat and maize (Hao et al. 2017). The assay based on SNP-based markers was specific and sensitive in detecting and differentiating *Colletotrichum gossypii* infecting upland cotton from other related fungal pathogens (Nawaz et al. 2018). The *ITS, GAPDH, CHS-1*, and *TUB2* genes of *Colletotrichum acutatum* and *C. gloeosporioides* species complex were amplified and sequenced. *C. acutatum* species complex contained 114 isolates, whereas 93 isolates were assigned to *C. gloeospoioides* species complex. Phylogenetic analysis of *Colletotrichum* species causing apple Glomerella leaf spot disease revealed the involvement of several pathogen species in the disease (Moreira et al. 2019). Phylogenetic analysis of sequences of multiple genes of *Colletotrichum* spp. causing anthracnose in strawberry, was performed. Sequences of actin, internal transcribed spacer, calmodulin, glyceraldehydes-3-phosphate dehydrogenase, and chitin synthase were analyzed. All *Colletotrichum* spp. belonged to *C. gloeosporioides* (*Cg*) complex. *C. fructicola* was the most abundant species causing anthracnose in China (Chen et al. 2020).

2.3.4.7 Single Nucleotide Polymorphism–Based Assay

It is possible to sequence and align multiple genomes to identify informative SNPs. The SNP-based assays provide additional information for other molecular assays such as SSR analysis that could be used for typing of lineages of *Phytophthora infestans*. However, SSR analysis offers an advantage over SNP-based assays of being able to identify novel alleles, whereas SNP-based assay can be applied only to known alleles. HRM analysis is a technique capable of differentiating genotypes based on the presence of SNPs. Isolates of *Phytophthora infestans* from potato and tomato were differentiated into US lineages. The HRM analysis and locked nucleic acids (LNAs) technique were evaluated for differentiating the US-23 lineage from other US lineages of *P. infestans*. Whole *P. infestans* genomes were aligned and SNPs were identified as targets. Informative

SNPs were employed in HRM assays and LNA probes. Two specific SNPs were present at the expected locations for all isolates of US-3 lineage. Two differentially labeled primers targeting one to three SNPs were able to differentiate US-23 lineage from US-8, US-11, and US-24 using HRM analysis. Another HRM primer pair targeted a highly variable genomic region containing nine polymorphism within 63-bp. These primers differentiated US-23, US-11 and US-8 plus US-24 into three groups following HRM analysis (Hansen et al. 2016a). An HRM assay was developed for simultaneous identification and differentiation of *Colletotrichum* spp., causing strawberry anthracnose disease. HRM marker was designed based on ITS region of DNAs of *C. acutatum* and *C. gloeosporioides* from strawberry. The procedure detected a SNP in the cytochrome *b* (*cytb*) gene of *C. acutatum* and *C. gloeosporioides* associated with resistance to QoI fungicides. An end-point SNP genotyping assays efficiently detected both pathogens (Forcelini et al. 2018).

2.3.4.8 Next-Generation Sequencing Technique

The next-generation sequencing (NGS) technique provides the advantage of sequencing genomes of microbial pathogens at high speed and at low cost over the existing procedures. The NGS methods are particularly useful especially in the case of obligate pathogens causing downy mildews, powdery mildews, and rust diseases in several crops cultivated in different agroecosystems. *Pseudoperonospora cubensis* has an unusually broad host range covering cucurbits such as cucumber, cantaloupe, watermelon, squash, and pumpkin. *P. cubensis* is closely related to *P. humuli* (hop downy mildew). The RNA sequence data permitted identification of diagnostic candidates with differential expression between *P. cubensis* and *P. humuli*, but conserved within each species and robust across isolates of these indistinguishable pathogens. The presence or absence of an exon or a gene or a complete gene in *P. humuli* could be confirmed by NGS (Withers et al. 2016). The NGS procedure was applied for large-scale pathogen diagnosis in soybean during the survey undertaken in Manitoba, Canada. *Peronospora manshurica* (soybean downy mildew) was efficiently detected. Other pathogens of soybean viz., *Pseudomonas savastanoi* pv. *glycinea* and *Septoria glycines* were also detected by the NGS method (Díaz-Cruz et al. 2019).

2.3.4.9 Heteroduplex Mobility Analysis

Heteroduplex mobility (HMA) is based on the delay in the rate of migration of a DNA heteroduplex in comparison with a DNA homoduplex to identify mismatches or deletions in DNA sequences. The efficacy of HMA of internal transcribed spacer (ITS) regions of *Colletotrichum* spp. DNA, as a biochemical tool, was investigated. The ITS regions of *C. acutatum, C. musae, C. graminicola, C. capsici, C. dematium, C. lindemuthianum*, and three unidentified *Colletotrichum* spp. were amplified by PCR. Analysis of ITS sequences of *Colletotrichum* isolates revealed greater DNA divergence within ITS1 region than that existing within ITS2. Phylogenetic analysis of ITS regions of *Colletotrichum* isolates divided them into five distinctive species groups. The results showed that HMA of ITS regions might be a relatively rapid and convenient procedure to identify *Colletotrichum* spp. (Huang et al. 2010).

2.3.4.10 Nanodiagnostic Techniques

Nanoscale materials with remarkable biospecificity of engineered molecular recognition are promising candidates of detection of microbial plant pathogens and diagnosis of diseases induced by them. Nanodiagnostic tools useful for agricultural applications may be placed into two categories: (i) nanostructured devices and tools like microneedle patch, nanopore, and wearable devices and (ii) nanoparticle-based nanosensors such as gold nanoparticles, quantum dot, and array-based nanosensors. Direct methods of detection involve analysis of microbial pathogens or biomolecular markers (nucleic acids, proteins, and carbohydrates) isolated from infected plant tissues. Indirect methods recognize the pathogen presence through changes in physiological or histological indices. Integration of nanotechnology with standard biomolecular analyses may improve sensitivity, portability, and versatility of traditional diagnostics.

A polymer-based microneedle patch in combination with real-time PCR (qPCR) was effective for rapid and sensitive detection of *Phytophthora infestans* (late blight disease of potato and tomato) (Paul et al. 2019; Paul et al. 2020). A prototype microneedle made of polyvinyl alcohol (PVA) was developed that was sufficiently strong to break the cell walls of plant tissues. The isolated DNA could be used for amplification. Integration of the microneedle patch technique with miniaturized DNA amplification assay was useful for detection of fungal pathogens (Koo et al. 2013). Inorganic nanomaterials such as Ag, Au, Si, and several metal oxide nanoparticles are employed commonly as nanosensors that use recognition ligands (antibodies or DNA oligonucleotides) and signal transducers to detect and quantify molecular targets of interest. Ag nanorods have been employed as sensor substrates for identification of plant pathogens or toxins produced by them using surface-enhanced Raman spectroscopy (SERS) (Yüksel et al. 2015). Fluorescent silica nanoparticles doped with a Ru (II) complex and conjugated with a secondary antibody was effective for detection of *X. campestris*, causing nightshade bacterial leaf spot disease. Quantum dots (QDs) are semiconductor nanocrystals with high potential as nanosensors. QDs are efficient as biosensors for plant imaging and identification of plant diseases. Profiling the volatile chemical signature of diseased plants is an important indirect method of detection of plant pathogens. Infection by microbial pathogens may result in release or change in the composition of characteristic volatile organic compounds (VOCs) that are indicative of biotic or abiotic stresses of host plants. Electronic noses made of metal oxidases or conductive polymer coatings have been used as multiplexed gas sensors for tracing VOC biomarkers. Detection of ethylene, a phytohormone, associated with infection of plant tissues by pathogens was effectively performed (Fang et al. 2014). Vapor analyzers embedded with gas chromatography (GC), tandem mass spectroscopy (MS), or ion mobility spectroscopy IMS) are effective mobile tools for on-site analysis of plant health status that could assess the abundance of each emitted VOC species (Sharma et al. 2019).

Early detection and precise identification of plant pathogens are required as a basic step for development of effective disease management systems. The development of advanced DNA sequencing technologies enabled assessment of total NA contents in biological materials. The single-molecule sequencing platform of Oxford Nanopore used as a general method for diagnosis of crop diseases was evaluated. Sequencing DNA or RNA isolated from tissues with symptoms from plants inoculated with microbial plant pathogens was performed. In addition, samples containing 1 of 200 seeds infected by two or three pathogens as well as samples with symptoms induced by unidentified pathogen were also included for testing. Sequencing results were analyzed with nanopore data analysis tools. Pathogens were detected and identified in all inoculated plants in real time within 1–2 h of running the nanopore sequences, and they were identified up to the genus or species levels. DNA sequencing or direct RNA sequencing of samples with unidentified pathogens were validated by conventional PCR, ELISA, and Koch's postulates tests, the results of which corroborated those of nanopore sequencing. Rapidity, portability, and low cost are the primary advantages offered by the nanopore sequencing platform for identification of microbial plant pathogens simultaneously (Chalupowicz et al. 2019).

REFERENCES

Abang MM, Winter S, Green KR, Hoffmann P, Mignouna HD et al. (2002) Molecular identification of *Colletotrichum gloeosporioides* causing anthracnose of yam in Nigeria. Plant Pathol 51: 63–71.

Abdelrazig AO, Suwannarat S, Raijiravanich P, Surareungchai W (2020) Identification and detection of chilli anthracnose using three new species-specific PCR primers. Eur J Plant Pathol 158: 571–582.

Aboukhaddour R, Strelkov SE (2016) Exploring de no specificity: the *Pyrenophora tritici-repentis*-barley interaction. Plant Pathol 65: 1342–1357.

Aboukhaddour R, Turkington TK, Strelkov SE (2013) Race structure of *Pyrenophora tritici-repentis* (tan spot of wheat) in Alberta, Canada. Canad J Plant Pathol 35: 256–268.

Aggarwal R, Sharma S, Gupta S, Manjunatha C, Singh VK et al. (2017a) Gene-based analysis of *Puccinia* species and development of PCR-based marker to detect *Puccinia striiformis* f.sp. *tritici* causing yellow rust of wheat. J Gen Plant Pathol 83: 205–215.

Aggarwal R, Sharma S, Manjunatha C, Gupta S, Singh VK (2017b) Development and validation of loop-mediated isothermal amplification-based detection assay of *Puccinia striiformis* f.sp. *tritici* causing wheat stripe rust of wheat. Austr Plant Pathol 46: 577–583.

Agrios GN (2005) *Plant Pathology*, fifth edition. Academic Press/ Elsevier, Amsterdam.

Agusti-Brisach C, Leon M, Garcia-Jimenez J, Armengol J (2015) Detection of grapevine fungal trunk pathogens on pruning shears and evaluation of their potential for speed of infection. Plant Dis 99: 976–981.

Ahmed Y, Hubert J, Fourrier-Jeandel C, Dewdney MM, Aguayo J et al. (2019) A set of conventional and multiplex real-time PCR assays for direct detection of *Elsinoe fawcetti, E. australis* and *Pseudocercospora angolensis* in citrus fruits. Plant Dis103: 345–356.

Akhavan A, Turkington TK, Kebede B, Xi K, Kuma K et al. (2016) Genetic structure of *Pyrenophora teres* f.sp. *teres* and *P. teres* f.sp. *maculata* populations from Western Canada. Eur J Plant Pathol 146: 325–335.

Alleyne AT, Gilkes JM, Briggs G (2015) Early detection of super elongation disease in *Manihot esculenta* Crantz (cassava) using molecular markers for gibberellic acid biosynthesis. Eur J Plant Pathol 141: 27–34.

Amour MS, Bilodeau GL, Tremblay DM, Van der Heyden H, Yaseen T et al. (2017) Development of real-time isothermal amplification assays for on-site detection of *Phytophthora infestans* in potato leaves. Plant Dis 101: 1269–1277.

Anderson JM, Aitken EAB, Dann EK, Coates LM (2013) Morphological and molecular diversity of *Colletotrichum* spp. causing pepper spot and lychee (*Litchi sinensis*) in Australia. Plant Pathol 62: 279–288.

Bahri BA, Saadani M, Mechichi G, Rouissi W (2019) Genetic diversity of *Colletotrichum gloeosporioides* species complex associated with citrus wither-tip of twigs in Tunisia using microsatellite markers. J Phytopathol 167: 351–362.

Barbedo JGA (2014) An automatic method to detect and measure leaf disease symptoms using digital image processing. Plant Dis 98: 1709–1716.

Barnes CW, Szabo LJ (2007) Detection and identification of four common rust pathogens of cereals and grasses using real-time polymerase chain reaction. Phytopathology 97: 717–727.

Baysal-Gurel F, Lewis Ivey ML, Dorrance A, Luster D, Frederick R et al. (2008) An immunofluorescence assay to detect urediniospores of *Phakopsora pachyrhizi*. Plant Dis 92: 1387–1393.

Bello JC, Sakalidis ML, Perla DE, Hausbeck MK (2021) Detection of airborne sporangia of *Pseudoperonospora cubensis* and *P. humuli* in Michigan using Burkard spore trap coupled with quantitative PCR. Plant Dis 105: 1373–1381.

Billones-Baaijens R, Urbez-Torres JR, Ayres M, Sosnowski M (2018) Molecular methods to detect and quantify Botryosphaeriaceae inocula associated with grapevine dieback in Australia. Plant Dis 102: 1489–1499.

Blanco-Meneses M, Ristaino JB (2011) Detection and quantification of *Peronospora tabacina* using a real-time polymerase chain reaction assay. Plant Dis 95: 673–682.

Böhm J, Hahn A, Schubert R, Bahnweg G, Adler N et al. (1999) Real-time quantitative PCR: DNA determination in inoculated spores of mycorrhizal fungus *Glomus mosseae* and monitoring *Phytophthora infestans* and *Phytophthora citricola* in their respective host plants. J Phytopathol 147: 404–416.

Bohnenkamp D, Kuska MT, Mahlein AK, Behmann J (2019) Hyperspectral signal decomposition and symptom detection of wheat rust disease at the leaf scale using pure fungal spore spectra as reference. Plant Pathol 68: 1188–1195.

Bossi R, Dewey F (1992) Development of a monoclonal antibody-based immunodetection assay for *Botrytis cinerea*. Plant Pathol 41: 472–482.

Bouterige S, Robert R, Bouchara JP, Marot-Leblond A, Molinero V et al. (2000) Production and characterization of two monoclonal antibodies specific for *Plasmopara halstedii*. Appl Environ Microbiol 66: 3277–3282.

Boyle B, Hamelin RC, Seguin A (2005) In vivo monitoring of obligate biotrophic pathogen growth by kinetic-PCR. Appl Environ Microbiol 71: 1546–1552.

Brewer MT, Frenkel O, Milgroom MG (2012) Linkage disequilibrium and spatial aggregation of genotypes in sexually reproducing populations of *Erysiphe necator*. Phytopathology 102: 997–1005.

Brylinska M, Sobkowiak S, Stefanczyk E, Sliwka J (2018) Evaluation of PCR markers for *Phytophthora infestans* mating type determination. Eur J Plant Pathol 152: 33–44.

Burdon JJ, Roelfs AP, Brown AHD (1986) The genetic basis of isozyme variation in the wheat stem rust fungus (*Puccinia graminis tritici*). Canad J Genet Cytol 28: 171.

Burlakoti RR, Estrada R, Rivera VV, Boddeta A, Secor GA et al. (2007) Real-time PCR quantification and mycotoxins production of *Fusarium graminearum* in wheat inoculated with isolates collected from potato, sugar beet and wheat. Phytopathology 97: 835–841.

Burlakotti RR, Neate SM, Adhikari TB, Gyawali S, Salas B et al. (2011) Trichothecene profiling and population genetic analysis of *Gibberella zeae* from barley in North Dakota and Minnesota. Phytopathology 101: 687–695.

Bussaban B, Lumyong S, Lumyong P, Seelanan T, Park DC et al. (2005) Molecular and morphological characterization of *Pyricularia* and allied genera. Mycologia 97: 1002–1011.

Cabral PGC, Maciel—Zambolim E, Oliveira SAS, Caixeta ET, Zambolim L (2016) Genetic diversity and structure of *Hemileia vastatrix* populations in *Coffea* spp. Plant Pathol 65: 194–204.

Caiazzo R, Tarantino P, Porrone G, Lahoz E (2006) Detection and early diagnosis of *Peronospora tabacina* Adam in tobacco plant with systemic infection. J Phytopathol 154: 432–435.

Camele I, Maricone C, Cristinzio G (2005) Detection and identification of *Phytophthora* species in Southern Italy by RFLP and sequence analysis of PCR amplified nuclear ribosomal DNA. Eur J Plant Pathol 113: 1–14.

Cao X, Yao D, Zhou Y, West JS, Xu X et al. (2016) Detection and quantification of airborne inoculum of *Blumeria graminis* f.sp. *tritici*. Eur J Plant Pathol 146: 225–229.

Carrieri R, Lahoz E, Tarantino P, Cermola M, Morra L (2017) DNA analysis of a tobacco herbarium specimen revels the presence of *Peronospora tabacina* in Ital in 1934. Eur J Plant Pathol 148: 219–221.

Carzaniga R, Fiocca D, Bowger P, O'Connell RJ (2002) Localization of melanin in conidia of *Alternaria alternata* using phage antibodies. Molec Plant-Microbe Interact 15: 216–224.

Chalupowicz L, Dombrovsky A, Gabi V, Luria N, Reaven M et al. (2019) Diagnosis of plant diseases using nanopore sequencing platform. Plant Pathol 68: 229–238.

Chavhan RL, Hinge VR, Chincole MB, Chakraborty PK, Patade VY et al. (2015) Rapid, specific and sensitive molecular detection assay for *Alternaria helianthi* that causes leaf blight in sunflower. Eur J Plant Pathol 143: 663–675.

Chen R-S, Chu C, Cheng C-W, Chen W-Y, Tsay J-G (2008) Differentiation of two powdery mildews of sunflower (*Helianthus annuus*) by a PCR-mediated method based on ITS sequences. Eur J Plant Pathol 121: 1–8.

Chen S, Cao Y-Y, Li T-Y, Wu W-X (2015) Simultaneous detection of three wheat pathogenic fungal species by multiplex PCR. Phytoparasitica 43: 449–460.

Chen XY, Dai DJ, Zhao SF, Shen Y, Wang HD et al. (2020) Genetic diversity of *Colletotrichum* spp. causing strawberry anthracnose in Zhejiang, China. Plant Dis 104: 1351–1357.

Chen YY, Conner RL, Gillard CL, McLaren DL, Boland GJ et al. (2013) A quantitative real time PCR assay for detection of *Colletotrichum lindemuthianum* in many bean seeds. Plant Pathol 62: 900–907.

Chethana CS, Chowdappa P, Biju CN, Praveena R, Sujatha AM (2016) Molecular and phenotypic characterization revealed six *Colletotrichum* species responsible for anthracnose disease of small cardamom. Eur J Plant Pathol 146: 465–481.

Ciampi-Guillardi M, Ramiro J, de Moraes MHD, Barbieri MCG, Massola NS Jr (2020) Multiplex qPCR assay for direct detection and quantification of *Colletotrichum truncatum*, *Corynespora cassiicola*, and *Sclerotinia sclerotiorum* in soybean seed. Plant Dis 104: 3002–3009.

Conde-Ferráez L, Grijalva-Arango R, Canto-Canché BB, Manzo-Sanchez G, Canul-Salazar MI et al. (2010) The development of mating type-specific primers for *Mycosphaerella fijiensis*, the causal agent of black Sigatoka of banana and analysis of the frequency of idiomorph types in Mexican populations. Austr Plant Pathol 39: 217–225.

Crespo M, Cazorla FM, de Vicente A, Arrebola E, Torres J et al. (2016) Analysis of genetic diversity of *Fusarium tupiense*, the main causal agent of mango malformation disease in southern Spain. Plant Dis 100: 276–286.

Cumagun CJR, Rabenstein F, Miedaner T (2004) Genetic variation and covariation for aggressiveness, deoxy-nivalenol production and fungal colonization among progeny of *Gibberella zeae* in wheat. Plant Pathol 53: 446–453.

Dai T, Chang X, Hu Z, Li L, Sun M et al. (2019) Untargeted metabolomics based on GC-MS and chemometrics: a new tool for the early diagnosis of strawberry anthracnose caused by *Colletotrichum theobromicola*. Plant Dis 103: 2541–2547.

Dai Y, Gan L, Ruan H, Shi N, Du Y et al. (2020) Characterization of natural isolates of *Bipolaris maydis* associated with mating types, genetic diversity and pathogenicity in Fujian Province, China. Plant Dis 104: 323–329.

De Baker M, Bonants P, Pedley KF, Maes M, Roldan-Ruiz I et al. (2013) Genetic relationships in an international collection of *Puccinia horiana* isolates based on newly identified molecular markers and demonstration of recombination. Phytopathology 103: 1169–1179.

De Silva DD, Ades PK, Crous PW, Taylor PWJ (2017) *Colletotrichum* species associated with chilli anthracnose in Australia. Plant Pathol 66: 254–267.

Deahl KL, Jones RW, Perez FM, Shaw DS, Cooke LS (2006) Characterization of isolates of *Phytophthora infestans* from four solanaceous hosts growing in association with late blight-affected commercial potato crops. HortScience 41: 1635–1639.

Demers JE, Crouch JA, Castlebury LA (2015) A multiplex real-time PCR assay for the detection of *Puccinia horiana* and *P. chrysanthemi* on chrysanthemum. Plant Dis 99: 195–200.

Díaz-Cruz GA, Smith CM, Wiebe KF, Villanueva SM, Klonowski AR et al. (2019) Applications of next-generation sequencing for large-scale pathogen diagnoses in soybean. Plant Dis 103: 1075–1083.

Dowling ME, Bridges WC, Cox BM, Sroka T, Wilson JR et al. (2019) Preservation of *Monilina fructicola* genotype diversity within fungal cankers. Plant Dis 103: 526–530.

Dreiseitl A (2014) Pathogenic divergence of Central European and Australian populations of *Bluemeria graminis* f.sp. *hordei*. Ann Appl Biol 165: 364–372.

Dushnicky LG, Balance GM, Summer MJ, MacGregor AW (1998) Detection of infection and host responses on susceptible and resistant wheat cultivars to toxin-producing isolate of *Pyrenophora tritici-repentis*. Canad J Plant Pathol 20: 19–27.

Dyer RB, Kendra DF, Brown DW (2006) Real-time PCR assay to quantify *Fusarium graminearum* wild-type and recombinant mutant DNA in plant material. J Microbiol Meth 67: 534–542.

Edwards SG, Seddon B (2001) Selective media for the specific isolation and enumeration of *Botrytis cinerea* conidia. Lett Appl Microbiol 32: 63–66.

Ekefan EJ, Simons SA, Nwankii AO, Peters JC (2000) Semiselective medium for isolation of *Colletotrichum gloeosporioides* from soil. Experi Agric 35: 313–321.

Elverson TR, Kontz BJ, Markell SG, Harveson RM, Mathew FM (2020) Quantitative PCR assays developed for *Diaporthe helianthi* and *Diaporthe gulyae* for Phomopsis stem canker diagnosis and germplasm screening in sunflower (*Helianthus annuus*). Plant Dis 104: 793–800.

Fang U, Umashankar Y, Ramasamy RP (2014) Electrochemical detection of p-ethylguaiacol, a fungi infected fruit volatile using metal oxide nanoparticles. Analyst 139: 3804–3810.

Firpo PDO, Newcomb M, Flath K, Sommerfeldt-Impe N, Szabo LJ et al. (2017) Characterization of *Puccinia graminis* f.sp. *tritici* isolates derived from an unusual wheat stem rust outbreak in Germany in 2013. Plant Pathol 66: 1258–1266.

Flor HH (1946) Genetics of pathogenicity in *Melampsora lini*. J Agric Res 73: 335–337.

Flor HH (1956) Current status of the gene-for-gene concept. Adv Genet 8: 29–54.

Forcelini BB, Lee S, Oliveira MS, Peres NA (2018) Development of high throughput SNP genotyping assays for rapid detection of strawberry *Colletotrichum* species and the G143A mutation. Phytopathology 108: 1501–1508.

Fountaine JM, Shaw MW, Napier B, Ward E, Fraaije BA (2007) Application of real-time and multiplex polymerase chain reaction to study leaf blotch epidemics in barley. Phytopathology 97: 297–303.

Francis SA, Roden BC, Adams MJ, Weiland J, Michael A (2007) Comparison of ITS sequences in UK and North America powdery mildews and the designation of *Erysiphe betae*. Mycol Res 111: 204–212.

Gagkaeva TV, Yli-Mattila T (2004) Genetic diversity of *Fusarium graminearum* in Europe and Asia. Eur J Plant Pathol 110: 551–562.

Gagnon M-C, Kawchuk L, Tremblay DM, Carrise O, Danies G et al. (2016) Identification of the dormant genotypes of *Phytophthora infestans* in Canada using real-time PCR with ASO-PCR assays. Plant Dis 100: 1482–1491.

Gannibal PB, Orina AS, Mironenko NV, Levitin MM (2014) Differentiation of the closely related species *Alternaria solani* and *A. tomatophila* by molecular and morphological features and aggressiveness. Eur J Plant Patho 139: 609–623.

Gao L, Yu HX, Kang XH, Shen HM, Li C et al. (2016) Development of SCAR markers and an SYBR Green assay to detect *Puccinia striiformis* f.sp. *tritici* in infected wheat leaves. Plant Dis 100: 1840–1847.

Garcia-Benitez C, Melgarejo P, De Cal A (2017) Detection of latent *Monilinia* infections in nectarine flowers and fruit by qPCR. Plant Dis 101: 1002–1008.

Garrido C, Carbu M, Fernandez-Acero FJ, Boonham N, Colyer A et al. (2009) Development of protocols for detection of *Colletotrichum acutatum* and monitoring of strawberry anthracnose using real-time PCR. Plant Pathol 58: 43–51.

Gnocato FS, Dracatos PM, Karaoglu H, Zhang P, Berlin A et al. (2018) Development, characterization and application of genomic SSR markers for the oat stem rust pathogen *Puccinia graminis* f.sp. *avenae*. Plant Pathol 67: 457–466.

Gomes LIS, Douhan GW, Lehner MS, Bibiano LBJ, Mizubuki ESG (2018) Yellow Sigatoka epidemics caused by a panmictic population of *Mycosphaerella musicola* in Brazil. Plant Pathol 67: 295–302.

Gonzáles E, Sutton TB, Correll JC (2006) Clarification of the etiology of Glomerella leaf spot and bitter rot apple caused by *Colletotrichum* spp. based on morphology and genetic, molecular and pathogenicity tests. Phytopathology 96: 982–992.

Goodwin SB, Schneider RF, Fry WE (1995) Use of cellulose acetate electrophoresis for rapid identification of allozyme genotypes of *Phytophthora infestans*. Plant Dis 79: 1181–1185.

Grasso V, Siertozki H, Garibaldi A, Gissi U (2006) Relatedness among agronomically important rusts based on mitochondrial cytochrome *b* gene and ribosomal ITS sequences. J Phytopathol 154: 110–118.

Griffin DW, Kellog CA, Peak KK, Shinn EA (2002) A rapid and efficient assay for extracting DNA from fungi. Lett Appl Microbiol 34: 210–214.

Grinbergs D, Chilian J, Carrasco-Fernandez J, France A, Moya-Elizondo E et al. (2020) A PCR based method for the rapid detection of *Chondrostereum purpureum* in apple. Plant Dis 104: 702–707.

Gubis I, Hudcovicoá M, Klcová L, Ćervená V, Bojnanská K et al. (2004) Detection of leaf blotches causal agents in barley leaves and grains. Czech J Genet Plant Breed 40: 111–117.

Guillemete T, Iacomi-Vasilescu B, Simoneau P (2004) Conventional and real-time PCR-based assay for detecting pathogenic *Alternaria brassicae* in cruciferous seed. Plant Dis 88: 490–496.

Gultyaeva EI, Shaydayuk EL, Kazartsev IA, Akhmetova A, Kosman E (2018) Microsatellite analysis of *Puccinia triticina* from *Triticum* and *Aegilops* hosts. Austr Plant Pathol 47: 163–170.

Guo D, Jing J, Hu W, Navi SS, Jing L (2016) Internal transcribed spacer sequence analysis of *Puccinia helianthi* Schw. and its application in detection of sunflower rust. J Phytopathol 164: 141–146.

Guo J-R, Schneider F, Beyer M, Verreet J-A (2005) Rapid detection of *Mycosphaerella graminicola* in wheat using reverse transcription-PCR assay. J Phytopathol 153: 674–679.

Gurung S, Goodwin SB, Kabbage M, Bockus WW, Adhikari TB (2011) Genetic differentiation at microsatellite loci among populations of *Mycosphaerella graminicola* from California, Indian, Kansas and North Dakota. Phytopathology 101: 1251–1259.

Gurung S, Mahto BN, Gyawali S, Adhikari TB (2013) Phenotypic and molecular diversity of *Cochliobolus sativus* from wheat. Plant Dis 97: 62–73.

Gusberti M, Patocchi A, Gessler C, Broggini GAL (2012) Quantification of *Venturia inaequalis* growth in *Malus* x *domestica* with quantitative real-time polymerase chain reaction. Plant Dis 96: 1791–1797.

Ha Y, Fessehaie A, Ling KS, Wechter WP, Keinath AP et al. (2009) Simultaneous detection of *Acidovorax avenae* subsp. *citrulli* and *Didymella bryoniae* in cucurbit seed lots using magnetic capture hybridization and real-time polymerase chain reaction. Phytopathology 99: 666–678.

Hafez M, Abdelmagid A, Adam LR, Daayf F (2020) Specific detection and identification of *Fusarium graminearum* sensu stricto using a PCR-RFLP tool and specific primers targeting the translational elongation factor 1a gene. Plant Dis 104: 1076–1086.

Hallan L, Neumann M, Klatt B, Kleinhenz B, Klein T et al. (2018) Automated identification of sugar beet diseases using smart phones. Plant Pathol 67: 399–410.

Han SS, Ra DS, Nelson RJ (1995) Relationship between DNA fragments and virulence of *Pyricularia grisea* from rice and new hosts in Korea. Internatl Rice News Notes 20(1): 26–27.

Hansen ZR, Knaus BJ, Tabima JF, Press CM, Judelson HS et al. (2016a) SNP-based differentiation of *Phytophthora infesatans* clonal lineages using locked nucleic acid probes and high resolution melt analysis. Plant Dis 100: 1297–1306.

Hansen ZR, Knaus BJ, Tabima JF, Press CM, Judelson HS et al. (2016b) Loop-mediated isothermal amplification for detection of the tomato and potato late blight pathogen, *Phytophthora infestans*. J Appl Microbiol 120: 1010–1020.

Hao JJ, Xie SN, Sun J, Yang GQ, Liu JZ et al. (2017) Analysis of *Fusarium graminearum* species complex from wheat-maize rotation region in Henan in China. Plant Dis 101: 720–725.

Haudenshield JS, Hartman GL (2011) Exogenous controls increase negative call veracity in multiplexed quantitative PCR assays for *Phakopsora pachyrhizi*. Plant Dis 95: 343–352.

Hayden KJ, Ivors K, Wilkinson C, Garbelotto M (2006) TaqMan chemistry for *Phytophthora ramorum* detection and quantification with comparison of diagnostic methods. Phytopathology 96: 846–854.

Hu J, Johnson EG, Wang N-Y, Davoglio T, Dewdney MD (2014) qPCR quantification of pathogenic *Guignardia citricarpa* and nonpathogen *G. mangiferae* in citrus. Plant Dis 98: 112–120.

Huang T-P, Yeh Y, Tzeng DD (2010) Heteroduplex mobility assay for identification and phylogenetic analysis of anthracnose fungi. J Phyotpathol 158: 46–55.

Hussain S, Lees AK, Duncan JM, Cooke DEL (2005) Development of a species-specific and sensitive detection assay for *Phytophthora infestans* and its application for monitoring of inoculum in tubers and soil. Plant Pathol 54: 373–382.

Hyun JW, Peres NA, Yi S-Y, Timmer LW, Kim KS et al. (2007) Development of PCR assay for the identification of species and pathotypes of Elsinoe causing scab on citrus. Plant Dis 91: 865–870.

Iori A, Scala V, Cesare D, Pinzari F, D'Egidio MGD et al. (2015) Hyperspectral and molecular analysis of Stagonospora nodorum blotch disease in durum wheat. Eur J Plant Pathol 141: 689–702.

Jackson EW, Avant JB, Overturf KE, Bonman JM (2006) A quantitative assay of *Puccinia coronata* f.sp. *avenae* DNA in *Avina sativa*. Plant Dis 90: 692–696.

Jaimes YY, Gonzalez C, Rojus J, Cornejo OE, Mideros MF et al. (2016) Geographic differentiation and population genetic structure of *Moniliophthora roreri* in the principal cocoa production areas in Colombia. Plant Dis 100: 1548–1558.

Johnson RD, Kohmoto K, Otani H, Lane CR, Kodama M (2000) A polymerase chain reaction based method to specifically detect *Alternaria alternata* apple pathotype (*A. mali*), the causal agent of Alternaria blotch of apple. Phytopathology 90: 993–976.

Kaczmarek AN, King M, West JS, Stevens M, Sparkes D et al. (2019) A loop-mediated isothermal amplification (LAMP) assay for rapid and specific detection of airborne inoculum of *Uromyces betae* (sugar beet rust). Plant Dis 103: 417–421.

Kandan A, Akhtar J, Singh B, Dixit D, Chand D et al. (2015) Molecular diversity of *Bipolaris oryzae* infecting *Oryza sativa* in India. Phytoparasitica 43: 5–14.

Karimi K, Arzanlou M, Pertot I (2020) Development of novel species-specific primers for the specific identification of *Colletotrichum nymphaeae* based on conventional PCR and LAMP techniques. Eur J Plant Pathol 156: 463–475.

Karolewski Z, Fitt BDL, Latunde-Dada AO, Foster SJ, Todd AD et al. (2006) Visual and PCR assessment of light leaf spot (*Pyrenopeziza brassicae*) on winter oilseed rape (*Brassica napus*) cultivars. Plant Pathol 55: 387–400.

Karpovich-Tate N, Spanu P, Dewey FM (1998) Use of monoclonal antibodies to determine biomass of *Cladosporium fulvum* in infected tomato leaves. Molec Plant-Microbe Interact 11: 710–716.

Kaufmann PJ, Weidemann GJ (1996) Isozyme analysis of *Colletotrichum gloeosporioides* from five host genera. Plant Dis 80: 1289–1293.

Kennedy R, Wakeham AJ, Cullington JE (1999) Production and immunodetection of ascospores of *Mycosphaerella brassicola*: ringspot of vegetable crucifers. Plant Pathol 48: 297–307.

Kgatle MG, Truster M, Ramusi TM, Flett B, Aveling TAS (2018) *Alternaria alternata*, the causal agent of leaf blight of sunflower in South Africa. Eur J Plant Pathol 151: 677–688.

Kiss L, Takamatsu S, Cunnington JH (2005) Molecular identification of *Oidium neolycopersici* as the causal agent of the recent powdery mildew epidemics in North America. Plant Dis 89: 491–496.

Koch G, Kohler W (1990) Isozyme variation and genetic distances of *Erysiphe graminis* DC formae speciales. J Phytopathol 129: 89.

Kolmer JA (2015) Collections of *Puccinia triticina* in different provinces of China are highly related for virulence and molecular genotype. Phytopathology 105: 700–706.

Kolmer JA, Hanzalov A, Goyeau H, Bayles R, Morgounov AC (2013) Genetic differentiation of the wheat leaf rust fungus *Puccinia triticina* in Europe. Eur J Plant Pathol 62: 21–31.

Komjati H, Bakomyi J, Spring O, Viranyi F (2008) Isozyme analysis of *Plasmopara halstedii* using cellulase acetate gel electrophoresis. Plant Pathol 57: 57–63.

Kontz B, Adhikari S, Subramanian S, Mathew FM (2016) Optimization and application of a quantitative polymerase chain reaction assay to detect *Diaporthe* species in soybean plant tissue. Plant Dis 100: 1669–1676.

Koo C, Malapi-Wight M, Kim HS, Cifci OS, Vaughn-Diaz V et al. (2013) Development of a real-time microchip PCR system for portable plant disease diagnosis. PLoS One 8: e82704.

Kosaraju B, Sannasi S, Mishra MK, Subramani D, Bychappa M (2017) Assessment of genetic diversity of coffee rust pathogen *Hemileia vastatrix* using SRAP markers. J Phytopathol 165: 486–493.

Krátká J, Pekárova-Kynérová B, Kudlikova I, Slováček J, Zemankova M (2002) Utilization of immunochemical methods for detection and quantification of rice blast fungus *Pyricularia oryzae*. Plant Protect Sci 38: 56–63.

Kuan CP, Wu M-T, Huang HC, Chang H (2011) Rapid detection of *Colletotrichum lagenarium*, causal agent of anthracnose of cucurbitaceous crops by PCR and real-time PCR. J Phytopathol 159: 276–282.

Kunjeti SG, Anchieta A, Martin FN, Choi Y-J, Thines M et al. (2016a) Detection and quantification of *Bremia lactucae* by spore trapping and quantitative PCR. Phytopathology 106: 1426–1437.

Kunjeti SG, Anchieta A, Subbarao KV, Koike ST, Klosterman SJ (2016b) Plasmolysis and vital staining reveal viable oospores of *Peronospora effusa* in spinach seed lots. Plant Dis 100: 59–65.

Ladhalakshmi D, Vijayasamundeeswari A, Paranidharan V, Samiyappan R, Velazhahan R (2009) Molecular identification of isolates of *Peronosclerospora sorghi* from maize using PCR based SCAR markers. World J Microbiol Biotechnol 25: 2129–2135.

Langrell SRH, Barbara DJ (2001) Magnetic capture hybridization for improved PCR detection of *Nectria galligena* from lignified apple extracts. Plant Molec Biol 19: 5–11.

Larkin RP, Groves CL (2003) Identification and characterization of isolates of *Phytophthora infestans* using fatty acid methyl ester (FAME) profiles. Plant Dis 87: 1233–1243.

Leconte P, Peros JP, Blancard D, Bastien N, Delye C (2000) PCR assays that identify the grapevine dieback fungus *Eutypa lata*. Appl Environ Microbiol 66: 4475–4480.

Lees AK, Sullivan L, Lynott JS, Cullen DW (2012) Development of a quantitative real-time PCR assay for *Phytophthora infestans* and its applicability to leaf, tuber and soil samples. Plant Pathol 61: 867–876.

Leišova L, Minaríkova L, Ovesná J (2005) Genetic diversity of *Pyrenophora teres* isolates detected by AFLP analysis. J Phytopathol 153: 569–578.

Leisova L, Minarikova V, Kucera L, Ovesna J (2006) Quantification of *Pyrenophora teres* in infected barley leaves using real-time PCR. J Microbiol Meth 67: 446–455.

Lesuthu P, Mostert L, Spies CFJ, Moyo P, Regnier T (2019) *Diaporthe nebulae* sp. nov. and first report of *D. cynaroids, D. novem*, and *D. serafiniae* on grapevines in South Africa. Plant Dis 103: 808–817.

Leung H, Williams PH (1986) Enzyme polymorphism and genetic differentiation among geographic isolates of the rice blast fungus. Phytopathology 76: 778.

Levy M, Romano J, Marchetti MA, Hamer JE (1999) DNA fingerprinting with a dispersed repeated sequence resolves pathotype diversity in rice blast fungus. Plant Cell 3: 95–102.

Li L, Zhang SY, Zhang C-Q (2019) Establishment of a rapid detection method for rice blast fungus based on one-step loop-mediated isothermal amplification (LAMP). Plant Dis 103: 1967–1973.

Liang J, Wan Q, Luo Y, Ma Z (2013) Population genetic structures of *Puccinia striiformis* in Ningxia and Gansu Provinces. Plant Dis 97: 501–509.

Liao C-Y, Chen M-Y, Chen Y-K, Wang T-C, Kuo K-C et al. (2012) Characterization of three *Colletotrichum acutatum* isolates from *Capsicum* spp. Eur J Plant Pathol 133: 599–608.

Lievens B, Thomma BPHJ (2005) Recent developments in pathogen detection arrays: implications for fungal plant pathogens and use in practice. Phytopathology 95: 1374–1380.

Lima CS, Monteiro JHA, Crespo NC, Costa SC, Leslie JF et al. (2009) VCG and AFLP analyses identify the same groups in the causal agents of mango malformation in Brazil. Eur J Plant Pathol 123: 17–26.

Liu B, Louws FJ, Sutton TB, Correll TC (2012) A rapid quantitative molecular method for identification of *Colletotrichum acutatum* and *C. gloeosporioides*. Eur J Plant Pathol 132: 593–607.

Liu N, Lei Y, Zhang M, Zheng W, Shi Y et al. (2019) Latent infection of powdery mildew on volunteer wheat in Sichuan Province, China. Plant Dis 103: 1084–1091.

López-Moral A, Reya-Orlega MC, Agusti-Brisach C, Roca LP, Lovera M et al. (2017) Morphological pathogenic and molecular characterization of *Colletotrichum acutatum* isolates causing almond anthracnose in Spain. Plant Dis 101: 2034–2045.

Lu L, Wang Q, Jia Y, Bi YQ, Li CY et al. (2019) Selection and mutation of the avirulence gene *AVR-Pii* of the rice blast fungus *Magnaporthe oryzae*. Plant Pathol 68: 127–134.

Lu S, Platz GJ, Edwards MC, Friesen TL (2010) Mating type locus-specific polymerase chain reaction markers for differentiation of *Pyrenophora teres* f.sp. *teres* and *P. teres* f.sp. *maculata*, the causal agents of barley net blotch. Phytopathology 100: 1298–1306.

Lu TH, Groth JV (1987) Isozyme detection and variation *Uromyces appendiculatus*. Canad J Bot 66: 885.

Lu X, Zheng Y, Zhang F, Yu J, Dai T et al. (2020) A rapid equipment-free method for detecting *Phytophthora infestans* in the field using a lateral flow strip-based recombinase polymerase amplification assay. Plant Dis 104: 2774–2778.

Lukman A, Afifuddin A, Lubberstedt T (2016) Tracing the signature of *Peronosclerospora maydis* in maize seeds. Austr Plant Pathol 45: 73–82.

Mahlein AK (2016) Plant disease detection by imaging—parallels and specific demands for precision agriculture and phenotyping. Plant Dis 100: 241–251.

Manzo-Sanchez G, Orozco-Santos M, Islas-Flores I, Martinez-Bolaños L, Guzman-Gonzalez S et al.(2019) Genetic variability of *Pseudocercosporella fijiensis*, the black Sigatoka pathogen of banana (*Musa* spp.) in Mexico. Plant Pathol 68: 513–522.

Marimon N, Eduardo I, Leon M, Beerbegal M, Armengol J, Luque J (2020) A qPCR-based method for the detection and quantification of the peach powdery mildew (*Podosphaera pannosa*) in epidemiological studies. Eur J Plant Pathol 158: 1005–1016.

Markovic VL, Stummer BE, Hill AS (2002) Immunodetection and characterization of antigens expressed by *Uncinula necator*. J Phytopathol 150: 667–673.

Matsuda Y, Sameshima T, Moiura N, Inoue K, Nomomura T et al. (2005) Identification of individual powdery mildew fungi infecting leaves and direct detection of gene expression of single conidium by polymerase chain reaction. Phytopathology 95: 1137–1143.

Medina R, López SMY, Franco MEE, Rollan C, Ronco BL et al. (2015) A survey on occurrence of *Cladosporium fulvum* identifies race 0 and race 2 in tomato growing areas of Argentina. Plant Dis 99: 1732–1737.

Mills PR, Sreenivasaprasad S, Muthumeenakshi S (1998) Assessing diversity in *Colletotrichum* and *Trichoderma* species using molecular markers. In: Bridge P, Couteaudier Y, Carlson J (eds.), *Molecular Variability of Fungal Pathogens*. CAB International, Willingford, UK, pp. 108–120.

Mitchell MN, Ocamb CM, Grunwald NJ, Mancino LE, Gent DH (2011) Genetic and pathogenic relatedness of *Pseudoperonospora cubensis* and *P. humuli*. Phytopathology 101: 805–818.

Moreira RR, Peres NA, May De Mio LL (2019) *Colletotrichum acutatum* and *C. gloeosporioides* species complex associated with apple. Plant Dis 103: 268–275.

Moretti C, Quaglia M, Cerri M, Nicosa DE, Buonaurio R (2015) A real-time PCR assay for detection and quantification of *Botrytis cinerea* in *Pelargonium* x *hortorum* plants and its use for evaluation of plant resistance. Eur J Plant Pathol 43: 159–171.

Mumford R, Boonham N, Tomlinson J, Barker I (2006) Advances in molecular phytodiagnostics—new solutions for old problems. Eur J Plant Pathol 116: 1–19.

Mutiga SK, Rotich F, Ganeshan VD, Mwongera DT, Mgonja E et al. (2017) Assessment of the virulence spectrum and its association with genetic diversity in *Magnaporthe oryzae* population form sub-Saharan Africa. Phytopathology 107: 852–863.

Naegele RP, Quesada-Ocampo LM, Kurjan JD, Saude C, Hausbeck MK (2016) Regional and temporal population structure of *Pseudoperonospora cubensis* in Michigan and Ontario. Phytopathology 106: 372–379.

Nannapeneni R, Gergerich RC, Lee FN (2000) Technology for rapid detection, identification and quantification of rice blast fungus Pyricularia oryzae. Ark Agric Exp Sta Res (476): 480–485.

Narayanasamy P (2002) *Microbial Plant Pathogens and Crop Disease Management*. Science Publishers, Enfield, the United States of America.

Narayanasamy P (2005) *Immunology in Plant Health and Its Impact on Food Safety*. The Haworth Press, New York.

Narayanasamy P (2008) *Molecular Biology of Plant Pathogenesis and Disease Management, Volume 1, Microbial Plant Pathogens*. Springer Science, Heidelberg, Germany.

Narayanasamy P (2011) *Microbial Plant Pathogens—Detection and Disease Diagnosis: Fungal Pathogens*, volume 1. Springer Science + Business Media B. V., Heidelberg, Germany.

Narayanasamy P (2017) *Microbial Plant Pathogens—Detection and Disease Management in Seeds and Propagules*. Wiley-Blackwell, Chichester, the United Kingdom.

Narayanasamy P (2020) *Soilborne Microbial Plant Pathogens: Detection and Disease Management*. CRC Press, Florida, The United States of America.

Nawaz HH, Anam U, Rajaofera MJ, He Q, Liu W et al. (2018) Development of SNP-based markers to identify *Colletotrichum gossypii* in upland cotton. Plant Dis 102: 1426–1433.

Newcomb M, Olivero PD, Rouse MN, Szabo LJ, Johnson J et al. (2016) Kenyan isolates of *Puccinia graminis* f.sp. *tritici* from 2008 to 2014: virulence to *SrTmp* in the Ug99 race group and implications for breeding programs. Phytopathology 106: 729–736.

Nicholson P, Rezanoor HN, Hollins TN (1994) The identification of a pathotype specific DNA probe for the R-type of *Pseudocercosporella herpotrichoides*. Plant Pathol 43: 694–700.

Ordoñez ME, German SE, Kolmer JA (2010) Genetic differentiation within the *Puccinia triticina* population in South America and comparison with the North American population suggests common ancestry and international migration. Phytopathology 100: 376–383.

Ortega SF, Bustos Lopes MP, Nari L, Boonham N, Gullino ML et al. (2019) Rapid detection of *Monilinia fructicola* and *Monilinia laxa* on peach and nectarine using loop-mediated isothermal amplification. Plant Dis 103: 2305–2314.

Otero AJ, Saracent J, Hernández H, Sánchez M, Muirragui D, Villamar M, Moreta D, Jiménez M, Perez L, Maribona H (2007) Monoclonal antibody-based TAS-ELISA for quantitative detection of *Mycosphaerella fijiensis*. J Phytopathol 155: 713–719.

Passey TAJ, Shaw MW, Xu X-M (2016) Differentiation in population of the apple scab fungus *Venturia inaequalis* on cultivars in a mixed orchard remain over time. Plant Pathol 65: 1133–1141.

Patzak J (2003) PCR detection of hop fungal pathogens. Proceedings of the International Hop Growers Conference, Dobrna-Zalec, Slovania, pp. 12–16.

Paul R, Ostermann E, Gu Z, Ristaino JB, Wei Q (2020) DNA extraction from plant leaves using a microneedle patch. Curr Protoc Plant Biol 5: e-20104.

Paul R, Saville AC, Hansel JC, Ye Y, Ball C et al. (2019) Extraction of plant DNA by microneedle patch for rapid detection of plant diseases. ACS Nano 13: 6540–6549.

Peres NA, Harakava R, Carrot GC, Adaskaveg JE, Timmer LW (2007) Comparison of molecular procedures for detection and identification of *Guignardia citricarpa* and *G. mangiferae*. Plant Dis 91: 525–531.

Perez-Hernandez O, Nam MH, Gleason ML, Kim HG (2008) Development of a nested polymerase chain reaction assay for detection of *Colletotrichum acutatum* on symptomless strawberry leaves. Plant Dis 92: 1665–1671.

Pieck ML, Ruck A, Farman ML, Peterson QL, Stack JP et al. (2017) Genomics based marker discovery and diagnostic assay development for wheat blast. Plant Dis 101: 103–109.

Pilet F, Pelle R, Ellisséche D, Andrivon D (2005) Efficacy of the *R2* resistance gene as a component for durable management of potato late blight in France. Plant Pathol 54: 723–732.

Puri KD, Saucedo ES, Zhong S (2012) Molecular characterization of Fusarium head blight pathogens sampled from a naturally infected disease nursery used for wheat breeding programs in China. Plant Dis 96: 1280–1285.

Qu B, Li HP, Zheng JB, Xu YB, Huang T et al. (2008) Geographic distribution and genetic diversity of *Fusarium graminearum* and *Fusarium asiaticum* on wheat spikes throughout China. Plant Pathol 57: 15–24.

Rampersad SN (2013) Genetic structure of *Colletotrichum gloeosporioides* sensu lato isolates infecting papaya inferred by multilocus ISSR markers. Phytopathology 103: 182–189.

Rinaldi PA, Paffetti D, Comparini C, Broggini GAL, Gessler C et al. (2017) Genetic variability of *Phyllosticta ampelicida*, the agent of black rot disease of grapevine. Phytopathology 107: 1406–1416.

Ristaino JB, Saville AC, Paul R, Cooper DC, Wei Q (2020) Detection of *Phytophthora infestans* by loop-mediated by isothermal amplification, real-time LAMP and droplet digital PCR. Plant Dis 104: 708–716.

Rojas A, Kirk WW (2016) Phenotypic and genotypic variation in Michigan populations of *Phytophthora infestans*. Plant Pathol 65: 1022–1033.

Rush TA, Golan J, Mc Taggart A, Kane C, Schneider RW et al. (2019) Variation in the internal transcribed spacer region of *Phakospsora pachyrhizi* and implications for molecular diagnostic assays. Plant Dis 103: 2237–2245.

Salinas J, Schots A (1994) Monoclonal antibodies-based immunofluorescence test for detection of conidia of *Botrytis cinerea* on cut flowers. Phytopathology 84: 351–356.

Samuelian SK, Greer LA, Savocchia S, Steel CC (2011) Detection and monitoring *Greeneria uvicola* and *Colletotrichum acutatum* development in grapevines by real-time PCR. Plant Dis 95: 298–303.

Schaad NW, Frederick RD, Shaw J, Schneider WL, Hickson A et al. (2003) Advances in molecular based diagnostics in meeting crop biosecurity and phytosanitary issues. Annu Rev Phytopathol 41: 305–324.

Schena L, Hughes KJ, Cooke DES (2006) Detection and quantification of *Phytophthora ramorum, P. kernoviae, P. citricola* and *P. quercina* in symptomatic leaves by multiplex real-time PCR. Molec Plant Pathol 7: 365–379.

Scott Jr DL, Clark CW, Tooley PW, Carras MM, Maas JL (2002) The use of biomagnetic separation to recover DNA suitable for PCR from *Claviceps* species. Lett Appl Microbiol 31: 95–99.

See PT, Moffat CS, Morina J, Oliver RP (2016) Evaluation of a multilocus Indel DNA region for the detection of wheat tan spot pathogen *Pyrenophora tritici-repentis*. Plant Dis 100: 2215–2225.

Sharma R, Zhou M, Hunter MD, Fan X (2019) Rapid in situ analysis of plant emission for disease diagnosis using a portable gas chromatography device. J Agric Food Chem 67: 7530–7537.

Shimomoto Y, Sato T, Hojo H, Morita Y, Takeuchi S et al. (2011) Pathogenic and genetic variation among isolates of *Corynespora cassiicola*. Plant Pathol 60: 1253–1260.

Sippell DN, Hall R (1995) Glucose phosphate isomerase polymorphism distinguish weakly virulent from highly virulent strains of *Leptosphaeria maculans*. Canad J Plant Pathol 17: 1–6.

Sireesha Y, Velazhahan R (2018) Rapid and specific detection of *Peronosclerospora sorghi* in maize seeds by conventional and real-time PCR. Eur J Plant Pathol 150: 521–526.

Skottrup P, Frokiaer H, Hearty S, O'Kennedy R, Hejgaard et al. (2007a) Monoclonal antibodies for the detection of *Puccinia striiformis* urediniospores. Mycol Res 111: 332–358.

Skottrup P, Nicolaissen M, Justesen AF (2007b) Rapid determination of *Phytophthora infestans* sporangia using a surface Plasmon resonance immunosensor. J Microbiol Meth 68: 507–515.

Sowley ENK, Dewey FM, Shaw MW (2010) Persistent, symptomless, systemic and seedborne infection of lettuce by *Botrytis cinerea*. Eur J Plant Pathol 26: 61–71.

Srinivasan M, Satya VK, Paranidharan V, Velazhahan R (2014) Development of conventional and real-time PCR protocols for specific and sensitive detection of *Colletotrichum capsici* in chilli (*Capsicum annuum* L.) Phytoparasitica 42: 437–444.

Srivastava DA, Yakubov M, Feldbaum R, Tish N, Shoughet H et al. (2018) Multiparametric analysis of diversity in *Botrytis cinerea* from Israel. Phytoparasitica 46: 569–581.

Stace-Smith R, Bowler DS, Mc Kenzie DJ, Ellis P (1993) Monoclonal antibodies differentiate the weekly virulent from highly virulent strain of *Leptosphaeria maculans*, the organism causing black leg of canola. Canad J Bot 15: 127–133.

Strandberg JO (2002) A selective medium for the detection of *Alternaria dauci* and *A. radicina*. Phytoparasitica 30: 269–284.

Stroud JA, Shaw DS, Hale MD, Steele KA (2016) SSR assessment of *Phytophthora infestans* populations on tomato and potato in British gardens demonstrates high diversity but no evidence for host specialization. Plant Pathol 65: 334–341.

Subbarao CS, Anchieta A, Ochoa L, Dhar N, Kunjeti SG et al. (2015) Detection of latent *Peronospora effusa* infections in spinach. Plant Dis 102: 1766–1771.

Sumabat LG, Kemerait RC Jr, Brewer MT (2018) Phylogenetic diversity and host specialization of *Corynespora cassiicola* responsible for emerging target spot disease of cotton and other crops in the southeastern United States. Phytopathology 108: 892–900.

Sun G, Liu J, Li G, Zhang X, Chen T et al. (2015) Quick and accurate detection and quantification of *Magnaporthe oryzae* in rice using real-time quantitative polymerase chain reaction. Plant Dis 99: 219–224.

Svircev AM, Smith RJ, Zhou T, Day AW (2000) Localization of fungal fimbriae by immunochemistry in pathogenic and nonpathogenic isolates of *Venturia inaequalis*. Canad J Microbiol 46: 800–808.

Thangavel T, Jones S, Scott JB, Livermore M, Wilson CR (2018) Detection of two *Peronospora* spp. responsible for downy mildew in opium poppy seed. Plant Dis 102: 2277–2284.

Thierry M, Gladieum P, Fournier E, Tharreau D, Ioos R (2020) A genome approach to develop a new qPCR test enabling detection of *Pyricularia oryzae* lineage causing wheat blast. Plant Dis 104: 60–70.

Thiessen LD, Keune JA, Neill TM, Turechek WW, Grove GG et al. (2016) Development of a grower-conducted inoculum detection assay for management of grape powdery mildew. Plant Pathol 65: 234–249.

Thuan NTN, Bigirimana J, Roumeen E, vander Stracton D, Hofte M (2006) Molecular and pathotype analysis of the rice blast fungus in North Vietnam. Eur J Plant Pathol 114: 381–386.

Tian Q, Lu C, Wang S, Xiong Q, Zhang H et al. (2017) Rapid diagnosis of soybean anthracnose caused by *Colletotrichum truncatum* using a loop-mediated isothermal amplification (LAMP) assay. Eur J Plant Pathol 148: 785–793.

Tian Y, Zhan G, Chen X, Tungruentragoon A, Lu X et al. (2016) Virulence and simple sequence repeat marker in a *Puccinia stiiformis* f.sp. *tritici* population produced by selling a Chinese isolate on *Berberis shensiana*. Phytopathology 106: 185–191.

Tillessen A, Menkhaus J, Verreet J-A (2018) Development of specific PCR primers for diagnosis and qualitative detection of the fungal maize pathogen *Kabatiella zeae*. Eur J Plant Pathol 152: 503–506.

Tooley PW, Carras MM, Lambert DM (1998) Application of a PCR-based test for detection of potato late blight and pink rot in tubers. Amer J Potato Res 75: 187–194.

Tóth B, Mesterchazy A, Nicholson P, Teren J, Varga J (2004) Mycotoxin production and molecular variability of European and American isolates of *Fusarium culmorum*. Eur J Plant Pathol 110: 587–599.

Tymon LS, Peever TL, Johnson DA (2016) Identification and enumeration of small-spored *Alternaria* species associated with potato in the U. S. Plant Dis 100: 465–472.

Ueng PP, Cunfer BM, Alano AS, Youmans JD, Chen W (1995b) Correlation between molecular and biological characteristics in identifying the wheat and barley biotypes of *Stagonospora nodorum*. Phytopathology 85: 44–52.

Úrbez-Torres JR, Haag P, Bowen P, Lowery T, O'Gorman DTO (2015) Development of a DNA macroarray for the detection and identification of fungal pathogens causing decline of young grapevines. Phytopathology 105: 1373–1388.

Vandecasteele M, Landscoot S, Carrete J, Verwaeren J, Hofte M et al. (2018) Species prevalence and disease progression studies demonstrate a seasonal shift on the *Alternaria* population in composition on potato. Plant Pathol 67: 325–336.

Vandemark GJ, Barker BM (2003) Quantifying *Phytophthora medicaginis* in susceptible and resistant alfalfa with real-time fluorescent assay. J Phytopathol 151: 577–583.

Verreet SA, Klink H, Hoffmann GM (2000) Regional monitoring for disease prediction and optimization of plant protection measures: the IPM wheat model. Plant Dis 84: 816–826.

Vidic M, Petrovic K, Dordevic C, Riccioni L (2013) Occurrence of *Phomopsis longicolla* B conidia in naturally infected soybean. J Phytopathol 161: 470–477.

Vilanova L, Usall J, Teixido N, Torres R (2017) Assessment of viable conidia of *Monilinia fructicola* in flower and stone fruit combining propidium monoazide (PMA) and qPCR. Plant Pathol 66: 1276–1287.

Villafana RT, Ramadass AC, Rampersad SN (2019) Development of a new technology for the detection of *Colletotrichum truncatum* and *Fusarium* sp. in bell pepper seed. Phytoparasitica 47: 543–555.

Vittal R, Haudenshield JS, Hartman GL (2012) A multiplexed immunofluorescence method identifies *Phakopsora pachyrhizi* urediniospores and determines their viability. Phytopathology 102: 1143–1152.

Vöhringer G, Sander G (2001) Comparison of antibodies in chicken egg yolk (IgY) and rabbit IgG for quantitative strain detection of *Colletotrichum falcatum* and *Fusarium subglutinans*. J Plant Dis Protect 108: 39–48.

Vujanovic V, Vidovic S, Fernandez MA, Daida P, Korber D (2009) Whole cell proteins and ITS rDNA profiles as diagnostic tools to discriminate *Fusarium avenaceum* intraspecific variability and associated virulence. Canad J Microbiol 55: 117–125.

Wakeham AJ, Kane G, Kennedy R (2016) Field evaluation of a competitive lateral-flow assay for detection of *Alternaria brassicae* in vegetable Brassica crops. Plant Dis 100: 1831–1839.

Wang N-Y, Zhang K, Hugnet-Tapia JC, Rollins JA, Dewdney MM (2016) Mating type and simple sequence repeat markers indicate a clonal population of *Phyllosticta citricarpa* in Florida. Phytopathology 106: 1300–1310.

Wang S, Chu B, Liu Q, Luo Y, Ma Z (2017a) Development of a sequence-characterized amplified region marker using inter-simple sequence repeats for detection of *Puccinia striiformis* f.sp. *tritici*. J Phytopathol 165: 442–447.

Wang S, Ye W, Tian Q, Dong S, Zheng X (2017b) Rapid detection of *Colletotrichum gloeosporioides* using a loop-mediated amplification assay. Austr Plant Pathol 46: 493–498.

Wang W, Tang JH, Wang YC (2008) Molecular detection of *Colletotrichum lindemuthianum* by duplex PCR. J Phytopathol 156: 431–437.

Wen D-M, Chen M-X, Zhao L, Li M, Yang X-T (2019) Use of thermal imaging and Fourier transform infrared spectroscopy for presymptomatic detection of cucumber downy mildew. Eur J Plant Pathol 155: 405–416.

Wieczorek TM, Jorgensen LN, Hansen AL, Munk L, Justesen AF (2014) Early detection sugar beet pathogen *Ramularia beticola* in leaf and air samples using qPCR. Eur J Plant Pathol 138: 775–785.

Wijekoon CP, Pageni BB, Kalischuk ML, Lupwayi NZ, Kawachuk LM (2017) Amplification of the *Phytophthora infestans* RG57 loci facilitates in planta T-RFLP identification of late blight genotypes. Amer J Pot Res 94: 251–257.

Winneker M, Pham KTK, Woudenberg JHC, Thomma PHJ (2019) Identification of *Alternaria* sp. is causal agent of dead flower buds disease of pear (*Pyrus communis*) in the Netherlands and methods of disease control. Eur J Plant Pathol 155: 967–981.

Withers S, Gongora-Castillo E, Thomas GA, Ojimbo PS, Quesada-Ocampo LM (2016) Using next-generation sequencing to develop molecular diagnostics for *Pseudoperonospora cubensis*, the cucurbit downy mildew pathogen. Phytopathology 106: 1105–1116.

Wong M-H, Henderson J, Drenth A (2013) Identification and differentiation of *Phytllosticta* species causing freckle disease of banana using high resolution melting (HRM) analysis. Plant Pathol 62: 1285–1306.

Wong M-Y, Smart CD (2012) A new application using a chromogenic assay in a plant pathogen DNA macroarray detection system. Plant Dis 96: 1365–1371.

Wu WS, Chen TW (1999) Development of a new selective medium for detecting *Alternaria brassicola* in cruciferous seeds. Seed Sci Technol 27: 397–409.

Wyenandt CA, Maimone LR, Homa K, Kadeiras AM, Wick RL et al. (2018) Detection of the downy mildew pathogen on seed of basil following field infection in Southern New Jersey. Hortic Technol 28: 637–641.

Xiaojie W, Chunlei T, Jimlong C, Buchaneur H, Jie Z et al. (2009) Detection of *Puccinia*-infected wheat leaves by nested polymerase chain reaction. J Phytopathol https://doi.org/10.1111/j.1439–0434.2008.01521.x

Xing J, Jia Y, Correll JC, Lee FN, Cartwright R et al. (2013) Analysis of genetic and molecular identity among isolates of the rice blast fungus with an international differential system, Rep-PCR and DNA sequencing. Plant Dis 97: 491–495.

Xu M, He R, Peng Y, Zeng C-B, Liu Y et al. (2017) Isolation and molecular identification of *Colletotrichum gloesoporioides* causing brown spot disease in *Camellia oleifera* in Hainan, China. J Phytopathol 165: 380–386.

Xu X, Harvey N, Roberts A, Barbara D (2013) Population variation of apple scab (*Venturia inaequalis*) within mixed orchards in the UK. Eur J Plant Pathol 135: 97–104.

Yan I, Zhang C, Ding L, Ma Z (2008) Development of a real-time PCR assay for the detection of *Cladosporium fulvum* in tomato leaves. J Appl Microbiol 104: 1417–1424.

Yan J, Luo Y, Chen T, Huang C, Ma Z (2012) Field distribution of wheat stripe rust latent infection using real-time PCR. Plant Dis 96: 544–551.

Yang H-C, Hartman GL (2015) Multiplex real-time PCR detection and differentiation of *Colletotrichum* species infecting soybean. Plant Dis 99: 1559–1568.

Yang X, Al-Attala MN, Zhang Y, Zhang A-F, Zhang H-Y et al. (2018) Rapid detection of *Ustilaginoidea virens* from rice using loop-mediated isothermal amplification assay. Plant Dis 102: 1741–1747.

Yao J, Lan C, Huang P, Yu D (2018) PCR detection of *Colletotrichum gloeosporioides* in *Psidium guajava*. Austr Plant Pathol 47: 95–100.

Yasuhara-Bell J, Pedley KF, Farman M, Valent B, Stack JP (2018) Specific detection o the wheat blast pathogen (*Magnaporthe oryzae* Triticum) by loop-mediated isothermal amplification. Plant Dis 102: 2550–2559.

Yoshida S, Tsukiboshi T, Shinohara H, Koitabashi M, Tsushima S (2007) Occurrence and development of *Colletotrichum acutatum* on symptomless blueberry bushes. Plant Pathol 56: 871–877.

Yüksel S, Schwenkbier L, Pollok S, Weber K, Cialla-May D et al. (2015) Label—free detection of *Phytophthora ramorum* using surface–enhanced Raman spectroscopy, Analyst 140: 7254–7262.

Zang R, Yin Z, Ke X, Wang X, Li Z et al. (2012) A nested PCR assay for detecting *Valsa mali* var. *mali* in different tissues of apple trees. Plant Dis 96: 1645–1652.

Zelinger E, Hawves CR, Gurr SJ, Dewey FM (2004) An immunochemical and ultra-structural study of the extracellular matrix produced by germinating spores of *Stagonospora nodorum* on natural and artificial surfaces. Physiol Molec Plant Pathol 65: 123–135.

Zeng X, Luo Y, Zheng Y, Duan X, Zhou Y (2010) Detection of latent infection of wheat leaves caused by *Blumeria graminis* f.sp. *tritici* using nested PCR. J Phytopathol 158: 227–235.

Zhang J, Zhang P, Karaoglu H, Park RF (2017) Molecular characterization of Australian isolates of *Puccinia graminis* f.sp. *tritici* supports long-term clonality but also reveals cryptic genetic variation. Phytopathology 107: 1032–1038.

Zhang Y, Li X, Shen F, Xu H, Li Y et al. (2018) Characterization of *Botryis cinerea* isolates from grape vineyards in China. Plant Dis 102: 40–48.

Zhao J, Wang XJ, Chen CQ, Huang LL, Kang ZS (2007) A PCR-based assay for detection of *Puccinia striiformis* f.sp. *tritici* in wheat. Plant Dis 91: 1669–1674.

Zheng Z, Nonomura T, Boka K, Matsud Y, Visser RGF et al. (2013) Detection and quantification of *Leveillula taurica* growth in pepper leaves. Phytopathology 103: 623–632.

Zou Z, Zhang X, Wannakuwattewaduge FGD (2018) Distribution of mating-type alleles and genetic variability in field populations of *Leptosphaeria maculans* in western Canada. J Phytopathol 156: 438–447.

Additional References

Ali S, Adhikari TB (2008) Variation in aggressiveness of *Stagonospora nodorum* isolates in North Dakota. J Phytopathol 156: 140–145.

Angelini RMD, Rotolo C, Pollastro S, Faretra F (2016) Molecular analysis of the mating type (MAT1) locus in strains of the heterothallic ascomycete *Botrytis cinerea*. Plant Pathol 65: 1321–1332.

Armengol J, Vicent A, Torné CL, García-Figueves G, García-Jimenez T (2001) Fungi associated with esca and grapevine decline in Spain: a three-year survey. Phytopathol Mediterr 40: 325–329.

Arzanlou M, Abeln ECA, Kema GHJ, Waalwijk C, Carlier J et al. (2007) Molecular diagnostics for the Sigatoka disease complex of banana. Phytopathology 97: 1112–1118.

Chung WH, Ishii H, Nishimura K, Fukaya M, Yano Y et al. (2006) Fungicide sensitivity and phylogenetic relationship of anthracnose fungi isolated from various fruit crops in Japan. Plant Dis 90: 506–512.

Danies G, Small IM, Myers K, Childers R, Fry WE (2013) Phenotypic characterization of recent clonal lineages of *Phytophthora infestans* in the United States. Plant Dis 97: 873–881.

Delwart ELH, Sheppard HW, Walker BC, Goudsmit J, Mullins I (1994) Human immunodeficiency virus type I evolution in vivo tracked by DNA heteroduplex mobility assays. J Virol 68: 6672–6683.

Falconi CE, Visser RG, van Heusden AW (2013) Phenotypic molecular and pathological characterization of *Colletotrichum acutatum* associated with Andean lupine and tamarillo in the Ecuadorian Andes. Plant Dis 97: 819–827.

Fraser DE, Shoemaker PB, Ristaino JB (1999) Characterization of isolates of *Phyophthora infestans* from tomato and potato in North Carolina from 1993 to 1995. Plant Dis 83: 633–638.

Frenkel D, Portillo I, Brewer MT, Peros JP, Cadle-Davidson L et al. (2012) Development of microsatellite markers from the transcriptome of *Erysiphe necator* of analyzing population structure in North America and Europe. Plant Pathol 61: 106–119.

Guinet C, Fourrier-Jeandel C, Cerf-Wendling I, Ioos R (2016) One-step detection of *Monilinia fructicola, M. fructigena*, and *M. laxa* on *Prunus* and *Malus* by a multiplex real-time PCR assay. Plant Dis 100: 2465–2474.

Guo L, Zhu X-Q, Hu C-H, Ristaino JB (2010) Genetic structure of *Phytophthora infestans* populations in China indicates multiple migration events. Phytopathology 100: 997–1006.

Hayden KJ, Rizozo D, Tse J, Garbelotto M (2004) Detection and quantification of *Phytophthora ramorum* from California forests using a real-time polymerase chain reaction assay. Phytopathology 94: 1075–1083.

Hu M-J, Grabke A, Schnabel G (2015) Investigation of the *Colletotrichum gloeosporioides* species complex causing anthracnose in South Carolina. Plant Dis 99: 797–805.

Ishiie H (2006) Impact of fungicide resistance in plant pathogens on crop disease control and agricultural environment. Japan Agric Res Quart 40: 205–211.

Ishiie H, Tanoue J, Oshima M, Chang W-H, Nishimura K et al. (2008) First application of PCR-Luminex system for molecular diagnosis of fungicide resistance and species identification of fungal pathogens. J Gen Plant Pathol 74: 409–416.

Ishiie H, Yano K, Date H, Furuta A, Sagehashi Y et al. (2007) Molecular characterization and diagnosis of QoI resistance in cucumber and eggplant fungal pathogens. Phytopathology 97: 1458–1466.

Keeper FJ, Capio E, Grcic M, Wallwork H (2007) Development of sequence-tagged microsatellites for the barley net blotch pathogen *Pyrenophora teres*. Molec Ecol Notes 7: 664–666.

Keeper FJ, Grcic M, Capio E, Wallwork H (2008) Diagnostic microsatellite markers for the barley net blotch pathogens *Pyrenophora teres* f.sp. *maculata* and *Pyrenophora teres* f.sp. *teres*. Austr Plant Pathol 37: 428–430.

Keiper FJ, Hayden MJ, Wallwork H (2006) Development of sequence-tagged microsatellites for the barley scald pathogen *Rhynchosporium secalis*. Molec Ecol Notes 6: 543–546.

Kendall JJ, Hollomon DW, Sellely A (1998) Immunodiagnosis as an aid to timing of fungicide sprays for the control of *Mycosphaerella graminicola* on winter wheat in UK. Brighton Crop Protect Conf 2: 701–706.

Kim DH, Martyn RD, Magil CW (1993) Mitochondrial DNA (mtDNA) relatedness among formae speciales of *Fusarium oxysporum* in Cucurbitaceae. Phytopathology 83: 91–97.

Kim H, Newell AD, Cost-Sieckmeyer RG, Rupe JC, Fakhoury AM et al. (2013) Mating-type distribution and genetic diversity of *Cercospora sojina* populations on soybean from Arkansas: evidence for potential sexual reproduction. Phytopathology 103: 1045–1054.

Kubota R, Vine BG, Alvarez AM, Jenkins DM (2008) Detection of *Ralstonia solanacearum* by loop-mediated isothermal amplification of DNA. Phytopathology 98: 1045–1051.

Kuhnen PR, Spotti P, Del Ponte EM, Cummings JA, Bergstrom GC (2015) Trichothecene genotype composition of *Fusarium graminearum* not differentiated among isolates from maize stubble maize ears, wheat spikes and atmosphere in New York. Phytopathology 105: 695–699.

Leach JE, White FE (1991) Molecular probes for disease diagnosis and monitoring. In: Khush GS, Toennissen GH (eds.), *Rice Biotechnology*. CAB International UK and International Rice Research Institute, Philippines.

Lehmensiek A, Bester-Vander Merwe AE, Sutherland MW, Platz G, Kriel WM et al. (2010) Population structure of South African and Australian *Pyrenophora teres* isolates. Plant Pathol 59: 504–515.

Liu ZH, Zhong S, Stasko AK, Edwards MC, Friesen TL (2012) Virulence profile and genetic structure of a South Dakota population of *Pyrenophora teres* f.sp. *teres*, the causal agent of net form net blotch of barley. Phytopathology 102: 539–546.

Lucciardello G, Grasso FM, Bella P, Cirvilleri G, Girimaldi V et al. (2006) Identification and detection of *Phoma tracheiphila*, causal agent of citrus mal seco disease by real-time PCR reaction. Plant Dis 90: 1523–1530.

Luna MPR, Aime C, Chilvers MI, Wise KA (2017) Genetic diversity of *Stenocarpella maydis* in the major corn production areas of the United States. Plant Dis 101: 2020–2026.

Luo QQ, Xu JQ, Hou YP, Chen CJ, Wang JX et al. (2009) PIRA-PCR for detection of *Fusarium graminearum* genotypes with moderate resistance to carbendazim. Plant Pathol 58: 882–887.

Ma L, Michailides Tj (2003) Nested PCR assay for detection of *Monilinia fructicola* in stone fruit orchards and *Botryosphaeria dothidea* from pistachios in California. J Phytopathol 151: 312–331.

Malihipour A, Gillert J, Piercey-Normore M, Cloutier S (2012) Molecular phylogenetic analysis, trichothecene chemotype patterns and variation in aggressiveness of Fusarium isolates causing head blight in wheat. Plant Dis 96: 1016–1025.

Martos S, Torres E, El Bakali MA, Raposo R, Gramaje D et al. (2011) Cooperational PCR coupled with dot blot hybridization for the detection of *Phaeomoniella chlamydospora* on infected grapevine wood. J Phytopathol 159: 247–254.

McMaugh SJ, Lyon BR (2003) Real-time quantitative RT-PCR assay of gene expression in plant roots during fungal pathogenesis. Biotechniques 34: 982–986.

Mills PR, Sreenivasaprasad J, Brown AF (1992) Detection and differentiation of *Colletotrichum gloeosporioides* isolates using PCR. FEMS Microbiol Lett 98: 137–143.

Mirzaei S, Goltapeh EM, Shams-Bakhsh M, Safaie N (2008) Identification of *Botrytis* spp. on plants grown in Iran. J Phytopathol 156: 21–28.

Mori T, Nagamine K, Tomita N, Notomi T (2001) Detection of loop-mediated isothermal amplification reaction by turbidity derived from magnesium pyrophosphate formation. Biochem Biophys Res Comm 289: 150–154.

Muñoz M, Faust JE, Schanabel G (2017) Characterization of *Botrytis cinerea* from commercial cut flower roses. Plant Dis 103: 1577–1583.

Narayanasamy P (2001) *Plant Pathogen Detection and Disease Diagnosis*, second edition. Marcel Dekker Inc., New York.

Nasehi A, Kadir J-B, Nasr-Esfahani M, Abed-Ashtiani F, Wong M-Y et al. (2014) Analysis of genetic and virulence variability of *Stemphylium lycopersici* associated with leaf spots of vegetable crops. Eur J Plant Pathol 140: 261–273.

O'Garman DT, Sholberg PL, Stokes SC, Ginns J (2008) DNA sequence analysis of herbarium specimen facilitates the revival of *B. mali*, a postharvest pathogen of apple. Mycologia 100: 227–235.

Oliver RP, Farman ML, Jones JDG, Harmon-Kosack KE (1993) Use of fungal transformants expressing β-glucuronidase activity to detect infection and measure hyphal biomass in infected plant tissues. Molec Plant-Microbe Interact 6: 521–525.

Otero-Colina G, Rodriguez-Alvarado G, Fernández-Pavia S, Maymon M, Ploetz RC et al. (2010) Identification and characterization of a novel etiological agent of mango malformation disease in Mexico, *Fusarium mexicanum* sp. nov. Phytopathology 100: 1176–1184.

Peraza-Echeverria L, Rodriguez-Garcia CM, Zapata-Salazar DM (2008) A rapid, effective method for profuse in vitro conidial production of *Mycosphaerella fijiensis*. Austr Plant Pathol 37: 460–463.

Peters RD, Platt (Bud) HW, Hall R (1999) Use of allozyme markers to determine genotype to *Phytophthora infestans* in Canada. Canad J Plant Pathol 21: 144–153.

Pinto JMA, Pereira R, Mota SF, Ishikawa FH, Souza EA (2012) Investigating phenotype variability in *Colletotrichum lindemuthianum* populations. Phytopathology 102: 490–497.

Rispail N, De Matteis L, Santos R, Miguel AJ, Custardoy L et al. (2014) Quantum dot and supermagnetic nanoparticles interaction with pathogenic fungi: internalization and toxicity profile. Appl Mater Interfaces 6: 9100–9110.

Rohenkamp D, Kuska MT, Mahlein AK, Behmann J (2019) Hyperspectral signal decomposition and symptom detection of wheat rust disease at the leaf scale using pure fungal spore spectra as reference. Plant Pathol 68: 1188–1195.

Romero MP, Wise KA (2015) Development of molecular assays for detection of *Stenocarpella maydis* and *Stenocarpella macrocarpa* in corn. Plant Dis 99: 761–769.

Ruan C-P, Wu M-T, Huang C, Chang H (2011) Rapid detection of *Colletotrichum lagenarium*, causal agent of anthracnose of cucurbitaceous crops by PCR and real-time PCR. J Phytopathol 159: 276–282.

Ruiz E, Ruffner HP (2002) Immunodetection of *Botrytis*-specific invertase in infected grapes. J Phytopathol 150: 76–85.

Schenk PM, Kazan K, Manners JM, Anderson JP, Simpson RS et al. (2003) Systemic gene expression in *Arabidopsis* during an incompatible interaction with *Alternaria brassicola*. Plant Physiol 132: 999–1010.

Somai BM, Keinath AP, Dean RA (2002) Development of PCR-ELISA detection and differentiation of *Didymella bryoniae* and *Phoma* species. Plant Dis 86: 710–716.

Svircev AM, Gardiner RB, Mc Keen WE, Day AW, Smith RJ (1986) Detection by protein A- gold of antigen of *Botrytis cinerea* in cytoplasm of infected *Vicia faba*. Phytopathology 76: 622–626.

Ueng PP, Cunfer BM, Alano AS, Youmans JD, Chen W (1995) Correlation between molecular and biological characters in identifying the wheat and barley pathotypes of *Stagonospora nodorum*. Phytopathology 85: 882–887.

Úrbes-Torres JR, Leavitt GM, Guerrero JC, Guevara WD (2005) Identification and pathogenicity of *Lasidiplodia theobromae* and *Diplodia seriata*, the causal agents of Bot canker disease of grapevine in Mexico. Plant Dis 92: 519–529.

Vanderplank JE (1963) *Plant Diseases: Epidemics and Control*. Academic Press, New York.

Vanderplank JE (1978) *Genetic and Molecular Basis of Plant Pathogenesis*. Springer-Verlag, Berlin.

Walcott RR, Gitaitis RD, Langston DB (2004) Detection of *Botrytis aclada* in onion seed using magnetic capture hybridization and polymerase chain reaction. Seed Sci Technol 32: 425–438.

Wang X, Zheng W, Buchenauer H, Zhao J, Han Q et al. (2008) The development of a PCR-based method for detecting *Puccinia striiformis* latent infections in wheat leaves. Eur J Plant Pathol 120: 241–247.

Weier HU. Grau JW (1988) A programmable system to perform polymerase chain reaction. DNA 7: 441–447.

Wennecker M, de Jong PF, Joosten NN, Goedhart PW, Thomma BPHJ (2017) Development of a method for detection of latent European fruit tree canker (*Neonectria ditissima*) infection in apple and pear nurseries. Eur J Plant Pathol 148: 631–635.

Zhang AW, Harman GL, Riccioni L, Chen WD, Ma RZ et al. (1997) Using PCR to distinguish *Diaporthe phaseolorum* and *Phomopsis longicolla* from other soybean fungal pathogens and to detect them in soybean tissues. Plant Dis 81: 1143–1149.

Zheng Y, Luo Y, Zhou Y, Zang X, Duan X et al. (2013) Real-time PCR quantification of latent infection of wheat powdery mildew in the field. Eur J Plant Pathol 136: 565–575.

3 Biology of Phyllosphere Fungal Plant Pathogens

3.1 INTRODUCTION

Phyllosphere fungal pathogens have a different lifestyle and induce several different kinds of symptoms characteristic of their interaction with susceptible host plants. They adopt a biotrophic or necrotrophic lifestyle by producing various kinds of enzymes, toxins, or other metabolites that facilitate their entry into the plant tissue and proliferation in the initially infected cell(s). Successful infection of plants by fungal pathogens depends on the completion of different stages of pathogenesis. Different compounds known as effectors produced by the pathogens interact with preformed and postinfectional compounds produced due to the response of plants to the presence of pathogens. Inability of susceptible plants to recognize the pathogen-associated molecular patterns results in progression of pathogen development. On the other hand, resistant plants produce several compounds that counteract the adverse effects of effectors produced by the fungal pathogens leading to suppression of disease development. Knowledge of the biology of fungal pathogens may be useful to develop appropriate tactics that may arrest pathogen development in different crops.

Phyllosphere fungal pathogens have access to different epigeal (above-ground) plant organs such as stem, leaves, and inflorescence. Various diseases such as downy mildews, powdery mildews, rusts, leaf spots, blasts, gray molds, stem blight, and bunch rots are induced by them, resulting in varying degrees of loss. Fungal pathogens vary in their virulence, host range, mode of dispersal, overwintering capacity on alternative hosts in the absence of crops, survival under adverse environmental conditions, and response to chemicals applied to protect plants. Various aspects of fungal pathogen interaction with their host plants resulting in establishment of infection and consequent development of disease symptom in susceptible plants or suppression of disease development in resistant plants have been elucidated. The knowledge of the biology of the fungal pathogens will be useful for applying measures to reduce disease incidence and spread (Narayanasamy 2002, 2017).

3.2 BIOLOGICAL CHARACTERISTICS OF FUNGAL PATHOGENS

3.2.1 Morphological Characteristics

Among the microbial plant pathogens, fungus-like and fungal pathogens are highly evolved and they belong to the Eukaryotes, which enclose plants and animals. They have a well-developed thallus (body) comprising innumerable filamentous thread-like hyphae weaving into mycelium. Fungal populations increase rapidly by means of both asexual and sexual reproduction. Fungal pathogens produce different types of asexual and sexual spores. Taxonomy, nomenclature, and phylogeny are different aspects of fungal systematics which are considered in the study of biological diversity. Nomenclature of fungi is governed by the International Code of Botanical Nomenclature. True fungi represent a distinct kingdom. Molecular sequence data indicated that true fungi are more closely related to animals than to plants. Oomycetes are considered to be related to the yellow-brown algae based on the DNA sequences that form the basis for determining phylogenetic relationships. Oomycetes share a common ancestor with other members of the heterokont algae. Differences in the sexual reproduction, nuclear state of vegetative mycelium, cell wall composition, type of flagellae, and mitochondria distinguish Oomycota from true fungi. All major groups of true fungi are placed in Chytridiomycota, Zygomycota, Ascomycota, and Basidiomycota. The fungi that do not produce sexual spores are included in the phylum Deuteromycota (Webster and Weber 2007).

3.2.2 Plant Disease Cycle

The ability to induce disease in susceptible host plant species is known as pathogenicity, and this characteristic differentiates other fungi lacking pathogenicity, which are known as saprophytes, capable of growing on dead organic matter. Pathogens may differ in their levels of pathogenic potential (virulence), which is assessed based on severity of disease symptoms induced by the pathogen concerned. Disease severity induced by isolates of a fungal pathogen is scored using a standard scale of 1–9, reflecting ascending order of disease severity. Different cultivars or genotypes of a crop species called differentials are used to determine variations in the levels of virulence of the isolates from different geographical regions/host species. Physiologic races of the pathogen are identified based on the infection grades/types. The number of cycles of spore production and amount spores produced are factors of epidemiological importance. The ways in which a pathogen survives between two cropping seasons or during the periods when unfavorable environment prevails and the modes of dispersal of the pathogen from infected plant to healthy plants are important aspects of plant disease cycles. Fungal pathogens initiate infection following germination of infective propagules (sporangia, conidia, or sclerotia) by penetrating and colonizing different host tissues. *Fusarium veriticillioides*,

DOI: 10.1201/9781003456827-3

infecting maize ears, produces mycotoxins in grains. Two methods of inoculating the ears with *F. veritcillioides* were employed to investigate ear symptomatology and morphological resistance mechanisms in maize inbred lines. Kernel bulk density was predictive of fumonisin levels. The inundative inoculation method resulted in greater disease severity and fumonisin contamination than point inoculation. The inoculation methods facilitated infection of different ear tissues by *F. verticillioides*. Cob morphology was a more important component of resistance for point inoculation method (Morales et al. 2018).

3.2.2.1 Reproduction of Fungal Pathogens

Increase in population of fungal pathogens occurs in asexual and sexual phases. In the anamorphic (asexual phase) of life cycle, asexual spores such as sporangia or conidia are produced in large numbers in specialized structures known as pycnidia, acervuli, or synnemata. Sexual spores are formed in the teleomorphic phase, and they are resistant to adverse environmental conditions, helping the fungal pathogens in overwintering between two crop seasons. The importance of infection of wheat basal leaves by blast pathogen *Magnaporthe oryzae* as source of inoculum was assessed. *Magnaporthe oryzae* could produce millions of conidia on small amounts of wheat basal leaves. A mean of 169,000 conidia were produced on 1 gram of dry basal leaves of a highly susceptible wheat cultivar under optimum conditions. Wheat basal leaves infected by *M. oryzae* could function as potential sources of inoculum (Cruz et al. 2015). Field isolates of *M. oryzae* were analyzed for mating type (MAT), DNA fingerprinting, and disease severity on differential rice lines. All MAT1-1 and MAT1-2 isolates were virulent on lines harboring 22 major resistance genes used as differentials. Populations of *M. oryzae* could exclusively remain in the asexual phase in rice fields during different seasons in China (Li et al. 2016). The homothallic nature of *Pseudoperonospora humuli* was revealed by inoculating hop leaves with diverse monosporangial isolates, and oospores were readily formed within hop leaves. However, the oospores did not germinate in planta (Gent et al. 2017).

Asexual reproduction of the *Podosphaeria xanthii* KMP-6N strain was studied using melon plants grown hydroponically. Cycles of conidial release were repeated on several melon leaves. At maturity, conidia were released without causing growth and septation of generative cells (Takikawa et al. 2015). *Zymoseptoria tritici* (Septoria tritici blotch) produced ascospores (sexual) and pycnidiospores (asexual) constituting primary and secondary inoculum, respectively. A two-stage inoculation method was applied under controlled conditions to compare the virulence of ascospores and pycnidiospores. Neither lesion size nor pycnidial density was influenced by the spore forms (Morais et al. 2015). Optimal conidial production of *Elsinoe ampelina* (grapevine anthracnose), occurred at 21° C for 24 h in the dark under controlled conditions. Approximately 5×10^6 conidia were produced under these conditions, and conidial production ceased at 25° C in the dark (Li et al. 2018). In another investigation, conidial production by *E. ampelina* of ten isolates from Australia and seven isolates from Brazil was induced. The results indicated the variability in conidial production capacity of Australian and Brazilian isolates from different locations (dos Santos et al. 2018). Members of *Colletotrichum acutatum* species complex remain dormant in asymptomatic plant tissues. No relationship could be established between conidial masses of *C. acutatum* in leaves and extent of fruit infection. However, the asymptomatic apple leaves might be potential sources of inoculum for fruit infection (Børve and Stensvand 2017).

Several cycles of asexual propagule production by *Phytophthora infestans* (potato late blight disease) may occur in a crop season when favorable conditions exist. *P. infestans* is heterothallic, requiring two mating types (A1 and A2) for sexual reproduction. Sexual reproduction was found to be a consistent component of the life history of *P. infestans* in central Mexico (Grünwald and Flier 2005). The British and Irish populations of *P. infestans* have been displaced due to immigration of novel strains from other locations. Reemergence of the Ib mitochondrial DNA haplotype in the British and Irish *P. infestans* populations was associated with the 6_A1 genotype. Appearance of the 6_A1 genotype in these populations might be due to migration of isolates. Use of different markers for pathogen monitoring may offer advantages over employing a single approach (Kildea et al. 2013).

Primary propagation of downy mildew pathogen *Pseudoperonospora cubensis* is clonally by sporangia. Formation of oospores was observed under natural conditions, but not under laboratory conditions. The heterothallic nature of *P. cubensis* required two opposite mating types, A1 and A2, for sexual reproduction. Oospore production was abundant on cucumber and melon. (Cohen and Rubin 2012). Asexual and sexual phases constitute the disease cycles of *Peronospora effusa* (spinach downy mildew disease). Sporangiophores, emerging from stomata primarily on the abaxial surface of leaves, produced sporangia carried by wind or rain splash. Sporangia caused infection on foliage of susceptible plants at the cotyledon stage and progressed rapidly in true leaves. The infection may remain asymptomatic during a latent period, and symptoms may appear during the postharvest stage (Figure 3.1) (Subbarao et al. 2018). Two different mating types (P1 and P2) are essential for sexual reproduction in *P. effusa*. Thick-walled oospores are formed within infected leaves and seeds. Sexual recombination in the heterothallic pathogen can generate extensive variation in virulence resulting in emergence of new isolates with higher levels of virulence (Kandel et al. 2019). The viability of oospores of *P. effusa* present on spinach seeds was assessed. Spinach seeds could carry both asexual and sexual spore to long distances and introduce the pathogen into new locations (Kunjeti et al. 2016). Occurrence and distribution of *Pseudoperonospora cubensis* (cucurbit downy mildew) in cucurbits in the United States was investigated. Mating type was significantly associated with host type. All isolates from cucumber belonged to the A1 mating type,

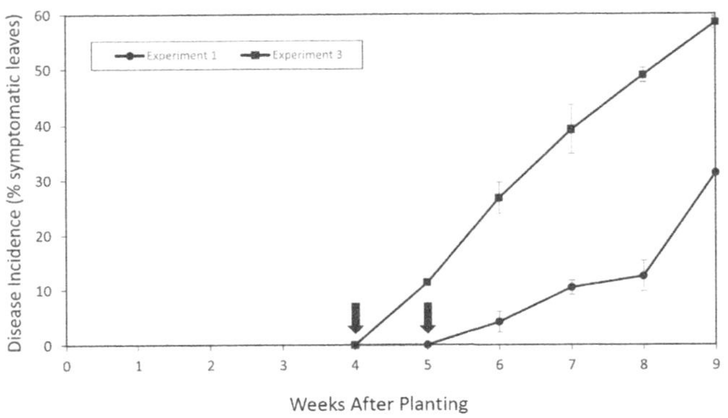

FIGURE 3.1 Detection of *Peronospora effusa* in spinach leaves by symptom expression and pathogen DNA by PCR assay. Pathogen DNA could be detected during latent period, though symptoms appeared at 5 weeks after planting.

Source: (Courtesy of Subbarao et al. 2018 and with kind permission of the American Phytopathological Society, MN, USA)

whereas isolates from squash and watermelon were of the A2 mating type (Thomas et al. 2017a). *Pseudoperonospora humuli* (hop downy mildew) formed oospores readily in large numbers in diseased hop tissues, as the pathogen was homothallic. Single zoospore isolates had similar ability to produce oospores within hop leaves as the sporangial isolates. Oospores derived from monosporangial isolate of *P. humuli* after overwintering under field conditions failed to be pathogenic (Gent et al. 2017).

The process of ascocarp development, ascospore maturation, and dispersal of ascospores of *Erysiphe necator* (grapevine powdery mildew) were studied. Ascospore discharge was assessed using silicon-coated slides. Vine trunk sections contained chasmothecia (ascocarps), which had mature ascospores or immature ascospores, and some were empty (Rossi et al. 2010). A period of superficial growth of colonies of *Erysiphe necator* preceded initiation of asexual spore production. Quorum-sensing signal in *E. necator* commenced at the center of colony and stimulated conidiation throughout the colony. In addition a signal repression or another signal to cease conidiation might operate once pairing occurred and initiation of the ascigerous stage for overwintering commences (Gadoury et al. 2012). The conidia of *Blumeria graminis* f.sp. *tritici* are produced in large numbers that are disseminated by wind, and they are important for repeated infections in the same crop or other crops away from the fields. Sexually produced chasmothecia contain asci which enclose ascospores that are ejected out and transported by wind to different distances. Chasmothecia were embedded in a dense mat of secondary hyphae. Ascospores were differentiated in most chasmothecia following submersion in water. They contained asci filled with protoplasm appeared within asci in most chasmothecia submerged in water. At 72 h, most of the protoplasm disappeared from asci, and at this stage asci contained only wall-differentiated ascospores. The ascospores, under moisture conditions, germinate by producing two germ tubes one at each end. The germ tubes penetrate epidermal cells and form powdery mildew colonies on wheat leaves after 5–7 days after ascosporic infection. Observation on wheat leaves revealed formation of one to four germ tubes from each ascospore on wheat seedlings. Septate hyphae were formed from the germ tubes. This was followed by production of penetration peg and haustorium inside the epidermal cells. At 3–5 days after germination of ascospores on young wheat leaves, conidiophore initials and young conidiophores with undifferentiated conidia were formed from the powdery mildew colonies (Jankovics et al. 2015).

Production of single-celled conidia by *Leveilulla taurica* (pepper powdery mildew) at the end of conidiophores which emerge from internal mycelium is the characteristic feature, whereas *Erysiphe* spp. and *Blumeria* spp. produce conidia in chains on conidiophores that are produced from epiphytic mycelium. The presence of intercellular mycelium in mesophyll was detected by light and transmission electron microscope (TEM) (Zheng et al. 2013). *Venturia inaequalis* (apple scab) induces primary lesions on leaves due to infection by spores released at bud burst from overwintering ascocarps present in leaf litter. The relative importance of conidia or ascospores as primary inoculum was investigated in the orchard in the United Kingdom. The results indicated that conidia might be a minor contributor as primary inoculum for infection (Passey et al. 2017).

Barberry and Mahonia plants have been identified as alternate hosts of *Puccinia striiformis* f.sp. *tritici (Pst)* under controlled conditions. Single-uredium isolates were characterized by virulence tests and simple sequence repeat (SSR) marker analysis. Cluster analysis with SSR marker revealed existence of two genetic groups which were significantly correlated to the two virulence groups. The results suggested that asexual reproduction was predominant in the *P. striiformis* f.sp. *tritici* population in the Pacific Northwest (PNW) as well as in other regions of the United States (Cheng and Chen 2014). *Phakopsora pachyrhizi* produces only three spore types viz., urediniospores, teliospores, and basidiospores. Urediniospores formed during asexual phase germinated, forming a single germ tube and an appressorium. The appressorial cone initiated the penetration or infection hypha to enter the epidermal cell by turgor pressure independently of melanin accumulation (Chang et al. 2014). In the susceptible plants, primary haustoria are produced in the mesophyll cells, and secondary haustoria are formed within 12 days and production of dome-shaped uredinia begins to generate numerous urediniospores. Infected leaves have a rusty appearance due to the presence of large number of uredinia that break open the epidermal layer to release urediniospores into the atmosphere. In the incompatible (resistant) interaction between soybean and pathogen, restriction of hyphal development in the mesophyll cell tissue was observed resulting in necrosis of infected cells (Vittal et al. 2014).

Phyllosticta citricarpa (citrus black spot), requiring the presence of isolates of different MAT idiomorphs and mating type locus, was characterized (Tran et al. 2017b). Mating type genes, controlling sexual reproduction, were located in the *MAT* locus. The sequence of DNA lyase gene was identified by genome sequence analysis of two individual *P. citricarpa* isolates. *P. citricarpa* had complete and apparently functional copies of *MAT* genes, containing α-1 and HMG domains present in different isolates. *P. citricarpa* mating idiomorphs were differentiated by multiplex polymerase chain reaction (PCR) assay (Amorin et al. 2017). Conidiophore production dynamics of *Guignardia bidwellii* (grapevine black rot) were determined under conditions of 25° C, 100% relative humidity. The production of conidia declined over washings and was almost completely depleted after four washings. The decline in conidial production was faster at 29° C than at 20° C (Onesti et al. 2017).

Hop powdery mildew pathogen, *Podosphaeria macularis,* is heterothallic, and it produces ascocarps (chasmothecia) under natural conditions in some locations. Genomic sequencing of a MAT1 isolate revealed the structure of the MAT1 locus and presence of MAT1-1-3 showing similarities to other powdery mildew pathogens. Absence of chasmothecia formation by *P. macularis* in the PNW was primarily due to the absence of the opposite mating idiomorph MAT1-2 (Wolfenberger et al. 2015). *Botrytis cinerea* has a heterothallic, bipolar mating type system with occasional of occurrence of homothallic sexual reproduction. MAT1 genes and flanking regions in the reference strains SAS50 (MAT1-1) and SAS405 (MAT1-2) and their monoascospore progeny were analyzed. Constitutive expression of all the four mating type genes was confirmed by real-time (RT)-PCR assays (Angelini et al. 2016). *Cochliobolus lunatus* (maize foliar spot disease) has a single mating type (*MAT*) locus with two idiomorphs controlling sexual reproduction. A *MAT1-1* or *MAT1-2* idiomorph was identified in single isolate, and sexual reproduction occurred heterothallic mode in vitro (Lu et al. 2018). *Pyrenophora teres* f. *teres (Ptt)* and *P. teres* f. *maculata (Ptm)* (net and spot forms). respectively. of barley net blotch disease, could be differentiated based on MAT locus-specific single nucleotide polymorphisms (SNPs). Primers designed based on the identified SNP patterns were employed to differentiate each atypical isolate from highly virulent forms. Genetic variation at the MAT locus correlated with variation in the form-level differentiation in *P. teres* (Lu et al. 2013). Mating-type loci and population genetic diversity of *Venturia carpophila* (peach scab) were investigated. Mating-type loci structures were consistent with those of *V. effusa* and *V. inaequalis.* Among 81 isolates, 69 multilocus genotypes were identified, indicating its genetic diversity (Bock et al. 2021).

3.2.2.1.1 Modes of Overwintering of Fungal Pathogens

Infested rice plant tissues stored indoors form the primary substrate for overwintering of *Magnaporthe oryzae*. The overwintering location of *M. oryzae* infecting wild, green and giant foxtails was studied. *M. oryzae* survived in blast lesions on leaves and seeds. After harvesting the rice, the withered standing plants might be available for infection of wild grasses which have higher winter survival frequency under field conditions (Akase and Kusaba 2017). Leaf blast incidence was significantly higher in the presence of infected rice residues. Survival of *M. oryzae* decreased rapidly on stems buried in the soil. Sporulation of the pathogen was observed in stems left on the mulch for up to 18 months (Raveloson et al. 2018). Saprophytic survival of the wheat blast disease pathogen *M. oryzae* in wheat residues was investigated. Inoculated air-dried plant parts and spikes obtained from field were examined for sporulation. Survival of blast pathogen decreased rapidly on the rachis compared with stems and leaves. Sporulation continued for up to 5 months (Pizolotto et al. 2019). Overwintering and non-overwintering regions of *Puccinia striiformis* f.sp. *tritici (Pst)* (wheat stripe rust) were identified. Active sporulation, indicating successful overwintering, was observed in samples from most sites. Increasing altitude led to decreasing incidence of visible *P. striiformis* symptoms and sporulation. *P. striiformis* f.sp. *tritici* could overwinter in young green leaves as latent infection established in autumn (Ma et al. 2016). *Botrytis cinerea* (grapevine bunch rot and grey mold) produced conidia from infected debris left on the ground in vineyards and in the canopy in the season following harvest. Necrotic overwintering grapevine debris might produce *B. cinerea* conidia throughout the following growing season, contributing to infection potential for bunch rot development (Jaspers et al. 2013). *Podosphaeria macularis* (hop powdery mildew) did not produce ascospores in the PNW. Bud infection appeared to occur over an extended period of time in the previous season, which might allow the pathogen to perennate. Preventive measures have to be taken to reduce bud infection by *P. macularis* by eliminating infected crown buds (Gent et al. 2018).

3.2.2.1.2 Role of Host Range of Fungal Pathogens

Fungal pathogens may have a wide or restricted host range. The fungal pathogens with a wide host range may have continuous infection chains in the absence of crop plants. Two strains of *Peronospora cubensis* and *P. humuli* reproduced on two wild cucurbit relatives *Bryonia dioca* and *Sicyos angulatus.* Ability of *P. cubensis* to infect the perennial *Bryonia dioca* and *Humulus tupulus* might favor overwintering to some extent in temperate regions (Runge and Thines 2012). As rice is cultivated all year round in tropical regions, susceptible plants are available for continuous infection by *M. oryzae*. Furthermore, different weed species could function as alternative hosts providing inoculum for the rice crops. Self-sown volunteer plants and wild relatives of crop plant species also may serve as important sources of inoculum for crops sown or planted in the next season. *M. oryzae* populations from nonrice hosts such as *Digitaria sanguinalis, Eleusine indica, Lolium boucheanum, L. multiflorum*, and *Festuca elatior* could serve as sources of inoculum for rice crops (Han et al. 1995). Phylogenetic analyses based on SNPs distributed across entire genomes indicated that *Magnaporthe* isolates associated with *Triticum* and *Avena* form a single monophyletic clade clearly distinct from the *P. oryzae* clade (Yoshida et al. 2016). The ability of *Fusarium graminearum* (wheat Fusarium head blight [FHB] disease) to infect other crop plant species was assessed, using artificially inoculated wheat florets as inoculum. Barley, rye, triticale, canary seed, flax, canola (*Brassica napus* and *B. rapa*), mustard, bean, field pea, lentil, and chickpea crops could be alternative sources of inoculum of the pathogen to wheat crops (Chongo et al. 2001). *F. graminearum* could infect noncereal crops such as canola, potato, soybean, and sugar beet. Inoculum produced in these crops might contribute significantly to the development of FHB epidemics in wheat (Burlakoti et al. 2007). Of the plant species tested, 54 weed species were identified as alternative hosts of *F. graminearum*. Strains belonging to 15-diacetyldeoxynivalenol (15-ADON) and nivalenol (NIV) chemotypes were differentiated among isolates obtained from weed plant species, which could function as alternative host plants capable of serving as sources of inoculum for wheat crops (Mourelos et al. 2014). In a later investigation, weed samples from ten weed species in China, yielded seven *Fusarium* spp. *F. asiaticum* was the most abundant, followed by *F. fujikuroi*, *F. proliferatum*, *F. graminearum*, *F. tricinctum*, *F. acuminatum*, and *F. sporotrichioides*. Genotype and mycotoxin analyses showed that some of the *F. asiaticum* isolates produced deoxynivalenol (DON) with 3-acetyldeoxynivalenol (3-ADON), while other isolates produced NIV. Both 3-ADON and NIV chemotypes of *F. asiaticum* were pathogenic to rice causing FHB disease. NIV chemotypes were significantly more aggressive than 3-ADON chemotypes. Mycotoxins DON, NIV, and zearalenone were present at low concentrations in naturally infected weed samples (Dong et al. 2020).

Hairy nightshade (*Solanum physalifolium*) was susceptible to *Phytophthora infestans* (potato late blight). Single-lesion leaves infected by *P. infestans* were collected from potato and hairy nightshade plants to determine phenotypic and genotypic population differentiation of *P. infestans*. The odds of infection for isolates from hairy nightshade showed a significantly shorter latency period and higher sporulation capacity compared with potato isolates. The weed host might function as a filter for selection of populations with higher virulence level (Grönberg et al. 2012). In another investigation, the host range of three clonal lineages, US-22, US-23, and US-24, was determined under controlled conditions on detached leaves of cultivated solanaceous plants and solanaceous weed species. Hairy nightshade plants supported abundant sporulation and symptom expression indicating the possibility of this weed becoming potential sources of inoculum for potato and tomato for these lineages of *P. infestans* (Johnson and Gevens 2014). *Stemphylium* sp. (sugar beet yellow spot) could also infect potato, white mustard (*Sinapsis alba*), red beet (*Beta vulgaris*), spinach, and fat hen (*Chenopodium album*), and these host plant species might be sources of inoculum for sugar beet (Hanse et al. 2015). The sunflower powdery mildew pathogen *Golovinomyces orontii* infected weeds and cucurbits present in and around sunflower fields. Common bean, *Zinnia elegans*, and *Luffa acutangula* showed infection by *G. orontii* (Mulpuri et al. 2016). Weed species present in strawberry fields were infected by *Colletotrichum fructicola* (strawberry anthracnose). *C. fructicola* produced acervuli on leaves of *Amaranthus blitum*, *Digitaria ciliaris*, and *Soloiago oleraceus* (Hirayama et al. 2018). *Colletotrichum acutatum*, causing citrus postbloom fruit drop disease, could survive on citrus leaf surfaces after producing appressoria. The weed species allowed survival of *C. acutatum* for different periods after infection (Frare et al. 2016). *Cercospora beticola* (Cercospora leaf spot, CLS) was pathogenic to *Brassica kaber*, *Chenopodium album*, *Carthamus tinctorium*, *Rumex obtusifolius* and *Spinacia oleracea* (Knight et al. 2019). Sweet basil (*Ocimum basilicum*) was seriously affected by downy mildew disease caused by *Peronospora belbahrii (Pb)*. *Salvia eigii*, *S. pinnata*, *S. fructicosa*, *Rosmarinus officinalis*, *Nepeta curviflora*, *Microneria fructiossa*, and *Agastache* spp. developed disease symptoms with sporulation by the pathogen (Naim et al. 2019).

New physiologic races of wheat stem rust pathogen, *Puccinia graminis* f.sp. *tritici (Pgt)*, may be formed due to genetic recombination following spermatization during sexual reproduction in the alternate host plant (barberry). Eradication of barberry was advocated to prevent formation of new races that can infect wheat cultivars with resistance to prevailing races of *P. graminis* f.sp. *tritici* as a disease management strategy (Narayanasamy 2002). Severe rust infection by *Puccinia graminis* f.sp. *tritici (Pgt)* on *Barberis shensiana*, *B. brachypoda*, *B. potanenii*, *B. souliean*, and *B. aggregata* were observed during field surveys. Wheat seedlings with the *Sr31* resistance gene were not infected by aeciospores from naturally infected barberry plants. (Zhao et al. 2015). Grass species were inoculated with aeciospores of *Puccinia striiformis* f.sp. *tritici (Pst)* (stripe rust) from naturally infected *Berberis chinensis* and *B. koreana*, resulting in production of uredinia on *Poa pratensis*, containing typical urediniospores of *Pst*. *Berberis* spp. was confirmed as alternate hosts for *P. striiformis* f.sp. *tritici*, facilitating production of new physiologic races of stripe rust in China (Jin et al. 2010b). Inability of *Puccinia striiformis* f.sp. *tritici* (*Pst*) to infect barberry in the United States in the PNW was investigated. Optimal conditions of prolonged leaf wetness conditions was required for *P. striiformis* f.sp. *tritici* to infect barberry plants in the region (Wang and Chen 2015). The 'Kranich' race of *Puccinia striiformis* f.sp. *tritici* (*Pst*) contributed to the replacement of existing *Pst* populations in several European countries, as this race had high capacity for teliospore production and showed similarity to populations in the center of diversity of *Pst* in the near-Himalayan region. SSR markers were employed to randomly selected 23 progeny isolates. Eight heterozygous markers in the parental isolate revealed segregation in the progenies leading to novel multilocus genotypes, confirming recombination following sexual reproduction. The results indicated a high sexual capacity of the Kranich race, highlighting risks associated with sexual reproduction under suitable natural conditions for rust development (Rodriguez-Algaba et al. 2020). *Puccinia sorghi* infects maize (primary host), and *Oxalis* spp., a perennial weed species, functions as an alternate host. *Oxalis conorrhiza* plants were inoculated with the basidiospores, and formation of spermagonia with spermatia and aecia with aeciospores on *O. conorrhiza* plants was observed. Maize (corn) plants were inoculated with aeciospores, which infected corn plants producing uredinia and telia, confirming the role of *O. conorrhiza* as the alternate host for *P. sorghi* in Argentina (Guerra et al. 2019). The host range of two Australian isolates of *Uromyces viciae-fabae* was investigated by inoculating field pea, chickpea, lupin, lentil, and mungbean genotypes. These two isolates exhibited race specificity on *Vicia faba* (faba bean), and two faba bean genotypes showed partial resistance. They could colonize field pea seedlings and produced uredinia in the greenhouse. This pathogen induced variable infections in other leguminous crops tested, ranging from small infection sites with rare haustoria formation in mungbean to more extensive growth and development of potential uredinial structures in chickpea (Ijaz et al. 2020).

3.2.2.1.3 Role of Infected Seeds and Propagules

Fungal pathogens may be present on the surface and in deeper tissues of seeds and vegetatively multiplied propagules. These infected plant tissues may play an important role in pathogen survival and dispersal to short and long distances. *M. oryzae* could sporulate on infested seeds left on the soil surface under field conditions and dispersal of conidia by wind or rain-splash could initiate infection on rice seedlings (Long et al. 2001). *Fusarium graminearum* (wheat FHB) produces macroconidia in sporodochia on infected heads and debris in the asexual stage. Survival of *F. graminearum* in Fusarium-damaged kernels (FDKs) was assessed under field conditions. The pathogen could be recovered from FDKs left on the soil surface or buried at 5 or 10 cm depth. Ascospores were produced from the perithecia present on FDKs left on the soil surface. The results revealed the role of FDKs as means of survival and potential sources of primary inoculum for infection of wheat earheads (Inch and Gilgert 2003). Isolates of *F. graminearum* from infected potato and sugar beet might serve as sources of inoculum for wheat crops and also facilitate survival of the pathogen (Burlakoti et al. 2007). As *Cladosporium variabile* and *Stemphylium botryosum* are seedborne, spinach seed lots were evaluated for the extent of seed infection. Pathogenic isolates of *S. botryosum* were detected more frequently than *C. variabile*. These pathogens could invade seed embryos, and it was difficult to eliminate them, and infected seeds could help their survival (Hernández-Perez and du Toit 2006). *Ramularia collo-cygni* DNA, representing pathogen presence, could be detected in several harvested barley samples (Havis et al. 2014).

Passage or transmission of *Phytophthora infestans* from potato tubers to emerging sprouts was investigated. Transmission of *P. infestans* from the base of below-ground potato stems to above-ground shoots was assessed, using stem inoculation assay. Isolates BP05 (US-8), WA10.1 (US-24), and 110B (US-11) were inoculated and sporangia developed above-ground stems. Sporangial production was greater in cv. Russet Norkotah than in cv. Umatilla Russet, when inoculated near the point of seed piece attachment. Age of stem did not seem to have any effect on the transmission of *P. infestans* from the base of below-ground stems to shoots (Johnson and Cummings 2013). Survival of *P. infestans* (potato tuber blight and late blight) as sporangia was investigated under ambient, relatively dry atmosphere conditions. The sporangia might survive longer when cool, cloudy days and rainy days were available (Sunseri et al. 2002). Survival of *P. infestans* infecting potato tubers and tomato fruits was determined. Tuber-bearing solanaceous plants offer conditions for survival of asexual phase of *P. infestans*. Under natural conditions, tubers of solanaceous plants are dormant until the next season, and the tubers provide safe conditions for the pathogen until favorable conditions become available (Fry et al. 2013). Among the isolates of the four mitochondrial haplotypes of *P. infestans*, only the Ia mtDNA haplotype was identified in infected potato leaves from the famine era (historical Irish Famine) herbarium specimens collected in England, Ireland, and Europe in the 19th century. Detection and identification of the same Ia mtDNA haplotype of *P. infestans* in three diseased tubers indicated the presence of the founder Ia mtDNA haplotype in potato tubers. The results showed that the founder Ia mtDNA haplotype could survive in potato tubers after 1846 (Ristaino et al. 2013). The mode of survival of *Phytophthora infestans*, in winter in Wisconsin, the United States, was investigated, since relatively new clonal lineages, US-22, US-23, and US-24, persisted within production region. *P. infestans* could survive over the winter season in tomato seed, indicating the need for use of disease-free tomato seeds to restrict tomato late blight disease incidence and spread in the next season (Frost et al. 2016).

3.2.2.1.4 Role of Abiotic Carriers in Pathogen Survival

The occurrence of epidemics, even in the absence of alternate or green bridge hosts, necessitated investigating the possibility of urediniospores of *Puccinia graminis* f.sp. *tritici (Pgt)* (wheat stem rust) surviving on abiotic carriers. Viability of urediniospores was retained at 36/14° C day/night for a maximum of 300 days on denim and jute. Urediniospores recovered even after 300 or 365 days from lower temperature regimes could initiate infections on wheat seedlings. The results showed that inert materials had the potential for functioning as carriers of urediniospores for long periods (Barua et al. 2018).

3.2.2.1.5 Dispersal of Fungal Pathogens

Fungal pathogens may be dispersed in different ways which may facilitate their short- or long-distance dissemination.

Fungal pathogens infecting foliar organs of host plants are generally disseminated via air (wind), and some of them may have multiple modes of dispersal. Various techniques such as microscopy, immunoassays, and DNA-based methods have been effective in detection and quantification of specific fungal species in air samples. Microtiter immunospore trapping (MTIST) device was used to monitor airborne inoculum of *Mycosphaerella brassicae* and *Botrytis cinerea*, and a relationship was established between airborne inoculum and severity of disease on Brassica in the field (Kennedy et al. 2000). *Ramularia beticola* infecting sugar beet was detected in air samples using qPCR assay, 14 days ahead of the appearance of the first visible symptom on sugar beet plants (Wieczorek et al. 2013). The TaqMan-based qPCR format was applied to quantify populations of *Botrytis squamosa* in air, and the potential as a risk predictor for onion was revealed. Periods of high risk of grape powdery mildew infection were indicated by estimating the numbers of conidia in air (Carisse et al. 2009a, 2009b). Fungal spores are dispersed in dust by air currents from few meters to hundreds of kilometers, depending on air turbulence, velocity, UV radiation, and desiccation tolerance. Long distance dispersal via 'wind highways' was reported in some pathogens, as in the case of coffee rust pathogen *Hemileia vastatrix*, which was considered to have been dispersed from Africa to Brazil. Fungal phylotypes were identified in dust samples and 386 different species of putative fungal pathogens were identified across all samples in the United States. By limiting the geographical distribution of multiple plant pathogens across the United States, it was possible to delimit five different sets of fungal pathogens. *Stemphylium solani*, the causal agent of tomato gray leaf spot was found to be restricted to California, based on molecular signatures in dust samples, although the pathogen has worldwide distribution (Dietzel et al. 2019). *Phytophthora infestans* has several mechanisms of dispersal. Short- and long-distance dispersal may be achieved by rain-splash of sporangia during irrigation events and wind. Sporangia may be dislodged from sporulating lesions by wind and deposit them on plant surfaces. Farm equipment, animals, and clothing may carry them to other locations. Dispersal over long distances may require human activities also. A highly efficient method of transporting plant pathogens is through seeds and propagules to different states within the country or intercontinental locations of diseased seed materials. *P. infestans* could sporulate on discarded infected plant materials and become sources of inoculum for neighboring crops (Narayanasamy 2017).

Disease distribution of dispersal distances (known as dispersal kernel) for *Mycosphaerella fijiensis* (banana black Sigatoka disease) was studied. Disease gradients induced by sexually and asexually produced spores of wind-dispersal were measured. Although ascospores distributed evenly in all directions, average dispersal distance was greater in two directions without obvious correlation with wind patterns. For conidia, few dispersal events occurred outside the source plot (Rieux et al. 2014). Spore trapping using the selective medium was employed for obtaining airborne spores of *Microdochium nivale* (winter wheat scab) under field conditions in wheat fields in Hokkaido, Japan. The production of characteristic pinkish colonies of the pathogen was observed on the selective medium (Hayashi et al. 2014). Monitoring pathogen spore concentrations of air around *Brassica* crops was taken up. DNA-based techniques were employed to quantify *Leptoshphaeria maculans* (Phoma stem canker) and *Pyrenophora brassicae* (light leaf spot) in air samples. The nested PCR format was highly effective in detecting and quantifying populations in the air of two important fungal pathogens infecting Brassica crops (Caledron et al. 2002). Spread of *L. maculans* in areas colonized earlier by *L. biglobosa* was assessed by PCR assays using species-specific primers. The rates of spread per year across spring oilseed rape and 70 km per year across winter oilseed rape in China were estimated (Zhang et al. 2014). Nested PCR assay was developed based on microsatellite regions of the pathogen DNA for detection of *Monilinia fructicola* (blossom blight and brown rot of stone fruits). The density of spores in the air trapped in the Burkard spore trap was determined using the nested PCR assay. Spore density threshold as a basis for taking a decision for fungicide application to manage the disease caused by *M. fructicola* was determined (Ma et al. 2003). In a further study, real-time PCR assay amplified only the DNA from isolates (ten) of *M. fructicola* resulting in production of the expected 390-bp PCR product. High spore densities in the air were detected in March and again in July, decreasing during April–June (Luo et al. 2007).

A PCR-based assay involving use of species-specific primers (uncin 144 and uncin 511) was employed to detect *E. necator* in the air samples. Under field conditions, the PCR procedure detected the pathogen in air samples within hours of sample rod collection and prior to disease onset. The initial PCR detection of *E. necator* of the season occurred during seasonal ascospore releases caused by precipitation. Detection of *E. necator* ascospores was possible several days before the appearance of visible symptoms of powdery mildew disease (Falacy et al. 2007). A PCR assay was developed for specific detection of *Pseudoperonospora humuli* and closely related to *P. humuli* (hop downy mildew), infecting cucurbits and to monitor airborne inoculum in the hop orchards. During eight years of monitoring, *P. humuli* could be detected in air samples not later than 8 days before the first appearance of basal (Gent et al. 2009). *Peronospora effusa* (spinach downy mildew) is a polycyclic pathogen with several cycles of asexual reproduction in a crop season. Sporangia from infected lesions became airborne and dispersed locally within the field or across different fields. The presence of airborne spores of *P. effusa* was detected continuously. Increase in detectable DNA contents of the pathogen during major spinach growing season was recorded in Salinas Valley (Choudhury et al. 2016). A combination of spore traps and quantitative amplification of pathogen DNA revealed a strong correlation between physical spore counts and the corresponding DNA copy number amplified by the quantitative PCR assay (Klosterman et al. 2014). Dispersal of *P. effusa* through spinach seeds has been studied. Oospore formation in the susceptible cultivar Viroflay leaf tissue was routinely recorded in the Salinas Valley, indicating that both mating types *P. effusa* were present. Oospore formation occurred both in seed crops and spinach (Kandel et al. 2019).

Wind-dispersed urediniospores of rust pathogens can be disseminated over long distances following prevailing winds and infect wheat crops across regions and even continents. Spread of *Sr31* virulent races of *Puccinia graminis* f.sp. *tritici* in the Ug99 race group and other races through wind-disseminated urediniospores caused severe epidemics and localized outbreaks in Europe and Central Asia (Bhattacharya 2017; Shamanin et al. 2018). Long-distance transport of *P. striiformis* f.sp. *tritici* urediniospores after oversummering and overwintering in Chinese provinces was studied. The results suggested that the pathogen via wind could interchange frequently among the provinces. Long-distance transport of inoculum might be responsible for the occurrence of stripe rust in the spring (Wang et al. 2010). The physical characteristics of urediniospores of *P. pachyrhizi* are such that the energy in a turbulent atmosphere is often able to lift large numbers of spores from a plant canopy and keep them airborne for long periods of time. Hence, most urediniospores are released when the atmosphere is turbulent and thus conducive for transport to other locations. Urediniospores remain viable for many days after exposure to a range of temperature and humidity conditions typically existing in the lower atmosphere during soybean growing seasons. Dark pigmented urediniospores could provide protection against adverse effects of ultraviolet radiation (Isard et al. 2006).

Glume infections by *Parastagonospora nodorum* (wheat Septoria nodorum blotch, SNB) resulted in seed infections and then seedling infection after sowing infected seeds. Infected wheat debris is an important inoculum source, as the pathogen is dispersed by splash-dispersal of conidia in rain droplets or through wind-spread ascospores from ascocarps overwintering on the wheat debris. Spore production occurred repeatedly in the same pycnidia Ascospores produced in pseudothecia could be spread over long distances with air currents. Ascospores could be trapped continuously during a season and ascospores were discharged in the presence of free water for at least 8 months from the pseudothecia formed in wheat debris. Seedborne inoculum is primarily responsible for long-distance dispersal of the pathogen and also for the spread to wheat crops planted in the subsequent seasons (Ficke et al. 2018). The effect of a combination of rain and wind on the spread of conidia of *Ascochyta rabei* to different distances was assessed by placing chickpea traps. Trap plants were transferred to a greenhouse (20° C) to allow lesion development. Trap plants placed at all distances were infected in all sprinkler irrigation and rainfall events. Number of lesions reached the maximum on trap plants placed closest to infected plots, with lesion number decreasing with increase in distance from infected plots (Khaliq et al. 2020).

Potato crops are commonly affected by two foliar pathogens, *Alternaria solani* (early blight) and *A. alternata* (brown spot). They share similar disease cycles, but vary in conidial morphology and host range. Conidial concentrations of *A. alternata* were greater and more variable than those of *A. solani*. The atmospheric concentrations of *A. alternata* were higher than those of *A. solani*, indicating greater potential of *A. alternata* for long-distance dispersal via wind. The results indicated that potato fields might be the source of conidial inoculum for *A. solani*, but *A. alternata* conidial inoculum might reach potato crops, not only from potato but also from other sources (Ding et al. 2019a, 2019b). Buds were found to be more important source of inoculum for *Colletotrichum acutatum* (bitter rot of apple) than twigs. Summer inoculum was primarily splashed up from ground. In spring, inoculum was recovered in similar amounts from all heights up to a meter, suggesting that splash dispersal might occur from the canopy as well as from the ground (Everett et al. 2018). Importation of wheat seeds by different countries in large quantities facilitated rapid spread of the blast disease primarily caused by *Pyricularia graminis-tritici (Pygt) Triticum* pathotype, indicating the pathogen dispersal occurred through seeds. Wheat blast was considered to have emerged via host shift from a *Pyricularia* population infecting *Lolium* sp. (Ceresini et al. 2019). The role of honey bees (*Apis mellifera*) as dispersal agents of *Colletotrichum acutatum* and *C. gloeosporioides* (citrus postbloom fruit drop, PFD) was investigated. Under screenhouse conditions, citrus plants were protected in insect-proof cages or left open. The source of inoculum was placed on one side of the screenhouse and a honeybee hive was placed on the opposite side. All uncaged plants showed symptoms of PFD and none of the caged plants exhibited disease symptoms. Conidium-like structures of *Colletotrichum* spp. attached to the bodies of honeybees were visualized under scanning microscope (SEM). The results indicated that honeybees might be transmitters of *Colletotrichum* spp. causing PFD in citrus (Gasparoto et al. 2017).

3.2.2.2 Vegetative Compatibility Groups of Fungal Pathogens

The ability to form heterokaryons by fusion between genetically different strains within a morphologic species may be useful to identify different strains. Vegetative compatibility may be governed by genetic systems. Strains that are identical at a particular loci and are capable of forming stable heterokaryons are placed in one group, whereas the second group includes the strains that differ at any of these loci and are unable to form vegetatively stable heterokaryons. This type of interaction is known as allelic compatibility. Generally, chlorate-sensitive strains can reduce nitrate to nitrite, whereas chlorate-resistant strains cannot do so. The *nit* mutants unable to utilize nitrate but able to use nitrite, ammonium, hypoxanthine, and uric acid were named *Nit1* mutants. These Nit mutants are incapable of using nitrate and hypoxanthine, but are capable of using the remaining three nitrogen sources. They are called *NitM* (Correll et al. 1987). The vegetative compatibility groups (VCGs) of *Botrytis cinerea* were identified. Selenate-resistant mutants (*sel*) were recovered from *B. cinerea* strains grown in the presence of sodium selenate. The majority of *sel* mutants were also sulfate non-utilizing (*sul*) mutants (Beever and Parkes 2003). The *sul* mutants were divided into two complementary groups: resistant and sensitive to chlorate. Based on compatibility reactions, chlorate-resistant and chlorate-sensitive *sul* mutants of *B. cinerea* strains were found to be compatible only within the group, and they were classified into different VCGs. (Korolev et al. 2008). The results suggested that the parasexual cycle might be an alternative source of genetic variability in *C. acutatum* isolates, since viable heterokaryons might be formed among vegetative compatible isolates of *C. acutatum* (Franco et al. 2011).

3.3 GENETIC CHARACTERISTICS OF FUNGAL PATHOGENS

Crop production has been seriously limited by fungal pathogens like *Phytophthora infestans, Pyricularia oryzae*, and *Puccinia graminis* f.sp. *tritici*. Microbial plant pathogens differ from the saprophytic microbes in their ability to infect plants and cause diseases with characteristic symptoms due to the presence of unique genes required for their pathogenicity. On the other hand, plants are endowed with genes for resistance to counter the adverse effects of pathogen during pathogenesis. Pathogenicity genes of fungal pathogens regulate production of effectors such as enzymes, toxins, and factors for inducing susceptibility of plants, facilitating infection and invasion of various host plant tissues.

3.3.1 Variability in Pathogen Virulence

Variations in physiological activities that contribute to the growth, reproduction, and aggressiveness are observed. Furthermore, the isolates of fungal pathogens exhibit variations in their sensitivity to fungicides and other environmental conditions. Fungal pathogens possess specific genes for adhesion to plant surface (receptors), production of toxins and enzymes, resistance to oxidative stress, degradation of plant cell walls, and breaching defense strategies of plant hosts. Pathogen factors capable of inducing susceptibility in host plants like selective toxins and small molecular suppressors produced from fungal pathogens have been characterized. Susceptibility or resistance of plants to a pathogen is observed as result of expression of virulence and avirulence gene(s), respectively (Narayanasamy 2008).

Wheat stem rust disease caused by *Puccinia graminis* f.sp. *tritici* was found to be a reemerging disease. Understanding the sources and origins of genetic variations in pathogen populations globally would be useful for the development of effective disease management strategies. High stem rust infection in wheat crops was observed in Georgia, the United States. In addition, common barberry (*Berberis vulgaris*) was widely distributed in the main wheat-growing areas. Unique combinations of virulence phenotypes including Sr22 + Sr24 and Sr13b + Sr35 + Sr37 were detected in wheat stem rust samples. Using a custom SNP Chip, 61 isolates were genotyped and 17 genotypes were identified. Higher levels of virulence and genotype diversity observed in 2,013 samples strongly indicated that sexual recombination could have occurred in *P. graminis* f.sp. *tritici* population (Olivera et al. 2019). Spread of races of wheat yellow rust pathogen *Puccinia striiformis* from wheat and triticale in Europe was analyzed using virulence phenotype data of isolates. Microsatellite markers were employed to assess a subset of 239 isolates. These non-European races, called Warrior, Kranich, and Tritical aggressive, were identified among pathogen populations. The presence of at least six multilocus genotypes of the Warrior race and five genotypes of the Kranch race and across large areas indicated the possibility of aerial spread from a genetically diverse source of populations (Hovmøller et al. 2016). Race phenotyping of cereal rust pathogens is based on pathogenicity of the field isolates on a set of selected differential host genotypes/cultivars. The emergence of new races of *Puccinia graminis* f.sp. *tritici (Pgt)* including the Ug99 race necessitated surveillance as an essential component of a disease management system. Isolates of *Pgt* were classified into five race groups (21C3, 34C0, 34C1, 34C3, and 34C6). The stem rust resistance gene *Sr31* was effective against all isolates (Cao et al. 2019). A strong correlation between seedling infection type and field response occurred for differential lines carrying *Sr5, 6, 7b, 8a, 8b, 9a, 9b, 9d, 9e* (Vernstein), *10, 11, 24, 27, 31, 36, 38, 44, McN, Kw*, and *Satu* (Boshoff et al. 2019). A new set of differentials of wheat genotypes with 18 stripe rust (yellow rust) single gene (*Yr*) lines was used to differentiate races of *P. striiformis* f.sp. *tritici (Pst)*, which were called race PSTv in distinction from PST races. An octal system was used to describe virulence and avirulence patterns of PSTv races. Races PSTv-1 to PSTv-41 were named following inoculation on 18 *Yr* single-gene differentials. Race PSTv-37 was well-distributed throughout the United States (Wan and Chen 2014). Virulence of isolates of *P. striiformis* f.sp. *tritici* were characterized and the races were identified by testing the samples on a set of 18 *Yr* single-gene differentials. The races occurring predominantly in different regions in the United States were identified. Races with virulence to *Yr6, Yr7, Yr8, Yr9, Yr17, Yr27, Yr44 and YrExp2* were detected at high frequencies, whereas virulence to *Yr5* and *Yr15* were not detected, indicating the effectiveness of these genes against *Pst* in the United States (Wan et al. 2016).

M. oryzae strains were analyzed using a Chinese differential system (CDS), repetitive element-based PCR/rep-PCR assay, and presence and absence of AVR-genes by PCR amplification with gene-specific primers. A total of 28 races of *M. oryzae* were identified with CDS and classified into 20 distinct groups with rep-PCR assay (Xing et al. 2017). *Pyricularia graminis-tritici (Pygt)*, incitant of wheat blast disease, existed in the form of several races with varying levels of virulence on wheat differentials. *Lolium*-derived isolates of *Pygt* carrying the *Ao* virulence allele at the *PWT3* locus infected susceptible cultivar carrying the *rwt3* allele conditioning susceptibility. Later selection on the less common Rwt3 wheat varieties favored the emergence of pathogen strains with nonfunctional *PWT3* alleles (Inoue et al. 2017). The *Pygt* populations infecting wheat and other hosts showed significantly higher genetic variations than the populations of *P. oryzae* (rice blast disease) in Brazil. Mutations in the avirulence genes *PWT3* and *PWT4* are the genetic factors that have been associated with adaptation of *Pygt* to wheat allowing the pathogen to escape recognition (Castrogulin et al. 2017). Rice blast isolates from Japan showed wide variation in virulence. Cluster analysis using the reaction patterns of differential varieties (DVs) classified isolates into three groups. Group I isolates were confined to northern Japan and group IIb to central Japan, whereas group IIa was well distributed throughout Japan. Based on the reactions on DVs, 123 races of *P. oryzae* were identified and called using a new designation system (Kawasaki-Tanaka et al. 2016).

The population density of pathogens and mechanisms of compatible and incompatible reactions using different host genotypes were investigated. Discrimination of strains with different levels of virulence is basically required for differentiating pathotypes of pathogen populations. A bioassay method was developed for assessing virulence based on sporulation density. Necrotic reactions of host genotypes were also determined simultaneously to identify the defense strategy of the host. Comparison of symptoms induced by *Plasmopara viticola* (grapevine downy mildew) on 12 grapevine genotypes after inoculation with different single sporangium strains indicated that resistance diversity was high in cultivars. *Vitis riparia* showed the strongest resistance among four North American species tested. Necrotic reaction in *V. amurensis* indicated a different mechanism of resistance, when compared to *Ampelpsis japonica* which did not show any necrosis (incompatible interaction) (Gómez-Zeledón et al. 2017). *Phialophora gregata* f.sp. *sojae (Pgs)* isolates have been grouped into genotype A and B. The genotype B isolates preferentially colonized many soybean accessions described as resistant to brown stem rot (BSR), and these isolates were considered less virulent than genotype A isolates, due to induction of minor or no foliar symptoms characteristic of BSR on soybean. Monoculture of a resistant or susceptible soybean might favor an increase in the virulence of isolates of *Pgs* genotype B (Hughes and Grau 2016). Host specialization of *Botrytis cinerea* on tomato and lettuce was investigated by phenotypic and genotypic approaches. Both genetic diversity and allelic richness were significantly higher in lettuce strains than in tomato strains. Tomato strains were significantly more aggressive than lettuce strains when inoculated on tomatoes. Strains of *B. cinerea* did not exhibit significant differences in virulence on lettuce or on apple. The results did not reveal a clear host specialization of *B. cinerea* on tomato and lettuce (Leyronas et al. 2015).

Generation of races of *Puccinia striiformis* f.sp. *tritici (Pst)* after sexual reproduction in alternate host *Barberis* spp. was examined. During a large-scale survey in China, single-acervuli (SA) samples were inoculated on susceptible wheat cultivar. The results provided evidence for production of new races that have virulence on different resistance genes reflected by responses of wheat differentials to *P. striiformis* f.sp. *tritici* (Wang et al. 2016). Changes in races and virulence genes in *P. striiformis* f.sp. *tritici (Pst)*, occurring in the United States during 1968–2009 were assessed. New races with more virulence genes were detected in greater numbers after 2000 (Liu et al. 2017). Segregating a population of 119 isolates of *Puccinia striiformis* f.sp. *tritici (Pst)* was obtained by self-fertilizing *Pst* isolate 08–220 (race PSTv-11) on barberry leaves under controlled greenhouse conditions to determine inheritance of virulence and to map virulence genes. The results of studies in virulence genes and their inheritance were useful for selection of effective resistance genes or gene combinations for development of cultivars resistant to wheat stripe rust disease (Yuan et al. 2018). The extent of relationship of *Puccinia striiformis* causing stripe rust diseases of cereals and grasses was examined. Isolates of *P. striiformis* from wheat, barley, triticale, rye and grasses were characterized by virulence tests and SSR markers. The SSR markers differentiated 44 multilocus genotypes (MLGs) and clustered them into three major molecular groups (MGs). Grasses could harbor more diverse isolates of *P. striiformis* than cereals (Cheng and Chen 2014). Of the 71 virulence phenotypes, the most common virulence phenotypes of *Puccinia triticina*, occurring across the United States, were MBTNB, MBDSD, and TNBJJ. The phenotype MBTNB was virulent to *Lr11*, and was the most common in the soft red winter wheat region of the United States. Phenotype MBDSD with virulence to *Lr17* and *Lr39* was the most commonly infecting hard red winter wheat. Hard red winter wheat cultivars showed infection by the TNBJJ frequently. Isolates of *P. triticina* could be classified into two major regional groups of virulence phenotypes (Kolmer and Hughes 2017). The isolates of *Phakopsora pachyrhizi* (soybean rust) were tested for virulence on a differential set of soybean genotypes that included six genotypes with known resistance genes (*Rpp*) and one each of resistant and one susceptible genotype. Three pathotypes were identified. Six aggressiveness groups were differentiated based on sporulating-uredinia production on differentials (Twizeyimana and Hartman 2012). Determination of virulence phenotypes is important for effective resistance gene deployment against the sunflower rust pathogen *Puccinia helianthi*. The pathogen isolates were obtained from cultivated and wild *Helianthus* spp. and volunteer plants. Virulence phenotypes of 23 single-pustule isolates were identified using a differential set of sunflower genotypes. Among the 29 races identified, races 300, 304, and 324 occurred more frequently than others. Race 777 was virulent on all different genotypes (Friskop et al. 2015).

Aggressiveness of single-lesion isolates of *Phytophthora infestans* on detached leaflets of four potato cultivars with differing levels of nonspecific resistance was compared to determine selection of pathogen populations. In leaves inoculated with a mixture of isolates of all genotypes of *P. infestans*, only a single genotype was isolated from one cultivar, but different genotypes were recovered from different cultivars. Isolates of *P. infestans* from Northern Ireland exhibited significant variation for foliar aggressiveness, and pathogen genotypes showed differential aggressiveness to partially resistant cultivar and interacted competitively in genotype selection (Young et al. 2018). The population structure of *P. infestans* existing in a potato germplasm nursery in northwestern China was studied. Isolates of *P. infestans* from 63 potato varieties were genotyped by mitochondrial DNA haplotypes and a set of SSR markers. The *P. infestans* population showed a low level of genetic diversity and had a high percentage of individuals with virulence on all 11 *R* genes. The presence of both A1 and A2 mating types as well as self-fertile isolates was detected. Migration and asexual reproduction could be the possible factors influencing *P. infestans* population structure in the germplasm nursery examined (Tian et al. 2015).

Virulence variation in the isolates of *Peronospora effusa* (spinach downy mildew) was investigated using samples collected from spinach production areas. Two new races (15 and 16) and eight novel strains were identified. Several loci (RPF1, 2, 4, and 6) were effective in preventing infection by Pfs15. Race Pfs16 could overcome several resistance loci (RPF2, 5, 9, and 10). One novel strain, UA1014, could overcome resistance of spinach resistant loci RPF1 to RPF7, but could infect only cotyledons and not true leaves of some cultivars. None of the germplasm entries of USDA accessions was resistant to Pfs 16 or UA1014 (Feng et al. 2018). Diversity in virulence and genetic structure in the isolates of *Plasmopara viticola (Pv)* (grapevine downy mildew) was investigated. Seven pairs of SSR primers were used to differentiate the strains in China (Li et al. 2016). Virulence variability of isolates of *P. cubensis* was investigated, using 22 isolates from the United States. Virulence structure was evaluated by inoculating a set of 12 differential genotypes from eight genera of cucurbits. All isolates were pathogenic to *Cucumis sativus*–susceptible cultivar, whereas virulence of the isolate was much lower on *Luffa cylindrica* and *Momordica charantia*. Based on the compatible interaction with differentials, five pathotypes, 1, 3, 4, 5, and 6 were identified. Virulence within the pathogen population was expressed based on virulence factors, virulence phenotypes, and virulence complexity (Thomas et al. 2017b). The spatial distribution of pathotypes of wheat powdery mildew pathogen, *Blumeria graminis* f.sp. *tritici (Bgt)*, was assessed using isolates derived from three separated plants. Several introgressed powdery mildew resistance genes from wild wheat relatives (*Pm37, Pm53, MlAG12, NcAG13,* and *MlUM15*) exhibited complete or nearly complete resistance in all local *Bgt* populations from across the sampled area. *B. graminis* f.sp. *tritici* populations in the United States had different virulence profiles in the hard and soft wheat regions which might be maintained by host selection, isolation by distance, and west-to-east gene flow (Cowger et al. 2018).

A set of differential genotypes was used to identify 17 apple scab (*Venturia inaequalis)* resistance genes in an updated system for defining gene-for-gene relationship (GFG) in a *V. inaequalis-Malus* pathosystem. A reference set of 23 *V. inaequalis* isolates was evaluated on 14 differential apple hosts carrying major resistance genes under controlled conditions. Six different virulent patterns were observed. Overall, the pathogen isolates carried one to three kinds of virulence, but some of them were more complex, carrying up to six kinds of virulence (Caffier et al. 2015). The potato pathogens *Alternaria solani*, causing early blight, and *A. alternata,* causing brown spot, were isolated from foliar lesions at three commercial locations. One genotype of *A. solani* and five genotypes of *A. alternata* were identified. *A. alternata* isolates induced less disease severity than *A. solani*. *A. alternata* caused little disease on breeding line 24–24–12, while *A. solani* induced fewer disease symptoms on 24–24–12 than on cv. Russet Burbank or Atlantic. Isolates of *A. alternata* inducing chlorosis caused larger lesions on cv. Russet Burbank, compared to isolates that did not induce chlorosis (Ding et al. 2019a). Population structure and genetic diversity of isolates of *Zymoseptoria tritici* (wheat Septoria tritici blotch) disease were analyzed using molecular markers. Isolates from Denmark and Sweden were grouped together, whereas isolates from the Baltic countries and Finland formed another group. Significant differences were detected between isolates sampled in different years in Denmark and for isolates sampled from specific cultivars in different years (Vagndorf et al. 2018).

3.4 PROCESS OF INFECTION AND DISEASE DEVELOPMENT

The ability to discriminate between 'self' and 'nonself' is considered the key step in the evolution of eukaryotic immune systems. Plants have variable receptor families for recognition of nonself structures/organism-mediated compounds. Plants have a repertoire of receptors that function as a radar system to detect the pathogen-derived nonself molecules. When there is a match between a strain-specific pathogen effector and its corresponding host plant resistance (*R*) gene product, the host plant may be able to recognize the pathogen presence and to express an effective resistance response. Such recognition of the pathogen effector by a plant R receptor results in rapid necrosis of plant cells at sites of attempted penetration/invasion by the pathogen. In contrast, if the plant is unable to recognize a pathogen effector, infection by the pathogen may be initiated. Further, the course of pathogenesis depends on the ability of the pathogen to overcome the effects of inhibitory compounds present prior to infection or formed following infection to arrest pathogen development and symptom expression. Five phases of the process of infection and disease development or pathogenesis have been recognized: (i) attachment of fungal pathogen to plant surfaces, (ii) germination of spores or infection (pathogenic) unit, (iii) penetration/entry into the host tissues, (iv) colonization of host plant tissues, and (v) symptom expression after completion of latent (incubation period). Biochemical, histological, and molecular investigations have been performed to have an insight into the pathogen interactions with susceptible and resistant plants (Narayanasamy 2002, 2008).

3.4.1 Attachment of Fungal Pathogens to Plant Surfaces

Fungal pathogens including oomycetes, after dispersal from infected plants, land on different aerial organs of the host plants such as leaves, stems or inflorescence. The spore tip mucilage (STM) of *Magnaporthe grisea* was involved in adhesion of conidia to the cuticle through hydrophobic action (Hamer et al. 1988). In a later investigation, hydrophobins present on the outer surface of conidia and hyphae were found to be involved in adhesion. The effects of two hydrophobins, class I *Mpg1* and class II *Mhp1*, on adhesion, germling differentiation, and pathogenicity of wheat isolate of *M. oryzae* were evaluated. Hydrophobin mutants were generated by gene knockdown procedure. *Mpg1* expression in the *Mpg1* silenced mutants was correlated with adhesion, germling differentiation, and pathogenicity. On the other hand, *mhp1* null mutants had full pathogenicity on wheat and barley leaves (Inoue et al. 2016).

Organ specificity is exhibited by some pathogens that can initiate infection only on reaching the specific host tissue or organ. An extreme case of organ specificity is observed in citrus-*Colletotrichum acutatum* interaction resulting in PFD disease. The pathogen can infect only the flower petals of most citrus cultivars (Fagan 1979; Timmer and Brown 2000). A fungal gene was required for pathogenicity on Key lime leaves and the gene *KLAP1* (**K**ey **l**ime **a**nthracnose **p**athogenicity). The putative Cys2His2 transcription activator required for pathogenicity of *C. acutatum* was identified in the isolates causing anthracnose disease (Chen et al. 2005). The first step in the initiation of infection by fungal pathogen is attachment of spores or other infection unit to the plant surface. The extracellular enzymes secreted by fungal pathogens might facilitate adhesion of spores to plant surfaces. Cutinases and esterases may have a role in adhesion of infection structures of *Colletotrichum graminicola* and *Uromyces viciae-fabae* (Deising et al. 1992; Pascholati et al. 1993). Conidia and cell walls of *Colletotrichum lindemuthianum* contained a fibrillar spore coat that aided adhesion to aerial plant surfaces for initiation of infection. Attachment of conidia was inhibited by treatment of conidia with the MAb UB20 IgG to hydrophobic surface in an antigen-specific manner, suggesting a role for specific glycoproteins in the initial attachment of conidia. The presence of specific glycoproteins on conidia, germ tubes, appressoria, and primary hyphae of *C. lindemuthianum* was detected. None of the glycoproteins produced in the early stage of infection could be detected during necrotrophic growth of the pathogen, indicating modification of fungal cell surface during this stage (Perfect et al. 2001).

Peronospora parasitica sporangia produced germ tube and appressoria on which a fibrillar matrix containing β-1,3-glucans confined to germling-substrate interface and another matrix consisting of protein spreading beyond the contact interface as a thin film (Carzaniga et al. 2001). The behavior of conidia of *Botryosphaeria dothidea* and other fungal pathogens associated with grapevine Botryosphaeria dieback disease complex was investigated at the pre-infection stage on substrata with different affinities for water. As conidia adhered well to all artificial substrates tested, it seemed that conidia attachment was nonspecific. Treatment of conidia of *Neofusicoccum luteum* with a protease entirely prevented adhesion of conidia, indicating spore surface proteins might have a role in the adhesion process. Mucopolysaccharide and protein production might be involved in a second phase of adhesion of conidia of *N. luteum* causing Botryosphaeria dieback in grapevine (Sammonds et al. 2016). Common sage (a herbal plant) was infected by the downy mildew pathogen *Peronospora salvia-officinalis* (*Pso*). The process of infection of sage plants by *Pso* was monitored using confocal laser scanning microscopy and autofluorescence techniques at different stages of infection from adhesion of conidia on leaf surface to de novo sporulation. The improved method of staining involved use of a 1:1 mixture of aniline blue and trypan blue. The staining method allowed discrimination of pathogen and host plant structures and visualization of plant immune responses (Hoffmeister et al. 2020).

3.4.2 Spore Germination and Penetration of Host Plant Surfaces

Phytophthora infestans (late blight disease) produces sporangia on infected leaves, and they are dispersed by wind. Three genes of *P. infestans* are activated just before germination, and they, as a cluster, code for novel proteins know as Car (**c**yst-germination specific **a**cidic **r**epeat) proteins (Görnhardt et al. 2020). Fungal pathogens produce soluble molecules and enzymes that could inhibit development of hypersensitive response (HR)-based plant defense mechanisms. Soluble glucans were detected in spore germination fluids produced by *P. infestans*, and the glucans may suppress the oxidative burst and HR in potato. Sporangia formed by oomycetes are dispersed by wind, and they initiate infection on the same or other plants. They are hydrated and physiologically active, enabling them to respond to germination-conducive conditions. Exposing sporangia to water triggers zoosporogenesis changes within minutes of the stimulus (Tani et al. 2004). Oomycete hyphae lack septa and their cytoplasm flows into sporangial initials during sporulation. Therefore, proteins and mRNAs synthesized during colonization may be deposited in sporangia, where they might contribute to pathogenesis later (Ah-Fong et al.

2017). Functional equivalence of sporangia from artificial media (rye grain-based medium) and lesions from infected plants was assessed. Transcriptions and infection potential of sporangia from rye-sucrose media, potato and tomato leaflets, and potato tubers were compared. When sporangia of similar maturity were used, those sourced from media and plant lesions had similar rates of zoospore release and cyst germination. Further, infection rates or virulence (aggressiveness) on leaflets based on single-spore inoculation assay were also similar, indicating the functional similarity of sporangia from different sources (Leesutthiponchai and Judelson 2019).

Penetration of plant surfaces by fungal pathogens, following spore germination, occurs by mechanical force as a result of melanized appressoria accumulating high turgor pressure. Different types of melanin, including dihydroxynaphthalene (DHN) melanin and L-3,4-dihydroxyphenylalanine (DOPA) melanin, are synthesized by different fungi. Infection of rice plants by *M. oryzae* showed the requirement of melanized appressoria for penetration of leaf surface. Treatment with tricyclazole inhibited DHN melanin biosynthesis in a wild-type strain. *M. oryzae* accumulated glycerol as an osmolyte inside appressoria that allowed inflow of surrounding water and the melanin layer prevented the outflow of osmolytes, resulting in high turgor pressure inside appressoria. The MPG1 is expressed at initiation of pathogenesis in rice inoculated with blast pathogen. *MPG1* codes for a gene product that forms a component of the host surface perception process, in addition to its requirement for efficient appressoria formation and cellular differentiation. In addition, *MPG1* encodes a fungal hydrophobin secreted by *M. grisea* on the rice leaf surface. Dramatic reduction in virulence following abolition of MPG1 function was observed, indicating its role as a virulence gene required for pathogenicity (Narayanasamy 2008).

The role of mitogen-activated protein kinase (MAPK) gene *PMK1* in *Magnaporthe grisea* in pathogenesis was investigated. Generation of hydrophobic turgor by accumulating molar concentrations of glycerol was also required along with *PMK1* for pathogenicity. Degradation of glycogen occurred rapidly during conidial germination. A mutant defective in conidiogenesis and appressorial functions showed reduced pathogenicity on susceptible rice cultivars compared with wild parent strain (Balhadere et al. 1999). The proximal carnitine acetyl transferase (CAT) activity in *M. grisea* was essential for appressorium function, especially elaboration of primary penetration hyphae. The pathogenicity gene *PTH2* in *M. grisea* encodes for the major CAT activity which registers enhanced expression in response to acetate and lipid. The nonpathogenic mutants obtained by target deletion of *PTH2*, lacking CAT activity, was unable to utilize a range of lipid substrates. A delay in lipid reserve mobilization from the germ tubes into developing infection structures might be due to impairment of appressorium function in Δ*pth2* (Bhambara et al. 2006). Gene expression profile during appressorium development process of *Magnaporthe grisea*, (rice blast) revealed upregulation and down-regulation of number of genes was almost consistent during appressorium development. More genes were differentially expressed during appressorium maturation than during appressorium induction and formation during appressorium maturation. Genes generally appeared to be most actively expressed (Jin et al. 2010a). Appressorium differentiation was strongly coordinated with cell cycle. Deletion mutants did not exhibit any deficiency in vegetative growth compared to wild-type and complimented strains, even on media containing DNA-damaging agents. However, these null mutants had abnormal appressoria and formed more appressoria per conidium than the wild-type strain and they were unable to penetrate the epidermis of rice leaves. The eGPP-fused Mop53BPI and ART-PCR analyses revealed that *Mop53BPI* was expressed during the first hour of appressorium formation. In addition, overexpression mutants, Mop53BPI localized to nuclei during all stages of appressorium maturation and penetration. The mutants were resistant to the nucleotide inhibitor benomyl, suggesting that MopBPI might be a nuclear protein (Ohara et al. 2018).

The *CMK1* gene of *M. grisea* could complement appressorium formation of the *PMK1-MAP* mutant of *M. grisea*. Loss of *CMK1* resulted in reduction in conidiation and complete lack of pathogenicity. Mutants of *M. grisea* defective in pathogenicity on leaves or roots of cereals, could also be detected (Dufrense and Osbourn 2001). The *PMK1* gene of *M. grisea* is homologous to the *FUS3* and *KSS1* genes that regulate the transcription factor STE12 in yeast. The *MST12* gene of *M. grisea* that is homologous to yeast *STE12* was identified. The *mst12* mutant was nonpathogenic on rice and barley leaves. The appressoria produced by *pmk1* mutants could not penetrate the onion epidermal cells. Further, *mst12* mutants when inoculated, did not form spreading lesions and appeared to be defective in infectious growth, suggesting that transcription factor(s) other than *MST12* might be present in *M. grisea* (Park et al. 2002). The MAPKinase-PMK1 gene in *M. griseas* controls appressorium formation and infectious growth. Two PAKinase genes, *HM1* and *MST20* in *M. grisea* were characterized. Reduction in growth of aerial hyphae and conidiation was observed in mutants disrupted in *MST20*. In *chm1* mutants all these parameters were reduced. CHM1 appeared to play a critical role in appressorium formation and penetration of plant surface, whereas *MST20* was not required for infection. CHM1 and MST20 might have redundant role in *M. grisea* without activating PMK1 MAPKinase pathway during early infection stage (Li et al. 2004). The MAP kinase (MAPK) Mst7 and MEK kinase (MEKK) Mst11 that activate *Pmk* in *Magnaporthe grisea* were identified. Like the *pmk1* mutants, *mst11* and *mst7* deletion mutants were defective in appressorium formation and were nonpathogenic, indicating that the Mst11-Mst7-Pmk1 MAP kinase cascade may be conserved for regulation of infection-dependent morphogenesis. In the later investigation, regulation of development of appressoria of *M. grisea* by the Mst11-Mst7-Pmk1 MAP kinase cascade was observed. The *MST50* gene directly interacting with both *Mst11* and *Mst7* was characterized. The *mst50* mutant was also defective in appressorium formation, sensitive to osmotic stresses and was nonpathogenic (Park et al. 2006). Interaction between *M. oryzae* conidia and the leaf surface of wild rice species *Oryza australiensis* was investigated. Significant differences in conidial development between two isolates of *P. oryzae* with different kinds of virulence were observed. Conidial germination at 6 h postinoculation (hpi) for virulent isolate was significantly delayed. Appressorial development for the virulent isolate was significantly faster ($P < 0.001$) at its later growth stages of 12 and 18 hpi, when approximately 100% of germ tubes had formed appressoria. Appressoria formation on *O. australiensis* by *P. oryzae* was significantly greater compared with *O. sativa* (rice) for both virulent and avirulent isolates, indicating the influence of host plant species on pathogen development (Pak et al. 2018).

Appressorium, a specialized infection structure, is differentiated from the tip of conidial germ tube of *M. oryzae* and regulated by the cyclic-3′-5′-adenosine monophosphate/protein kinase A (cAMP/PKA) signaling pathway. High turgor pressure is generated in the appressorium due to accumulation of osmolytes such as glycerol, forcing penetration peg into host cells and primary infection hyphae develop with sequential colonization. The target of rapamycin (TOR) kinase complex coordinates nutrients and energy to regulate cell proliferation, cell cycle, cell wall integrity, and cell metabolic activities by interacting with different subunits (Qian et al. 2020). The GATA-type transcription factor Asd4 required for glutamine hydrolysis, promoted appressorium formation by preventing suppression of the cAMP/PKA signaling pathway by TOR. Existence of a feed-forward regulatory mechanism between TOR and cAMP/PKA signaling pathway controlling appressorium formation in *M. oryzae* was observed (Sun et al. 2019). MoWhi2, the homolog of *Saccharomyces cerevisiae* Whi2 (Whisky2), an important regulator controlling appressorium formation was identified in *M. oryzae*. A putative phosphatase, MoPsrl interacted with MoWhi2. Knockout mutant Δ*Mossrl* displayed similar phenotypes to the Δ*Mowhi2* strain. Further, expression levels of MoTOR and the MoTOR signaling activation marker gene *MoRS3* showed increases, suggesting that inappropriate activation of the MoTor signaling pathway might be a reason for defects in the *dMowhi2* and *dMopsrl* strains (Shi et al. 2021)

Botrytis cinerea knockout mutant in the *bmp3* gene encoding region was detected and replaced by the hygromycin resistance cassette. Deletion of *bmp3* resulted in reduced vegetative growth on various media, strongly impaired conidiation and loss of sclerotia formation in the Δ*bmp3* mutant. After penetration, progress of necrotic lesion induced by Δ*bmp3* was retarded by genetic complementation of the mutant with wild-type *bmp3* gene (Rui and Hahn 2007). A cDNA library was produced from *B. cinerea* grown in axenic culture. From this library, expressed sequence tags (ESTs) were generated. Only *BCPG1* was present in the ESTs, and the *BCPG1* gene appeared highly expressed (37 ESTs). A signaling pathway depending on calcium and calcineurin was identified in the EST set. This pathway is involved in fungal morphology and virulence. The EST set also provided the *BCPLS1* gene, a homolog of *MgPLS1* that is necessary for appressorium-mediated penetration of *M. grisea* into host plants. *BCPLS1* and *MgPLS1* null mutants had similar penetration defects (Viaud et al. 2005). Null mutants of *BOS1* were found to be essential of normal macroconidiation and full virulence in *B. cinerea* (Viaud et al. 2005). Infection of wheat spikelets by *Bipolaris sorokiniana* was monitored after inoculation of the surfaces of different spike tissues. Extracellular sheaths were seen on germ tubes and appressoria attached to the substrates of lemma, palea, and seeds. Appressoria, frequently present over grooves, formed penetration hyphae invading the epidermal walls. Fungal growth was predominantly observed on the cell walls of epidermal and parenchyma cells. Hyphae also spread intercellularly as well as intracellularly. Secretion of cell wall hydrolytic enzymes at the apex of penetrating hyphae might facilitate pathogen spread (Han et al. 2010).

The *CMK1* gene encoding a MAP kinase has a critical role in conidial germination, formation of melanized appressoria, penetration of the plant surface and subsequent colonization of tissues by *C. lagenarium*, causing cucumber anthracnose. Loss of *CMK1* resulted in conidiation and complete lack of pathogenicity on cucumber, since the conidia did not germinate and form appressoria (Takano et al. 2000). Darkly melanized appressoria formed at the end of germ tubes from conidia of *C. lagenarium* penetrate host plant surface. The *CMK1* gene, a homolog of the yeast *FUS3/KSS1/MAPK* genes, regulated conidial germination, appressorium development and invasive growth of pathogen hyphae. The *cst* Delta strains were nonpathogenic on cucumber leaves but could form lesions when wound-inoculated. Infection hyphae were not produced from the appressoria of *Cst1* Delta strains, suggesting that *CST1* was essential for penetration of host surface by infection hyphae (Tsugi et al. 2003). The role of cAMP signaling was evaluated by functional characterization of the regulatory subunit gene of the cAMP-dependent protein kinase (PEA). The *RPK1* gene encoding mutants of *Colletotrichum lagenarium (Cl)* had reduced vegetative growth, and conidiation and were nonpathogenic on cucumber. However, the ability to form lesions when wound-inoculated was retained (Takano et al. 2000). The gene ClaPEX6 (PEX6) encoding protein was involved in peroxisome biosynthesis in *C l.* Requirement of *PEX6* gene for pathogenicity of the pathogen was demonstrated by knockout analysis of this gene. *ICL1* encoded isocitrate lyase involved in the glyoxylate cycle in peroxisomes. The *icl1* mutants could not utilize fatty acids and acetate for growth. These mutants formed highly pigmented appressoria, but had severely reduced virulence and penetration hyphae failed to develop in planta. Exogenous supply of glucose partially restored virulence of *icl1* mutants. *PEXb* was required for appressorial melanization. The peroxisomal pathway might have a functional role in appressorial melanization and host invasion at later stages of infection by *C. lagenarium* (Asakura et al. 2006).

Severity of symptoms induced by ascospores of *Colletotrichum lindemuthianum* was milder compared with severity of symptoms induced by conidial infections. The majority of ascospores produced germ tubes and appressoria and formation of hyphae and production of germ tubes and appressoria were high later. Ascospore germination and appressorium formation were similar on both resistant and susceptible cultivars (Ishikawa et al. 2010). The infection process of *Elsinoe ampelina* (grapevine anthracnose) was investigated by inoculating potted seedlings of cv. Red Globe under greenhouse conditions during a period of 6 days after inoculation. The first symptom on leaves appeared at 3 days postinoculation (dpi). Conidial germination and early penetration were observed in epidermal cells with or without appressorial formation at 2 dpi. The pathogen enlarged in intracellular spaces. New conidia were produced in the acervuli on leaf surface from 6 dpi. Defense responses of grapevine plants such as collapse of cell walls and chloroplasts, changes in the number of mitochondria, plasmolysis, and cell death were also revealed by ultrastructural observations (Li et al. 2019).

Colletotrichum acutatum (citrus PFD) induced necrotic lesions on petals and stigmas of citrus flowers and subsequently abscission of young fruits. Infection of citrus floral parts by *C. acutatum* was monitored. The pathogen could penetrate and colonize pollen grains of citrus at 24 h after inoculation. The germ tubes from conidia either penetrated the pollen sporodermis directly or passed through pollen germ pores. A single hypha could colonize more than one pollen grain. The production of conidia was observed on the surface of stigma (Marques et al. 2013). *C. acutatum* penetrated the cuticle layer of pepper fruit and often formed a dendroid structure localized in the cuticle layer in pepper petals, mango leaves or fruits of tomato and aubergine. When inoculated on resistant pepper fruits these structures turned dark brown and there was no further infection in the epidermal cells (Liao et al. 2012). Soybean anthracnose pathogen *Colletotrichum truncatum* invaded susceptible plants by forming appressoria. MAPK genes of the fungal pathogen were involved in various infection processes such as spore germination, melanized appressorium formation and penetration, and subsequent host tissue invasion. The *Fus3/Kss1*-related MAPK gene, *CtPMK1* was identified in *C. truncatum* and it was related to PMK1 in *M. oryzae*. Disruption of *CtPMK1* resulted in a mutant with slightly reduced mycelial growth, reduced melanization. Conidia of the *Ctpmk1* mutant germinated normally on onion epidermal surfaces. Further, the *Ctpmk1* mutant did not form appressoria or hyphopodia, subsequently resulting in complete loss of pathogenicity on soybean plants (Xiong et al. 2015).

Bulk degradation of cellular components in lysosome/vacuoles is mediated by autophagy, which has an important role in impacting the growth, morphology, and development of fungal pathogens. The structure of autophagy in infection by *Puccinia triticina* (wheat leaf rust disease) was visualized using TEM. The results suggested that autophagy could be strongly induced in haustorium mother cells (HMC), but did not occur in other special infection structures such as appressorium, substomatal vesicle, and haustoria. Application of autophagy inhibitor 3-methyladenine (3-MA), arrested mycelial growth and delayed infection (Liu et al. 2017). Several in planta induced genes (PIGs) have been identified in the haustoria of *Uromyces fabae* (cowpea rust disease). Some of the PIGs were highly expressed and were present as a single or low copy number genes in the rust genome. By screening a cDNA library constructed from haustorial DNA of *U. fabae*, the PIG2 gene encoding a protein with high homologies to fungal amino acid exporters was isolated. The putative amino acid exporter protein, formed following expression of PIG2 mRNA was localized to the plasma membrane of haustoria. Modification of cell wall components during penetration of cowpea epidermal cells by *U. fabae* was studied. The immunoassay indicated that density of pectin and xyloglucan epitopes were reduced, suggesting that *U. fabae* could degrade the cell wall at the penetration point by secreting the appropriate enzyme(s) (Narayanasamy 2008). Haustoria may also be actively involved in establishing and maintaining a biotrophic relationship with host plant cells. A rust-transferred protein1 from *Uromyces fabae* (*Uf*-RTP1p) was detected in the host-parasite interface, the extrahaustorial matrix, and inside the infected plant cells. A homolog of *Uf*-RTP1p was identified in *U. striatus* (*Us*-RTP1p). The presence of these two proteins inside infected plant cell was detected by using specific polyclonal antibodies (PAbs). *Us*-RTP1p could be detected in the host cell nucleus later. Putative nuclear localization signals (SNLs) could be observed in the predicted RTP1p sequences. Western blot analysis indicated that *Uf*-RTP1p and *Us*-RTP1p might gain entry into host cells as N-glucosylated proteins (Kemen et al. 2005). The genes encoding haustorially expressed secreted proteins (HESPs) were identified by screening a flax rust pathogen (*Melampsora lini*) haustorium-specific cDNA library. One HESP corresponded to the *AvrL567* gene (Catanzariti et al. 2006).

The majority of germ tubes of *Puccinia striiformis* f.sp. *tritici* (*Pst*) pathotype CY32 formed on the surface of spike organs of susceptible cv. Suwon 11 penetrated through stomatal pore, and only a few germ tubes produced small appressoria over stomata. The pattern of pathogen spread differed depending on floral tissues (Wang et al. 2011). Changes in the proteome profile during urediniospore germination of *Pst* and germ tubes (germlings) were studied. Proteins of spores and germlings were either upregulated or downregulated during urediniospore germination. Proteins involved in catabolic process, energy production and transport specially accumulated in germlings, which indicated the metabolic transition from dormancy to germination.

(Zhao et al. 2016). The structure of autophagy in infection by *Puccinia triticina* (wheat leaf rust disease) was visualized using TEM. Application of autophagy inhibitor 3-methyladenine (3-MA), arrested mycelial growth and delayed infection (Liu et al. 2017). The contribution of haustoria of *Puccinia striiformis* f.sp. *tritici* (*Pst*) to the interactions with host plants was studied by isolating haustoria from susceptible wheat leaves infected with *Pst* race CYR31 and sequencing their transcriptome, as well as those of urediniospores and germ tubes. The upregulated genes from haustoria were related to thiamine biosynthesis, glycolysis, and lipid metabolic processes. Silencing seven of these genes resulted in reduction of growth of *Pst* in wheat. Further, haustoria-secreted proteins (HASPs) in large number were detected in haustoria. In addition, HASPs suppressed Bax-triggered programmed cell death (PCD) in tobacco and also reduced callose deposition in wheat significantly employing the type III secretion system (Xu et al. 2020).

Phakopsora pachyrhizi (soybean rust disease) is unusual in that the urediniospores penetrated the leaf surface directly through epidermal cuticle, using mechanical force, and subsequently through epidermal cell wall by enzymatic digestion. A fibril-like matrix developed between lower appressorial wall and the host epidermal cuticle. The area on the lower side of the appressorium could be recognized by the presence of a dark-staining material (DSM), which had arm-like extensions radiating into the appressorial protoplast. Then the DSM-infused appressorial wall disappeared, allowing the appressorial plasma membrane to bulge into the area vacated by the appressorial wall and established contact with host cuticle (Figure 3.2). Penetration of the host cell wall began with a central point of the plasma membrane bulge penetrating the cuticle and pushing into the epidermal cell wall. The penetration hypha formed as it passed through the penetration pore in the cuticle and outer epidermal cell wall subsequently. It expanded as it passed through papillar material and traversed the host epidermal cell to reach the lower epidermal cell wall (Figure 3.3). Penetration hypha then grew along the inside of the wall between the protoplast and mesophyll cell wall. The primary hypha produced a bleb-like haustorial mother cell which was pressed tightly to host mesophyll cell. Then a small penetration peg formed and began digesting a hole through the mesophyll cell wall. The penetration peg invaginated into the mesophyll protoplast, pushing the plasma membrane ahead of the peg and finally the peg grew into a haustorial lobe. A characteristic neck ring in the haustoria was observed, as in other rust pathogens (Figure 3.4) (Edwards and Bonde 2011).

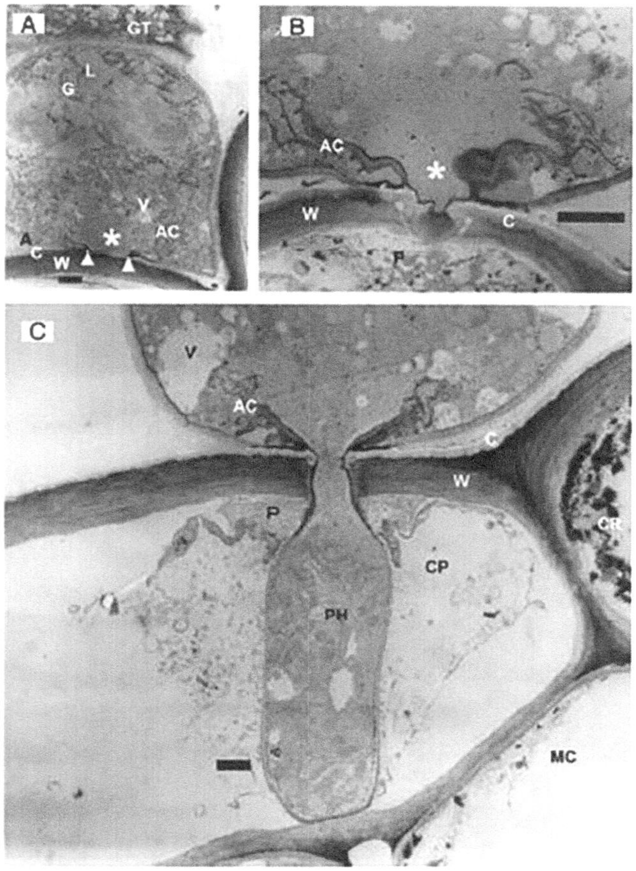

FIGURE 3.2 Visualization of penetration of soybean leaf epidermis by *Phakopsora pachyrhizi* using transmission electron microscopy (TEM). Penetration of epidermal cuticle with appressorial membrane, making direct contact with the host plant cuticle (A), followed by penetration of host cell wall (B), and penetration hypha traversing host epidermal cell. A-appressorial wall; AC-appressorial cone; C-cuticle; CP-cytoplasm; CR-crystalline material; G-granules; GT-germ tube; L-lipid; MC-mesophyll cell; P-papilla; penetration hypha; V-vacuole; and W-host cell wall.

Source: (Courtesy of Edwards and Bonde 2011 and with kind permission of the American Phytopathological Society, MN, USA)

The infection processes of *Phakopsora pachyrhizi* and *Colletotrichum truncatum* infecting soybean were compared. Appressoria of *P. pachyrhizi* were hyaline with no pigmentation on the appressorial wall, whereas appressoria of *C. truncatum* were opaque and melanized. Appressoria of *P. pachyrhizi*, after treatment with tricyclazole and niacin, were visually identical to untreated control appressoria. On the other hand, appressoria of *C. truncatum* treated with tricyclazole blocked melanin synthesis and turned the intermediate products into reddish compounds, whereas niacin did not eliminate the melanin, indicating the opaque color was due to DHN melanin. Soybean leaflets inoculated with conidia of *C. truncatum* mixed with tricyclazole produced fewer necrotic spots than conidia mixed with niacin or in Tween 20 control. Soybean leaflets inoculated with urediniospores of *P. pachyrhizi* suspended in tricyclazole or niacin were visually indistinguishable from control based

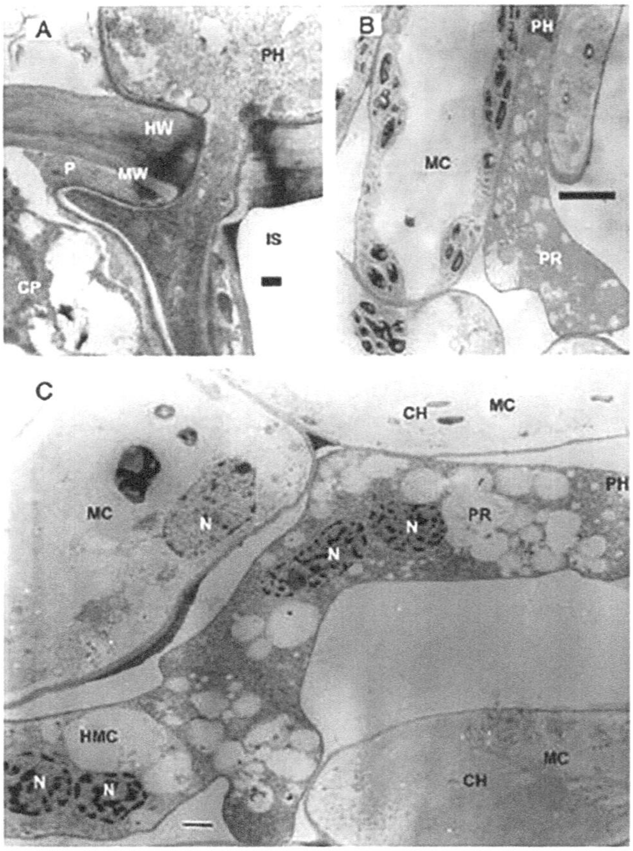

FIGURE 3.3 Colonization of epidermal and mesophyll cells of host plant by *Phakopsora pachyrhizi*. Penetration hypha passing through epidermal and mesophyll cell walls (A), penetration hypha and primary hypha growing between mesophyll cells (B), and penetration hypha and haustorial mother cell in the mesophyll cell (C). CH-chloroplast; CP-cytoplasm; HMC-haustorial mother cell; HW-epidermal cell lower wall; N-nucleus; P-papilla; PH-penetration hypha; PR-primary hypha.

Source: (Courtesy of Edwards and Bonde 2011 and with kind permission of the American Phytopathological Society, MN, USA)

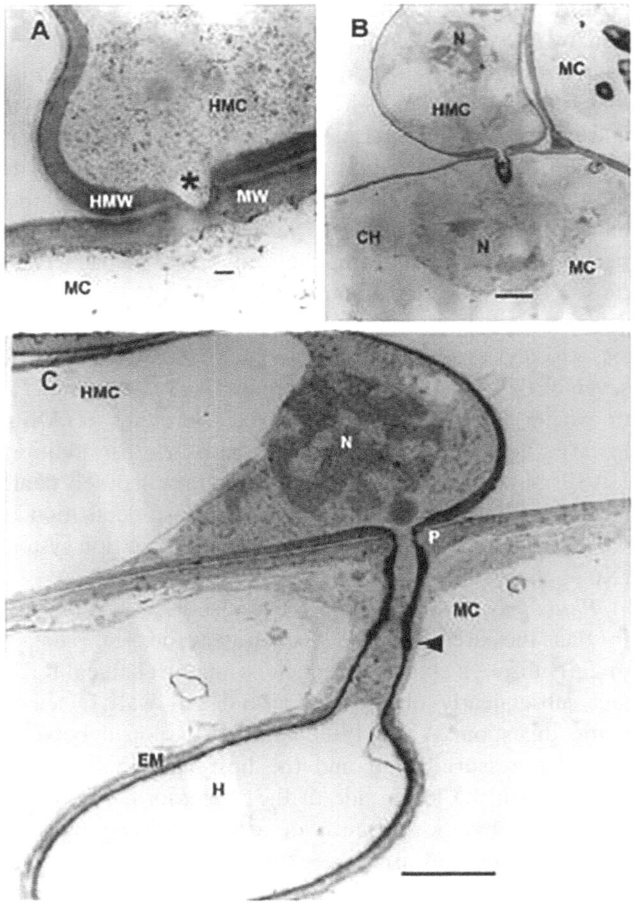

FIGURE 3.4 Formation of haustorium by *Phakopsora pachyrhizi* in infected soybean leaves. Initiation of haustorial penetration by forming haustorial peg (A), penetration of mesophyll cell by penetration peg (B), and formation of well-developed haustorium in mesophyll cell (C) C-chloroplast; EM-extrahaustorial membrane; H-haustorium; HMC-haustorial mother cell; N-nucleus; P-papilla.

Source: (Courtesy of Edwards and Bonde 2011 and with kind permission of the American Phytopathological Society, MN, USA)

on appearance of sporulating uredinia (Figure 3.5). TEM observations showed that appressoria did not retain their spherical shape during penetration. Soybean cells formed papillae in response to infection to counteract pathogen invasion at penetration site. An appressorial cone was formed at initiation of penetration. The penetration hyphae invaded through the epidermis and developed into primary hyphae which colonized the leaf tissue progressively. However, no melanin layer was observed in the cell walls of appressoria in all these stages, although a gray waxy layer was present locally. A dark material was associated with the appressorial cone near the penetration site. But accumulation of this dark material did not follow appressorial wall (Figure 3.6) (Chang et al. 2014).

Transcriptome analyses were employed for identifying differentially expressed genes in soybean ring infection by *Phakopsora pachyrhizi* in compatible and incompatible interactions. *P. pachyrhizi* secreted proteins from haustoria that could be transferred into plant cells to promote virulence. One candidate, *P. pachyrhizi* effector protein, was characterized to understand the mechanism of suppression of plant defenses by the pathogen. Effector protein Phapa7431740 suppressed pathogen-associated molecular pattern-triggered immunity (PTI). The effector protein interacted with a soybean glucan endo-1,3-ß-glucosidase (GmßGLU), a PR protein (PR-2 family). Transcriptional profiling during infection time course showed that GmßGLU mRNA was highly induced during initial hours after infection, coinciding with peak of expression of Phapa7431740 which might suppress PTI by interfering with glucan endo-1,3-ß-glucosidase activity (Bueno et al. 2022).

FIGURE 3.5 Effects of melanin inhibitors on the pathogenicity of *Phakopsora pachyrhizi (Pp)* and *Colletotrichum truncatum (Ct)* on soybean leaves. (A) Treatment with tricyclazole reducing necrotic area induced by *Ct*. (B) Treatment with tricyclazole or macin suppressing development of *Pp* entirely.

Source: (Courtesy of Chang et al. 2014 and with kind permission of the American Phytopathological Society, MN, USA)

A micro-needle technique was employed to remove cellular contents from single cells. Isolated conidia were subjected to PCR amplification of the 5.8S rDNA and its adjacent ITS sequences, followed by nested PCR assay to attain sensitivity high enough to amplify target nucleotide sequences. Transcripts expressed in single conidia could be amplified. Conidia at pre- and post-germination were tested by this procedure. The β-tubulin homolog *TUB2-ol* was expressed at pre- and post-germination stages of *Blumeria graminis* f.sp. *hordei (Bgh)* in inoculated barley. Cutinase homolog *CUT1-ol* was expressed at post-germination stage only (Matsuda et al. 2005). High density cDNA microarrays were employed to analyze transcript profiles of *B. graminis* f.sp. *hordei (Bgh)* to cover the entire asexual life cycle and development of infection by *Bgh*. The presence of a global switch in the expression between the pre- and post-penetration stages was observed. Accumulation of RNA-encoding protein biosynthesis genes was recorded in the later stages. Other functional clusters such as virulence-related genes and sterol metabolism genes were found to be upregulated in pre- and post-penetration stages, respectively. Expression of *cap20* was related to abundance of a group of RNAs, and they are likely to be pathogenicity factors in *Bgh* (Both et al. 2005). The role of cuticle-derived molecules in the prepenetration process of *Blumeria graminis* f.sp. *hordei (Bgh)* on plant leaf in vivo was investigated. In maize, conidial germination and appressorial differentiation of *B. graminis* in mutant *glossy11* were strongly impeded by wax. Spraying mutant leaf surface with aldehyde-containing wild-type wax or pure n-hexacosanal (C26-aldehyde) fully restored fungal penetration, whereas maize wild-type leaf surfaces coated with n-docosanoic acid showed reduced conidial germination rates and only a few conidia produced appressoria (Hansjakob et al. 2011). The infection process of *B. graminis* f.sp. *tritici (Bgt)* was investigated using ascospores as inoculum. All germinated ascospores produced a single germ tube type both in vitro and on wheat plant surfaces and penetrated epidermal cells subsequently producing haustoria as in conidial infections. Ascosporic infections occurred frequently under field conditions, when suitable environmental conditions prevailed (Jankovics et al. 2015).

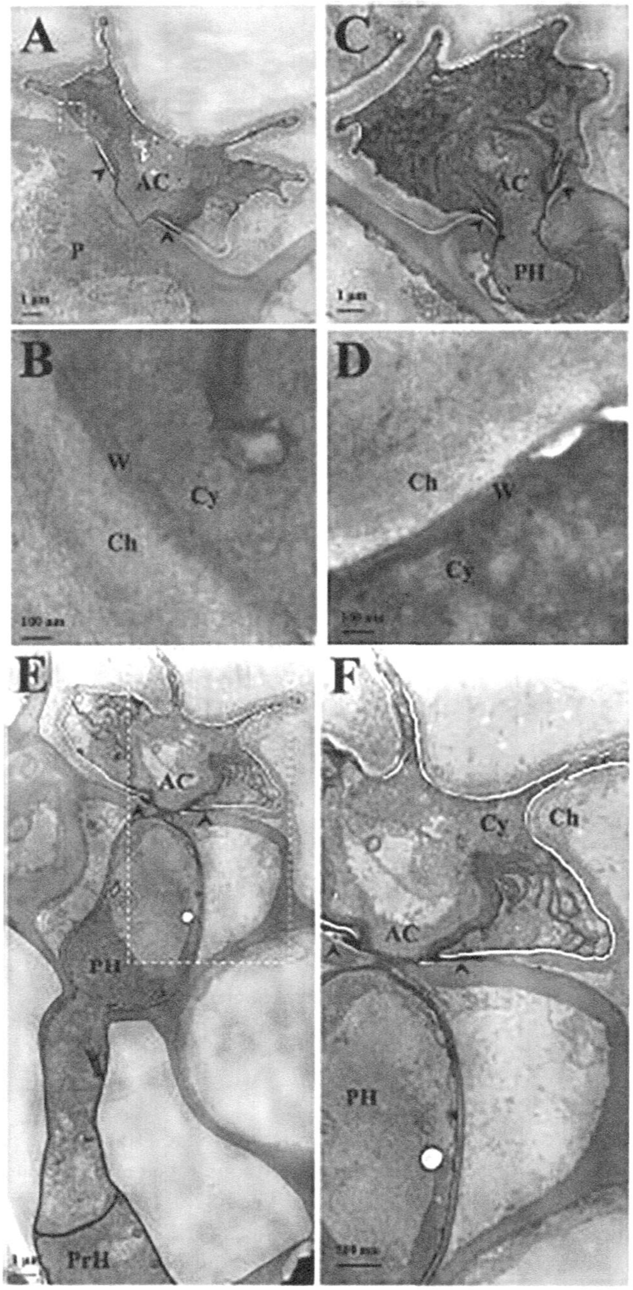

FIGURE 3.6 Process of penetration of soybean leaf by *Phakopsora pachyrhizi (Pp)* as revealed by transmission electron microscopy. (A) Papillae formation to prevent invasion of *Pp* penetration hyphae; (B) enlarged view of marked section of area between chitin cell wall and cytosol (Cy); (C) invasion of epidermal cell by penetration hypha; (D) enlarged view of marked section of gray waxy layer; (E) development of primary hypha from penetration hypha for further invasion through epidermis; (F) enlarged view of the marked section (white dash) showing dark material at penetration site.

Source: (Courtesy of Chang et al. 2014 and with kind permission of the American Phytopathological Society, MN, USA)

Genetic regulation of conidiation in grapevine powdery mildew pathogen *Erysiphe necator* was studied. Obligate biotrophic lifestyle of *E. necator* might have some novel gene regulation. The cDNA amplified fragment length polymorphism (AFLP) analysis with 45 selective primer combinations produced 1,600 transcript-derived fragments (TDFs). Differential gene expression during morphogenesis in *E. necator* and candidate genes involved in conidiation were identified (Wakefield et al. 2011). Individual trichome (Type I) cells of two tomato genotypes were inoculated with pathogenic and nonpathogenic powdery mildew *Pseudoidium neolycopersici*, *Erysiphe trifoliorum*, and *Podosphaeria xanthii*. On trichome cells of tomato, hyphae of pathogenic *Pseudoidium neolycopersici* (KTP-03 and KTP-04 isolates) grew vigorously. In contrast, the nonpathogenic *E. trifoliarum* and *P. xanthii* failed to develop after appressorium formation, which was associated with papilla formation and hypersensitive death respectively. A similar pattern of infection processes of the powdery mildew species occurred on epidermal cells of tomato. Both isolates KTP-03 and KTP04 did not produce conidiophores on the hyphae elongating on trichomes of cv. Moneymaker. Similarly, conidiophores were not formed on the hyphae elongating on trichomes of *Solanum peruvianum*, which was resistant to KTP-03 and susceptible to KTP-04. Further, delayed death occurred in LA2172 (wild tomato species) epidermal cells attacked by KTP03 hyphae elongating from trichomes and conidiophores were formed on new hyphae growing from leaf epidermal cells (Suzuki et al. 2018).

Host–*Pseudocercosporella capsellae* interaction at early stages of development on cotyledons of resistant and susceptible *Brassica carinata, B. juncea*, and *B. napus* were studied. Unique brown structures, especially in susceptible genotypes, were seen. Two morphologically distinct form were observed as thread-like structures within cortical tissues. These brown structures were found to be most prevalent in highly susceptible *B. juncea*. Susceptible *B. napus* also showed high incidence in cortical tissue. The presence of smaller thread-like structures forming highly branched networks and ropy strand-like structures in *B. juncea* and *B. napus* was recorded. There were fewer brown structures and fine thread-like structures that formed networks in *B. carinata*. Resistant genotypes showed brown structures at very low frequency or even absent. Fine thread-like structures were present in resistant *B. juncea* cv. Dune and *B. napus* cv. Hyola 42. Both internal ropy strands within cortical tissues and ropy strand extrusions contained the toxin cercosporin produced by *P. capsellae* (Gunasinghe et al. 2017). Infection process of *Pseudocercospora musae* (banana yellow Sigatoka disease) was studied. Conidia germinated and penetrated through stomata followed by colonization of the spongy parenchyma and palisade parenchyma. Later sporulation commenced in adaxial surface of leaves and conidiophores bearing conidia emerged through stomata (Freitas et al. 2017). Conidial germination in *Phyllosticta maculata* on banana leaf surface commenced within 3 h postinoculation (*hpi*) by producing a long and slender germ tube from which appressoria were produced. A layer of an extracellular matrix surrounded the appressoria at pathogen-host cell interface. On the other hand, conidia germinated on fruit surface producing predominantly swollen germ tubes which functioned as lateral appressoria together with some slender ones. Direct penetration through cuticle with or without formation of appressoria was observed. Cuticular degradation on the leaf surface was frequently noted with a circular, darkened area around point of penetration by hyphae or appressoria (Wong et al. 2013). Differences in the dynamics of *Alternaria brassicola* development on leaves of cabbage cultivars were determined. Greater percentages of germination on Savoy leaves was observed, compared to that on red and white cabbage leaves at 4 h postinoculation (*hpi*). Penetration of the leaf surface occurred mainly through appressoria, rarely through stomata. Formation of infection cushions was observed only on red cabbage leaves. Conidiogenesis occurred first directly from mature conidia, and conidia were produced on conidiophores later (Maciozek et al. 2018).

3.4.3 Colonization of Host Plant Tissues

3.4.3.1 Histopathology of Infected Plants

The infection process of *Diplocarpon mali* (apple leaf blotch) was investigated, using fluorescence and electron microscopy techniques. Conidia became attached to the leaf surface apparently by mucilage and could germinate on both surfaces of the leaves. The pathogen penetrated the cuticle by infection pegs produced either from germ tubes or appressoria, and then haustoria were formed in host epidermal cells and mesophyll cells accompanied by extension of subcuticular and intercellular hyphae. Subcuticular hyphal strands (SHS) were produced simultaneously as a means for fast invasion and reproduction. Acervuli were formed on inoculated leaves (Zhao et al. 2013a). Early events of colonization of wheat floral tissues by *Fusarium graminarum (Fg)* were investigated by employing the strain constitutively expressing green fluorescent protein (GFP). Wheat cultivars Rubin (susceptible) and Sumai 3 (resistant) were point-inoculated at anthesis with a spore suspension of transformant under greenhouse conditions. *F. graminearum (Fg)* hyphae were visible inside the florets of both cultivars. The pathogen showed distinct affinity for pollen and anthers of both cultivars. Abundant growth of *Fg* was seen in the ovaries and stigmas of both cultivars at different intervals after inoculation. The pathogen later spread preferentially to the softer tissues of ovaries and stigmas and reached the spikes spreading internally through rachis and florets. Spread of *Fg* through the parenchyma and vascular tissues was slower and less extensive in resistant cultivar, compared to susceptible cultivar (Miller et al. 2004). In a later study, the infection pathway in wheat spikes by *F. graminearum* and the role of trichodiene synthase gene in pathogenicity were investigated by employing a GFP-labeled strain and its knockout mutants lacking trichodiene synthesis. Trichodienes did not have any role as a virulence factor during infection through fruit coat. However, *Fg* was blocked in the absence of trichodienes by development of heavy cell wall thickening in the rachis node of Nandu wheat, a defense strategy, inhibited by the mycotoxin. In contrast, hyphal development in barley was inhibited at the rachis node and rachilla, limiting infection of adjacent florets through the phloem and along the surface of rachis (Jansen et al. 2005).

Adult plants of wheat cv. Falat and the line ERN9204 were inoculated at the anthesis stage by injecting spore suspension of *Fusarium graminearum* (*Fg*) in medial spikelets using the point-inoculation method. Hyphal colonization and germination of spores in line ERN9204 spikelets were greater than in Falat spikelets. LM and TEM observations revealed that mycelium penetrated into cells through stomata and spread to other cells via cell wall or plasmodesmata. The pathogen showed biotrophic and necrotrophic lifestyles at 24–84 and 120 *hpi*, respectively. Accumulation of lignin displayed at a different pattern during activation of plant defense responses and brownish areas could be seen on the infected florets at 72 and 120 *hpi*. Lignin accumulation was recorded from 72 *hpi* and continued up to 120 *hpi* in the cultivar and genotype examined (Kheiri et al. 2019). In another investigation, *F. graminearum (Fg)* development was restricted in resistant cultivar, although the process of infection by *Fg* was similar in both resistant and susceptible cultivars. In addition, formation of thick layered appositions and papillae was essentially more pronounced in infected tissues of resistant cultivar compared to the susceptible variety. The presence of β-1,3-glucan was detected in the appositions and papillae. The immunogold labeling technique revealed that labeling densities for lignin, thionins, and hydroxyproline-rice glycoproteins (HRGPs) over the cell walls of infected tissues of susceptible cultivar increase only slightly. On the other hand, high accumulation of these compounds in cell walls of infected resistant cultivars was observed. A similar greater increase in the plant hydrolase, β-1,3-gluganase and chitinase were found in infected spikes of resistant cultivar. The mycotoxin DON concentrations produced by *F. graminearum* in the spikes of resistant cultivar was significantly less than in susceptible cultivar, as revealed by immunogold labeling technique (Kang et al. 2008).

Infection process of *Fusarium graminearum* (*Fg*), commenced as conidia of adhered to the surface of wheat spikelets with the help of hydrophobin proteins (Quarantin et al. 2019). Then *Fg* grew epiphytically on plant surface and differentiated into specialized running hyphae (RH) colonizing floral surface. RH differentiated into infection cushion (IC) that complex appressoria comprising agglomerated hyphae. IC facilitated several penetrating events and colonization of wheat epidermal cells. RH and IC formed on the surface of wheat paleae were isolated by laser capture microdissection technique. RNA-seq-based transcriptome analyses were performed on RH and IC and compared to mycelium grown in complete medium (MY). A high number of infection upregulated genes encoding carbohydrate-active enzymes (CAZymes), putative effectors (PE), or secondary metabolism gene clusters (SMC) was identified. RH specifically upregulated one SMC that corresponded to aurofusarin biosynthesis, representing an antibiotic activity. Likewise, IC specifically upregulated 248 genes encoding mostly putative virulence factors such as 7 SMC, including DON and fusaoctaxin A, 33 PE, and 42 CAZymes. Among the infection-regulated genes, 634 encoded secreted proteins, of which 306 were infection downregulated, while 328 were infection-upregulated genes. The highest transcriptional changes on infection-regulated secreted proteins were on CAZymes, reactive oxygen species (ROS), and PE. A greater number of the infection-related nonsecreted proteins upregulated, while some were infection-downregulated (Mentges et al. 2020).

High levels of zoosporangial germination in *Hyaloperonospora parasitica* causing crucifer downy mildew occurred on tissues of ten cruciferous hosts including resistant and susceptible genotypes. Formation of haustoria was the key infection phase for establishment of biotrophic pathogens. Initiation of haustoria occurred inside epidermal cells across all tested species. But haustorial development seemed to abort in penetrated epidermal cells of *Raphanus raphanistrum, Triticum aestivum, Lupinus angustifolius,* or *Trifolium subterraneum.* Haustorial development in epidermal mesophyll cells of *Sinapsis alba* and *Pisum sativum* proceeded. But further hyphal growth and continued haustorial formation could be seen rarely. Hypersensitive reaction was the key resistance response displayed among host and nonhost resistant species in the initial stages of interaction with *H. parasitica* (Li et al. 2011). *H. brassicae* isolates P501, P005, and P006 were inoculated on cotyledons of *Brassica oleracea* genotypes KB207/00 (resistant), KB14/00 (partially resistant) and CrGC3.4 (susceptible). The susceptible genotype CrGR3.4 showed no necrosis, but abundant sporulation was observed in the infected area of cotyledons. Hyphal length and number of haustoria per infection site were lower in the resistant and partially resistant genotypes. On the other hand, susceptible genotype was densely colonized without any recognizable inhibition of host defense mechanism. Restricted pathogen growth in the resistant genotype was associated with autofluorescence and/or browning of host cells at sites of penetration and encasement of haustoria (Baheevandziev et al. 2015).

Leveillula spp. are endophytic, producing haustoria in the mesophyll cells, and conidiophores emerge through stomata, mainly on the abaxial leaf surface, producing primary and secondary conidia which differ in their morphology. Following inoculation of the abaxial pepper leaf surface, conidia of *Leveillula taurica* germinated and formed primary adhesion bodies at the ends of the first germ tubes after inoculation. Infection hyphae were formed from primary adhesion bodies and they rapidly penetrated the leaf tissues through stomata on the leaf surface. Hyphae and conidiophores were projected from internal mycelium through stomata at 14–16 days after inoculation (Figure 3.7). After entry into the leaf tissues, the hyphae continued their growth intracellularly in the spongy parenchyma tissue. Intracellular hyphae were detected in the veins also among the cells of vascular tissue (Figure 3.8). Penetration pegs were surrounded by papillae consisting of thinner electron-dense and much thicker electron-lucent, callose-like materials. Penetration pegs produced haustoria inside the mesophyll cells. Fully the developed haustorium consisted of a central body and several lobes embedded in a haustorial matrix and surrounded by an extrahaustorial membrane (Zheng et al. 2013).

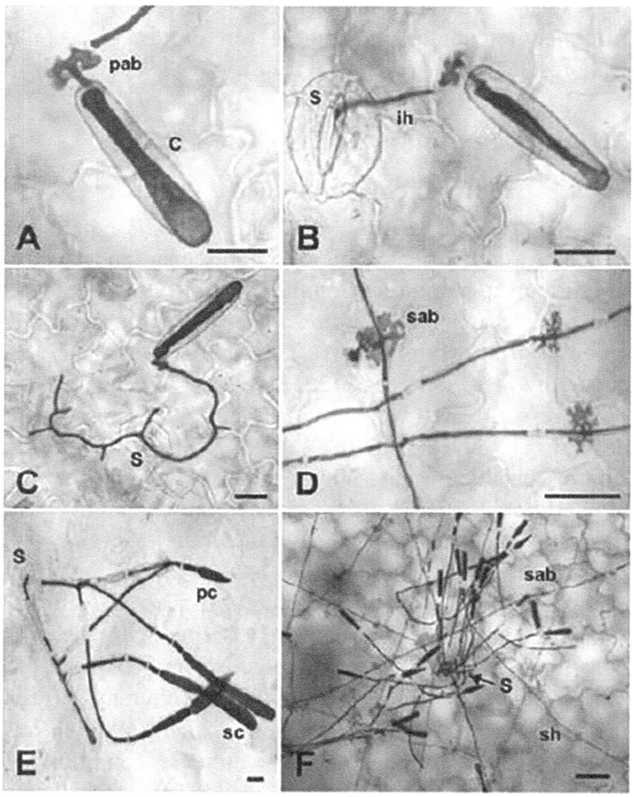

FIGURE 3.7 Emergence of conidiophores with terminal conidia of *Leveilulla taurica* through stomatal opening on the abaxial leaf surface of pepper. (A) Production of lobed primary adhesion bodies (*pab*) from germinated conidia; (B) initiation of infection by *pab*; (C) epiphytic growth of hyphae near stoma; (D) production of secondary adhesion bodies by superficial hyphae; (E) emergence of conidiophores at 2–3 weeks after inoculation; (F) appearance of superficial hyphae produced from internal mycelium colonizing mesophyll cells.

Source: (Courtesy of Zheng et al. 2013 and with kind permission of the American Phytopathological Society, MN, USA)

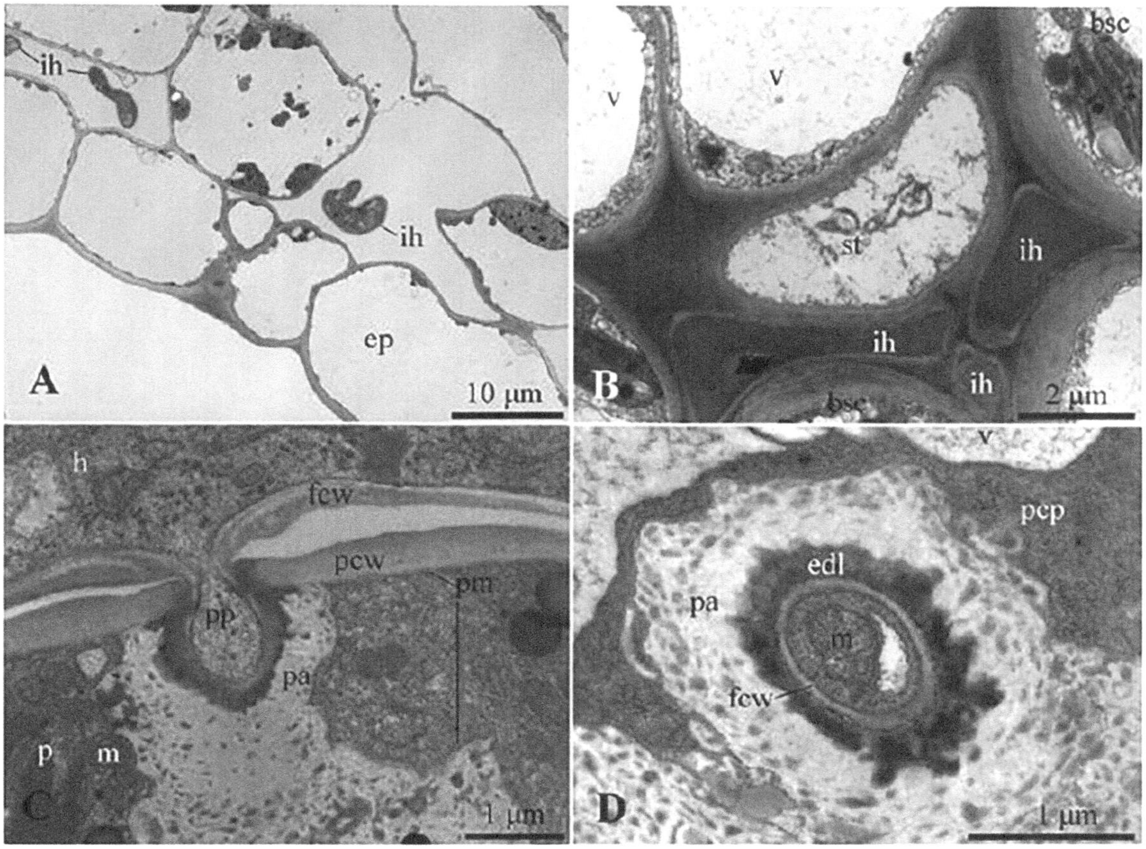

FIGURE 3.8 Progress of development of *Leveilulla taurica* visualized through transmission electron microscopy (TEM). (A) Presence of intercellular hypha (ih) in substomatal cavity and intercellular spaces below epidermis; (B) invasion of leaf vein and sieve tube (st) by intercellular hyphae (ih) vascular bundle sheath cells (bsc); (C) LS through a penetration peg (pp) emerged from intercellular hypha (h) in the mesophyll surrounded by a papilla (pa), fungal cell wall (fcw), mitochondria (m), plastid, plant cell wall (pcw), plasma membrane (pm); (D) TS through a penetration peg surrounded by a papilla (pa).

Source: (Courtesy of Chang et al. 2014 and with kind permission of the American Phytopathological Society, MN, USA)

Grapevine leaves inoculated with *Phakopsora euvitis* (grapevine rust) were examined using SEM. Formation of pustules by the pathogen from stomata and presence of hyphae of the pathogen in the substomatic chamber leading to protrusion of stomata were observed. In the mesophyll adjacent to the water-soaked halo area, the pathogen grew vigorously in the intercellular spaces of the palisade. Mesophyll cells, particularly spongy parenchyma, in the water-soaked halo showed hypertrophy, accumulation of phenolic compounds, additional deposition of pectin and cellulose in cell walls, and pectin in intercellular spaces. In the vascular bundles near pustules, hyphae were seen only in the injured area, but not in the water-soaked halo. The vascular bundle sheath in cv. Mosato Giallo restricted the spread of *P. euvitis* to vascular tissues which showed pectic parietal thickening in cells. The grapevine plants exhibited induced defense responses following infection by *P. euvitis*. The water-soaked halo along with changes in the vascular bundle sheath cells and delimited injuries acted as structural barriers on fungal colonization. Accumulation of pectic substances and the parietal thickenings in the cells of the cellulosic halo and vascular bundle sheath restricted fungal colonization which led to reduction in pustule size on leaves (Navarro et al. 2019).

Puccinia horiana, causing white rust of chrysanthemum, produces two spore types. Basidiospore is the primary infectious propagule, produced from germinating teliospores embedded in infected leaves (Figure 3.9). Basidiospores germinated on leaf surface by forming germ tubes. At the site of penetration, the tip of the germ tube appeared to be attached to the leaf surface by microfibrils. The germ tube penetrated the leaf directly through cuticle. *P. horiana* hyphae grew both intercellularly and intracellularly. An electron-dense material formed at the interface between fungal and mesophyll cells. Penetration of mesophyll cells by penetration (infection) hyphae might be by enzymatic digestion of host cell wall. The pathogen could grow along the inside surface of the xylem vessels within leaf veins. Pathogen hyphae could be commonly seen in mature xylem tracheid cell wall contents, often replacing

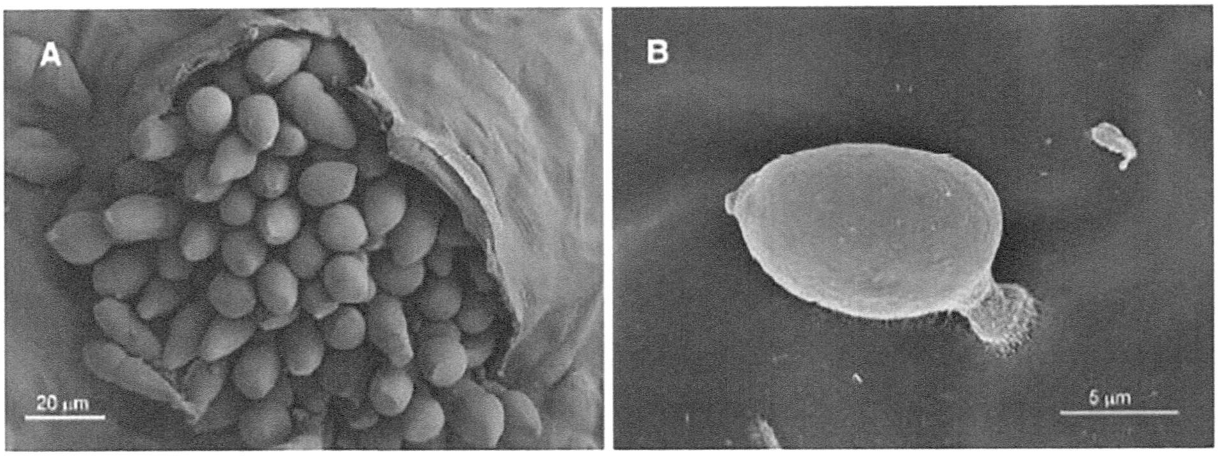

FIGURE 3.9 (A) Teliospores of *Puccinia horiana* embedded in telium that breaks open the chrysanthemum leaf surface. (B) On germination, teliospores produce basidiospores which initiate infection on leaves of susceptible plants by producing germ tube with terminal microfibrils for attachment on leaf surface.

Source: (Courtesy of Bonde et al. 2015 and with kind permission of the American Phytopathological Society, MN, USA)

most of the tracheid cell wall contents. The pathogen frequently moved from one tracheid cell to adjacent tracheid cell, either through pit pair or through nonpitted area of the tracheid cell walls (Figure 3.10). The pathogen developed a septum at or new point of entry into tracheid cell wall (Bonde et al. 2015).

Infection behavior of scab pathogen, *Elsinoe fawcetti* on young leaves of Satsuma mandarin was studied. Of the several conidia inoculated and adhered, only a few germinated and induced characteristic symptoms. Globose appressorium-like structures (APLs) were formed at their tip. APLs were present on both leaf surfaces and symptoms induced even in their absence. Cuticle and mesophyll cells including epidermis were degraded, resulting in cellular necrosis. Adjacent cells of the necrotic region were enlarged and gradually covered leaving a groove-shaped necrotic center, which was the first visible symptom of infection at 3 *dpi*. Hyphae spread through necrotic epidermis and colonized mesophyll tissue intracellularly and intercellularly. Also the lamellated cell wall was formed as the boundary between the infected and uninfected regions in the symptomatic leaves. Lamellation in the cell simultaneously occurred forming a compaction of host cells and a semi-circular arc of lamellated cells was formed in between affected and unaffected areas in the symptomatic leaves (Paudyal et al. 2019).

Hyphae of *Taphrina deformans* (leafcurl disease) penetrate through the stoma or into cuticle. Anatomical structures were similar in both tolerant peach tree (TPT) and very susceptible peach tree (VSPT). However, degradation of mesophyll cells was seen in leaves of VSPT. Cells were free of organelles and middle lamella was expanded and invaded by pathogen hyphae. Some of the samples from TPT showed deformed epidermal cells, loss of some spongy cells and increase of intercellular spaces and division of palisade cells. In VSPT, stimulation of cell division and pathogen proliferation was evident. In addition, increased number of plasmodesmata and loss of differentiation of palisade and spongy cells were discernible. In later stages of disease development in TPT, abundance of fungal hyphae and

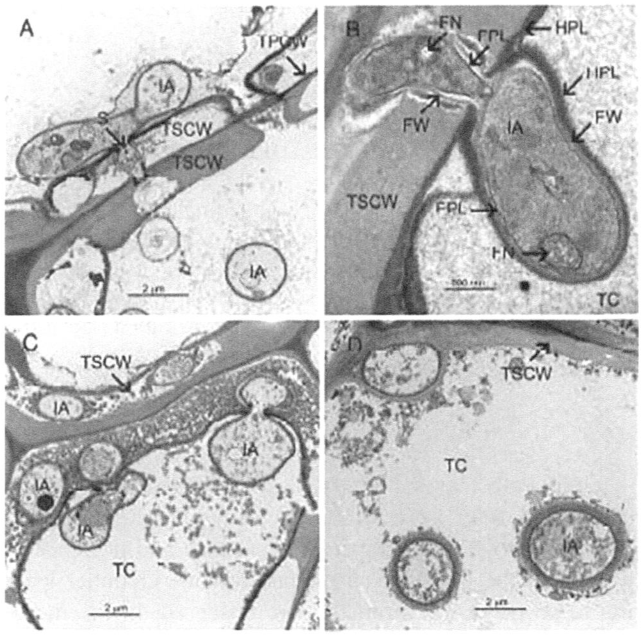

FIGURE 3.10 Colonization of tracheid cells of chrysanthemum by *Puccinia horiana.* (A) Invasion of vascular elements through plasmodesmata from one tracheid to an adjacent tracheid cell; (B) penetration of secondary cell wall of a tracheid cell by *P. horiana*; (C) presence of spherical pathogen cells in infected crown cell of chrysanthemum; (D) presence of isolated pathogen cells in a tracheid cell of stem formed from infected chrysanthemum crown.

Source: (Courtesy of Bonde et al. 2015 and with kind permission of the American Phytopathological Society, MN, USA)

presence of numerous endoplasmic reticulum of rough type (RER) in the cytoplasm marked the destructive changes due to infection of susceptible plants by *T. deformans*. Leaves of VSPT showed empty mesophyll with collapsed cytoplasm and adaxial epidermis covered with fungal fructifications (Giordani et al. 2013). *Colletotrichum acutatum* and *C. gloeosporioides* (citrus PFD) induced necrotic lesions in petals and stigmas of citrus flowers. In petals, penetration by the pathogen occurred intracellularly and intercellularly and also entry through stomata was also observed. Distinct tissues of petals were colonized, including vascular system, particularly xylem. Acervuli were produced on both sides of petals. Increase in contents of phenolic compounds in epidermal cells of stigma was observed, although the pathogen did not penetrate through these cells. A protective layer rich in lipophilic and phenolic compounds was produced under necrotic sites and crystals of oxalate were associated with the sites where pathogen was present (Marques et al. 2013). Infection and colonization processes of *Elsinoë ampelina* were investigated using light, transmission and scanning electron microscopic techniques. Conidia of *E. ampelina* produce one or several germ tubes, and areas under the germ tube showed degraded cuticle during adhesion process. A dense mycelium was formed between trichomes on the abaxial leaf surface. At this stage visible lesions as brown necrotic spots surrounded by a chlorotic halo could be seen. Phenolic compounds accumulated in the epidermal and parenchyma cells of chlorotic areas of the leaf lesions. Hyphae occupied intercellular spaces between cuticle and adaxial epidermis and also between epidermal cells in the mesophyll spaces, where vesicle-like structures were produced. Cellulose levels in cell walls of infected cells were reduced. Elongated lesions were produced along veins, showing depression at the epidermis with formation of conidigenous cells. Ascus-like structures were observed in the collapsed cells. As necrosis progressed, parenchyma inside the vascular system also exhibited cell collapse. Hyphae were present in the lumen of treachery elements (Braga et al. 2019).

3.4.3.2 Role of Enzymes in Disease Development

Fungal pathogens produce various kinds of enzymes and toxins at different stages of pathogenesis to degrade host cells or denature cellular functions to gain entry into host cells and invasion of other host tissues. The enzymes produced by the fungal pathogens degrade pectin by depolymerization. *Botrytis cinerea* degraded pectin and the endoPG genes (*Bcpg*) were expressed during invasion of host plant tissues. EndoPGs were required for virulence of *B. cinerea* (Kars et al. 2005). Purified glycoprotein from *B. cinerea* strain T4, identified as endoPG1 (T4BcPG1) is a virulence and avirulence factor that could activate defense responses in grapevine cv. Gamay (Poinssot et al. 2003). The role of PGIP-dependent generation of oligogalacturonic acids (OGAs) in activating plant defense responses to pathogen in vivo, was investigated. Tomato plants segregating for a mutation in the jasmonate receptor CORONATINE INSENSITIVITY (coi1) and for overexpression of pear PGIP (pPGIP) was challenged with *Botrytis cinerea*. The *coi1* mutant was hypersensitive to *B. cinerea*, but overexpression of *pPGPIP* in *coi1* mutant background reduced pathogen susceptibility. Furthermore, pPGIP overexpression suppressed pathogen induction of salicylic acid (SA) in *coi1* mutant and activated expression of acidic ß-1,3-glucanase independently of the coi1 mutation. However, expression of proteinase inhibitor II (PIN II) in pPGIP overexpressing tomato plants was dependent on COI1 (Akagi et al. 2010). Alternative oxidase (AOX) is involved in fungal mitochondrial oxidative phosphorylation. The role of AOX in *B. cinerea* was studied by generating BcAOX deletion mutants. The mutants showed defects in mycelial growth, sporulation, spore germination, and virulence. The results revealed involvement of BcAOX in regulation of vegetative development and adaptation to environmental stress and virulence of *B. cinerea* (Lin et al. 2019). A gel-based proteomics approach was employed to identify secretome of *Fusarium graminearum (Fg)*, (wheat FHB disease) into the culture medium. Proteins secreted by *Fg* were primarily involved in fungal cell-wall remodeling and the degradation of plant cell walls and starch. In planta expression of corresponding *Fg* genes was confirmed by qRT-PCR assay in barley and wheat spikelets. In addition, a significant difference in the accumulation of pathogen biomass and extent of pathogen-induced proteolysis of plant β-amylase was observed in barley and wheat (Yang et al. 2012).

Colonization of plant tissues by *B. cinerea* was accompanied by rapid killing of plant cells. All pathogenic strains of *B. cinerea* secreted a nonspecific phytotoxin: the sesquiterpene botrydial. The functional characterization of bcbot1, encoding a P450 monooxygenase, revealed that it was involved in botrydial pathway. The botrydial biosynthetic gene *bcbot1* was expressed in planta. Its expression in vitro and in planta could be controlled by an α-subunit of a heterotrimeric GTP-binding protein BCG1. Deletion of *bcbot1* resulted in reduced virulence only in the strain T4 among the strains tested, indicating that the effect on virulence of *B. cinerea* was strain-dependent (Siewers et al. 2005). Involvement of fungal enzymes that degrade cellulose was studied in the interaction between *Botryits cinerea* and tomato. The gene *cel5A* appeared to be a single copy gene. The levels of *cel5A* transcripts and extracellular β-1,4-glucanase activity were regulated by carbon source carboxy methylcellulose (CMC)–induced formation of *cel5A* transcripts and enzyme activity, whereas glucose repressed both functions. The *cel5A* mRNA was detected during infection of tomato leaves by *B. cinerea*. No significant difference between mutants and the wild-type strain either in lesion size or the rate of disease development was discernible. Likewise, no distinct differences in the extracellular β-1,4 glucanase activity and growth rate of the mutants in comparison with wild-type strain was observed (Espino et al. 2005). The gene *xyn11A* coding for an endo β-1,4-xylanase belonging to the family of 11 of glycosylhydrolases was expressed in plant in infected by

Botrytis cinerea. Deletion of the gene *xyn11A* significantly reduced virulence. Complementation of the gene restored virulence in the mutant to the level equal to that of the wild-type strain (Brito et al. 2006). *Colletotrichum gloeosporioides* (anthracnose disease of *Camellia oleifera*) secreted enzymes to degrade the cuticle of aerial plant parts. Cloned cDNA of *CglCUT1* encoded a cutinase in the pathogen. The entire *CglCUT1* gene was knocked out, and the knockout mutant (Δ*CgCUT39*) was subsequently complemented using *Agrobacterium tumefaciens*–mediated transformation. The knockout mutants had significantly reduced cutinase activity, but the virulence remained unaffected and the level was similar to that of the wild-type strain. The complemented transformants of the disrupted transformant Δ*CglCUT39* showed significant increase in cutinase activity and virulence compared to disrupted transformant Δ*CglCUT39* (Wang et al. 2017).

Extracellular hydrolytic enzymes encoded by multiple fungal genes are required to degrade the physical barriers present in plant surfaces for successfully infecting and subsequently colonizing host tissues. The *pelB* gene of *C. gloeosporioides* was disrupted via homologous recombination. The mutants did not show any reduction in growth on glucose. They did not secrete pectate lyase B (PLB) and had a lower PLB and PG activities, compared with the wild-type strain. Lesion size induced by mutants on avocado fruits was reduced. PLB appeared to be an important virulence factor, and probably it might have a role in induction of host defense (Yakoby et al. 2001). Accumulation of ammonia and consequent tissue alkalinization predisposed avocado fruits to infection by *C. gloeosporioides*. The transcript level of *pelB* increased parallel as a function of pH linked to PL secretion and virulence of *C. gloeosporioides* (Prusky et al. 2003). *C. acutatum* secreted seven isozymes of PG. The genes *capg1* and *capg2* encoding two of these enzymes were differentially regulated with respect to pH and carbon sources. Expression pattern of *capg1* was positively correlated with total PG activity (McEvoy et al. 2003). Molecular analysis of the pathogenesis of *Alternaria alternata*–citrus interaction was performed. The *aapk1* gene encoding the cAMP-dependent protein kinase catalytic subunit was cloned and sequenced. The *aapk1* deletion mutant had reduced vegetative growth and was less toxic than the wild-type strain sd1. Deletion of *aapk1* also delayed symptom development on detached tobacco leaves (Xu et al. 2011). Protein kinase HOG1 is one of the mitogen-activated protein kinases, playing an important role in fungal cellular responses to external stress and stimuli. AHOG1, a yeast HOG1 homolog, was isolated from *Alternaria longipes* infecting tobacco and disrupted mutants were generated using homologous recombination. AlHOG1 was involved in multiple physiological processes in *A. longipes*, including mycelia development, melanin production, nutrients sensing, and multistress responses (Yin et al. 2017).

The molecular mechanisms of pathogenesis of *Colletotrichum gloeosporioides (Cg)* were investigated. The T-DNA flanking sequences of the transformants showed homology with the pathogenesis-associated *CAP20* gene. At the onset of appressorium formation in *Colletotrichum* spp., the level of glycerol decreased rapidly, but rapidly rose during turgor pressure generation. The *CAP20* gene in *C. gloeosporioides* was cloned and CAP20 protein was characterized. Most of the CAP20 protein was localized in the lipid droplets in hyphae and conidia. Subcellular localization of CAP20 was visualized fusing a coding sequence of CAP20 and its terminus with GFP and introduced into a constricted Δ*CAP20–16* mutant. The conidia from CAP20 to 16-GFP were examined for the presence of GFP and NR signals, using epifluorescence microscopy. Transformed conidia were treated with GFP, showing the presence of CAP20 localization to the lipid droplets in the germ tube. Perilipin protein acted as a lipid shield which prevented lipid degradation by cytosolic lipases, and it was located at the periphery of lipid droplets. The role of CAP20 on virulence of *Cg*, a target gene deletion, was carried out and null mutants were confirmed by a Southern blot analysis and sequencing. Mutants lacking CAP20 had thinner spores and smaller appressoria. CAP20 could induce virulence in *C. gloeosporioides* (Lin et al. 2007). Regulation of defense pathways in beans against the anthracnose pathogen *Colletotrichum lindemuthianum* (*Cl*) was studied. Hydrogen peroxide (H_2O_2) might have a differential role in establishing incompatible and compatible interactions. Free radicals (H_2O_2) might be elevated at 24 *hpi* in incompatible interaction to arrest pathogen growth, whereas H_2O_2 was promoted at a late stage (72 *hpi*) in compatible interactions, facilitating pathogen invasion. Expression of six defense-related genes was differentially regulated depending on the type of interactions (Shams et al. 2020).

Compatible interactions between *Z. tritici* and both bread wheat (BW) and durum wheat (DW) species were compared at cytological, biochemical, and molecular levels. Spore germination of and leaf penetration by the pathogen were quite similar in BW and DW species. During necrotrophic phase, disease severity and sporulation levels were associated in both interactions with increases of endo β-1,4 xylanase and endo β-1,3-glucanase as well as protease. During the first 5 days postinoculation, induction of *GLUC*, *Chit4*, *POX*, and *PAL* genes and repression of *LOX* gene expression associated with defense responses were detected in both wheat species. Further, activities of peroxidase, catalase, glucanase, phenylalanine-ammonia-lyase, and lipoxygenase were induced in both wheat species. In general, interactions between *Z. tritici* and wheat species proceeded in a similar pattern in infection and host plant responses (Somai-Jemmali et al. 2017). *Zymoseptoria tritici* enters wheat leaf through stomata and establishes a hyphal network in the mesophyll and remains asymptomatic without inducing any visible symptoms in the early stage. Necrosis and production of pycnidia containing pycnidiospores were observed later. Pycnidiospores were dispersed via leaf-to-leaf contact or rain-splashes to neighboring plants. *Z. septoria* possessed a heterothallic mating system and ascospores were disseminated by wind. Expression profiles of *Z. tritici* infection on compatible and incompatible interactions with wheat (*Triticum aestivum*) and *Brachypodium distachyon* (nonhost grass) were compared. Infection of nonhost grass was blocked early after stomatal penetration. During the abortive infections, the pathogen exhibited markedly different gene expression patterns. MoCOD1, a member of fungal Zn(II)1Cys6 transcription factor family, could directly affect pathogenicity in the rice blast pathogen *M. oryzae*. Hence, the role of the putative transcription factor Zt107320, a homolog of MoCOD1, during infection of compatible and incompatible hosts by *Z. tritici* was explored. Zt107320 was differentially expressed in host versus nonhost infections, and the lower expression corresponded to an incompatible infection of nonhost (Habig et al. 2020).

Plant pathogens have to counteract the effects of preformed protective compounds and antimicrobial substances formed in response to initiation of infection. Infection with *Venturia inaequalis* enhanced the metabolism of phenolic compounds in scab spot on fruits and in healthy peel of cv. Golden Delicious. Scab spot had higher enzyme activity for hydroxy cinnamic acids and dihydrochalcones. On contrast, healthy peel had higher concentrations of flavonoids that were present in scab spot (Slatnar et al. 2010). *Curvularia lunata*, infecting maize, produced 1,8-dihydroxynaphthalene (DHN) melanin. The *Brn1* gene involved in melanin production and pathogenicity of *C. lunata* was cloned and silenced. Specific proteins associated with *Brn1* expression in *C. lunata* wild-type CX-3 and *Brn1*-silenced mutant T5 were detected. One upregulated and seven downregulated proteins were identified in T5 mutant. Elongation factor 3 facilitated the synthesis of important proteins required for production of melanin and stress tolerance-related proteins (HSP30 and HSP70) and catalase (CAT) contributed to pathogenicity of *C. lunata*. On the other hand, transaldolase, NADP-dependent mannitol dehydrogenase, formate dehydrogenase, and UTP-glucose-1-phosphate uridylyltransferase were involved in energy metabolism (Gao et al. 2015). Laccase-like multicopper oxidases (LMCOs) play important physiological roles in growth and pathogenicity of fungal pathogens. Nine LMCOs of *Setosphaeria turcica* (maize northern leaf blight) were detected. The genes *StLAC6* and *StLAC8* were highly expressed in mycelium and expression of *StLAC2* was significant in nongerminated conidia. During infection, expression of *StLAC8* was high during appressorium formation and expression of *StLAC6* was high during formation of the penetration peg. When treated with laccase inducer Cu2+, the laccase activity significantly increased. Expression of all genes was significantly increased, except that of *StLAC7*. In the presence of the phytoalexin resveratrol, laccase activity did not increase, but expression levels of *StLAC2, StLAC4,* and *StLAC5* were upregulated (Liu et al. 2019).

Plant class III chitinases known as pathogenesis-related (PR)-8 proteins include rice xylanase RIXI, in the GH18 family. GH11 xylanase was inhibited to the maximum, while rice endogenous xylanase (OSX) was the least inhibited. Xylanase inhibitor gene *RIXI* was upregulated in rice seedlings infected by *Magnaporthe grisea*. Increased *RIXI* expression was also accompanied by significantly elevated expression of PR-protein gene *PR-1* and class III chitinase gene (*chit-3*) (Hou et al. 2014). The role of *Sas3* gene, encoding a histone acetyltransferase (HAT) ortholog MoSAS3 and its histone acetylation activity in the regulation of growth, development and pathogenicity of *M. oryzae*, was studied by combining genetic and transcriptomic approaches. Cellular localization of MoSAS3 was examined by expressing MoSAS3 protein fused to GFP at its C-terminal. MoSAS3 was localized to pathogen nuclei. However, a faint and diffuse fluorescence signal was detected in the cytoplasm also. Deletion of MoSAS3 resulted in substantial reduction in H3K14ac, a histone substrate. Further, more deletion of *MoSaS3* gene significantly impaired vegetative mycelial growth, compared to the wild-type strain. The production of conidiophores in mutant was reduced significantly. Complementation of *Mosas3* gene in the mutant restored the capacity to produce conidia normally. Both conidial germination and appressorium production were also significantly reduced in the Δ*Mosas3* mutant. The wild-type KJ201 and MoSAS3c strains induced typical blast lesions, dark brown spots, and yellowing of leaves. In contrast, the mutant Δ*Mosas*3-noculated leaves did not produce any lesions, indicating the loss of pathogenicity (Figure 3.3). *MoSAS3* gene was required for growth, spore germination, appressorium formation, and pathogenicity of *M. oryzae* (Dubey et al. 2019).

3.4.3.3 Role of Toxins and Other Metabolites in Disease Development

The classical example for host-specific toxin is victorin produced by *Cochliobolus victoriae* (oat Victoria blight disease), and it is the primary pathogenicity determinant. Isolates of *C. victoriae* producing victorin are pathogenic, whereas isolates lacking ability to produce victorin are nonpathogenic. Victorin could function not only as toxin but also as a specific elicitor of phytoalexin accumulation and stress-responding genes in victorin-sensitive oat lines, when applied at a low concentration. During apoptotic cell death induced by victorin, rRNA and mRNA were specifically degraded in oat cells. The rRNA species from cytosol, mitochondria, and chloroplasts were degraded via specific degradation intermediates during victorin-induced apoptotic cell death. Specific cleavage of mRNA of house-keeping genes, such as actin and ubiquitin, during victorin-induced cell death was observed (Hoat et al. 2006). Southern corn blight pathogen *Cochliobolus heterostrophus* race T, infecting maize lines with Texas male sterile (tms) cytoplasm, produced a host-specific toxin (HST), T-toxin. The production of T-toxin was under the control of two unlinked loci, *TOX1A* and *Tox1B*, carried on DNA. The T-toxin production depended on *PKS1*, a polyketide synthase (PKS) encoding gene *TOX1A,* and *DEC1*, a carboxylase-encoding gene at *TOX1B*. The requirement of PKS2, a second PKS-encoding gene that maps at *TOX1A,* for T-toxin biosynthesis and high virulence to maize revealed by ligation specificity-based expression analysis display (LEAD) (Baker et al. 2006). The genome of *C. heterostrophus* contains 12 NRPS-encoding genes (NRPSs). NRPS6 was involved in virulence of *C. heterostrophus* to maize. Deletion of NRPS6 resulted in concomitant reduction in virulence and enhancement of sensitivity to H_2O_2 (Lee et al. 2005).

Alternaria alternata strains produce different host-selective (specific) toxins (HST) that cause necrosis on different hosts and they are called 'pathotypes' of *A. alternata*. *A. alternata* pathotype causes Alternaria brown spot in tangerines (*Citrus reticulata*), whereas the pathotype infecting rough lemon (*C. jambhiri)* induces Alternaria leaf spot disease. Distinct host-specific (selective) toxins ACT- and ACR-toxin are produced by these pathotypes, but they have different host ranges (Akimitsu et al. 2003). ACTT genes involved in the biosynthesis of ACT-toxin were identified (Akimitsu et al. 2003; Matsunaka et al. 2005). *A. alternata* strawberry pathotype produces ACT-toxin involved in disease induction. Four AFT genes, *AFT-1, AFT3–1, AFTR-1,* and *AFTS-1* were located in 2-kb region of *pcAFT-1* (Hatta et al. 2006). *ACTTS2* gene was present only in ACT-toxin producers. ACT-toxin production and pathogenicity were reduced, following targeted gene disruption of ACTTS2. Further, transcriptional knockdown of ACTTS2, using RNA silencing resulted in complete loss of ACT-toxin production and pathogenicity. The results revealed the requirement of ACTTS2 for ACT-toxin biosynthesis in the tangerine pathotype of *A. alternata* as well as for pathogenicity on tangerines (Ajiro et al. 2010). The rough lemon pathotype of *A. alternata* infecting rootstock species of rough lemon (*Citrus jambhiri*) and Rangpur lime (*C. limonia*) produces host-selective ACR-toxin. Genes governing toxin production were detected in the chromosome carrying the ACR-toxin biosynthesis gene cluster (ACRT) in the genome of rough lemon pathotype. Three copies of *ACRTS* genes were present and disruption of two copies of the gene significantly reduced ACR-toxin production and pathogenicity as well. The RNA-silenced transformants did not produce detectable ACR-toxin and were not pathogenic. The *ACRTS1* gene was required for ACR-toxin biosynthesis and also full virulence to rough lemon plants (Izumi et al. 2012). *LaeA* global regulator homologs including methyl transferase were identified in tomato, strawberry, and apple pathotypes of *A. alternata* and they were called *AtLAE-1, AsLAE1,* and *AaLAE1,* respectively. In the *AtLAE*-deleted mutant of the tomato pathotype, expression of AAL-toxin biosynthetic gene *ALT1* was lower than that in the wild-type strain. Consequently AAL-toxin production and virulence of the mutant were significantly lower and sporulation of the mutant was also reduced significantly. Likewise, the *AsLAE1*-deleted mutant of strawberry pathotype (Δ*AsLAE1*) and the *AaLAE1*-deleted mutant of apple pathotype (Δ*AaLAE1*) produced lower concentration of respective toxins (AF- and AM-toxin). Virulence of the mutant hyphal growth and sporulation were affected. The results indicated the global regulator gene *LaeA* positively regulated HST biosynthesis, pathogenicity, growth, and differentiation in *A. alternata* pathotypes (Takao et al. 2016).

Pyrenophora tritici-repentis (Ptr) produces combinations of three host-specific toxins, ToxA, ToxB, and ToxC. Each toxin interacted with a host sensitive locus, respectively, *Tsn1, Tsc2,* and *Tsc1*. The virulence of an isolate was partially connected with the presence of these toxins and resistance in the host was associated with the absence of sensitivity loci. The presence of the gene for ToxA was detected in all isolates, whereas the gene for ToxB was absent in all isolates of *Ptr*. (Antoni et al. 2010). *P. tritici-repentis* produces multiple necrotrophic effectors contributing to virulence. The proteinaceous effector ToxA induced necrosis in wheat genotypes possessing the *Tsn1* gene. In general, ToxA sensitivity was associated with susceptibility to disease. The results indicated that ToxA might not be the major determinant in tan spot disease development. Infection by mutant *toxa* in three varieties resulted in enhancement of chlorosis. The results showed that *P. tritic-repentis* seemed to interact with wheat cultivars in a complex manner, leading to a variety of disease reactions that might be dependent or independent of *ToxA-Tsn1* interaction (See et al. 2018). *Cercospora kikuchii* infecting soybean elaborates cercosporin. Targeted disruption of the genomic copy of the mRNA transcription of one of the cDNA clones, called cercosporin facilitator protein (CFP), resulted in reduction in virulence on soybean. Cercosporin plays a pivotal role in pathogenicity of *C. nicotianae* and lesions formation on tobacco. A polyketide synthase gene (CTB1) was functionally identified and it played a key role in cercosporin biosynthesis by *C. nicotianae* (tobacco frog eye leaf spot). Targeted disruption of CTB1 resulted in loss of both CTB1 transcript and cercosporin biosynthesis in *C. nicotianae*. Pathogenicity could be restored to null mutants by complementation with a full-length CTB1 clone, in addition to production of cercosporin. (Choquer et al. 2005). Another *CTB3* gene was essential for biosynthesis of cercosporin, Expression of eight genes (*CTB1–CTB8*) in a cluster was coordinately induced (Chen et al. 2007). *Exserohilum turcicum* infecting maize produces the host-specific HT-toxin. Differential expression of *Ht*-related genes between maize near-isogenic lines (NILs) was analyzed by cDNA-AFLP procedure, following treatment with crude extract of HT-toxin. Transcript-derived fragments (TDFs) (21), called H1 to H21, were specifically expressed or upregulated in NIL, HZSHt2, following exposure to the HT-toxin (Wang et al. 2010). The presence of the *ToxA* gene in wheat spot blotch pathogen, *Bipolaris sorokiniana,* was identified. The *ToxA* gene was present in *B. sorokiniana* populations infecting winter wheat. Duster wheat leaves showing necrotic lesions contained isolates harboring *BsToxA* gene and they secreted ToxA in culture. Sensitivity assays conducted with differential line BG261 containing the dominant ToxA sensitivity gene Tsn1 showed the presence of toxin. Culture filtrates of the isolate lacking BsToxA did not induce necrosis on any of the lines. Culture filtrates

(CFs) of the four *BsToxA*-harboring isolates induced necrosis on the wild-type *Tsn1* lines, but not on the corresponding *tsn1* mutant lines. All isolates harbored the same *BsToxA* nucleotide sequence, which was identical to one of the two haplotypes identified in Australia (Friesen et al. 2018).

The purple pigment produced by *Pseudocercosporella capsellae (Ps)* (Brassica white leaf spot disease) is due to the toxin cercosporin. CFs containing cercosporin produced by *Ps* were phytotoxic to *Brassic juncea, B. napus*, and *Raphanus raphanistrum*. Sensitivity of the *Brassica* spp. to cercosporin was analogous to the susceptibility to the pathogen. The presence of cercosporin in the inoculum significantly increased disease severity on the highly cercosporin-sensitive *B. juncea* plants. Cercosporin was a pathogenicity factor, since it reproduced white spot lesions on susceptible plants and enhanced virulence of *P. capsellae* (Gunasinghe et al. 2016). The effects of virulence, incubation time and temperature on production of cercosporin by *Cercospora coffeicola* were assessed. Isolates LFP12 and LFP43 produced the highest and lowest concentrations of cercosporin, respectively, indicating the variability in toxin production by pathogen isolates. The production of toxin ceased in all isolates after 15 days of incubation at 20–23° C (Vale et al. 2019). *Curvularia lunatus* maize pathotype (maize Curvularia leaf spot) produces a nonhost-specific toxin methyl 5-(hydroxymethyl) furan-2-carboxylate (M5HF2C). *Agrobacterium tumefaciens*-mediated transformation (ATMT) was employed to tag the gene for specific toxin biosynthesis. Southern blot analysis showed that the gene was *Clt-1* was essential for both toxin production and pathogenicity, as shown by gene disruption experiments (Gao et al. 2014). In the further study, five upregulated and two downregulated proteins were identified in Δ*Clt-1*. The succinate dehydrogenase flavoprotein subunit (Sdh1p) functioned as a coenzyme of dehydrogenase in the anabolism of toxin. Eukaryotic elongation factor 3 (EF-3) facilitated synthesis of proteins required for toxin production in *C. lunatus*. Melanin synthesis-related protein scytalone dehydratase (SCD) and stress tolerance-related proteins (HSP30 and HSP70) contributed to pathogenicity of *C. lunatus*. *Scd* gene exhibited similar expression pattern at transcriptional level, corresponding to the protein detected by 2-DE (Gao et al. 2015). The presence of genes related to melanin production and toxin production or pathogenicity was investigated using a transcriptome approach, based on the RNA-Seq technique. Specific genes were associated with *pks18* expression in *pks18* deletion mutant Δ*PKS18* and wild-type CX-3 (Gao and Chen 2018).

3.4.3.4 Role of Auxins in Disease Development

Rice transformants containing an auxin-inducible reporter construct exhibited an auxin response around the apex of infection hyphae of *M. oryzae* following invasion of living tissue. The results suggested that the pathogen might produce small amount of auxin during its biotrophic phase of invaded rice tissues. The auxin was likely to be involved in the biotrophic phase of infection hyphae (Tanaka et al. 2011). *Colletotrichum acutatum* infects citrus petals and induces premature fruit drop and formation of persistent calyces. Differential gene expression, in addition to accumulation of hormones and other growth regulators were investigated. Ethylene evolution showed 3-fold increase, whereas IAA accumulation was very high. Northern blot analysis revealed that the genes encoding ACC oxidase or ACC synthase and 12-oxo-phytodienoic acid (12-oxo-PDA) reductase were highly expressed in affected flowers. The gene encoding auxin-related proteins were upregulated (Lahey et al. 2004). The effects of ethylene on *Botrytis cinerea*, incitant of gray mold diseases were assessed. Exposure of *B. cinerea* to ethylene inhibited mycelial growth and induced transcriptional changes in large number of genes. The G-protein signaling pathway partially mediated the response of *B. cinerea* to ethylene and ethylene-induced host resistance might involve effects of plant ethylene on both plant and pathogen (Chagué et al. 2006). Spore germination and hyphal growth of *A. alternata* were promoted by ethylene production. Strain AO induced higher ethylene production and formation of larger lesions. Isolates from infected berries induced formation of ethylene on inoculation onto berries. *A. alternata* produced ethylene via the KMBA pathway and utilized it for enhanced expression of virulence during pathogenesis (Zhu et al. 2017). Host response to microbial pathogens may be influenced by plant signaling hormones such as ethylene (ET). Resistant responses to necrotrophic fungal pathogens are often mediated via synergistic interactions of ET with jasmonate signaling pathway. In addition, ET can also function as an inducer of senescence and cell death. The effects of ET on development of *Fusarium graminearum* (*Fg*) (wheat FHB) were investigated. ET-inhibitor treatments broke down resistance to both initial infection and disease spread in three resistant genotypes, whereas ET-enhancer treatments resulted in reduced susceptibility in three susceptible genotypes (Foroud et al. 2019).

3.4.3.5 Role of Mycotoxins in Disease Development

Accumulation of mycotoxins produced by fungal pathogens in seeds makes them unfit for consumption by human and animals. The nature of fungal pathogens, crop genotype, and chemical applied at the heading stage may influence infection of spikes and consequent accumulation of mycotoxins in the grains. FHB of wheat is caused by different *Fusarium* spp. *F. graminearum* occurs predominantly in most wheat growing countries. Various *Fusarium* spp. producing different mycotoxins are differentiated based on the nature of the mycotoxins produced by them. Isolates/strains of a morphologic species producing mycotoxins are grouped into one chemotype, and there may be different chemotypes within one morphologic species of *Fusarium*. *F. graminearum (Fg)* strains produce DON, 15-acetyldeoxynivalenol (15-ADON) and 3-acetyldeoxynivalenol (3-ADON) (Narayanasamy 2017).

Early events of colonization of wheat floral tissues by *F. graminearum (Fg)* were investigated by employing a *gfp*-expressing strain of *Fg*. A strain of *F. graminearum (Fg)* was transformed with green fluorescent protein (*gfp*) gene to visualize infection patterns in barley and *Arabidopsis* model system to determine primary site of infection by *Fg*. The rapid route of infection involved the brush hairs at the extruded seed tip. Colonization of seed tissue was followed by a pattern of a rapid basipetal growth along the pericarp epithelium, followed by a slower growth inward through pericarp and testa (Skadsen and Hohn 2004). Expression of core deoxynivalenol (DON) biosynthetic gene *Tri5* during infection of wheat head by *F. graminearum (Fg)* of susceptible and resistant cultivars treated with azoxystrobin (strobilurin fungicide) was studied. As infection progressed, earliest-infected kernels showed diminished relative *Tri5* expression, but its expression did not cease during the experimentation period of 21 days. DON levels were negligible in wheat at 8 days after inoculation (*dai*) with detectable DON at later sampling time. *Tri5* was detected even in fully senesced kernels at 21 *dai*. *Tri5* transcripts were present in susceptible cultivar over a much longer time period. The ability of the pathogen to rapidly resume toxin biosynthesis in dried grain might be a possible mechanism for accumulation of DON to reach high levels in asymptomatic grain (Hallen-Adams et al. 2011). Pathogen development and host responses to infection by *F. graminearum* (*Fg*) on wheat spikes of resistant Sumai 3 and susceptible Xiaogyan were investigated. Pathogen spread was restricted in resistant variety, although the process of infection of *Fg* was similar in both resistant and susceptible varieties. Further, production of thick layered appositions and papillae was distinctly more pronounced in tissues of Sumai 3, compared to Xiaoyan. The presence of β-1,3 glucan was detected in the appositions and papillae. Densities of labeling for lignin, thionins and hydroxyproline-rich glycoproteins (HRGPs) over cell walls of infected tissues of susceptible cultivar increased only slightly. On the other hand, high accumulation of these compounds in cell walls of infected resistant cultivars was observed. Similar increases in concentrations of plant hydrolases, β-1,3-glucanase and chitinase were detected in infected spikes of resistant cultivar. DON contents produced by *F. graminearum* in spikes of resistant cultivar was significantly less than in susceptible cultivar, as revealed by immunogold labeling technique (Kang et al. 2008).

Progression of DON and NIV accumulation in wheat grain infected by *F. graminearum (Fg)* and influence of time of infection on mycotoxin accumulation were studied. Four wheat cultivars were spray-inoculated at three different stages with a mixture of DON and NIV chemotypes of *Fg*. High levels of DON and NIV were produced beyond 20 days after anthesis (DAA), even in early infection of spikes under greenhouse and field conditions. Late infection under greenhouse conditions at as late as 20 DAA resulted in grain contamination with toxins, even without clear disease symptoms on the spike (Yoshida and Nakajima 2010). *Fusarium graminearum* (maize seedling blight) occurs in the form of several chemotypes capable of producing different forms of DON and NIV. The roles of genotype-determined mycotoxins in disease induction were studied. These isolates produced both DON and 15-ADON in either rice culture medium or maize seedling roots. The 15-ADON was predominant initially, whereas DON predominated later. The biomass of *F. graminearum* doubled and was positively correlated with virulence of the isolates. Both mycotoxins contributed to symptom expression of seedling blight, but 15-ADON was more destructive than DON (Meng et al. 2017).

3.4.4 Molecular Bases of Infection of Fungal Pathogens

3.4.4.1 Role of Effectors in Fungal Pathogenesis

Plant cells (tissues) deploy extracellular and intracellular receptors to detect invasion patterns of the pathogens and to recognize the presence or activities of pathogens and to initiate immune responses. On the other hand, pathogens secrete effector proteins to deregulate host responses. However, effectors themselves can become invasion patterns, when they are recognized by plant immune receptors that trigger immune responses, resulting in incompatible interaction between the pathogen and host plants. In order to avoid recognition by plant immune system, pathogens in turn evolve novel effectors or modify or lose existing ones (Cook et al. 2015). Effectors secreted by pathogens exert their activity either in host's extracellular environment (apoplastic effectors) or inside host cells (cytoplasmic effectors). Effectors are generally secreted via the endoplasmic reticulum (ER) Golgi route, which requires a hydrophobic N-terminal signal peptide. Oomycetes-secreted effectors contained such an N-terminal secretion signal (Lo Presti et al. 2015). Effectors may subsequently be translocated from extracellular space into the host cytoplasm. In oomycetes, two distinct classes of cytoplasmic effectors, the RXLRs and crinklers (CRNs) have been differentiated. Both classes have conserved amino acid motifs in the N-terminal region downstream of the signal peptide (Schornack et al. 2010). Effector genes in *Leptosphaeria maculans* may be located on distinct chromosomes that are not required for growth under specific conditions (Balesdent et al. 2013). Genome comparisons may be useful to identify effectors and avirulence effectors that are recognized by the plant immune system. Comparisons of avirulent and virulent strains of a pathogen may lead to identification of SNPs in protein-encoding regions of effectors or reveal the presence/absence of polymorphisms in genomic regions that are present exclusively in avirulent or virulent strains. Likewise, sequencing and subsequent analyses of transcriptomes might lead to the identification of effectors (Mesarich et al. 2014). Identification of effectors from genomic sequences is crucial for enhancing host plant resistance. Existing effector predicting methods are unable to identify hundreds of candidates and may result in high false-positive rates. Fungal effector proteins relied on filtering secretomes for proteins of small size and high cysteine content. Such user-driven selection is generally of low accuracy. A contrasting data-driven approach is machine learning that can automatically identify relevant patterns in data. EffectorP versions 1.0 and 2.0 had a major impact on plant-pathogen community and introduced a shift in methodology from user-driven selection to data-driven selection of effectors (Sperschneider et al. 2018). Differences in apoplastic and cytoplasmic environments that allow for accurate prediction of pathogen proteins localizing in these compartments were utilized in the EffectorP 3.0 system. This system could predict whether a secreted protein is an apoplastic, cytoplasmic, or noneffector. EffectorP 3.0 accurately predicted known apoplastic and cytoplasmic effectors in fungal and oomycete secretomes with low estimated false-positive rates of 3% and 8%, respectively (Sperschneider and Dodds 2022).

'Effectoromics' is a large-scale screening approach to utilize effector candidates to identify host resistance (R) genes. This approach is based on transient expression of candidate effectors in plant leaves by agro-infiltration (*Agrobacterium tumefaciens* transient transformation assay) and/or a virus vector like *Potato virus X*. Then plants are screened for development of hypersensitive cell death responses (HR), indicating recognition of the effector by matching plant immune receptor. Pathogen effectors may be employed to develop protection strategies against pathogens, since they might reduce pathogen development. Host-induced gene silencing (HIGS) is based on exploring an RNA silencing mechanism to target a selected pathogen gene (effector gene) via host plant. Plants may be transformed with harpin RNA constructs that target selected pathogen effectors to suppress their expression. HIGS in barley expressing ds RNA targeting the effector gene *Avra10* from powdery mildew pathogen *Blumeria graminis* led to reduced disease incidence (Nowara et al. 2010). To avoid pathogen evasion by loss of the targeted effector conserved genes that are unlikely to be lost by the pathogen may target HIGS in barley. Targeting the cytochrome P450 gene *CYP51* of the wheat FHB pathogen *Fusarium graminearum* resulted in complete plant immunity (Koch et al. 2013).

Although plants lack an adaptive immune system, they have a sophisticated innate immune system that is able to circumvent infections by most microbial pathogens. In addition to physical barriers, cuticle, and cell walls, plant apoplast is equipped with various hydrolases, protease inhibitors and antimicrobial compounds to impede infection by pathogens. Soon after the physical barriers are breached, plant cell surface receptors are activated to sense invading microbial pathogens via detection of molecular patterns derived from microbial pathogens, known as microbe-associated molecular patterns (MAMPs) or pathogen-associated molecular patterns (MAMPs), which ultimately activate pattern-triggered immunity (PTI), including reactive oxygen species (ROS) production, ion fluxes, callose depositions, defense-related gene expression, and secondary metabolite production (Boller and Felix 2009). Plant pathogens secrete effectors to undermine plant immunity leading to successful infection. The apoplastic effectors, acting on the apoplast and cytoplasmic effectors, secrete inside plant cells fulfill virulence functions. Cytoplasmic effectors are translocated into different cell compartments and interfere with various plant physiological processes, including protein synthesis or secretion, ROS production, secondary metabolites synthesis, MAPK activation, as well as epigenetic modifications. The initial interactions between plants and pathogens are initiated in plant apoplast, which hosts extracellular pathogens as well as feeding structures of different fungal pathogens (Lo Presti et al. 2015). Some of the apoplastic effectors may be detected by plants as elicitors inducing defense responses. The effectors functioning in apoplast are highly conserved and induce defense responses among microbial taxa or race-specific avirulence proteins secreted by extracellular fungal pathogens. The necrosis- and ethylene-inducing peptide-like proteins (NLPs) present in a broad spectrum of pathogens are cytotoxic effectors that trigger necrosis in multiple dicotyledonous plants. Tomato leaf mold pathogen *Cladosporium fulvum* could release dozens of small secreted effectors into apoplast during infection. Several race-specific avirulence effectors such as Avr2, Avr4E, Avr5, and Avr9 were recognized by tomato lines carrying cognate *C. fulvum* resistance genes (de Wit 2016). In a later investigation, a survey of 63 putative apoplastic effectors secreted by *Zymoseptoria tritici* (wheat Septoria tritici blotch) showed that 13 effectors could be recognized by nonhost *Nicotiana benthamiana* and trigger cell death (Kettles et al. 2017). All immune receptors sense the secreted apoplastic effectors belonged to the superfamily of receptor-like proteins with membrane-localized leucine-rich repeat (LRR) receptors in their extracellular domains (Wang and Wang 2018).

Adapted pathogens have evolved virulence effectors to downregulate pattern triggered immunity (PTI) basal responses in plants. Modulation of chitin-induced plant immunity by fungal apoplastic effectors has been investigated. Plants secrete apoplastic hydrolytic enzymes such as chitinases, which target plant cell wall components to release soluble chitin oligosaccharides and damage fungal cell wall integrity. Chitin oligosaccharides are perceived as microbe-associated molecular patterns (MAMPs) by plant membrane receptors to initiate PTI. In response, fungal pathogens secrete effectors interfering with plant recognition of chitin and blocking chitin-induced immunity. *Cladosporium fulvum* effector Avr4 contains a chitin-binding domain that binds chitin and protect the cell walls of chitinous fungal pathogens from hydrolysis by plant chitinases (Van den Burg et al. 2006). Chitin-binding activity of Avr4 inhibited the release of soluble chitin oligomers, preventing activation of chitin-induced plant immunity. Ectopic expression of *Avr4* increased *Arabidopsis* susceptibility to chitinous fungal pathogens, but not to bacterial or oomycete pathogens lacking chitinous cell walls, indicating the ability of Avr4 to promote virulence via countering chitin-induced plant defense (Van Esse et al. 2007). Extracellular protein 6 (Ecp6), another protein secreted by *Cladosporium fulvum*, was required for full virulence in tomato. The action of Ecp 6 was different from that of Avr4, and it sequestered chitin oligosaccharides released from cell walls of invading fungi, thereby preventing activation of chitin-induced plant immunity (de Jonge et al. 2010). *Stagonospora nodorum* (wheat SNB) produces several necrotrophic effectors (NE, also termed host selective toxins). Host genotypes have corresponding sensitivity genes (*Snn*). The production of NEs was detected in isolates. Of the 39 isolates, 17 produced NEs different from those previously characterized in the Upper Great Plains region. The novel NEs might correspond to unidentified *Snn* genes in the differential wheat cultivars. Only two cultivars appeared to contain known *Snn* genes. Effector sensitivity was more frequent in SNB-susceptible cultivars than in moderately resistant (MR) cultivars, although some susceptible cultivars were not sensitive to NEs produced by isolates tested (Crook et al. 2012). Pathogen-derived, small secreted cysteine-rich proteins (SSCPs) are the common sources of fungal effectors that trigger susceptibility/resistance, depending on host-pathogen interactions. Comprehensive analyses of SSCPs encoded in the genome of *Fusarium graminearum (Fg)*, were performed using proteomics and sequence transcriptional approaches to select candidate effector proteins. Transcriptional analysis showed that at least 34 SSCPs (including 23 detected in the in vitro secretome) were expressed in infected wheat heads. About half of them were upregulated with expression patterns correlating with development of FHB (Lu and Edwards 2016).

In *M. oryzae*, cytoplasmic effectors preferentially accumulated in the biotrophic interfacial complex (Khang et al. 2010), whereas apoplastic effectors followed the conventional secretory pathway (Giraldo et al. 2013). The apoplastic LySM effector Slp1, secreted by *M. oryzae*, was functional in a manner similar to Ecp 6 of *Cladosporium fulvum*. Slp1 bound specifically to chitin and competed for chitin-binding with plant chitin receptors to suppress chitin-induced immunity (Mentlak et al. 2012). Regulatory relationships between transcription factors and putative effectors proteins are important to understand the course of rice blast disease development. The period of 36 h postinoculation, termed the 'early infection' stage, is crucial for deploying pathogen effectors to overcome the defenses of the host resulting in symptom expression. During the early infection stage, the pathogen was confined to the first host cell and secreted many effectors to weaken the host immune responses (Kim et al. 2020). Processes triggered during early infection stage are essential for survival of *M. oryzae*. Transcription factors are essential for regulating gene expression and cell development. During *M. oryzae* infection, of the 6,000 expressed genes detected, >800 were putative effectors. The roles of two **e**arly **i**nfection-induced **t**ranscription **f**actors (MoEITF1 and MoETIF2) in infection process were analyzed. The mutants Δ*moeitf1* and Δ*moeitf2* were defective only for infection progression. Most of the effector genes were downregulated in mutants employing a potential regulation by the transcription factors. Overexpression of two highly downregulated effectors, T1REP and T2REP, in the mutants partially restored the pathogenicity of Δ*moeitf1* and Δ*moeitf2* respectively, indicating that these were directly regulated. T1REP and T2REP showed similar BIC (*M. oryzae*-specific plant-derived structure) accumulation, in addition to a general localization in the plant apoplast. Mutants of Moeitf1 and Moeitf2 were defective in coping with host ROS stress development and strongly downregulated a set of putative effectors. Deletion of T2REP severely reduced fungal pathogenicity. Both T1REP and T2REP are critical effectors required for infection of rice by *M. oryzae* (Cao et al. 2022).

Phytophthora infestans secreted several RXLR class effectors involved in diverse functions and translocated them into host cells, facilitating infection of susceptible plants. The candidate effector Pi14054 functioned as a suppressor of RNA silencing in *N. benthamiana*. Pi14054 complemented TCV-deficiency for a viral suppressor of RNA silencing. Transcripts encoding Pi14054 were most abundant during early stages, suggesting that expression of this gene was specifically induced in contact with host tissue to modulate plant defenses. Pi14054 was detected in many European *P. infestans* isolates (Vetukuri et al. 2017). The mode of adaptation of newly emerging isolates (strains) of *Phytophthora infestans* resulting in resistance breakdown was investigated. Genome and transcription sequencing of HB1501 and HN1602 field isolates from different regions of China showed extensive polymorphisms in these isolates. Effector-encoding genes which contribute to virulence showed distinct expression landscapes in HB1501 and HN1602 isolates. Polymorphisms at multiple effectors required for recognition (Avr loci) enabled these isolates to overcome corresponding *R* gene-based resistance. However, several *R* genes, *R8*, *RB*, and *Rpi-vnt1.1*, continued to be resistant against these isolates of *P. infestans*, which evolved multiple strategies to evade recognition by other resistance genes of potato cultivars (Zhang et al. 2021).

Phylogenetically related *Plasmopara halstedii* and *P. viticola* causing downy mildew disease in sunflower and grapevine crops, respectively, have different pathological profiles. In oomycetes, two classes of effectors were translocated into host cytoplasm, the RXLRs and CRNs. About 50 putative RXLRs and 60 CRNs were identified for each pathogen. Sequence analysis revealed the presence of 55 RXLR families, 12 of them shared by both *Plasmopara* spp. and 19 showing amino acid conservation. The analyses confirmed the presence of eight CRN C-terminal domains described for *Phytophthora infestans* and identified CRN effectors showing similarity to serine proteases. Species-specific effectors were also identified, and they may be involved in host specificity, as well as effectors conserved among oomycetes (Mestre et al. 2016). *Plasmopara viticola* secreted several effectors to manipulate host immunity.

The role of cell death-inducing PvRXLR111 in manipulating host immunity was investigated. PvRXLR-induced cell death was prevented when coexpressed with other PvRXLR effectors. PvRXLR111 interacted with *Vitis vinifera* putative WRKY transcription factor 40 (VvWRKY40) which increased VvWRKY40 stability. The results suggested that VvWRKY40 might function as a negative regulator in plant immunity and PvRXLR111 might suppress host immunity by stabilizing VvWRKY40 (Ma et al. 2021). Apoplastic effectors include secreted hydrolytic enzymes such as proteases, lipases, and glycosylases that degrade plant tissue and protease inhibitors that protect oomycetes from host defense enzymes, necrosis- and ethylene-inducing peptide-like proteins (NLPs) and PcF-like small cysteine-rich proteins (SCRs) (Kamoun 2006). On the other hand, RXLR domain-containing proteins and crinklers are cytoplasmic effectors produced only by plant pathogenic oomycetes (Morgan and Kamoun 2007). A gene family encoding jacalin-like lectin domain-containing proteins (JRLs) was identified earlier. In the further study, JRL from *Sclerospora graminicola* (SgJRL) was found to function as an apoplastic virulence factor. The presence of eight SgJRLs in the extracellular fluid from Sg-inoculated foxtail millet leaves was detected by mass spectrometry analysis. Expression of SgJRL was induced by *S. graminicola* infection. The results suggested that JRLs were novel apoplastic effectors contributing to pathogenicity of oomycetes by suppressing plant defense responses (Kobayashi et al. 2022).

Secretion of effector proteins by rust pathogens *Uromyces appendiculatus* and *Phakopsora pachyrhizi* infecting common bean and soybean, respectively, was investigated. Comparative analysis between the two pathogens allowed differentiation of a set of 24 *U. appendiculatus* proteins targeted for secretions that were specifically detected in infected beans and a set of 34 *U. appendiculatus* proteins targeted for secretion that were present in germinated spores in infected bean. Proteins specific to infected beans included family 26 and family 76 glycosidic hydrolysis that might be involved in degradation of plant cell walls. Many of the *P. pachyrhizi* proteins showed similarity to those secreted by *U. appendiculatus* in infected common bean (Cooper et al. 2016). The host-induced gene silencing (HIGS) approach was applied to identify genes involved in early fungal aggressiveness. Secreted effector proteins (CSEPs) of *Blumeria graminis* f.sp. *hordei (Bgh)* were screened. Expression data revealed that these genes involved in virulence were upregulated at initial stages of infection, suggesting their involvement in early fungal virulence (Aguilar et al. 2016). When expressed in planta, a matured effector candidate from *Puccinia striiformis* f.sp. *tritici* (*Pst*) localized to chloroplasts. This effector had no conserved detectable transit signal region. N-terminal tagged red fluorescent protein had no effect on chloroplast localization of *Pst* CTE1. Entry of the candidate effector into the chloroplast was observed even with the construct having the intact signal peptide in the N-terminal of the transit peptide region. Accumulation of the effector in the endoplasmic reticulum (ER) might be due to overexpression of the protein in *Nicotiana benthamiana*. The *Pst* CTE1 might either escape from the secretory pathway by retrograde transport or translation might occur at an alternative site (Andac et al. 2020).

3.4.4.2 Expression of Genes during Pathogenesis

Fungal pathogens possess virulence and avirulence genes that are expressed during their interaction with their hosts with different levels of susceptibility/resistance. Avirulence gene *AVR-Pita1* in Japanese isolates of *Magnaporthe oryzae* was analyzed to determine the manner of gaining virulence toward rice cultivars containing the *Pita* resistance gene. An avirulent isolate OS99-G-7a (G7a) contained two paralogs of AVR-Pita1, called AVR-Pita1^{JA} and AVR-Pita1^{JB}. Analysis of virulent independent mutant derived from G7a, a single avirulent progenitor strain, indicated that AVR-Pita1^{JA} was functional, but AVR-Pita1^{JB} was nonfunctional. The most frequent mutation was loss of AVR-Pita1^{JA}. Most of the genes of AVR-Pita1 carried by Japanese isolates were identical to AVR-Pita1^{JA} or AVR-Pita1^{JB} (Takahashi et al. 2010). Inoculation of rice plants with spores of nonpathogen *Puccinia triticina* (wheat leaf rust disease) resulted in formation of infection hyphae and appressoria on rice leaf surfaces cv. Nipponbare. The cultivar responded to inoculation with brown localized discoloration of tissue and fragmentation of rust infection hyphae and appressoria. A microarray gene expression analyses of the host transcriptional response was performed at 24 h after inoculation, revealing rice genes that were upregulated or downregulated following the interaction. Rice has an active transcriptional and defense priming reactions in response to *P. triticina* and other nonpathogens (Li et al. 2017).

Hypersensitivity response (HR) against *Blumeria graminis* f.sp. *hordei (Bgh)* infection in barley was associated with stomata 'lockup' leading to increased leaf water conductance (gl). Unique spatiotemporal patterns of HR formation was observed in barley with *Mla1, Mla3* or *MlLa R* genes challenged with *Bgh*. With *Mla1*, a rapid HR, limited to epidermal cells arrested pathogen growth before colonies initiated secondary attacks. With *Mla3*, mesophyll HR preceded in epidermal cells whose initial survival supported secondary infections. Guard cell autofluorescence was detected only after drying epidermal strips and closure of stomata suggesting that locked open stomata were visible. The results indicated the link of stomatal lock-up to HR-associated cell death in barley (Prats et al. 2010). The function of HvSRP72 protein in barley powdery mildew resistance/susceptibility responses was studied. HvSRP72 was silenced by *Barley stripe mosaic virus* (BSMV)-induced gene silencing. The level of HvSRP72 mRNA in silenced plant samples showed substantial reduction. Under susceptibility conditions, HvSRP72-silenced samples of spores germinated approximately four times more efficiently and *Bgh* development also increased twice compared to control plants. In contrast, under resistant conditions, no difference was observed in HvSRP72-silenced samples compared to non-silenced control plants. The results showed that silencing HvSRP72 could enhance susceptibility to *B. graminis* f.sp. *hordei*, but no change in hyphal growth was evident during resistance response (Bedassa et al. 2019).

Role of microRNAs (miRNAs) in response of wheat to infection by *Blumeria graminis* f.sp. *tritici (Bgt)* (powdery mildew disease) was studied. The miRNA expression pattern was investigated to identify *Bgt*-responsive miRNAs in wheat leaves using plant miRNA microarray platform. Seven of 79 miRNAs were involved in wheat powdery mildew response. A negative correlation with the seven miRNAs in mock- and *Bgt*-infected samples was observed (Wu et al. 2015). Genome-wide gene expression profiles of Chinese cabbage resistant line 318R and susceptible line 361S in response to *Hyaloperonospora brassicae* (downy mildew) infection were determined by employing high-throughput RNA sequencing technique. Number of transcripts identified was greater in *H. brassicae*-infected samples of both lines, especially in the susceptible line 361S. Differentially expressed genes were primarily involved in plant-pathogen interaction pathway. They were annotated to defense response genes, translocation factors and cell wall-related genes functioning during protein processing in endoplasmic reticulum pathway or metabolism pathways (Li et al. 2018). Mechanism of defense in Shine Muscat, a hybrid grapevine cultivar of *Vitis vinifera* and *V. labrascana* to *Plasmopara viticola*, (downy mildew) was studied by transcriptome analyses. Shine Muscat might recognize *P. viticola* invasion at cell membrane level with subsequent transmission of signals into cells to initiate a response. Further, *P. viticola*-responsive paralogous genes were distributed among 8 clusters located on chromosomes 2, 5, 6, 9, 12, and 16 in the *V. vinifera* genome. These gene clusters encode PR5, PR10, dirigent protein, pleiotropic drug resistance protein, an uncharacterized protein, phenylalanine ammonia lyase (PAL), ET-responsive transcription factor, and stilbene synthase. The gene cluster encoding ethylene-responsive transcription factor was downregulated by *P. viticola* infection, whereas other gene clusters were upregulated. Synchronized transcription of these gene clusters might play an important role in the grapevine defense mechanism against the downy mildew pathogen (Shimizu and Suzaki 2020).

The genes involved in the transition of biotrophic phase to necrotrophic phase during infection process of *C. lindemuthianum* on common bean were studied. Suppression subtractive hybridization (SSH) in combination with qRT-PCR assay allowed identification of genes that were differentially expressed at each development stage of the pathogen in the plant. Among the expressed sequence tags (ESTs) sequences encoding proteins essential for anthracnose disease development as well as sequences encoding proteins related to primary and secondary metabolism and other functions were identified (Fontonelle et al. 2017). *Colletotrichum higginsianum* infecting crucifers passes through different stages of pathogenesis during infection. The *ChRgf* gene encoding a Ras guanine-nucleotide exchange factor protein was identified. Targeted gene deletion of *ChRgf* resulted in a null mutant (*ΔChrgf-42*) with defects in vegetative growth, hyphal morphology, conidiation, and poor surface attachment and low germination on hydrophobic surfaces. However, no difference in appressorial turgor pressure was observed between the wild-type and mutant strains. Conidia of the mutant did not germinate on attached leaves of *A. thaliana* and induce any disease symptom. The results suggested that *ChRgf* might be a key regulator in response to salt and osmotic stresses in *C. higginsianum* and might be involved in fungal pathogenicity (Gu et al. 2017). In order to understand early challenges that SA-mediated defense responses during strawberry–*Colletotrichum fructicola* interaction, dynamic changes of SA levels were monitored. A very early burst of SA followed by a fast quenching was recorded. The genes related to the salicylic acid (SA) pathway were differentially expressed, although they were all induced at different time points. Transcriptional characterization of host genes suggested that de novo SA-synthesis and the free SA release from methyl salicylate might contribute equally to the intricate control of SA homeostasis in strawberry. *C. fructicola* manipulation of SA-dependent resistance in strawberry might be closely related to multihormonal interplay among SA, jasmonic acid, abscisic acid, and cytokinins (He et al. 2019).

The survival motor neuron (SMN) protein has essential roles in cell viability and stress responses in higher eukaryotes. The role of SMN-like (*Mos*) gene in *M. oryzae* (rice blast disease) was investigated. Bioinformatics and Southern blot analysis confirmed that *Mos* was a single-copy gene located on *M. oryzae* chromosome. But it contained three transcripts isoforms. Expression patterns of these transcripts revealed that *Mos* was associated with stress response and conidiogenesis process in *M. oryzae*. Disruption mutant Δ*Mos* was sensitive to oxidative, osmotic and nitrogen- or carbon-starved stresses, suggesting that *Mos* was required for environmental adaptation and basic metabolism. Further, pathogenicity assay showed that *Mos* contributed to the development of rice blast disease and it was a pathogenicity gene (Liang et al. 2015). *A> alternata* existed in the form of different pathotypes with pathogenic specificity to crop plant species. G-protein-coupled receptors (GPCR)–encoding genes, *AaGPR1, AaGPR2*, and *AaGPR3,* were identified in the genomic data of *A. alternata* tomato pathotype, which produced the host-specific AAL-toxin. Each *AaGPR1, AaGPR2*, and *AaGPR3* encodes a protein containing a seven-transmembrane domain, characteristic of GPCRs. Deletion mutants *AaGPR1, 2*, and *3* had similar colony morphology or AAL-toxin-producing ability as the wild-type strain. However, one deletion strain of Δ*AaGPR3* showed aberrant conidial morphology including decreased conidial length and beak formation. This deletion mutant strain also had reduced ability to induce formation of necrotic lesions on susceptible leaves, indicating the adverse effect on pathogenicity/virulence of the mutant strain. The G-protein signal transduction pathway might have a role in conidial development and virulence of the *A. alternata* tomato pathotype (Takao et al. 2016). Necrotic infections by *Botrytis cinerea* on detached lettuce leaves had similar gene expression patterns to necrotic infections on leaves of seedlings. Asymptomatic infections on leaves that expanded after inoculation of young seedlings had similar pathogen gene expression patterns which distinctly differed from those in necrotizing infections. The results suggested that the physiological state of *B. cinerea* during asymptomatic infection was distinct from necrotizing infections (Emmanuel et al. 2018).

3.4.5 Pathophysiology of Infected Plants

Similarities in the pathways involved in the defense of plants against biotrophic and necrotrophic pathogens have been indicated by different investigations. The jasmonate/ethylene pathway is more important for plants for defense against necrotrophic pathogens, while salicylic acid (SA)–dependent responses of plants are more effective against biotrophic pathogens. Microbial plant pathogens may be placed into two groups as those infecting above-ground plant organs (epigeal) and those infecting below-ground organs (hypogeal). Above-ground tissues are green, assimilate-producing tissues (like leaves) or assimilate-importing tissues (like flowers). Pathogens infecting source tissues may have to encounter different conditions related to primary metabolism as well as to defense responses, compared to those pathogens infecting sink or assimilate importing tissues such as roots, leaves, and sink leaves (Narayanasamy 2008). *Botrytis cinerea* secreted a small protein, BcIEB, which is a PAMP recognized by plants. The activity is mapped to a highly conserved region of 35 amino acids, the peptide ieb35. The protein and peptide secreted by the pathogen-induced systemic resistance to *B. cinerea*, following infiltration into tobacco leaves. Tobacco seedlings sprayed, infiltrated or treated with the protein or peptide exhibited similar reduction in lesion size induced by *B. cinerea* (about 30%). Defense-related genes *PR1a*, *NPR1*, and osmotin *ap24* were induced by both BcIEB and ieb35, indicating the possible use of the pathogen-secreted PAMP for protecting tobacco, as an environmentally friendly treatment (Pérez-Hernández et al. 2020).

3.4.5.1 Influence of Infection on the Photosynthetic Process

Infection of plants by microbial pathogens results in alterations in primary metabolism of plants and initiation of defense activities requiring high demand for assimilates in plants. Fungal pathogens manipulate plant carbohydrate metabolism to satisfy their needs. The effect on photosynthesis could be analyzed by monitoring in vivo chlorophyll fluorescence which responds to the changes in energy conversion at photosystem II reaction centers. The usefulness of chlorophyll fluorescence analysis was improved by chlorophyll fluorescence imaging, which showed that generally changes in photosynthesis following infection were localized. In tomato plants infected by *Botrytis cinerea*, a ring of enhanced photosynthesis was detected in areas surrounding the area with decreased photosynthesis at the infection site (Berger et al. 2007). In compatible and incompatible interactions between barley and *Blumeria graminis* causing powdery mildew, photosynthesis was reduced. Quantitative imaging of chlorophyll fluorescence showed that photosynthesis was reduced both in cells directly below the powdery mildew colonies and in adjacent cells during compatible interaction (Swarbrick et al. 2006). Different lifestyles are adopted by fungal pathogens to obtain available nutrients from host plant tissues. Genomes of the obligate biotrophs *Blumeria graminis* and *Puccinia graminis* are smaller, compared to necrotrophs for genes encoding proteins involved in primary metabolism and nitrogen assimilation as well as for genes encoding carbohydrate active enzymes (CAZys) that might degrade cell wall components (Spanu et al. 2010). Loss of CAZy genes from the genomes of the biotrophs indicated that aggressive degradation of the host cell wall is required for the biotrophic lifestyle. On the other hand, necrotrophs like *B. cinerea* do not need a specialized feeding structure and instead secrete toxins and lytic enzymes that degrade the plant cell membrane. *B. cinerea* carries genes for assimilating nitrate, as well as high number of genes that encode putative cell wall-degrading enzymes (CWDEs/CAZys) (Amselem et al. 2011).

The effects of *M. oryzae* on photosynthetic process in infected wheat plants were assessed using a susceptible (BR18) and a partially resistant cultivar (BR229). The net carbon assimilation rate (*A*), stomata conductance (*gs*), and transpiration rate were significantly reduced in both cultivars following infection by *M. oryzae*, but to a lesser level in BR229. Photosynthesis was impaired in asymptomatic leaf tissues, indicating blast severity did not represent a reliable indicator for predicting *M. oryzae*-induced reactions in *A*. The proportionally larger decreases in *A* than in *gs*, in parallel with increases in internal CO_2 concentration (*Ci*), suggested that the lower influx of CO_2 into the diseased leaves might be caused by stomatal closure was not a prominent factor associated with the reduction in *A*. Both the maximum rate of carboxylation and the maximum rate of electron transport were significantly depressed at advanced stages of *P. oryzae* infection, particularly in susceptible BR18, although the reduction in *A* was not closely related to the decrease in electron transport rate (Debona et al. 2014). Changes associated with photoassimilate production and partitioning in the source-sink relationship of flag leaves and spikes of wheat plants inoculated with *P. oryzae* at 10 and 20 days after anthesis (*daa*) were assessed. Impairment of the photosynthetic performance in both infected flag leaves and spikes, coupled with reduced concentrations of chlorophyll $a + b$ and carotenoids was observed. In the infected flag leaves and grains concentrations of soluble sugars showed decrease, whereas hexose-to-sucrose ratio registered increase in infected flag leaves only. Infected flag leaves had lower sucrose phosphate synthase (SPS) activity and lower expression of sucrose synthesis (*SuSy*) gene, whereas higher expression and activity of acid invertase also was observed. ADP-glucose phosphorylase activity and expression of *ADP-glucose phosphorylase* genes and downregulation of ß and α-amylase expression occurred at advanced stages of infection in flag leaves and spikes (Rios et al. 2017). The impact on photosynthesis by soybean target spot caused by *Corynespora cassiicola* was investigated using cultivars Fundacep 59 (moderately resistant) and TMG 132 (susceptible). Area under disease progress curve (AUDPC) decreased in susceptible (S) plants. The concentration of total Chl*a*+*b* significantly increased and Chl*a*+*b*/carotenoids ratio significantly decreased in S plants. With progress of disease, the values of maximal photosystem II quantum yield (Fv/Fm), effective photosystem II quantum yield [YC(II)], quantum yield of regulated energy dissipation [Y(NPQ)] and quantum yield of nonregulated energy dissipation [Y(NO)] increased, especially in susceptible plants. Positive correlations between AUDPC and both Y(NPQ) and Y(NO) were observed in both cultivars. In susceptible plants a negative correlations between AUDPC and electron transport was recorded. The results indicated the potential of target spot to impact the functionality of photosynthetic apparatus that could assist plants to invest in metabolic pathways involved in host defense responses (Forutunato et al. 2018).

Phakopsora euvitis (grapevine rust disease) induced early defoliation and a consequent reduction in photosynthesis and photoassimilate supply and storage. The effects of different intensities of rust disease in leaves of cv. Moscato Giallo and Niagara Rosada on photosynthetic activity were assessed. Average lesion size in *V. vinifera* was greater than in *V. labrusca*. A water-soaked halo around the lesion was seen only in *V. vinifera*. Rust pathogen had similar effects on photosynthesis in both *Vitis* spp. with increasing disease severity. Stomatal conductance was slightly affected by rust infection, whereas significant decreases were observed in effective quantum efficiency of photosystem II and instantaneous carboxylation efficiency of both species with increasing disease severity. β-value (photosynthetic rate) of instantaneous carboxylation efficiency was higher in *V. vinifera* than in *V. labrusca* (Navarro et al. 2019). Application of silicon (Si) mitigated physiological derangements in soybean infected by *Cercospora sojina* causing soybean frog eye spot. No changes in physiological parameters of noninoculated +Si plants were observed. Susceptible plants showed lower values for leaf gas exchange parameters, photochemical efficiency and concentration of photosynthetic pigments, though their susceptibility was increased by Si supply. Impairments due to infection on the physiology of resistant plants were governed primarily by reduction in stomatal conductance regardless of Si supply. Increase in hexose content in inoculated plants of both cultivars was associated with their resistance to frog eye spot disease. Increase in susceptibility of resistant cultivar by Si supply did not lead to physiological derangements following infection by *C. sojina* (Nascimoto et al. 2018).

3.4.5.2 Effects of Infection on Carbohydrate Metabolism

The effect of infection on sugar levels varies significantly depending on the host-pathogen combination. In tomato–*Botrytis cinerea* (grey mold disease) interaction, levels of sucrose decreased more than the levels of hexoses, leading to an increase in the hexose-to-sucrose ratio (Berger et al. 2004). Increases in the sucrose and hexose levels in apoplast and invertase activity in tobacco infected by *Phytophthora nicotianae* were observed (Scharte et al. 2005). Soluble sugars, glucose, mannose, arabinose, and xylose significantly increased in shoots inoculated with *Lasiodiplodia theobromae* (peach gummosis) compared with noninoculated control shoots at different intervals after inoculation. Contents of key transcripts in the synthesis pathway of uredine diphosphate (UDP)-D-glucuronate, UDP-D-galactose, and UDP-D-arabinose increased, but the synthesis of L-galactose and guanosine diphosphate-L-galactose were inhibited. After inoculation, the transcript levels of sugar transport-related genes (*SUT*, *SOT*, *GMT*, and *UGT*) were induced. Changes in sugar content and gene expression were directly associated with peach gum polysaccharide formation and may induce gummosis symptoms in peach (Li et al. 2014). Sugars have dual roles as nutrients required for growth, respiration and accumulation of storage compounds and also as signals that can regulate gene expression. Hexoses released due to increased invertase activity function as signaling molecules and repress photosynthetic genes. Downregulation of photosynthetic genes, in turn, decreases the net photosynthesis rates. Several factors contribute to the complexity of reaction between carbohydrate status and development of disease or resistance. Carbohydrate status affects defense as well as general metabolism of the plant. Sugars function also as signals for host plant as well as for pathogen. The presence of extracellular sucrolytic enzymes such as invertases, fructoexohydrolases, and levensucrases has been reported in some pathogens. Expression of these pathogen enzymes might change hexose and sucrose levels in the apoplast, where several pathogens develop (Berger et al. 2007). Plants of two rice cultivars, BRS Querencia and InovCL, exposed to 700 ppm of atmospheric CO_2 exhibited changes in anatomical characteristics, photosynthetic parameters, and carbohydrate accumulation when compared to plants exposed to 400 ppm CO_2. Changes in morphological characteristics of the leaf and accumulation of carbohydrates, which were stimulated in the rice plants by enhanced level of CO_2 were associated with reduction in brown spot severity induced by *Bipolaris oryzae* (Dorneles et al. 2020).

3.4.5.3 Effects on Nitrogen Metabolism

Nitrogen metabolism is regulated by wide-domain, positive-acting GATA family transcription factors such as Area (of *Aspergillus nidulans*) and Nit1 (of *Neurospora crassa*). Orthologous Area/Nit2-encoding genes have been identified in the genomes of necrotrophic and biotrophic fungal pathogens. *Colletotrichum lindemuthianum* (common bean anthracnose) required functional Area/Nit2 ortholog, *Clnr1* for infection. The mutant Δ*clmr1* strains became nonpathogenic, but in infected plants produced primary hyphae in biotrophic phase. However, necrotrophic development was inhibited and only a few secondary hyphae were produced, compared with the wild-type strain. The results revealed the requirement of *Clnr1* for biotrophic to necrotrophic switch and also for ammonium and L-glutamine metabolism of *C. lindemuthianum* (Pellier et al. 2003). The ortholog Area/Nit2(*Nrf1*) of rice blast pathogen *M. oryzae* was dispensable for pathogenicity although the mutant strains Δ*nut1* could produce reduced lesions compared with wild-type strains. *Nrf1* was required for expressing a virulence determinant, MPG1 under nitrogen-limiting conditions. Nut1 was not required for metabolizing several amino acids by *M. oryzae*. The pathogen functioned as a biotroph for very short time, before progressing into necrotrophic phase. As the time spent for deriving nutrient from host apoplast increased the ortholog Area/Nit2 might become more important (Divon and Fluhr 2007). The nitrogen regulators, Npr1 and NPr2, are required for growth on nitrate and they are needed also for pathogenicity. The NIRA/NIT4 ortholog in NIR1 in *M. oryzae* which encodes a pathway-specific activator required for growth on nitrate and nitrite, is dispensable for pathogenicity. The role of NirA/Nit4 orthologs in plant disease development appeared to be lifestyle-dependent (Fernandez et al. 2014).

Acquisition of nutrients from host plant is a key component in the lifestyle of a microbial pathogen and it is crucial for maintaining growth in plants. After entry into the host, the transcriptome of the pathogen responds to the environment and many effectors are produced specifically at different stages of infection process. Fungal pathogens prefer readily assimilated nitrogen sources such as ammonium, glutamine and glutamate. Among the global nitrogen regulators needed for utilization of secondary nitrogen sources, GATA family transcription factors *AreA* and *Nit2* might be required. *AreA* transcript was degraded in response to primary nitrogen sources at transcriptional level. GATA-containing genes might be involved in *AreA/Nit2*-mediated gene induction. The GATA transcription factor *clnr1* from *Colletotrichum lindemuthianum* was a pathogenic factor. The *Clnr1* mutants produced appressoria, infection vesicles, and primary hyphae at the same rate as the wild-type strain. However, development of mutants was arrested at the onset of necrotrophic stage of the infection process (Pellier et al. 2003). Protein encoded by *Nir1* gene of *Colletotrichum acutatum* was required for N metabolism, in addition to GATA transcription factors,. Mutants of *Nir1* were unable to grow on medium containing nitrate as sole nitrogen source and they were nonpathogenic to strawberry (Horrowitz et al. 2006). Wheat apoplastic amino acid and sugar contents during infection by *Mycosphaerella graminicola* were assessed. At the phase of symptomless growth of the pathogen, no significant differences could be observed in pathogen-inoculated and mock-inoculated wheat plants. Later significant increases in several sources of carbohydrates and nitrogen were recorded in pathogen-inoculated plants compared to mock-inoculated plants. This might be due to pathogen-induced programmed cell death (PCD) (Keon et al. 2007). Infection by *Colletotrichum lindemuthianum* increased amino acid concentration by almost 5-fold at 12 dpi. Glutamine and glutamate, the two primary nitrogen sources were most abundant at this stage of infection. Glutamine increased significantly and correlated with bean glutamine synthetases (GS1a) mRNA accumulation, an enzyme that catalyzed the formation of glutamine on glutamate and ammonia (Tavernier et al. 2007). The amino acids cysteine and tryptophane were not detected in the tomato–*Cladosporium fulvum* interaction, whereas methionine, arginine, histidine, and proline were in very low concentrations during infection. On the other hand, alanine, asparagine, glutamine, glycine, serine, and GABA as well as nitrate were present in mM concentrations in the tomato apoplast following infection. Fungal pathogens might use amino acids preferentially during infection (Solomon and Oliver 2001; Bolton and Thomma 2008).

A metabolomics-based approach was applied to investigate rice-blast pathogen interaction with cv. Hawacheong and compatible KJ201 and incompatible KJ401 strains of *Magnaporthe oryzae*. No significant differences in the metabolic response could be detected in plants inoculated with both strains until 24 *hpi*. Alanine, along with other metabolites, malate, glutamine, proline, cinnamate and an unknown sugar, showed good correlation between time of pathogen penetration into leaf and the divergence of metabolic profiles in each interaction (Oliver et al. 2011). Nut 1 is dispensable for pathogenicity, but it is required for production of fully expanded necrotic lesions. Among the major sensing proteins, Tps1 is essential for establishing infection. Tps1 integrates carbon metabolism through control of carbon catabolite repression (CCR) and nitrogen metabolism via control of Nut1 in response to available glucose-6-phosphate (G6P) (Fernandez and Wilson 2012). Two pathogenicity genes of *M. oryzae*, *SPM1* which encoded a serine protease, were expressed under nitrogen starvation conditions. These genes *SPM1* and *PTH1* were under Nut 1 control and expressed in response to glucose regardless of the nitrogen source in a Tps1-dependent manner (Fernandez et al. 2014).

The molecular mechanism underlying nitrogen-induced susceptibility (NIS) in interaction between *M. oryzae* and rice was investigated. A dual RNA-seq experiment was performed using infected tissues of rice plants grown under two nitrogen fertilization regimes. Enhanced susceptibility of rice plants due to a high nitrogen level was observed, in spite of overinduction of defense gene expression by infection. On the other hand, *M. oryzae* expressed a high concentration of effectors, and pathogenicity-related genes were also induced under the high nitrogen regime. Modifications in glutamine synthetase OsGS1–2 expression and fungal pathogenicity led to nitrogen-induced susceptibility to blast disease. Some rice genes implicated in nitrogen recycling were also highly induced during NIS. OsGS1–2 glutamine synthase gene increased plant resistance to *M. oryzae* and counteracted NIS (Huang et al. 2017). Leaf proteome changes in wheat cultivars susceptible and resistant to infection by *Blumeria graminis* f.sp. *tritici (Bgt)* (wheat powdery mildew) were compared. Protein spots reproducibly detected in leaves of susceptible and resistant cultivars were almost the same. Most of the spot changes were unique in each genotype. Responses to *Bgt* infection at the translational level were different in near isogenic lines. Differentially expressed proteins included functional categories such as transcription and translation, energy and metabolism, signal transduction, disease, and defense as well as unclassified proteins (Wang et al. 2012). Proteome profiles during compatible interaction of wheat with stripe rust pathogen (*Puccinia graminis* f.sp. *striiformis*) were examined. Many wheat proteins, such as antioxidants, pathogenesis-related (PR) proteins, and cold-responsive proteins, detected in susceptible plants, could be implicated in defense and stress tolerance. Proteins of the pathogen origin included pathogenicity proteins and those that might suppress host defense. In addition, nine proteins with signal sequence and some with effector function were also present (Demirici et al. 2016).

3.4.5.4 Effect on Mineral Metabolism

The role of magnesium (Mg), essentially required for the growth of wheat, in the development of resistance to rice blast caused by *M. oryzae* was investigated. Foliar Mg content increased with increase in Mg applied, whereas calcium (Ca) decreased. Blast disease severity increased as the applied Mg increased. Photosynthesis, stomatal conductance, transpiration, and photosynthetic pigment concentration decreased in inoculated plants compared to noninoculated plants regardless of the Mg concentration. A higher internal CO_2 concentration, invertase activity and malondialdehyde concentrations was recorded in inoculated plants grown with lower concentration of Mg compared to those grown with higher level of Mg. Mg-induced susceptibility to leaf blast seemed to be responsible for the photosynthetic impairment (Debona et al. 2016). The effects of manganese (Mn), applied as foliar spray, on amelioration of negative effects of *Puccinia kuehnii* (orange rust disease) on sugarcane production were assessed. The percentage of diseased leaf area was reduced by Mn application. An increase in biomass production was observed, following a decrease in disease severity. Mn treatment was responsible for direct damage to pathogen spores and improved lignin deposition in the mesophyll. In the presence of orange rust infection, Mn-sprayed leaves exhibited lesser levels of oxidative stress, in addition to improved structural organization of xylem and phloem vessels, compared to untreated control. The negative effects of orange rust on gas exchange and photochemistry were also ameliorated by Mn application (Mesquita et al. 2019). The effects of different concentrations of Fe^{3+} on pathogenic characteristics of *Curvularia lunata*, (maize Curvularia leaf spot) were assessed. Iron (Fe3+) showed significant enhanced mycelial biomass, conidiation and conidial germination. Exogenous application of Fe^{3+} enhanced pathogenicity of *C. lunata* CX-3 strain on maize inbred line. Polygalacturonase (PG) activity was improved in early stages of disease development. Activities of cellulase (Cx), pectin methylgalacturonase (PMG), polygalacturonase and transeliminase (PGTE) and pectin methyl transeliminase (PMTE) were enhanced after inoculation of maize (Wang et al. 2020).

3.4.5.5 Effect of Infection on Host Enzymes

Effects of infection by *M. oryzae* on activities of various enzymes involved in different metabolic functions of wheat susceptible (BR18) and partially resistant (BRS229) plants were studied. Blast severity in BRS229 was less than that in BR18. Activities of superoxide dismutase (SOD), peroxidase (POX), ascorbate peroxidase (APX) and glutathione-*S*-transferase (GST) increased in inoculated plants of both cultivars. But increases in the activities of these enzymes were more pronounced in BRS229 than in BR18. Catalase (CAT) increased only in inoculated plants of BRS229 at 96 hai. In BR18, CAT activity was reduced in inoculated plants. Further, inoculated plants of BR18 exhibited increased activity of glutathione peroxidase (GPX). Infection by *M. oryzae* enhanced concentrations of superoxide, hydrogen peroxide and malondialdehyde (MDA) and increased electrolyte leakage (EL) (Debona et al. 2012). The effects of *Puccinia coronata* f.sp. *avenae* (*Pca*) (oat crown rust) on oxidative stress in oat cultivar URS21 (partially resistant), Legett (race-specific resistant), URS22, and Clintland 64 (susceptible) were assessed. After inoculation, Legget and URS21 did not exhibit any increase in peroxidase levels. The susceptible cultivars showed increased ROS and ascorbate peroxidase activities. Clintland also had enhancement of catalase activity, whereas URS22 showed increases in glutathione reductase activity and expression of genes encoding antioxidant enzymes. Antioxidant enzyme was not induced in URS21, but it showed increased expression of genes encoding lipoxygenase and peroxidase, soon after inoculation with *Pca* (Figueiro et al. 2015).

Higher peroxidase activity and total phenol content in potato showed a positive relationship with levels of resistance to early blight disease induced by *Alternaria solani* isolates in cv. Diamond. Variability in responses of potato cultivar to isolates of *A. solani* suggested that genotypic diversity among isolates might result in phenotypic diversity for pathogen aggressiveness (virulence) (Shahbazi et al. 2010). The relationship between cellular damage induced by the infection by *Corynespora cassiicola* (soybean target spot) and activation of antioxidant system in infected plants was investigated. In general the antioxidant enzyme activities of superoxide dismutase, peroxidase, ascorbate peroxidase (AsA), hydrogen peroxide (H_2O_2), superoxide, and malondialdehyde (MDA) increased significantly in response to *C. cassiicola* in both cultivars, but higher levels were attained in cv. Fundacep. Concentration of MDA, H_2O_2, and superoxide increased in cv. TMG132, while activity of AsA increased in cv. Fundacep. The results indicated the requirement of an efficient antioxidant system to scavenge ROS generated in soybean plants for development of resistance against *C. cassiicola* (Fortunata et al. 2015).

REFERENCES

Aguilar GB, Pedersen C, Thordal-Christensen H (2016) Identification of eight effector candidate genes involved in early aggressiveness of the barley powdery mildew fungus. Plant Pathol 65: 953–958.

Ah-Fong AM, Kim KS, Judelson HS (2017) RNA seq of life stages of the oomycete *Phytophthora infestans* reveals dynamic changes in metabolic, signal transduction, and pathogenesis genes and a major role for calcium signaling in development. BMC Genomics 18: 198.

Ajiro N, Miyamoto Y, Masunaka A, Tsuge T, Yamamoto M et al. (2010) Role of host-selective ACT-toxin synthesis gene ACTTS2 encoding an emoyl-reductase in pathogenicity of the tangerine pathotype of *Alternaria alternata*. Phytopathology 100: 120–126.

Akagi A, Engelberth J, Stotz HO (2010) Interaction between polygalacturonase-inhibiting protein and jasmonic acid during defence activation in tomato against *Botrytis cinerea*. Eur J Plant Pathol 128: 423–428.

Akase K, Kusaba M (2017) Overwintering of *Pyricularia oryzae* in wild infected foxtails. J Gen Plant Pathol 83: 197–204.

Akimitsu K, Peever TL, Timmer LW (2003) Molecular, ecological, and evolutionary approaches to understanding Alternaria plant pathogens known to produce host-specific toxins. Curr Genet 33: 642–656.

Amorin R, Savi DC, Ferreira-Maba L, Aluizo R, Gadin EH et al. (2017) MAT gene idiomorphs suggest a heterothallic sexual cycle in the citrus pathogen *Phyllostica citricarpa*. Eur J Plant Pathol 147: 325–337.

Amselem J, Cuomo CA, van Kan JAL, Viaud M, Benito EP et al. (2011) Genomic analysis of necrotrophic fungus pathogen *Sclerotinia sclerotiorum* and *Botrytis cinerea*. PLoS Genet 7: e1002230.

Andac A, Ozketen A, Dagvadorj B, Akkaya MS (2020) An effector of *Puccinia striiformis* f.sp. *tritici* targets chloroplasts with a novel and robust targeting signal. Eur J Plant Pathol 157: 751–765.

Angelini RMDM, Rotolo C, Pollastro S, Faretra F (2016) Molecular analysis of the mating-type (*MAT1*) locus in strains of the heterothallic ascomycete *Botrytis cinerea*. Plant Pathol 65: 1321–1332.

Antoni EA, Rybak K, Tucker MP, Hane JK, Solomon PS et al. (2010) Ubiquity of ToxA and absence of ToxB in Australian populations of *Pyrenophora tritici-repentis*. Austr Plant Pathol 39: 63–68.

Asakura M, Okuno T, Takano Y (2006) Multiple contributions of peroxisomal metabolic function to fungal pathogenicity in *Colletotrichum lagenarium*. Appl Environ Microbiol 72: 6345–6354.

Baheevandziev K, Silva MC, Monteiro AA (2015) Histological characterization of downy mildew (*Hyaloperonospora brassicae*) resistance in Portugese cabbage (*Brassica oleracea* var, *tronchuda*). Eur J Plant Pathol 141: 199–207.

Baker SE, Koken S, Inderbitizin P, Asvarak T, Li BY et al. (2006) Two polyketide synthase-encoding genes are required for biosynthesis of polyketide virulence factor, T-toxin by *Cochliobolus heterostrophus*. Molec Plant-Micro Interact 19: 139–149.

Balesdent M-H, Fudal I, Ollivier B, Bally P, Grandaubert J et al. (2013) The dispensable chromosome of *Leptosphaeria maculans* shelters an effector gene conferring avirulence towards *Brassica rapa*. New Phytol 198: 887–898.

Balhadere PV, Foster AJ, Talbot NJ (1999) Identification of pathogenicity mutants of rice blast fungus *Magnaporthe grisea* by insertional mutagenesis. Molec Plant-Micro Interact 12: 129–143.

Barua P, You MP, Bayliss KL, Lanoiselet V, Barbetti MJ (2018) Extended survival of *Puccinia graminis* f.sp. *tritici* urediniospores: implication for biosecurity and on-farm management. Plant Pathol 67: 799–809.

Bedassa SB, Akkaya MS, Ersoy F (2019) HvSRP72 silencing enhanced *Blumeria graminis* f.sp. *hordei* growth in compatible interaction with barley. J Plant Pathol 101: 91–96.

Beever RE, Parkes SL (2003) Use of nitrate nonutilizing (Nit) mutants to determine vegetative compatibility in *Botrytinia fuckeliana (Botrytis cinerea)*. Eur J Plant Pathol 109: 607–613.

Berger S, Papadopoulos M, Schreiber U, Kaiser W, Roitsch T (2004) Complex regulation of gene expression, photosynthetic and sugar levels by pathogen infection in tomato. Physiol Plantar 122: 419–428.

Berger S, Sinha AK, Roitsch T (2007) Plant physiology meets phytopathology: plant primary metabolism and plant-pathogens interactions. J Exptl Bot 58: 4019–4026.

Bhambara KG, Wang ZY, Soanes DM, Wakley GE, Talbot NJ (2006) Peroxisomal carnitineacetyl transferase is required for elaboration of penetration hyphae during plant infection by *Magnaporthe grisea*. Molec Microbiol 61: 46–60.

Bhattacharya S (2017) Deadly new wheat disease threatens Europe's crops. Nature 542: 145–146.

Bock CH, Young CA, Zhang M, Chen C, Brannen PM et al. (2021) Mating type idiomorph, heterothallism, and high genetic diversity in *Venturia carpophila*, cause of peach scab. Phytopathology 111: 408–424.

Boller VT, Felix G (2009) A renaissance of elicitors: Perception of microbe-associated molecular pattern recognition receptors. Annu Rev Plant Biol 60: 379–406.

Bolton MD, Thomma BPHJ (2008) The complexity of nitrogen-regulated gene expression on plant pathogenic fungi. Physiol Molec Plant Pathol 72: 104–110.

Bonde MR, Murphy CA, Bauchan GR, Luster DG, Palmer CL et al. (2015) Evidence for synthetic infection by *Puccinia horiana*, causal agent of chrysanthemum white rust in chrysanthemum. Phytopathology 105: 91–98.

Børve J, Stensvand A (2017) *Colletotrichum acutatum* occurs asymptomatically on apple leaves. Eur J Plant Pathol 147: 943–948.

Boshoff WHP, Bender CM, Pretorium ZA (2019) The value of field ratings of differential lines for pathotyping *Puccinia graminis* f.sp. *tritici*. Eur J Plant Pathol 155: 349–352.

Both M, Eckert SE, Csukai M, Miller E, Dimopoulos G et al. (2005) Transport profiles of *Blumeria graminis* development during infection reveal a cluster of genes that are potential virulence determinants. Molec Plant-Micro Interact 18: 125–133.

Braga ZV, dos Santos RF, Amorim L, Appezzato-da-Glória B (2019) Histopathology of infection and colonization of *Elsinoë ampelina* on grapevine leaves. Eur J Plant Pathol 154: 1009–1019.

Brito N, Espino JJ, Gonzalez C (2006) The endo-beta-4 xylanase xyn11A, is required for virulence in *Botrytis cinerea*. Molec Plant-Micro Interact 19: 25–32.

Bueno TV, Fontes PP, Abe VY, Utiyama AS, Senra RL et al. (2022) A *Phakopsora pachyrhizi* effector suppresses PAMP-triggered immunity and interacts with a soybean glucan endo-1,3-ß-glucosidase to promote virulence. Molec Plant-Micro Interact 35: 779–790.

Burlakoti RR, Estrada R Jr, Rivera VV, Boddeda A, Secor GA et al. (2007) Real-time PCR quantification of mycotoxin production by Fusarium graminearum in wheat inoculated with isolates of Fusarium graminearum collected from potato, sugar beet, and wheat. Phytopathology 97: 835–841.

Caffier V, Patocchi A, Expert P, Bellanger MN, Durel CE et al. (2015) Virulence characterization of *Venturia inaequalis* reference isolates on differential set of *Malus* hosts. Plant Dis 99: 370–375.

Caledron C, Ward E, Freeman J, Fraser SJ, McCartney HA (2002) Detection of airborne inoculum of *Leptosphaeria maculans* and *Pyrenopeziza brassicae* in oilseed rape crops by polymerase chain reaction (PCR) assay. Plant Pathol 51: 303–310.

Cao Y, Chen J, Xie X, Liu S, Jiang Y et al. (2022) Characterization of two infection-induced transcription factors of Magnaporthe oryzae reveals their roles in regulating early infectin and effector expression. Molec Plant Pathol 23: 1200–1213.

Cao Y, Si B, Zhu S, Xu X, Li W et al. (2019) Race and virulence of asexual and sexual populations of *Puccinia graminis* f.sp. *tritici* in China from 2009 to 2015. Eur J Plant Pathol 153: 545–555.

Carisse O, Bacon R, Lefebvre A (2009a) Grape powdery mildew (*Erysiphe necator*) risk assessment based on airborne conidium concentration. Crop Protect 28: 1036–1044.

Carisse O, Tremblay DM, Levesque CA, Gindro K, Ward P et al. (2009b) Development of a TaqMan real-time PCR assay for quantification of airborne conidia of *Botrytis squamosa* and management of Botrytis leaf blight of onion. Phytopathology 99: 1273–1280.

Carzaniga R, Bowyer P, O'Connell RJ (2001) Production of extracellular matrices during development of infection structures by the downy mildew *Peronospora parasitica*. New Phytol 149: 83–93.

Castrogulin VL. Danielli A, Moreira SI, Reyes JJA, Carvalho G et al. (2017) The wheat blast pathogen *Pyricularia graminis-tritici* complex origin and a disease cycle spanning multiple grass hosts. bioRxiv 203455. https://doi.org/10.1101/203455

Catanzariti AM. Dodds PN, Lawrence GJ, Aylifffe MH, Ellis JG (2006) Haustorially expressed secreted proteins from flax rust are highly enriched for avirulence factors. Plant Cell 18: 243–256.

Ceresini PC, Castroagudin VL, Rodrigues FA, Rios JA, Aucique-Perez CE, Moreira SI, Croll D, AlvesI, DeCarvalho G, Maicel JLN, Mc Donald BA (2019) Wheat blast: from its origin in South America to its emergence as a global threat. Molec Plant Pathol 20: 155–172.

Chagué JC, Danit LV, Siewers V, Gronover CS, Tudzynski B et al. (2006) Ethylene sensing and gene activation in *Botrytis cinerea*: a missing link in ethylene regulation of fungus-plant interactions. Molec Plant-Micro Interact 19: 33–42.

Chang H-X, Miller LA, Harman GL (2014) Melanin-independent accumulation of turgor pressure in appressoria of *Phakopsora pachyrhizi*. Phytopathology 104: 977–984.

Chen HQ, Cao L, Dekkers KL, Rollins JA, Ko NJ et al. (2005) A gene with domains related to transcripts regulation is required for pathogenicity in *Colletotrichum acutatum* causing Key lime anthracnose. Molec Plant Pathol 6: 513–525.

Chen HQ, Lee MH, Daub ME, Chung KR (2007) Molecular analysis of the cercosporin biosynthetic gene cluster in *Cercospora nicotianae*. Molec Microbiol 64: 755–770.

Cheng P, Chen XM (2014) Virulence and molecular analyses support asexual reproduction of *Puccinia striiformis* f.sp. *tritici* in the U. S. Pacific Northwest. Phytopathology 104: 1208–1220.

Chongo G, Goosen BD, Kutcher HR, Gilbert J, Turkington TK et al. (2001) Reaction of seedling roots of 14 crop species to *Fusarium graminearum* from wheat heads. Canad J Plant Pathol 23: 132–137.

Choquer M, Dekkers KL, Chen HQ, Cao L, Ueng PP et al. (2005) The CTB1 gene encoding a fungal polyketide synthase is required for cercosporin biosynthesis and fungal virulence of *Cercospora nicotianae*. Molec Plant-Micro Interact 18: 468–476.

Choudhury RA, Koike ST, Fox AD, Anchieta A, Subbarao KV et al. (2016) Season-long dynamics of spinach downy mildew determined by spore-trapping and disease incidence. Phytopathology 106: 1311–1318.

Cohen Y, Rubin AE (2012) Mating type and sexual reproduction of *Pseudoperonospora cubensis*, the downy mildew agent of cucurbits. Eur J Plant Pathol 132: 577–592.

Cook DE, Mesarich CH, Thomma BPHJ (2015) Understanding plant immunity as a surveillance system to detect invasion. Annu Rev Phytopathol 53: 541–563.

Cooper B, Campbell KB, Beard HS, Garrett WM, Islam N (2016) Putative rust effector proteins in infected bean and soybean leaves. Phytopathology 106: 491–499.

Correll JC, Klittich CJR, Leslie JF (1987) Nitrate nonutilizing mutants of *Fusarium oxysporum* and their use in vegetative compatibility tests. Phytopathology 77: 1640–1646.

Cowger C, Mehra L, Arellano C, Meyers E, Murphy JP (2018) Virulence differences in *Blumeria graminis* f.sp. *tritici* the central and eastern United States. Phytopathology 108: 402–411.

Crook AD, Friesen TL, Liu ZH, Ojiamabo PS, Cowger C (2012) Novel necrotrophic effectors from *Stagonospora nodorum* and corresponding host sensitivities to winter wheat germplasm in the southeastern United States. Phytopathology 102: 498–505.

Cruz CD, Kiyund J, Bockus WW, Todd TC, Stack JP et al. (2015) *Magnaporthe oryzae* conidia on basal wheat leaves as a potential source of wheat blast inoculum. Plant Pathol 64: 1491–1498.

de Jonge R, van Esse HP, Kombrink A, Shinya T, Desaki Y et al. (2010) Conserved fungal LysM effector Ecp6 prevents chitin-triggered immunity in plants. Science 329: 953–955.

de Wit PJGM (2016) *Cladosporium fulvum* effects: weapons in the arms race with tomato. Annu Rev Phytopathol 54: 1–23.

Debona D, Rios JA, Nascimento KJT, Silva LC, Rodrigues FA (2016) Influence of magnesium on physiological responses of wheat infected by *Pyricularia oryzae*. Plant Pathol 65: 114–123.

Debona D, Rodrigues FA, Rios JA, Martins SCV, Pereira LF et al. (2014) Limitations to photosynthesis in leaves of wheat plants infected by *Pyricularia oryzae*. Phytopathology 104: 34–39.

Debona D, Rodrigues FA, Rios JA, Nascimento KJT (2012) Biochemical changes in the leaves of wheat plants infected by *Pyricularia oryzae*. Phytopathology 102: 1121–1129.

Deising H, Nicholson RL, Haug M (1992) Adhesion pad formation and the involvement of cutinase and esterases in the attachment of uredospore to the host cuticle. Plant Cell 4: 1001–1011.

Demirici YE, Inan C, Gunel A, Maytalman D, Mert Z et al. (2016) Proteome profiling of the compatible interaction between wheat and stripe rust. Eur J Plant Pathol 145: 941–962.

Dietzel K, Valle D, Fierer N, U'Ren JM, Barberan A (2019) Geographical distribution of fungal plant pathogens in dust across the United States. Front Ecol Evol 7: 304, https://doi.org/10.3389/fevo.2019.00304

Ding S, Meinholz K, Cleveland K, Jordan S, Gebebs AJ (2019a) Diversity and virulence of *Alternaria* spp. causing potato early blight and brown spot in Wisconsin. Phytopathology 109: 436–445.

Ding S, Rouse DI, Meinholz K, Gevens AJ (2019b) Aerial concentration of pathogens causing early blight and brown spot within susceptible potato fields. Phytopathology 109: 1425–1432.
Divon HH, Fluhr R (2007) Nutrition acquisition and strategies during fungal infection of plants. FEMS Microbiol Lett 266: 65–74.
Dong F, Xu J, Zhang X, Wang S, Xing Y et al. (2020) Graminaceous weeds near paddy fields are alternative hosts for *Fusarium graminearum* species complex that causes Fusarium head blight in rice. Plant Pathol 69: 433–441.
Dorneles KR, Martins AC, Fernando JA, do Amarante L, de Avila LA et al. (2020) Increased atmospheric CO2 concentration cuased modification of physiological, biochemical and histological characteristics that affect rice-*Bipolaris oryzae*. Eur J Plant Pathol 157: 29–38.
dos Santos RF, Spósito MB, Ayres M, Sosnowski M (2018) In vitro production of conidia of *Elsinoe ampelina*, the causal fungus of grapevine anthracnose. Eur J Plant Pathol 152: 815–821.
Dubey A, Lee J, Kwon S, Lee Y-H. Jeon J (2019) A MYST family histone acetyl transferase, MOSAS3, is required for development and pathogenicity in the rice blast fungus. Molec Plant Pathol 20: 1491–1505.
Dufrense M, Osbourn AE (2001) Definition of tissue-specific and general requirement of plant infection in phytopathogenic fungus. Molec Plant-Micro Interact 14: 300–307.
Edwards HH, Bonde MR (2011) Penetration and establishment of *Phakopsora pachyrhizi* in soybean leaves as observed in transmission electron microscopy. Phytopathology 101: 894–900.
Emmanuel CJ, van Kan JAL, Shaw MW (2018) Differences in the gene transcription state of *Botrytis cinerea* between necrotic and symptomless infections of lettuce and *Arabidopsis thaliana*. Plant Pathol 67: 1865–1873.
Espino JJ, Brito N, Noda J, Gonzalez C (2005) *Botrytis cinerea* endo β-1,4-glucanase Cel5A is expressed during infection but not required for pathogenesis. Physiol Molec Plant Pathol 66: 213–221.
Everett KR, Pusparajah IPS, Timudo CE, Ah Chee A, Scheper RWA et al. (2018) Infection criteria, inoculum sources and splash dispersal pattern of *Colletotrichum acutatum* causing bitter rot apple in New Zealand. Eur J Plant Pathol 152: 367–383.
Fagan H (1979) Post bloom fruit drop of citrus associated with a form of *Colletotrichum gloeosporioides*. Ann Appl Biol 91: 13–20.
Falacy JS, Grove GG, Mahaffee WF, Galloway H, Galwe DA et al. (2007) Detection of *Erysiphe necator* in air samples using polymerase chain reaction and species specific primers. Phytopathology 97: 1290–1297.
Feng C, Saito K, Liu B, Manley A, Kammeijer K et al. (2018) New races and novel strains of the spinach downy mildew pathogen *Peronospora effusa*. Plant Dis 102: 613–618.
Fernandez J, Marroquin-Guzman M, Wilson RA (2014) Mechanism of nutrient acquisition and civilization during fungal infections of leaves. Annu Rev Phytopathol 52: 155–174.
Fernandez J, Wilson RA (2012) Why no feeding freznzy? Mechanism of nutrient acquisition and utilization during infection by rice blast fungus *Magnaporthe oryzae*. Molec Plant-Micro Interact 25: 1286–1293.
Ficke A, Cowger C, Bergstrom G, Brodal G (2018) Understanding yield loss and pathogen biology to improve disease management: septoria nodorum blotch—a case study in wheat. Plant Dis 102: 696–707.
Figueiro AA, Reese N, Hernandez JLG, Pacheco MT, Martinelli JA et al. (2015) Reactive oxygen species are not increased in resistant oat genotypes challenged by crown rust isolates. J Phytopathol 163: 795–806.
Fontonelle MR, Santana MF, Cnossen A, Bazzolli DMS, Bromonschenkel SH et al. (2017) Differential expression of genes during the interaction between *Colletotrichum lindemuthianum* and *Phaseolu vulgaris*. Eur J Plant Pathol 147: 653–670.
Foroud NA, Pordel R, Goyal RK, Ryabova D, Eranthodi A et al. (2019) chemical activation of ethylene signaling pathway, promotes *Fusarium graminearum* resistance in detached wheat heads. Phytopathology 109: 769–809.
Fortunata OAA, Debona D, Bernardeli AMA, Rodrigues FA (2015) Changes in the antioxidant system in soybean leaves infected by *Corynespora casiicol*. Phytopathology: 1050–1058.
Forutunato AA, Debona D, Aucique-Pérez CE, Correa EF, Rodrigues FA (2018) Chlorphyll *a* fluorescence imaging of soybean leaflets infected by *Corynespora cassiicola*. J Phytopathol 166: 782–789.
Franco CCS, Anna JRS, Rosada IJ, Kaneshima EW, Stangarlin JR et al. (2011) Vegetative compatibility groups and parasexual segregation in *Colletotrichum acutatum* isolates from different hosts. Phytopathology 101: 923–928.
Frare GF, Couto HTZ, Ciampi-Guillardi M, Amorim L (2016) The causal agent of citrus postbloom fruit drop *Colletotrichum acutatum* can survive on weeds. Austr Plant Pathol 45: 339–346.
Freitas AS, Pozza EA, Moreira SI, Gilmara M, Soares O et al. (2017) Infection process of *Pseudocercospora musae* on banana leaf. Phytoparasitica 45: 317–324.
Friesen TL, Holmes DJ, Bowden RL, Faris JD (2018) ToxA is present in the U. S. *Bipolaris sorokiniana* population and is a significant virulence factor on wheat harboring *Tsn1*. Plant Dis 102: 2446–2452.
Friskop AJ, Gulya TJ, Harveson RM, Humann RM, Acevedo M et al. (2015) Phenotypic diversity of *Puccinia helanthii* (Sunflower rust) in the United States from 2011 and 2012. Plant Dis 99: 1604–1609.
Frost KE, Johnson ACS, Gevens AJ (2016) Survival of isolates of the US-22, US-23, and US-24 clonal lineages of *Phytophthora infestans* by asexual means in tomato seed at cold temperatures. Plant Dis 100: 180–187.
Fry WE, Myers K, Roberts P, McGrath MT, Everts K et al. (2013) The 2009 late blight pandemic in the eastern United States—causes and results. Plant Dis 97: 296–306.
Gadoury DM, Wakefield LM, Cadle-Davidson L, Dry IB, Seem RC (2012) Effect of prior vegetative growth inoculum density, light, and mating on conidiation of *Erysiphe necator*. Phytopathology 102: 65–72.
Gao J-X, Chen J (2018) Transcriptions analysis identifies candidate genes associated with melanin and toxin biosynthesis and pathogenicity of the maize pathogen, *Curvularia lunata*. J Phytopathol 166: 223–241.
Gao J-X, Jung J, Liu T, Chen J (2015) Identification of Clt-1 regulated proteins associated with the production of nonhost-specific toxin and pathogenicity in *Cochliobolus lunatus*. J Phytopathol 163: 33–41.
Gao J-X, Liu T, Chen J (2014) Insertional metagenesis and cloning of the gene required for the biosynthesis of the nonhost specific toxin in *Cochliobolus lunatus* that causes maize leaf spot. Phytopathology 104: 332–339.
Gasparoto MCG, Lourenço SA, Tanaka FAO, Spósito MB, Marchini LC et al. (2017) Honeybees can spread *Colletotrichum acutatum* and *C. gloeosporioides* among citrus plants. Plant Pathol 66: 777–782.

Gent DH, Claassen BJ, Twomey MC, Wolfenbarger SN, Woods JL (2018) Susceptibility of host crown buds to powdery mildew and its relation to perennation of *Podosphaeria macularis*. Plant Dis 102: 1316–1325.

Gent DH, Cohen Y, Runge F (2017) Homothallism in *Pseudoperonospora humuli*. Plant Pathol 66: 1508–1516.

Gent DH, Nelson ME, Farnsworth JL, Grove GG (2009) PCR detection of *Pseudoperonospora humuli* in air samples from hop yards. Plant Pathol 58: 1081–1091.

Giordani E, Padula G, Radice S (2013) Compared anatomy of young leaves of *Prunus persica* (L.) Batsch. with different degrees of susceptibility to *Taphrina deformans* (Berk.) Tul. J Phytopathol 161: 190–196.

Giraldo MC, Degdes YF, Gupta YK, Mentlak TA, Yi M et al. (2013) Two distinct secretion system facilitate tissue invasion by rice blast fungus Magnaporthe oryzae. Nat Commun 4: 1996.

Gómez-Zeledón I. Kaiser M, Spring O (2017) Exploring host-pathogen combinations for compatible and incompatible reactions in grapevine downy mildew. Eur J Plant Pathol 149: 1–10.

Görnhardt B, Rouchara I, Schmelzer E (2020) Cyst germination proteins of the potato pathogen *Phytophthora infestans* share homology with human mucins. Molec Plant-Micro Interact 13: 32–42.

Grönberg L, Anderson B, Yuen J (2012) Can weed hosts increase aggressiveness of *Phytophthora infestans* on potato? Phytopathology 102: 429–433.

Grünwald NJ, Flier WG (2005) The biology of *Phytophthora infestans* at its center of origin. Annu Rev Phytopathol 43: 171–190.

Gu Q, Chen M, Huang J, Wei Y, Hsiang T et al. (2017) Multifaceted roles of the Ras guanine-nucleotide exchange factor *ChRgf* in development, pathogenesis, and stress responses of *Colletotrichum hogginsianum*. Phytopathology 107: 433–443.

Guerra FA, De Rossi RL, Brücher E, Vuletic E, Plazas MC et al. (2019) Occurrence of the complete cycle of *Puccinia sorghi* Schw. in Argentina and implications on the common corn rust epidemiology Eur J Plant Pathol 154: 171–177.

Gunasinghe N, You MP, Cawthray GR, Barbetti MJ (2016) Cercosporin from *Pseduocercosporella capsellae* and its critical role in white leaf spot development. Plant Dis 100: 1521–1531.

Gunasinghe N, You MP, Clode P, Cawthray GR, Barbetti MJ (2017) Unique infection structures produced by *Pseudocercosporella capsellae* on oilseed crops *Brassica carinata, B. juncea* and *B. napus* in Western Australia. Plant Pathol 66: 304–315.

Habig M, Bahena-Garrido SM, Barkmann F, Haueisen J, Stukenbrock Eh (2020) The transcription factor Zt107320 affects the dimorphic switch, growth and virulence of the fungal pathogen *Zymoseptoria tritici*. Molec Plant Pathol 21: 124–138.

Hallen-Adams H, Wenner N, Kuldau GA, Trail F (2011) Deoxynivalenol biosynthesis-related gene expression during wheat kernel colonization by *Fusarium graminearum*. Phytopathology 101: 1091–1096.

Hamer JE, Howard RJ, Chunley FG, Valent B (1988) A mechanism for surface attachment in spores of a fungal plant pathogen. Science 239: 288–290.

Han Q, Huang L, Buchanauer H, Wang C, Kang Z (2010) Cytological study of wheat spike infection by *Bipolaris sorokiniana*. J Phytopathol 158: 22–29.

Han SS, Ra DS, Nelson RJ (1995) Relationship between DNA fingerprints and virulence of *Pyricularia grisea* from rice and new hosts in Korea. Internatl Rice News Notes 20(1): 26–27.

Hanse B, Raaijmakers EEM, Schoone AHL, vaw Oorschot PMS (2015) *Stemphylium* sp., the cause of yellow leaf spot disease in sugar beet (*Beta vulgaris* L.) in the Netherlands. Eur J Plant Pathol 142: 319–330.

Hansjakob A, Riederer M, Hildebrandt U (2011) Wax matters: absence of very long-chain aldehydes from leaf cuticular wax of the glossy 11 mutant of maize compromises the prepenetration processes of *Blumeria graminis*. Plant Pathol 60: 1151–1161.

Hatta R, Shinjo A, Ruswandi S, Kitani K, Yamamoto M et al. (2006) DNA transposon fossils present on the conditionally dispensable chromosome controlling AF-toxin biosynthesis and pathogenicity of *Alternaria alternata*. J Gen Plant Pathol 72: 210–219.

Havis ND, Nyman M, Oxley SJP (2014) Evidence for seed transmission and symptomless growth of *Ramularia cello-cygni* in barley (*Hordeum vulgare*). Plant Pathol 63: 929–936.

Hayashi Y, Kozawa T, Aiuchi D, Tani M, Koike M (2014) A selective medium to isolate airborne spores of *Microdochium nivale*, causing winter wheat scab. Eur J Plant Pathol 138: 247–256.

He C, Duan K, Zhang L, Zhang L, Song L et al. (2019) Fast quenching the burst of host salicylic acid is common in early strawberry/*Colletotrichum fructicola* interaction. Phytopathology 109: 531–541.

Hernández-Perez P, du Toit LJC (2006) Seedborne *Cladosporium variabile* and *Stemphylium botryosum* in spinach. Plant Dis 90: 137–145.

Hirayama Y, Asano S, Okayama K, Okhi T, Tojo M (2018) Weeds as the potential inoculum source of *Colletotrichum fructicola* responsible for strawberry anthracnose in Nara, Japan. J Gen Plant Pathol 84: 12–19.

Hoat TX, Nakayashiki H, Tosa Y, Mayama S (2006) Specific cleavage of ribosomal RNA during victorin-induced apoptotic cell death in oat. Plant J 46: 922–933.

Hoffmeister M, Maier W, Thines M, Becker Y (2020) Tracking host infection and reproduction of *Peronospora salvia-officinalis* using an improved method for confocal laser scanning microscopy. Plant Pathol 69: 922–931.

Horrowitz S, Freeman S, Zveibil A, Yarden O (2006) A defect in *nir1*, a *nirA*-like transcription factor confers morphological abnormalities and loss of pathogenicity in *Colletotrichum acutatum*. Molec Plant Pathol 7: 341–534.

Hou C-X, Zhan Y-H, Jiang De-An, Weng X-Y (2014) Functional characterization of a new pathogen-induced xylanase inhibitor (RIXI) from rice. Eur J Plant Pathol 138: 405–414.

Hovmøller MS, Walter S, Bayles RA, Hubbard A, Flath K et al. (2016) Replacement of the European wheat yellow rust population by new races from the center of diversity in the near-Himalayan region. Plant Pathol 65: 402–411.

Huang H, Thi Thu TN, He X, Gravot A, Bernillon S et al. (2017) Increase of fungal pathogenicity and role of glutamine in nitrogen-induced susceptibility (NIS) to rice blast. Front Plant Sci. https://doi.org/10.3389/fpls.2017.00265

Hughes TJ, Grau CR (2016) Influence of soybean monoculture on *Phialophora gregata* f.sp. *sojae* IGS-genotype B isolate aggressiveness. Plant Dis 94: 69–74.

Ijaz U, Ayliffe M, Adhikari K, Bansal HBU (2020) Australian *Uromyces viciae-fabae*: host and nonhost interactions among cultivated grain legumes. Plant Pathol 69: 1227–1236.

Inch SA, Gilgert J (2003) Survival of *Gibberella zeae* in Fusarium damaged wheat kernels. Plant Dis 87: 282–287.

Inoue K, Kitaoka H, Park P, Ikeda K (2016) Novel aspects of hydrophobins in wheat isolate of *Magnaporthe oryzae*: Mpg1, but not Mhp1, is essential for adhesion and pathogenicity. J Gen Plant Pathol 82: 18–28.

Inoue Y, Vy TTP, Yoshida K, Asano H, Mitsuoka et al. (2017) Evolution of the wheat blast fungus through functional losses in a host specificity determinant. Science 357: 80–83.

Isard SA, Dufault NS, Miles MR, Harman GL, Rasso JM et al. (2006) The effect of solar irradiance on the mortality of *Phakopsora pachyrhizi* urediniospores. Plant Dis 90: 941–945.

Ishikawa FJ, Barcelos QL, Alves E, Camargo Jr AC, De Souza EA (2010) Symptoms and penetration events associated with the infection of common bean by the anamorph and teleomorph of *Glomerella cingulata* f.sp. *phaseoli*. J Phyotpathol 158: 270–277.

Izumi Y, Kamel E, Miyamoto Y, Ohtani K, Masunaka A et al. (2012) Role of pathotype-specific ACRTS1 gene encoding a hydrolase involved in the biosynthesis of host-selective ACR-toxin in the rough lemon pathotype of *Alternaria alternata*. Phytopathology 102: 741–748.

Jankovics T, Komaromi J, Fabian A, Jager K, Vida G et al. (2015) New insights into life cycle of the wheat powdery mildew: direct observation of ascosporic infection in *Blumeria graminis* f.sp. *tritici*. Phytopathology 105: 797–804.

Jansen C, von Wettstein D, Schafer W, Kogel K-H, Felk A, Maier FJ (2005) Infection patterns in barley and wheat spikes inoculated with wild-type and trichodiene-disrupted *Fusarium graminearum*. Proc Natl Acad Sci USA, 16892–16897.

Jaspers MV, Seyb AM, Trought MCT, Balasubramanian R (2013) Overwintering grapevine debris as an important source of *Botrytis cinerea*. Plant Pathol 62: 132–138.

Jin Q, Chen B, Li Y, Peng Y, Deng Y et al. (2010a) Expression profile during the development of appressoria induced by hydrophobic surfaces in *Magnaporthe griesea* Y34. J Phytopathol 158: 1143–1153.

Jin Y, Szabo LJ, Carson M (2010b) Century-old mystery of *Puccinia striiformis* life history solved with the identification of *Berberis* as an alternate host. Phytopathology 100: 432–435.

Johnson ACS, Gevens AJ (2014) Investigating the host range of the US-22, US-23, and US-24 clonal lineages of *Phytophthora infestans* on solanaceous cultivated plants and weeds. Plant Dis 98: 754–760.

Johnson DA, Cummings TF (2013) A plant stem inoculation assay for assessing transmission of *Phytophthora infestans* from potato seed tubers to emerged shoots. Plant Dis 97: 183–188.

Kamoun S (2006) A catalogue of the effector secretome of plant pathogenic oomycetes. Annu Rev Phytopathol 44: 41–60.

Kandel SL, Mou B, Shiskoff N, Shi A, Subbarao KV et al. (2019) Spinach downy mildew: advances in our understanding of the disease cycle and prospects for disease management. Plant Dis 103: 791–803.

Kang Z, Buchenauer H, Huang L, Haw Q, Zhang H (2008) Cytological and immunological studies on responses of wheat spikes of resistant Chinese cv. Sumai 3 and the susceptible cv. Xiaoyan to infection by *Fusarium graminearum*. Eur J Plant Pathol 120: 383–396.

Kars I, Kroshof GH, Wagemakers L, Joosten R, Benen JA et al. (2005) Necrotizing activity of five *Botrytis cinerea* endopolygalacturonases produced by *Pichia pastoris*. Plant J 43: 213–225.

Kawasaki-Tanaka A, Hayashi N, Yanagihara S, Fukuta Y (2016) Diversity and distribution of rice blast (*Pyricularia oryzae*, Cavara) races in Japan. Plant Dis 100: 816–823.

Kemen E, Kemen AC, Rafiqi M, Hempel U, Mendgen K et al. (2005) Identification of a protein from rust fungi transferred from haustoria into infected plant cells. Molec Plant-Micro Interact 18: 1130–1139.

Kennedy R, Wakeham AJ, Bryne KG, Meyer UM, Dewey FM (2000) A new method to monitor airborne inoculum of fungal plant pathogens *Mycosphaerella brassicola* and *Botrytis cinerea*. Appl Environ Microbiol 66: 2996–3003.

Keon J, Antoniw J, Carzniga R, Deller S, Ward JL et al. (2007) Transcriptional adaptation of *Mycosphaerella graminis* to programmed cell death (PCD) of its susceptible wheat host. Molec Plant-Micro Interact 20: 178–193.

Kettles GJ, Bayon C, Canning G, Rudd JJ, Kanyuka K (2017) Apoplastic recognition of multiple candidate effectors from the wheat pathogen *Zymoseptoria tritici* in nonhost plant *Nicotiana benthamiana*. New Phytol 213: 338–350.

Khaliq I, Fanning J, Melloy P, Galloway J, Moore K et al. (2020) The role of conidia in the dispersal of *Ascochyta rabei*. Eur J Plant Pathol 158: 911–924.

Khang CH, Berruyer R, Giraldo MC, Kankanala P, Park SY et al. (2010) Translocation of *Magnaporthe oryzae* effectors into rice cells and their subsequent cell-to-cell movement. Plant Cell 22: 1388–1403.

Kheiri A, Jorf AM, Malihapour A (2019) Infection process and wheat response to Fusarium head blight caused by *Fusarium graminearum*. Eur J Plant Pathol 153: 489–502.

Kildea S, Quinn L, Mehenni-Ciz J, Cooke DEL, Perez FM et al. (2013) Reemergence of the Ib mitochondrial haplotype within the British and Irish *Phytophthora infestans* populations. Eur J Plant Pathol 135: 237–242.

Kim S, Kim C-Y, Park S-Y, Kim K-T, Jeon J et al. (2020) Two nucelar effectors of the rice blast fungus modulate host immunity via transcriptional reprogramming. Nature Commun 11: 5845.

Klosterman SJ, Anchieta A, Mc Roberts N, Koike ST, Subbarao KV et al. (2014) Coupling spore traps and quantitative PCR assays for detection of the downy mildew pathogens of spinach (*Peronospora effusa*) and beet (*P. schactii*). Phytopathology 104: 1349–1359.

Knight NL, Vaghefi N, Kikkert JR, Pethybridge SJ (2019) Alternative hosts of *Cercospora beticola* in field surveys and inoculation trials. Plant Dis 103: 1983–1990.

Kobayashi M, Utsushi H, Fujisaki K, Takeda T, Yamashita T et al. (2022) A jacalin-like lectin domain-containing protein of *Sclerospora graminicola* acts as an apoplastic virulence effector in plant-oomycete interactions. Molec Plant Pathol 23: 845–854.

Koch A, Kumar N, Weber L, Keller H, Imani J et al. (2013) Host-induced gene silencing of cytochrome P450 lanosterol C14α-demethylase-encoding genes confers strong resistance to *Fusarium* species. Proc Nat Acad Sci USA 110: 19324–19329.

Kolmer JA, Hughes ME (2017) Physiologic specialization of *Puccinia triticina* on wheat in the United States in 2015. Plant Dis 101: 1968–1973.

Korolev N, Elad Y, Katan T (2008) Vegetative compatibility grouping *in Botrytis cinerea* using sulphate nonutilizing mutants. Eur J Plant Pathol 122: 369–383.

Kunjeti SG, Anchieta A, Subbarao KV (2016) Plasmolysis and vital staining reveal viable oospores of *Peronospora effusa* in spinach seed lots. Plant Dis 100: 59–65.

Lahey KA, Yuan R, Burns JK, Ueng PP, Timmer LW et al. (2004) Induction of phytohormones and differential gene expression in citrus flowers infected by *Colletotrichum acutatum*. Molec Plant-Micro Interact 17: 1394–1401.

Lee BN, Kroken S, Chou DYT, Robbertse B, Yoder OC (2005) Functional analysis of all nonribosomal peptide synthetases in *Cochliobolus heterostrophus* reveals a factor NSP6, involved in virulence and resistance to oxidative stress. Eukary Cell 4: 545–555.

Leesutthiponchai W, Judelson HS (2019) *Phytophthora infestans* sporangia produced in artificial media and plant lesions have subtly divergent transcription profiles, but equivalent infection potential and aggressiveness. Molec Plant-Micro Interact 32: 1077–1087.

Leyronas C, Bryone F, Duffaud M, Troulet C, Nicot PC (2015) Assessing host specialization of *Botrytis cinerea* on lettuce and tomato by genotypic and phenotypic characterization. Plant Pathol 64: 119–127.

Li H, Ge X, Han S, Sivasithamparam K, Barbetti MJ (2011) Histological responses of host and nonhost plants to *Hyaloperonospora parasitica*. Eur J Plant Pathol 129: 221–232.

Li H, Mahmood T, Antony G, Lu N, Pumphreys et al. (2017) The nonhost pathogen *Puccinia triticina* elicits an active transcriptional response in rice. Eur J Plant Pathol 147: 553–569.

Li J, Lu L, Jia Y, Wang Q, Fukuta Y et al. (2016) Characterization of field isolates of *Magnaporthe oryzae* with mating type, DNA fingerprinting, and pathogenicity assays. Plant Dis 100: 298–303.

Li L, Xue C, Bruno K, Nishimura M, Xu R (2004) Two PAK kinase genes, *CHM1* and *MST20* have distinct functions in *Magnaporthe grisea*. Molec Plant-Micro Interact 17: 547–556.

Li Z, Dang H, Yuan X, He J, Hu Z et al. (2018) Morphological characterization and optimization of conditions for conidial production of *Elsinoe ampelina*, the causal organism of grapevine anthracnose. J Phytopathol 166: 420–428.

Li Z, Gao R, Wang T, Zhu W, Ye JL, Li GH (2014) Carbohydrate metabolism changes in *Prunus persica* gummosis infected with *Lasidiplodia theobromae*. Phytopathology 104: 445–452.

Li Z, Zhang S, Han R, Zhang H, Li K et al. (2019) Infection process and host responses to *Elsinoe ampelina*, the causal organism of grapevine anthracnose. Eur J Plant Pathol 155: 571–582.

Liang X-L, Liu J-L, Liu S-S, Liang X-N, Zhang S-H (2015) Alternatively spliced SMN orthologue in *Magnaporthe oryzae* is required for stress resistance and disease development. Eur J Plant Pathol 142: 427–439.

Liao C-Y, Chen M-Y, Chen Y-K, Kuo K-C, Chung K-R et al. (2012) Formation of highly branched hyphae by *Colletotrichum acutatum* within the fruit cuticles of *Capsicum* spp. Plant Pathol 61: 427–434.

Lin C, Liu X, Shi T, Li C, Huang G (2007) The *Colletotrichum gloeosporioides* periplipin homologue CAP20 regulates functional appressorial formation and fungal virulence. J Phytopathol 11: 216–225.

Lin Z, Wu J, Jamieson PA, Zhang C (2019) Alternative oxidase is involved in the pathogenicity, development, and oxygen stress response in *Botrytis cinerea*. Phytopathology 109: 1677–1678.

Liu N, Cao Z, Cao K, Ma S, Gong X et al. (2019) Identification of laccase multicopper oxidases from the pathogenic fungus *Setosphaeria turcica* and their expression pattern during growth and infection. Eur J Plant Pathol 153: 1149–1163.

Liu T, Wan A, Liu D, Chen X (2017) Changes in races and virulence genes in *Puccinia striiformis* f.sp. *tritici*, the wheat stripe rust pathogen in the United States from 1968–2009. Plant Dis 101: 1522–1532.

Lo Presti L, Lanver D, Schweizer G, Tanaka S, Liang L et al. (2015) Fungal effectors and plant susceptibility. Annu Rev Plant Biol 66: 513–545.

Long DH, Correll JC, Lee FN, Te Beest DO (2001) Rice blast epidemics initiated by infested rice grain on the soil surface. Plant Dis 85: 612–616.

Lu S, Edwards MC (2016) Genome-wide analysis of small-secreted cysteine-rich proteins identifies candidate effector proteins potentially involved in *Fusarium graminearum*-wheat interactions. Phytopathology 106: 166–176.

Lu S, Edwards MC, Friesen TL (2013) Genetic variation of single nucleotide polymorphisms identified at the mating-type locus correlates with form-specific disease phenotype in the barley net blotch fungus *Pyrenophora teres*. Eur J Plant Pathol 135: 49–65.

Lu YY, Lin KX, Li GF, Yan LB, Xiao SQ et al. (2018) Identification of mating-type locus structure and distribution of *Cochliobolus lunatus* in China. Eur J Plant Pathol 151: 487–500.

Luo Y, Ma Z, Reyes HC, Morgan D, Michailides TJ (2007) Quantification of airborne spores of *Monilinia fructicola* in stone fruit orchards in California using real-time PCR. Eur J Plant Pathol 118: 145–154.

Ma L, Kong X, Qiao J, Fei A, Hu X et al. (2016) Overwintering of *Puccinia striiformis* on winter wheat at varying altitudes in Gansu and Qinghai Provinces. Plant Dis 100: 1138–1145.

Ma T, Chen S, Liu J, Fu P, Wu W et al. (2021) *Plasmopara viticola* effector PvRXLR111 stabilizes VvWRKY40 to promote virulence. Molec Plant Pathol 22: 231–242.

Ma Z, Luo Y, Michailides TJ (2003) Nested PCR assays for detection of *Monilina fructicola* in stone fruit orchards and *Botryosphaeria dothidea* in pistachios in California. J Phytopathol 151: 312–322.

Maciozek VK, Lawrence CB, Kononowicz AK (2018) Infection cycle of *Alternaria brassicola* on *Brassic oleracea* leaves under growth room condition. Plant Pathol 67: 1088–1096.

Marques JPR, Amorin L, Sposito MB, Marin D, Glona BA (2013) Infection of citrus pollen grains by *Colletotrichum acutatum*. Eur J Plant Pathol 136: 35–50.

Matsuda Y, Sameshima T, Moriura N, Inoue K, Nonomura T et al. (2005) Identification of individual powdery mildew fungi infecting leaves and direct detection of gene expression by single conidium polymerase chain reaction. Phytopathology 95: 1137–1143.

Matsunaka A, Ohtani K, Peever TL, Timmer LW, Tsuge T et al. (2005) An isolate of *Alternaria alternata* that is pathogenic to both tangerines and rough lemon and produces two selective toxins ACT-toxin and ACR-toxins. Phytopathology 95: 241–247.

McEvoy JL, Park E, Conway WS (2003) Virulence factors involved in postharvest apple decay by fungal pathogens. Phytopathology 93: S105 (Abst).

Meng Y, Hao J, Mayfield D, Luo L, Munkvold GP et al. (2017) Roles of genotype-determined mycotoxins maize seedling blight caused by *Fusarium graminearum*. Plant Dis 101: 1103–1112.

Mentges M, Glasenpp A, Boenisch M, Malz S, Henrissat B et al. (2020) Infection cushions of *Fusarium graminearum* are fungal arsenals for wheat infection. Molec Plant Pathol 21: 1070–1087.

Mentlak TA, Kombrink A, Shinya T, Ryder LS, Otomo I et al. (2012) Effector-mediated suppression of chitin-triggered immunity by *Magnaporthe oryzae* is necessary for rice blast disease. Plant Cell 24: 322–335.

Mesarich CH, Griffiths SA, van der Burgt A, Okmen B, Beeman HG et al. (2014) Transcriptome sequencing uncovers the *Avr5* avirulence gene of the tomato leaf mold pathogen *Cladosporium fulvum*. Molec Plant-Micro Interact 27: 846–857.

Mesquita GL, Tanaka FAO, Zambrosi FCB, Chapola R, Cursi D et al. (2019) Foliar application of manganese increases sugarcane resistance to orange rust. Plant Pathol 68: 1296–1307.

Mestre P, Carrere S, Gouzy J, Piron M-C, de Labrohe DT et al. (2016) Comparative analysis of expressed CRN and RxLR effectors from two *Plasmopara* species causing grapevine and sunflower downy mildew. Plant Pathol 65: 767–781.

Miller SA, Chabor DMP, Oullet T, Harris LJ, Fedak G (2004) Use of a *Fusarium graminearum* strain transformed with green fluorescent protein to study infection in wheat (*Triticum aestivum*). Canad J Plant Pathol 26: 453–463.

Morais D, Laval V, Sache I, Suffert F (2015) Comparative pathogenicity of sexual and asexual spores of *Zymoseptoria tritici* (septoria tritici blotch) on wheat leaves. Plant Pahthol 64: 1429–1439.

Morales L, Marino TP, Fouts JQ, Holland JB, Nelson RJ (2018) Dissecting symptomatology and fumonisin contamination produced by *Fusarium verticillioides* in maize. Phytopathology 108: 1475–1485.

Morgan W, Kamoun S (2007) RXLR effectors of plant pathogenic oomycetes. Curr Opin Microbiol 10: 332–338.

Mourelos RR, Malbran I, Balatti PA, Ghiringhelli PD, Lori GA (2014) Graminaceous and nongraminaceous hosts of *Fusarium graminearum*, causal agent of Fusarium head blight of wheat in Argentina. Crop Protect 65: 100–104.

Mulpuri S, Soni PK, Gonela SK (2016) Morphological and molecular characterization of powdery mildew on sunflower (*Helinanthus annuus* L.), alternative hosts and weeds commonly found in and around sunflower fields in India. Phytoparasitca 44: 353–367.

Naim YB, Falach-Block L, Ben-Daniel B-H, Cohen Y (2019) Host range of *Peronospora belbahrii*, causal agent of basil downy mildew in Israel. Eur J Plant Pathol 155: 789–799.

Narayanasamy P (2002) *Microbial Plant Pathogens and Crop Disease Management*. Science Publishers, Enfield, USA.

Narayanasamy P (2008) *Molecular Biology in Plant Pathogenesis and Disease Management*, volume 2. Springer Science + Business Media B.V., Heidelberg, Germany.

Narayanasamy P (2017) *Microbial Plant Pathogens—Detection and Management in Seeds and Propagules*. Wiley-Blackwell, Oxford, Chichester, UK.

Nascimoto KJT, Debona D, Rezende D, DaMatta FM, Rodrigues FA (2018) Changes in leaf gas exchange and chlorophyll *a* fluorescence on soybean plants supplied with silicon and infected by *Cercospora sojina*. J Phytopathol 166: 747–760.

Navarro BL, Marques JPR, Appezato-da-Gloria B, Sposito MB (2019) Histopathology of *Phakopsora euvitis* in *Vitis vinifera*. Eur J Plant Pathol 154: 1185–1193.

Nowara D, Gay A, Lacomme C, Shaw J, Ridout C et al. (2010) HIGS: host-induced gene silencing in obligate biotrophic fungal pathogen *Blumeria graminis*. Plant Cell 22L: 3130–3141.

Ohara A, Tashika Y, Abe A, Sone T (2018) The DNA damage signal transducer orthology Mop53BP1 is required for proper appressorium differentiation and pathogenicity in *Pyricularia oryzae*. J Gen Plant Pathol 84: 176–188.

Oliver AH, Maguire ML, Giffin JL, Jung Y-H, Shoboto J et al. (2011) Using metabolic profiling to assess plant-pathogen interactions: in example using rice (*Oryza sativa*) and the blast pathogen *Magnaporthe grisea*. Eur J Plant Pathol 129: 539–554.

Olivera PD, Sikharulidze Z, Dambadze R, Szabo LJ, Newcomb M et al. (2019) Presence of a sexual population of *Puccinia graminis* f.sp. *tritici* in Georgia provides a hotspot for genotypic and phenotypic diversity. Phytopathology 109: 2152–2160.

Onesti G, Gonzalez-Dominguez E, Rossi V (2017) Production of *Guignardia bidwellii* conidia on grapevine leaf lesions is influenced by repeated washing events by alternation of dry and wet periods. Eur J Plant Pathol 147: 949–953.

Pak D, You MP, Lanoiselet V, Barbetti MJ (2018) Comparative colonization by virulent versus avirulent *Pyricularia oryzae* on wild *Oryza australiensis*. Eur J Plant Pathol 151: 927–936.

Park G, Xue C, Zhao X, Kim Y, Orbach M et al. (2006) Multiple upstream signals coverage on the adaptor protein Mst50 in *Magnaporthe grisea*. Plant Cell 18: 2822–2935.

Park G, Xue C, Zheng L, Lam S, Xu JR (2002) MST12 regulates infectious growth but not appressorium formation in rice blast fungus *Magnaporthe grisea*. Molec Plant-Micro Interact 15: 183–192.

Pascholati SF, Deising H, Leite B, Anderson D, Nicholson RL (1993) Cutinase and nonspecific esterase activities in the conidial mucilage of *Colletotrichum graminicola*. Physiol Molec Plant Pathol 42: 37–51.

Passey TAJ, Robinson JD, Shaw MM, Xu X-M (2017) The relative importance of conidia and ascospores as primary inoculum of *Venturia inaequalis* in a southeast England orchard. Plant Pathol 66: 1445–1451.

Paudyal DP, Hyun J-W, Hwang R-Y (2019) Infection and symptom development by citrus scab pathogen *Elsinoë fawcettii* on leaves of Satsuma mandarin. Eur J Plant Pathol 148: 807–816.

Pellier AL, Lauge R, Veneault-Fourrey C, Langam T (2003) CLNR1, the AREA/NIT2-like global nitrogen regulator of plant fungal pathogen *Colletotrichum lindemuthianum* is required for the infection cycle. Molec Microbiol 48: 639–655.

Pérez-Hernández A, González M, Gonzalez C, Brito N (2020) The elicitor protein BcIEB1 and the derived peptide ieb35 provide long-term plant protection. Plant Pathol 69: 807–817.

Perfect SE, Green JR, O'Connell RJ (2001) Surface characteristics of necrotrophic secondary hyphae produced by the bean anthracnose fungus *Colletotrichum lindemuthianum*. Eur J Plant Pathol 107: 813–819.

Pizolotto CA, Maciel JLN, Fernandes JMC, Boller W (2019) Saprophytic survival of *Magnaporthe oryzae* in infested residues. Eur J Plant Pathol 153: 327–339.

Poinssot B, Vandelle E, Bentejac M, Adrian M, Levis C et al. (2003) The endopolygalacturonase 1 from *Botrytis cinerea* activates grapevine defense reactions unrelated to its enzymatic activity. Molec Plant-Micro Interact 16: 553–564.

Prats E, Gay AP, Roberts PC, Thomas BJ, Sanderson R et al. (2010) *Blumeria graminis* interactions with barley conditioned by different single R genes demonstrate and temporal and spatial relationship between stomatal dysfunction and cell death. Phytopathology 100: 21–32.

Prusky D, Dori N, Kramer-Haimovich H, Rollins I (2003) External pH and nitrogen source affect secretion of pectate lyase by *Colletotrichum gloeosporioides*. Phytopathology 93: S106 (Abst).

Qian B, Liu X, Ye Z, Zhou Q, Liu P et al. (2020) Phosphatase-associated protein MoTip41 interacts with phosphatase MoPpe1 to mediate cross-talk between TOR and cell wall integrity signaling during infection by rice blast fungus *Magnaporthe oryzae*. Environ Microbiol 23: 791–809.

Quarantin A, Hadeler B, Kroger C, Schaefer W, Favron F et al. (2019) Different hydrophobins of *Fusarium graminearum* are involved in hyphal growth, attachment, water-air interface penetration and plant infection. Front Microbiol 10: 751.

Raveloson H, Ramonta IR, Tharreau D, Sester M (2018) Long-term survival of blast pathogen in infected rice residues as major source of primary inoculum in high altitude upland ecology. Plant Pathol 67: 610–618.

Rieux A, Soubeyrand S, Bennot F, Klein EK, Ngando JE (2014) Long-distance wind-dispersal of spores in a fungal plant pathogen: estimation of an anisotropic dispersal kernels from an extensive field experiment. PLoS ONE 9(8): e103225. https://doi.org/10.1371/journal.pone.0103225

Rios JA, Rios VS, Aucique-Perez CE, Cruz MFA, Morai LE et al. (2017) Alteration of photosynthetic performance and source-link relationships in wheat plants infected by *Pyricularia oryzae*. Plant Pathol 66: 1496–1507.

Ristaino JA, Hu CH, Fitt BDL (2013) Evidence for the presence of the founder Ib mt DNA haplotype of *Phytophthora infestans* in 19th century potato tubers from the Rothamsted archives. Plant Pathol 62: 492–500.

Rodriguez-Algaba J, Hovmoller MS, Justesen AF (2020) Sexual recombination with the "Kranich" race of the yellow rust fungus *Puccinia striiformis* f.sp. *tritici* on *Berberis vulgaris*. Eur J Plant Pathol 156: 1169–1173.

Rossi V, Caffi T, Legler SEC (2010) Dynamics of ascospore maturation and discharge in *Erysiphe necator*, the causal agent of grape powdery mildew. Phytopathology 100: 1321–1329.

Rui O, Hahn M (2007) The Slt2-type MAP kinase Bmp3 of *Botrytis cinerea* is required for normal saprophytic growth, conidiation, plant surface sensing and host tissue colonization. Molec Plant Pathol 8: 173–184.

Runge F, Thines M (2012) Reevaluation of host specificity of the closely related species *Pseudoperonospora humuli* and *P. cubensis*. Plant Dis 96: 55–61.

Sammonds J, Jaspers MV, Jones EE (2016) Preinfection process of *Botryosphaeria* spp.: adhesion of conidia to different substrates. Plant Pathol 65: 1142–1152.

Scharte J, Schörn H, Weis E (2005) Photosynthesis and carbohydrate metabolism in tobacco leaves during incompatible interaction with *Phytophthora nicotianae*. Plant Cell Environ 28: 1421–1435.

Schornack S, van Damme M, Bozkurr TO, Cano LM, Smoker M et al. (2010) Ancient class of translocated oomycete effectors targets the host nucleus. Proc Nat Acad Sci USA 107: 17421–17426.

See PT, Marathamuthu KA, Iagallo EM, Oliver RP, Moffat CS (2018) Evaluating the importance of the tan spot *ToxA-Tsn1* interaction in Australian wheat varieties. Plant Pathol 67: 1066–1075.

Shahbazi H, Aminian H, Shebani N, Alterman D (2010) Biochemical evaluation of resistance responses of potato to different isolates of *Alternaria solani*. Phytopathology 100: 454–459.

Shamanin V, Salina E, Zelenskiv Y, Kohkmetova A, Patpour M et al. (2018) Large scale wheat stem rust outbreaks in Western and Siberia Northern Kazakhstan in 2015–2017. www.globalrust.org/content/large-scale-wheat-stem-rust-outbreaks-western-siberia-northern kazkhstan-2015-2017

Shams E, Javan-Nikkhah M, Gohari AM (2020) Dissecting molecular events and gene expression signatures involved in *Colletotrichum lindemuthianum-Phaseolus vulgaris* pathosystem in compatible and incompatible interactions. Eur J Plant Pathol 156: 925–937.

Shi H, Meng S, Qiu J, Wang C, Shi Y et al. (2021) MoWhi2 regulates appressorium formation and pathogenicity via the MoTor signalling pathway in *Magnaporthe grisea*. Molec Plant Pathol 22: 969–983.

Shimizu T, Suzaki K (2020) Multiple gene clusters responsive to *Plasmopara viticola* infection in grapevines. Eur J Plant Pathol 158: 681–691.

Siewers V, Viaud M, Jimenez-Teja D, Collado JG, Schulze-Gronover CH et al. (2005) Functional analysis of cytochrome P450 monooxygenase gene *cbot1* of *Botrytis cinerea* indicates that botrydial is a strain-specific virulence factor. Molec Plant-Micro Interact 18: 602–612.

Skadsen RW, Hohn TM (2004) Use of *Fusarium graminearum* transformed with *gfp* to follow infection patterns in barley and *Arabidopsis*. Physiol Molec Plant Pathol 64: 45–53.

Slatnar A, Petkovsek M, Halbwirth H, Stampar F, Stich K et al. (2010) Enzyme activity of the phenylpropanoid pathway as a response to apple scab infection. Ann Appl Biol 156: 449–456.

Solomon PS, Oliver RP (2001) The nitrogen content of tomato leaf apoplast increases during infection by *Cladosporium fulvum*. Planta 213: 241–249.

Somai-Jemmali L, Randoux B, Siah A, Magnin-Robert M, Halama P et al. (2017) Similar infection process and induced defense patterns during compatible interactions between *Zymoseptoria tritici* and both bread and durum wheat. Eur J Plant Pathol 147: 787–801.

Spanu PD, Abbott JC, Amselem J, Burgla TA, Soanes DM et al. (2010) Genome expansion and gene loss in powdery mildew fungi reveal tradeoffs in extreme and transcriptome analyses. Nature Genet 44: 1060–1065.

Sperschneider J, Dodds PN (2022) Effector 3.0: prediction of apoplastic and cytoplasmic effectors in fungi and oomycetes. Molec Plant-Micro Interact 35: 146–156.

Sperschneider J, Dodds PN, Gardiner DM, Singh KB, Taylor JM (2018) Improved prediction of fungal effector proteins from secretomes with Effector 2.0. Molec Plant Pathol 19: 2094–2110.

Subbarao CS, Anchieta A, Ochoa L, Dhar N, Kunjeti SG et al. (2018) Detection of *Peronospora effusa* infection in spinach. Plant Dis 102: 1766–1771.

Sun G, Qi X, Wilson RA (2019) A feed-forward subnetwork emerging from integrated TOR- and cAMP/PKA-siganlling architecture reinforces *Magnaporthe oryzae* appressorium morphogenesis. Molec Plant-Micro Interact 32: 593–607.

Sunseri MH, Johnson DA, Dasgupta JN (2002) Survival of detached sporangia of *Phytophthora infestans* exposed to ambient relatively dry atmospheric conditions. Amer J Potato Res 79: 443–450.

Suzuki T, Nishimura S, Yagi K, Nakamura R, Takikawa Y et al. (2018) Effect of light quality on conidiophore formation of the melon powdery mildew pathogen *Podosphaera xanthii*. Phytoparasitica 46: 31–43.

Swarbrick PJ, Schulze-Lefert P, Scholes JD (2006) Metabolic consequences of susceptibility and resistance in barley leaves challenged with powdery mildew. Plant Cell Environ 29: 1061–1076.

Takahashi M, Ashizawa T, Hirayae K, Moriwaki J, Sone T et al. (2010) One of two paralogs of AVR-Pita1 is functional in Japanese rice blast isolates. Phytopathology 100: 612–618.

Takano Y, Kikuchi T, Kubo Y, Hamer JE, Mise K et al. (2000) The *Colletotrichum lagenarium* MAP kinase gene *CMK1* regulates diverse aspects of fungal pathogenesis. Molec Plant-Micro Interact 13: 374–383.

Takao K, Akagi Y, Tsuge T, Harimoto Y, Yamamoto M et al. (2016a) The global regulator LacA controls biosynthesis of host-specific toxins, pathogenicity and development of *Alternaria alternata* pathotypes. J Gen Plant Pathol 82: 121–131.

Takikawa Y, Nonomura T, Miyamoto S, Okamoto N, Murakami T et al. (2015) Digital microscope analyses of conidiogenesis of powdery mildew pathogens isolated from melon leaves. Phytoparasitica 43: 517–530.

Tanaka E, Koga H, Mori M, Mori M (2011) Auxin production by the rice blast fungus and its localization in host tissue. J Phytopathol 159: 522–530.

Tani S, Yatzkan E, Judelson HS (2004) Multiple pathways regulate the induction of genes during zoosporogenesis in *Phytophthora infestans*. Molec Plant-Micro Interact 17: 330–337.

Tavernier V, Cadiou S, Pageau K, Lauge R, Reisdorf-Cren M et al. (2007) The plant nitrogen mobilization promoted by *Colletotrichum lindemuthianum* in *Phaseolus* leaves depends on fungus pathogenicity. J Exp Botany 58: 3351–3360.

Thomas A, Carbone I, Cohen Y, Ojiambo PS (2017a) Occurrence and distribution of mating types of *Pseudoperonospora cubensis* in the United States. Phytopathology 107: 313–321.

Thomas A, Carbone I, Lebeda A, Ojamb PS (2017b) Virulence structure within populations of *Pseudoperonospora cubensis* in the United States. Phytopathology 107: 777–785.

Tian Y, Yin J, Sun J, Ma h, Ma Y, Quan J, Shan W (2015) Population structure of the late blight pathogen *Phytophthora infestans* in potato germplasm nursery in two consecutive years. Phytopathology 105: 1771–777.

Timmer LW, Brown GE (2000) Biology and control of anthracnose disease of citrus. In: Prusky D, Freeman S, Dickman MB (eds.), *Colletotrichum: Host Specificity, Pathology, and Host-Pathogen Interaction*, APS Press, St Paul, MN, USA, pp. 300–316.

Tran NT, Miles AK, Dietzgen RG, Dewdney MM, Zhang K et al. (2017b) Sexual reproduction on citrus black spot pathogen, *Phyllosticta citricarpa*. Phytopathology 107: 732–739.

Tsugi G, Tsuge S, Shiraishi T, Kubo Y (2003) Expression pattern of melanin biosynthesis enzymes during infectious morphogenesis of *Colletotrichum lagenarium*. J Gen Plant Pathol 69: 169–175.

Twizeyimana M, Hartman GL (2012) Pathogenic variation of *Phakopsora pachyrhizi* isolates on soybean in the United States from 2006 to 2009. Plant Dis 96: 75–81.

Vagndorf N, Heick TM, Justesen AJ, Andersen JR, Jahoor A et al. (2018) Population structure and frequency differences of CYP51 mutations in *Zymoseptoria tritici* populations in the Nordic and Baltic regions. Eur J Plant Pathol 152: 327–341.

Vale PAS, de Resende MLV, Botelho DMS, Pozza EA, Ogoshi C et al. (2019) Temperature, incubation time and virulence of *Cercospora coffeicola* in the production of cercosporin. J Phytopathol 167: 371–379.

Van den Burg HA, Harrison SJ, Joosten MHAJ, Vervoort J, de Wit PJGM (2006) *Cladosporium fulvum* Avr4 protects fungal cell walls against hydrolases by plant chitinases accumulating during infection. Molec Plant-Micro Interact 19: 1420–1430.

Van Esse HP, Bolton MD, Stergiopoulos I, de Wit PJGM, Thomma BPHJ (2007) The chitin-binding *Cladosporium fulvum* effector protein Avr4, is a virulence factor. Molec Plant-Micro Interact 20:1092–1101.

Vetukuri RR, Whisson SC, Grenville-Briggs LJ (2017) *Phytophthora infestans* effector Pi14054 is a novel candidate suppressor of host silencing mechanisms. Eur J Plant Pathol 149: 771–777.

Viaud MC, Legeai F, Fradier JM, Brygo Y, Batton F et al. (2005) Expressed sequence tags from phytopathogenic fungus *Botrytis cinerea*. Eur J Plant Pathol 111: 139–146.

Vittal R, Paul C, Hill CB, Hartman GL (2014) Characterization and quantification of fungal colonization of *Phakopsora pachyrhizi* in soybean genotypes. Phytopathology 104: 86–94.

Wakefield L, Gadoury DM, Seem RC, Milgroom MG, Sun Q et al. (2011) Differential gene expression during conidiation in the grape powdery mildew pathogen *Erysiphe necator*. Phytopathology 101: 839–846.

Wan A, Chen X (2014) Virulence characterization of *Puccinia striiformis* f.sp. *tritici* using a new set of Yr dingle-gene line differentiates in the United States in 2010. Plant Dis 98: 1534–1542.

Wan A, Chen X, Yuen J (2016) Races of *Puccinia striiformis* f.sp. *tritici*, the cause of wheat stripe rust in the United States in 2011 and 2012 and comparison with races in 2010. Plant Dis 100: 966–975.

Wang B, Xie C, Xin M, Song N, Ni Z et al. (2012) Comparative proteomic analysis of wheat response to powdery mildew infection in wheat *Pm30* near-isogenic lines. J Phytopathol 160: 229–236.

Wang C, Zhang H, Buchenauer H, Kang Z (2011) Microscopic observation on development of *Puccinia striiformis* in the spike and stems of wheat. J Phytopathol 159: 314–316.

Wang F, Lu YY, Liu MM, Xiao SQ, Gao YB et al. (2020) Effects of iron on the asexual reproduction and major virulence factors of *Curvularia lunata*. Eur J Plant Pathol 157: 497–507.

Wang H, Yang XB, Ma Z (2010) Long distance spore transport of wheat stripe rust pathogen from Sichuan, Yunnan and Guizhou. Plant Dis 94: 873–880.

Wang MN, Chen XM (2015) Barberry does not function as an alternate host for *Puccinia striiformis* f.sp. *tritici* in the U. S. Pacific Northwest due to teliospore degradation and barberry phenology. Plant Dis 99: 1500–1506.

Wang Y, Chen J, Li D-W, Lu Z, Huang J (2017) *cglCUT1* gene required for cutinase activity and pathogenicity of *Colletotrichum gloeosporioides* causing anthracnose of *Camellia oleifera*. Eur J Plant Pathol 147: 103–114.

Wang Y, Wang Y (2018) Trick or treat: microbial pathogens evolved apoplastic effectors modulating plant susceptibility to infection. Molec Plant-Micro Interact 31: 6–12.

Wang Z, Zhao J, Chen X, Peng Y, Ji J et al. (2016) virulence variations of *Puccinia striiformis* f.sp. *tritici* isolates collected from *Berberis* spp. in China. Plant Dis 100: 131–138.

Webster J, Weber R (2007) *Introduction to Fungi*, third edition. Cambridge University Press, Cambridge, UK.

Wieczorek TM, Jorgensen LN, Hansen AL, Munk L, Justesen AF (2013) Early detection of sugar beet pathogen *Ramularia beticola* in leaf and air samples using qPCR. Eur J Plant Pathol 138: 775–785.

Wolfenberger SN, Twomey MC, Gadoury DM, Knaus BJ, Grunwald NJ et al. (2015) Identification and distribution of mating-type idiomorph in population of *Podosphaeria maculans* and development of chasmothecia of the fungus. Plant Pathol 64: 1094–1102.

Wong M-H, Henderson J, Aitken EAB, Drenth A (2013) Mode of infection of *Phyllosticta maculata* on banana as revealed by scanning microscopy. J Phytopathol 161: 135–141.

Wu F, Guo Q, Zhang W, Jin W (2015) Identification and analysis of powdery mildew-responsive miRNAs in wheat. J Phytopathol 163: 264–270.

Xing J, Jia Y, Peng Z, Shi Y, He Q et al. (2017) Characterization of molecular identity and pathogenicity of rice blast fungus in Hunan province in China. Plant Dis 101: 557–561.

Xiong Q, Xu Y, Zhao Y, Wang K (2015) *CtPMK1*, a mitogen-activated protein kinase gene is required for conidiation, appressorium formation and pathogenicity of *Colletotrichum truncatum* on soybean. Ann Appl Biol 167: 63–74.

Xu HX, Wang Y, Zhao P, Zhang Y, Xu R et al. (2011) A cAMP-dependent protein kinase gene *aapk1* is required for mycelia growth, toxicity, and pathogenicity of *Alternaria alternata* on tobacco. J Phytopathol 159: 208–216.

Xu Q, Tang C, Wang L, Zhao C, Kang Z et al. (2020) Haustoria—arsenals during the interaction between wheat and *Puccinia striiformis* f.sp. *tritici*. Molec Plant Pathol 21: 83–94.

Yakoby N, Beno-Moualen D, Keen NT, Dinoor A, Pines O et al. (2001) *Colletotrichum gloeosporioides pelB* is an important virulence factor in avocado fruit-fungus interaction. Molec Plant-Micro Interact 14: 988–995.

Yang E, Jensen JD, Svensson B, Jorgensen HJ, Collingge DB, Finnie C (2012) Secretomics identifies *Fusarium graminearum* proteins involved in the interaction with barley and wheat. Molec Plant Pathol 13: 445–453.

Yin Z, Bi W, Mi Q, Kang Z, Liu C et al. (2017) Conserved and divergent roles of the HOG1 kinase of *Alternaria longipes* in mycelial and conidial development multi-stress responses, melanin production and pathogenicity. Eur J Plant Pathol 147: 415–430.

Yoshida K, Saunders DG, Mitsuoka C, Natsume S, Kosugi S et al. (2016) Host specialization of the blast fungus *Magnaporthe oryzae* is associated with dynamic gain and loss of genes linked to transposable elements. BMC Genetics 18: 370.

Yoshida M, Nakajima T (2010) Deoxynivalenol and nivalenol accumulation in wheat infected with *Fusarium graminearum* during grain development. Phytopathology 100: 460–471.

Young GK, Cooke LR, Watson S, Kirk WW, Perez FM et al. (2018) The role of aggressiveness and competition in the selection of *Phytophthora infestans* populations. Plant Pathol 67: 1539–1551.

Yuan C, Wang M, Skinner DZ, See DR, Xia C et al. (2018) Inheritance of virulence, construction of a lineage map, and mapping dominant virulence genes in *Puccinia striiformis* f.sp. *tritici* through characterization of a sexual population with genotyping by sequencing. Phytopathology 108: 133–141.

Zhang F, Chen H, Zhang X, Gao C, Huang J et al. (2021) Genome analysis of two newly emerged potato late blight isolates sheds light on pathogen adaptation and provides tools for disease management. Phytopathology 111: 96–107.

Zhang X, White RP, Demir E, Jedryczka M, Lange RM et al. (2014) *Leptosphaeria* spp., Phoma stem canker and potential spread of *L. maculans* on oil seed rape crops in China. Plant Pathol 63: 598–612.

Zhao H, Han Q, Wang J, Gao X, Xiao C-L et al. (2013a) Cytology of infection of apple leaves by *Diplocarpon mali*. Eur J Plant Pathol 136: 41–49.

Zhao J, Zhao S, Chen X, Wang Z, Wang L (2015) Determination of role of *Berberis* spp. in wheat stem rust in China. Plant Dis 99: 1113–1117.

Zhao J, Zhuang H, Zhan G, Huang L, Kang Z (2016) Proteomic analyses of *Puccinia striiformis* f.sp. *tritici* (Pst) during uredospore germination. Eur J Plant Pathol 144: 121–132.

Zheng Z, Nonomura T, Bókak K, Matsuda Y, Vissar RGF et al. (2013) Detection and quantification of *Leveillula taurica* growth on pepper leaves. Phytopathology 103: 623–632.

Zhu P, Xu Z, Zhang Z, Xu L (2017) Ethylene production by *Alternaria alternata* and its association with virulence on inoculated grape berries. Phytoparasitica 45: 273–279.

Additional References

Ahn JH, Walton JD (1997) A fatty acid synthase gene in *Cochliobolus carbonum* required for production of HC toxin cyclo (D-prolyl-L-alanyl-D-alanyl L-2-amino-9, 10 epoxi-8-oxodecanoyl). Molec Plant-Micro Interact 10: 207–214.

Augusti Brisach C, Gramje D, León M, Garcia-Jiménez J, Armengol J (2011) Evaluation of vineyard weeds as potential hosts of black-foot and Petri disease pathogens. Plant Dis 95: 803–810.

Babu TK, Sharma R, Thakur RP, Upadhyaya HD, Reddy PN et al. (2015) Selection of host differentials for elucidating pathogenic variation in *Magnaporthe grisea* populations adapted to finger millet (*Eleusine corocana*). Plant Dis 99: 1784–1799.

Bai GH, Desjardins AE, Plattner RD (2002) Deoxynivalenol nonproducing *Fusarium graminearum* causes initial infection but does not cause disease spread in wheat spikes. Mycopathologia 153: 91–98.

Bailey A, Mueller E, Bowyer P (2000) Ornithine decarboxylase of *Stagonospora (Septoria) nodorum* is required for virulence toward wheat. J Biol Chem 275: 14242–14247.

Bormann J, Lebrun M, Schafer W, Maritnez-Rocha AL (2020) Infection cushions of *Fusarium graminearum* are fungal arsenals for wheat infection. Molec Plant Pathol 21: 1070–1087.

Boshoff WHP, Pretorius ZA, Terefe TG, Bender CM, Heselman L et al. (2018) Phenotypic and genotypic description of *Puccinia graminis* f.sp. *tritici* race 2SA55 in South Africa. Eur J Plant Pathol 152: 783–789.

Brar GS, Kutcher HR (2016) Race characterization of *Puccinia striiformis* f.sp. *tritici*, the cause of wheat stripe rust, in Saskatchewan and southern Alberta, Canada and virulence comparison with races from the United States. Plant Dis 100: 1744–1753.

Burge HA, Solomon WR (1987) Sampling and analysis of biological aerosols. Atmos Environ 21: 451–456.

Caires NP, Rodrigues FA, Furtado GQ (2015) Infection process of *Botrytis cinerea* on eucalyptus leaves. J Phytopathol 163: 604–611.

Callahan TM, Rose MS, Meade MJ, Ehrenshaft M, Upchurch RG (1999) CFP the putative cercospora transporting *Cercospora kikuchii* is required for wild-type cercosporin production, resistance and virulence in soybean. Mole Plant-Micro Interact 12: 901–910.

Ceresini PC. Castrpagudin VL, Rodrigues FA, Rios JA, Aucique-Perez CE et al. (2015) Wheat blast: past, present and future. Annu Rev Phytopathol 56: 427–456.

Champouret N, Bouwmeester K, Rietman H, van der Lee T, Malipaard C et al. (2009) *Phytophthora infestans* isolates lacking class I ipiO variants are virulent on Rpi-blb1 potato. Molecular Plant-Micro Interact 22: 1535–1545.

Cheng YQ, Ahn JH, Waltson JD (1999) A putative branched-amino acid transaminase gene required for HC-toxin biosynthesis and pathogenicity in Cochliobolus carbonum. Microbiology (Reading) 145: 3539–3546.

Cohen Y, Belausov E, Maymon M, Elazar M, Shulman I et al. (2017) *Fusarium mangiferae* localization in planta during initiation and development of mango malformation disease. Plant Pathol 66: 924–933.

de Vallvieille-Pope C, Ali S, Leconte M, Eryalbert J, Delos M et al. (2012) Virulence dynamics and regional structuring of *Puccinia striiformis* f.sp.tritici in France between 1984 and 2009. Plant Dis 96: 131–140.

Dickman MB, Ha YS, Yang Z, Adams B, Huang C (2003) A protein kinase from *Colletotrichum trifolii* is induced by plant cutin and is required for appressorium formation. Molec Plant-Micro Interact 16: 411–421.

Divilov K, Wiesner-Hanks T, Barba P, Cadle-Davidson L, Reisch BI (2017) Computer vision for high throughput quantitative phenotyping: a case study of grapevine downy mildew sporulation and leaf trichomes. Phytopathology 107: 1549–1555.

Farman M, Peterson G, Chen L, Bachi P, Murdock L et al. (2017) The *Lolium* pathotype of *Magnaporthe oryzae* recovered from a single blasted wheat plant in the United States. Plant Dis 101: 684–692.

Feng C, Correll JC, Kammeijer KE, Koike ST (2014) Identification of new races and deviating strains of the spinach downy mildew pathogen *Peronospora farinosa* fsp. *spinaciae*. Plant Dis 98: 145–152.

Feng J, Chang KF, Hwang SF, Strelkov SE, Conner RL et al. (2012) Analysis of expressed sequence tags derived from pea leaves infected by *Peronospora viciae* f.sp. *pisi*. Ann Appl Biol 161: 214–222.

Fernandez-Pavia SP, Grünwald NJ, Diaz-Valasis M, Cadena-Hinojosa MA, Fry WE (2004) Soilborne oospores of *Phytophthora infestans* in Central Mexico survive winter fall and infect potato plants in the field. Plant Dis 88: 29–33.

Flor HH (1942) Inheritance of pathogenicity in *Melampsora lini*. Phytopathology 32: 653–669.

Flor HH (1946) Genetics of pathogenicity in *Melampsora lini*. J Agric Res 73: 335–357.

Funnell-Harris DL, Graybosch RA, O'Neill PM, Durray ZT, Wegulo SN (2019) Amylose-free ("waxy") wheat colonization by *Fusarium* spp. and response to Fusarium head blight. Plant Dis 103: 972–989.

Gao J-X, Jing J, Liu T, Yu C-J, Li Y-Q et al. (2015) Identification of proteins associated with production of melanin and with pathogenicity in maize pathogen *Curvularia luanta*. Austr Plant Pathol 44: 597–603.

Garcia-Brugger A, Lamotte O, Vandelle E, Bourque S, Lecourieux D et al. (2006) Early signaling events induced by elicitors of plant defenses. Molec Plant-Micro Interact 19: 711–724.

Geddes H, Eudes F, Tucker JR, Legge WG, Selinger B (2005) Evaluation of inoculation methods on infection and deoxynivalenol production by *Fusarium graminearum* on barley. Canad J Plant Pathol 33: 66–73.

Glenn DL, Phipps PM, Stipes RJ (2003) Incidence and survival of *Cylindrocladium parasiticum* in peanut seed. Plant Dis 87: 867–871.

Hahn M, Mendgen K (1997) Characterization of in planta induced rust genes isolated from a haustorium-specific cDNA library. Molec Plant-Micro Interact 10: 427–434.

Hu M, Luo L, Wang S, Liu Y, Li J (2014) Infection processes of *Ustilaginoidea virens* during artificial inoculation of rice panicles. Eur J Plant Pathol 139: 67–77.

Hughes HB, Carzaniga P, Rawlings SL, Green JR, O'Connell RJ (1999) Spore surfaces glycoproteins of *Colletotrichum lindemuthianum* are recognized by a monoclonal antibody which inhibits adhesion to polystyrene. Microbiology 145: 1927–1936.

Jones JD, Dangl JL (2006) The plant immune system. Nature 444: 323–329.

Kalischuk ML, Peters RD, Howard RJ, Platt (Bud) HW, Kawchuk LM (2012) Genetic composition of *Phytophthora infestans* in Canada reveals migration and increased diversity. Plant Dis 96: 1729–1735.

Kars I, van Kan JAL (2004) Extracellular enzymes and metabolites involved in pathogenesis of *Botrytis*. In: Elad Y, Williamson B, Tudzynski P, Nelson N (eds.), *Botrytis, Biology, Pathology and Control*. Kluwer Academic Publishers, Dordrecht, The Netherlands, pp. 99–118.

Kato H, Yamamoto M, Yamaguchi-Ozaki T, Kadouchi H, Iwamoto Y et al. (2000) Pathogenicity, mating ability, DNA restriction fragment length polymorphisms of *Pyricularia* populations isolated from Gramineae, Bambusideae, and Zingiberaceae plants. J Gen Plant Pathol 66: 30–47.

Kaur P, Sivasithamparam K, Barbetti MJ (2013) Host range and phylogenetic relationships of *Albugo candida* from cruciferous hosts in western Australia with special reference to *Brassica juncea*. Plant Dis 95: 712–718.

Kitner M, Lebeda A, Sharma R, Runge F, Dvorak P et al. (2015) Coincidence of virulence shifts and population genetic changes of *Pseudoperonospora cubensis* in the Czech Republic. Plant Pathol 64: 1461–1470.

Leplat J, Mangin P, Falchetto L, Heralld C, Gautheron E et al. (2018) Visual assessment and compute assisted image analysis of Fusarium head blight in the field to predict mycotoxin accumulation in wheat grains. Eur J Plant Pathol 150: 1065–1081.

Li J, Ding Q, Wang F, Li H, Zhang Y et al. (2018) Genome-wide gene expression profiles in response to downy mildew in Chinese cabbage (*Brassica rapa* ssp. *pekinensis*). Eur J Plant Pathol 151: 861–873.

Li J, Liang X, Wei Y, Liu J, Liri F et al. (2015) An ATP-dependent protease homolog ensures basic standards of survival and pathogenicity for *Magnaporthe oryzae*. Eur J Plant Pathol 141: 703–716.

Li X, Yin L, Ma L, Zhang Y, An Y et al. (2016) Pathogenicity variation and population genetic structure of *Plasmopara viticola*. J Phyotpathol 104: 863–873.

Liu G, Tian D, Shi C, Wang D-M (2017) Autophagy is induced in haustorial mother cells of *Puccinia triticina* and is necessary for plant infection. Eur J Plant Pathol 147: 833–843.

Marques JPR, Amorin L, Sposito MB, Appezzato-da-Gloria B (2013) Histopathology of post bloom fruit drop caused by *Colletotrichum* in citrus flowers. Eur J Plant Pathol 135: 783–790.

Martinez JP, Oesch NW, Ciuffetti LMC (2004) Characterization of the multiple copy host selective toxin gene *ToxB* in pathogenic and nonpathogenic isolates of *Pyrenophora tritici-repentis*. Molec Plant-Micro Interact 17: 467–474.

Metzer ML (2010) Sequencing technologies—The next generation. Nat Rev Genet 11: 31–46.

Navarro BL, Junior AFN, Ribeiro RV, Sposito MB (2019) Photosynthetic damage caused by grapevine rust (*Phakopsora euvitis*) in *Vitis vinifera* and *Vitis labrusca*. Austr Plant Pathol 48: 509–518.

Ou SH (1972) *Rice Diseases*. Commonwealth Mycological Institute, Kew, Surrey, England.

Pilet F, Pelle R, Ellisseche D, Andrivon D (2005) Efficacy of the *R2* resistance genes as a component for durable management of potato late blight in France. Plant Pathol 54: 723–732.

Pryce-Jones E, Carver T, Gurr SJ (1999) The role of cellulose enzymes and mechanical force in host penetration by *Erysiphe graminis* f.sp. *hordei*. Physiol Molec Plant Pathol 55: 175–182.

Puri KD, Sancedo ES, Zhong S (2012) Molecular characterization of Fusarium head blight pathogens sampled from a naturally infected disease nursery for wheat breeding programs in China. Plant Dis 96: 1280–1285.

Rodrigues FA, Silva IT, Cruz MFA, Carrie-Missio V (2014) The infection process of *Pestalotiopsis longisetula* leaf spot on strawberry leaves. J Phytopathol 162: 690–692.

Rogers LM, Kim YK, Guo WJ, Gonzalez-Candelas L, Li DX et al. (2000) Requirement of either a host or pectin-induced pectate lyase for infection of *Pisum sativum* by *Nectria haematococca*. Proc Natl Acad Sci USA 97: 9813–1818.

Sedlářova M, Petřivalský M, Piterková J, Luhová L, Kočiřova J et al. (2011) Influence of nitric oxide and reactive oxygen species on development of lettuce downy mildew in *Lactuca* spp. Eur J Plant Pathol 129: 267–280.

Singh RP, Hodson DP, Jin Y, Lagudah ES, Ayliffe MA et al. (2015) Emergence and spread of new races of security and prospects of genetic control. Phytopathology 105: 872–884.

Soares MA, Nogueira GB, Bazzolli DMS, de Arujo EF, Langin T et al. (2014) PacCl, a pH-response transcriptional regulator is essential in the pathogenicity of *Colletotrichum lindemuthianum*, causal agent of anthracnose in bean plants. Eur J Plant Pathol 140: 769–785.

Solomon PS, Oliver RP (2002) Evidence that r-aminobutyric acid is a major nitrogen source during *Cladosporium fulvum* infection of tomato. Planta 214: 414–420.

Stergiopoulos I, van den Burg HA, Okmen B, Beenen HG, van Liere S et al. (2010) Tomato Cf resistance proteins mediate recognition of cognate homologous effectors from fungal pathogens on dicots and monocots. Proc Natl Acad Sci USA 107: 7610–7615.

Suzuki T, Nakamura K, Takagi N, Takikawa Y, Kakutani K et al. (2019) Quantitative analysis of the lifelong production of conidia released from single colonies of *Podosphaera xanthi* on melon leaves using electrostatic techniques. Austr Plant Pathol 48: 297–307.

Tada Y, Hata S, Nakayashiki H, Tosa Y, Mayama S (2000) Signal mediators for phytoalexin production in defense response of oats elicited by victorin as a specific elicitor. J Gen Plant Pathol 66: 185–190.

Takao K, Akagi Y, Tsuge T, Kodama M (2016) Functional characterization of putative G- protein-coupled receptors in the tomato pathotype of *Alternaria alternata*. J Gen Plant Pathol 82: 82–88.

Tang YX, Jin J, Hu D-W, Yong ML, Xu Y et al. (2013) Elucidation of infection process of *Ustilaginoidea virens* (teleomorph-*Villosiclava virens*) in rice spikelets. Plant Pathol 62: 1–8.

Thines W, Webber RWS, Talbot NJ (2000) MAP-kinase and protein kinase A-dependent mobilization of tracylglyceros and glycogen during appressorium turgor generation by *Magnaporthe grisea*. Plant Cell 12: 1703–1718.

Thomma BPHJ, Bolton MD, Clergeot PH, deWit PJ (2006) Nitrogen controls in planta expression of *Cladosporium fulvum* Avr9, but not other effector genes. Molec Plant Pathol 7: 125–130.

Thomma BPHJ, Penninckx IA, Broekaert WF, Cammune BP (2001) The complexity of disease signaling in *Arabidopsis*. Curr Opin Immunol 13: 63–68.

Tran HS, You MP, Khan TN, Barbetti MJ (2017a) Infection process of *Phoma koolunga* on stem and leaf tissues of resistant and susceptible field pea (*Pisum sativum*). Plant Pathol 66: 212–222.

Tsuda M, Ueyama A (1976) Distribution of two mating types of *Cochliobolus miyabeanus* in field and a laboratory attempt to produce hybrids between isolates of *Helminthosporium oryzae* from the USA and Japan. Ann Phytopath Soc Jpn 42: 7–11.

Turgeon BG (1998) Application of mating type gene technology to problems in fungi biology. Annu Rev Phytopathol 36: 115–137.

Vaquero S, Patriarca A, Cabrera G, Pinto VF (2017) Temperature and water activity influence on simultaneous production of AAL toxins by *Alternaria arborescens* on tomato medium. Eur J Plant Pathol 148: 1003–1009.

Venturini G, Toffolatti SL, Assante G, Babazadeh L, Campia P (2015) The influence of flavanoids in maize pericarp on Fusarium ear rot symptoms and fumonisin accumulation under field conditions. Plant Pathol 64: 671–679.

Viaud MC, Brunet-Simon A, Brygoo Y, Pradier JM, Levis C (2002) Cyclophilin A and calcineurin functions investigated by gene inactivation, cyclosporin A inhibition and cDNA arrays approaches in the phytopathogenic fungus *Botrytis cinerea*. Molec Microbiol 50: 1451–1465.

Wang H, Li H, Zhu Z, Wang X (2010) Expression of Ht2-related genes in response to the HT-toxin of *Exserohilum turcicum* in maize. Ann Appl Biol 156: 111–120.

Widmark AK, Andersson B, Sandström M, Yuen JE (2011) Tracking *Phytophthora infestans* with SSR marker within and between seasons—a field study in Sweden. Plant Pathol 60: 938–945.

Winter M, Koopmann B, Doll K, Karlovsky P, Kropf U et al. (2013) Mechanisms regulating grain contamination with trichothecenes translocated from the stem base of wheat (*Triticum aestivum*) infected with *Fusarium culmorum*. Phytopathology 103: 682–689.

Xiao CL, Kim YK, Boal RJ (2014) Sources and availability of inoculum and seasonal survival of *Sphaeropsis pyripurescens* in apple orchards. Plant Dis 98: 1043–1049.

Xu HX, Mendgen K (1997) Targeted cell wall degradation at the penetration site of cowpea rust basidiosporelings. Molec Plant-Micro Interact 10: 87–94.

Xu X, Ma L, Hu X (2019) Overwintering of wheat stripe rust under field conditions in northwestern regions of China. Plant Dis 103: 638–644.

Zhan F, Xie Y, Zhu W, Sun D, Mc Donald BA et al. (2016) Linear correlation analysis of *Zymoseptoria tritici* aggressiveness with in vitro growth rate. Phytopathology 106: 1255–1261.

Zhao J, Wang L, Wang Z, Chen X, Zhang H et al. (2013b) Identification of eighteen *Berberis* species as alternate host of *Puccinia striiformis* f.sp. *tritici* and virulence variation in the pathogen isolates from natural infection of barberry plants in China. Phytopathology 103: 927–934.

Zheng L, Campbell M, Murphy L, Lam S, Xu JR (2000) The *BMP1* gene is essential for pathogenicity in the gray mold fungus *Botrytis cinerea*. Molec Plant-Micro Interact 13: 724–732.

4 Detection and Identification of Phyllosphere Bacterial and Phytoplasmal Pathogens

4.1 INTRODUCTION

Bacteria and phytoplasmas are prokaryotes with less defined morphological characteristics and primitive organization of genomic materials. They cause several economically important diseases in various crops and induce different kinds of symptoms. As the morphological characteristics of bacteria are insufficient to identify them precisely, biochemical, immunological, and genetic characteristics are used for their detection and identification in the infected plant materials. Biochemical tests are time-consuming and the results are often inconclusive. Immunoassays require polyclonal and monoclonal antibodies, the nature of which may vary depending on the cultural conditions in which the bacteria are grown. On the other hand, genetic characteristics of the bacteria remain constant, regardless of the environment to which they are exposed. Polymerase chain reaction (PCR)–based techniques are more frequently employed, because of their higher level of sensitivity, reliability, rapidity, and accuracy of the results for detection, identification, differentiation, and quantification of the bacterial pathogens in infected seeds, plants, and propagules, which have an important role in dispersal of the pathogens. Phytoplasmal pathogens are cell wall-less, obligate pathogens causing important diseases commonly found in several horticultural crops. Molecular techniques have been applied to detect and differentiate them to determine the extent of infection in various crops. Techniques have to be selected based on the cost-effectiveness and reliability of the results.

Bacterial plant pathogens are smaller in size with different shapes and have less defined structural characteristics, compared to the fungal plant pathogens. Fungi are included in eukaryotes, whereas bacteria are primitive organisms placed under prokaryotes which enclose Mollicutes containing the cell wall-less phytoplasmal pathogens also. Among 1,600 bacterial species recorded, about 100 are known to be pathogenic species capable of inducing different kinds of diseases in various crops (Agrios 2005). Most of the bacteria have a saprophytic lifestyle, helping with decomposition of large quantities of organic wastes produced by industrial establishments or by dead plants and animals. Many bacterial species are beneficial to humans, because of their role in building up nutritional levels of soils and as biological control agents (BCAs) that have potential for suppressing disease development, as well as promoting the growth of crop plants.

4.2 DETECTION AND IDENTIFICATION OF BACTERIAL PATHOGENS

Phyllosphere bacterial plant pathogens infecting aerial plant parts can be detected early and identified precisely, based on their biological, biochemical, physiological, immunological, and genetic characteristics. Guidelines for naming and describing bacterial species are described in *Bergey's Manual of Systematic Bacteriology* (Kreig and Holt 1984). The names proposed for newly discovered species should be in accordance with the International Code of Nomenclature of Bacteria. Variations in morphological characteristics of bacteria are insufficient for differentiating different species. Hence, physiological and biochemical properties of bacteria are taken into consideration for classifying them into class, family, and genus. The International Society of Plant Pathology established criteria for applying the term 'pathovar' for those bacterial pathogens that did not satisfy criteria for species designation. The nomenclature 'pathovar' is applied at infra-specific level of bacteria distinguished primarily based on difference in pathogenicity on a set of plant species/cultivars used as differentials. The predominantly occurring strain is known as type-strain, whereas other strains of the same species may vary from the type-strain to varying magnitude.

4.2.1 Morphological Characteristics of Bacterial Pathogens

All plant pathogenic bacterial species are rod-shaped, except the filamentous *Streptomyces* spp. Bacterial species are placed into two groups based on their ability to retain Gram stain (developed by Hans Christian Gram) as gram-positive and those that are unable to retain stain as gram-negative. Asexual reproduction by binary fission, is the most common and predominant method resulting in production of huge populations in a short period. The phenomenon of conjugation considered to be similar to sexual reproduction in eukaryotic organisms, involves side by side contact between two physiologically opposite cells. Three phenomena viz., mutation, transformation, and transduction, account for most variability in genetic characteristics observed in different isolates of bacterial species (Narayanasamy 2011, 2017).

DOI: 10.1201/9781003456827-4

4.2.2 Isolation-Dependent Methods

4.2.2.1 Culturing Bacteria on Artificial Media

Bacterial plant pathogens infecting aerial plant organs are isolated on appropriate nutrient media that favor their development. A general-purpose, nonselective, but differential medium containing tetrazolium chloride is helpful for distinguishing target pathogen from saprophytes (Schaad et al. 2002a). Nutrient broth and potato-peptone-glucose-agar (PPGA) medium may be useful for initial isolation for isolation of specific bacterial pathogens. A seed extraction buffer (SEB) was used along with antibiotic vancomycin to arrest development of saprophytic bacteria to isolate *Acidovorax avenae* subsp. *citrulli* (*Aac*) from seeds and other plant tissues (Zhao et al. 2009). The semiselective bacterial canker of tomato (BCT) medium was employed to isolate *Clavibacter michiganensis* subsp. *michiganensis* (*Cmm*). The BCT medium was efficient for isolation of 30 strains of *Cmm* associated with tomato seeds and symptomless plantlets to eliminate the pathogen from planting materials, while other bacterial species were inhibited to the extent of 98–100% (Ftayeh et al. 2011). Injection-infiltration bioassay and culturing on artificial media were evaluated for their efficacy in isolating *Xanthomonas citri* subsp. *citri* (*Xcc*) (citrus canker disease). Washates of lesions taken from fruit, leaves and shoots in a commercial orchard were tested. Lesions from shoots yielded higher concentration of *Xcc* cells than lesions from fruit or leaves. Culturing was found to be a reliable method to detect and quantify *Xcc*, compared with injection-infiltration assay (Bock et al. 2014). '*Candidatus* Liberibacter asiaticus' (CLas) associated with huanglongbing (HLB) disease of citrus is a phloem-limited α-proteobacterium, and it is primarily transmitted by Asian citrus psyllid (ACP) *Diphorina citri*. Liber A medium containing citrus vein extract and a growth factor of '*Candidatus* Liberibacter' was used to isolate three species '*Ca.* Liberibacter asiaticus', '*Ca.* Liberibacter africanus', and '*Ca.* Liberibacter americanus' (Sechler et al. 2009). Culturing CLas was difficult and transient cultures were produced. A new species of Liberibacter was axenically cultured from diseased mountain papaya and named '*Candidatus* Liberibacter crescens' strain BT-1 (Merfa et al. 2019).

4.2.2.2 Phage Typing

Bacteriophages infecting plant pathogenic bacterial species have been isolated from infected plant tissues, soil and irrigation water. Rice bacterial leaf blight (BLB) pathogen *Xanthomonas campestris* pv. *oryzae* (*Xco*) (= *X. oryzae* pv. *oryzae*) was susceptible to several phages. Sensitivity of the pathogen strains to phages did not exhibit any relationship to their virulence and immunological characteristics (Ou 1985). On the other hand, phage sensitivity of *X. campestris* pv. *malvacearum* (*Xcm*) was related to virulence of two races causing bacterial blight of cotton (Freigoun et al. 1994). The genome of OP_2 phage showed similarity to the phage Bccp1 which lysed *Burkholderia cepacia* infecting onion (Inoue et al. 2006). These two phages could lyse both *Burkholderia glumae* and *B. plantarii* infecting rice and suppressed disease development. Seed treatment with phage was effective in reducing seedborne infection by *B. glumae* and *B. plantarii* (Adachi et al. 2012). Lysotypes of bacterial species are differentiated, depending on their levels of sensitivity to bacteriophage. Citrus canker pathogen *Xanthomonas axonopodis* pv. *citri* (*Xac*), could be detected on overwintering canker lesions by employing the phage specific to *Xac*. Phage population assessed on symptomless citrus leaves was closely related to the disease incidence. Isolates of *Xac* were divided into two groups, lysotype I and lysotype II, based on their sensitivity to the phage CPK. Phage technique provided a distinct advantage over the enzyme-linked immunoassay (ELISA) test, due to its ability to detect and quantify only living pathogen cells. The phage technique is rapid, simple and inexpensive, as the lysis zones (plaques) could be counted after 18–24 h of incubation. The phage technique was applied practically in Korea for detection of *Xac* on Satsuma mandarin fruits meant for export (Myung et al. 2001, 2006). Bacteriophage isolates from naturally infected citrus canker lesions from diverse sources and locations in Florida and Argentina were characterized for host range. *Xanthomonas citri* pv. *citri* (*Xcc*)-infecting phages from Argentina had identical host ranges (group ARG). Phages from Florida were placed in two groups (FLA and FLB). Citrus-associated xanthomoads could be classified based on phage typing. Majority of *Xcc* strains from different countries were typical 'A' pathotype strains (Balogh et al. 2013).

4.2.3 Microscopy

Histological changes induced by bacterial pathogens were useful to identify the pathogens causing some diseases. Electron microscopic observations of plants infected by *Xylella fastidiosa* (*Xf*), inducing grapevine Pierce's disease (PD), citrus variegated chlorosis (CVC), and leaf scorch disease in plum, pear, mulberry, almond, and coffee, revealed presence of aggregates of bacteria attached to the walls of vessels by extracellular strands produced by the pathogen cells (Machado 2009). *Xf* inducing CVC caused severe symptoms in sweet orange plants, while lime and mandarin were resistant to infection. Differences in responses of these *Citrus* spp. were assessed using light microscope (Garcia et al. 2012).

4.2.4 Biochemical Methods

Bacterial plant pathogens are broadly classified into gram-positive and gram-negative bacteria. Coryneform bacteria including the genera, *Arthrobacter, Clavibacter, Curtobacterium,* and *Rhodococcus* are gram-positive, whereas those included in other genera are gram-negative. Other characteristics useful for identification of bacteria are presence of food reserve materials (volutin, fat, glycogen, and iogen); ability to reduce nitrates; production of hydrogen sulfide, ammonia, and indole; utilization of carbon and nitrogen compounds; starch hydrolysis; lipolytic activity; action on litmus milk; and gelatin liquefaction. In addition, tests for determining the activities of Kovoc's oxidases, Thornley's arginine dehydrolase, and tyrosine are also recommended for gathering information useful for identification of bacterial pathogens (Narayanasamy 2001).

4.2.4.1 Metabolic Fingerprinting

This technique is based on differential utilization of various carbon sources as a basis of for differentiating bacterial species/strains. The API system depending on visualization of growth of target bacterial species was applied to identify pathovars of *Pseudomonas syringae* (Gardan et al. 1984) and to distinguish pathotypes of *X. campestris* (Verniere et al. 1991). A system of identifying gram-negative bacteria on the basis of metabolic fingerprints was developed by Biolog, Inc., USA. The Biolog automated identification system is based on the differential utilization of selected carbon sources by different bacterial species/strain/isolates. Biolog's identification system was applied to identify American Type Culture Collection reference taxa (Klinger et al. 1992). Metabolic fingerprints of 148 strains of *X. axonopodis* pv. *citri (Xac)* from 24 countries associated with citrus bacterial canker disease (CBCD) were obtained by employing Biolog substrate utilization system (Verniere et al. 1993).

4.2.4.2 Isozyme Analysis

Isozyme profiles were used for identification of *Pseudomonas syringae* pv. *pisi (Pss)*. The patterns of enzyme esterase (EST) and superoxide dismutase (SOD) of *Pss* indicated that two EST zymotypes specific for *Pss* could be employed for identification of this pathovar. Zymotype1 corresponded to races 2, 3, 4, and 6, whereas zymotype 2 was correlated to races 1, 5, and 7 (Malandrin et al. 1996).

4.2.4.3 Fluorescent Markers for Bacterial Pathogens

The green fluorescent protein (GPF) assay is effective in detecting and monitoring distribution of bacterial pathogens in plant tissues. A bioluminescent strain of *Xanthomonas axonopodis* pv. *diffenbachiae (Xad)* (anthurium bacterial blight) was employed to detect latent infection using autoradiography technique (Fukui et al. 1998). The GFP assay was applied to detect the presence and track the movement of the pathogen in the inoculated anthurium plants. Detection of latent infection of anthurium by *Xad* using GFP assay might be useful for elimination of infected plants (Elibox and Umaharan 2007).

4.2.5 Physical Methods

4.2.5.1 Direct Colony Thin-Layer Chromatography

Thin-layer chromatography (TLC) was performed by applying bacterial colony (cell) suspension directly on a silica gel thin-layer chromatography plate. Chloroform-methanol was used as the solvent. Several species of *Clavibacter, Erwinia, Pseudomonas*, and *Xanthomonas* were identified using TLC procedure. TLC profiles were characteristic at the genus or species level and the results were highly reproducible. The profiles of pathovars of *Pseudomonas syringae* was identical. On the other hand, distinct diversity was exhibited in the profiles of *X. campestris* and *X. oryzae* (Matsuyama et al. 2009).

4.2.5.2 Polyacrylamide Gel Electrophoresis

The DNA isolated from target pathogen(s) are treated with restriction enzymes to generate fragments small enough to resolve by electrophoresis in agarose or acrylamide. Two major phenotypic groups (α and β) groups of *Xanthomonas campestris* pv. *vesicatoria (Xcv)* were identified using polyacrylamide gel electrophoresis (PAGE) and silver staining of sodium dodecylsulfate (SDS)-lysed bacterial cells (Bouzar et al. 1994). Pulsed-field gel electrophoresis (PFGE) can be used for electrophoretic separation of larger molecules in agar gels. Strains of *Erwinia amylovora* (apple fire blight disease) from Austria and Czechia could be differentiated as group Pt1, and strains from Eastern Europe and Mediterranean region as group Pt2 (Zhang et al. 1998). *Acidovorax avenae* subsp. *citrulli* (*Aac*) isolates from cucurbits inducing bacterial fruit blotch (BFB) were placed in two groups within *Aac* based on pathogenicity tests and DNA fingerprinting using PFGE analysis. Group II included strains associated more with watermelon, whereas group I isolates contained strains associated usually with non-watermelon cucurbits (Burdman et al. 2005).

4.2.5.3 Spectral Signatures of Infected Plant Materials

'Candidatus Liberibacter solanacearum' (Lso) causing zebra chip (ZC) disease in potato tubers, exhibited characteristic bands. Visible and near-infrared spectrometry and infrared thermal imaging were employed to differentiate ZC-affected tubers. Near infrared (NIR) and infrared (IR) thermal imaging failed to differentiate tubers based on Lso infection status after storage (Zhao et al. 2018).

4.2.6 Isolation-Independent Methods

4.2.6.1 Immunoassays for Bacterial Pathogens

Bacteria may have various immunodeterminants of capsular polysaccharide antigens, lipopolysaccharides (LPS) O and K antigens and murine lipoproteins, which are less characterized in the case of bacterial pathogens. Characterization of the inner and outer membrane fractions of *Xanthomonas campestris* pv. *campestris (Xcc)* showed that a 44-kDa peptide of the outer membrane of *Xcc* was species-specific (Dianese and Schaad 1986). Antiserum prepared against purified 125-, 100-, and 44-kDa antigenic determinant of membrane peptides could be employed to identify membrane proteins of *Xcc* by employing Ouchterlony double-diffusion tests (Thaveechai and Schaad 1986). Pathovar-specific MAbs were used in ELISA tests to identify *Xanthomonas oryzae* pv. *oryzae (Xoo)* (rice bacterial leaf blight, BLB) outbreak in the United States (Jones et al. 1989). MAbs raised against two German isolates of *Xanthomonas campestris* pv. *campestris (Xcc)* were employed to subdivide the *Xcc* isolates into three subgroups S1, S2, and S3 (Rabenstein et al. 1999). The antibodies raised against flagella from strains representing most pathovars of *Pseudomonas syringae* were used to differentiate flagella serotypes H1 or H2 (Malandrin et al. 1996). Many *Pseudomonas* strains elaborate phytotoxic lipodepsipeptides such as syringomycin, syringopeptins and related compounds. The MAbs generated against the toxin coronatine (COR) produced by *P. syringae* pv. *glycinea* infecting soybean were employed to detect the presence of toxin and quantitatively estimate its concentration (Jones et al. 2001).

4.2.6.1.1 Enzyme-Linked Immunosorbent Assay

Polyclonal antibodies (PAbs) and monoclonal antibodies (MAbs) have been widely employed for detection, identification, and quantification of bacterial pathogens using different formats of ELISA. The ELISA procedure is amenable for automation and effective in providing reproducible results in large-scale testing programs. Commercial detection kits could detect *Clavibacter michiganensis* subsp. *michiganensis* and *Xanthomonas axonopodis* pv. *vesicatoria* simultaneously in plant samples by ELISA test (Alvarez 2004). *Xanthomonas oryzae* pv. *oryzae (Xoo)* (rice bacterial blight) could be detected in leaves and seeds of infected plants. The MAB (Xco-1) specific for *Xoo* reacted positively with all 178 strains from diverse geographical locations in modified ELISA test, but not with *X. oryzae* pv. *oryzicola (Xoc)* (rice bacterial leaf streak, BLS) or other xanthomonads. The strains of *Xoo* were classified into four groups, I, II, III, and IV on the basis of reactions with 4 MAbs (Benedict et al. 1989). *Xoo* strains were classified into nine reaction types consisting of four serovars and seven subserovars, based on the serological reactivity with MAbs and PAbs (Huang et al. 1993). ELISA test was applied to detect *Xanthomonas campestris* pv. *citri (Xcc)* (citrus canker) in fruits). *Xac* could be more efficiently detected by ELISA than bacteriophage test (BPT) (Jin et al. 2001). The strains of *Xac* and *Xac-A*^{w} (Wellington strain) *Xac-A* (Asiatic strain) were differentiated reliably by the ELISA test (Sun et al. 2004). *X. campestris* pv. *vesicatoria (Xcv)* (pepper bacterial spot) could be detected in artificially inoculated and field-infected plants (Tsuchiya and d'Ursel 2004). The indirect ELISA format using PAb provided reliable detection and unambiguous differentiation of eight strains of *Xanthomonas axonopodis* pv. *vesicatoria* (*Xcv*). The ELISA assay specifically detected *X. campestris* pv. *malvacearum* infecting cotton (Mitrev and Kovacĕvic 2006). The plate-trapped antigen (PTA)–ELISA format was applied using PAbs to detect *Erwinia amylovora* (fire blight disease of apple) in plant materials (Kokošková et al. 2007).

Xanthomonas fastidiosa (Xf), causing almond leaf scorch, could be detected in asymptomatic almond trees by ELISA (Groves et al. 2005). The ELISA test detected *Xf* in xylem fluids more efficiently than in whole tissue samples from asymptomatic plants. The ELISA method was as efficient as the PCR assay (Bextine and Miller 2004). The ELISA test was employed to determine the extent of incidence of grapevine Pierce's disease (PD) within American hybrid vineyards. In addition, the presence of *Xf* in six additional plant species was detected using the ELISA test (Buzomobo et al. 2006). Transmission of *X. fastidiosa* causal agent of grapevine PD by smoke sharp shooter nymphs and apterae Cicada adults, was confirmed by ELISA tests (Krell et al. 2007). Both acropetal and basipetal movement of *Xf* in infected grapevine plants was revealed by the ELISA test. More acropetal movement was detected earlier than basipetal movement (Krell et al. 2008). *Acidovorax citrulli (A. avenae* subsp. *citrulli*) infecting watermelon and other cucurbits was detected using MAb11E5 in the ELISA test. The MAb11E5 efficiently detected a strain of *A. citrulli* that could not be detected by commercially available ELISA-based kit and commercial immunographic strip test. MAb11E5 reacted with an *A. citrulli* protein of >170 kDa. The MAb-captured-sandwich ELISA (Mc-sELISA) was effective also in detecting *A. citrulli* in naturally infected cucurbit leaves, fruit and seed, and it showed potential for use in routine field inspection (Himananto et al. 2011). The antibodies produced against Omp protein of '*Candidatus* Liberibacter asiaticus' (Las) positively reacted with Las in the ELISA test in infected citrus samples. The antibodies produced against Omp protein was effective in tracking the incidence and monitoring distribution of Las causing citrus greening disease (Lu et al. 2013).

4.2.6.1.2 Dot Immunobinding Assay

The antigen is immobilized on a nitrocellulose or nylon-based membrane in the dot immunobinding assay (DIBA), which is similar to the ELISA test where microtiter plates are used in place of membrane. Immunoassay based on the use of *Spiroplasma citri,* causing citrus stubborn disease (CSD), secreting protein as a detection marker was applied. Using mass spectrometry analysis, a unique secreted protein from *S. citri* was identified and ScCCPP1, an antibody generated against this protein, distinguished *S. citri*-infected citrus and periwinkle from healthy plants. Further, the antiserum could be used to detect CSD using direct tissue print assay under field conditions (Shi et al. 2014).

4.2.6.1.3 Immunofluorescence Test

Fluorescence emitted from the labeled antibodies reacting with epitopes present in the antigen or antigenic sites are detected by the immunofluorescence test (IF). PAb conjugated with an antirabbit fluorescein isothiocyanate (FITC) stain in immunofluorescence antibody staining (IFAS) technique detected *Erwinia amylovora* (apple fire blight disease) very efficiently even at low concentrations of the pathogen in plant samples (Kokošková et al. 2007) *Xanthomonas campestris* pv. *campestris (Xcc*), (cabbage black rot) was chromosomally tagged with *gfp* gene encoding green fluorescent protein (GFP). The *gfp* labeled strain enabled visualization of the process of infection of cabbage by *Xcc* in early stages of disease development (So et al. 2002). *X. campestris* pv. *vesicatoria (Xcv)* and *Clavibacter michiganensis* subsp. *michiganensis (Cmm)* were detected by immunofluorescence (IF) technique in tomato seeds. In contrast, immunofluorescence colony staining (IFC) method could not detect *Xcv* in seeds (Németh and van Vuurde 2006). *Xylella fastidiosa (Xf)* (grapevine Pierce's disease) was detected by performing immunofluorescence (IF) methods. Fluorescence microscopy was effective for detection and identification of *Xf* in thin cross-sections of the petioles from infected plants at 3 weeks after inoculation. Further, observations under fluorescence microscopy could be made at different periods of time after inoculation to monitor pathogen multiplication and invasion by *Xf* of xylem vessels (Buzkan et al. 2005). Indirect immunofluorescence (IF) microscopy procedure was adopted to detect *Xf* in American hybrid vineyards in the Texas Gulf Coast. *X. fastidiosa* was detected by indirect IF method in grapevine (Buzomobo et al. 2006). Rapid immunoassay strip test involving the use of fluorescein isothiocyanate (FITC) was developed for detection of *Acidovorax citrulli* (Ac) (=*A. avenae* subsp. *citrulli*), infecting watermelon and other cucurbits. IFS test detected strains of *Ac* in crude extract from field-infected plants. IFS reduced detection time significantly and simplified protocols compared with ELISA and PCR procedures (Zeng et al. 2018).

4.2.6.1.4 Immunolabeling and Electron Microscopy

Visualization of immunological reactions on grids is possible by applying immunolabeling and electron microscopy techniques. The procedures are also useful to observe localization and distribution of various proteins of bacterial pathogen/host plant origin in different pathosystems. Enzyme-labeled antibodies are applied after embedding and thin sectioning of plant tissues, or they may be allowed to diffuse inside the fixed cells and to interact with antigenic sites prior to thin sectioning. Vertical transmission of '*Candidatus* Liberibacter asiaticus', the causal agent of citrus huanglongbing (known earlier as citrus greening), to seedlings was investigated. Transmission electron microscopy (TEM) was employed to detect pathogen cells in phloem sieve elements of seed coat vascular bundle. Pleomorphic, round and elongated bacillus-like cells characteristic of '*Ca.* L. asiaticus' were seen. In addition, many round bacterial cells, sometimes in large numbers were also present. The bacterial DNA was most frequently detected in the seed coat. Electron microscopic observations also indicated the presence of intact pathogen cells in the sieve elements in the seed coat vascular bundle (Figure 4.1). The results indicated that the seed coats could be used as a source of relatively pure pathogen cells for culturing (Hilf et al. 2013). Polyclonal antiserum was raised against bacterial effector hopz5 protein of *Pseudomonas syringae* pv. *actinidiae (Psa)*, causing kiwifruit canker. The PAb hop35 detected *Psa* in culture and infected plant samples from field also (Chen et al. 2018).

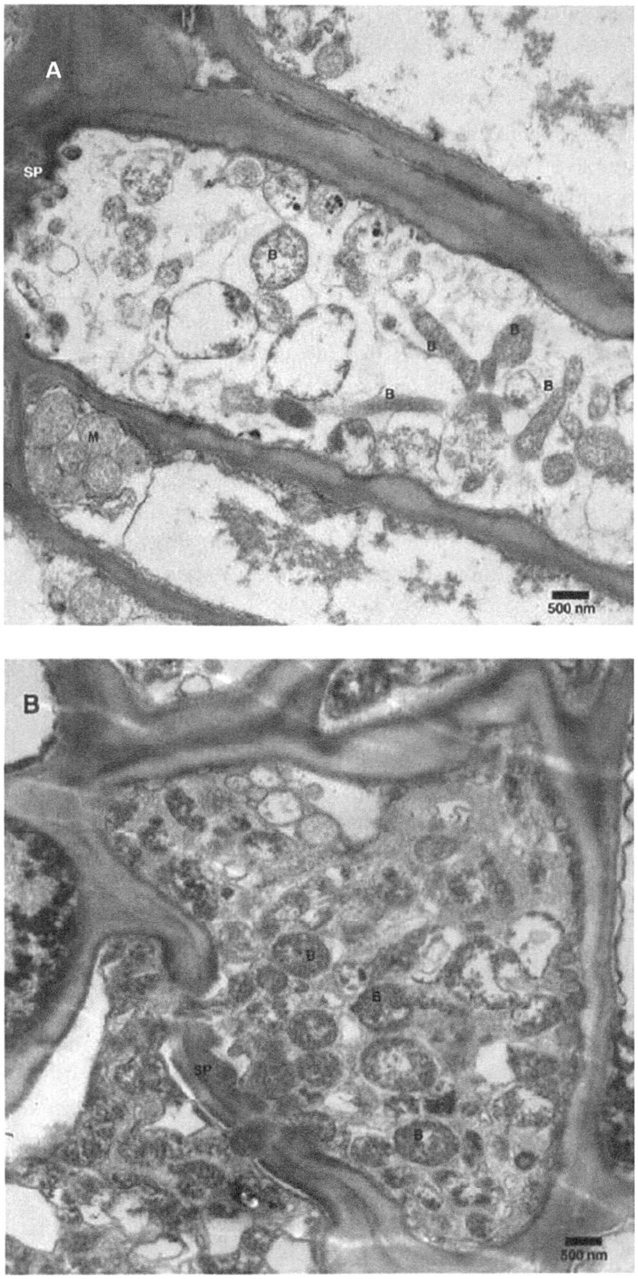

FIGURE 4.1 A. Visualization of pleomorphic cells of '*Candidatus* Liberibacter asiaticus' (Las) in sieve elements of seed coat of grapefruit cv. Connors. B. Presence of bacterial cells in the cross section of a sieve element in the grapefruit seed coat infected by Las. B-bacterial cells; SP- sieve plate; M-mitochondria.

Source: (Courtesy of Hilf et al. 2013 and with kind permission of the American Phytopathological Society, MN, USA)

4.2.6.1.5 Immunomagnetic Separation of Bacterial Pathogens

Immunomagnetic separation (IMS) involves coating of immunomagnetic beads (IMBs) with antibodies specific to target bacterial pathogen followed by selective trapping of the pathogen cells from the samples. IMS procedure was used for the detection of *Acidovorax avenae* subsp. *citrulli*

(Aac), causing bacterial fruit blotch (BFB) in watermelon (Walcott and Gitaitis 2000).

4.2.6.1.6 Flow Cytometry

Detection and identification of bacterial pathogens could be achieved by flow cytometry (FCM), as they move through a sensor in a liquid system. Fluorescent antibodies conjugated to specific antibodies are useful for identification of bacterial pathogens. Fluorescent probes calcein AM, cFDA and PI were combined with the FCM procedure to assess the viability of cells of *Clavibacter michiganensis* subsp. *michiganensis* (*Cmm*) (tomato canker disease). The ratio of viable/nonviable cells of *Cmm* was used as the basis of assessment to determine pathogen population capable of spreading the disease. Heat-killed (nonviable) and viable (untreated) *Cmm* cells labeled with fluorescent probes could be differentiated based on the intensity of fluorescence in FCM analyses. Viable cells of *Cmm* showed relatively high green fluorescence levels due to staining with either calcein AM or cFDA. In contrast, *Cmm* cells damaged by heat treatment showed high red fluorescence levels due to staining with PI. FCM analysis of calcein AM stained cells was superior to the conventional plate count, as the FCM required short assay time (Chitaara et al. 2006).

4.2.6.2 Nucleic Acid–Based Techniques

Among the diagnostic techniques applied for detection, identification, differentiation, and quantification of bacterial pathogens, nucleic acid–based techniques are more sensitive, reliable, rapid, and precise. The extent of relatedness between bacterial pathogens is determined based on nucleotide sequence similarity. Species/subspecies/pathovars/strains share greater nucleotide sequence similarity than those that are distantly related. A highly specific nucleotide sequence in the DNA of target bacterial species can be identified. Probes or primers based on the sequences of nucleotides present in the DNA fragment may be designed and used in various NA-based techniques. The PCR assay or its variants are the most widely employed technique for detection and identification of bacterial pathogens at species/pathovar/subspecies levels. However, NA-based techniques require use of hazardous chemicals and expensive equipment and are unable to differentiate live and dead bacterial cells. Nevertheless, these techniques particularly PCR assays either alone or in combination with other techniques are extensively applied in various investigations. With the development of probes labeled with nonradioactive digoxigenin, NA-based techniques are preferred, though they are relatively more expensive (Narayanasamy 2011).

Detection of plant pathogenic bacteria depends on accurate design of oligonucleotides and probes. Conserved regions for each target sequence(s) may be identified using the Basic Local Alignment Search Tool (BLAST). The PCR assay and its variants have the ability to detect a single/multiple targets in complex mixtures and potential to detect some unculturable bacterial species and phytoplasmas. The earlier method of visualizing the amplified product through agarose gel electrophoresis with ethidium bromide (EtBr) staining has been replaced by less toxic and more sensitive SYBR Green detection under UV irradiation. Reliability and specificity of PCR assay are improved by employing dye-quenched probes (Narayanasamy 2011).

4.2.6.2.1 Nucleic Acid Hybridization Methods

Specific DNA probes have been used to rapidly detect, identify, and quantify bacterial pathogens belonging to different species, subspecies, pathovars or strains, infecting various crops. The DNA sequences of 16S rRNA of different bacterial species are compared for selecting probes specific at genus/species level. A DNA probe for comparing partial sequences of 16S rRNA from 52 strains of bacteria, including *Xanthomonas oryzae* pv. *oryzae (Xoo)* (rice bacterial leaf blight, BLB) was developed. DNA sequence of the 16S rRNA hybridizing only with plant pathogenic *Xoo* pathovar could be identified (De Parasis and Roth 1990). *X. oyzae* pv. *oryzicola*, (rice bacterial leaf streak, BLS) was also detected by nucleic acid hybridization technique (Cottyn et al. 1994). *Xanthomonas axonopodis* pv. *vesicatoria* (bacterial leaf spot disease in tomato and pepper) was detected by employing a 1.75-kb fragment that preferentially hybridized to the pathogen isolates infecting both hosts (Kuflu and Cuppels 1999). Probes based on the sequences of the 16S rRNA gene effectively detected and quantified cells of *Xylella fastidiosa* in infected citrus plants (Rodrigues et al. 2006). Citrus greening (huanglongbing disease) pathogen '*Candidatus* Liberibacter asiaticus' was detected in mandarins, tangerines, oranges, and pummelos by employing pathogen-specific DNA fragment (Hung et al. 1999). Seed-to-seedling transmission of '*Ca.* L. asiaticus' (Las) was confirmed by applying fluorescence in situ hybridization (FISH) technique. Probes complementary to the Las 16S rRNA were employed to detect the pathogen in the vascular tissues of seed coats of grapefruit and pummelo and also in fragmented vascular bundles excised from grapefruit seed coats (Hilf et al. 2013). In a later investigation, in situ tissue print (ISTP) method was employed for detection of '*Ca.* L. asiaticus' (CLas) in sweet orange infected by HLB disease. The corresponding outer membrane genes of representative isolates were cloned and sequenced. The procedure efficiently differentiated both CLas and '*Ca.* L. africanus (CLaf) in phloem cells of different citrus tissues (Ding et al. 2020).

4.2.6.2.2 Restriction Fragment Length Polymorphism Technique

Restriction endonucleases cleave the pathogen DNA at specific sites and generate banding patterns unique to the target bacterial pathogen, which can be visualized after gel electrophoresis and it is useful for detection, identification and differentiation of some bacterial pathogens and their strains. The RFLP technique was applied for detection and differentiation of genomic DNAs of strains of *Xanthomonas axonopodis* pv. *citri (Xac)*, *X. axonopodis* isolated from Florida citrus nurseries and five pathovars of *X. axonopodis*.

Differentiation of *Xac* and *Xa* strains was achieved, indicating that the *Xa* strains occurring in citrus nurseries were not a form of strain of *Xac* (Hartung and Civerolo 1989). The cosmid clone PXCF 13–38 was employed as a probe for RFLP analysis of xanthomonads. Based on RFLP analysis, a dendrogram different from the ones based on biochemical properties was generated and it was useful for identification of various xanthomonads (Kanamori et al. 1999).

The genetic diversity of *X. oryzae* pv. *oryzae* strains present in 18 locations in India was investigated. Strains of *Xoo* belonging to a single lineage was distributed in most locations sampled. All strains with this linkage belonged to the pathotype 1b, as indicated by pathotyping analysis (Yashitola et al. 1997). The population structure of *Xoo* from the principal rice-cultivating region in China was studied, using RFLP and virulence analyses. Strains of *Xoo* were classified into 61 pathotypes which formed three clusters. Each distinct RFLP banding pattern of each strain was considered a haplotype. Forty-two haplotypes revealed by the probe were grouped into four lineages (Li et al. 2009).

4.2.6.2.3 Polymerase Chain Reaction–Based Assays

PCR is performed to amplify a target DNA sequence to reach sufficient concentrations, facilitating detection of the target pathogen efficiently and reliably. It is possible to use the PCR amplicons as targets for hybridization or for direct sequencing of DNA to determine variations in strain or as specific probes. PCR-based assays provide certain advantages over the other diagnostic methods. The purified or extract of the DNA is enough for performing PCR assay without the requirement of axenic culture of the pathogen to be detected. Using appropriate primers, single or multiple pathogens can be detected from the same sample. Another advantage of employing PCR assays is the possibility of detecting the target pathogen in samples contaminated with saprophytic bacteria. The primer pairs based on sequences of genes encoding virulence, enzyme or toxin production may be employed to detect bacterial pathogens infecting various crops. The primer pair amplifying a 222-bp DNA fragment of *Xanthomonas axonopodis* pv. *citri* (*Xac*) was used for diagnosing field outbreaks of citrus canker disease (Gillings et al. 1995). Sequence analysis of the 16S and 23S ribosomal genes indicated that *Xanthomonas*-like bacteria belonged to *X. campestris* subgroup and they were found to be common epiphytes on citrus leaf surfaces (Taylor et al. 2002). A rapid and sensitive PCR format was developed for detection of *Erwinia amylovora* (fire blight disease of apple). The assay was based on the amplification of 187–bp DNA fragment and could detect *E. amylovora* in apple flowers before symptom expression indicated its potential for use in the presymptomatic management of apple fire blight disease (Taylor et al. 2001). PCR assay using the RED Extract-Amp Plant PCR Kit for amplification of the *Pst*I fragment of the common *Erwinia amylovora (Ea)* plasmid pEA29 could detect *Ea* in greater number of asymptomatic plants than the standard PCR format (Stoger et al. 2006). The *Ea* DNA extracted using DNeasy Plant Mini Kit (Qiagen) from cultures and plant tissues was more specific and efficient than the PTA-ELISA or indirect fluorescent antibody staining methods in detecting the apple fire blight pathogen (Kokošková et al. 2007). The genetic variability within Spanish populations of *Pseudomonas syringae* pv. *phaseolicola* (*Psp*) and *P. syringae* pv. *glycinea (Psg)* (isolates infecting soybean) were assessed. The Tox$^+$ and Tox– isolates showed genomic divergence that included in ERIC-PCR and arbitrarily primed-profiles. The Tox$^+$ isolates showed distinct patterns of IS801 genomic insertions that contained a chromosomal IS801 insertion that was absent in Tox– isolates. The Tox– isolates also lacked the characteristic plasmid that contains a pathogenicity island. The results provided the basis for separation of the Tox$^+$ and Tox– isolates of *P. syringae* pv. *phaseolicola* into two genetic lineages, called Pph1 and Pph2, respectively (Oguiza et al. 2004). The strain of tomato bacterial speck pathogen *Pseudomonas syringae* pv. *tomato* NYS-T1 responsible for the outbreak was highly virulent on tomato. Primers developed for *avrA* and *hopW* were diagnostic for more virulent strains of *Pst*, whereas primers for *hopN* were specific for *P. syringae* pv. *tomato* DC 3000 and related *P. syringae* pv. *maulicola* strains. Primers designed against *hopR* distinguished both subgroups, high virulence T1-like subgroups and low virulence DC 3000/*P. syringae* pv. *maulicola* subgroup within *P. syringae* phylogenetic Group I (Jones et al. 2015).

The PCR-based method was developed for the detection of *Xanthomonas campestris* pv. *campestris (Xcc)* infecting cruciferous crops, employing primers designed to specifically amplify a 619-bp fragment of the *hrpF* gene from *X. campestris*. The primer pair DLH120 and DLH125 from the *hrpF* gene from *X. campestris* (*Xc*) amplified at the 3′ end of the *hrpF* gene and not from other *Xanthomonas* spp., indicating the primer specificity. This format consistently detected *Xcc* in infected plants, and it was more sensitive and specific than the conventional isolation methods or a commercially available ELISA format (Berg et al. 2005). The primer pair prepared from the nucleotide sequences of the *rhs* family gene was effective in detecting and identifying *X. campestris* pv. *vesicatoria (Xcv)* (bacterial leaf spot of pepper and tomato) by PCR assay. The 20-mer oligonucleotide XcvF and XcvR amplified the expected 517-bp fragment from all six strains of *Xcv* tested (Park et al. 2009). A triplex PCR assay was developed for rapid and reliable detection of *Xanthomonas oryzae* pv. *oryzae (Xoo)* employing three primer sets targeting internal regions of two genes (*hpaA*) and *Xor11*. It was possible to detect *Xoo* in asymptomatic rice (Song et al. 2012). Pathovars of *Xoo* and *X. oryzae* pv. *oryzicola (Xoc)* (rice BLS) could be differentiated by applying the SYBR Green real-time PCR assay to target a putative membrane protein gene of *X. oryzae* pv. *oryzicola* strains. A single band of the expected size was amplified by the primer set only from *Xoc*, but not from *Xoo* strains (Kang et al. 2012).

PCR assay detected *Xylella fastidiosa (Xf)* in citrus plants infected by citrus variegated chlorosis (CVC) disease in all main sweet orange fruit vascular bundles and seeds, which remained asymptomatic (Li et al. 2003). PCR-based procedure detected *Xf* strains in grapevine, almond and oleander using different sets of primers. The assay performed with primers XF1968-L and XF1968-R amplified a fragment of 638-bp from OLS strain infecting oleander, whereas PCR with primers XF2542-L/XF2542-R amplified a specific product from PD strains causing Pierce's disease in grapevine. The PCR primers ALM1 and ALM2 specifically amplified a fragment of 521-bp from ALS strains infecting almond. (Hernandez-Martinez et al. 2006). The primer set C from HLB pathogen amplifying a PCR product of 451-bp and Klen Taq polymerase were effective for detection of the pathogen in citrus in India (Gouda et al. 2006). Specific prokaryote PCR primers in combination with universal 16S rRNA primers were employed. Sequence analysis of the 16S rRNA gene, the 16S/23S rRNA spacer region and the rplKAJL-rpoBC operon revealed that the bacterium shared a strong identity with '*Candidatus* Liberibacter' species. The new bacterial pathogen infecting tomato and pepper in New Zealand was called '*Candidatus* Liberibacter solanacearum' (Liefting et al. 2009).

4.2.6.2.3.1 Targeting Specific Genes of Bacterial Pathogens Primers designed on the sequences of specific gene(s) involved in pathogenesis like adhesion, production of phytotoxins, resistance to oxidative stress, plant cell degradation, secretory systems and interference, and/r suppression of host defenses have been employed for detection of bacterial pathogens. *Pseudomonas syringae* pathovars possess a large transmissible plasmid containing coronatine (COR)-producing genes. The sequences of the gene cluster have been used for designing primers for their amplification by PCR assays which detected only strains of *P. syringae* possessing COR-producing genes (Bereswill et al. 1994). *P. syringae* pv. *maulicola* infecting leaf crucifers was detected by employing primers to amplify *cfl* gene sequences from the gene cluster encoding coronatine toxin (Zhao et al. 2002). PCR assay for detection of pathovars of *P. syringae(Ps)*-producing toxins have been applied based on the sequences of the *tblA* and *tabA* genes involved in the biosynthesis of tabtoxin. The product of 829-bp was amplified with *tblA* primers in reactions that contained cells from tabtoxin- producing strains of *P. syringae* pv. *tabaci* and *P. syringae* BR2R. On the other hand, the product of 1,020-bp was amplified in reactions that contained cells from all the tabtoxin-producing strains of *P. syringae* tested (Lydons and Patterson 2001). The specificity of *Xylella fastidiosa (Xf)* 16S rRNA and DNA gyrase (*gyrB*)-targeted primers was confirmed by amplification of specific products from *Xf* DNA only (Rodrigues et al. 2003). Primers based on DNA sequences related to *hrp* genes were employed for specific amplification of DNA of *X. campestris* pv. *vesicatoria (Xcv)* (Leite et al. 1994). The primer pair DLH120 and DLH125 from the *hrp* gene from *X. campestris (Xc)* amplified the 3′ end of the *hrpF* gene. Primers amplified a 619-bp fragment of the *hrpF* gene only from *Xc*, indicating the specificity of the primer pair. By including primers targeting a 360-bp section of the ITS region from *Brassica* spp. in a multiplex PCR, *X. campestris* pv. *campestris* was detected in seed lots (Berg et al. 2005). PCR assay using primers XCF and XCR designed based on the sequences of *hrpF* gene of *Xanthomonas campestris* pv. *campestris (Xcc)*, amplified a specific fragment of 525-bp from cabbage plants infected by *Xcc* (Park et al. 2004). The primers XACF and XACR derived from the sequences of *hrp*^{w} of citrus canker pathogen *X. axonopodis* pv. *citri (Xac)*, amplified a 561-bp only from leaves and fruits infected by *Xac*. The PCR format was effective in detecting *Xac* in naturally infected or artificially inoculated citrus leaves (Park et al. 2006).

Sequences of the *hrp* region of *Xanthomonas campestris* pv. *vesicatoria (Xcv)* infecting pepper, were used for designing primer pair RST2/RST3 for detection of *Xcv*. The primer pair amplified a unique fragment of 840-bp from the genomic DNA of the strains of *Xcv*. All strains infecting pepper produced the expected amplicon of 840-bp, indicating the reliability of the assay for detection and identification of *Xcv* strains pathogenic to pepper (Mitrev and Kovac̆evic 2006). *X. campestris* pv. *vesicatoria (Xcv)* and *X. axonopodis* pv. *vesicatoria (Xav)* induced indistinguishable bacterial spot diseases in pepper and tomato. The 20-mer oligonucleotide primer pair XCVF and XCVR based on the sequences of *rhs* family gene homolog in *Xanthomonas* strains was confirmed by dot blot analysis. *Xcv* strains exhibited high hybridization signal to the probe, whereas other *Xanthomonas* strains showed negative results in the dot blot analysis. The primer pair XCVF and XCVR amplified a 517-bp fragment from all six strains of *Xcv*, indicating the specificity of the PCR assay using XCVF and XCVR primer pair for detection and identification of *Xcv* (Park et al. 2009). Primers based on the partial sequence of *hrpB* gene of *X. campestris* pv. *viticola (Xcv)* (grapevine bacterial canker) were useful for specific detection of pathogen detection. Primer pairs Xcv1F/Xcv3R and RST2/Xcv3R were specific to amplify 243-bp and 340-bp fragments respectively from DNA of *Xcv* strains and also from the DNAs of *X. campestris* pv. *mangiferae-indicae* and *X. axonopodis*. PCR amplicons showed distinct differences from *Xcv* strains (Trindadae et al. 2007). Primers G1 and L1 employed for amplification of the 16S – 23S rRNA intergenic spacer region of *Xylella fastidiosa (Xf)* (almond leaf scorch disease, ALSD and grapevine Pierce's disease, PD) were useful for detection of the pathogen in many host plant species (Costa et al. 2004). Strains of ALSD collected from two locations in California were identified as two genotypically distinct types of *Xf* strains. The single nucleotide polymorphism (SNP)–based PCR format was applied to identify and differentiate *Xf* strains collected from ALSD-affected almond. When genotype-specific SNPs were employed to design primers for multiplex PCR assays of early passage of cultures, two genetically distinct types of *Xf* strains were identified. G-type and A-type strains were detected in the same almond orchards. RFLP analysis of different genetic loci RST31 – RST33, confirmed the results of multiplex PCR assays (Chen et al. 2005).

'*Candidatus* Liberibacter' (citrus greening/huanglongbing [HLB]) disease was detected using primers OI1 and OI2c designed based on 16S rDNA sequences in the PCR (Colletto-Filho et al. 2004). Primers LSg2f/LSg2r developed using the consensus sequences of 16S rDNA fragments were able to detect '*Ca.* Liberibacter' in samples that did not amplify with primers OI1 and OI2c employed earlier. In addition, the Asian '*Ca.* Liberibacter' was also detected (Colletto-Filho et al. 2005). A signature oligonucleotide sequence (OI1) in the 16S RNA gene (*rrs* with three genomic copies) was identified in the genome of '*Candidatus* Liberibacter asiaticus' (CLas). OI1 contains single nucleotide polymorphism (SNPs) differentiating CLas from non-CLas species. The primer HLBas was designed on the SNPs for CLas detection using TaqMan PCR system (HLBas_PCR). CLas whole genome sequence strains (WGSs) were identified by employing next-generation sequencing (NGS) technology. The sequence integrity of OI1/HLBas was reevaluated using WGSs information and a missing nucleotide G in the two primers were confirmed. A 4CP-PCR system based on a four-copy genomic locus was developed to enhance sensitivity of detection of CLas at low concentrations. The 4CP-PCR assay was more sensitive in detecting HLB in citrus and psyllid vector samples than HLBas-PCR format, and it exhibited similar sensitivity as the RNR-PCR assay (Bao et al. 2020).

Prophages are important genetic elements involved in lateral gene transfer, pathogenicity, environmental adaptations, and interstrain genetic variability. Sequence of a prophage terminase gene of '*Candidatus* Liberibacter asiaticus' (citrus HLB) was employed as a molecular biomarker. The frequency of prophage terminase gene was 15.8%, and it differed depending on locations/altitude. Partial prophage terminase gene sequences of strains were identical or highly similar (Liu et al. 2011). The prophage/phage region of '*Ca.* Liberibacter asiaticus' contained a mosaic genomic region exhibiting inter- and intra-continental variations in '*Ca.* L. asiaticus'. Pathogen isolates (267) present in China were analyzed with a primer set flanking the mosaic region plus downstream sequence. Most amplicons had sizes ranging from 1,400 to 2,000-bp. The amplicon of 550-bp (S550) was detected in 14 samples from south western China. The unique easy-to-detect S550 amplicon showed potential to serve as a molecular marker for detection and assessment of variations in the populations of '*Ca.* L. asiaticus' (Wang et al. 2015b).

4.2.6.2.3.2 Variants of the PCR Assay The repetitive sequence-based (rep)–PCR genome fingerprinting technique is more frequently employed for detection of bacterial pathogens. Rep-PCR is based on PCR-mediated amplification of DNA sequences located between specific interspersed repeated sequences in prokaryotic genomes. These repeated sequences are called REP (repetitive extragenic palindromic sequences), BOX (DNA sequences of the BOX A subunit of the Box element of *Streptococcus pneumoniae*), and ERIC (enterobacterial repetitive intergenic consensus) elements. Isolates of *Xanthomonas campestris* (Brassica leaf spot) and closely related pathovars could be detected and differentiated by employing rep-DNA PCR-based fingerprinting procedure. The leaf isolates were clustered separately from *X. campestris* pv. *campestris* isolates. Hence, it was proposed that *X. campestris* isolates causing nonvascular leaf spot disease of *Brassica* spp. should be called *X. campestris* pv. *raphani* and not as pv. *armoraciae* (Vicente et al. 2006). ERIC- and rep-PCR assays were effective in differentiating the isolates *Acidovorax avenae* subsp. *citrulli* (*Aac*) (watermelon BFB). Isolates from melon (M) and watermelon (W) were classified into group I and II, respectively (Burdman et al. 2005). The rep-PCR assay with ERIC and BOX primer sets was applied to assess phylogenetic diversity of *Burkholderia glumae* (rice bacterial seedling rot and grain rot disease). Ten strains from Japan, in which GyrA83 was isoleucine (Ile), were classified into two groups based on band patterns in rep-PCR analysis. However, the nucleotide sequences of *gyrB* and *rpoD* were identical among strains (Maeda et al. 2007). Genetic analysis of *X. campestris* pv. *phaseoli* (*Xcp*) and *X. campestris* pv. *phaseoli* var. *fuscans* (bean common bacterial blight) was performed by rep-PCR and random amplified polymorphic DNA (RAPD) methods. Relatively little genetic diversity was observed among spinach strains of *Xcp* and these strains were grouped with the New World strains into a large cluster. Strains of *X. campestris* pv. *phaseolicola* var. *fuscans* were more closely related and genetically distinct from *Xcp* strains (Lopez et al. 2006). Rep-PCR with ERIC and BOX primer sets was applied to analyze global genetic diversity of *X. campestris* pv. *malvacearum* (*Xcm*) (angular leaf spot, black arm and boll rot of cotton) as well as highly virulent strains (HVSs). The strains with lower moderate virulence clustered in one genogroup. All HVSs collected from Africa were placed in another genogroup. The results suggested that the virulence of *X. campestris* pv. *malvacearum* might be highly related to the genotype and/or geoclimatic origin of strains (Zhai et al. 2010).

A nested PCR procedure using two oligonucleotide primers was designed based on the unique fragment of the plasmid pEA29 of *Erwinia amylovora* (*Ea*) (apple fire blight disease). The nested PCR was more sensitive than conventional PCR format. The pathogen could be detected in different organs of infected trees such as leaves, axillary buds and mature fruit calyx samples (McManus and Jones 1996). *Xanthomonas axonopodis* pv. *diffenbachiae* (*Xad*) was detected by nested PCR format using a sequence characterized amplified region (SCAR) marker. One of the SCAR markers was used for detection of *Xad* by nested PCR assay in pure cultures and anthurium tissue extracts. The internal primer pair detected the expected products of 785-bp for all 70 strains of *Xad* pathogenic to anthurium and for isolates originating from syngonium and not pathogenic to anthurium. There was a high level of relatedness between strains from anthurium and syngonium. The nested PCR protocol could be used for screening propagation stock plant materials and monitoring international movement of

X. axonopodis pv. *diffenbachiae* (Roéne-Soustrade et al. 2006). Conventional and nested PCR formats detected *Pseudomonas syringae* pv. *actinidiae (Psa)*, (kiwifruit bacterial canker disease) at very low concentrations from plant and pollen extracts. RFLP analysis with *Bcl*I on PCR and nested PCR amplicons confirmed the specificity of the assays. The nested PCR format detected *Psa* in more numbers of samples, than the conventional PCR assay, indicating the greater sensitivity of nested PCR assay for detection of *P. syringae* pv. *actinidiae* in kiwifruit orchards (Biondi et al. 2013).

Competitive PCR assay with potential to eliminate the risk of false negatives, has been applied for detection of bacterial pathogens. '*Candidatus* Liberibacter (CL)' (citrus HLB; exists as Asian, African, and American forms which are named '*Ca.* L. asiaticus' [Las], '*Ca.* L. africanus' [Laf], and '*Ca.* L. americanus' [Lam], respectively). The competitive PCR format was developed for detection and quantification of the pathogen in infected citrus tissues. Competitive DNA was generated from a DNA fragment amplified by rep-PCR with DNA extracted from infected citrus plants. The concentration of CL varied between citrus varieties as determined by competitive PCR format (Kawabe et al. 2006). The tandem repeat-based polymerase chain displacement reaction (TR-PCDR) was developed for sensitive and reliable detection of Huanglongbing (HLB) pathogen '*Candidatus* Liberibacter asiaticus' (Las). A uniquely designed primer set (TR2-PCDR-F/TR2-PCDR-1R) and a thermostable *Taq*DNA polymerase mutant with strand displacement activity were employed for TR-PCDR amplification. The assay produced more than two amplicons after each amplification cycle in a regular thermal cycler. Sensitivity of the developed TR-PCDR was ten copies of the target DNA fragment. The efficiency of the TR-PCDR technique was maintained at a level better than conventional PCR and real-time PCR formats when tested using field-infected citrus samples. No false-positive TR-PCDR amplification was recorded from healthy citrus samples and water controls (Lou et al. 2018).

A multiplex PCR assay was employed for detection of *Pseudomonas syringae* pv. *actinidiae* (*Psa*), infecting kiwifruit. The assay could detect the pathogen in both artificially and naturally infected plant tissues. The multiplex PCR procedure could be employed for reliable detection and discrimination of haplotypes of *P. syringae* pv. *actinidiae* (Balestra et al. 2013). Infection of rice plants by multiple pathogens occurred frequently necessitating the development of a multiplex PCR method for detection of two or more bacterial pathogens coinfecting rice plants simultaneously. The multiplex PCR format detected *P. fuscovaginae, X. oryzae, B. glumae, B. gladioli, Sphingomonas* sp., and *Pantoea* spp. simultaneously (Bangratz et al. 2020). Recombinase polymerase amplification (RPA), an isothermal technique that is more tolerant to inhibitors compared with the PCR assay, was applied for detection of *Xanthomonas* spp., causing tomato bacterial spot disease. Novel RPA probes and primers were employed to amplify regions of the *hrcN* genes of *Xanthomonas* spp. RPA format was specific to *X. gardneri* and detected *X. gardneri* in lesions from naturally or artificially infected plants. *X. euvesicatoria* RPA detected *X. euvesicatoria* and *X. perforans* in plants artificially infected or naturally infected with either *X. euvesicatoria* or *X. perforans* RPAs could be efficiently applied for detection of the bacterial spot pathogens under field conditions (Strayer-Scherer et al. 2019).

4.2.6.2.3.3 Quantitative Real-Time PCR Technique
Real-time PCR, compared with the standard PCR format, is simpler, less labor intensive, and amenable for field testing. The quantitative real-time (QRT)-PCR procedure was developed by using point in time, when amplification of the target was first detected to estimate bacterial concentration. Fluorescence can be generated by nonspecific dyes such as SYBR green that binds to ds-DNA or fluorogenic probes. The cycle at which fluorescent signal exceeds a certain background fluorescent signal exceeds known as the threshold (Ct), is directly proportional to the amount of target DNA present in the sample. Primers for sensitive and reliable SYBR Green real-time PCR assay designed to amplify conserved regions of an essential pathogenicity gene were developed for detection of *Xanthomonas axonopodis* pv. *citri (Xac)* and its strains present in field-collected samples. A portable, field-hardened RAPID 7200 system was employed for performing real-time PCR assay. The real-time PCR assay was highly reproducible, quantitative and very sensitive, requiring minimum labor and time. The high level of sensitivity of real-time PCR procedure could be inferred from positive detection of *X. axonopodis* pv. *citri* in a herbarium sample from a 1912-canker outbreak in Florida (Mavrodieva et al. 2004). A QRT-PCR format was applied for detection, identification, and enumeration of bacteria in citrus plant samples infected by *Xac* and pv. *citrumelo.* Three sets of primers based on the pathogenicity gene *pth* in *Xac*, a ribosomal gene in *X. axonopodis* pv. *citrumelo,* and the leucine-responsive regulatory protein (*lrp*) gene in both pathovars were combined with TaqMan probes and applied for specific strain detection and quantification. QRT-PCR assay enhanced reliability of detection and differentiation of strains of *Xanthomonas* (Cubero and Graham 2005).

A direct real-time PCR assay was employed for detection of *Xylella fastidiosa (Xf)* (grapevine PD) in plant sap and samples of macerated chips of secondary xylem from trunks of grapevines without extraction of DNA. The procedure detected *Xf* in asymptomatic vines just coming out of dormancy, indicating the sensitivity and reliability of the real-time PCR assay (Schaad et al. 2002b). *X. fastidiosa* (CVC) was detected by applying conventional PCR and real-time PCR formats. No significant differences in populations of *X. fastidiosa* could be seen in three sweet orange cultivars tested. *X. fastidiosa* was detected and recovered from seedlings grown from infected seeds, indicating seed transmission of *X. fastidiosa* to the next generation in citrus (Li et al. 2003). Real-time PCR assay indicated the presence/absence of *X. fastidiosa* in the processed samples of plant tissues (Modesti et al. 2017). One-step protocol was developed for the detection of *Clavibacter michiganensis* subsp. *michiganensis (Cmm)* (tomato bacterial canker) in symptomatic seeds directly using a TaqMan probe (Zhao et al. 2007). As the PCR-based assays do not discriminate viable and nonviable cells of bacteria, quantitative real-time PCR format involving treatment with DNA-binding dye ethidium monoazide prior to real-time PCR was applied for detection and quantification of *C. michiganensis* subsp. *michiganensis* in tomato seeds. Viable cells of *Cmm* could be selectively quantified by EMA-real-time PCR format. Viable cells of *Cmm* could be discriminated from dead cells (Luo et al. 2008). Quantitative real-time PCR assay was employed to detect *Acidovorax avenae* subsp. *citrulli* (*Aac*) in watermelon seed lots generated by inoculating stigmas (pistil inoculation) or ovary pericarp (pericarp inoculation) with *Aac* strain AAC00–1. *Aac* populations from pericarp- and pistil-inoculated seeds were enumerated to determine *Aac* recovery from seed washes and seed macerates by real-time PCR assay. Use of macerates resulted in greater level of detection of *Aac*, irrespective of pathogen location in watermelon seeds (Dutta et al. 2012). A TaqMan probe-based insulated isothermal PCR (TiiPCR) assay was developed to detect *Acidovorax citrulli* (= *A. avenae* subsp. *citrulli*). TiiPCR format could detect *A. citrulli* cells in watermelon seeds efficiently (Wu et al. 2017).

A bio-polymerase chain reaction (BIO-PCR) assay was applied for specific and rapid detection of *Xanthomonas oryzae* pv. *oryzae* (*Xoo*) (rice bacterial blight, BLB) disease, and the SYBR Green real-time PCR format was useful for disease diagnosis and monitoring (Cho et al. 2011). An enrichment step of enhancing populations of *Xanthomonas campestris* pv. *campestris* (*Xcc*), (cabbage black rot) associated with infected seeds by seed germination followed by real-time TaqMan PCR format after extraction of target DNA was adopted (Laala et al. 2015). A novel, sensitive TaqMan-based qPCR format was developed to detect *Xanthomonas citri* pv. *malvacearum (Xcm)* in cotton plant tissues. The qPCR assay was effective for rapid and detection and quantification of *Xcm* populations in cotton plant tissues and also for monitoring disease incidence and spread (Wang et al. 2019). The SYBR Green real-time primer set was designed based on *rhs* family gene (YD repeat protein) for detection of bacterial grain rot pathogen *Burkholderia glumae* BGR1. The assay detected at least one genome equivalent of cloned amplified target DNA (Kim et al. 2012). Biomarkers specific to *Xylella fastidiosa (Xf)* DNA associated with grapevine PD and *Vitis vinifera* gene expression were identified for early diagnosis of the disease. Four genes upregulated in response to PD infection were individually multiplexed with an *Xf* marker RNA and scored using either real-time RT-PCR or gel-based conventional RT-PCR techniques. The system was sensitive enough to detect both host gene transcript and pathogen RNA in asymptomatically infected plants. Biomarker genes could be effectively used for early detection of infection by bacterial pathogens like *X. fastidisoa* (Choi et al. 2011). Novel primers were employed for detection of *Xylella fastidiosa* (*Xf*) in different peach tissue types sampled at different periods, in conventional (c) PCR and qPCR formats. The results indicated effectiveness of the primers employed to detect *X. fastidiosa* in root tissues (Chen et al. 2019).

Real-time quantitative PCR assays with SYBR Green primers and TaqMan probe and primers derived from albicidin toxin biosynthetic gene cluster efficiently and reproducibly amplified *Xanthomonas albilinians* (sugarcane leaf scald disease). The qPCR format was more sensitive than standard PCR. The qPCR assay could be employed for detection of and assessment of levels of resistance of germplasm to leaf scald caused by *X. albilinians* (Graces et al. 2014). The real-time PCR format was applied to detect *Leifsonia xyli* subsp. *xyli (Lxx)* (ratoon stunting disease) in the fully expanded youngest leaves of sugarcane cultivars at biweekly intervals from field nurseries. The qPCR and PCR formats detected *Lxx* in samples from different sugarcane cultivars throughout the year in Australia (Young et al. 2016). *Pantoea stewartii* subsp. *stewartii (Pss)* could not be differentiated from *P. stewartii* subsp. *indologenes (Psi)*, a nonpathogen occasionally detected on corn seeds along with *Pss*. A real-time PCR assay targeting *CpsAB* intergenic sequence was employed to specifically detect *Pss* and differentiate it from *Psi*. The assay efficiently detected *P. stewartii* subsp. *stewarti*i only (Pal et al. 2019). Dissemination of *Xanthomonas fragariae* (*Xf*), (strawberry angular spot) primarily occurs through contaminated planting stock. Propidium monoazide (PMA) quantitative PCR procedure was applied for specific detection of viable cells of *Xf*, employing two different primer pairs. The assay efficiently improved differentiation of viable from dead bacterial cells (Wang and Turechek 2020).

Citrus huanglongbing (HLB) is associated with three different *'Candidatus Liberibacter'* species, viz., '*Ca.* L. asiaticus', '*Ca.* L. americanus', and '*Ca.* L. africanus', depending on the geographical occurrence of the disease. A single assay that could detect all species of '*Ca.* Liberibacter' at the genus level was developed, providing convenience and saving time. Molecular analyses of '*Ca.* Liberibacter' species suggested that the *rpoB* gene (encoding ß-subunit of RNA polymerase) might provide an alternative target for identification of bacteria. A single pair of degenerate primers and a hybridization of probe corresponding to *rpoB* region were applied for detection of '*Ca.* Liberibacter' at the genus level. Further, species-specific primers and probes based on *rplJ/rplK* genes were designed and used for detection at species level in a multiplex format. Both the genus- and species-specific assays were validated in both SYBR Green and TaqMan formats, which could detect the pathogen in plant and insect vector extracts. Furthermore, the degenerate genus-specific primers and probe detected '*Ca.* Liberibacter solanacearum', in tomato, potato, and carrot (Ananthakrishnan et al. 2013). Combination of conventional PCR and real-time PCR format targeting respectively the putative DNA polymerase and 16S rDNA sequences of '*Ca.* Liberibacter asiaticus' (*Las*) was adopted for huanglongbing (HLB) pathogen detection. Quantification of *Las* showed that the pathogen was unevenly distributed in plants. A relatively high concentration of Las was recorded in fruit peduncles (Tatineni et al. 2008). Propidium monoazide (PMA), a DNA-binding dye, was used for determining the live and dead cells of '*Ca.* Liberibacter asiaticus'. Rapid PMA-qPCR was more precise in determining pathogen populations in host plants (Hu et al. 2013). Dynamics of population of '*Candidatus* Liberibacter asiaticus' (*Las*) (HLB) disease in different citrus hosts and each instar of Asiatic citrus psyllid (ACP) were studied together with seasonal growth and distribution of *Las* in different tissues, using conventional and TaqMan real-time PCR assay. Different levels of susceptibility/tolerance to HLB were recorded, resulting in varying degrees of symptom severity and growth effects on hosts and rootstocks. Mature and old leaves contained high populations of pathogen in cool temperatures in autumn and spring. *Las* was detected earlier through psyllid transmission than through graft inoculation, and the amounts of *Las* (AOL) varied in different hosts (Lin et al. 2017). Quantitative real-time PCR and conventional PCR formats could detect '*Ca.* L. asiaticus' in the Asian citrus psyllid (ACP) (Pandey and Wang 2019).

Most *Citrus* spp. are susceptible to stubborn disease caused by *Spiroplasma citri*. *S. citri* could be successfully cultured on cell-free liquid medium. The presence of helical structure and its mobility differentiated *S. citri* from phytoplasmas (Lee and Davis 1983). Primers were designed based on the sequences of *P89* or *P58* genes, which were multicopy genes present in *S. citri*. *P89* is a putative adhesion protein gene present on a plasmid as well as in the *S. citri* genome and *P58* is a putative adhesion multigene. The real-time PCR format was applied to quantify *S. citri* using both P58-base/primer pairs. Two distinct groups of *S. citri* were present, as revealed by conventional or real-time PCR formats. Primers based on sequences of *P58* or *P89* were effective for detection and quantification of *S. citri* in field samples from citrus plants (Yokomi et al. 2008). In a later investigation, two primer sets Php-orf1 and Php-orf3 targeting multicopy prophage genes developed from conserved prophage sequences in the *S. citri* genome were employed in SYBR Green-based quantitative PCR (qPCR) assays. Prophage primer set Php-orf1 increased detection sensitivity compared with housekeeping genes primers for spiralin and P58 putative adhesion gene. The results showed that *S. citri* prophage sequences enhanced the sensitivity of detection by qPCR in infected trees and reduced the false-negative results (Wang et al. 2015a). A field-deployable real-time PCR protocol was developed for rapid, specific, and sensitive detection of *S. kunkelii*, using the primers designed based on the sequences of adhesion-like gene and fluorigenic probe. The real-time PCR format could detect *S. kunkelii* DNA in the corn plant and vector tissue extracts. The sensitivity of the real-time PCR assay was significantly greater than the standard PCR format (Wei et al. 2016).

4.2.6.2.3.4 Loop-Mediated Isothermal Amplification Technique Loop-mediated isothermal amplification (LAMP) involves autocycling strand displacement DNA synthesis by a DNA polymerase which has high strand displacement activity. Amplification is completed within 30 min using a simple affordable water bath which is maintained at 65° C constantly. A thermal asymmetric interlaced (TAIL)-PCR assay was performed to amplify the uncharacterized regions adjacent to the *nusG-rplKAJL-rpoB* gene cluster of '*Ca.* Liberibacter' (CL) isolates. Conventional PCR was employed to amplify the internal *nusG-rplKAJL-rpoB* gene cluster of CL isolates. The LAMP technique based on the conserved sequence of the *nusG-rplKAJL-rpoB* gene cluster was employed for detection of '*Candidatus* Liberibacter' (CL) isolates. The LAMP product was detected on nylon membranes by staining with Azur B. Detection of colored amplicons on nylon membrane was simple and rapid (Okuda et al. 2005). The LAMP technique was employed to detect '*Ca.* Liberibacter solanacearum' (potato zebra chip) in potato plants and psyllid vector. The assay reliably detected the pathogen and also closely related species '*Ca.* Liberibacter asiaticus' (citrus huanglongbing, HLB) in plant DNA extracts. But LAMP assay was not as sensitive as quantitative real-time PCR format (Ravindran et al. 2012).

LAMP and real-time PCR assays were developed to enhance diagnostic efficiency under field conditions. Sequences of *rimM* gene of *Xylella fastidiosa*, were targeted. LAMP and real-time PCR assays could detect copies of the *rimG* gene (Harper et al. 2010). A rapid method of detection of *E. amylovora (Ea)* (pear and apple fire blight) was developed to improve disease risk assessment during flowering. LAMP assay could detect a single flower colonized epiphytically by *Ea* in samples of 100-flower clusters (Temple and Johnson 2011). The LAMP assay was applied to identify specific pathovars of *Pseudomonas syringae* such as *phaseolicola*, causing bean halo blight disease. *P. syringae* pv. *phaseolicola (Psp)* could be differentiated from all other fluorescent *Pseudomonas* spp. Three strains of *Psp* could be detected under standard LAMP reaction conditions. Addition of SYBR Green I dye to the reaction mixture after completion of LAMP reaction enabled the test result to be visualized under UV-light (Li et al. 2009). Strains of *Pseudomonas fuscovaginae* (rice sheath brown rot) were differentiated reliably, using a primer combination called Pf8 (Ash et al. 2014). The LAMP procedure was developed for detection of *P. syringae* pv. *tomato* (*Pst*) (tomato bacterial speck) based on the *hrpZ* gene. The presence of *Pst* was detected in artificially and naturally infected tomato leaf and stem samples. Early onset of disease could be detected under field conditions by using the LAMP assay (Chen et al. 2020).

Clavibacter michiganensis subsp. *michiganensis (Cmm)* (tomato bacterial canker) was detected by employing LAMP method, using primers targeting *mica*, encoding a type II antibiotic, michiganin A. *Cmm* and non-*Cmm* strains from a worldwide collection were tested to determine the specificity of LAMP assay. LAMP assay efficiently identified all *Cmm* strains and discriminated them from other *C. michiganensis* subspecies and non-*Clavibacter* bacteria. LAMP reaction was an easy, one-step test for rapid detection and differentiation of *C. michiganensis* subsp. *michiganensis* strains (Yashuhara-Bell et al. 2013). The LAMP assay with a fluorometer detected *Burkholderia gladioli* pv. *alliicola (Bga)* (onions and tulip rot) reliably and rapidly. Considerable variability was revealed in 16S – 23S rRNA intergenic transcribed spacer (ITS) sequences among *Burkholderia* species, and *B. gladioli* pathovars. A set of LAMP primers was designed based on ITS region. Primers were specific for *B. gladioli* pv. *alliicola*. Real-time LAMP assay was more sensitive than the conventional PCR format (Kuang et al. 2017). A unique marker was identified for reliable and sensitive detection of *Xanthomonas fragariae* (strawberry angular leaf spot) by LAMP assay. The pathogen was detected in crude leaf extract containing a population of 10^2 CFU/ml. Further, the assay did not need any advanced analytical knowledge or fully equipped laboratory, providing the results rapidly (Gétaz et al. 2017). LAMP assay primers were designed from specific targeting of non-ribosomal peptide synthetases (NRPs) genes from *Acidovorax citrulli* (watermelon BFB). The LAMP assay was highly specific to *A. citrulli* without cross-reaction with other bacterial pathogens tested. LAMP assay detected *A. citrulli* in greater number of cantaloupe seed lots, than the conventional PCR format could detect the pathogen (Yan et al. 2019).

The LAMP method was employed for sensitive and specific detection of *Xanthomonas phaseoli* pv. *phaseoli* (*Xpp*) and *X. citri* pv. *fuscans* (*Xcf*) (common bean bacterial blight). *Xpp* could be detected by LAMP assay in inoculated bean leaves even prior to appearance of disease symptoms. The presence of bacterial pathogens was detected in all seedlots obtained from infected plants. The LAMP assay detected the bacterial blight pathogens in common bean plants and seeds efficiently and rapidly (de Paiva et al. 2020). The Cas12a-based DNA endonuclease-targeted CRISPR transreporter (DETECTR) assay was highly specific and sensitive in detecting huanglongbing pathogen '*Candidatus* Liberibacter asiaticus' (CLas) nucleic acids in infected citrus samples. The DETECTR assay targeted the 5-copy *nrdB* gene specific to CLas. Coupled to isothermal amplification with Cas12a transcleavage of a fluorescent reporter oligonucleotide and enabled detection of CLas nucleic acids at the attomolar level. The pathogen was detected in samples from infected citrus, periwinkle, and vector psyllids. As it was compatible with lateral flow assay, DETECTR technology was considered to be a suitable tool for early detection of CLas in infected orchards/regions (Wheatley et al. 2021).

4.2.6.2.3.5 Combination of PCR Assay and Other Detection Techniques The sensitivity and specificity of detection of bacterial plant pathogens may be improved by integration of PCR with other detection techniques. Immuno-enzymatic detection of PCR amplified products may be accomplished by applying PCR assay in tandem with immunoassay. Sensitivity of the assay could be enhanced by immunocapture (IC)-PCR format by concentrating *Xanthomonas axonopodis* pv. *citrumelo (Xac)* from dilute plant extracts (Hartung et al. 1996). *Erwinia amylovora* (apple fire blight) was detected reliably, using PCR-ELISA combination. PCR primers were designed to amplify sequences of the ubiquitously present cryptic plasmid pEA29. Amplicons were labeled with digoxigenin (DIG)-dUTP during amplification reaction, captured by hybridization to a biotinylated oligonucleotide in streptavidin-coated ELISA microplates, followed by detection with anti-DIG-Fab-peroxidase conjugated antibodies. Strains of *Ea* from various host plants and geographical regions were detected by PCR-ELISA procedure, confirming the specificity and reliability of the technique (Merighi et al. 2000). *Xylella fastidisoa* (grapevine Pierce's disease (PD) isolated from xylem fluids on PD3 solid medium was purified using immunocapture (IC) procedure prior to PCR amplification. *Xf* suspension (mixed pellet of *Xf* and antibodies) was used as templates. IC-PCR protocol could detect *X. fastidiosa* in the xylem exudates from plants before symptom expression (Guo and Lu 2004). In another investigation, *X. fastidiosa* was detected by the IC-PCR procedure, with detection efficiencies of IC-PCR and ELISA techniques being similar in detecting *Xf* from naturally infected grapevine plants (Costa et al. 2004). The IC-PCR technique was effective in detecting *X. axonopodis* pv. *diffenbachiae (Xad)* (anthurium bacterial blight). The presence of *Xad* could be detected in both symptomatic and asymptomatic plants. The IC-PCR format was more sensitive than both conventional PCR assay and indirect ELISA test (Khoodoo et al. 2005). IC-PCR technique was also reported to be effective for detection of *Acidovorax avenae* subsp. *citrulli* (*Aac*) (watermelon BFB). Positive amplification of the expected 360-bp DNA fragment occurred only from four strains of *A. avenae* subsp. *citrulli* (Xiao et al. 2007).

The BIO-PCR procedure involved the enrichment of the causal bacterial species using an appropriate medium, followed by enzymatic amplification of PCR targets by employing specific primers (Schaad et al. 1995). *Pseudomonas syringae* pv. *phaseoli (Psp)* was detected on washing of leaves and seeds of bean (*Phaseolus vulgaris*) by BIO-PCR technique. The primers were designed from the sequences of a *tox-argK* chromosomal cluster of *Psp*-specific phaseolotoxin gene. The membrane real-time BIO-PCR format detected *Psp* at a lower concentration of 80 CFU/ml of bacterial cell suspension. The sensitivity of detection from leaf washings was lower compared with seed washings (Schaad et al. 2007). *Clavibacter michiganensis* subsp. *michiganensis (Cmm)* (tomato bacterial canker disease) was detected in seed samples only by BIO-PCR assay, but not by isolation or direct PCR methods, indicating higher level of sensitivity to BIO-PCR assay (Hadas et al. 2005). *Acidovorax avenae* subsp. *avenae (Aaa)* (rice bacterial stripe disease) was detected in contaminated rice seeds by isolation, followed by PCR amplification. *A. avenae* subsp. *avenae (Aaa)* was enriched prior to PCR amplification, by incubating washings of naturally contaminated rice seeds. BIO-PCR assay was more sensitive in detecting *Aaa* than the conventional PCR format (Song et al. 2004). In a later investigation showed that direct real-time PCR and real-time BIO-PCR and the semiselective medium could be employed successfully for pathogen detection in naturally infected seed lots (Zhao et al. 2009). BIO-PCR format was developed to detect *Xanthomonas campestris* pv. *campestris (Xcc)* (black rot of crucifers). The primer set Dhrp Xacc_F/Dhrp_Xcc_R was employed to amplify a specific 769-bp fragment of *Xcc*. The assay was highly sensitive in detecting bacterial DNA. The results indicated the greater level of sensitivity of BIO-PCR format in detecting *X. campestris* pv. *campestris* in seeds of cauliflower (Singh et al. 2014).

Efficiency of template DNA preparation could be enhanced by immunomagnetic separation (IMS), a process in which small supermagnetic polystyrene beads coated with specific antibodies are used. Antibodies covalently attached to magnetic beads bind target cells by specific antibody-antigen interactions during incubation with cell suspensions. Inhibitory compounds present in the immunomagnetic bead (IMB)-bacteria could be washed off by rinsing. PCR assay was performed for detection of *Acidovorax avenae* subsp. *citrulli (Aac)* in watermelon seed. IMS yielded a 10-fold increase in recovery of *Aac* over direct spread-plating on King's Medium B. Synthetic oligonucleotide primers were designed based on the 16S rRNA gene of *Aac* strain and tested for specific DNA amplification by PCR. The primers generated amplicons from DNAs of closely related bacteria. IMS-PCR assay showed 100-fold increase in sensitivity in detecting *Aac*, compared with the direct PCR format. In addition, the adverse effect of inhibitors of PCR present in watermelon seed was absent (Walcott and Gitaitis 2000). In a later investigation, the IMS-PCR assay was applied for detection of *Acidovorax avenae* subsp. *citrulli (Aac)* in watermelon seeds. Specific primers were designed based on PCR fragments obtained in ERIC and Box profiles of *Aac* strains. IMS-PCR assay did not amplify DNAs from closely related species and subspecies. The results obtained with the BX-S set were comparable with those of the *Aac* set, while no amplification occurred in control seed lots. The BX-S set was more sensitive than the *Aac* primers (Bahar et al. 2008). Specific primer pairs Xcv1F/Xcv3R and RST2/Xcv3R were employed to detect and identify *Xanthomonas campestris* pv. *viticola*, (grapevine bacterial canker). By performing the enrichment step prior to PCR amplification, bacterial colonies were recovered from symptomatic petioles of inoculated grapevine plants. The PCR assay was performed with aliquots of plate wash obtained from symptomatic tissues and amplification occurred in pathogen samples (Trindadae et al. 2007).

Multiplex real-time PCR assay and magnetic capture hybridization were combined for detection of *Acidovorax avenae* subsp. *citrulli (Aac)* and *Didymella bryoniae (Db)* (gummy stem blight) in cucurbit seeds. Comparative frequencies of detection of *Aac* for and Magnetic capture hybridization (MCH)-real-time PCR was more sensitive than direct real-time PCR assay in detecting Aac. A combination of MCH and real-time PCR assay was effective in detecting the pathogens present in watermelon and melon seed samples infested each with *A. avenae* subsp. *citrulli* and *D. bryoniae* (Ha et al. 2009). *Clavibacter michiganensis* subp. *michiganensis (Cmm)* and *Pepino mosaic virus* (PepMV) are seedborne. The magnetic capture hybridization (MCH) step to concentrate and purify target nucleic acids of the pathogen was applied prior to real-time PCR assay. Detection sensitivity of the combined methods was substantially enhanced. The results indicated that MCH could facilitate simultaneous detection of *C. michiganensis* subsp. *michiganensis* and *Pepino mosaic virus* by multiplex real-time PCR format (Johnson and Walcott 2012). A diagnostic method based on padlock probes (PLPs) was developed to improve sensitivity of detection and reduce the cost and time required for diagnosis. PLPs are long oligonucleotides (about 100-bp), whose ends are complementary to sequences adjacent to the site targeted for PCR amplification. *Acidovorax citrulli* was detected by combining PLP with dot blot hybridization in cucurbit seed extracts. The PLPs were designed based on the 16S – 23S internal transcribed spacer ribosomal DNA sequence of *A. citrulli*. The pathogen was detected in all artificially infested seed lots (100%) with 0.1% infestation or greater. Further, presence of the pathogen in melon and watermelon seed lots collected from different locations in China was detected using PLP assay (Tian et al. 2013). In order to increase the reliability and sensitivity of detection of *Xanthomonas oryzae* pv. *oryzae (Xoo)* and *X. oryzae* pv. *oryzicola* (*Xoc*) (rice bacterial leaf blight and bacterial leaf streak), two padlock probes (PLPs) P-Xoo and P-Xoc were developed based on 16S – 23S ITS rDNA sequences, respectively. The PLPs were combined with dot blot hybridization to detect the pathogens. PLP assays could detect *Xoo* in seed lots infested naturally. Both pathogens were detected simultaneously in two seed lots employing PLPs P-Xoo and P-Xooc combined with reverse blot hybridization. The results showed the effectiveness of a combination of PLPs and dot blot hybridization for the detection of multiple pathogens present in rice seed and plant samples (Tian et al. 2014).

4.2.7 Assessment of Variability in Bacterial Pathogens

4.2.7.1 Biological Characteristics

A set of cultivars or genotypes of a crop is used as differentials to distinguish different strains of species or subspecies. Pathovars of a bacterial species are distinguished primarily based on the ability or inability of isolates to infect a set of host plant species or cultivars. Based on visual assessment of BFB severity induced, *Acidovorax citrulli* strains were classified into two groups. Strains belonging to group I were moderately to highly virulent on several cucurbit hosts, whereas group II strains were predominantly found to be highly virulent on watermelon, but showed low virulence on other cucurbits (Zivanovic and Walcott 2017). An immature detached melon fruit assay based on cv. Joaquin Gold melon was developed to differentiate the host specificity between group I and II of *A. citrulli* strains. Ability to induce water-soaked lesions could be used as a basis for differentiating these two groups of strains of *A. citrulli* (Yan et al. 2017). Genetic diversity and population structure of *A. citrulli* strains occurring in Brazil were investigated employing pulsed-field gel electrophoresis (PFGE), multilocus sequence analysis (MLSA) of housekeeping genes and virulence-associated genes and pathogenicity tests were the methods employed to differentiate the strains of *A. citrulli*. MLSA efficiently differentiated *A. citrulli* strains into two clades, and PFGE analysis also indicated the presence of two groups of strains (Silva et al. 2016). Genetic differences between the virulent and less virulent strains of *Pseudomonas syringae* pv. *tomato* (tomato bacterial speck disease) were identified in toxin biosynthetic genes and elicitors of host innate immunity (Jones et al. 2015). Using a set of differentials, six isolates of *Pseudomonas syringae* pv. *phaseolicola* (Psp) (halo blight disease) were found to infect all differential plants and matched the compatibility pattern of race 6 (Felix-Gastelum et al. 2016). Pathovars of *Xanthomonas translucens* (cereal leaf streak, CLS), *undulosa*, *translucens*, *hordei*, and *secalis* were distinguished from noncereal pathovars using differentials (Langlois et al. 2017). Variations in biochemical and physiological characteristics have been used as bases for differentiation of some bacterial pathogens. The genomic DNAs of 63 Spanish strains of *Erwinia amylovora (Ea)* (apple fire blight) were analyzed by PFGE procedure, after digestion with *XbaI* restriction enzyme. Three distinct patterns, Pt1, Pt3, and Pt4, were identified among strains of *Ea*. All strains of *Ea* detected in imported plant materials were identified as Pt3. PFGE pattern of each strain could be reproduced in all cases (Dona et al. 2007). The identity of isolates of *Clavibacter michiganensis* subsp. *michiganensis (Cmm)*, causing tomato bacterial canker was established through gas chromatography-fatty acid methyl esters (GC-FAME) and PCR assay, using primers CMM5 and CMM6. The isolates were separated into 12 groups by PFGE analysis (Basim and Basim 2018).

Effects of the number of bacteriliferous psyllids on populations of '*Candidatus* Liberibacter solanacearum' Lso (potato zebra chip, ZC) and levels of amino acids, carbohydrates, phenolics, and subsequently symptom severity in potato tubers were assessed. Increasing the number of psyllids resulted in enhanced titer of '*Ca*. L. solancearum' and symptom severity in plants. Phenolic compounds showed a consistent increase, following infection and this response was positively correlated with this numbers of infective psyllids in most cases (Rashed et al. 2016). Strains of Lso strains could be differentiated into two haplotypes (A and B). ZC symptom severity in Lso A-infected plants was lower than in plants infected by either Lso B or Lso AB. ZC symptom severity in plants inoculated with psyllids coinfected with haplotypes A and B was similar to the symptoms associated with Lso haplotype B (Harrison et al. 2019).

4.2.7.2 Variability in Genetic Characteristics

Variability in genetic characteristics depends on differences in nucleic acid characteristics of bacterial pathogens. Nucleic acid (NA)–based techniques have been applied to detect and differentiate and quantify several bacterial pathogens more reliably and accurately. Restriction fragment length polymorphism (RFLP), PCR either alone or in combination with random amplified polymorphic DNA (RAPD) and amplified fragment length polymorphism (AFLP) techniques have been demonstrated to be effective for differentiation of strains of bacterial plant pathogens. The genetic variability of strains of *Erwinia amylovora (Ea)*, (apple fire blight) investigated using RFLP analyses. *Ea* strains collected from Maloideae host species and *Rubus* spp. were differentiated by RFLP analysis (Giorgi and Scortichini 2005). *X. albineans* (sugarcane leaf scald) produces the toxin albicidin production, and it is governed by the 40-kb albicidin biosynthetic gene. Fourteen haplotypes and two major genetic groups ALB-RFLPA and ALB-RFLPB of *X. albilineans* were identified by RFLP procedure. Genetic diversity of albicidin biosynthetic gene was very similar to the genetic diversity of the pathogen (Champoiseau et al. 2006). In a later investigation, genetic diversity of *X. albilineans* and pathogenicity variability were assessed based on short sequence repeat (SSR) loci. Of the 16 SSR loci, 12 were polymorphic and generated 54 polymorphic alleles. No correlation could be established between aggressiveness and groups of isolates formed based on SSR markers (Tardiani et al. 2014).

PCR-based techniques have been employed to differentiate strains of bacterial pathogens with varying degrees of effectiveness. Repetitive sequence based rep-PCR procedures are more frequently employed to fingerprint plant pathogenic bacteria. Rep-PCR assays are employed using primer sets that can amplify the specific conserved repetitive sequences, leading to the differentiation of bacterial strains within a species. Interrelatedness, genetic diversity, and geographical distribution of the common bean bacterial blight (CBB) pathogen *Xanthomonas axonopodis* pv. *phaseoli (Xap)* and *X. axonopodis* pv. *phaseoli* var. *fuscans* were investigated by employing RFLP analysis. The presence of 300- and 400-bp fragments in *Xap* isolates and their absence and presence of another fragment of 700-bp in *X. axonopodis* pv. *phaseoli* var. *fuscans* differentiated them (Mahuku et al. 2006). Genetic diversity of *Pseudomonas syringae* pv. *syringae (Pss)* (pear bud necrosis and leaf spot) was assessed by employing rep-PCR format with BOX A1 primer. Four genomic patterns, A, B, C, and D were recognized by BOX-PCR analysis (Natalini et al. 2006). Genetic diversity of *Xanthomonas campestris* pv. *campestris* (*Xcc*) isolates was investigated by employing rep-PCR, PFGE, and AFLP techniques. Rep-PCR, AFLP, and PFGE differentiated genotypes differently. No clear relationship between genotypes determined by molecular methods and virulence was observed by inoculation on differential host plant species (Valverde et al. 2007).

Multiple loci variable numbers of tandem repeat analyses (MLVAs) provide advantages over traditional fingerprinting methods for analysis of genetic diversity of bacterial pathogens. MLVA, based on a set of nine variable number tandem repeat loci, was employed to type 46 *E. amylovora* isolates collected from different host plants in different countries. Nine polymorphic loci showed high discriminatory ability and enhancement of resolution of the MLVA. Seven clonal complexes contained 38 clusters of haplotypes. Different MLVA types were identified among Italian strains only, indicating multiple introduction of the disease. MLVA showed potential for application as an inexpensive, rapid, and simple tool to track *E. amylovora* infection sources to gain insight into geographic diversity and to understand the dynamic evolution of *E. amylovora* (Alnaasan et al. 2017). Genetic diversity of *Xanthomonas phaseoli* pv. *manihotis*, causal agent of cassava bacterial blight disease was assessed using the MLVA scheme. The MLVA15 scheme consisted of 15 variable numbers of tandem repeat loci grouped into four multiplex PCR pools. The MLVA-15 scheme could be employed specifically to differentiate *X. phaseoli* from other *Xanthomonas* spp., using up to eight tandem repeat loci. Strains originating from the same field and isolates at the same time could be differentiated using the MLVA-15 scheme, demonstrating the advantages of the scheme targeting 6- or 7-bp repeats (Rache et al. 2019).

Phylogenetic relationships among *E. amylovora* bv. 4 and other biovars of *E. amylovora* and *E. pyrifoliae* were examined on the basis of nucleotide sequences of the genes for 16S rRNA, DNA gyrase B subunit (*gyrB*) and *rpoD*. These genes are required for survival of bacterial cells. Isolates of *E. amylovora* and *E. pyrifoliae* could be divided into two major groups. Phylogenetic analysis based on the *gyrB* and *rpoD* sequences might provide higher resolution of relationship between pathogen isolates than when sequences of 16S rRNA were used (Matsura et al. 2007). Strains of *Erwinia amylovora (Ea)* were screened for the presence of the plasmid pEA29, which carries genes involved in thiamine biosynthesis required for colonization of plant tissues. Strains with pEA29 were virulent in assays with immature pear or apple seedlings, but showed reduced growth in minimal medium without amino acids and thiamine. However, strains lacking pEA29 were detected rarely in different countries, where fire blight disease incidence was observed (Mohammadi et al. 2009).

Genetic diversity of strains of *Xanthomonas translucens* pv. *undulosa* (wheat bacterial leaf streak, BLS) strains was assessed using repetitive sequence based rep-PCR and insertion sequence (IS)-based PCR. The 16S rDNA sequences of the strains were identical and distinct from those of other *Xanthomonas* pathovars (Adhikari et al. 2012). Multilocus sequence typing (MLST) and analysis (MLSA) of four housekeeping genes of *ropD, dnaK, fyuA,* and *gyrB* were employed to evaluate the genetic diversity of strains of *X. translucens*. All strains were related, as indicated by MLSA, to *X. translucens* pv. *undulosa* and *X. translucens* pv. *translucens*. Phylogenetic analysis of the four housekeeping genes correlated with host origination of isolates (Curland et al. 2018). Genetic diversity of highly virulent strains (HVS) of *Xanthomonas campestris* pv. *malvacearum* (*Xcm*) (cotton black arm and boll rot) was assessed by repetitive sequence (rep)-based PCR with ERIC and BOX primer sets. The strains with low or moderate virulence were clustered into one genogroup, whereas all HVSs isolated in Africa were segregated in another group. The results suggested the virulence of *Xcm* was highly related to genotype and/or geoclimatic origin of strains (Zhai et al. 2010). Repetitive PCR markers have been employed to determine genetic relationships between *Xanthomonas* spp., including the species infecting *Citrus* spp. BOX-, ERIC-, and rep-PCR assays were performed using purified total DNA of the 42 strains of *X. citri* and *Xanthomonas* reference strains. Rep-PCR assay amplified fragments from complex genomic fingerprinting patterns with multiple and reproducible bands of distinct intensities. The average number of bands/strain varied, and the molecular size of the bands were useful for differentiating them from *X. aurantifolii* B and C strains. (Chiesa et al. 2013). Genetic diversity of the strains of *Xanthomonas campestris* pv. *campestris (Xcc)* (cabbage black rot) was analyzed using rep-PCR assay including primers for repetitive extragenic palindromic (REP) primers, ERIC and BOX elements. The isolates were grouped into 56 different DNA clusters with a 75% similarity coefficient (Singh et al. 2016). Rep-PCR assay confirmed the high diversity of *X. campestris* and used to generate 15 unique fingerprints from isolates of *X. campestris* (Lange et al. 2016).

Strains of *Xanthomonas campestris* pv. *vesicatoria (Xcv)* could be identified based on their strain-specific sequences. Pyrosequencing-based detection and typing of *Xcv* strains were rapid and highly specific and sensitive (Gan et al. 2014). Identification of strains of *Xanthomonas* genospecies causing bacterial spot of tomato at species level was consistently was possible using AFLP and MLSA. *X. euvesicatoria* was predominantly prevalent (Hamza et al. 2010). In a later investigation, the region of *hrpB* operon was evaluated to detect and identify *Xanthomonas* spp. involved in the bacterial spot disease affecting tomato and pepper by real-time PCR assay. The optimized multiplex real-time TaqMan PCR assay was highly specific to *X. euvesicatoria, X. vesicatoria, X. perforans*, and *X. gardneri*, the four target pathogen species (Strayer et al. 2016). Pigmented isolates of *Xanthomonas campestris* pv. *viticola*, were pathogenic to cashew. The MLSA method, average nucleotide identity (ANI) values, and tetranucleotide frequency correlation efficient (TETRA) were employed to analyze the phylogenetic relationship of isolates. *X. campestris* pv. *viticola* isolates were closely related to *X. citri* subsp. *citri* (MLSA and ANI). Pigmented isolates from cashew were closely related to *X. citri* pv. *anacardii* (rep-PCR, MLSA, ANI, TETRA) (Da Gama et al. 2018). Average nucleotide identity (ANI) analysis approach was applied to assess variability in 34 phytopathogenic strain genomes of *Pseudomonas syringae* species complex considered related species including many pathotypes strains. Strains of *Pseudomonas syringae* pv. *syringae (Pss)* and *P. syringae* pv. *morsprunorum (Psm)* associated with bacterial canker disease was characterized using biochemical and multilocus sequence analyses (MLSA). The isolates induced hypersensitive response (HR) in tobacco leaves and pathogenic to immature cherry fruits and cherry trees. Phenotype assays and MLSA using four genes, *cts, gapA, gyrB,* and *ropD* showed phenotypic and genetic identity with *P. syringae* pv. *syringae*, indicating that *P. syringae* pv. *syringae* could be the cause of cherry bacterial canker disease in South Africa (Otto et al. 2018).

Phylogenetic analyses were performed by comparing the sequences of 13 strains of *Xylella fastidiosa* (grapevine PD) of 16S rRNA gene and the 16–23S rRNA ITS region. The results showed that the PD strains were closely related to the American *Xf* grapevine strains, but not to the pear strains, suggesting their evolution independently from each other (Su et al. 2013). *X. fastidiosa* is taxonomically classified into subspecies with phylogenetic clusters within subspecies that are host specific. *X. fastidiosa* subsp. *pauca* restricted to South Africa was studied to determine its genetic diversity. MLST identified six sequence types (STs) among plant samples collected in Brazil. In a later survey of *Xf* genetic diversity (MLST-based) seven new STs in subspecies *pauca* and two new STs in subspecies multiplex were identified. The presence of subsp. *multiplex* in South America was considered to be consequence of a single introduction from its native range in North America. Different phylogenetic approaches clustered South American STs into four groups, with strains infecting citrus (subsp. *pauca*), coffee and olive (subsp. *pauca*), coffee, hibiscus, and plum (subsp. *pauca*) and plum (subsp. *multiplex*) (Coletto-Filho et al. 2017). *Xanthomonas* spp. isolates associated with tomato bacterial spot were characterized and differentiated using the BOX repetitive element PCR assay. Fingerprint patterns revealed the presence of four haplotypes among the strains. The multiplex real-time quantitative PCR identified *Xanthomonas* strains from tomato as *X. perforans* (Adhikari et al. 2019). Genetic diversity of strains of *P. syringae* pv. *actinidiae (Psa)* (kiwifruit bacterial canker) was investigated using rep-PCR, RAPD-PCR, and multilocus satellite analysis (MLSA) techniques. Molecular fingerprinting obtained through rep-PCR and RAPD-PCR analysis revealed a high level of variability in the population of different strains of *Psa* occurring in northern Italy (Prencipe et al. 2018). The biovars of *P. syringae* pv. *actinidiae (Psa)* exhibited differences in virulence and composition of pathogenicity-related genes, such as toxin biosynthesis and type III effector genes. In addition to the *tox* island, and various plasmids, several mobile genetic elements were also detected in the genome of *P. syringae* pv. *actinidiae* (Sawada and Fujikawa 2019).

The BOX-PCR technique differentiated four fingerprint types, A, B, C, and D, among strains of *Clavibacter michiganensis* subsp. *michiganensis (Cmm)*. PCR types detected all strains with BOX-PCR and *Cmm* populations in Michigan tomato fields were geographically structured (Quesada-Ocampo et al. 2012). *Cmm* strains were characterized by PCR targeting virulence genes, fingerprinting techniques, metabolic profiles, and virulence and chromosomal genes encoding virulence determinants *tomA, ppaA,* and *chpC*. Plasmid-encoded genes *pat-1* and *celA* were detected by PCR in all strains tested, except for four Sicilian *Cmm* strains from which the *pat-1* gene was not amplified. *Cmm* strains were differentiated into 13 haplotypes using BOX-PCR technique and their clonal populations were identified. The AFLP protocol based on amplification of *Apa*I/*Mse*I fragments could differentiate *C. michiganensis* subsp. *michiganensis* (Ialacci et al. 2016). Strains of *Clavibacter michiganensis* subsp. *michiganensis (Cmm)* were analyzed by rep-PCR assay employing either ERIC or BOX primers. Strains of *Cmm* were placed in four groups, A, B, C, and D. PCR-based detection of three genes located on different sites of pathogenicity island (PAI) revealed that all tested pathogenic strains of *Cmm* were positive for the presence of *ppaA, chpC*, but not *tomA*, indicating that *ppA* and *chpC* might be essential for virulence of *C. michiganensis* subsp. *michiganensis* (Kleitman et al. 2008). The strains of *Cmm* were differentiated into 21 haplotypes. The MLSA consisted of five housekeeping genes, *kdpA, sdhA, dnaA, ligA*, and *gyr B*, and putative pathogenicity genes, *celA, tomA*, and *nagA*. Rep-PCR with BOX-AIR primer confirmed the high diversity of *Cmm* isolates (Tancos et al. 2015). The 16S rRNA and multilocus sequence analyses showed that strains of *Pantoea ananatis, P. agglomerans,* and *P. allii* causing onion center rot disease were more closely related to *P. stewartii* subsp. *stewartii* and *P. stewartii* subsp. *indologenes*. Phylogenetic analysis (16S rRNA and atpD, gyrB, infB, and rpoB genes) revealed that these strains formed a distinct cluster (Stumpf et al. 2018).

BLB of cantaloupe (*Cucumis melo*) was considered to be due to *Pseudomonas syringae* pv. *lachrymans (Psl)*. But putative Psl isolated from cantaloupe did not react positively in *P. syringae* pv. *lachrymans*-specific assays, suggesting that *Psl*-like strains from cantaloupe might be distinct from those obtained from cucumber. In order to investigate the differences between *Psl*-like strains from cantaloupe and cucumber (13 strains from China and 1 from Zimbabwe), seven additional *P. syringae* reference strains were analyzed by catabolic profiling, phylogenetic analysis by MLSA, and pathogenicity tests on cantaloupe leaflets. Catabolic profiling and MLSA based on ten housekeeping genes and two HR and pathogenicity (*hrp*) genes allowed differentiation of strains isolated from cantaloupe and cucumber. *P. syringae* pv. *lachrymans* strains isolated from cucumber clustered with genomospecies 2, and 13 *P. syringae* strains isolated from cantaloupe belonged to genomospecies 1. Pathogenicity tests showed that strains of *P. syringae* pv. *lachrymans* were pathogenic only to their respective hosts, cantaloupe or cucumber. Strains from cantaloupe represented a novel phylotype of *P. syringae* pv. *lachrymans* (Tian et al. 2017). Genome analysis to determine genetic diversity within pathovar *lachrymans* was performed. Comparisons of sequences of other pathovars of *P. syringae* revealed that three unique plasmids and 24 type III effector proteins (TTEs) were present in strain 814/98. Phylogenetic analyses of MLST loci and TTEs revealed the existence of two distinct clusters of strains within *lachrymans*, which were grouped into either Phylogroup 1 or 3 (Slomnicka et al. 2018).

Putative isolates of *Pseudomonas syringae* pv. *syringae* (*Pss*) (grapevine bacterial inflorescence rot) from different vineyards were genotyped using RNA polymerase ß-subunit (*rpoB*) and MLST using primers for glyceraldehyde-3-phosphate dehydrogenase (*gapA*), citrate synthase (*gltA*), DNA gyrase B (*gyrB*), and σ factor 70 (*rpoD*). Two clades were recognized based on MLST data, each containing a mixture of pathogenic and nonpathogenic grapevine isolates, indicating that *Pss* in Australian vineyards was genetically diverse. Each clade also contained strains of *Pss* from non-grape hosts pathogenic to grapevine revealing absence of host specificity and possible potential cross-infection of grapevine and other horticultural crops (Hall et al. 2016). In a further study, genotypes and phenotypes of *P. syringae* pv. *syringae* were characterized using molecular multilocus sequence typing (MLST) of housekeeping genes and biochemical testing. The isolates of *Pss* formed a new clade within subgroup 2 (PG02), putatively called PG02f. some genetic differences between BIR isolates and the general vineyard *Pss* populations were detected. Phylogroup PG02f contained the epidemic BIR isolates, as well as nonpathogenic isolates of *P. syringae* pv. *syringae* present in grapevine (Hall et al. 2019). Association of multiple pathovars with cucurbit angular leaf spot necessitated performance of tests to assess genetic diversity of *Pseudomonas syringae*. Strains of *P. syringae* (27) involved in angular leaf spot disease were collected from watermelon, cantaloupe, and squash. *P. syringaegae* pv. *lachrymans* was included as a member of genomospecies 2. However, the genetic profiles obtained through both MLSA (*gyrB, rpoD, gapA*, and *gltA*) and BOX-PCR (BOXAIR) identified strains of *P. syringae* to be distributed among 3 clades within genomospecies 1 and phylogenetically distinct from genomospecies 2 member *P. syringae* pv. *lachrymans*. A novel MLSA haplotype of *P. syringae* pv. *lachrymans* present in cucurbit hosts was identified. The genetic diversity among *P. syringae* strains infecting cucurbits reflected larger ecological diversity of pathogen populations from genomospecies 1 (Newberry et al. 2016). Rep-PCR analysis showed genetic diversity of *Serratia marcescens* (cucurbit yellow vine disease, CYVD) which was confirmed by MLSA analysis. CYVD-inducing strains clustered into one group, separated from non-plant pathogenic *S. marcescens* strains. (Besler and Little 2017). Pathogenicity of *Xanthomonas* strains associated with bacterial leaf spot on poinsettia was analyzed by MLSA using six housekeeping genes. These strains were divided into three distinct classes based on MLSA. Clade I contained *X. axonopodis* pv. *poinsetticola* strain (LMG84.9^{PT}). Clade II was more closely related to *X. hortorum* pv. *pelargonii* (LMG7314^{PT}), and Clade III contained *X. codiaei* type strain (LMG8678^{PT}). Strains from three clades could not be differentiated by symptoms or virulence on poinsettia plants (Rockey et al. 2015). Bacterial strains isolated from leaf lesions on Araceae plants were identified either as *Pantoea ananatis* or *P. agglomerans*, based on morphological, physiological, biochemical, and molecular characteristics. Further, 16S rRNA gene sequencing showed that the strains were phylogenetically closely related to the genus *Pantoea*. MLSA of concatenated partial *atpD, gyrB*, and *rpoB* gene sequences of the six strains exhibited high similarity to three strains (B24, J3Y, and SY) of *P. ananatis* and another three strains (E60Y, E5MM, and E684) to *P. agglomerans*. Phylogenetic analysis confirmed these results (Yazdani et al. 2018).

Conventional PCR and real-time PCR formats were applied to detect '*Candidatus* Liberibacter' solanacearum' and '*Ca*. L. psyllaurous' in potato infected by Zebra disease complex, using pathogen-specific primers and probes. '*Ca*. Liberibacter' species were grouped into two clusters showing high similarity in their partial 16S rRNA gene sequences and in their partial intergenic spacer region ISR-23S RNA gene sequences. These two species were very closely related, as shown by sequence analysis (Wen et al. 2009). Five haplotypes were recognized among strains of '*Ca*. L. solanacearum' LsoA and LsoB infecting solanaceous crops and they were transmitted by tomato/potato psyllid *Bactericera cockerelli*. LsoC affected carrot, and it is transmitted by the carrot psyllid *Trioza apicalis*. Haplotypes LsoD- and LsoE-infected carrot and celery, and the psyllid *Bactericera trigonica* transmitted the pathogen (Alfaro-Fernandez et al. 2017). Infection of potato tuber sample by '*Ca*. L. solanacearum' was due to a new haplotype of the pathogen. Sequence analysis showed that the sequences of the pathogen were different from the six haplotypes present already. The new haplotype was called haplotype F of '*Ca*. L. solanacearum' (Grimm and Gaczynski 2019).

'*Candidatus* Liberibacter solanacearum' has two haplotypes A and B that can infect solanaceous crops. Comparison of genomes showed differences in prophage location and sequence both within and between haplotypes. Putative coding sequences present in haplotype A, were absent in haplotype B. Eight loci assays showed consistent A-B differentiation (Thompson et al. 2015). In order to identify other sources of inoculum and vectors involved in its transmission by *Trioza apicalis*, wild plants within and near carrot fields, psyllids feeding on these plants and parsnips growing next to carrots and carrot seeds were examined. The MLST procedure was applied to detect and identify CLso haplotype C. The pathogen was detected in 11% of *Trioza apicalis* samples and in *Anthriscus sylvestris* plants showing symptoms. The results of MLST showed that CLso in *T. anthrisci* and most *A. sylvestris* plants contained different strains. CLso haplotype D was detected in carrot seed lots tested, but was not detected in the plants grown from the seeds. Another haplotype of CLso, most closely related to A and D, was detected in psyllid *T. urticae* and stinging nettle (*Urtica dioica*, Urticaceae) and it was called haplotype U (Haapalainen et al. 2018). Dynamic populations of '*Ca*. L. asiaticus' (CLas) in different citrus hosts and each instar of *D. citri* were assessed and seasonal growth and distribution of CLas in different tissues was estimated, using conventional and TaqMan real-time PCR formats. CLas could be detected earlier in plants inoculated by psyllid vector than in graft-inoculated plants. But the titer reached varied in different hosts (Lin et al. 2017). Seven haplotypes of '*Ca*. Liberibacter solanacearum' (Lso) (zebra chip disease) in potato and tomato have been identified. Lso is transmitted by potato psyllid *Bactericera cockerelli* under field conditions. Significant changes in occurrence of Lso haplotype in Idaho, USA were observed. Prevalence of pathogen population changed exclusively from haplotype A in 2012 season to an almost equal distribution of haplotype A and B during 2016 (Dahan et al. 2019).

'*Candidatus* Liberibacter asiaticus' (CLas) contains prophages, the lysogenic form of bacteriophages, as an important genetic entity. Two classes of prophages, SC1 (Type 1) and SC2 (Type 2) are involved in the lytic cycle and lysogenic cycles of bacteriophage respectively. The presence of a new Type 3 prophage in the samples of CLas was detected. Typed by prophage-specific PCR, the CLas strains in southern China contained all combinations of three prophage types, with the exception of a Type 2–Type 3 combination (Zheng et al. 2018). Next-generation sequencing technique (Illumina MiSeq) was employed to characterize '*Candidatus* Liberibacter asiaticus' (CLas) strains occurring in the United States. Two prophages (P-HHCA1–2 and P-SGCA5–1) were identified by the MiSeq read mapping technique referenced to two known Florida CLas prophage sequences SC1 and SC2. Phylogenetic analysis revealed that P-HHCA1–2 was part of an Asiatic lineage within the Type 2 prophage group. Similarly, P-SGCA5–1 was part of an Asiatic lineage within Type 1 prophage group (Zheng et al. 2017). Two hypervariable prophage and phage-related loci, bacteriophage repressor protein C1 (CLIBASIA_01645 locus) and prophage terminase gene (CLIBASIA_5610 locus) were used to determine the diversity and characterization of CLas strains associated with HLB samples. Analyses of the CLIBASIA_01645 locus, characteristics of variable tandem repeat numbers (VTRNS) revealed the homogeneity of isolates occurring in Iran. Two distinct genotypes (TRN<10) and (TRN>10) of CLas could be differentiated (Saberi et al. 2018).'*Ca*. Liberibacter – asiaticus, africanus, and americanus – are involved in citrus huanglongbing (HLB) disease in different regions/countries. Three real-time PCR variants recommended by EPPO and FAO were evaluated for their efficiencies and compared with the conventional duplex PCR format. Interlaboratory reproducibility was high, indicating applicability of the real-time PCR methods for detection and differentiation of HLB pathogens in the early stages of infection (Cellier et al. 2020). Combination of conventional PCR amplification and Sanger sequencing of the 16S recombinant DNA was employed to detect and identify '*Candidatus* Liberibacter solanacearum' in black (*Fraxinus nigra*), and mancana (*F. mandshurica*) and cottony ash psyllid (*Psyllopsis discrepans*). The bacteria showed 99–100% similarity to a '*Ca*. L. solanacearum' sequence isolated from the Nearctic psyllid (*Bactericera maculipennis*). '*Ca*. L. solanacearum' was detected in psyllids collected from ash trees and specifically in *P. discrepans*. Dieback symptoms associated with psyllid feed were consistent with disease symptoms induced by '*Ca*. Liberibacter' pathogens (Wamonje et al. 2021).

4.8 DETECTION AND IDENTIFICATION OF PHYTOPLASMAL PATHOGENS

4.8.1 Detection of Phytoplasmal Pathogens

Phytoplasmas, earlier known as mycoplasma-like organisms (MLOs), are placed in the class Mollicutes, which include minute, cell wall-less organisms, and they are present primarily in phloem sieve elements of infected plants. Aster yellows disease described by Kunkel (1926) was considered to be due to a virus, because of the similarities in the modes of transmission through grafting and leafhopper vectors. But later, these phytopathogenic organisms were recognized as a distinct group of mycoplasma-like organisms, following investigations on yellows type of plant diseases by Doi et al. (1967) and Ishiie et al. (1967). The trivial name 'phytoplasma' was officially adopted to replace the designation 'mycoplasma-like organism'. Phytoplasmas associated with plant diseases are primarily transmitted by leafhoppers, planthoppers, and psyllids. Phytoplasmas are restricted to phloem tissues of infected plants, whereas they are present in hemolymph of infective leafhopper vectors and move to salivary glands, before transmission to healthy plants. It has not been possible to culture the phytoplasmas in artificial medium and consequently their cultural characteristics are not known. Molecular techniques were applied to determine the differences in the genetic structure of these plant pathogens. Designation of a new taxon for these pathogens, was proposed as '*Candidatus* Phytoplasma' (IRPCM 2004). Investigations on the phylogeny and taxonomy of phytoplasmas provided the reliable basis of classification based on the sequences of 16S rDNA (IRPCM 2004). Phytoplasmas have variable size and shape (pleomorphic). They lack a cell wall, but have a triple-layered membrane. They are insensitive to penicillin, but sensitive to tetracyclines. In contrast, bacteria possess distinct cell wall and are sensitive to penicillins. The absence of a rigid cell wall enables the phytoplasmas to pass through smaller pores, even though the viable phytoplasmal cell may be larger (Musetti and Favelli 2004). Scanning electron microscopy (SEM) was employed to visualize the causal agent of Australian papaya yellow crinkle (PYC) disease occurring in Mexico. Phytoplasmal bodies were detected in different organs of infected plants, dry and germinated seeds within the fruit, suggesting possible seed transmission of the pathogen. Asymptomatic tomato plants from field also showed the presence of phytoplasma cells at low titers in the phloem tissues (Lebsky et al. 2010, 2011).

Cloned phytoplasma DNA probes and especially phytoplasma-specific universal (generic) or phytoplasma group-specific PCR primers designed based on the sequences of highly conserved 16S rRNA genes were widely employed for detection and identification of large number of phytoplasmas. RFLP analysis of PCR-amplified 16S rDNA sequences using different restriction endonucleases such as *Alu*I, *Hae*III or *Rsa*I was employed to identify and differentiate phytoplasmas. This resulted in distinct RFLP patterns unique to different strains which could be resolved by agarose or acrylamide gel electrophoresis. Phytoplasma groups identified based on comprehensive RFLP analyses with restriction enzymes have been considered with phylogenetic groups delineated based on sequence data. Based on extensive RFLP analysis of phytoplasma 16S rDNA sequences, phytoplasmas were classified into 14 groups and 38 subgroups (Lee et al. 1998). However, this method did not provide clear resolution of all 16Sr groups (Wei et al. 2007). The current '*Candidatus* Phytoplasma' species classification is based on the 16S rRNA gene sequences and a threshold of 97.5% similarity with any previously described species is used as the basis to propose any new species of phytoplasma. The number of phytoplasma strains reported worldwide increased exponentially with lapse of time. In order to classify phytoplasmas more rapidly, an alternative approach, an online tool was developed. The iPhyClassfier, computerized system simulates an RFLP analysis and a new sequence of phytoplasma strain could be affiliated to a 16Sr group/subgroup using this system. More than 30 groups and 100 subgroups have been listed in this system (Wei et al. 2007).

4.8.1.1 Histochemical Methods

Histochemical methods have been employed to obtain the preliminary evidence to establish the causal nature of phytoplasma in plants. A fluorescent DNA-binding dye 4,6-diamidino-2-phenylindole-2HCl (DAPI) was employed for detection of phytoplasmal infections in plants. The presence of aster yellows (AY) in aster, blueberry stunt phytoplasma and *Spiroplasma citri* in periwinkle was detected by applying DAPI fluorescence stain (Dale 1988). DAPI fluorescence test was effective in detecting ash yellow in white ash trees and the test was as consistent as dot blot hybridization assay in detecting the pathogen (Sinclair et al. 1992). Later a DAPI fluorescence test was applied to detect mulberry dwarf phytoplasma in tissue-culture generated plants grown under greenhouse conditions. The results of DAPI fluorescence test were comparable to that of PCR assay (Dai et al. 1997). The presence of the phytoplasma in Arizona velvet ash (*Fraxinus velutina*) cv. Modesta was detected using DAPI fluorescence test (Bricker and Stutz 2004). The causal nature of coffee crispiness ('Crespera') disease occurring in Colombia was established by observations under TEM. The presence of phytoplasma in leaf petiole sections and localization of pleomorphic bodies in the phloem cells of petiole tissues provided strong evidence for infection of coffee plants by phytoplasma (Galvis et al. 2007).

4.8.1.2 Immunoassays

Immunoassays employing polyclonal or monoclonal antibodies provide reliable and sensitive detection of phytoplasmas in plants and establish relationships among phytoplasmas. ELISA formats have been more frequently employed for detection of phytoplasmas infecting different crops. Polyclonal antibodies (PAbs) produced against aster yellows (AY) phytoplasma recognized a specific protein (23-kDa) in infected, but not in healthy plants. The monospecific PAbs reacted positively with AY isolates from carrots and lettuce, indicating the specificity of the PAbs (Errampalli and Fletcher 1993). Using specific PAbs in ELISA tests, *Spiroplasma citri* (citrus stubborn) could be detected in citrus plants and its vector *Circulifer haematoceps* (Najar et al. 1998). Involvement of *S. kunkellii* as one of the causative agents of corn stunt was demonstrated by employing specific F(ab')2 in protein A ELISA format (Henriquez et al. 1999). The presence of a surface protein p89 in *S. citri* was detected by electrophoresis and Western blotting. ELISA was applied to detect *S. citri* in cells of infective leafhopper vector, *Circulifer tenellus* transmitting the disease to healthy plants. The surface protein might have a role in the adhesion of *S. citri* to vector cells (Yu et al. 2000). The specificity and sensitivity of immunoassays depend on primarily on the purity of the immunogens. Double antibody sandwich (DAS)-ELISA, PCR, and reverse transcription (RT)-PCR techniques were applied to detect causal agents of sugarcane yellow leaf syndrome. The results showed the disease was due to combined infection of sugarcane by sugarcane yellow leaf phytoplasma and *Sugarcane yellow leaf virus* (ScYLV) (Aljanabi et al. 2001). Apple proliferation (AP) phytoplasma could be detected in apple roots, stems, and leaves by employing two MAbs in ELISA and IF tests (Loi et al. 2002). PCR-based detection methods did not provide reliable detection due to uneven distribution of the pathogen in infected plants. A novel diagnostic technique was developed, based on the presence of *Spiroplasma citri* CSD–secreted protein as a detection marker. Of the six secreted proteins present in the phloem extract, the protein CAR98563 was detected as a CSD marker. The direct tissue–print immunoblot technique was effective in detecting *S. citri* in young branches and petioles of citrus plants. Use of a specific protein secreted by *S. citri* as a marker for reliable detection of the pathogen was revealed by this investigation (Shi et al. 2014).

4.8.1.3 Nucleic Acid–Based Techniques

NA-based techniques are more specific, sensitive, rapid, and reliable in detecting the phytoplasmas in a wide range of plant hosts and their vectors than immunoassays. Among the NA-based techniques, PCR-based methods have been the most widely employed techniques for detection, identification, quantification, and differentiation of phytoplasmas and their strains. Several methods have been developed to extract the phytoplasmas from infected plants or inoculative insects and detected using specific DNA probe in hybridization techniques and/or primers in PCR-based techniques.

4.8.1.3.1 Nucleic Acid Hybridization Methods

The taxonomic classification of phytoplasmas is based mainly on the sequences of their 16S rDNA. Phytoplasma-specific DNA probes have been designed from chromosomal or plasmid DNA of the phytoplasma pathogens. These probes may be labeled with either radioactive ^{32}P or nonradioactive biotin or digoxigenin. Labeled oligonucleotide probes have been effectively employed as strain-specific probes in the dot blot hybridization of PCR-amplified phytoplasma DNA and in the localization of phytoplasma in plant cells using electron microscopy. Probes based on extra-chromosomal DNA may give a stronger signal, as they are present in multiple copies in many phytoplasmas. Nonradioactive labels are preferred, due to hazards associated with radioactive labels. Biotinylated cloned DNA probes were employed for detection of aster yellows (AY) and ash yellows phytoplasmas in the plant hosts and leafhopper vectors, whereas digoxigenin-labeled DNA probes were effective in detecting sweet potato witches' broom, peanut witches' broom, and rice yellow dwarf phytoplasmas (Narayanasamy 2011). Nonradioactive nucleic acid probes were employed for efficient detection of papaya bunchy top disease (BTD) caused by '*Candidatus* Phytoplasma aurantifolia'. Very well-defined and strong hybridization signals were observed in symptomatic and asymptomatic papaya plant samples and *Empoasca papayae* samples (Arocha et al. 2008). Argentinian alfalfa witches' broom (ArAWB) phytoplasma could be detected by RFLP analysis of partial 16S rRNA gene of two ArAWB isolates (Conci et al. 2005). Tomato big bud (TBB) phytoplasma was detected by RFLP analysis using endonucleases *Hha*I and *Rsa*I (Mello et al. 2006).

4.8.1.3.2 Polymerase Chain Reaction–Based Assays

General and specific primers located in the 16S rDNA intergenic spacer (IGS) and the 23S rDNA regions of phytoplasma genomes have been employed frequently. In general, PCR-based assays are more sensitive than hybridization-based assays. High-quality phytoplasma DNA, free of PCR inhibitors of plant or insect origin, is basically needed for PCR assays. The suppression subtractive hybridization (SSH) method was employed to detect the stolbur phytoplasma efficiently (Cimmerman et al. 2006). Symptoms of cassava frogskin disease (FSD) became visible only after harvest. The phytoplasma associated with CFSD was identified as a member of the 16SrIII-A group by RFLP sequences and RFLP in silico. Association of both phytoplasma and a ds-RNA virus with cassava FSD disease was revealed by this investigation (de Souza et al. 2014). The multiplex real-time PCR amplified simultaneously a fragment of apple proliferation (AP) phytoplasma 16S rRNA gene and the host *Malus domestica* chloroplast gene coding for tRNA leucine used as internal analytical control. The primer/probe allowed specific detection of all AP strains. The primers and the probe matched the sequences in the variable region of the 16S rRNA gene. The real-time TaqMan PCR assay was the most precise, rapid, and reliable testing method for apple proliferation phytoplasma (Baric and Dalla-Via 2004; Baric et al. 2006).

Different TaqMan MGB- or SYBR Green-based real-time PCR procedures were effective in detecting '*Candidatus* Phytoplasm mali', causing apple proliferation (AP) and periwinkle in crude extracts of infected tissues. The presence of '*Ca*. P. mali' could be detected in greenhouse-grown and naturally infected apple shoots and periwinkle plants (Aldaghi et al. 2009). PCR assay and RFLP analyses of 16S rRNA gene and nonribosomal DNA fragment were applied for detection of apple proliferation (AP) phytoplasma. The presence of three AP subtypes (AP-1, AP-2, and AT-15) was detected symptomatic apple trees. AP-1 subtype was predominantly detected in the northeastern Italy. In silico SNP analysis based on the 16S rDNA sequence could be used to differentiate AP-1 subtype from AP-2 and AP-15 subtypes. Isothermal recombinase polymerase amplification (RPA) assay specifically detected '*Ca*. P. mali' (Valasevitch et al. 2017). The *nusA* gene sequences of '*Candidatus* Phytoplasma asteris' (clover phyllody [CPh] phytoplasma) related strains, coding NusA protein, were amplified in PCR, followed by RFLP analysis of PCR products for detection and differentiation of phytoplasma lineages with the group 16SrI. The *nusA* gene sequences might assist in detection of '*Ca*. P. asteris'-related strains. RFLP analysis of PCR products using *Mse*I and Isp5091 restriction enzymes differentiated subgroups 16SrI-A, 16SrI-B, and 16SrI-C (Shao et al. 2006). DNA extracted from oat corn plants exhibiting stunt symptoms was used as template. Amplification and sequencing of 16 S rRNA encoding sequences and chaperonin 60 universal target (*cpn60*UT) sequences followed by in vitro RFLP and phylogenetic analyses revealed that the phytoplasma causing maize bushy stunt (MBS) symptoms was due to '*Candidatus* Phytoplasma asteris', belonging to the subgroup 16 SrI-B (Pérez-López et al. 2016). A PCR format was developed for the detection of rice orange leaf disease (ROLD) using a set of primers based on ROLP genome sequence to amplify the housekeeping gene FtsH-1of rice and leafhopper vector (*Recilia dorsalis* and *Nephotettix cincticeps*) samples. The PCR format minimized false positives occurring in the nested PCR assay (Wang et al. 2020).

A nested PCR assay using a combination of two phytoplasma-specific universal primer pairs detected a soybean green stem syndrome pathogen. RFLP analysis of nested PCR products indicated that only two of these samples contained phytoplasma DNA. Phytoplasmas belonging to aster yellows (AY) phytoplasma group 16SrI, subgroup 16SrI-A and a newly established subgroup 16SrI-O were present in these two samples. Phylogenetic analysis using partial 16S rDNA sequences (1.2-kb) from representative phytoplasma strains clustered soybean phytoplasmas into two distinct phylogenetic lineages (Lee et al. 2002). Black pepper (*Piper nigrum*) phyllody phytoplasma was detected by applying the nested PCR format. A 1.20-kb DNA fragment was consistently amplified. Black pepper phyllody phytoplasma was identified as a member of

AY group based on sequence homology and phylogenetic studies (Bhat et al. 2006). Direct and nested PCR formats were applied to detect the sesame phyllody phytoplasma in sesame plants and weed *Sclerocarpus africanus*. BLAST analysis of 1.25-kb 16S rDNA partial sequences of nested PCR products obtained from symptomatic sesame plants, weed species and leafhopper revealed 99% sequence identities among themselves and 99% identity with other reported strains of '*Candidatus* Phytoplasma asteris' (16SrI group) (Nabi et al. 2015). Phytoplasma causing sesame phyllody disease in India, was investigated to establish its identity. The phytoplasma had sequence relatedness with similarity coefficiency of 1.00 to '*Ca. P. aurantifolia*', which represented 16SrII-D group (Pamei and Makandar 2016). RFLP analysis of PCR-amplified 16S rDNA sequences revealed that carrot plants exhibiting dark purple or yellow purple leaf symptoms tested positive for *Spiroplasma citri*. Clover proliferation phytoplasma (16SrVI, subgroup 16SrVI-A) and AY group (16SrI, subgroup 16SrI-A were detected in infected carrot plants (Lee et al. 2006). In a later investigation, PCR assay followed by RFLP analysis and/or sequencing of another carrot-infecting phytoplasma rDNA and ribosomal protein genes *l22* and *ss33*, *tuf*, putative akinase plus ribosomal recycling factor genes and DNA helicase gene effectively identified and differentiated phytoplasmas infecting carrots. Use of several molecular markers provided unambiguous identification and differentiation of phytoplasma strains infecting carrots (Duduk et al. 2009). TEM revealed the presence of pleomorphic phytoplasma-like bodies of tomato big bud phytoplasma in the phloem sieve tube elements in symptomatic tomato plants in China. The product specifically amplified by PCR using universal primers R16F2n/R16R2 exhibited 16S rRNA gene sequence similarity with members of '*Candidatus* Phytoplasma aurantifolia' (16SrII group). Phylogenetic and RFLP analyses of the 16S rRNA gene sequences confirmed the phytoplasma as a member of the 16SrII group (Xu et al. 2013). The phytoplasma inducing malformation of flowers (phyllody) in cauliflower was detected based on 16S rRNA gene sequences in DNA from infected plants. Phylogenetic and virtual RFLP analyses of the 16S rRNA gene sequence indicated that the phytoplasma was a strain of '*Candiatus* Phytoplasma aurantifolia' Molecular markers present in the *secY* and *rp* loci distinguished their phytoplasma from other members of the subgroup 16SrII-A (Cai et al. 2016).

Multiplex-PCR assay was developed using 36 primers based on the genome sequence of '*Candidatus* Phytoplasma asteris'(aster yellows phytoplasma) to amplify 18 single copy genes covering a wide range of pathogen genome. Nine genes could be amplified in a single PCR assay. The multiplex PCR format efficiently detected 10 phytoplasma strains belonging to the AY group, and different amplification patterns generated could be used to differentiate these strains (Kakizawa and Kamagata 2014). The phytoplasma 16S rDNA gene primers were used in the nested PCR format for detection of the pathogen in the leafhopper vector. The assay was performed with universal primers P1/P7 and R16F2n/R16R2. The presence of AYp in leafhoppers was detected by nested PCR and TaqMan qPCR assay. The qPCR assay was more rapid than nested PCR assay (Demeuse et al. 2016). DNA extracts from symptomatic and asymptomatic potato plants showing rosette symptoms, were subjected to PCR with primer pair P1/P7, followed by nested PCR with P1A/P7A, P1A/16S-SR or R16F1n/R16R2n for amplification of 16S rRNA of phytoplasmas. Nested PCR products were cloned and sequenced. Phytoplasmas infecting potatoes shared high similarity with '*Candidatus* Phytoplasma fragariae' (16SrXII-E7) and other strains in 16SrXII subgroups. Five strains belonged to 16SrXII-E and 11 strains were called a new 16SrXII-I subgroup (Cheng et al. 2015). Phytoplasma associated with HLB-infected grapefruit (*Citrus paradisi*) plants was characterized. Phylogenetic analysis and virtual RFLP analysis confirmed the phytoplasma was a variant (16SrII-A*) of phytoplasma subgroup 16SrII-A (Lou et al. 2014). HLB samples that tested negative for '*Ca.* L. asiaticus', '*Ca.* L. americanus' and 16SrIX group phytoplasma, were infected by 16SrIII phytoplasmas (Wulff et al. 2019).

Real-time PCR assays were employed for specific diagnosis of Flavescence doree (FD), Bois noir (BN), and apple proliferation (AP) phytoplasmas and universal primer pairs for detection of phytoplasmas belonging to groups 16SrV, 16SrX, and 16SrXII were employed. The group-specific assays provided highly efficient detection comparable to nested PCR format (Galetto et al. 2005). In a later study, grapevine yellows (GY) phytoplasma was detected using real-time PCR format with TaqMan minor groove binder probes (Hren et al. 2007). GY phytoplasma infection occurred along with molecularly indistinguishable phytoplasmas, causing FD and BN diseases. By employing generic primers for phytoplasmas, positive amplification of expected pathogen DNA fragment was accomplished. In the second round amplification using specific primers for FD and BN, presence of BN (stolbur) was detected in three samples. Infection of grapevine by stolbur phytoplasma could be confirmed by applying a multiplex nested PCR format. The technique was reliable for testing mother plants during crop season and selecting healthy stocks (Avramov et al. 2008). TaqMan probes and primers were designed based on the 16S rRNA gene sequences of FD, BN and AY phytoplasma genomes. Another TaqMan assay targeting a grapevine gene encoding chloroplast chaperonin 21 was employed to ascertain the DNA quality and to infer absence of PCR inhibitors. Two TaqMan assays and nested PCR format were equally efficient in detecting the phytoplasmas in grapevine, periwinkle, and other plant species tested. Real-time PCR assay showed potential for analyzing a large number of samples, suitable for certification and quarantine programs (Angelini et al. 2007).

North American grapevine yellows (NAGY) disease-infected plants contain phytoplasmas belonging to different groups. The strains were subjected to genotyping based on analysis of 16S rRNA and *secY* genes and to in silico three-dimensional modeling on the SecY protein.

The results from genotyping and protein modeling might reflect ongoing evolutionary divergence of NAGY1 strains from related strains in subgroup 16SrIB (Davis et al. 2018). '*Candidatus* Phytoplasma solani' inducing Bois noir phytoplasma (BNp) disease in grapevine was detected by PCR and RFLP analysis of 16S rDNA, followed by multiple gene typing analysis (*vmp1* and *stamp* genes) (Quaglino et al. 2016). '*Ca.* P. solani' exists in the form of several strains included in different taxonomic subgroups 16SrXII-A, -F, and -G. Existence of extensive intraspecies diversity among strains of '*Ca.* P. solani' infecting grapevines was recorded (Quaglino et al. 2017). Genetic diversity among strains of '*Ca.* P. solani' was determined by nucleotide sequence analysis of *vmp1* and *stamp* genes and 12 *vmp1* and 16 *stamp* sequence variants, showing an overall positive selection for such genes were identified. The genotype Vm43/St10 showed significant differences in their distribution in grapevines exhibiting distinct symptom severity (Pierro et al. 2018). Immunodominant membrane protein or Imp of apple proliferation group phytoplasmas showed high inter- and intra-species gene variability. Discrimination of '*Ca.* Phytoplama pyri' genotypes could be achieved by using a RFLP-PCR analysis of genotypes (Bohunicka et al. 2018). Recombinase polymerase amplification (RPA) method was developed to detect rapidly '*Ca.* P. pruni' causing Western X (WX) disease of sweet cherry, for removal and replacement of the infected trees in time. RPA assay targeting putative immunodominant protein coding regions was efficient in detecting '*Ca.* P. pruni' in crude sap of sweet cherry tissues. Multilocus sequence analyses using immunodominant protein A (*idpA*), *imp, ropE, secY*, and 16S rRNA regions from several isolates from WX-infected trees showed that this novel strain represented a new subgroup within the 16SrIII group (Villamor and Eastwell 2019).

Phytoplasma inducing yellow symptoms in pomegranate was detected by extracting total nucleic acids from fresh leaf midribs and phloem tissues from young branches of ten symptomatic and five asymptomatic plants. Nested PCR assay using universal phytoplasma specific 16S rRNA and *tuf* gene primers generated unique amplicons. Restriction enzymes *Tru*1I, *Tsp*5091I and *Hha*I were used to digest amplicons. The phytoplasma profiles were identical to each other and to aster yellow (16SrI-B) strain (Gazel et al. 2016). Primrose (*Primula acaulis*) plants showed stunting, leaf-yellowing, and virescence, indicating possible infection by phytoplasma belonging to the aster yellows group, ribosomal subgroup 16SrI-B, tufI-B, groELIB-III and SecY-IB subgroups. Phylogeny of the 16Sr rRNA gene sequences as well as sequence analysis of several chromosomal regions confirmed its affiliation with '*Candidatus* Phytoplasma asteris' species (Fránová et al. 2016).

Sugarcane plants may be simultaneously infected by sugarcane yellows phytoplasma (SCYP) and sugarcane yellow leaf virus (SCYLV). A nested PCR assay with different primers and RT-PCR assay were employed to detect SCYP and SCYLV, respectively. Sequencing and RFLP analysis showed that all phytoplasma isolates belonged to the 16SrXI (rice yellow dwarf phytoplasma) group (El Sayed et al. 2016). Infection of coconut by a phytoplasma of 16SrIV group resulted in lethal yellowing (LY) symptoms in Mexico. The *GroEL* gene was employed for detection of subgroup 16SrIV-A and 16SrIV-D. Sensitive, rapid, and reliable singleplex and duplex assays were developed based on real-time PCR targeting both *GroEL* gene and 16S rRNA. The duplex TaqMan probe assay had the potential for simultaneous detection and quantification of LY phytoplasmas of 16Sr subgroups -A and -D (Córdova et al. 2019). Infection of poinsettia by poinsettia branch-inducing (PoiBI) phytoplasma caused variable infection patterns with ornamental value. PCR and sequence analysis were performed to detect the pathogen in cv. Luv U Pink. Infection densities of PoiBI tend to be higher in source leaves located in the lower parts of the plant as indicated by qPCR format. Densities were lower consistently at the top of plants, regardless of tissue types. PoiBI infection density and extent of branching showed positive relationship in cutting-propagated plants. Uneven distribution of PoiBI within among stock plants might lead to the production of cuttings with variable pathogen densities which might influence phenotypic uniformity of plants (Lee et al. 2020).

4.8.1.3.3 Loop-Mediated Isothermal Amplification Assays

LAMP assay, when combined with rapid DNA extraction procedure, has the potential to provide rapid in-field detection and identification of phytoplasmas. Primers for LAMP assays were designed against a range of ribosomal groups, 16SrI, 16SrII, 16SrIII, 16SrV, 16SrXI, and 16SrXXII. The primers were group-specific, and they amplified only DNA samples from the group which they were designed to amplify. Samples of phytoplasmas belonging to 16SrXXII group were amplified within 20 min (Hodgetts et al. 2011). Phytoplasma detection protocol based on LAMP assay was applied by employing primers targeting *groEL* gene and 16S rDNA to detect '*Candidatus* Phytoplasma asteris'. Pathogen DNA was efficiently amplified by one of the primer sets targeting the *groEL* gene. Sensitivity of LAMP assay was enhanced by 10-fold, and *groEL* gene could be effectively amplified from several strains of '*Ca.* P. asteris' using this primer set (Sugawara et al. 2012). A real-time isothermal LAMP assay was developed for detection of Flavescence dorée phytoplasma (FDp) (grapevine yellows disease). The assay detected FDp in all leaves and crude homogenate of grapevine leaf vein, flower or berry samples from symptomatic plants as template under field conditions (Kogovsek et al. 2015).

Fluorescence based real-time LAMP assay was optimized for rapid and sensitive detection of '*Candidatus* Phytoplasma solani, (Bois noir phytoplasma, BNp). Grapevine leaf samples were tested with real-time LAMP assay and the assay was also employed for detection of Flavescence dorée phytoplasma (FDp). Both phytoplasmas could be rapidly detected by LAMP assay. LAMP assay provided the results in less than 35 minutes as effectively as qPCR assay (Kogovsek et al. 2017). For the detection of phytoplasma infections in *Malus, Pyrus*, and *Prunus*, the three principal hosts for apple proliferation (AP), pear decline (PD), and European stone fruit yellows (ESFY), the LAMP-based method was applied. The LAMP technique was preferred because of its high specificity and sensitivity, requirement of less stringent need for DNA purification, and short analysis time and inexpensive equipment required for the assays. LAMP primer set for the ribosomal group 16SrX, containing AP, PD, and ESFY phytoplasmas were designed and validated for detection of these phytoplasmas by LAMP procedure, which showed potential for on-site diagnosis of the three phytoplasmas (De Jonghe et al. 2017). LAMP assay was developed for detection of '*Ca.* Phytoplasma pyri' (pear decline, PD). A field-applicable LAMP assay was developed for detection of '*Ca.* P. pyri', targeting the phytoplasmal 16S rDNA sequence. Ease of application, cost, and rapidity of the LAMP assay are the advantages over PCR-based assays (Siemonsmeier et al. 2019). An efficient LAMP detection system was developed based on two sets of primers, 16SrDNA-2 and 16SrDNA-3 of the genome of areca palm yellow leaf (AYL) phytoplasma (16SrI group). AYL phytoplasma could be detected in all diseased arecanut seedlings tested (Yu et al. 2020). On-site applicable LAMP assay was developed for detection of sugarcane white leaf disease (SCWLD) phytoplasma for rapid and efficient detection of the phytoplasma belonging to 16SrXI group. Plant cytochrome oxidase (cox) LAMP primers that amplify housekeeping gene in plants were employed as controls (Quoc et al. 2021).

4.8.2 Variability in Phytoplasmal Pathogens

Phytoplasma DNA extracted from infected plant or vector tissues were analyzed to determine the extent of variations in sequences of genes governing pathogenicity, vector transmissibility, and other characteristics. Genetic diversity of isolates of phytoplasmas was assessed by different molecular typing methods. Stolbur phytoplasmas belonging to 16SrXII-A subgroup, were detected in some of the symptomatic plants, based on 16S rDNA amplification and RFLP analysis. The genotype of the isolates was determined based on the restriction patterns or sequence analysis of two nonribosomal housekeeping genes (*tuf* and *secY*) and polymorphic gene *vmp1*. The *tuf* genotyping placed all isolates in the tufAY-b genotype (Fialova et al. 2009). Stolbur phytoplasma causes Bois noir (BN) in grapevines. Two nonribosomal genes *tuf* and *vmp1* sequences were used as the basis for differentiation of stolbur isolates following their characterization by PCR/RFLP techniques. The gene *vmp1* exhibited higher polymorphism in isolates of grapevines than in those of herbaceous plants (Murolo et al. 2010). Isolates of Bois noir (BN) phytoplasma inducing grapevine yellows, '*Candidatus* Phytoplasma solani', were characterized by MLSA, and their *secY* and *stamp* gene sequences were compared with strains from other plants. Genotypes were determined for two nonribosomal housekeeping genes *tuf* and *secY*, as well as two variable markers namely *stamp* and *mleP1* genes. The BN phytoplasma isolates corresponded to three *tufB* and *secY* genotypes. Five different *mleP1* genetic variants could be identified. Further, *stamp* allowed discrimination of four new genotypes in grapevine (Balakishiyeva et al. 2018).

'*Candidatus* Phytoplasma prunorum' causes ESFY in trees and shrubs of the genus *Prunus*, whereas '*Ca.* P. mali' induces AP in plants belonging to *Malus* and '*Ca.* P. pyris' is responsible for PD in *Pyrus*. These phytoplasmas belong to the same phylogenetic cluster 16SrX group and they are transmitted by psyllids of the genus *Cacopsylla*. The presence of hypovirulent isolates of '*Ca.* P. prunorum' indicated the existence of variability in virulence, suggesting the possible use of hypovirulent strains for cross-protection against virulent strains. The genetic diversity of the three temperate fruit tree phytoplasmas was investigated by multilocus sequence typing (MLST) approach. Among the four genetic loci used, the genes *imp* and *aceF* differentiated 30 and 24 genotypes, respectively, and exhibited the highest degree of genetic variability. The percentage of substitution varied between 9% and 12% for *aceF*, where variation was between 5% and 6% for *imp* and *secY* genes. The most prevalent *aceF* genotype of '*Ca.* P. prunorum' was detected in both plant and insect vectors, confirming propagation of the phytoplasma in insects. Hypovirulent isolates had the same *aceF* sequence, indicating a possible monophyletic origin. Among 34 haplotypes of '*Ca.* P. prunorum' four hypovirulent isolates were grouped together in the same clade. Genotyping of some Spanish and Azerbaijanian '*Ca.* P. pyri' isolates showed that they shared some alleles with '*Ca.* P. prunorum' indicating the existence of interspecies recombination between these two species (Danet et al. 2011). Genetic variability of phytoplasmas present in selected vineyards showing grapevine yellows in Northern Italy was investigated. The presence of flavescence dorée (FD) (16SrV-C/D), as well as '*Ca.* P. solani' causing bois noir (BN) (16SrVII-A), '*Ca.* P. asteris'-related aster yellow (16SrI-B), '*Ca.* P. prunorum'-related European stone fruit yellows, and '*Ca.* P. ulmi'-related yellows (16SrV-A) was detected. Direct sequences of selected amplicons obtained from *16S rRNA*, *rp*, and *tuf* genes from grapevine and insect samples confirmed the identity of the phytoplasmas occurring in test location (Zambon et al. 2018).

Variability in genotypes of phytoplasma involved in the GY pathosystem was assessed by employing an MLST approach. Samples from grapevine, weed species, and insect vectors were analyzed. Species-specific *stamp* and *vmp1* genes were amplified from BN strains, whereas the genes *secY, map*, and *uvrB-degV* in FD strains were analyzed. MLST analysis showed a diversity of BN genotypes. One genotype was present grapevine as well as in the vector *H. obsoletus*. The presence of distinct BN strains in bindweed and two grapevine samples indicated the existence of different BN pathosystems. Combined infection of grapevine by BN and FD phytoplasmas in the same vineyard was detected. Genotyping of FD strains from both grapevine plants and its vector *Scapholetus titanus* revealed the presence of two distinct FD genotypes at two different locations (Plavec et al. 2015). Two Flavescence dorée phytoplasma (FDp) taxonomic groups, 16SV-C and 16SV-D, were differentiated based on the sequence and RFLP analyses of the 16S rRNA and 16S-23S intergenic spacers. Further, sequencing of two nonribosomal loci, *secY* and *rpsC* allowed the identification of three genetic clusters with FDp populations present in France and Italy. Two new genetic markers for FDp characterization, *dnaK* and *malG*, along with *vmpA* locus were used to genetically characterize the FDp populations. The markers *dnaK* and *malG* were highly efficient in discriminating FD-C and FD-D strains due to robustness and variability (Rossi et al. 2019).

Aster yellows phytoplasma (AYp) '*Candidatus* Phytoplasma asteris' (AYp) can infect several commercially important crops such as lettuce, celery, carrot, wheat and oats. Transmission of AYp occurs primarily horizontally by aster leafhopper (ALH) *Macrosteles quadrilineatus* to plants and infected plants to insect vector. Analysis of 16S rRNA region showed that AYp subclade is composed of multiple subgroups. AYp was detected in ALH using primers for amplification of the conserved 16Sr RNA region of AYp genome. The isolates of AYp belonged to either subgroup 16SrI-A or 16SrI-B. Some of the ALHs appeared to be coinfected by both subgroups. Although similar AYp strains were detected in geographically distinct locations, the proportions of each genotype varied over time (Clements et al. 2020). '*Candidatus* Phytoplasma ulmi' (elm yellows) belonging to 16SrV group was subdivided into five subgroups A to E, based on RFLP analyses, although the members of this group are homogenous in respect to the 16S rRNA sequences. A close relationship between American and European strains was revealed by Southern blot hybridization and RFLP and sequence analyses. Based on sequence homology of 16S rRNA in plant and insect vectors, the novel taxon '*Ca.* P. ulmi' was proposed. Phylogenetic analyses of *groEL* gene clustered all '*Ca.* P. ulmi' strains present in Germany and separated them from related phytoplasmas of the 16SrV group (Schneider et al. 2020).

REFERENCES

Adachi N, Tsukamoto S, Inoue Y, Azegami K (2012) Control of bacterial seedling rot and seedling blight of rice by bacteriophages. Plant Dis 96: 1033–1036.

Adhikari P, Adhikari TB, Timilsina S, Meadows I, Jones JB et al. (2019) Phenotypic and genetic diversity of *Xanthomonas perforans* populations from tomato in North Carolina. Phytopathology 109: 1533–1543.

Adhikari TB, Gurung S, Hansen JM, Bonman JM (2012) Pathogenic and genetic diversity of *Xanthomonas translucens* pv. *undulosa* in North Dakota. Phytopathology 102: 390–402.

Agrios GN (2005) *Plant Pathology*, fifth edition. Elsevier/Academic Press, Amsterdam, Netherlands.

Aldaghi M, Massart S, Dutrecq O, Bertaccini A, Jijakli MH (2009) A simple and rapid protocol of crude DNA extraction from apple trees for PCR and real-time PCR detection of '*Candidatus* Phytoplasma mali'. J Virol Meth 156: 96–101.

Alfaro-Fernandez A, Hernandez-Lopez D, Font MI (2017) Haplotypes of '*Candidatus* Liberibacter solanacearum' identified in umbelliferous crops in Spain. Eur J Plant Pathol 149: 127–131.

Aljanabi SM, Paramessure Y, Moutia Y, Saumtally A, Dookun A (2001) Further evidence of the association of a phytoplasma and a virus with yellows syndrome in sugarcane. Plant Pathol 50: 628–636.

Alnaasan Y, Valentini F, Balestra GM, Mazzaglia A, D'Onghia AM et al. (2017) Modification of a multiple locus variable number tandem repeat analysis (MLVA) for typing isolates of *Erwinia amylovora*. Plant Pathol 66: 1075–1080.

Alvarez AM (2004) Integrated approaches for detection of plant pathogenic bacteria and diagnosis of bacterial diseases. Annu Rev Phytopathol 42: 339–366.

Ananthakrishnan G, Choudhary N, Roy A, Sengoda VG, Postnikova E et al. (2013) Development of primers and probes for genes and species-specific detection of '*Candidatus* Liberibacter' species by real-time PCR. Plant Dis 97: 1235–1243.

Angelini E, Bianchi GL, Filippin L, Morassutti C, Borgo M (2007) A new TaqMan method for the identification of phytoplasmas associated with grapevine yellows by real-time PCR assay. J Microbiol Meth 68: 613–622.

Arocha Y, Almeida R, Jones P (2008) Using a ribosomal probe for the detection of phytoplasma associated with bunchy top symptom of papaya (BTS) in plants and insect hosts. Rev Prot Veg 23: 16–20.

Ash GJ, Lang JM, Triplett LR, Stodert BJ, Verdier V et al. (2014) Development of a genomic-based LAMP (loop-mediated isothermal amplification) assay for detection of *Pseudomonas fuscovaginae* from rice. Plant Dis 98: 909–915.

Avramov Z, Gille J, Langinova M (2008) First detection of stolbur phytoplasma in grapevine (*Vitis vinifera* cv. Merlot) affected with grapevine yellows in Bulgaria. J Phytopathol 156: 112–114.

Bahar D, Efrat E, Hadar E, Dutta B, Walcott RR et al. (2008) New subspecies-specific polymerase chain reaction based assay for the detection of *Acidovorax avenae* subsp. *citrulli*. Plant Pathol 57: 754–763.

Balakishiyeva G, Bayramova I, Mamamadov A, Salar P, Danet J-L et al. (2018) Important genetic diversity of '*Candidatus* Phytoplasma solani-related strain associated with bois noir grapevine yellow and planthoppers in Azerbaijan. Eur J Plant Pathol 151: 937–946.

Balestra GM, Taratufolo MC, Vinatzer BA, Mazzaglia A (2013) A multiplex PCR assay for detection of *Pseudomonas syringae* pv. *actinidiae* and differentiation of populations with different geographic origin. Plant Dis 97: 472–478.

Balogh B, Dickstein ER, Jones JB, Canteros BI (2013) Narrow host range phages associated with citrus canker lesions in Florida and Argentina. Eur J Plant Pathol 135: 253–264.

Bangratz M, Wonni I, Kini K, Sondo M, Brugidou C et al. (2020) Design of a new multiplex PCR assay for rice pathogenic bacteria detection and its application to infer disease incidence and detect coinfection in rice fields in Burkina Faso. PLoS ONE: https://doi.org/10.1371/journal.pone 0232115

Bao M, Zhang Z, Sun X, Chen J, Deng X (2020) Enhancing PCR capacity to detect '*Candidatus* Liberibacter asiaticus' utilizing whole genome sequence information. Plant Dis 104: 527–532.

Baric S, Dalla-Via J (2004) A new approach to apple proliferation detection: a highly sensitive real-time PCR assay. J Microbiol Meth 57: 135–145.

Baric S, Kerschbamer C, Via JD (2006) TaqMan real-time PCR versus four conventional PCR assays for detection of apple proliferation phytoplasma. Plant Molec Biol Rep 24: 169–184.

Basim H, Basim E (2018) Phenotypic and genotypic characterization of *Clavibacter michiganensis* subsp. *michiganensis*, causing tomato bacterial canker and wilt disease in Turkey. Eur J Plant Pathol 151: 355–369.

Benedict A, Alvarez AM, Berestecky J, Imanaka W, Mizumoto C, Pollard L, Mew T, Gonzalez C (1989) Pathovar-specific monoclonal antibodies for *Xanthomonas campestris* pv. *oryzae*. Phytopathology 79: 322–328.

Bereswill S, Bugert P, Volksch B, Ulrich M, Bender CL (1994) Identification and relatedness of coronatine-producing *Pseudomonas syringae* pathovars by PCR analysis and sequence determination of amplification products. Appl Environ Microbiol 60: 2924–2930.

Berg T, Tesoriero I, Hailstones DL (2005) PCR-based detection *of Xanthomonas campestris* pv. *campestris* pathovars in Brassica. Plant Pathol 54: 416–426.

Besler KR, Little EL (2017) Diversity of *Serratia marcescens* strains associated with cucurbit yellow vine disease in Georgia. Plant Dis 101: 129–136.

Bextine BR, Miller TA (2004) Comparison of whole-tissue and xylem fluid collection techniques to detect *Xylella fastidiosa* in grapevine and oleander. Plant Dis 88: 600–604.

Bhat AI, Madupriya R, Hareesh PS, Anandaraj M (2006) Detection of characterization of the phytoplasma associated with a phyllody disease of black pepper (*Piper nigrum* L.) in India. Sci Hortic 107: 200–204.

Biondi E, Galeone A, Kuzmanovic N, Ardizzi S, Lucchese C et al. (2013) *Pseudomonas syringae* pv. *actidiniae* detection in kiwifruit plant tissue and bleeding sap. Ann Appl Biol 162: 60–70.

Bock CH, Gottwald TR, Graham JH (2014) A comparison of pathogen isolation in culture and injection-infiltration bioassay of citrus leaves for detecting *Xanthomonas citri* subsp. *citri*. J Phytopathol 162: 291–301.

Bohunicka M, Valentova L, Sucha J, Necas T, Eichneier A et al. (2018) Identification of 17 '*Candidatus* Phytoplasma pyri' genotypes based on the diversity of the *imp* gene sequence. Plant Pathol 67: 971–977.

Bouzar H, Jones JB, Minsavage GV, Stall RE, Scott JE (1994) Proteins unique to phenotypically distinct groups of *Xanthomonas campestris* pv. *vesicatoria* revealed by silver staining. Phytopathology 84: 39–44.

Bricker JS, Stutz JC (2004) Phytoplasmas associated with ash decline. J Arboric 30: 193–199.

Burdman S, Kots N, Kritzman G, Kopelowitz J (2005) Molecular, physiological and host range characterization of *Acidovorax avenae* subsp. *citrulli* isolates from watermelon and melon in Israel. Plant Dis 89: 1339–1347.

Buzkan N, Kocsis L, Walker MA (2005) Detection of *Xylella fastidiosa* from resistant and susceptible grapevine by tissue sectioning and membrane entrapment immunofluorescence. Microbiol Res 160: 225–231.

Buzomobo P, Jaimes J, Lam V, Centrell K, Harkness M et al. (2006) An American hybrid vineyard in Texas Gulf Coast: analysis within a Pierce's hot zone. Amer J Enol Vitic 57: 347–355.

Cai H, Wang L, Mu W, Wan Q, Wei W et al. (2016) Multilocus genotyping of a '*Candidatus* Phytoplasma aurantifolia'-related strain associated with cauliflower phyllody disease in China. Ann Appl Biol 169: 64–74.

Cellier G, Redondo C, Cubero J, Rosello M, de Andarade E et al. (2020) Comparison of the performance of the main real-time and conventional PCR detection tests for '*Candidatus* Liberibacter' spp., plant pathogenic bacteria causing the huanglongbing disease in *Citrus* spp. Eur J Plant Pathol 157: 919–941.

Champoiseau P, Daugrois J-H, Girard D-C, Royer M, Rott PC (2006) Variation in albicidin biosynthesis genes and in pathogenicity of *Xanthomonas albilineans*, the sugarcane scald pathogen. Phytopathology 96: 33–45.

Chen C, Bock CH, Brannen PM (2019) Novel primers and sampling for PCR detection of *Xylella fastidiosa* in peach. Phytopathology 109: 307–317.

Chen J, Groves R, Civerolo EL, Viveros M, Freeman M et al. (2005) Two *Xylella fastidiosa* genotyping associated with almond leaf scorch disease on the same location in California. Phytopathology 95: 708–714.

Chen J, Hu Y, Qin K, Yang X, Jia Z et al. (2018) A serological approach for the identification of the effector *hopz5* of *Pseudomonas syringae* pv. *actinidiae*, a tool for the rapid immunodetection of kiwifruit bacterial canker. J Plant Pathol 100: 171–177.

Chen Z-D, Kang H-J, Chai A-L, Shi Y-X, Xie X-W et al. (2020) Development of a loop-mediated isothermal amplification (LAMP) for rapid detection of *Pseuodomonas syringae* pv. *tomato* in planta. Eur J Plant Pathol 156: 739–750.

Cheng M, Dong J, Bottner-Parker KD, Zhao Y, Davis RE et al. (2015) Group 16SrXII phytoplasma strains, including subgroup 16SrXII-E ('*Candidatus* Phytoplasma fragariae') and a new subgroup 16SrXII-I are associated with diseased potatoes (*Solanum tuberosum*) in theYunnan and Inner Mongolia regions of China. Eur J Plant Pathol 142: 305–318.

Chiesa MA, Sicialiano MF, Ornella L, Roeschlin RA, Favro MA et al. (2013) Characterization of a variant of *Xanthomonas citri* subsp. *citri* that triggers a host specific disease response. Phytopathology 103: 555–564.

Chitaara LG, Breeuwer P, Abee T, van den Bulk RW (2006) The use of fluorescent probes to assess viability of the plant pathogenic bacterium *Clavibacter michiganensis* subsp. *michiganensis* by flow cytometry. Fitpatol Bras 31: 349–356.

Cho MS, Kang MJ, Kim CK, Seol Y-J, Hahn JH et al. (2011) Sensitive and specific detection of *Xanthomonas oryzae* pv. *oryzae* by real-time BIO-PCR using pathovar-specific primers based on *rhs* family gene. Plant Dis 95: 589–594.

Choi H-K, da Silva FG, Lim H-J, Iandolino LA, Seo YS et al. (2011) Diagnosis of Pierce's disease using biomarkers specific to *Xylella fastidiosa* RNA and *Vitis vinifera* gene expression. Phytopathology 100: 1089–1099.

Cimmerman A, Arnaud G, Foissac X (2006) Stolbur phytoplasma genome survey achieved using a suppression subtractive hybridization approach with high specificity. Appl Environ Microbiol 72: 3274–3283.

Clements J, Garcia M, Bradford B, Crubaugh L, Piper S et al. (2020) Genetic variations among geographically disparate isolates of aster yellows phytoplasma in the contiguous United States. J Econ Entomol 113: 604–611.

Coletto-Filho HD, Francisco CS, Lopes JRS, Muller C, Almeida RPP (2017) Homologous recommendation and *Xylella fastidiosa* host-pathogen associations in South America. Phytopathology 107: 305–312.

Colletto-Filho HD, Takita MA, Targon MLPN, Machado MA (2005) Analysis of 16S rDNA sequences from citrus huanglongbing bacteria reveal a different '*Candidatus* Liberibacter' strain associated with citrus disease in Sao Paulo. Plant Dis 88: 848–852.

Colletto-Filho HD, Targon MLPN, Takita MA, de Negri JD, Pompedu J Jr et al. (2004) First report of the causal agent of huanglongbing ('*Candidatus* Liberibacter asiaticus') in Brazil. Plant Dis 88: 1382.

Conci L, Meneguzzi N, Galdeano E, Torres L, Nome C et al. (2005) Detection and molecular characterization of an alfalfa phytoplasma in Argentina that represents a new group in the 16S rDNA ash yellow group ('*Candidatus* Phytoplasma fraxini'). Eur J Plant Pathol 113: 255–265.

Córdova I, Oropeza C, Harrison N, Ku-Rodríguez S, Puch-Hau C et al. (2019) Simultaneous detection of coconut lethal yellowing phytoplasmas (group 16SrIV)) by real-time PCR assays using 16Sr- and *GroEL*-based TaqMan probes. J Plant Pathol 101: 609–619.

Costa HS, Raetz E, Pinckard TR, Gaspert C, Hernandez-Martinez R et al. (2004) Plant hosts of *Xylella fastidiosa* in and near southern California vineyards. Plant Dis 88: 1255–1261.

Cottyn B, Bautista AT, Nelson RJ, Leach LE, Swings J et al. (1994) Polymerase chain reaction amplification of DNA from bacterial pathogens of rice using specific oligonucleotide primers. Internatl Rice Res Notes 19: 30–32.

Cubero J, Graham JH (2005) Quantitative real-time polymerase chain reaction for bacterial enumeration and allelic discrimination to differentiate *Xanthomonas* strains on citrus. Phytopathology 95: 1333–1340.

Curland RD, Gao L, Bull CT, Vinatzer BA, Bill-Macky et al. (2018) Genetic diversity and virulence of wheat and barley strains of *Xanthomonas translucens* from the upper Midwestern United States. Phytopathology 108: 443–453.

Da Gama MAS, Mariano RLM, da Silva WJ Jr, de Farias ARG, Barbosa MAG et al. (2018) Taxonomic respositioning of *Xanthomonas campestris* pv. *viticola* (Nayudu 1972) as *Xanthomonas citri* pv. *viticola* (Nayudu 1972) Dye 1978 comb. nov and emendation of the description of *Xanthomonas citri* pv. *anacardi* to include pigmented isolates pathogenic to cashew plant. Phytopathology 108: 1143–1153.

Dahan J, Wenninger EJ, Thompson BD, Eid S, Olsen N et al. (2019) Prevalence of '*Candidatus* Liberibacter solanacearum' haplotypes in potato and psyllid vectors in Idaho from 2012 to 2018. Plant Dis 103: 2587–2591.

Dai Q, He FT, Lue PY (1997) Elimination of phytoplasma by stem culture from mulberry plants (*Morus alba*) with dwarf disease. Plant Pathol 46: 56–61.

Dale JL (1988) rapid comparison technique for detecting mycoplasma-like organisms in leaf midrib sieve tubes by fluorescence microscopy. Phytopathology 78: 118–120.

Danet JL, Balakrishiyeva G, Cemerman A, Sauvion N, Marie-Jeanne V et al. (2011) Multilocus sequence analysis reveals the genetic diversity of European fruit tree phytoplasmas and supports the existence of inter-species recombination. Microbiology 157: 438–450.

Davis RE, Dally EL, Zhao Y, Wolf TK, Smith AH (2018) Genotyping points to divergent evolution of '*Candidatus* Phytoplasma asteris' strains causing North American grapevine yellows and strains causing aster yellows. Plant Dis 102: 1697–1702.

De Jonghe K, De Roo I, Maes M (2017) Fast and sensitive onsite isothermal assay (LAMP) for diagnosis and detection of three tree phytoplasmas. Eur J Plant Pathol 147: 749–759.

de Paiva BAR, Wendland A, Teixeira NC, Ferreira MASV (2020) Rapid detection of *Xanthomonas citri* pv. *fuscans* and *Xanthomonas phaseoli* pv. *phaseolicola* in common bean loop-mediated isothermal amplification. Plant Dis 104: 198–203.

De Parasis J, Roth DA (1990) Nucleic acid probes for identification of genus specific 16S rRNA sequences. Phytopathology 80: 618.

de Souza AN, da Silva FN, Bedendo IP, Carvalho CM (2014) A phytoplasma belonging to a 16SrIII-A subgroup and ds-RNA virus associated cassava frogskin disease in Brazil. Plant Dis 98: 771–779.

Demeuse KL, Grode AS, Szendrei Z (2016) Comparing qPCR and nested PCR diagnostic methods for aster yellows phytoplasma in aster leafhoppers. Plant Dis 100: 2513–2519.

Dianese JC, Schaad NW (1986) Immunochemical characterization of inner and outer membranes of *Xanthomonas campestris* pv. *campestris*. Phytopathology 76: 148–153.

Ding F, Peng S-A, Hartung JS (2020) Enhanced serologically based detection of Liberibacters associated with citrus huanglongbing. Plant Dis 104: 1584–1588.

Doi Y, Ternaka M, Yora K, Asuyama H (1967) Mycoplasma and PLT group-like microorganisms in plants infected with mulberry dwarf, potato witches' broom, aster yellows and paulownia witches' broom. Ann Phytopathol Soc Jpn 33: 259–266.

Dona C, Biosca EG, Penalver J, Lopez MM (2007) Exploring diversity among Spanish strains of *Erwinia amylovora* and possible infection sources. J Appl Microbiol 103: 1639–1649.

Duduk B, Calari A, Paltrinieri S, Duduk N, Bertaccini A (2009) Multigenic analysis for differentiation of aster yellows phytoplasmas infecting carrots in Serbia. Ann Appl Biol 154: 219–229.

Dutta B, Vernaiz MAC, Castro-Sparks AC, Scherm H, Walcott RR (2012) Location of *Acidovorax avenae* subsp. *citrulli* in watermelon seeds affects efficiency of pathogen detection by seed health testing. Seed Sci Technol 40: 309–319.

El Sayed AZ, Soufi Z, Wahdan KM, Komor E (2016) Detection and characterization of phytoplasma and *Sugarcane yellow leaf virus* associated with leaf yellowing of sugarcane. J Phytopathol 164: 217–225.

Elibox W, Umaharan P (2007) A green fluorescent protein-based screening method for identification of resistance in anthurium to systemic infection by *Xanthomonas axonopodis* pv. *diffenbachiae*. Plant Pathol56: 819–827.

Errampalli D, Fletcher J (1993) Production of monospecific polyclonal antibodies against aster yellows mycoplasma-like organisms associated antigen. Phytopathology 83: 1279–1282.

Felix-Gastelum R, Maldonado-Mendoza IE, Navarrete-Maya R, Olivas-Peraza NG, Brito-Vega H et al. (2016) Identification of *Pseudomonas syringae* pv. *phaseolicola* as the causal agent of halo blight in yellow beans in northern Sinsloa, Mexco. Phytoparasitica 44: 369–378.

Fialova R, Valova P, Balakrishiyeva G, Danet JL, Safarova D et al. (2009) Genetic variability of stolbur phytoplasma in annual crop and wild plant species in south Moravia. J Plant Pathol 91: 411–416.

Fránová J, Přibylová J, Koloniuk J, Podrábska K, Špak J (2016) Multigene sequence analysis of aster yellows phytoplasma associated with primrose yellows. J Phytopathol 164: 166–176.

Freigoun SO, El Faki HI, Gelie B, Schirmer M, Lemattre M (1994) Phage sensitivity in relation to pathogenicity and virulence of cotton bacterial blight pathogen of Sudan. Plant Pathol 43: 493–497.

Ftayeh RM, von Tiedemann A, Rudolph KWE (2011) A new selective median for isolation of *Clavibacter michiganensis* subsp. *michiganensis* from tomato plants and seed. Phytopathology 101: 1355–1364.

Fukui H, Alvarez AM, Fukui R (1998) Differential susceptibility of anthurium cultivars to bacterial blight in foliar and systemic infection phases. Plant Dis 82: 800–806.

Galetto L, Bosco D, Marzachi C (2005) Universal and group-specific real-time PCR diagnosis of flavescence dorée

(16SrV), bois noir (16SrXII) and apple proliferation (16SrXII) phytoplasmas from field-collected plant hosts and insect vectors. Ann Appl Biol 147: 191–201.

Galvis CA, Leguizamon JE, Gaitan AL, Meija JF, Alvarez E et al. (2007) Detection and identification of a group 16SrIII-related phytoplasma associated with coffee crispiness disease in Colombia. Plant Dis 91: 248–252.

Gan Q, Lu Y, Shao X, Sun T, Li Y et al. (2014) Rapid detection and typing of *Xanthomonas campestris* pv. *vesicatoria* by pyrosequencing. J Phytopathol 162: 190–194.

Garcia AL, Torres SCZ, Heredia M, Lopes SA (2012) Citrus responses to *Xylella fastidiosa* infection. Plant Dis 96: 1245–1249.

Gardan L, Moreno B, Guinebretierre JP (1984) The use of a micromethod of auxanogram and data analysis for identification of Pseudomonas. Proceedings of the 2nd Workshop Group, Athens, Greece, pp. 37–39.

Gazel MC, Caglayan K, Baspinar H, Mejia JF, Paltrinieri S et al. (2016) Detection and identification of phytoplasmas in pomegranate trees with yellows symptoms. J Phytopathol 164: 136–140.

Gétaz M, Buhlmann A, Schneeberger PHH, Van Malderghem C, Duffy B et al. (2017) A diagnostic tool for improved detection of *Xanthomonas fragariae* using a rapid and highly specific LAMP assay designed with coronatine genomics. Plant Pathol 66: 1094–1102.

Gillings MR, Fahy PC, Broadbent P, Barnes D (1995) Rapid identification of a second outbreak of Asiatic canker in the Northern Territory using polymerase chain reaction and genomic fingerprinting. Austr Plant Pathol 24: 104–111.

Giorgi S, Scortichini M (2005) Molecular characterization of *Erwinia amylovora* strains from different host plants through RFLP analysis and sequencing of *hrpN* and *dspA/E* genes. Plant Pathol 54: 789–798.

Gouda KA, Baranwal VK, Ahlawat YS (2006) Simplified DNA extraction and improved PCR-based detection of greening bacterium in citrus. J Plant Biochem Biotechnol 15: 117–121.

Graces FF, Gutierrez A, Hoy JW (2014) Detection and quantification of *Xanthomonas albilineans* by qPCR and potential characterization of sugarcane resistance to leaf scald. Plant Dis 98: 121–126.

Grimm, KDS, Gaczynski SF (2019) Identification of a new haplotype of '*Candidatus* Liberibacter solanacearum' in *Solanum tuberosum*. Plant Dis 103: 468–474.

Groves PL, Chen J, Civerolo EL, Freeman MW, Viveros MA (2005) Spatial analysis of almond leaf scorch disease in the San Joaquim Valley of California factors affecting pathogen distribution and spread. Plant Dis 89: 581–587.

Guo X, Lu J (2004) Use of a pressure chamber to isolate and detect *Xylella fastidiosa* in xylem exudates of grapevines. Amer J Enol Vitic 52: 202–204.

Ha Y, Fessehaie A, Ling KS, Wechter WP, Keinath AB et al. (2009) Simultaneous detection of *Acidovorax avenae* subsp. *citrulli* and *Didymella bryoniae* in cucurbit seed lots using magnetic capture hybridization and real-time polymerase chain reaction. Phytopathology 99: 666–678.

Haapalainen M, Wang J, Latvala S, Lehtonen MT, Pirhonen M et al. (2018) Genetic variation of '*Candidatus* Liberibacter solanacearum' haplotype C and identification of novel haplotype from *Troza urticae* and stinging nettle. Phytopathology 108: 925–934.

Hadas R, Kritzman G, Kliatman F, Gefew T, Manulis S (2005) Comparison of extraction procedures and determination of the detection threshold for *Clavibacter michiganensis* subsp. *michiganensis* in tomato seeds. Plant Pathol 54: 643–649.

Hall SJ, Dry IB, Blanchard CL, Whiteleaw-Weckert MA (2016) Phylogenetic relationships of *Pseudomonas syringae* pv. *syringae* isolates associated with bacterial inflorescence rot in grapevine. Plant Dis 100: 607–616.

Hall SJ, Dry IB, Gopurenko D, Whitelaw-Wecker MA (2019) *Pseudomonas syringae* pv. *syringae* from cool climate Australian grapevine vineyards with bacterial inflorescence rot. Plant Pathol 68: 312–322.

Hamza AA, Robéne-Soustrade I, Jouen E, Gagnevin L, Lefeuvre P et al. (2010) Genetic and pathological diversity among *Xanthomonas* strains responsible for bacterial spot on tomato and pepper in the southwest Indian Ocean Region. Plant Dis 94: 993–999.

Harper SJ, Ward LI, Clover GRG (2010) Development of LAMP and real-time PCR methods for rapid detection of *Xylella fastidiosa* for quarantine and field applications. Phytopathology 100: 1282–1288.

Harrison K, Tamborigndeguy C, Schevring DC, Herrera AM, Silva A et al. (2019) Differences in zebra chip severity between '*Candidatus* Liberibacter solanacearum' haplotypes in Texas. Amer J Potato Res 96: 86–93.

Hartung JS, Civerolo EL (1989) Restriction fragment length polymorphisms distinguish *Xanthomonas campestris* strains isolated from Florida citrus nurseries from *X. campestris* pv. *citri*. Phytopathology 79: 793–799.

Hartung JS, Pruvost OP, Villeniot I, Alvarez A (1996) Rapid and sensitive colorimetric detection of *Xanthomonas axonopodis* pv. *citri* by immunocapture and nested polymerase chain reaction assay. Phytopathology 86: 95–101.

Henriquez NP, Kanyone L, Quiroz L (1999) Corn stunt complex Mollicutes in Belize. Plant Dis 83: 77.

Hernandez-Martinez R, Costa HS, Dumenyo CK, Cooksey DA (2006) Differentiation of strains of *Xylella fastidiosa* infecting grape, almonds, and oleander using multiprimer PCR assay. Plant Dis 90: 1382–1388.

Hilf ME, Sims KR, Folimonova SY, Achor DS (2013) Visualization of '*Candidatus* Liberibacter asiaticus' cells in the vascular bundles of citrus seed coats with fluorescence in situ hybridization and transmission electron microscopy. Phytopathology 103: 545–554.

Himananto P, Thummabenjapone P, Luxananil P, Kumpoosiri M, Hongprayoon R et al. (2011) Novel and highly specific monoclonal antibody to *Acidovorax citrulli* and development of ELISA base detection in cucurbit leaves and seed. Plant Dis 95: 1172–1178.

Hodgetts J, Tomlinson J, Boonham N, Gonzalez-Martin I, Nikoli P et al. (2011) Development of rapid in-field loop-mediated isothermal amplification (LAMP) assays for phytoplasmas. Bull Insectol 64(Suppl) S41–S42.

Hren M, Boben J, Rotter A, Kralj P, Gruden K et al. (2007) Real-time PCR detection systems for flavescence dorée and bois noir phytoplasmas in grapevine: comparison with conventional PCR detection and application in diagnostics. Plant Pathol 56: 785–796.

Hu H, Davis MJ, Brlansky RH (2013) Quantification of live '*Candidatus* Liberibacter asiaticus' populations using real-time PCR and propidium monoazide. Plant Dis 97: 1158–1167.

Huang TH, Wu ML, Su HJ (1993) Detection of fastidious bacteria citrus greening disease by nonradioactive DNA probes. Ann Phytopathol Soc Jpn 65: 140–146.

Hung TH, Wu ML, Su HJ (1999) Detection of fastidious bacteria citrus greening disease by nonradioactive DNA probes. Ann Phytopathol Soc Jpn 65: 140–146.

Ialacci GM, Bella P, Licciardello G, Strano CP, Eichenlaub R et al. (2016) Clonal populations of *Clavibacter michiganensis* subsp. *michiganensis* are responsible for the outbreaks of bacterial canker in greenhouse tomatoes in Italy. Plant Pathol 65: 484–495.

Inoue Y, Matsura T, Ohara T, Azegami K (2006) Sequence analysis of the genome of P2, a lytic bacteriophage of *Xanthomonas oryzae* pv. *oryzae*. J Gen Plant Pathol 72: 104–110.

International Research Program on Comparative Mycoplasmology (IRPCM) (2004) '*Candidatus* Phytoplasma' a taxon for the wall-less, nonhelical prokaryotes that colonize plant phloem and insects. Internatl J System Evol Microbiol 54: 1243–1255.

Ishiie T, Doi Y, Yora K, Asuyama H (1967) Suppressive effects of an antibiotic of tetracycline group on symptom development in mulberry dwarf disease. Ann Phytopathol Soc Jpn 33: 267–275.

Jin K-S, Kang I-H, Ko K-I, Lee E-S, Heo L-Y et al. (2001) Detection of *Xanthomonas axonopodis* pv. *citri* on citrus fruits using enzyme-linked immunosorbent assay. Plant Pathol J 17: 62–66.

Johnson KL, Walcott RR (2012) Progress towards a real-time PCR assay for the simultaneous detection of *Clavibacter michiganensis* subsp. *michiganensis* and *Pepino mosaic virus* in tomato seed. J Phytopathol 160: 353–363.

Jones LA, Saha S, Collmer A, Smart CD, Lindeberg M (2015) Genome-assisted development of a diagnostic protocol for distinguishing high virulence *Pseudomonas syringae* pv. *tomato* strains. Plant Dis 99: 527–534.

Jones R, Barnes L, Gonzalez C, Leach J, Alvarez A et al. (1989) Identification of low virulence strains of *Xanthomonas campestris* pv. *oryzae* from rice in the United States. Phytopathology 79: 984–990.

Jones WT, Harvey D, Zhao YF, Mitchell RF, Bender CL et al. (2001) Monoclonal antibody-based immunoassays for the phytotoxin coronatine. Food Agric Immunol 13: 19–32.

Kakizawa S, Kamagata Y (2014) A multiplex PCR method for strain identification and detailed phylogenetic analysis of AY-group phytoplasmas. Plant Dis 98: 299–305.

Kanamori H, Sugimoto H, Ochiar K, Kaku H, Tsuyumu S (1999) Isolation of *hrp* cluster from *Xanthomonas campestris* pv. *citri* and its application for RFLP analyses of xanthomonads. Ann Phytopathol Soc Jpn 65: 110–115.

Kang MJ, Kim MH, Hwang DJ, Cho MS, Seol Y (2012) Quantitative in planta PCR assay for specific detection of *Xanthomonas oryzae* pv. *oryzicola* using putative membrane protein-based primer set. Crop Protect 40: 22–27.

Kawabe K, Truc NTN, Lan BTN, Hong LTT, Onuki M (2006) Quantification of DNA of citrus huanglongbing pathogen in diseased leaves using competitive PCR. J Gen Plant Pathol 72: 355–359.

Khoodoo MHR. Sahin F, Jaufeerally-Fakin Y (2005) Sensitive detection of *Xanthomonas axonopodis* pv. *diffenbachiae* by immunocapture-PCR (IC-PCR) using primers designed from sequence characterized amplified regions (SCAR) of the blight pathogen. Eur J Plant Pathol 112: 379–390.

Kim BK, Cho MS, Kim MH, Choi HJ, Kang MJ et al. (2012) Rapid and specific detection of *Burkholderia glumae* in rice seed by real-time BIO-PCR using species-specific primers based on an *rhs* family gene. Plant Dis 96: 577–580.

Kleitman F, Barash I, Burger A, Iraki N, Falah Y et al. (2008) Characterization of a *Clavibacter michiganensis* subsp. *michiganensis* population in Israel. Eur J Plant Pathol 121: 464–475.

Klinger JM, Stowe RP, Obenhuber DC, Groves TO, Mishra SK et al. (1992) Evaluation of the Biolog automated microbial identification system. Appl Environ Microbiol 58: 2089–2092.

Kogovsek P, Hodgetts J, Hall J, Prezelj N, Nikoli P et al. (2015) LAMP assay and rapid sample preparation method for on-site detection of flavescence dorée phytoplasma in grapevine. Plant Pathol 64: 286–296.

Kogovsek P, Mehle N, Pugelj A, Jakomin T, Schroers H et al. (2017) Rapid loop-mediated isothermal amplification assays for grapevine yellows phytoplasmas in crude leaf vein homogenate has the same performance as qPCR. Eur J Plant Pathol 148: 75–84.

Kokošková B, Mraz I, Hyblova J (2007) Comparison of specificity and sensitivity of immunochemical and molecular techniques for reliable detection of *Erwinia amylovora*. Folia Microbiol 52: 175–182.

Kreig NY, Holt JG (eds.) (1984) *Bergey's Manual of Systematic Bacteriology*, volume 1. Williams & Wilkins, Baltimore, MD, USA.

Krell RK, Boyd EA, Nay JE, Park Y-L, Perring TM (2007) Mechanical and insect transmission of *Xylella fastidiosa* to *Vitis vinifera*. Amer J Enol Vitic 58: 211–216.

Krell RK, Perring TM, Hashim-Buckey JM, Pinckard TR (2008) Susceptibility of *Vitis vinifera* L. cv. Redglobe and Thompson Seedless to Pierce's disease. Amer J Enol Vitic 59: 61–66.

Kuang W, Luo L, Gao W, Lei Y, Iv Q et al. (2017) Development of a real-time fluorescence loop-mediated isothermal amplification assay for detection of *Burkholderia gladioli* pv. *alliicola*. J Phytopathol 165: 82–90.

Kuflu KM, Cuppels DA (1999) Development of diagnostic cDNA probe for xanthomonads causing bacterial leaf spot of peppers and tomatoes. Appl Environ Microbiol 63: 4462–4470.

Kunkel LO (1926) Studies on aster yellows. Amer J Bot 23: 646–705.

Laala S, Bouznad Z, Manceau C (2015) Development of a new technique to detect living cells of *Xanthomonas campestris* pv. *campestris* in crucifers seeds: the seed-qPCR. Eur J Plant Pathol 141: 637–646.

Lange HW, Tancos MA, Carlson MO, Smart CD (2016) Diversity of *Xanthomonas campestris* isolates from symptomatic crucifers in New York State. Phytopathology 106: 113–122.

Langlois PA, Snelling J, Hamilton JP, Bragard C, Koebnik R et al. (2017) Characterization of the *Xanthomonas translucens* complex using draft genomes comparative genomics, phylogenetic analysis and diagnostic LAMP assays. Phytopathology 107: 519–527.

Lebsky V, Hernandez-Gonzalez J, Arguello-Astorga G, Gardenas-Conejo Y, Poghosyan A (2011) Detection of phytoplasmas in mixed infection with begomoviruses – a case study of tomato and pepper in Mexico. Bull Insectol (Suppl): 555–556.

Lebsky V, Poghosyan A, Silva-Rosales L (2010) Application of scanning electron microscopy for diagnosing phytoplasma in single and mixed (virus-phytoplasma) infection in papaya. Proceedings of the 21st International Conference Virus and Other Graft-Transmissible Diseases of Fruit Crops, Julius-Kuhn Arch 427, pp. 1–9.

Lee I-M, Bottner KD, Munyaneza JE, Davis RE, Crosslin JM et al. (2006) Carrot purple leaf: a new spiroplasmal disease associated with carrots in Washington State. Plant Dis 90: 989–993.

Lee I-M, Davis RE (1983) Phloem-limited prokaryotes in sieve elements isolated by enzyme treatment of diseased plant tissues. Phytopathology 73: 1540–1543.

Lee I-M, Gunderson-Rindal DE, Davis RE, Bartaszyk IM (1998) Revised classification scheme of phytoplasmas based on RFLP analysis of 16S rRNA and ribosomal protein gene sequences. Internatl J Syst Bacteriol 48: 1153–1169.

Lee ME, Grau CR, Lukaesko LA, Lee I-M (2002) Identification of aster yellows phytoplasma in Wisconsin based on RFLP analyses of PCR amplified products. Canad J Plant Pathol 24: 125–130.

Lee S, Chu C-Y, Chu C-C (2020) Variability of phytoplasma infection density in poinsettia and evaluation of its association with the level of branching in host plants. Plant Dis https://doi.org/10.1094/PDIS-10-20-2304-RE

Leite RP Jr, Minsavage GV, Bonas U, Stall RF (1994) Detection and identification of phytopathogenic *Xanthomonas campestris* pv. *vesicatoria*. Appl Environ Microbiol 60: 1068–1077.

Li G, Song C-F, Pang X-M, Wang J-S (2009) Analysis of pathotypic and genotypic diversity of *Xanthomonas oryze* pv. *oryzae* in China. J Phytopathol 157: 208–218.

Li WB, Pria WDJ, Laeava PM, Qin X, Hartung JS (2003) Presence of *Xylella fastidiosa* in sweet orange fruit and seeds and its transmission to seedlings. Phytopathology 93: 953–958.

Liefting LW, Sutherland PW. Ward LI, Paice KL, Weir BS et al. (2009) A new '*Candidatus* Liberibacter species associated with diseases of solanaceous crops. Plant Dis 93: 208–214.

Lin C-Y, Tsai C-H, Tien H-J, Wu M-L, Su H-J et al. (2017) Quantification and ecological study of '*Candidatus* Liberibacter asiaticus' in citrus hosts, rootstocks and the Asian citrus psyllid. Plant Pathol 66: 1555–1568.

Liu R, Zhang P, Pu X, Xiang X, Chen J et al. (2011) Analysis of a prophage gene frequency revealed population variation of *Candidatus* Liberibacter asiaticus' from two citrus growing provinces in China. Plant Dis 95: 431–435.

Loi N, Ermacora P, Carraro L, Osler R, Chen TA (2002) Production of monoclonal antibodies against apple proliferation phytoplasma and their use in serological detection. Eur J Plant Pathol 108: 81–86.

López R, Assensio C, Gilbertson RL (2006) Phenotypic and genetic diversity in strains of common blight bacteria (*Xanthomonas campestris* pv. *phaseoli* and *X. campestris* pv. *phaseoli* var. *fuscans*) in a secondary center of diversity of the common bean host suggests multiple introduction events. Phytopathology 96: 1204–1213.

Lou B, Song Y Roychowdhury M, Deng C, Niu Y et al. (2018) Development of a tandem repeat-based polymerase chain displacement reaction method for highly sensitive detection of '*Candidatus* Liberibacter asiaticus'. Phytopathology 108: 292–298.

Lou NA, Bai X, Bai Y, Deng C, Roychowdhury M et al. (2014) Detection and molecular characterization of a 16SrII-A* phytoplasma in grapefruit (*Citrus paradisi*) with huanglongbing-like symptoms in China. J Phytopathol 162: 387–395.

Lu L, Cheng B, Yao J. Peng A, Du D et al. (2013) A new diagnostic system for detection of '*Candidatus* Liberibacter asiaticus' infection in citrus. Plant Dis 97: 1295–1300.

Luo LX, Walters C, Bolkan H, Liu XL, Li JQ (2008) Quantification of viable cells of *Clavibacter michiganensis* subsp. *michiganensis* using a DNA binding dye and a real-time PCR assay. Plant Pathol 57: 332–337.

Lydons J, Patterson CD (2001) Detection of tabtoxin-producing strains of *Pseudomonas syringae* by PCR. Lett Appl Microbiol 32: 166–170.

Machado MA (2009) *Xylella fastidiosa* and the citrus variegated chlorosis (CVC), a new destructive citrus disease in Brazil. www.lbi.ic.unicampa.br/xylella_cvc-strain/home.mmachdo.html

Maeda Y, Horita M, Shinohara H, Kiba A, Ohnishi K et al. (2007) Analysis of sources of oxolinic acid-resistant field strains of *Burkholderia glumae* based on rep-PCR analysis and nucleotide sequences of *gyrB* and *rpoD*. J Gen Plant Pathol 73: 46–52.

Mahuku GS, Jara C, Henrique MA, Castellanos G, Cuasquer J (2006) Genotypic characterization of the common bean bacterial blight pathogen *Xanthomonas axonopodis* pv. *phaseoli* and *Xanthomonas axonopodis* pv. *phaseoli* var. *fuscans* by rep-PCR and PCR-RFLP of ribosomal genes. J Phytopathol 154: 35–44.

Malandrin L, Huard A, Samsun R (1996) Discrimination of envelope profiles between *Pseudomonas syringae* pv. *pisi* and *P. syringae* pv. *syringae*. FEMS Microbiol Lett 41: 11–17.

Matsura T, Shinohara H, Inouey Y, Azegami K, Tsushima S et al. (2007) *Erwinia* isolates from the bacterial shoot blight of pear in Japan are closely related to *Erwinia pyrifoliae* based on phylogenetic analysis of *gyrB* and *rpoD* genes. J Gen Plant Pathol 73: 53–58.

Matsuyama N, Daikohara M, Yoshimura K, Manabe K, Khan MAA et al. (2009) Presumptive differentiation of phytopathogenic and nonpathogenic bacteria by improved rapid-extraction TLC method. J Fac Agric Kyushu Univ 54: 1–11.

Mavrodieva V, Levy L, Gabriel DW (2004) Improved sampling methods for real-time polymerase chain reaction diagnosis of citrus canker from field samples. Phytopathology 94: 61–68.

McManus PS, Jones AC (1996) Detection of *Erwinia amylovora* by nested PCR and PCR-dot blot and reverse blot hybridization. Acta Hortic 411: 87–90.

Mello APOA, Bedendo IP, Camargo LEA (2006) Source homogeneity in the 16S rDNA of tomato big bud phytoplasma belonging to group 16SrII. J Phytopathol 154: 245–249.

Merfa MV, Perez-Lopez E, Narenjo E, Jain M, Gabriel DW et al. (2019) Progress and obstacles in culturing '*Candidatus* Liberibacter asiaticus' the bacterium associated with huanglongbing. Phytopathology 109: 1092–1101.

Merighi M, Sandrini A, Landini S, Ghini S, Girotti S et al. (2000) Chemiluminescent and colorimetric detection of *Erwinia amylovora* by immunoenzymatic determination of PCR amplicons from plasmid pEA29. Plant Dis 84: 49–54.

Mitrev S, Kovačevic B (2006) Characterization of *Xanthomonas axonopodis* pv. *vesicatoria* isolated from peppers in Macedonia. J Plant Pathol 88: 321–324.

Modesti V, Pucci N, Lucchesi S, Campus L, Loreti S (2017) Experience of Latium region (central Italy) as a pest-free area for monitoring of *Xylella fastidiosa* distinctive features of molecular diagnostic methods. Eur J Plant Pathol 148: 557–566.

Mohammadi M, Moltmann E, Zeller W, Geider K (2009) Characterization of naturally occurring *Erwinia amylovora* strains lacking the common plasmid pEA29 and their detection with real-time PCR. Eur J Plant Pathol 124: 293–302.

Murolo S, Marcone C, Prota V, Garau R, Foissac X et al. (2010) Genetic variability of the stolbur phytoplasma *vmp1* gene in grapevines, bindweeds and vegetables. J Appl Microbiol 109: 2049–2059.

Musetti R, Favelli MA (2004) Microscopy techniques applied to the study of phytoplasma diseases: traditional and innovative methods. www.formatex.org/microscopy2/papers/72–80.pdf

Myung I-S, Hyn J-W, Chow W-D (2006) Detection of *Xanthomonas axonopodis* pv. *citri* on Satsuma mandarin orange fruit using phage technique in Korea. Plant Pathol J 22: 314–317.

Myung I-S, Nam K-W, Kwon H-M (2001) Dispersal of citrus bacterial canker caused by *Xanthomonas axonopodis* pv. *citri* in nursery plots of Unshiu orange. Plant Pathol 19: 205–209.

Nabi S, Madhupriya, Dubey DK, Rao GP, Baranwal VK et al. (2015) Molecular characterization of '*Candidatus* Phytoplasma asteris' subgroup I-B associated with sesame phyllody disease and identification its natural vector and weed reservoir in India. Austr Plant Pathol 44: 289–297.

Najar A, Bouachem S, Danet JL, Saillard C, Garnier M et al. (1998) Presence of *Spiroplasma citri*, the pathogen responsible for citrus stubborn disease and its vector leafhopper *Circulifer haematoceps* in Tunisia: contamination of both *C. haematoceps* and *C. opasipennis*. Fruits (Paris) 53: 391–396.

Narayanasamy P (2001) *Plant Pathogen Detection and Disease Diagnosis*, second edition. Marcel Dekker Inc., NY, USA.

Narayanasamy P (2011) *Microbial Plant Pathogens – Detection and Disease Diagnosis: Bacterial and Phytoplasmal Pathogens*, volume 2. Springer Science + Business Media BV, Heidelberg, Germany.

Narayanasamy P (2017) *Microbial Plant Pathogens – Detection and Disease Management in Seeds and Propagules*, volume 1. Wiley-Blackwell, Chichester, UK.

Natalini E, Rossi MP, Barionovi D, Scortichini M (2006) Genetic and pathogenic diversity of *Pseudomonas syringae* pv. *syringae* isolates associated with bud necrosis and leaf spot of pear in a single orchard. J Plant Pathol 88: 219–223.

Németh J, van Vuurde JWL (2006) Development of immunofluorescence colony staining (IFC) for detection of *Xanthomonas campestris* pv. *vesicatoria* and *Clavibacter michiganensis* subsp. *michiganensis* in tomato seeds. Seed Sci Technol 34: 8–100.

Newberry EA, Jardini TM, Rubio I, Roberts PD, Koike ST et al. (2016) Angular leaf spot of cucurbits is associated with genetically diverse *Pseudomonas syringae* strains. Plant Dis 100: 1397–1404.

Oguiza JA, Rico A, Rivas LA, Sutra L, Vivian A et al. (2004) *Pseudomonas syringae* pv. *phaseolicola* can be separated into two genetic lineages distinguished by the possession of the phaseolotoxin biosynthetic cluster. Microbiology 150: 473–482.

Okuda M, Matsumoto M, Tanaka Y, Subandiyah S, Iwananu T (2005) Characterization of the *tuftB-secE-nusG-rplKAJL-rop* gene cluster of the citrus greening organism and detection by loop-mediated isothermal amplification. Plant Dis 89: 705–711.

Otto M, Petersen Y, Roux J, Wright J, Coutinho TA (2018) Bacterial canker of cherry trees, *Prunus avium* in South Africa. Eur J Plant Pathol 151: 427–438.

Ou SH (1985) *Rice Diseases*, second edition. Commonwealth Mycological Institute, Surrey, UK.

Pal N, Block CC, Gardner LAC (2019) A real-time PCR differentiating *Pantoea stewartii* subsp. *stewartii* from *P. stewartii* subsp. *indolgenes* in corn seed. Plant Dis 103: 1474–1486.

Pamei I, Makandar R (2016) Association of 16SrII-D phytoplasma with phyllody disease in sesame (Sesamum indicum) in Telengana from southern India. Plant Dis 100: 1774.

Pandey SS, Wang N (2019) Targeted early detection of citrus huanglongbing causal agent '*Candidatus* Liberibacter asiaticus' before symptom expression. Phytopathology 109: 952–959.

Park DS, Hyun JW, Park YJ, Kim JS, Kang HW et al. (2006) Sensitive and specific detection of *Xanthomonas axonopodis* pv. *citri* by PCR using pathovar specific primer based on *hrp*[w] gene sequences. Microbiol Res 161: 145–149.

Park DS, Shim JK, Kim JS, Lim CK, Shreshtha R et al. (2009) Sensitive and specific detection of *Xanthomonas campestris* pv. *vesicatoria* by PCR using pathovar-specific primers based on *rhs* family gene sequences. Microbiol Res 164: 36–42.

Park YJ, Lee BM, Ho-Hahn J, Lee GB, Park DS (2004) Sensitive and specific detection of *Xanthomonas campestris* pv. *campestris* by PCR using species-specific primers based on *hrpF* gene sequences. Microbiol Res 159: 419–423.

Pérez-López E, Olivier CY, Luna-Rodriguez M, Rodriguez Y, Iglesias LG (2016) Maize bushy stunt phytoplasma affects native corn at high elevations in southeast Mexico. Eur J Plant Pathol 45: 963–971.

Pierro R, Passera A, Panattoni A, Casati P, Luvisi A et al. (2018) Molecular typing of bois noir phytoplasma strains in Chianti Classio area (Tuscany, Central Italy) and their association with symptom severity in *Vitis vinifera* 'Sangivese'. Phytopathology 108: 362–373.

Plavec J, Križanac I, Budinšćak Ž, Škoric D, Musić MS (2015) A case study of FD and BN phytoplasma variability in Croatia: multigene sequence analysis approach. Eur J Plant Pathol 142: 591–601.

Prencipe S, Gullino ML, Spadaro D (2018) *Pseudomonas syringae* pv. *actinidiae* isolated from *Actinidia chinensis* var. *deliciosa* in northern Italy: genetic diversity and virulence. Eur J Plant Pathol 150: 191–204.

Quaglino F, Maghradze D, Casati P, Chkhaidze N, Lobjancdze M et al. (2016) Identification and characterization of new '*Candidatus* Phytoplasma solani' strains associated with bois noir disease in *Vitis* vinifera L. cultivars showing a range of symptoms severity in Georgia, the Caucasus Region. Plant Dis 100: 904–915.

Quaglino F, Murolo S, Zhao Y, Casati P, Durante G et al. (2017) Identification of new –J and –K 16SrXII subgroups and distinct single nucleotide polymorphism genetic lineages among '*Candidatus* Phytoplasma solani' strains associated with bois noir in Central Italy. Austr Plant Pathol 46: 31–34.

Quesada-Ocampo LM, Landers NA, Lebeis AC, Fulbright DW, Hausbeck MK (2012) Genetic structure of *Clavibacter michiganensis* subsp. *michiganensis* populations in Michigan commercial tomato fields. Plant Dis 96: 788–796.

Quoc NB, Xuan NTT, Phuong NDN, Trang HTT, Chau NNB et al. (2021) Development of loop-mediated isothermal amplification assays for the detection of sugarcane white leaf. Physiol Molec Plant Pathol 113: 101595. https://doi.org/10.1016/j.pmpp.2020.101595

Rabenstein F, Nachtigali M, Boll E (1999) Production and characterization of new monoclonal antibodies against *Xanthomonas campestris* pv. *campestris*. Beit Zucht Bunde Zucht Kultur 5: 4–6.

Rache L, Blondin L, Flores C, Trujillo C, Szurek B et al. (2019) An optimized microsatellite scheme for assessing populations of *Xanthomonas phaseoli* pv. *manihotis*. Phytopathology 109: 859–869.

Rashed A, Wallis CM, Workneh F, Paetzold L, Rush CM (2016) Variations in zebra chip disease expression and tuber biochemistry in response to vector density. Phytopathology 106: 854–860.

Ravindran A, Levy J, Pierson E, Gross DC (2012) Development of a loop-mediated isothermal amplification procedure as a sensitive and rapid method for detection of '*Candidatus* Liberibacter solanacearum' and potatoes and psyllids. Phytopathology 102: 899–907.

Rockey W, Potnis N, Timilsina S, Hong JC, Vallad GE et al. (2015) Multilocus sequence analysis reveals genetic diversity in xanthomonads associated with poinsettia production. Plant Dis 99: 874–882.

Rodrigues JLM, Silva-Stencio ME, de Souza AN, Lopes JRS, Tsai SM (2006) In situ probing of *Xylella fastidiosa* in honey dew of a xylem sap-feeding insect using 16S rRNA-targeted fluorescent oligonucleotides. Environ Microbiol 8: 747–754.

Rodrigues JLM, Silva-Stencio ME, Gomes JE, Lopes JRS, Tsai SM (2003) Detection and diversity assessment of *Xylella fastidiosa* in field-collected plant and insect samples by using 16S rRNA and *gyrB* sequences. Appl Environ Microbiol 69: 4249–4255.

Roéne-Soustrade I, Laurent P, Ganevin L, Jouen E, Pruvost O (2006) Specific detection of *Xanthomonas axonopodis* pv. *diffenbachiae* in anthurium (*Anthuruium andreanum*) tissues by nested PCR. Appl Environ Microbiol 72: 1072–1087.
Rossi M Pegoraro M, Ripamonti M, Abba S, Beal D et al. (2019) Genetic diversity of flavescence dorée phytoplasmas at the vineyard scale. Appl Environ Microbiol 85: e03123–18.
Saberi E, Mehdi S, Safaie N (2018) Diversity and characterization of '*Candidatus* Liberibacter asiaticus' strains causing huanglongbing disease in Iran, based on two prophage loci. J Phytopathol 166: 558–562.
Sawada H, Fujikawa T (2019) Genetic diversity of *Pseudomonas syringae* pv. *actinidiae*, pathogen of kiwifruit bacterial canker. Plant Pathol 68: 1235–1248.
Schaad NW, Berthier-Schaad Y, Knorr D (2007) A high throughput membrane BIO-PCR technique for ultra-sensitive detection of *Pseudomonas syringae* pv. *phaseolicola*. Plant Pathol 56: 1–8.
Schaad NW, Cheong SS, Tamaki E, Panapoulos NJ (1995) A combined biological and enzymatic amplification (BIO-PCR) technique to detect *Pseudomonas syringae* pv. *phaseolicola* in bean seed extracts. Phytopathology 85: 243–248.
Schaad NW, Jones JB, Chun W (2002a) *Laboratory Guide for Identification of Plant Pathogenic Bacteria*. The American Phytopathological Society, St Paul, MN, USA.
Schaad NW, Opgenorth D, Gaush P (2002b) Real-time polymerase chain reaction for one-hour on-site diagnosis of Pierce's disease of grape in the early season asymptomatic vines. Phytopathology 92: 721–728.
Schneider B, Huttel B, Zubert C, Kube M (2020) Genetic variation, phylogenetic relationship and spatial disruption of '*Candidatus* Phytoplasma ulmi' strains in Germany. Scientific Rep 10: 21864. https://doi.org/10.1038/s41598-020-78745-w
Sechler A, Schuenzel EL, Cooke S, Donnua, Thaveechai N, Postnikova E, Stone AL, Schneider WL, Damsteegt D, Schaad NW (2009) Cultivation of *Candidatus* Liberibacter asiaticus', '*Ca*. L. africanus' and '*Ca*. L. americanus' associated with huanglongbing. Phytopathology 99: 480–486.
Shao J, Jomantiene R, Dally EL, Zaho Y, Lee I-M et al. (2006) Phylogeny and characterization of phytoplasmal NusA and use of the *nusA* gene in detection of group 16 SrI strains. J Plant Pathol 88: 193–201.
Shi J, Pagliaccia D, Morgan R, Qiao Y, Pan S et al. (2014) Novel diagnosis for citrus stubborn disease by detection of *Spiroplasma citri*-secreted protein. Phytopathology 104: 188–195.
Siemonsmeier A, Hadersdorfer J, Neumuller M, Schwab W, Treutter D (2019) A LAMP protocol for the detection of '*Candidatus* Phytoplasma pyri', the causal agent of pear decline. Plant Dis 103: 1397–1404.
Silva GM, Souza RM, Yan L, Junior RS, Medeiros FHV et al. (2016) Strains of the group I lineage of *Acidovorax citrulli*, the causal agent of bacterial fruit blotch of cucurbitaceous crops and predominant in Brazil. Phytopathology 106: 1486–1494.
Sinclair QA, Griffiths HM, Davis RE, Lee I-M (1992) Detection of ash yellows mycoplasma-like organisms in different tree organs and in chemically preserved specimens by cDNA probe vs DAPI. Plant Dis 76: 154–158.
Singh D, Raghvendra BT, Rathaur PS, Singh H, Raguwanshi R et al. (2014) Detection of black rot disease pathogen *Xanthomonas campestris* pv. *campestris* by BIO-PCR from seeds and plant parts of cole crops. Seed Sci Technol 42: 36–46.
Singh D, Rathaur PS, Vicente JG (2016) Characterization of genetic diversity and distribution of *Xanthomonas campestris* pv. *campestris* races causing black rot disease in cruciferous crops in India. Plant Pathol 65: 1411–1418.
Slomnicka R, Olczak-Woltman H, Oskiera M, Schollenberger M, Niemirowicz-Szczytt K et al. (2018) Genome analysis of *Pseudomonas syringae* pv. *lachrymans*, strain 814198 indicates diversity within pathovar. Eur J Plant Pathol 151: 663–676.
So J-S, Lim HT, Oh E-T, Heo T-R, Koh S-C et al. (2002) Visualizing in the infection process of *Xanthomonas campestris* in cabbage using green fluorescent protein. World J Microbiol Biotechnol 18: 17–21.
Song E-S, Lee B-M, Lee C-S, Park Y-J (2012) PCR-based rapid assay for discriminative detection of latent infections of rice bacterial blight. J Phytopathol 160: 195–200.
Song WY, Kim HM, Hwang CY, Schaad NW (2004) Detection of *Acidovorax avenae* susbp. *avenae* in rice seeds using BIO-PCR. J Phytopathol 152: 667–676.
Stoger A, Schaffer J, Ruppitsch W (2006) A rapid and sensitive method for direct detection of *Erwinia amylovora* in symptomatic and asymptomatic plant tissues by polymerase chain reaction. J Phytopathol 154: 469–473.
Strayer AL, Jeyaparakash A, Minsavage GV, Timilsina S, Vallad GE et al. (2016) A multiplex real-time PCR assay differentiates four *Xanthomonas* species associated with bacterial spot of tomato. Plant Dis 100: 1660–1668.
Strayer-Scherer A, Jones JB, Paret ML (2019) Recombinase polymerase amplification assay for field detection of tomato bacterial spot pathogens. Phytopathology 109: 690–700.
Stumpf S, Kvitko B, Gitaitis R, Dutta B (2018) Isolation and characterization of novel *Pantoea stewarti* subsp. *indologenes* strains exhibiting center rot in onion. Plant Dis 102: 727–733.
Su C-C, Chang CJ, Chang C-M, Shih H-T, Tzeng K-C et al. (2013) Pierce's disease of grapevine in Taiwan: isolation, cultivation and pathogenicity of *Xylella fastidiosa*. J Phytopathol 161: 389–396.
Sugawara K, Himeno M, Keima T, Kitazawa Y, Maejima K et al. (2012) Rapid and reliable detection of phytoplasma by loop-mediated isothermal amplification targeting a housekeeping gene. J Gen Plant Pathol 78: 389–397.
Sun X, Stall RE, Jones JB, Cubero J, Gottwald TR et al. (2004) Detection and characterization of a new strain of citrus canker bacteria from Key/Mexican lime and Alemow in South Florida. Plant Dis 88: 1179–1188.
Tancos MA, Lange HW, Smart CD (2015) Characterizing the genetic diversity of the *Clavibacter michiganensis* subsp. *michiganensis* populations in New York. Phytopathology 105: 169–179.
Tardiani AC, Perecin D, Peixo-Junior RF, Sanguino A, Landelt MMG et al. (2014) Molecular and pathogenic diversity among Brazilian isolates of *Xanthomonas albilineans* assessed with SSR marker loci. Plant Dis 98: 540–546.
Tatineni S, Sagaram US, Gowda S, Robertson CJ, Dawson WO et al. (2008) In planta distribution of '*Candidatus* Liberibacter asiaticus' populations using real-time PCR and propidium monoazide. Plant Dis 97: 1158–1167.
Taylor RK, Guilford PJ, Clark RG, Hale CN, Forster RLS (2001) Detection of *Erwinia amylovora* in plant material using novel polymerase chain reaction (PCR) primers. NZ J Crop Hortic Sci 29: 35–43.
Taylor RK, Tyson JL, Fullerton RA, Hale CN (2002) Molecular detection of exotic phytopathogenic bacteria: a case study involving canker-like symptoms in citrus. NZ Plant Protect 55: 53–57.

Temple TN, Johnson KB (2011) Evaluation of loop-mediated isothermal amplification for rapid detection of *Erwinia amylovora* on pear and apple fruit flowers. Plant Dis 95: 423–430.

Thaveechai N, Schaad NW (1986) Immunochemical characterization of a subspecies-specific antigenic determinant of a membrane protein complex of *Xanthomonas campestris* pv. *campestris*. Phytopathology 76: 148–153.

Thompson SM, Johnson CP, Lu AY, Frampton RA, Sullivan KL et al. (2015) Genomes of '*Candidatus* Liberibacter solanacearum' haplotype A from New Zealand and the United States suggest significant genome plasticity in the species. Phytopathology 105: 863–871.

Tian Y, Zhao Y, Bai S, Walcott RR, Hu B et al. (2013) Reliable and sensitive detection of *Acidovorax citrulli* in cucurbit seed using padlock-probe based assay. Plant Dis 97: 961–966.

Tian Y, Zhao Y, Chen X, Dai Y, Zhao W et al. (2017) Evidence for a bacterial leaf blight of cantaloupe in China. Plant Dis 101: 1746–1752.

Tian Y, Zhao Y, Xu R, Liu F, Hu B et al. (2014) Simultaneous detection of *Xanthomonas oryzae* pv. *oryzae* and *X. oryzae* pv. *oryzicola* in rice seed using a padlock probe-based assay. Phytopathology 104: 1130–1137.

Trindadae LC, Marques E, Lopes DB, Ferreira MASV (2007) Development of a molecular method for detection and identification of *Xanthomonas campestris* pv. *viticola*. Summa Phytopathol 33: 16–23.

Tsuchiya K, d'Ursel CCM (2004) Development of a sensitive ELISA using three monoclonal antibodies against lipopolysaccharides to detect *Xanthomonas campestris* pv. *vesicatoria*, the causal agent of bacterial spot of tomato and pepper. J Gen Plant Pathol 70: 21–26.

Valasevitch N, Schneider B (2017) Rapid detection of '*Candidatus* Phytoplasma mali' by recombinase polymerase amplification. J Phytopathol 165: 762–770.

Valverde A, Hubert T, Stolov A, Dagar A, Kopelowitz J et al. (2007) Assessment of genetic diversity of *Xanthomonas campestris* pv. *campestris* isolates from Israel by various DNA fingerprinting techniques. Plant Pathol 56: 17–25.

Verniere C, Devaux M, Pruvost O, Couteau A, Luisetti J (1991) Studies on the biochemical and physiological variations among strains of *Xanthomonas campestris* pv. *citri*, the causal agent of citrus bacterial canker disease. Fruits 46: 162–170.

Verniere C, Pruvost O, Civerolo EL, Gambin O, Jaquemoud-Collet JP et al. (1993) Evaluation of the Biolog substrate utilization system to identify and assess metabolic variation among strains of *Xanthomonas campestris* pv. *citri*. Appl Environ Microbiol 59: 243–249.

Vicente JG, Everett B, Roberts SJ (2006) Identification of isolates that cause a leaf spot disease of Brassicas as *Xanthomonas campestris* pv. *raphani* and pathogenic and genetic comparison with related pathovars. Phytopathology 96: 735–745.

Villamor DEV, Eastwell KC (2019) Multilocus characterization, gene expression analysis of putative immunodominant protein coding regions, and development of recombinase, polymerase amplification assays for detection of '*Candidatus* Phytoplasma pruni' in *Prunus avium*. Phytopathology 109: 983–992.

Walcott RR, Gitaitis RD (2000) Detection f *Acidovorax avenae* susbp. *citrulli* in watermelon seed by immunomagnetic separation and polymerase chain reaction. Plant Dis 84: 470–474.

Wamonje FO, Zhou N, Bamrah R, Wist T, Prager SM (2021) Detection and identification of a '*Candidatus* Liberibacter solanacearum' species from ash tree-infesting psyllids. Phytopathology 111: 76–80.

Wang H, Turechek WW (2020) Detection of viable *Xanthomonas fragariae* cells in strawberry using propidium monoazide and long-amplicon quantitative PCR. Plant Dis 104: 1105–1112.

Wang X, Doddapaneni H, Chen J, Yokomi RK (2015a) Improved real-time PCR diagnosis of citrus stubborn disease by targeting prophage genes of *Spiroplasma citri*. Plant Dis 99: 149–154.

Wang X, Su H, Huang L, Deng X, Chen J et al. (2015b) Identification of a novel 1033-nucleotide deletion of polymorphism in prophage region of '*Candidatus* Liberibacter asiaticus': potential applications for bacterial epidemiology. J Phytopathol 163: 681–685.

Wang X-Q, Allen TW, Wang H, Peterson DG, Nichols RL et al. (2019) Development of a qPCR protocol to detect the cotton bacterial blight pathogen, *Xanthomonas citri* pv. *malvacearum* from cotton leaves and seeds. Plant Dis 103: 422–429.

Wang Z, Zhu Y, Li Z, Yang X, Zhang T et al. (2020) Development of a specific polymerase chain reaction system for the detection of rice orange leaf phytoplasma. Plant Dis 104: 521–526.

Wei W, Davis RE, Lee I-M, Zhao Y (2007) Computer simulated RFLP analysis of 16S rRNA genes: identification of ten new phytoplasma groups. Internatl J Syst Evol Microbiol 57: 1855–1867.

Wei W, Opgenorth DC, Davis RE, Chang CJ, Summers GG et al. (2016) Characterization of a novel adhesion-like gene and design of a real-time PCR for rapid sensitive and specific detection of *Spiroplasma kunkelii*. Plant Dis 90: 1233–1238.

Wen A, Mallik I, Alvarado VY, Pasche JS, Wang A et al. (2009) Detection, distribution and genetic variability of '*Candidatus* Liberibacter' species associated with zebra complex disease of potato in North America. Plant Dis 93: 1102–1115.

Wheatley MS, Duan Y-P, Yang Y (2021) Highly sensitive and rapid detection of citrus huanglongbing pathogen ('*Candidatus* Liberibacter asiaticus) using Cas12a-based methods. Phytopathology 111: 2375–2382.

Wu P-Y, Ho L-C, Chang J-J, Tzeng K-C, Deng W-L et al. (2017) Development of a TaqMan probe-based insulated isothermal PCR (TiiPCR) for detection of *Acidovorax citrulli*, the bacterial pathogen of watermelon fruit blotch. Eur J Plant Pathol 147: 869–875.

Wulff N, Fassini CG, Marques VV, Martins EC, Colletti DAB et al. (2019) Molecular characterization and detection of 16SrIII group phytoplasma associated with huanglongbing symptoms. Phytopathology 109: 366–374.

Xiao W, Le Z, Fu-Shou X, Li-Han Z, Guan-Lin X (2007) Immunocapture PCR method for detecting *Acidovorax avenae* susbp. *citrulli* from watermelon. China J Agric Biotechnol 4: 173–179.

Xu X, Mou H-Q, Zhu S-F, Liao X-L, Zhao W-J (2013) Detection and characterization of phytoplasma associated with big bud disease of tomato in China. J Phytopathol 161: 430–433.

Yan L, Hu B, Chen G, Zhao M, Walcott RR (2017) Further evidence of cucurbit host specificity among *Acidovorax citrulli* groups based on detached melon fruit pathogenicity assay. Phytopathology 107: 1305–1311.

Yan L, Zhao Y, Zhou J, Chen S, Bai S et al. (2019) Rapid and sensitive detection of *Acidovorax citrulli* on cucurbit seeds by visual loop-mediated isothermal amplification assay. J Phytopathol 167: 10–18.

Yashitola J, Krishnaveni D, Reddy APK, Sonti RV (1997) Genetic diversity within the population of *Xanthomonas oryzae* pv. *oryzae* in India. Phytopathology 87: 760–765.

Yashuhara-Bell J, Kubota R, Jenkins DM, Alvarez AM (2013) Loop-mediated amplification of the *Clavibacter michiganensis* subsp. *michiganensis*. Phytopathology 103: 1220–1226.

Yazdani R, Safaie N, Shams-Bakhsh M (2018) Association of *Pantoea ananatis* and *Pantoea agglomerans* with leaf spot disease on ornamental plants of Araceae family. Eur J Plant Pathol 150: 167–178.

Yokomi RK, Mello AFS, Saponari M, Fletcher J (2008) Polymerase chain reaction based detection of *Spiroplasma citri* associated with citrus stubborn disease. Plant Dis 92: 253–260.

Young AJ, Kawamata A, Ensbey MA, Lambley E, Nock CJ (2016) Efficient diagnosis of ratoon stunting disease of sugarcane by quantitative PCR on pooled leaf sheath biopsies. Plant Dis 100: 2492–2498.

Yu J, Wayadande AC, Fletcher J (2000) *Spiroplasma citri* surface protein p89 implicated in adhesion to cells of vector *Circulifer tenellus*. Phytopathology 90: 716–722.

Yu S-S, Che H-Y, Wang S-J, Cai C-L, Lin M-X et al. (2020) Rapid and efficient detection of 16SrI group areca pal yellow leaf phytoplasma in China by loop-mediated isothermal amplification. Plant Pathol 36: 459–467.

Zambon Y, Canel A, Bertacini A, Contaldo N (2018) Molecular diversity of phytoplasmas associated with grapevine yellows disease in Northern Italy. Phytopathology 108: 206–214.

Zeng H, Zhai X, Xie M, Liu Q (2018) Fluorescein isothiocyanate labeling antigen-based immunoassay strip for rapid detection of *Acidovorax citrulli*. Plant Dis 102: 527–532.

Zhai J, Luo Y, Zheng D, Huang X (2010) Evaluation of genetic diversity of highly virulent strains of *Xanthomonas campestris* pv. *malvacearum* by rep-PCR fingerprinting. J Phytopathol 158: 764–768.

Zhang Y, Merghi M, Bazzi C, Geider K (1998) Genomic analysis by pulsed-field gel electrophoresis of *Erwinia amylovora* from the Mediterranean region including Italy. J Plant Pathol 80: 225–232.

Zhao T, Feng J, Sechler A, Randhawa P, Li J et al. (2009) An improved assay for detection of *Acidovorax citrulli* in watermelon and melon seed. Seed Sci Technol 37: 337–349.

Zhao W-J, Chen H-Y, Zhu S-F, Xia M-X, Tan T-W (2007) One step detection of *Clavibacter michiganensis* subsp. *michiganensis* in symptomless tomato seeds using a TaqMan probe. J Plant Pathol 89: 349–351.

Zhao Y, Damicone JP, Bender CL (2002) Detection, survival and sources of inoculum for bacterial diseases of leaf crucifers in Oklahoma. Plant Dis 86: 883–888.

Zhao Z, Prager SM, Cruzado RK, Liang X, Cooper WR (2018) Characterizing zebra chip symptom severity and identifying spectral signatures associated with '*Candidatus* Liberibacter solanacearum''-infected potato tubers. Amer J Potato Res 95: 584–596.

Zheng Z, Bao M, Wu F, Van Hom C, Chen J (2018) A type 3 prophage of '*Candidatus* Liberibacter asiaticus' carrying a restriction-modification system. Phytopathology 108: 454–461.

Zheng Z, Wu F, Kumagai LB, Polak M, Deng X et al. (2017) Two '*Candidatus* Liberibacter asiaticus' strains recently found in California harbor different prophages. Phytopathology 107: 662–665.

Zivanovic M, Walcott RR (2017) Further characterization of genetically distinct groups of *Acidovorax citrulli* strains. Phytopathology 107: 29–35.

Additional References

Abeysinghe S, Abeysinghe PD, de-Silva CK, Udagama P, Warawichanee K et al. (2016) Refinement of the taxonomic structure of 16SrXI and 16SrXIV phytoplasmas of graminaceous plants using multilocus sequence typing. Plant Dis 100: 2001–2010.

Baharuddin B, Rudolph K, Niepold F (1994) Production of monospecific antiserum against the blood disease bacterium affecting banana and plantain. Phytopathology 84: 570–575.

Bekele B, Hodgetts J, Tomlinson J, Boonham N, Nikoli P et al. (2011) Use of a real-time LAMP isothermal assay for detecting 16SrII and XII phytoplasmas in fruit and weeds of Ethiopian Rift Valley. Plant Pathol 60: 345–355.

Bella P, Licciardello G, Sicilia A, Strano CP, Catara A et al. (2016) Collecting and preserving plant DNA for huanglongbing diagnosis in citrus samples from China. Eur J Plant Pathol 146: 829–836.

Bertolini E, Penyalver R, Garcia A, Olmos A, Quesada JM et al. (2003) Highly sensitive detection of *Pseudomonas savastanoi* pv. *savastanoi* in asymptomatic olive plants by nested PCR and single closed tube. J Microbiol Meth 52: 261–266.

Bhader BW, Helmick EE, Charkrabarti S, Osorio S, Soto N et al. (2018a) Disease progression of a lethal decline caused by the 16SrIV-D phytoplasma in Florida palms. Plant Pathol 67: 1821–1828.

Bhader BW, Helmick EE, Mou D-F, Harrison NA, Davis RE (2018b) Digital PCR technology for detection of palm-infecting phytoplasmas belonging to group 16SrIV that occur in Florida. Plant Dis 102: 1008–1014.

Biondi E, Campos-Vargas R, Meneses C, Perez S, Bertacini A et al. (2018) Draft whole genome sequence analyses on *Pseudomonas syringae* pv. *actidiniae* hypersensitive response negative strains detected from kiwifruit bleeding sap samples. Phytopathology 108: 552–560.

Bonants P, Griekspoor Y, Houwers I, Krijger M, van der Zouwen P et al. (2019) Development and evaluation of a triplex TaqMan assay and next-generation sequence analysis for improved detection of *Xylella* in plant material. Plant Dis 103: 645–655.

Casati P, Quaglino F, Tedeschi R, Spiga FM, Alma A et al. (2010) Identification and molecular characterization of '*Candiatus* Phytoplasma mali' isolates in north-western Italy. J Phytopathol 158: 81–87.

Chapman JR, Taylor RK, Weir BS, Romberg MK, Vanneste JL et al. (2012) Phylogenetic relationships among global populations of *Pseudomonas syringae* pv. *actinidiae*. Phytopathology 102: 1034–1044.

Davis RE, Dally EL, Zhao Y, Lee I-M, Wei W et al. (2015) Unraveling the etiology of North American grapevine yellows (NAGY) novel NAGY phytoplasma sequevars related to '*Candidatus* Phytoplasma pruni'. Plant Dis 99: 1087–1097.

Davis RE, Sinclair WA (1998) Phytoplasma identity and disease etiology. Phytopathology 88: 1372–1376.

Duduk B, Peric P, Mařuć D, Drobnjaković T, Picuau L et al. (2005) Phytoplasma in carrots: disease and potential vectors in Serbia. Bull Insectol 61: 327–331.

Fernandes C, Albuquerque P, Sousa R, Cruz, Tavares F (2017) Multiple DNA markers of identification of *Xanthomonas arboricola* pv. *juglandis* isolates and its direct detection on plant samples. Plant Dis 101: 858–865.

Franova J, Bertaccini A, Duduk B (2014) Molecular tools in COSTFAO807 Action. In: Bertaccini A (ed.), *Phytoplasmas and Phytoplasma Disease Management: How to Reduce Their Economic Impact*, pp. 179–184.

Gaitaitis R, Sumner D, Gay D, Smittle D, McDonald G et al. (1997) Bacterial streak and bulb rot of onion. I. A diagnostic medium for semiselective isolation and enumeration of *Pseudomonas viridiflava*. Plant Dis 87: 897–900.

Ghosh DK, Bhose S, Motghare M, Warghane A, Mukherjee K et al. (2015) Genetic diversity of the Indian populations of '*Candidatus* Liberibacter asiaticus' based on tandem repeat variability in a genomic locus. Phytopathology 105: 1043–1049.

Grisham MP, Pan YB, Richard EP Jr (2007) Early detection of *Leifsonia xyli* subsp. *xyli* in sugarcane leaves by real-time polymerase chain reaction. Plant Dis 91: 430–434.

Hodgetts J, Ball T, Boonham N, Mumford R, Dickinson M (2007) Use of terminal restriction fragment length polymorphism (T-RFLP) for identification of phytoplasmas in plants. Plant Pathol 56: 357–365.

Ji Z-Y, Zakria M, Zou L-F, Xion L, Li Z et al. (2014) Genetic diversity of transcriptional activator-like effector genes in Chinese isolates of *Xanthomonas oryzae* pv. *oryzicola*. Phytopathology 104: 672–682.

Jun-Hai N, Yue-Rong G, Jun-Mei Y, Qung-Yun L, Guang-Sui Y et al. (2015) Development and evaluation of a loop-mediated isothermal amplification assay for rapid detection of bacterial blight pathogen (*Xanthomonas axonopodis* pv. *diffenbachiae*) in anthurium. Eur J Plant Pathol 142: 801–813.

Kim WS, Hildebrand M, Jock S, Geider K (2001) Molecular comparison of pathogenic bacteria from pear trees in Japan and the fire blight pathogen *Erwinia amylovora*. Microbiology 147: 2951–2959.

Kubota R, Matsumoto M, Tanaka Y, Subandiyaj S, Iwanami T (2005) Characterization of the *tuftB, secE-nusG-rplKAJL-rpo* gene cluster of the citrus greening organism and detection by loop-mediated thermal amplification. Plant Dis 89: 705–711.

Lana UGP, Gomes EA, Silva DD, Costa RV, Cota LV et al. (2012) Detection and molecular diversity of *Pantoea ananatis* associated with white spot disease in maize, sorghum and crabgrass in Brazil. J Phytopathol 160: 441–448.

Lang JM, Ducharme E, Caballero JJ, Luna E, Hartman T et al. (2017) Detection and characterization of *Xanthomonas vasicola* pv. *vasiculorum* (Cobb1894) comb. nov, causing bacterial leaf streak of corn in the United States. Phytopathology 107: 1312–1321.

Lang JM, Gent DH, Schwartz HF (2007) Management of Xanthomonas leaf blight of onion with bacteriophage and plant activator. Plant Dis 91: 871–878.

Lee I-M, Zhao Y, Daivs RE, Wei W, Martini M (2007) Prospects of DNA-based systems for differentiation and classification of phytoplasmas. Bull Insectol 60: 239–244.

Li WB, Hartung JS, Levy L (2007) Evaluation of DNA amplification methods for improved detection of '*Candidatus* Liberibacter asiaticus' associated with citrus huanglongbing. Plant Dis 91: 51–58.

Li X, Nie J, Ward L, Madani M, Hsiang T et al. (2009) Comparative genomics-guided loop-mediated isothermal amplification for characterization of *Pseudomonas syringae* pv. *phaseolicola*. J Appl Microbiol 107: 717–726.

Lu S, Tian Q, Zhao W, Hu B (2017) Evaluation of the potential of five housekeeping genes for identification of quarantine *Pseudomonas syringae*. J Phytopathol 165: 73–81.

Lyon NF, Taylor JD (1990) Serological detection and identification of bacteria from plants by conjugated *Staphylococcus aureus* slide agglutination test. Plant Pathol 39: 584–590.

Marcelletti S, Scortichini M (2014) Definition of plant pathogenic *Pseudomonas* genomospecies of the *Pseudomonas syringae* complex through multiple comparative approaches. Phytopathology 104: 1274–1232.

Matsuyama N (1995) Trials for rapid identification of phytopathogenic bacteria by HPLC and direct colony TLC. J Fac Agric Kyushu Univ 40: 87–91.

Mazivele MOM, Nuaila V, Durante M, Colombo MM, Taviani E (2018) Promising pairs for detection of phytoplasma causing coconut lethal yellowing disease in Mozambique. Phytoparasitica 46: 301–308.

Mori M, Sogou K, Inoue Y (2019) Development of a selective medium and antisera for *Pseudomonas syringae* pv. *syringae* from seeds of barley and wheat. J Gen Plant Pathol 85: 211–220.

Mori T, Nagamine K, Tomita N, Notomi T (2001) Detection of loop-mediated isothermal amplification reaction by turbidity derived from magnesium pyrophosphate formation. Biochem Biophys Res Comm 289: 150–154.

Naumann L, Karl H, Zielke R, Schmidt A, Griesbach E (1988) Comparative analysis of bean seeds on the incidence of causal agent of halo blight *Pseudomonas syringae* pv. *phaseolicola* by different identification methods. Zentr Mikrobiol 143: 487–498.

Ozdemir Z (2005) Development of a multiplex PCR assay for concurrent detection of *Clavibacter michiganensis* subsp. *michiganensis* and *Xanthomonas axonpodis* pv. *vesicatoria*. Plant Pathol J 4: 133–137.

Pérez-López E, Wei W, Wang J, Davis RE, Luna-Rodríguez M (2017) Novel phytoplasma strains of X-disease group unveil genetic markers that distinguish North American and South American geographic lineages within subgroups 16SrIII-J and 16SrIII-U. Ann Appl Biol 171: 405–416.

Primo ED, Santoro MA, Giordano WB, Lisa AT (2019) Identification and genetic diversity analysis of strain *Pseudomonas syringae* S5: a pathogen responsible for bacteriosis in *Avena sativa* plants. J Phytopathol 167: 146–155.

Rao GP, Madhupriya, Kumar M, Tomar S, Maya B et al. (2018) Detection and identification of four 16Sr subgroups of phytoplasmas associated with different legume crops in India. Eur J Plant Pathol 150: 507–513.

Roberts R, Cook G, Grout TG, Khamis F, Rwomushana I et al. (2017) Resolution of identity of '*Candidatus* Liberibacter' species from huanglongbing affected citrus in East Africa. Plant Dis 101: 1481–1488.

Rosete YA, Diallo HA, Konankonan JL, Yankey N, Saleh M et al. (2017) Detection and differentiation of the coconut lethal yellowing phytoplasma in coconut-growing villages of Grand-Lahou, Coted'Ivoire. Ann Appl Biol 170: 333–347.

Roumagnac P, Gagnevin L, Pruvost O (2000) Detection of *Xanthomonas* sp., causal agent of onion bacterial blight in onion seeds using a newly developed semiselective isolation medium. Eur J Plant Pathol 106: 867–877.

Schaad NW, Azad H, Peet RC, Panopoulos NJ (1989) Identification of *Pseudomonas syringae* pv. *phaseolicola* by a DNA hybridization probe. Phytopathology 79: 903.

Streten C, Gibb KS (2005) Genetic variation in '*Candidatus* Phytoplasma australiense'. Plant Pathology 54: 8–14.

Teresani G, Bertolini E, Alfano-Fernández A, Martinez C, André F et al. (2014) Association of '*Candidatus* Liberibacter solanacearum' with a vegetative disorder of celery in Spain and development of a real-time PCR method for its detection. Phytopathology 104: 804–811.

Walcott RR, Fessehaie A, Castro AC (2004) Differences in pathogenicity between two genetically distinct groups of *Acidovorax avenae* subsp. *citrulli* on cucurbit hosts. J Phytopathol 52: 277–285.

Wang K, Hiruki C (2005) Distinction between phytoplasmas at subgroup level detected by heteroduplex mobility assay. Plant Pathol 54: 625–633.

Wang ZK, Luo HH, Yi SZ (1997) Application of DIA (dot immunobinding assay) for rapid detection of *Xanthomonas axonopodis* pv. *citri*. J Southwest Agric Univ 19: 529–532.

Weier HU, Fahy PC, Broadbent P, Barnas D (1995) Rapid identification of second outbreak of Asiatic citrus canker in the Northern Territory using polymerase chain reaction and genomic fingerprinting. Austr Plant Pathol 24: 104–111.

Wu WC, Lee ST, Kuo HP, Wang LY (1993) Use of phage for identifying the citrus canker bacterium *Xanthomonas campestris* pv. *citri* in Taiwan. Plant Pathol 42: 389–395.

Ye G, Hong N, Zou L-F, Zou H-S, Zakria M et al. (2013) Tale-based genetic diversity of Chinese isolates of the citrus canker pathogen *Xanthomonas citri* pv. *citri*. Plant Dis 97: 1187–1194.

Young AJ, Nock CJ (2017) Molecular detection of diverse *Leifsonia* strains associated with sugarcane. Plant Dis 101: 1422–1431.

Yuan X, Morano L, Bromley R, Spring-Pearson S, Stouthamer R (2010) Multilocus typing of *Xylella fastidiosa* causing Pierce's disease and oleander leaf scorch in the United States. Phytopathology 110: 601–611.

Zheng J, Lin Z-Y, Ma Y-M, Gao B-D, Liu H-Q et al. (2016) Rapid discrimination between groups I and II of *Acidovorax citrulli*, using a primer pair specific to a *pilL* gene. J Phytopathol 164: 558–562.

Zombre C, Sankara P, Ouedraogo SL, Wonni I, Boyer J et al. (2016) Natural infection of cashew (*Anacardium occidentale*) by *Xanthomonas citri* pv. *mangiferaeindicae* in Buskina Faso. Plant Dis 100: 718–723.

5 Biology of Phyllosphere Bacterial and Phytoplasmal Pathogens

5.1 INTRODUCTION

Plant bacterial pathogens are prokaryotic, single-celled primitive organisms, and they are classified as gram-negative or gram-positive based on their ability/inability to retain the Gram stain. They vary in the levels of virulence, host range, and sensitivity to environment and chemicals. During pathogenesis they elaborate different kinds of enzymes, toxins, and secondary metabolites which function as effectors facilitating entry into the plants and colonization of host tissues. Further, the pathogen-derived effectors are capable of manipulating host responses to derive nutrients for their multiplication and inducing various kinds of symptoms in the susceptible plants. On the other hand, plants are endowed with preformed structural compounds that counteract the effects of pathogen-derived effectors, arresting further progress of pathogen development. Resistant plants produce postinfectional defense-related compounds such as PR proteins and phenolic compounds that restrict colonization of host tissues at different stages of disease development. Investigations on the molecular basis of pathogenesis reveal the complex nature of interactions between different bacterial pathogens with their respective host plants resulting in either successful establishment of infection leading to disease development in susceptible plants or suppression of pathogen development in resistant plants.

Bacterial plant pathogens are primitive microorganisms and they are classified as prokaryotes, whereas fungal pathogens are classified as eukaryotes. Phytopathogenic bacteria are either gram-positive, classified within the Phylum Actinobacteria, or gram-negative, placed in the Phylum Proteobacteria. Bacteria as causal agents of plant diseases were implicated only in the late 19th century. *Erwinia amylovora* was indicated to be the incitant of apple fire blight disease by Burrill in 1878. *E. amylovora* was established as the causal agent of apple fire blight by Arthur (1885) based on conventional Koch's postulates. *Agrobacterium tumefaciens* was proved to induce crown gall diseases in several plant species (Smith and Townsend 1907). During the course of 20th century, many crop diseases were reported to be caused by different bacterial pathogens in various countries.

5.2 BIOLOGY OF BACTERIAL PATHOGENS

5.2.1 General Characteristics of Bacterial Pathogens

Bacterial pathogens infecting plants have simpler structural characteristics, compared to fungal plant pathogens. Single cells constituting a population formed in cultures show variations in shape and size, depending on the bacterial species. Most bacteria are strictly saprophytic in nature and are beneficial to humans, since they are involved in decomposition of organic wastes produced by industries and death of animals, plants and other organisms. In addition, they have an important role in building up nutritional levels of soils by fixing atmospheric nitrogen and breaking down complex organic matter into simpler forms that can be absorbed by plant root system. The cell walls of most bacterial species are enveloped by a viscous gummy material which may form as a mass around bacterial cells known as capsule or it may be present as a slime layer around the bacterial cells. Most of the phytopathogenic bacteria possess delicate, thread-like flagella, organs of locomotion, which are much longer than the bacterial cells themselves. Fimbriae or pili, smaller thread-like appendages may be present at multiple locations and they are believed to be helpful for attachment of bacteria to host surface. Chromosomes composed of DNA are coiled and plasmids are extra-chromosomal genomic entities coding for essential virulence factors required for infection of susceptible hosts. Storage granules are present in some bacterial species. Nuclear material of the bacteria is not separated from the cytoplasm by a nuclear membrane as in eukaryotes. Internal cell contents constitute the protoplast which contains a cytoplasmic or protoplasmic membrane. Cytoplasm contains a complex mixture of proteins, lipids, carbohydrates, and different kinds of other organic compounds, minerals, and water. The nuclear material consists of a large circular chromosome composed of DNA, which forms the primary body of genetic material of the bacterial species concerned. Plasmids in bacterial cells can replicate independently and govern some characteristics such as pathogenicity, tumor production, and resistance to chemicals. Plasmids can move or be moved between bacteria or even between bacteria and plants. During asexual reproduction, each bacterial cell divides into two daughter cells by a process known as binary fission. This process can be repeated indefinitely, leading to formation of large population of bacterial cells under optimal conditions.

Variability in the genetic characteristics of the microbial pathogens may be primarily due to sexual recombination or mutation. Variations in genetic characteristics may occur through other modes also. Mutation may occur in a very small percentage of bacterial cells which multiply giving rise to a population of bacterial cells with characteristics different from the parent/wild type cells. The process of transformation involves liberation of genetic material from one cell either by secretion or rupture of the cell wall.

DOI: 10.1201/9781003456827-5

A fragment of the DNA is liberated and gains entry into a genetically compatible bacterial cell of the same or closely related species. The recipient cell becomes genetically different. The phenomenon of transduction requires a bacteriophage to function as a vector for the transfer of a fragment of the DNA of one bacterial cell to another bacterial cell. Bacteria may harbor prophage, a stable heritable form of bacterial virus or remnants of prophage integrated into their characteristics. Different characters of bacteria may show variation due to mutation caused by environmental conditions and exposure to certain chemicals applied frequently/excessively as a disease management tactic to reduce disease incidence. In addition to these phenomena, horizontal transfer, passage of DNA from one bacterial cell to another may be effected under in vitro and it is considered to be an important mode leading to variability in genetic characteristics of bacterial pathogens (Narayanasamy 2001; Vidaver and Lambrecht 2004).

5.2.2 Disease Cycles of Bacterial Pathogens

Phyllosphere bacterial pathogens do not produce any special structure required for different functions. Bacterial cells do not change during binary fission which is the most common asexual method of multiplication. Investigations on different biological characteristics such as the sources of inoculum, host range, survival, modes of spread of the pathogen and sensitivity to changes in the environmental conditions and chemicals may be useful to develop effective management systems for containing bacterial diseases in different agroecosystems.

5.2.2.1 Pathological Characteristics

Phyllosphere bacterial pathogens induce leaf spots, streaks, cankers, blights, and seed rot diseases. *Pseudomonas syringae* pv. *tomato* (*Pst*) (tomato bacterial speck disease) induces necrotic lesions on leaves, stem, and fruit. Infection of fruits may lead to infestation of seeds. Seedlings growing from infected seeds may develop symptoms or remain asymptomatic. Formation of bacterial ooze from infected tissues is a characteristic of some diseases. *Xanthomonas oryzae* pv. *oryzae* (*Xoo*) induces bacterial leaf blight and vascular wilt resulting in drying up of foliage and plants may be killed in large number (Leach et al. 1989). *X. citri* subsp. *citri (Xcc)* (Asiatic citrus canker) occurs in the form of several strains with varying virulence levels and the strains have been grouped into different pathotypes (Koizumi 1985). *X. citri* subsp. *citri* grows initially epiphytotically on leaf surfaces and gains entry into the host through stomata or wounds followed by colonization of mesophyll parenchyma (Laia et al. 2000). *X. campestris* pv. *campestris* causes black rot disease in cabbage, cauliflower, lettuce, and other *Brassica* crops. The pathogen is transmitted through seeds and flies (*Calliphora vomitoria*) (Van der Wolf and Van der Zouwen 2010). Colonization of pepper blossoms by *X. euvesicatoria* posed a substantial risk after transmission through seeds (Dutta et al. 2014). *Acidovorax avenae* subsp. *citrulli (Aac)* induces bacterial fruit blotch in cucurbitaceous crops such as watermelon, and melon. Blossom colonization in female plants by the pathogen may result in seed infestation by *Aac* (Walcott et al. 2003).

Burkholderia glumae (bacterial grain rot and leaf sheath browning) could be detected consistently in infected rice leaf sheaths and grains. *B. gladioli* can infect orchidaceous gladiolus plants and onion bulbs (Ura et al. 2006). *Clavibacter michiganensis* subsp. *michiganensis (Cmm)* (bacterial canker) infects tomato stem and fruits and subsequently invades seeds externally and internally (Kaneshiro and Alvarez 2003). Infected seeds may carry infection to seedlings and positive correlation between seed infection and incidence of bacterial canker in seedlings may be high (Hadas et al. 2005). The pathogenicity of *X. arboricola* pv. *fragariae (Xaf)* isolated from strawberry plants exhibiting bacterial blight disease was examined. A spray leaf inoculation technique was developed to inoculate strawberry plants with pathotype strain PD2780. Cultivars Candonga, Sabrina, and Murano showed typical symptoms of infection by *Xaf* (Ferrante and Scortichini 2018). *Xanthomonas axonopodis* pv. *phaseoli* (*Xap*) (bean common bacterial blight) induces symptoms on stems, pods and seed in addition to leaves. Lesions appear as small water-soaked spots which enlarge and turn necrotic with a border of yellowish zone. Spots on pods are circular, yellow to brown in color, whereas seeds show brown spots and seed germination and vigor of seedlings are significantly reduced (Akhavan et al. 2009). Bacterial spot disease of pumpkin induced by *X. cucurbitae* showed variations in lesion types which were grouped into three classes. Infected seeds may be an important source of inoculum (Zhang and Babadoost 2018). Infection of onion leaves by *Pantoea ananatis* results in leaf blight disease. Leaf infection by *P. ananatis* led to bulb infection. Pathogenic strains of *P. ananatis* induced non-macerated yellow-brown discoloration in fleshy bulb scales following inoculation of bulbs. Subepidermal inoculation of onion leaves with *P. ananatis* resulted in gray-white foliar lesions that enlarged acropetally and basipetally from points of inoculation (Carr et al. 2013). Onion plants infected by *Pseudomonas syringae* pv. *allii (Psa)* exhibited irregular lesions with necrotic centers and water-soaked margins (Moloto et al. 2017).

'*Candidatus* Liberibacter' spp. (citrus huanglongbing, HLB), is a phloem-limited fastidious nonculturable α-proteobacterium. Three different species involved in HLB disease have been identified. Spread of the disease occurs through infected planting materials and psyllid vectors (Bové 2006). Citrus stubborn disease caused by *Spiroplasma citri* induces different degrees of severity of symptoms in infected plants. Fruits from severely symptomatic plants contained very high concentrations of spiroplasmas compared to plants with mild symptoms (Mello et al. 2010). Media incorporated with juice from infected grapefruit significantly prolonged '*Candidatus* Liberibacter spp.' cell viability, compared with K medium. In addition, a biofilm formed at the air-liquid interface of juice cultures contained '*Ca.* L. asiaticus' cells (Parker et al. 2014). '*Ca* Liberibacter solanacearum' (Zebra chip, ZC disease) induces stunting, chlorosis, swollen nodes, proliferated axillary buds, aerial tuber formation, browning of vascular system in stems, and leaf scorching. Symptoms on below-ground organs include enlarged lenticels of stem, collapsed stolon, brown discoloration of vascular ring, necrotic flecking in internal tuber tissues (Lin and Gudmestad 2013). Two '*Ca* L. solanacearum' haplotypes A and B (Lso A and Lso B) were differentiated. Plants infected with LsoB died prematurely, whereas plants infected with LsoA survived (Mendoza-Herrera et al. 2018).

5.2.2.2 Sources of Pathogen Inoculum

Infected seeds function as primary sources of inoculum for bacterial disease incidence and subsequent spread. The role of wild relatives of crops and weeds serving as sources of inoculum has been reported in many crops. The presence of weed species in the field not only leads to loss of nutrients applied to crops but also provides support to bacterial pathogen development and their survival in the absence of crops. Populations of *Xanthomonas campestris* pv. *vitians* (*Xcv*) (grapevine bacterial leaf spot) from different weed species varied significantly. *Lactuca* sp. supported pathogen development to the maximum extent. Other weed species either showed mild symptoms or remained asymptomatic (Toussaint et al. 2012). In a later investigation, assessment of epiphytic populations of *Xcv* in 13 weed species showed wide variations. Leaf spot-like symptoms developed on *Sonchus oleraceus*, which had significantly greater populations than those on other weed species (Fayette et al. 2018). The ability of cruciferous weeds and ornamentals to serve as sources of inoculum of *X. campestris* pv. *campestris (Xcc)* (black rot of *Brassica*) was investigated. *Xanthomonas* isolates from wallflower appeared to have potential for development of epidemics of black rot disease in crucifers (Kranthausen et al. 2018). Four representative strains of *Xanthomonas citri, X. phaseoli*, and *X. euvesicatoria* (bacterial leaf blight) were genome sequenced and they showed variable virulence-associated gene contents. Pathogenicity to aroids was correlated to the presence of three specific T3 effector genes and with a T6SS gene sequence (Constantin et al. 2017). Asiatic citrus canker pathogen *Xanthomonas citri* subsp. *citri (Xcc)* exists in the form of three pathotypes A, A*, and A^w. A* type strains have restricted host range to Mexican lime and they are classified into four subgroups (Jalali et al. 2017). *Xanthomonas* species infect several vegetable crops and large number of weed species. The weeds *Aeollanthus suaveolans*, *Amaranthus lividus,* and *Sida glomerata* exhibited lesions on leaves induced by *Xanthomonas* spp. and these isolates were tested for their pathogenicity. Phylogenetic analysis of the *gyrB* and *ropD* genes and PCR-specific assays were applied to identify *X. euvesicatoria* pv. *euvesicatoria* and *X. campestris* pv. *campestris.* All strains except CCRMXe03, were pathogenic. All *X. euvesicatoria* pv. *euvesicatoria* were pathogenic to leaves and fruits of tomato and bell pepper. Strain CCRMXcc 371 was pathogenic to cabbage, kale, cauliflower, and broccoli (Santos et al. 2020). The genomic relationships among pathovars of *Xanthomonas campestris* pv. *campestris* (*Xcc*), as well as their genetic determinants of host range and tissue specificity were investigated. Genomes of a collection of *X. campestris* strains isolated from cruciferous crops and weeds as well as strains from global collections were determined. Pathogenicity assays and genomic sequence comparison across the collections showed a correlation between pathovar and genomic relatedness, linking strain host range with type III effector repertoires identified. The presence of a copy of the cellobiosidase gene *cbsA* with coding sequence for a signal peptide was correlated with ability to infect vascular tissues (Dubrow et al. 2022).

Xylella fastidiosa, a xylem-limited gram-negative bacterial pathogen causing bacterial leaf scorch disease, has a wide host range that includes several crops such as grapevine, citrus, peach, pecan, and almond. Olive trees (*Olea europaea*) showing leaf scorch or branch dieback symptoms were examined for their ability to serve as inoculum source of *X. fastidiosa* capable of infecting several crops including grapevine. Pathogenicity tests with olive strains on grapevine and almond indicated that they induced symptoms in grapevine and almond typical of *X. fastidiosa* subsp. *multiplex*. Glassy-winged sharp shooters (*Homalodisca vitripennis*) could transmit strains of both subsp. *multiplex* and *fastidiosa* and green sharpshooter *Draculacephala minerva* transmitted olive strains (Krugner et al. 2014). '*Candidatus* Liberibacter solanacearum' (Lso) (potato zebra chip, ZC) is transmitted by potato/tomato psyllid *Bactericera cockerelli* to solanaceous plants and it depends on psyllids *Trioza apicalis* and *B. trigonica* for transmission to carrot plants. Several solanaceous weed species have been reported to be hosts for Lso. Isolates of Lso originating from different geographical regions were grouped into five haplotypes, LsoA, LsoB, LsoC, LsoD, and LsoE. The single nucleotide polymorphism (SNP) analysis revealed the presence of two distinct bacterial linkages that corresponded to haplotypes D and E. Phylogenetic analysis also indicated that isolates of Lso in France, clustered into two groups, are closely related to haplotypes LsoD and LsoE (Hajri et al. 2017). Strains of *Erwinia amylovora* (fire blight of rosaceous plants) are placed in two groups based on host range as Spiraeoideae and *Rubus* strains. Spiraeoideae strains have a wide host range, infecting many genera in the Rosaceae family, including apple and pear, under field conditions. *Rubus* strains infect plant belonging to the genus *Rubus* exclusively which includes raspberry and blackberry (Asselin et al. 2011). Infected plant parts or organs may be sources of inoculum for infection by bacterial pathogens and function as sites for overwintering of *Erwinia amylovora.* Cankers and symptomless infected apple shoots infected by *E. amylovora* might be sources of inoculum to blossoms of nonhost plants growing close to infected trees. Fruit mummies and buds might favor overwintering and functioning as primary sources of inoculum for dissemination of the pathogen in the early growing season. Nonhost blossoms might also serve as alternative inoculum source for the pathogen population buildup (Weißhaupt et al. 2016). Asiatic citrus canker pathogen *Xanthomonas citri* subsp. *citri (Xcc)* exists in the form of three pathotypes A, A*, and A^w. A* type strains have a restricted host range to the Mexican lime, and they are classified into four subgroups. The results indicated the differences in sequences might determine the host range of strains of *X. citri* subsp. *citri* (Jalali et al. 2017).

5.2.2.3 Survival of Bacterial Pathogens

Phyllosphere bacterial pathogens may survive in infected leaves, seeds/propagules, and plant debris left on the soil after harvest. Spread of *Xanthomonas oryzae* pv. *oryzae* (*Xoo*) (bacterial leaf blight, BLB) in rice leaves was more rapid and extensive with compatible than incompatible interactions, indicating that susceptible rice cultivars might support survival of *Xoo* in the leaves of infected plants (Leach et al. 1989). *Xanthomonas axonopodis* pv. *phaseoli (Xap)* (bean common bacterial blight), may be carried both externally and internally in seed, bean crop debris, and epiphytically on volunteer plants and perennial alternate host plants. *Xap* might survive in the seed coat for several years (Pernezny and Jones 2002). But *Xap* overwintered between crops in crop residues in Zimbabwe (Karavina et al. 2008). *Xap* could also survive and multiply as an epiphyte or resident on shoot surfaces of weed hosts, primarily on leguminous plant species, without exhibiting visible symptoms (Pernezny and Jones 2002). *Clavibacter michiganensis* subsp. *michiganensis (Cmm)* (tomato bacterial canker) was transmitted through infected seeds or transplants. *Cmm* survived primarily on host plant debris. Survival and half-life of *C. michiganensis* subsp. *michiganensis* in plant debris were relatively long and might be influenced by both tissue exposure and geographic location (Fatmi and Schaad 2002). *Xanthomonas campestris* pv. *campestris* (black rot of Brassica), and it could be transmitted to cauliflower blossoms via flies (*Calliphora vomitora).* Infestation of seed with *Xcc* by flies was confirmed by seedling grow-out test (van der Wolf and van der Zouwen 2010). *Xcc* may survive on plant debris only for a short period (Roberts et al. 2004).

The role of infected seeds, plant debris, and infested soil in the survival of *Xanthomonas campestris* pv. *vignicola* (*Xcv)* (cowpea bacterial blight) was investigated under in vitro and in vivo conditions. Seeds harbored the highest populations of *Xcv.* Infested seeds and plant debris left on the soil surface after harvest might function as important sources of inoculum (Okechukwu and Ekpo 2008). Survival duration of group I and II isolates of *Acidovorax avenae* subsp. *citrulli (Acc*) (bacterial fruit blotch) was similar. However, *Aac* survival was significantly greater in seed lots generated via pistil inoculation than via pericarp inoculation. The results showed that *A. avenae pv. citrulli* might survive or overwinter in citron melon seeds for several years (Dutta et al. 2014). Influence of tissue localization of *A. citrulli* on its survival was investigated. Bacteria localized in the embryo and endosperm survived effectively and were in higher populations compared to those localized in seed coat (Dutta et al. 2016). Distribution of *Xylophilus ampelinus* (grapevine bacterial blight) in infected plants was investigated. The presence of *Xa* in the axillary buds, but not in the epidermis and vascular bundle, was detected by PCR assay. *Xa* was detected in axillary buds and vascular bundles in naturally infected plants also. Detection frequency of *Xa* in whole axillary buds decreased only slightly even after overwintering. *X. ampelinus* might not survive on plant surfaces, but might be present on the underside surface of bract and bud wood, during winter in Hokkaido (Komatsu and Kondo 2015). The possibility of *Erwinia amylovora (Ea)* (fire blight disease of apple and pear) had the potential for survival as a necrotroph or seminecrotroph on host leaves, allowing overwintering of *E. amylovora* in dead apple leaves (Sobiczewski et al. 2017).

5.2.2.4 Dissemination of Bacterial Pathogens

5.2.2.4.1 Dissemination through Seeds

Several species of *Xanthomonas* and *Pseudomonas* are seedborne and disseminated through seed to different locations. *X. oryzae* pv. *oryzae* (in rice), *X. axonopodis* pv. *malvacearum* (in cotton), *X. campestris* pv. *campestris* (in crucifers) *X. campestris* pv. *vesicatoria* (in pepper), *X. campestris* pv. *vignicola* (in cowpea), *Pseudomonas syringae* pv. *phaseolicola* (in bean), *P. syringae* pv. *tomato* (in tomato), *Acidovorax avenae* susbp. *citrulli* (in cucurbits), and *Clavibacter michiganensis* subsp. *michiganensis* (in tomato) are some of the bacterial pathogens transmitted through seeds to different extents. Infected seeds and propagules may have triple roles as carrier of inoculum, agency of dissemination, and substrate for survival of bacterial pathogens (Narayanasamy 2017). *Xanthomonas euvesicatoria*–infected pepper blossoms and the relationship between the pathogen concentration in blossoms and seed disinfestation was investigated under artificial inoculated conditions. The pathogen colonized stylar and ovary tissues and reached high populations. Pathogen concentration in inoculated plants was positively correlated to percentage of infected seed lots. *A. citrulli* colonized pepper blossoms more effectively than *X. euvesicatoria.* Pepper blossoms might be a potential site of ingress of *X. euvesicatoria* into seed, leading to infestation of seed lots with bacterial pathogen (Dutta et al. 2014). The pathway of ingress of *Acidovorax citrulli* (bacterial fruit blotch, BFB) in watermelon seeds was studied. With stigma inoculation, *A. citrulli* colonized style and ovary tissues. The bacteria co-localized with pollen germ tubes in these tissues, with ovary pericarp inoculation. The results showed that stigma inoculation resulted in faster colonization of watermelon ovules by *A. citrulli* than pericarp inoculation (Dutta et al. 2012).

Infected or contaminated seed and transplants form the primary methods of dissemination of *Clavibacter michiganensis* susbp. *michiganensis (Cmm*) (bacterial canker disease in tomato). Bacterial canker of tomato (BCT) medium could be employed for detection of *Cmm* in tomato seed and asymptomatic tomato seedlings (Ftayeh et al. 2011). Disbudding and defoliation using scissors or by hand contributed significantly to the secondary spread of the disease in commercial greenhouses (Kawaguchi et al. 2010). *Cmm* dispersal may occur via contaminated tools to a limited extent. A large population of *Cmm* was exuded in gutation fluid which might result in buildup of epiphytic population and secondary spread of the disease (Sharabani et al. 2013). *Cmm* dispersed spatially from root-inoculated source seedlings and colonized leaf surfaces of surrounding seedlings to short distances (Frenkel et al. 2016). *Xanthomonas campestris* pv. *campestris (Xcc)* in the infested seed coat multiplied and emerged along with cotyledons serving as the primary inoculum source. The secondary inoculum is disseminated by rain drop splashes from infected leaves (Roberts et al. 2004). Seeds infected by *X. campestris* pv. *vignicola (Xcv)* and plant debris left on the soil surface after harvest might be a major source of primary inoculum for subsequent years for cowpea crops (Okechukwu and Ekpo 2008).

Transmission of *Xanthomonas campestris* pv. *campestris (Xcc)* from bean flowers to seeds and from seeds to seedlings was determined in compatible and incompatible interactions. Transmission of *Xcc* from seeds to seedlings was compared with *X. citri* pv. *phaseoli* var. *fuscans* (*Xcp*) in compatible interactions with bean. Colonization of germinating seeds and seedlings was considered a critical step in the establishment of bacterial diseases. Populations of *Xcc* ATC33913 increased in contaminated seed and seedling to high levels. Dynamics of pathogen population sizes during seedling development were analyzed by inoculating seeds by vacuum infiltration with *Xcc*. Population sizes did not vary initially in compatible and incompatible interactions. Later *Xcp* began to multiply at a faster rate than *Xcc* on bean seedlings. No visible symptoms appeared on seeds and seedling following seed contamination with *Xcc*, implying that seedling colonization might not require bacterial pathogenicity determinants. Colonization of seedling relied most on surface colonization and not on parasitic relationships (Darrasse et al. 2010). *Xanthomonas citri* pv. *citri (Xcc)* (citrus canker disease) in the sonicates from mature asymptomatic fruits could not be detected by PCR, indicating that the asymptomatic fruits might not be source of inoculum of *Xcc* for infection of other trees. PCR assay showed that no release of pathogen cells into rain water occurred from canker lesions present on (Shiotani et al. 2009). *Xanthomonas fragariae* (strawberry angular leaf spot) is transmitted through propagative plant materials. Risk of infection by *X. fragariae* in strawberry plants exposed to aerosolized inoculum has to be prevented to restrict incidence of angular leaf spot of strawberry (van der Wolf et al. 2018). *Pseudomonas syringae* pv. *phaseolicola (Psp)* (bean halo blight) in addition to seed transmission, may be disseminated from bacterial exudation by water-splash and wind during rainfall. *Psp* might persist on dry bean seeds and actively multiply after germination (Arnold et al. 2011).

5.2.2.4.2 Dissemination through Vectors

Hemipteran insects, leafhoppers, and psyllids are by far the most important vectors of bacterial pathogens. Specialized mouth parts, stylets, allow hemipterans to penetrate the epidermal tissues and reach the preferred tissues, phloem or xylem. *Xylella fastidiosa*, the only known xylem-limited bacterial pathogen, exhibits a semipersistent association with the vectors, *Homalodisca vitripennis* and *Graphocephala atropunctata* (Backus and Morgan 2011). Phloem-limited bacterial pathogens, *Candidatus* Phytoplasma spp., *Candidatus* Liberibacter spp., and *Spiroplasma citri*, have persistent relationships with their respective leafhopper vectors. *Xanthomonas citri* subsp. *citri (Xcc)* (Asiatic citrus canker disease) could be spread by larvae of leafminer *Phyllocnistis citrella* through wounds made while feeding, in addition to rain-splash and wind. Cauliflower blossom infection by *Xanthomonas campestris* pv. *campestris (Xcc)* was achieved by brush or pollination with blue bottle flies *Calliphora vomitoria* as a vector, resulting in seed infection by *Xcc*. Flies were fed on agar-grown bacteria prior to inoculation of cauliflower blossoms. Poly-tunnel experiments with fly-inoculation of cauliflower showed that some seed batches of a breeding line had high densities of the bacterial pathogen (van der Wolf and van der Zouwen 2010). *Erwinia tracheiphila* causing bacterial wilt of cucurbits was transmitted by striped (*Aaclymma vittatum*) and spotted (*Diabrotic undecimpuntata*) cucumber beetles (Rojas and Gleason 2012).

A wide range of vectors are involved in the transmission of *Xylella fastidiosa* from infected to healthy plants and *X. fastidiosa* can insect several plant species. The xylem-limited *X. fastidiosa* has non-circulative relationship with its vectors, whereas all phloem-colonizing '*Candidatus* Liberibacter spp.' and *Spiroplasma citri* exhibit a circulative type of relationship with their insect vectors. *Xylella fastidiosa* could propagate extracellularly within the vector and it is non-circulative, whereas all other insect-transmitted bacterial pathogens are either propagative or circulative in nature (Beanland et al. 2000; Ammar et al. 2011). *Xylella fastidiosa*, has larger genome than those of phloem-limited bacterial pathogens, and this may be required for the pathogen to survive in the nutrient-limited xylem tissue (Orlovskis et al. 2015). *X. fastidiosa* did not invade the epithelial gut, hemolymph, or salivary glands, and its cells were retained in the alimentary canal. *Xf* could be transmitted within a few minutes after acquisition feeding on infected source plant (Killiny et al. 2012). Lipopolysaccharide (LPS), the outermost layer of sugar polymers surrounding gram-negative bacteria, had a critical role in adhesion of bacteria to the vector tissue and subsequent transmission (Rapicavoli et al. 2015). *X. fastidiosa*, causing Pierce's disease (PD) in grapevine forms a biofilm on portions of the foregut of the vector *Graphiocephala atropuncta*, which is apparently the source of inoculated cells. Diffusible signal factor (DSF), an unsaturated fatty acid, accumulates in populations of *Xf* as cell numbers increase. A variety of genes encoding fimbrial and afimbrial adhesions are upregulated by DSF, while genes encoding type IV pili involved in twitching motility as well as those encoding extracellular enzymes such as polygalacturonases and endoglucosidases are repressed in the presence of DSF. The wild-type strain of *Xf* might restore the ability of *rfpF* mutants, deficient in rpf quorum-sensing system blocked in DSF production. The *rpfF* mutant was very poorly transmitted by vectors irrespective of whether they had also fed on plants infected with wild-type strain. A factor released into the xylem by *rpfF* mutants and to a lesser extent by the wild-type strain inhibited their attachment to and their transmission by sharpshooter vectors and may also enable them to move more readily through host plants (Baccari et al. 2014).

Many proteins required for attachment of *Spiroplasma citri* (citrus stubborn disease) to vector were identified, using *S. citri* mutants impaired in insect transmission and with *S. citri* strains that lost insect attachment properties after multiple in vitro cultivations (Mutaqin et al. 2011). Bacterial protein–insect interactions were investigated using leafhopper *Ciruculifer tenellus* monolayer cell culture assays with spiroplasma. Many candidate attachment proteins including P58, SARP, and plasmid-borne protein P32 were identified. Mutants of *S. citri* defective in spiralin, another membrane protein, production had reduced transmission efficiency of the vector *Circulifer haematoceps*, compared to the wild-type strain (Gasparich 2010). '*Candidatus* Liberibacter asiaticus' (Las) (HLB) is spread locally by citrus psyllid *Diaphorina citri* and also through contaminated scion wood (Hartung et al. 2010). Carrot seed samples were important sources of inoculum of '*Candidatus* Liberibacter solancearum' (Lso) (Monger and Jeffries 2018). All species of '*Candidatus* Liberibacter' associated with citrus HLB, viz., 'africanus', americanus', and 'asiaticus' and '*Ca.* L. solanacearum' and associated with potato zebra chip (ZC) disease and other diseases of Solanaceae and Apiaceae families are transmitted by psyllid vectors. Asian citrus psyllid *Diaphorina citri* and the African citrus psyllid *Trioza erytrea* are involved in the transmission of HLB disease affecting different *Citrus* spp. Both *D. citri* and HLB pathogen are well established in all citrus growing areas of Florida. *Murraya* spp. grown widely as ornamental hedges was colonized by *Diaphorina citri* commonly and it transmitted *Ca.* L. asiaticus to *M. paniculata M. exotica* plants very efficiently (Damsteegt et al. 2010). Infection process of '*Candidatus* Liberibacter asiaticus' transmitted by psyllid *Diaphorina citri* was investigated. '*Ca.* L. asiaticus' infection was a function of leaf age and number of psyllids. Higher infection rates were recorded on younger leaves inoculated with higher number of vectors and longer IAP and IP (Raiol-Junior et al. 2017). The Asian citrus psyllid (ACP), *Diaphorina citri*, transmitted Las in a persistent-propagative manner. Psyllids could acquire Las from infected seedlings and nursery plants. The results indicated that latency of Las in planta was shorter than incubation period required for symptom expression. The presence of HLB pathogen in presymptomatic stage of plants could be a critical factor in disease spread (Canale et al. 2020).

Xylella fastidiosa (*Xf*) (grapevine PD) is foregut-borne and does not circulate in hemolymph of the sharpshooter *Homalodisca vitripennis*. Established populations of *Xf* could disappear from vector foreguts over time (Backus and Morgan 2011). Cibarial biofilm from field-collected and greenhouse-reared sharpshooters *Homalodisca vitripennis* showed large fibrillar aggregations of wild type microorganisms (false-colored red) that often resembled those of Xf GFP (green) (Figure 5.5). Bright-greenish yellow Xf GFP was visible against darker-crimson red antifluorescence of the insect cuticle. Within the first 24 h, *X. fastidiosa* GFP was apparent throughout the cibarium, whereas the amount varied with location and acquisition access period (AAP). The amount of Xf GFP was lower consistently in all insects on day 5 AAP. *Xf* GFP density in cibarium, trough, basin, value, pit, and D-sensilla field differed. The cibarium had the highest *Xf* GFP density rank among foregut locations. Density of *Xf* GFP in the cibarium was significantly lower on day 1 than on other days, increased to a level that remained the same for every AAP thereafter. In most locations, the mean density rank increased for the first three AAP days, the ranks then rapidly declined in day 4 and 5 and they began to build up on day 6 (Backus and Morgan 2011). Concentrations of green fluorescent protein (GFP)-expressing *Xf* or fluorescent nanoparticles acquired from artificial diets by the vectors were monitored, as they were egested into simultaneously secreted saliva. *X. fastidiosa* or nanoparticles mixed with gelling saliva to form fluorescent deposits and salivary sheath on artificial diets, providing direct and conclusive evidence for egestion by the vectors. The results supported the egestion-salivation mechanisms of *Xf* inoculation (Figure 5.1) (Backus et al. 2015).

Pantoea ananatis (onion center rot) was transmitted by *Frankliniella fusca*, the tobacco thrips. The relationship of *P. agglomerans* also causing onion center rot with *Thrips tabaci* was investigated. *T. tabaci* could be an efficient vector of both *P. ananatis* and *P. agglomerans* (Dutta et al. 2014). In a later investigation, acquisition of *P. ananatis* by *F. fusca* and transstadial persistence of the pathogen were studied. Persistence of *P. ananatis* through larval, pupal and adult stages of *F. fusca* was observed. Immunofluorescence microscopic examination showed that *P. ananatis* was localized in the gut only. The results suggested that *P. ananatis* was not circulative in the vector body (Dutta et al. 2016).

Transmission efficiency of the *rpfF* mutant of *Xylella fastidiosa* by *Graphocephala atropuncta* was poor regardless of whether the vectors had fed on plants infected with the wild-type strain. On the other hand wild-type strain was not efficiently transmitted, if the vector had fed on plants infected with *rpfF* mutant. A factor released into xylem by *rpfF* mutants and to a lesser extent by wild-type strain inhibited their attachment to and thus transmission by vectors and might also enable them to move more readily through host plants (Baccari et al. 2014). Vector specificity and efficiency or different sequence types (STs) within

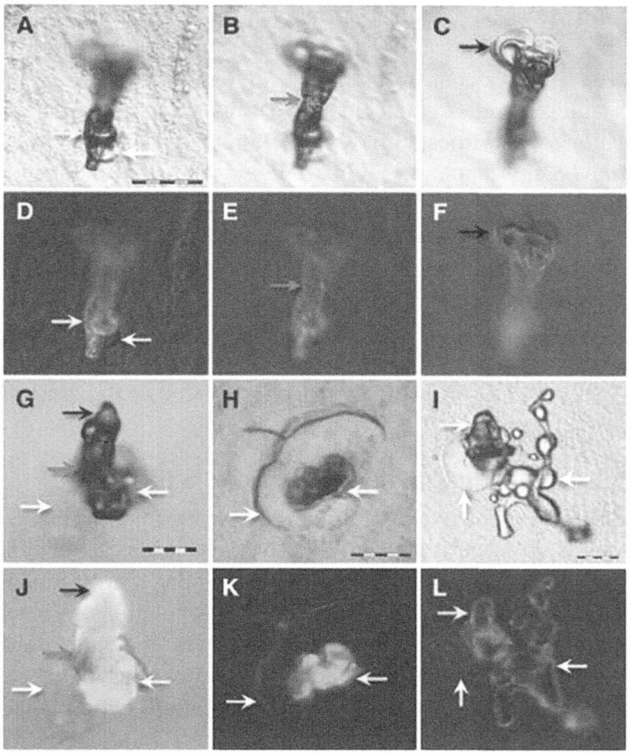

FIGURE 5.1 Acquisition of *Xylella fastidiosa* by *Homalodisca vitripennis* from artificial diet with green fluorescence nanoparticles. (A, B, and C) Bright field view; (D, E, and F) UV-excited filtered green fluorescence. Nanoparticles mix with gelling saliva to form fluorescence deposits and salivary sheaths on artificial diet for *H. vitripennis*; (G) green fluorescence only at the base of air-formed s sheath; (H) bright field; (I) bright field; (J) flange of the air-formed sheath as indicated by arrows; (K) green fluorescence of an aborted sheath; and (L) green filtered fluorescence view of a sheath control (minus nanoparticles).

Source: (Courtesy of Backus et al. 2015 and with kind permission of the American Phytopathological Society, MN, USA)

X. fastidiosa subp. *pauca (Xfp)*, a genetically diverse pathogen with many vector species involved in transmission were investigated. The artificial diet was effective in determining variations in acquisition and transmission of ST13 by four sharpshooter species. ST13 was more effectively transmitted by *B. xanthoplus* and *M. leucomelas*, when compared with *M. cavifrons* and *Sibovia sagata*. The differences in vector competence might suggest that ST-vector foregut or vector-plant interactions might influence bacterial acquisition, retention and inoculation by the insect (Esteves et al. 2019).

Transmission characteristics of '*Candidatus* Liberibacter asiaticus' (Las) by psyllid vector *Diaphorina citri* were studied. The median latent periods of 16.8 and 17.8 days for psyllids were required at nymphal, and adult stages respectively. A minimum latent period of 7–10 days was observed. Psyllids transmitted the pathogen up to 5 weeks. Long latent period and persistence of transmission provided indirect evidence for Las having a circulative relationship with *D. citri* (Canale et al. 2017). Acquisition success of '*Candidatus* Liberibacter solanacearum' (Lso) (potato zebra chip, ZC) by *Bactericera cockerelli* was dependent on feeding site on the plant. The highest acquisition success was attained, when whole plant was the source of inoculum. Progress of disease was not affected by bacterial population injected or psyllid numbers (Rashed et al. 2012). Seven-day fecundity and nymphal survival percentage were reduced in Lso-positive isofemale lines compared to Lso-negative insects. '*Ca.* L. solanacearum' might exert negative effects on vector population growth rate on tomato plants (Nachappa et al. 2012). Potato psyllid *B. cockerelli* (BC) could transmit '*Candidatus* Liberibacter solanacearum' (Lso) to several cultivated and wild solanaceous plant species. Lso is transmitted in a circulative-propagative mode of transmission by BC. Infected silver nightshade (SLN) plants could serve as potential sources of inoculum of Lso to potato crops and provide inoculum to BC adults colonizing potato in the next season (Thinakaran et al. 2015). Of the five haplotypes of Lso, only two haplotypes, hapA and hapB are known to infect potato, inducing zebra chip (ZC) disease. Both haplotypes, causing ZC disease, are transmitted by potato/tomato psyllid *Bactericera cockerelli* (Bc). A shift from hapA toward hapB population of Lso was revealed during four seasons, indicating possible evolution of Lso in Idaho fields (Dahan et al. 2017).

'*Ca.* L. solanacearum' (Lso) (zebra chip, ZC) is transmitted by circulatively and propagatively by potato psyllid *B. cockerelli*. Lso-specific DNA probes annealed to the bacterium that formed extensive, monocultural biofilm on gut, salivary gland and oral region tissues, confirming that it was one of the morphotype of the pathogen. Rod-shaped cells with variable length and rough granular cytosol were observed. Further, Lso prepared from shredded midgets and negatively stained for transmission electron microscopy (TEM) contained pili- and flagella-like surface appendages (Cicero et al. 2016). In adult midgets, Lso cells were present between the basal lamina and basal epithelial cell membranes as well as in basal laminar perforations, on the outer laminar surface, and in the ventricular lumen, epithelial cytosol, and filter chamber periventricular space. Lso cells were clustered abundantly in the salivary gland pericellular spaces and in the epidermal cell cytosol of the head. Intrusive, systemic invasion by Lso employing an endo/exocytosis-like mechanism was in operation of a propagative and circulative mode of transmission (Cicero et al. 2017). Lso females might inoculate higher Lso titers into plants than males. The proportion of individual that reached phloem tissues was similar for males and females. However, electrical penetration graph (EPG) also showed that females probed more frequently, ingested longer from phloem sieve elements and reached phloem tissues more frequently than the males (Antolinez et al. 2017).

5.2.2.5 Factors Influencing Bacterial Pathogen Development

Xanthomonas axonopodis pv. *phaseoli* (*Xap*) (bean common bacterial blight) is seedborne with an epiphytic phase on bean plants. Under field conditions, *Xap* epiphytic population consists of two major kinds: solitary cells and biofilms, and they are genetically identical as revealed by ERIC fingerprinting (Jacques et al. 2005). *X. campestris* pv. *campestris* also forms biofilms and there is a cell-to-cell signaling procedure. This system is coded by an *rpf* gene cluster with an important role in pathogenicity (Crossman and Dow 2004). Acyl homoserine lactones (AHLs) essentially required for biofilm formation in *Pseudomonas syringae* pv. *syringae* were not detected in another bean pathogen *Xanthomonas* spp. Butyrolactones may act as signal molecules in quorum-sensing-like systems in *Xanthomonas* (Jacques et al. 2005). Solitary cells are sensitive to desiccation stress, UV radiation, and chemicals, whereas bacteria in biofilms are protected against these environmental conditions (Monier and Lindow 2003; Jacques et al. 2005). Foliar plant pathogens exploit stomata as points of entry to gain access into the apoplast of leaves. Bacterial pathogens possess specific virulence factors such as coronatine (COR) and syringolin A toxins, with which the stomatal aperture could be manipulated. These toxins promoted stomatal opening to facilitate bacterial entry into the plant. The phytotoxin COR produced by *Pseudomonas syringae* pv. *tomato* (*Pst*) DC3000 may play multiple roles in pathogenesis, including promoting entry of *Pst* through stomata at initial stages of infection process. *P. syringae* pv. *glycinea* (*Psg*) COR-deficient mutants of *Psg* strain PG180 had reduced multiplication rate and symptom development, when inoculated by spraying (Budde and Ullrich 2000). An effector of *P. syringae* HopAM1 promoted stomatal closure in an abscisic acid (ABA)-dependent manner (Goel et al. 2008; Melotto et al. 2017). High atmospheric humidity triggers degradation of ABA by upregulating key enzymes involved in ABA catabolism, leading to stomatal opening (Okamoto et al. 2009). Bacteria-induced stomatal closure might be suppressed by high humidity due to early activation of jasmonic acid (JA) hormone signaling and suppression of SA defense signaling pathway within stomatal guard cells (Panchal et al. 2016).

Many bacterial pathogens induce water-soaked lesions as the early visible symptom in their respective host plant species. Water-soaking was a transient process occurring in the early stages of infection in many plants following bacterial infection. Water-soaked lesions induced by *Pseudomonas syringae* pv. *tomato (Pst)* DC3000 disappeared, when necrosis of infected tissue developed. Virulent *Pst* DC3000 strain could experience suitable water potential for its multiplication, whereas avirulent *Pst* DC3000 (*avrRpm1*) strain experienced a high level of water stress, resulting in inhibition of its growth (Wright and Beattie 2004). *Xanthomonas axonopodis* pv. *manihotis* (cassava bacterial blight) delivered TAL20$_{xam668}$ to regulate expression of the sugar transporter gene MeSWEET10a in cassava (Cohn et al. 2014). *X. citri* subsp. *malvacearum* (*Xcm*) (cotton bacterial blight) secreted Avrb6 to induce expression of the sugar exporter gene GhSWEET10 in cotton (Cox et al. 2017). *Xanthomonas euvesicatoria* (bacterial spot) is seedborne in pepper and tomato. Viable colonies of the bacteria could be recovered from them, indicating asymptomatic infection of pepper fruit might occur from infested blossom. Further, some seedlots were also infected due to transmission of *X. euvesicatoria* from infested blossom (Dutta et al. 2014). *Erwinia amylovora* (apple fire blight) could spread systemically from blighted blossoms and shoots to the central leader and rootstock of the tree, causing rootstock blight and eventually tree death. Examination of individual buds showed that *E. amylovora* was localized within the tissue beneath the bud scales containing meristem (Tancos et al. 2017).

5.2.3 Process of Infection of Plants by Bacterial Pathogens

5.2.3.1 Pathogenicity of Bacterial Pathogens

Several phytopathogenic bacterial species multiply or survive epiphytically on aerial plant parts without inducing any visible symptoms. The asymptomatic phase, among other consequences, allows bacterial populations to reach sizes permitting development of diseases under favorable conditions. The phenomenon of pathogenesis has been studied to have an insight into the interactions between bacterial pathogens and various plant species. Phytopathogenic bacteria have cell surface-anchored structures such as pili, flagella, lipopolysaccharides, exopolysaccharides slimy layers, and outer membrane proteins, which assist in ingression and survival within host plant apoplast environment. Secreted effector proteins of pathogenic origin play a central role in the interaction of bacterial pathogens with their host plants. Six different secretion systems have been identified in gram-negative bacterial pathogens (Mougous et al. 2006). Bacterial pathogens produce effectors secreted through different secretion systems required to promote bacterial growth in host plant apoplast by suppressing host defenses for releasing nutrients from plant cells. Host defenses operate at two levels. Basal defenses are triggered from outside of plant cells by nonpathogenic mutants, the type III secretion system (T3SS), heat-killed bacterial cells, and factors bearing pathogen-associated molecular patterns (PAMPs). These factors known as general elicitors, include lipopolysaccharides, flagellin, and cold shock proteins. Basal resistance to nonpathogenic bacteria is recognized by production of callose-rich papilla beneath plant cell wall at the site of bacterial contact, localized induction of reactive oxygen species (ROS) and enhancement of expression of defense-related genes. Basal defenses are triggered early in plant-pathogen interaction in response to the perception by plant pathogen recognition receptors (PRRs) of extracellular PAMPs. Plants have evolved PRRs that can recognize certain PAMPs which are important molecules required for the pathogen lifestyle. PAMPs contain a conserved structural feature that is recognized by a PRR. Recognition of a MAMP accelerates several early front line defenses. The second level of host defenses, the hypersensitive reaction (HR), in contrast, is elicited inside plant cells by many bacterial pathogen effectors during gene-for-gene interactions involving matching of *avr* and *R* genes. The programmed cell death (PCD) is the resultant of recognition and signal transduction events and the surrounding plant tissues that exhibit resistance following a complex physiological activities of the host tissues (Narayanasamy 2008).

Bacterial pathogens secrete exoenzymes and toxins through the type I pathway, which is sec-independent. Proteases produced by *Erwinia chrysanthemi* are transported from bacterial cytoplasm directly into the extracellular environment. '*Candidatus* Liberibacter solanacaearum' (zebra chip disease) has a relatively small, AT-rich genome that contains all components of type I secretion system, ABC transporters, and also ten bifunctional protein genes that encode proteins with two different enzymatic domains. Two of the bifunctional genes encode proteins similar to those described for pathogenicity and virulence determinants in other organisms (Gilkes et al. 2018). Pectinases and cellulases are secreted through a sec-dependent, two-step, type II pathway, also known as the general secretary pathway (GSP). Secretion of virulence factors through GSP in EC is known. As a first step the enzymes are exported to the periplasm, but they are unable to move through outer membrane. In the second step, the periplasmic form of these enzymes moves through the outer membrane and then into the extracellular environment. Secretion of major pectate lyases, polygalacturonases and cellulases is controlled by the *out* gene cluster and the number of genes may vary with pathogen species (Narayanasamy 2002).

Cell wall-degrading enzymes, cellulases, polygalacturonases, xylanases, and proteases secreted by T2SS contributed to the virulence of *X. euvesicatoria*. Xylanases and proteases under the control of *HrpG* and *HrpX* were secreted via the T2SS Xps system (Szczesny et al. 2010). The lipopolysaccharide (LPS) biosynthesis cluster (14–26 kb) region flanked by *etfA* and *metB*, was found to be variable in terms of identity and size across the sequenced xanthomonads. Variation in LPS cluster was not associated host or tissue specificity (Lu et al. 2018). Bacterial pathogens causing bacterial spot possess an identical LPS cluster. LPS O-antigen contributed to pepper pathogenicity by increasing *in planta* growth, when expressed with two other pepper pathogen-specific genes *Xcv1839* and *XopG* effector (Potnis et al. 2011). A secreted lipolytic enzyme encoded by *lipA* of *X. euvesicatoria* was involved in virulence during interaction of the pathogen with tomato (Tamir-Ariel et al. 2012). *Pseudomonas syringae* pv. *actinidiae (Psa)* (kiwifruit canker or leaf spot) genes involved in virulence of *Psa* were investigated by performing a genetic screen to identify transposon mutants altered in the lipolytic activity. A total of 58 *Tn5* transposon mutants with altered lipolytic activity were identified. Transposons were inserted in genes that play roles in cell division, chemotaxis metabolism, movement, recombination, regulation, signal transduction, and transport. *Tn5* mutants with functions affected in phosphomannomutase AlgC, lipidA biosynthesis acyltransferase, glutamate-cysteine ligase, and the type IV pilus protein PilI were identified (Patel et al. 2017).

Phytobacterial pathogens in the genera *Erwinia, Pseudomonas,* and *Xanthomonas* possess *hrp* (**h**ypersensitive **r**esponse and **p**athogenicity) genes. Functions of *hrp* genes are the characteristic traits of type III apparatus of secretion system (T3SS). Pathogenicity and elicitation of resistance by bacterial pathogens may be controlled by these genes. *P. syringae* pv. *syringae* has 25 *hrp* genes. Three biochemical functions viz., gene regulation, protein secretion, and induction of HR are attributed to *hrp* genes. Many *hrp* genes and the genes involved in secretions of proteinaceous virulence factors of pathogenic bacteria have significant similarities. Products of *hrp* genes constitute the T3SS machinery. The virulence/pathogenicity of bacterial pathogens is frequently abolished/reduced, following mutations affecting type III protein secretion. When bacterial pathogens come in contact with host cells, T3SS may accelerate leakage of plant nutrients in the extracellular space (apoplast) in plant tissues. Expression of a single gene may result in host-specific symptoms as in the case of citrus canker caused by *X. axonopodis* pv. *citri*. Division, enlargement, and death of host cells are controlled by the *pthA* gene. The Pth protein may be a specific plant signal with a site of action inside the plant cells that might elicit specific symptom (Narayanasamy 2002). The outer membrane vesicles (OMVs) secretory pathway may serve as an alternative route for extracellular enzymes secreted by type II secretion system. T2S xylanase secreted by *X. campestris* was produced through T2S system and OMVs, and it was required as virulence factor to degrade plant cell walls. Packaging of LesA, a lipase/esterase secreted by T2SS in *Xf* onto OMVs, promotes spread of disease symptoms. The results indicated that OMVs could be an alternative pathway for T3SS system or act in coordination with it (Solé et al. 2015; Nascimento et al. 2016). Many other virulence factors are required for induction of different symptoms in infected susceptible plants in addition to *hrp*-controlled pathogen factors. Water-soaking symptom associated with several bacterial diseases is due to the action of extracellular polysaccharides (EPS) produced by bacterial pathogens. Highly virulent strain of *Xanthomonas oryzae* pv. *oryzae (Xoo)* (rice bacterial leaf blight) produced greater amount of EPS than less virulent strain (Singh et al. 2006). The protein-protein interactions involving subunits, regulators, and substrates of the type III secretion system (T3SS) encoded by *hrp* cluster of *Xanthomonas axonopodis* pv. *citri (Xac)* were identified. Mutants for the operons *hrpB, hrpD,* and *hrpF* could not induce canker in citrus plants or HR in nonhost (cotton), indicating the need for the presence of operons *hrpB, hrpD,* and *hrpF* for pathogenicity and HR functions (Dunger et al. 2005). *Xanthomonas euvesicatoria* (bacterial spot) contain about 45 effectors as shown by computational analyses. The effectors *XopD, XopL,* and *XopN* were virulence factors that interfered with pathogen-associated molecular patterns (PAMP)-triggered immunity (PTI) (Kim et al. 2009; Singer et al. 2013).

Pseudomonas syringae includes model pathogens that differentiated into at least nine genomospecies and 60 pathovars (Lamichhane et al. 2015). Three mechanisms underlying the origin of DNA modifications may elucidate the evolution of genomic traits governing bacterial pathogenicity: (i) point mutations (small local sequence change), (ii) rearrangements of DNA segments, and (iii) acquisition of new DNA segments (gene duplication or insertion/deletion of entire or portions of genes) or acquisition of new DNA components from other organisms via horizontal gene transfer (HGT). The *hopZ1* effector gene of *P. syringae* strains known to promote infection of soybean, evolved into three functional and two nonfunctional forms as a consequence into more virulent variant (Ma et al. 2006). Similarly, point mutants in *avrBS1* of *X. axonopodis* pv. *vesicatoria* enabled evasion of recognition for some strains isolated from pepper field (Wichmann et al. 2005). *P. syringae* encountering the first layer of defense was the PRRs present in plasma membrane of plant cells. PRRs recognize conserved microbial features, the PAMPs that are present in the extracellular space. The flagellin-sensitive 2 (FLS2) receptor recognized the peptide flg22 derived from bacterial flagellin as well as the EF-TU receptor (EFR) that recognizes a peptide derived from bacterial EF-Tu protein. Perception of PAMPs by PRRs is triggered within minutes, and many cellular responses including the production of ROS, ion fluxes across membranes, activation of mitogen-activated protein kinase (MAPK) signaling pathways, and altered transcription of large number of genes follow later. Many diverse pathogens have evolved mechanisms to suppress the establishment of PTI-mediated defenses. *P. syringae* depends on T3SS to suppress PTI (Bütner and He. 2009). *P. syringae* delivered effector proteins into host cells and many of them could interfere directly with activation of PRRs. The role of *Hrp* pathogenicity island (PAI) conserved effector locus (CEL) was investigated. The Delta CEL mutation affected three putative or known effector genes, *avrE1, hopM1* and *hopAA1–1*. Comparison of genome sequences showed that only these effector genes are present in the CEL of *Pst*, *P. syringae* pv. *phaseolicola* (*Psp*) 1448H, and *P. syringae* pv. *syringae* (*Pss*) B728a. *P. syringae tomato* (*Pst*) DC3000 Delta avrE1 mutant was reduced in its ability to produce lesions, but not in its ability to grow in host tomato leaves. Deletion of both *avrE1* and *hopM1* resulted in significant reduction in growth and lesion phenotype of *Pst* DC3000 Delta CEL mutant. Further, both *avrE1/hopM1* and Delta CEL mutants were partially impaired in their ability to elicit the hypersensitive response (HR) in nonhost *Nicotiana benthamiana* leaves (Badel et al. 2006).

A link between nonprotein amino acid γ-aminobutyric acid (GABA) contents both in planta and in vitro and ability of *Pseudomonas syringae* pv. *tomato (Pst)* DC3000 to express virulence genes was observed. A higher concentration of GABA repressed expression of *hrp* genes that encode the T3SS (Park et al. 2010). High exogenous GABA levels repressed expression of T3SS in *Pst* DC3000, resulting in reduced elicitation of HR in the nonhost tobacco (McGraw et al. 2016). A method of coculturing *P. syringae* pv. *tomato* DC3000 and *Arabidopsis* suspension cells was developed. The T3SS-associated genes were inhibited when *Pst* DC3000 was cocultured with PAMP-treated plant suspension cells. The mechanism of decreased T3SS gene expression of was studied. Metabolites showing significant increases/decreases following PAMP treatment were identified. Citric acid, a known inducer of T3SS gene expression in *Pst* DC3000 and several other organic acids were in reduced concentrations in exudates from PAMP-treated cells, indicating that a portion of PAMP- induced defense in *Pst-Arabidopsis* system was reduced by extracellular release of this metabolite (Yan et al. 2019).

The outer membrane (OM) confers stomatal stability to bacterial cells and protects bacteria forming a restrictive permeability barrier from antimicrobial compounds, while facilitating material exchange with the environment. The contribution of OMVs to bacterial virulence was revealed in pathogens limited to vascular elements. In the xylem-inhabiting pathogen *Xylella fastidiosa*, quorum-sensing through production and perception of DSF-induced aggregation, surface attachment, and biofilm formation. Lack of DSF production in Δ*rpf* mutant of *X. fastidiosa* promoted a more virulent-free swimming form of bacteria that overproduced OMVs (Ionescu et al. 2014). LPS is an essential OM component and has a role in the interactions between bacterial cells and environment. Due to its outer most localization and hydrophilic properties, O-polysaccharide (OPS) may mediate adhesion of bacterial cells to host plant surfaces. One of the two known OPS biosynthesis clusters, the common polysaccharide antigen (CPA) gene cluster was conserved only in some strains of *P. fluorescens* group. Orthologs of the glucosyltransferase WbpL, required for initiation of the O-specific antigen (OSA) synthesis could be identified in genomes of *Pseudomonas*. Knockout mutants of *wbpL* ortholog in *P. syringae* pv. *tomato* DC3000 (Pst) and *P. cichorii (Pci)* resulted in strains lacking OPS. *Pcl*Δ*wbpL* caused reduced symptom severity, compared to the wild-type strain. In addition, swarming motility was reduced in Δ*wbpL* mutants of *Pst* and *Pci* (Kutschera et al. 2019). In *X. fastidiosa*, type IV pili are regulated by *pilG*, a response regulator protein putatively involved in chemotaxis-like operon sensing stimuli through signal pathways. The role of *pilG* on pathogenicity of *X. fastidiosa*, was investigated. The *pilG* deletion mutant *Xf*Δ*pliG* showed significant reduction in cell matrix adherence and biofilm production, compared with wild-type (WT) *X. fastidiosa* and *Xf*Δ*pilG-C*. The genes *pilE, pilU, pilT*, and *pilS* were downregulated in *Xf*Δ*pilG*, when compared with its complemented strain and wild-type *X. fastidiosa*. Grapevine plants inoculated with *Xf*Δ*pilG* failed to develop symptoms of PD, whereas WT strains could induce typical symptoms. Complemented *Xf*Δ*pilG* strain also induced symptoms as the wild-type strain. The results showed that *pilG* had a role in virulence of *X. fastidiosa* infecting grapevines (Shi and Lin 2018). *X. fastidiosa* released outer membrane vesicles (OMVs) known to play a role in its systemic invasion/dissemination. OMVs inhibited bacterial attachment to xylem wall and traffic lipases/esterases that degrade plant cell wall. The two CVC strains produced diffusible signaling factor 2(DSF2) and CVC-DSF. These compounds of the DSF (*X. fastidiosa* DSF) family are associated with OMV-enriched proteins (Feitosa-Junior et al. 2019).

Bacterial response to temperature is likely to play an important role in the pathogenesis of PD, as virulence-associated genes are upregulated in *Xylella fastidiosa* after exposure to heat and cold conditions. A crucial and highly conserved aspect of bacterial adaptation to cold temperatures is the production of cold-shock proteins (CSPs) which are strongly induced by temperature change. Bacterial genomes encode multiple CSPs, some of which are strongly induced following exposure to cold. A putative CSP homolog (Cspl) with RNA-binding activity was identified in *Xylella. fastidiosa*. Deletion mutant *cspl* (Δ*cspl*) had decreased survival rates following cold exposure and salt stress in vitro. The deletion mutant also was significantly less virulent in grapevine, as compared with wild-type strain in the absence of cold temperatures (Burbank et al. 2016). Expression of luciferase from pCR8-ECcspA-luc was induced after temperature drop and showed increasing levels over the course of 2 h at 15° C. In contrast, luciferase expression from pCR8-Xfcspl-luc was constitutive in relation to temperature. The role of CSP in virulence of *X. fastidiosa* Cspl in infection process was studied. Susceptible grapevine cc. Chardonnay plants were inoculated with wild-type *X. fastidiosa* Stag's Leap, Δ*cspl*, and Δ*cspr/cspl*$^+$ and development of symptoms in inoculated plants was recorded at 12 weeks postinoculation. The mean disease score of Δ*cspl* inoculated plants was significantly reduced compared with wild-type strain inoculated plants. The plants inoculated with Δ*cspl*/Δ*cspl*$^+$ showed symptoms similar to wild-type inoculated plants (Figure 5.2). The pathogen titer was significantly reduced in dcspl-inouclated plants, compared to wild-type strain-inoculated plants. The results indicated that *X. fastidiosa* Capl was a virulence factor required for colonization of plant in the absence of cold stress (Burbank and Stenger 2016).

Bacterial genes involved in virulence of *Pseudomonas syringae* pv. *actidiniae (Psa)*, (kiwifruit canker or leaf spot) were investigated by performing a genetic screen to identify transposon mutants altered in the lipolytic activity. A total of 58 Tn5 transposon mutants with altered lipolytic activity were identified. Mapping all these Tn5 mutants revealed that the transposons were inserted in genes that play roles in cell division, chemotaxis, metabolism, movement, recombination, regulation, signal transduction and transport (Patel et al. 2017). The loss of HR- a process used by *Xanthomonas citri* subsp. *citri* (*Xcc*) strains to evade plant immune defenses, was associated with non-synonymous mutations in the effector genes *avrXacE, pthA2* and *avrXacA3* (Trivedi and Wang 2016). The prevalence of type III secretion effector *avrBsT* was detected frequently among strains of *Xanthomonas perforans* (*Xp*) infecting tomato plants. It appeared that acquisition of *avrBsT* by *X. perforans* provided a fitness advantage. Contributions of *avrBsT* to fitness on tomato for representative strains GEV872 and

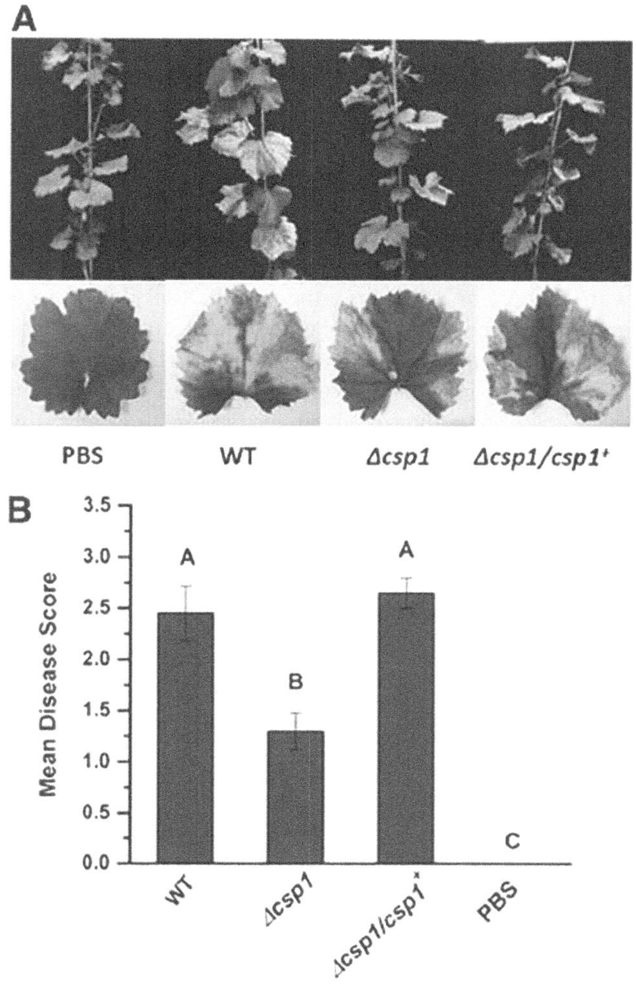

FIGURE 5.2 Effect of cold shock protein (CSP) on virulence of *Xylella fastidiosa* on grapevine cv. Chardonnay. (A) Development of symptoms on plants inoculated with wild type (WT), *csp* deletion (dcspl) mutant and complemented deletion mutant (dcsp/csp*) and control (C). (B) Disease severity score based on 0- to 5-point scale.

Source: (Courtesy of Burbank and Stenger 2016 and with kind permission of the American Phytopathological Society, MN, USA)

GEV1001 of two phylogenetic groups of *Xp*, were selected for generating *avrBsT* mutants. WT strain induced greater disease severity compared to *avrBsT* mutant strains. The effect of *avrBsT* on pathogen fitness was observed under field conditions. WT strains could be recovered more frequently than *avrBs*T mutant strains. WT strains could spread to longer distances across field plots compared to *avrBsT* mutant strains (Abrahamian et al. 2018).

Xanthomonas oryzae pv. *oryzae (Xoo)* (rice bacterial leaf blight) secretes effector proteins (T3Es) directly into rice cell through a *hrp*-encoded specialized type III secretory system (T3SS) to induce disease symptoms in rice. The function of XopF, an effector in *Xoo,* was investigated using a null mutant. Inducible, *hrp*-dependent expression pattern of *xopF* was studied. XopF was translocated in rice cytosol through T3SS using adenylate cyclase activity assay. XopF regulated in planta growth and suppressed PAMP-triggered immunity (PTI) response in rice. Wild-type strain produced intense blight lesions upon inoculation using leaf clipping method. In addition, *Xoo*Δ*xopF* exhibited significant reduction in planta colonization relative to the wild-type strain. The genes *PR10, OsWRKY13, OsRLCK16*, and *OsFLS2* were upregulated upon challenge inoculation with *Xoo* Δ*xopF*, indicating the role of XopF in suppressing PTI in rice. *Xoo* Δ*xopF* mutant induced more callose deposition in infected rice leaves. XopF might express basal PTI response in plants, favoring growth of *X. oryzae* pv. *oryzae* and its pathogenicity in rice (Mondal et al. 2016). Hypersensitive responses and pathogenicity (*hrp*) genes of *Xanthomonas oryzae pv. oryzae (Xoo)* encode components of T3SS and are essential for virulence. Expression of *hrp* genes is regulated by two key *hrp* regulators. HrpG and HrpX. HrpG regulated *hrpX* and *hrpA*, whereas HrpX regulated other *hrp* genes (*hrpB – hrpF*) operons. Quantitative regulation of HrpX occurred due to high *hrp* gene expression. The results suggested that inactivation of XyIR might result in greater xylan and xylose utilization and simultaneous higher accumulation of HrpX, followed by higher *hrp* gene expression in *Xoo* in the presence of xylose (Ikawa et al. 2018). Mechanisms of infection and induction of symptoms induced by *Xoo* were investigated. The twin-arginine translocation (Tat) pathway played a vital role in *Xoo* fitness and virulence. Potential Tat-dependent proteins were identified (TDTPs) in *Xoo* strain PX099A. Four TDTP were involved in virulence; three TDTPs contributed to colonization in planta; one TDTP had a key role in attachment to leaf surface; four TDTPs were involved in tolerance to multiple stresses and two TDTPs were required for full swarming motility (Wu et al. 2017). Two unique diffusible signal factor (DSF)-controlled virulence related gene *hshB* in RS105 strain located in the *nodB-rghB* locus was absent in rice bacterial leaf blight pathogen *Xoo* PX 099A strain. Further, two additional genes within this locus, *hshA* and *hshC*, unique to rice bacterial leaf streak (BLS) pathogen *Xanthomonas oryzae* pv. *oryzicola (Xoc)* XocRs105 not present in *Xoo*, were also identified. Transcription of these genes was regulated by DSF-signaling in *Xoc*. Mutation in these genes impaired virulence of the wild-type Rs105 strain. But mutation in *hshA* or *hshC* significantly reduced the in planta growth ability and epiphytic survival of *Xoc* cells (Song et al. 2015). Functional characterization of a 9-gene *his* operon in *Xoc* was performed. Two genes within this operon, *trpR* and *hisB* were essential for virulence and growth of the pathogen in planta. The *trpR* functioned as a transcriptional repressor and could negatively regulate the expression of *hisG, -D, -C, -B, -H, -A*, and *–F*. Two genes in histidine biosynthesis operon might play an important role in the pathogenicity of *Xoc* (Su et al. 2018).

Bacterial plant pathogens contain many small DNA-binding proteins abundantly, forming a group of proteins known as nucleoid-associated proteins (NAPs), capable of binding DNA contributing to the structure of bacterial nucleoid. Factor for inversion stimulation (Fis) influences DNA topology by directly binding DNA, whereas heat unstable (HU) binds to DNA nonspecifically. The role of NAPs in regulation of virulence of *Xanthomonas campestris* pv. *campestris* (*Xcc*) (black rot of crucifers) was investigated. *Xcc* genome encodes many virulence factors, such as T3SS-dependent effectors, cyclic glucans, lipopolysaccharides, EPSs, and a serious of enzymes, including amylase, endoglucanase, polygalacturonate lyase and protease (Ryan et al. 2011). *Xanthomonas campestris* pv. *campestris* (*Xcc*) infecting crucifers elaborated large number of effectors such as EPSs, lipopolysaccharides, type III secreted proteins, and extracellular enzymes which contribute to virulence of *Xcc* needed. *Xcc* uses the type II secretion system (T2SS) to export extracellular enzymes outside host cell. Xcc is equipped with two T2SSs (Xps-T2SS and Xcs-T2SS), but only Xps-T2SS is functional capable of secreting extracellular enzymes. Inactivation of genes encoding key components of Xps-T2SS, but not Xcs-T2SS resulted in abolition of secretion of enzymes and loss of virulence. Deletion of XC3487 (renamed $amyA_{xcc}$), a putative amylase-encoding gene in the genome of *Xcc* strain 8004, led to a complete loss of amylase activity and significant reduction in virulence. Extracellular amylase activity and virulence of the $amyA_{xcc}$ mutant could be restored to the level of wild-type strains by expressing $amyA_{xcc}$ in trans (Lin et al. 2021).

Complex regulatory networks, including quorum-sensing (QS) pathways, multiple two-component systems, and transcriptional regulators, may act to ensure appropriate expression of these virulence factors promoting successful invasion and proliferation during pathogenesis. The role of *XC_1355* was predicted to encode DNA-binding protein, HU-like protein (Hlp). Mutation of *XC_1355* in *Xcc* reduced virulence, extracellular polysaccharide production and cell motility. Transcriptome analysis also indicated that *hlp* was involved in virulence that had partially overlapping and complementing functions with *flp* in *Xcc*, although these two regulators exhibited opposite effects on expression of genes involved in T3SS (Su et al. 2021). Horizontal gene transfer (HGT) has a role in the emergence of phytopathogenic bacteria. Plant pathogenic bacterial genera, *Pseudomonas, Xanthomonas*, and *Erwinia* have evolved two different *hrp/hrc* gene cluster groups that encode for the central core of type III secretion system (T3SS) (Araki et al. 2006). In silico whole-genome analysis revealed that genes associated with pathogenicity of *X. campestris* pv. *campestris* and *X. axonopodis* pv. *citri* and suppression of nonhost immune defenses were acquired by HGT during evolutionary histories of these bacterial pathogens (Lima et al. 2008). Among the virulence factors of bacterial pathogens, effector molecule that can modify host cells structures facilitates infection and/or trigger host defense responses. In '*Candidatus* Liberibacter asiaticus' (Las), of the 14 candidates for secreted hypothetic proteins, one WSEC01 was predicted to carry a nuclear localization sequence (NLS). Putative effectors HLBSP04 and HLBSP21 were conserved in all genomics evaluated, whereas HLBSP01 and HLBSP19 and HLBSP30 were present exclusively in Las. Expression analysis of candidate effector genes was performed in citrus leaves and used as markers for selection of effectors for further functional analysis. Genes coding for candidate effectors HLBSP13, HLBSP20, HLBSP28, and NSEC01 exhibited high levels of expression in blotch mottle leaves, suggesting their usefulness for use as marker genes to detect the pathogen in trees showing HLB symptoms (Granato et al. 2020).

Pseudomonas spp. infecting plants produce effectors, toxins and cyclic lipopeptides required for disease development. The roles of *srf* genes in virulence of the pathogen were studied by knocking out each of *srfA*, *srfB*, *srfC*, and *srfD* from *srf* cluster of *P. cichorii*, infecting tomato JBC1. Infection of tomato was suppressed only when inoculated with *srfC*-defective Δ*srfC* mutants, compared with wild-type strain. However, when the Δ*srfC* mutant was inoculated into apoplast of tomato leaves directly by vacuum infiltration, disease symptoms developed similar to that of wild-type strain. Further, the Δ*srfC* showed defective swarming motility and biofilm formation, and the attachment of Δ*srfC* cells on the leaf surface was significantly reduced, compared to the wild-type strain. SrfC protein could play important roles in different stage of pathogenesis, resulting in development of symptoms characteristic of infection by *P. cichorii* in tomato (Nagendran and Lee 2019). Strains of race 0 expressed either or both of the type III effectors AvrPto or AvrPtoB which were recognized by tomato cultivar expressing the *Pto* resistance gene in *Pseudomonas syringae* pv. *tomato (Pst)*. The *Pto* encodes a protein kinase that activates host immune system and consequently inhibiting pathogen multiplication and disease development. Race 1 strains of *Pst* did not express these effectors and were virulent on all tomato varieties, regardless of expression of *Pto* by them. *Pst* strains encoded *avrPto* which encoded for *avrPto* which was transcribed and translated. Deletion of *avrPto* from a New York strain of *Pst* enabled this strain to reach high populations in *Pto*-expressing tomato cultivars, without compromising its virulence on susceptible tomato plants (Kraus et al. 2017). One of the important aspects of pathogenicity of microbial plant pathogens is the delivery of effectors into the host cells. Effectors can modulate host metabolism, shut down host defense signaling and suppress cell death (Toruño et al. 2016). Gram-negative foliar bacterial pathogens, like *Pseudomonas syringae*, employ the T3SS. A proteinaceous needle-like structure is used to directly deliver effectors inside the host cells (Chang et al. 2014). On the other hand, plants have evolved intracellular immune receptors with nucleotide-binding leucine-rich repeat (NLR) domain architecture that specifically recognize pathogen effectors, resulting in development of effector-triggered immunity (ETI). The hallmark of ETI is the HR, a specialized form of PCD (Chiang and Coaker 2015). Intracellular delivery of virulence protein VirE2 by *Agrobacterium tumefaciens (At)* could be visualized using the GFP strand through the type IV secretion system in tobacco (Yang et al. 2017). A GFP strand system was developed to visualize directly the delivery of bacterial effectors into plant cells through T3SS. Infection of plants expressing GFP-10 with bacteria delivering GFP11 tagged effectors enabled direct effector detection in planta. Temporal and spatial delivery of GFP11-tagged effectors during infection by *P. syringae* was studied (Henry et al. 2017).

Twitching mobility of *Xylella fastidiosa*, a means of flagellin-independent bacterial movement through extension, attachment, and retraction of polar type IV pili was investigated. The *pilB*, *pilQ*, and *pilR* mutants with defect in type IV pili and non-twitching phenotypes could induce only mild disease symptoms, suggesting that twitching mobility was required not only for long distance movement in plants and colonization but also virulence (Li et al. 2007). Type IV pili of *X. fastidiosa* were regulated by *pilG*, a response regulator protein putatively involved in chemotaxis-like operon sensing stimuli through signal transduction pathways. The *pilG*-deletion mutant *Xf*Δ*pilG* and complemented strain *Xf*Δ*pilG-C* exhibited significant reduction in cell matrix adherence and biofilm production compared with wild-type and complemented strains. Further, the mutant strain did not induce any symptom of PD in susceptible grapevine plants. The results indicated the requirement of *pilG* gene for virulence and pathogenicity of *X. fastidiosa* (Shi and Lin 2018). Harpin, a protein playing a major role in virulence of *Erwinia amylovora*, is exported via T3SS machinery. The extracellular localization of harpin is in agreement with the physiological effects induced by purified harpin applied as an exogenous elicitor. The *dsp* (disease specific) region located next to *hrp* gene cluster is essential for *E. amylovora*. The operon *dspEF* contains genes *dspE* and *dspF* and it is positively regulated by *hrpL* gene. Three proteins secreted by *E. amylovora* through functional Hrp-secretion pathway were identified. Harpin (HrpN), a glycine-rich, heat stable protein capable of eliciting HR on tobacco was identified. Examination of the mutants defective in *hrpN*, *dspA*, and *hrpW* showed that HrpN was a virulence factor, rather than a pathogenicity factor. DspA was an essential pathogenicity factor and HrpW was not required for pathogenicity of *E. amylovora* (Barny et al. 1999; Perino et al. 1999). The DspA/E protein, essential for pathogenicity of *E. amylovora* was secreted via T3SS in vitro. HrpN and DspA/E were located specifically along Hrp pili, indicating that HrpN and DspA/E were apparently secreted at Hrp pilus assembly site (Jin et al. 2001). Functions of the gene *dspA/E* encoding an essential pathogenicity factor of 198-kDa were determined. DspA/E interacted physically and specifically with four similar putative leucine-rich receptor (LRR)-like serine/threonine kinases (RLK) from apple. Genes encoding these four DspA/E interacting proteins of *Malus* x *domestica* (DIPM1–4) are conserved in a wide range of cultivars representing highly susceptible to highly resistant categories. All the four DIPMs were expressed constitutively in host plants (Meng et al. 2006).

Secretome of a mutant of *Erwinia amylovora* (*Ea*), deficient in production of two harpins HrpN and HrpW and mutant deficient in the production of HrpJ was studied. Six effector-like helper proteins [Eop1, Eop2, Eop3, Eop4 (*Erwinia* outer protein), HrpJ, and HropK] were identified. These proteins were identified earlier as proteins secreted by T3SS of *E. amylovora*. In addition, HrpJ was required for extracellular accumulation of harpins for pathogenicity and for wild-type elicitation of the HR. Four proteins, Eop1, Eop3, Eop4, and DspA/E, exhibited similarity to known effectors. HrpJ was essential for pathogenicity and also for elicitation of HR on tobacco. The wild-type strain *Ea* 273 induced ten proteins that were absent or extremely faint in the *hrp* secretion and *hrp* regulation mutants Ea273-hrcN: Tn10 and Ea273 hrpL: Tn10 respectively (Nissinen et al. 2007). Strains of *Ea* exhibit variations in their virulence. Amylovoran, exopolysaccharides, and T3SS are three major virulence factors of *Ea*. Among the four wild-type (WT) strains, Ea273 and Ea110 elicited higher severity of disease symptoms than that induced by strains Ea1189 and CFB1430 on apple plants. Molecular analysis indicated that amylovoran production and expression of both amylovoran (*amsG*) and T3SS (*dspE* and *hrpL*) genes were significantly higher in Ea273 and Ea110 than those of Ea1189 and CFBp 1430 (Wang et al. 2010). *E. amylovora (Ea)* possesses RcsC, a hybrid sensor kinase which contains a sensor domain, a histidine kinase domain, and a receive domain. The *rcsC* mutant of *Ea* produced more amylovoran than wild-type (WT) strain in vitro, and it was nonpathogenic on immature and apple plants. Expression of *Erwinia rcsC* gene suppressed amylovoran production in various amylovoran overproducing WT and mutant strains (Wang et al. 2011). The interaction between effector proteins of *E. amylovora* and resistance of *Malus* wild species were investigated. Responses of *Malus* wild species were assessed by inoculation with a set of effector deletion mutant strains and several highly virulent *Ea* strains. Deletion of the T3SS abolished virulence of *Ea*. Further, a new gene-for-gene relationship could be established between effector protein Eop1 and fire blight resistant ornamental apple cultivar Everest and the wild species *Malus floribunda* 821 (Wöhner et al. 2018).

Complete genome of strain MAFF31108 of *Xanthomonas oryzae* pv. *oryzae (Xoo)* (rice bacterial leaf blight) has been sequenced. The *hrp* genes encode T3SS that determines the pathogenicity of *Xoo*. The *hrp* gene cluster contains 27 genes extending from *hrpaZ* to *hrpaF*. Three novel genes were detected in the region between *hpaB* and *hpaF*. Many gram-negative bacterial pathogens have type IV fimbriae (also known as pili) for adhesion. Type IV fimbriae genes are distributed in several loci on the genome of *Xoo*. In addition, a nonfimbrial adhesion gene known as *XadA* detected in *Xoo*, may also have a role in virulence. Two *xadA* homologs at different loci in *Xoo* MAFF have been identified (Ochiai et al. 2005). *Xoo* produces AvrXa7 protein, the product of *Xoo* gene *avrXa7* and this protein is a virulence factor. The gene *avrXa7* is a member of the *avrBs* avirulence gene family which encodes proteins targeted to plant cells by the T3SS. The products of the *avrBs3* related genes are virulence factors targeted to the host cell nuclei and have the potential to interact with host DNA and transcriptional machinery (Yang et al. 2000). AvrXa7 is the only known T3SS effector from *Xoo* with a major contribution to pathogen growth and lesion formation in bacterial leaf blight disease. Growth of *Xoo* in plants was severely inhibited following inactivation of a single effector gene in representative strains (Yang and White 2004). The genes *avrXa27* and *avrXa10* from *Xoo* failed to elicit disease resistance-associated HR and to prevent development of *X. oryzae* pv. *oryzicola (Xoc)* in rice cultivars with corresponding *R* genes after introduction into *Xoc*, despite the ability of this pathovar to deliver an AvrXa10:Cya fusion protein into rice cells (Makino et al. 2006). Two species of *Xanthomonas* are involved in citrus canker. *X. citri* induces type A citrus canker (Xc-A), while *X. fuscans* pv. *aurantifolii* (*Xfa*) strains cause B and C types. The avirulence gene *avrGf2* in C strains induced HR in grapefruit. AvrGf2, a member of XopAG effector family, shared 45% identity at amino acid level with another of the same family, AvrGf1 from strain Xc-AW, which elicited an HR in grapefruit. AvrGf2 shared sequence identity with other XopAG effectors present in other bacterial pathogens. Mutagenesis of *avrGf2* in strain C led to a compatible reaction in grapefruit. Expression of *avrGf1* or *avrGf2* in Xc-A occurred following infiltration into grapefruit leaves, although the *avrGf2* transconjugant elicited a faster HR and lower populations than the transconjugant containing *avrGf1* (Gochez et al. 2015).

Erwinia amylovora employs hypersensitive response (HR) and pathogenicity (*hrp*)-type III secretion system (T3SS) to translocate effector proteins directly into plant cells and modulate cell functions during infection. The role of integration host factor (IHF) in regulating *hrpL* and T3SS gene expression was investigated. The growth of mutants was greatly reduced, and expression of *hrpL* and T3SS genes was significantly downregulated as compared to those of the wild-type strain. Further, expression of *ihfA*, but not *ihfB*, was under auto-suppression by IHF. In addition, both *ihfA* and *ihfB* mutants were hypermotile, due to significantly reduced expression of small RNA (sRNA) rsmB. IHF bound to the promoters of the *hrpL* and *ihfA* genes, as well as *rsmB* sRNA gene. The results indicated that IHF might be required for RpoN-dependent *hrpL* gene expression and virulence and controls motility in positively regulating the *rsmB* sRNA in *E. amylovora* (Lee and Zhao 2017). Plasmid-borne cellulase (*celA*) and serine protease (*pat-1*) and serine proteases chpC and ppa A residing on the *chp/tomA* PAI of *Clavibacter michiganensis* subsp. *michiganensis (Cmm)* were examined. The *celA* and *pat-1* were significantly induced in *Cmm*382 at initial 12 to 72 h, whereas *chpC* and *ppaA* were highly expressed only at 96 h after inoculation, as revealed by transcriptional analysis. Expression of chromosomal and of plasmid-located genes was revealed. Expression of *celA* and *pat-1* was substantially reduced in the absence of *chpq/tomA* in *tomA* PAI, whereas *chpC* and *ppA* expressions were reduced in the absence of virulence plasmids. Transcription of chromosomal genes involved in cell wall degradation was also observed at early stages of infection. Expression of the host defense, chitinase class II and pathogenesis related (PR) protein-5 isoform was induced in the absence of the PAI at early stages of infection. PAI-harbored genes might be involved in suppression of tomato basal defenses (Chalupowicz et al. 2010). Three *C. michiganensis* subsp. *insidiosus (Cmi)* stains varied in their gene content due to genome rearrangements. The core *C. michiganensis* genome contained 1,917 genes of which 379 genes were unique to *Cmi*. Unique features of *Cmi* were an operon for synthesis of the extracellular blue pigment indigoidine, enzymes for pectin degradation and an operon for inositol metabolism. Secreted serine proteases belonging to both *pat-1* and *ppa* families were present (Lu et al. 2018).

Pseudomonas syringae pv. *tomato* (*Pst*) DC3000 uses T3SS for delivery of the bacterial effector proteins directly into the host cells. Mutations in the T3SS caused loss of ability to induce disease, indicating that the effectors are essential agents of *P. syringae* pathogenesis. Many effectors have been identified based on their ability to trigger host immunity. The concept of plant immunity based on the gene-for-gene model is based on the recognition of a pathogen avirulence (Avr) effector protein by resistance (*R*) protein of the host plant (Cohn and Martin 2003). The tomato R protein Pto, a serine/threonine protein kinase recognizes and directly interacts with DC3000 effector protein AvrPto and AvrPtoB and uncharacterized signal mechanisms. Based on the genome sequence of *Pst*, more than 30 putative effector genes are inferred to be present in *Pst* (Abramovitch and Martin 2004). Deletion of the effector gene *hopPtoM* in *P. syringae* pv. *tomato* DC3000 resulted in decrease in symptom severity induced in tomato leaves without marked reduction in pathogen growth (Badel et al. 2003). In contrast, dramatic increase in disease symptoms occurred due to mutation of another effector gene *hopPtoN* in *Pst* DC3000 which did not affect pathogen growth (Lopez-Solanilla et al. 2004). Necrosis-associated disease increased when both AvrPtoA and AvrPtoB were delivered by T3SS of Pst. AvrPtoB acted inside the plant cell to inhibit PCD of host plant cells by the Pto and Cf disease resistance proteins (Lin and Martin 2005). The AvrPto protein from *Pseudomonas syringae* pv. *tomato (Pst)* DC3000, following delivery into plant cells by T3SS, either promoted host susceptibility or expressing the *Pto* kinase elicits disease resistance in tomato plants. When Ala was substituted for S149, the ability of AvrPto to enhance disease symptoms was reduced significantly, in addition to reduction in growth of *Pst* in susceptible leaves. Further, the avirulence activity of AvrPto in resistant tomato plants was reduced by S149A (Anderson et al. 2006).

Pathogenicity of *Xanthomonas oryzae* pv. *oryzae (Xoo)* depends on a functional T3SS, whereas members of the *avrBs3/pthA* gene family of T3SS substrate effectors are required for its ability to induce symptoms of BLB in rice. Products of *avrBs3/pthA* have characteristics of nuclear transcription activation factors and are named transcription activator-like (TAL) effectors. They contribute to the virulence of bacterial pathogens (Yang and White 2004; Schornack et al. 2006). The TAL effectors secreted through T3SS are targeted to host cell nuclei, and each effector has a C-terminal transcription activation-like domain that is essential for both virulence and avirulence activities. Expression of the rice *Os8N3*, a member of the MtN3 family from plants was enhanced upon infection by *Xoo* strainPX099, and it depended on the type III effector gene *PthX01*. The results showed that *Os8NS* is a host susceptibility gene for BLB targeted by the type III TAL effector PthX01 (Yang et al. 2006). Xa27 is a dominant gene whose expression was specifically elevated in the presence of the TAL effector AvrXa27 (Gu et al. 2005). Likewise, PthX01 is the second TAL effector demonstrated to be associated with enhanced expression of a host gene that has known consequences for the outcome of the host-pathogen interaction (Yang et al. 2006). Characteristics involved in the virulence of *Xanthomonas vesicatoria (Xv)*, member of a species complex causing bacterial spot on tomato, were investigated. The strain BNM208, the most virulent, exhibited the greatest swarming and twitching mobilities and developed a mature biofilm with presence of defined cell clusters, a homogenous and compact structure, and higher biomass and substratum coverage than the other two strains. BNM208 strain produced the most viscous exopolysaccharide that possibly had better characteristics of its biofilm. The results suggested that aggressiveness of *Xv* strains might be related to their ability to move by flagellum or type IV pili, adhere to leaves and form well developed biofilms, factors that might improve colonization of tomato phyllosphere (Felipe et al. 2018). The functional importance of global regulator Crp-like protein (Clp) in *Xanthomonas* pv. *glycines (Xag)* infecting soybean was assessed, since Clp was shown to be involved in the production of virulence factors in some *Xanthomonas* spp. Deletion of *clp* resulted in reduction in virulence of *Xag* in soybean, which was strongly correlated with drastic reduction in carbohydrates utilization, EPS production, biofilm formation, cell motility, and synthesis of cell wall-degrading enzymes (CWDEs). Integration of a single-copy *clp* gene of the wild-type strain into the *clp* mutant restored all the impaired characteristics. Overexpression of clp in the wild-type strain led to significant increases in cell motility and synthesis of CWDEs. The Clp might function as both activator and repressor in multiple process in *X. axonopodis* pv. *glycines* required for its full virulence (Guo et al. 2019).

Virulence factors introduced into the host cells render the host more hospitable for bacterial pathogen proliferation. The gene *avrRpt2* from *Pseudomonas syringae pv. tomato (Pst)* JL1065 strain promoted virulence of heterologous *P. syringae* strains on *Arabiopsis thaliana*. Virulence activity of *avrRpt2* could be detected in tomato lines that were defective in either ethylene perception or accumulation of salicylic acid (SA), but could not be detected on tomato mutant insensitive to jasmonic acid (JA) (Lim and Kunkel 2005). Involvement of two-arginine translocation (Tat) system in physiological functions of *Pseudomonas syringae* pv. *tomato* PstDC3000 was investigated. The Tat pathway operates in the inner membrane of many Gram-negative bacteria. The Tat system has a remarkable ability to transport proteins that have already folded in the cytoplasm. Inactivation of the Tat system in *Pst* DC 3000 led to multiple complex phenotypes, including loss of mobility on soft agar plates, deficiency in siderophore synthesis and ion acquisition, sensitivity to copper, loss of extracellular phospholipase activity and attenuation in tomato leaves. In addition, localization of a Tat-specific protein, the powerful periplasmic disulfide catalyst DSbA protein to the periplasm compared with wild-type *Pst* DC3000. Inactivation of either Tat or type III secretion system resulted in a significant attenuation of virulence (Bronstein et al. 2005). Bacterial plant pathogens employ a range of different secretion systems to deliver virulence factors into host cells. The type VI secretion system (T6SS) plays an important role in the virulence of plant bacterial pathogens, *Agrobacterium tumefaciens, Pectobacterium atsepticum, Xanthomonas oryzae,* and *Pseudomonas syringae* (Mattinen et al. 2007; Wu et al. 2012; Filloux et al. 2008).

In the sequenced genome of *Erwinia amylovora*, three T6SS gene clusters were identified (Mann et al. 2013). Deletion mutants within three T6SS clusters were generated from *E. amylovora* strain WCPPB1665. Of the 33 mutants, 19 exhibited reduced virulence on immature pear fruit, compared with wild-type strain. These mutants produced less amylovoran or levan or both. *E. amylovora* T6SS might play a role in competition with other bacteria and virulence probably by influencing exopolysaccharide production (Tian et al. 2017). The 17-gene cluster encoded signature T6SS proteins in *Acidovorax citrulli* AAC00–1 genome. The core T6SS genes were deleted individually to generate mutants in an *A. citrulli* xjl12 background. Four of these mutants, Δ*vasD*, Δ*impJ*, Δ*impK*, and Δ*impF* were impaired in seed to seedling transmission of BFB in melon and in biofilm formation. Further, these mutants were reduced in seed colonization. Other mutants did not show any significant differences in BFB seed-to-seed transmission, melon tissue colonization and biofilm formation, compared to the wild-type xjl12 strain of *A. citrulli* (Tian et al. 2015). *Acidovorax citrulli* employs the functional T3SS for its pathogenicity. An AraC-type transcriptional regulator, HrpX regulates expression of genes encoding T3SS components and type III secreted effectors (T3Es). Examination of a sequenced strain of *A. citrulli* revealed the presence of 11 T3E genes. Machine learning and transcriptomics approaches were employed to identify novel T3Es. All *A. citrulli* M6 genes encoded T3Es, as shown by machine learning procedure. RNA-seq revealed differential gene expression between wild-type M6 and a mutual defective in HrpX:159 and 28 genes showed significantly reduced and increased expression in the mutant relative to wild-type M6, respectively. Transient expression in *Nicotiana benthamiana* revealed that two of these T3Es localized to the cell nucleus and one interacts with endoplasmic reticulum (Jimenez-Guerrero et al. 2020).

Burkholderia glumae (rice grain rot) was more virulent than *B. gladioli* and caused greater yield reduction. Virulence factors of isolates of *B. glumae* and *B. gladioli*, characterized previously, were highly conserved between the two species. Virulence factors *of B. glumae and* *B. gladioli* were predicted to encode distinct groups of genes encoding type IV secretion systems, transcription regulators, and membrane-sensing proteins. The identified genotypic differences between *B. glumae* and *B. gladioli* might be useful to differentiate disease phenotypes (Fory et al. 2014). *Acidovorax avenae* subsp. *avenae (Aaa)*, (rice brown stripe disease) possesses the *clpB* gene that represents T3SS. Function of *clpB* and its contribution to bacterial pathogenesis were studied, using insertional deletion mutation and complementation approaches. Mutation in *clpB* significantly affected bacterial growth, virulence, exopolysaccharide production, biofilm formation and expression of other T6SS genes of *Aaa* Rs-1 strain. Change in the composition of EPS in the mutant might be also partially responsible for reduction in virulence. Structural difference was also discernible between ClpB and type IV pili (TFP) whose functions were found to be similar to ClpB. The results showed that *clpB* gene could play a key role in virulence of *A. avenae* subsp. *avenae* infecting rice (Zhang et al. 2017). Transcriptional regulators of the LuxR family has a role in quorum sensing (QS), which is required for communication with other bacterial cells. Further, LuxR-type proteins are involved in cellular signaling pathways. Several *luxR* genes are responsible for modulating expression of virulence factors, biofilm formation, and host immune responses (Santos et al. 2012). Regulatory role of the LuxR-type regulator, AcR in the group II strain Aac-5 of *Acidovorax citrulli* was investigated by mutational analysis and RNA sequencing. Deletion of *acR* affected flagellar biosynthesis, cell mobility, biofilm formation, and virulence of AacR strain of *A. citrulli.* Expression of many genes involved in various functions of Aac-5 strain, including flagellar assembly were altered in *acR* deletion mutant strain, suggesting that *acR* might function as a global transcriptional regulator, especially for flagella-related functions in *A. citrulli* (Guan et al. 2020).

5.2.3.2 Ectophytic Interactions of Bacterial Pathogens with Their Host Plants

Phyllosphere bacterial pathogens, after reaching plant surfaces, have to survive an initial ectophytic phase, during which they are likely to be exposed to adverse environmental conditions. Phyllosphere bacterial pathogens establish a stable biofilm population size prior to initiation of infection. Localization in protected sites and clustering in structures such as biofilms results in enhanced resistance to stress that may be due to adaptive traits of epiphytic bacteria. The most common sites of bacterial colonization of aerial plant surface are stomata, base of trichomes, and depressions along veins. Biofilms formed in these sites provide protection to the bacterial pathogens from adverse environmental conditions. *Xanthomonas axonopodis* pv. *phaseoli (Xap)* (common bean bacterial blight) could survive on leaf surfaces and also endophytically. Populations of *Xap* in biofilms remained stable throughout the growing season, while solitary population sizes were more abundant. Microbial epiphytic community of *Xap* after reaching a threshold enter the leaves through natural openings, like stomata or wounds and establish an endophytic population leading to development of disease (Jacques et al. 2005). Bacteria harbored in biofilm could resist any copper-based chemicals, since the EPS of the biofilm could bind the chemicals, while viability of most solitary cells could be affected significantly (Amino 2010).

Entry of bacteria into leaf tissue appears to be more complex and dynamic. *Arabidopsis* stomata responded to the presence of live bacteria and to treatment with purified bacteria-derived PAMPs. Treatment of epidermal peels of *Arabidopsis* with suspension of *P. syringae* pv. *tomato* DC3000 (Pst DC3000) resulted in significant reduction in the number of open stomata. The bacterial pathogens might encode a virulence factor to allow them to overcome stomatal closure and enter leaf tissues (Melotto et al. 2006). Plant-bacterial pathogen interactions commence with molecular recognition of each other, leading to signaling events resulting in resistance or susceptibility to pathogen. Plants have developed a mechanism to close stomata upon sensing the MAMPs. Guard cells surrounding the stomatal pore can sense the presence of microbial pathogens and close the pore as a function of stomatal defense. The process of entry into the host plant is not passive, but dynamic and complex, requiring active inducible responses of both the host and pathogen, as well as specific environmental conditions at the time of penetration. But bacterial pathogens can trigger stomata closure under bright daylight, suggesting that a signaling cascade that overrides the natural circadian rhythm of stomatal movement. *Xanthomonas campestris* pv. *campestris* interfered with stomatal closure induced by MAMP by inducing production of diffusible signal factor that is involved in bacterium-to-bacterium signaling. Likewise, *X. axonopodis* pv. *citri* produced a plant natriuretic peptide-like compound, capable of opening stomata during plant infection, which correlated with enhanced disease severity (Gotting et al. 2008).

A COR-deficient mutant of *Pseudomonas syringae* pv. *tomato (Pst)* caused less disease when inoculated on leaf surface than when inoculated directly into the apoplast of *Arabidopsis thaliana* or tomato (Zipfel et al. 2004). Immune signal SA is required for stomatal defense. SA-responsible genes ICS1, EDS1, and PAD4 were induced in guard cells within 1 h after exposure to fla22 (Zheng et al. 2015). The outcome of stomatal defense is a reduction of pathogen entry into the plant, which diminishes the severity of foliar diseases. High relative humidity negatively affected stomatal defense against *P. syringae* (Roy et al. 2013). Plants possess multiple cell surface pattern-recognition receptors (PRRs) that can recognize either buried or surface-exposed regions of flagellin. FLS2 is a PRR present in all plants, and it recognizes flagellin-derived 22-amino acid (aa) peptide, flg22, considered a microbe-associated molecular pattern (MAMP). FLS2, as well as PRRs, can recognize regions of flagellin. Solanaceous-specific FLS3 receptor recognized the 28-aa peptide, flgII-28. Tomato plants have both FLS2 and FLS3 with subtle differences (Roberts et al. 2020). Expression and biosynthesis of flagellin components are tightly regulated. Flagellin is encoded by the gene *fliC* which is a part of the conserved gene cluster. Flagellin proteins have conserved N-terminal and C-terminal domains that are located within the center of the flagellin structure, whereas regions that are more surface-exposed have greater variability between different bacterial species (Fleigmann and Felix 2016). Bacterial mobility contributes to bacterial pathogen virulence and chemotaxis. Variability in epitope regions of bacterial flagellin, including in regions harboring MAMPs flg22 and flgII-28, are recognized by PRRs FLS2 and FLSII-28, respectively. Variations in flg22 and flgII-28 epitopes allowed some *Xanthomonas* spp. to evade both FLS2- and FLS3-mediated oxidative burst responses. The presence of larger bacterial populations of strains with flagellin variants predicted not to be recognized by either FLS2 or FLS3 was observed (Malvino et al. 2022).

Several pathogens are able to overcome stomatal defense using secreted virulence factors. The production of phytotoxins and the T3SS appear to be important factors for overcoming stomatal response by bacterial pathogens. COR, a phytotoxin produced by several pathovars of *Pseudomonas syringae* including *Pst* DC3000, could effectively prevent PAMP-triggered stomatal closure (Panchal et al. 2016). Another virulence factor impacting stomatal defense is syringolin A, a peptide toxin produced by *P. syringae* pv. *syringae* (*Pss*), had a proteosome-inhibiting function. Syringolin A promoted stomatal opening. Syringolin A diffused from the site of infection and suppressed SA silencing, leading to decrease in immune responses in adjacent tissues. Syringolin A might inhibit proteosome-mediated turnover of NPR1, an important component of SA signaling to induce stomatal defense (Schellenberg et al. 2010). Recognition of PAMPs by cognate PRRs has been investigated in several host-pathogen interactions. Flagellin recognition plays an important role in stomatal defense in interaction of *P. syringae* pv. *tomato* DC3000 (*Pst* DC3000) (Zeng and He 2010). Effectors produced by type III secretion system of bacterial pathogens can overcome stomatal defense by either inhibiting PAMP-triggered stomatal closure or actively inducing stomatal opening. The T3SE HopM1 of *P. syringae* disrupted the function of a 14–3-3 protein GRF8, resulting in reduction in PAMP-triggered ROS burst and stomatal defense (Lozano-Duran et al. 2014). The *P. syringae* effector HopF2 inhibited flg22-induced ROS and stomatal defense (Hurley et al. 2014). *P. syringae* effector AvrB-induced stomatal opening depended on canconical jasmonic acid (JA) signaling pathway. The *avrB* appeared to regulate stomatal opening through manipulation of the SA-JA antagonism (Zhou et al. 2015). The COR-deficient mutant *Pst* DC3118 colonized leaf apoplast less efficiently in the darkness, compared to plant inoculated under light. The COR deficient mutant *Pst* DC3118 colonized leaf apoplast inoculated under light. COR-producing *Pst* DC 300, however, could open dark closed stomata and efficiently invaded leaf at night (Panchal et al. 2016).

The formation of biofilm may function as an important virulence factor. Functional variations due to targeted mutation in *flg K* and *pilD* genes in the wild-type strain KU-P-SW005 of *Xanthomonas axonopodis* pv. *glycinis* (soybean pustule disease) were studied. The *flgK* mutants lacked monopolar flagellum and lost swimming motility. On the other hand, the *pilD* mutant lacked type IV pili and was unable to move via twitching, a form of surface motility. Biofilm formation was altered due to mutation in both genes. In addition, *flgK* and *pilD* mutants induced reduced disease severity in susceptible soybean cultivar Spencer in comparison to wild-type strain. The flgK and pilD genes were required for swimming and twitching motility of *X. axonopodis* pv *glycines* (Athinuwat et al. 2018). The presence of a cell-to-cell signaling system was detected in *Xanthomonas campestris* pv. *campestris* (*Xcc*). This system is coded by *rpf* gene cluster, which is known to be important for pathogenicity also. These genes govern the production of diffusible factors like butyrolactones. Acyl homoserine lactones (AHLs) have an important role in biofilm formation by *P. syringae* pv. *syringae* (Crossman and Dow 2004). Butyrolactones may act as signal molecules in quorum-sensing-like systems in *Xanthomonas*, instead of AHL present in *Pseudomonas* spp. (Jacques et al. 2005). Scanning electron microscope (SEM) and confocal laser scanning microscopy (CLSM) observations revealed that all strains of *X. campestris* pv *viticola* (*Xcv*) (grapevine bacterial canker) adhered to polystyrene in the microtiter plates. They formed biofilms in all culture media tested at different levels. In glass tubes, only two strains, Xcv 229 and Xcv 158, formed biofilms. SEM of Xcv 229 and Xcv 158 showed typical biofilm architectures. CLSM showed that only Xcv 229 formed an initial matrix structure characteristic of biofilms (Guerra et al. 2020).

Infection process (pathogenesis) of *Xanthomonas citri* subsp. *citri (Xcc)* commences with formation of biofilm on susceptible leaf surfaces. It is a factor for epiphytic survival of *Xanthomonas citri* subsp. *citri (Xcc)* on leaves and infection initiation. N-acetylcysteine (NAC) reduced biofilm formation, production of EPS and bacterial stickiness. Motility of *Xcc* was also affected in the presence of NAC. Under epiphytic conditions, NAC increased the susceptibility of *Xcc* cells to copper by affecting biofilm formation (Picchi et al. 2016). Inhibition of biofilm formation by small molecules may prevent infection by *Xcc*. D-leucine and 3-indodylacetonitrile (IAN) prevented biofilm formation by *Xcc* on different biotic surfaces and host leaves at a concentration lower than the minimum inhibition concentration (MIC). Quantitative RT-PCR analysis indicated that IAN repressed expression of chemotaxis/mobility-related genes in *Xcc* (Li and Wang 2014). Biofilm formation is an important virulence factor with a crucial role in the early stages of infection by increasing persistence of the epiphytic population of *Xanthomonas citri* subsp. *citri (Xcc)* (citrus canker disease). Biofilm formation was suggested to be linked to *Xcc* colonization prior to and after ingress. This view was verified by using two strains of *Xcc* A^w and A^* with restricted host range and testing their behavior in citrus hosts and nonhosts. SEM observations revealed that biofilms on leaf and fruit surfaces differed in structure of bacterial aggregates with respect to the strain's host range. Further, swimming motility varied widely, depending on the host range of the strain concerned. Biofilm formation in vitro and in planta for strains *Xcc* and *Xac* was related to their host range, since these processes affected colonization at early stages of infection process (Sena-Vélez et al. 2015). *Xcc* required bacterial aggregation and biofilm formation to colonize citrus plants. Biofilms were dependent on environmental conditions and their structures differed depending on their host range of *Xcc* strains. The extracellular structures produced by *Xcc* strains with different host range were visualized at early stages of biofilm formation in mature biofilms and during planktonic growth. The presence of these structures was related to the transcription of quorum-sensing, flagellar, and fimbrial genes. Variation in biofilm formation between *Xcc* strains was associated with expression of fimbrial and flagellar genes which in some cases was dependent on a quorum-sensing system (Sena-Vélez et al. 2020).

Ability to perceive and respond to the presence of neighboring populations of the same species/strains is present in bacteria. This kind of density-dependent phenomenon is termed quorum-sensing (QS). This system is accomplished by extracellular accumulation of small, self-generated chemical signaling moieties. Microbially derived signaling molecules can be divided into two main classes: (i) amino acids and short peptide derivatives commonly utilized by gram-positive bacteria and (ii) fatty acid derivatives known as homoserine lactones (HSLs) frequently utilized by gram-negative bacteria. The bacterial density is determined based on the release of chemical signaling molecules or autoinducers. The autoinducer then establishes a communication between cells that gets reflected in the expression of luciferase gene (*lux*) (Hastings and Greenberg 1999). Gram-negative and -positive bacteria differ in the nature of their signal molecules. Gram-negative bacteria have the family of N-AHL as universal signal factor. An increasing number of gram-negative bacteria have genes similar to either *luxR* or *luxL*. Gram-negative bacteria employ luxR-type proteins for autoinduction, whereas gram-positive bacteria utilize two-component adaptive response proteins for detection of autoinducers. Inter-species communication through use of autoinducers might be a possible mechanism by which pathogenicity of some virulent bacterial species may be enhanced. N-acyl homoserine lactones (AHLs) function as mediators of cell-density-dependent communication among gram-negative bacteria, whereas 2-heptyl-3-hydroxy-4(1H)-quinone and autoinducer 2 (4,5-dihydroxy 2,3-pentanedione) are converted into a variety of extracellular molecules (Sperandio et al. 2003). Swarming motilities are regulated by QS in many bacterial phenotypes. QS might directly regulate the genes for flagellum biosynthesis (Daniels et al. 2004). Flagellum gene regulation may be of two distinct types based on the nature of master regulators. The master transcriptional FlhDC is used by the bacteria with peritrichous flagella for activation of the expression of flagellum. On the other hand, bacteria with polar flagellum system employ the б54 dependent on NtrC family of transcriptional activators for flagellum gene expression (Arora et al. 1997; Aldridge and Hughes 2002).

The mechanism of flagellum biosynthesis under control of QS in *Burkholderia glumae* (rice seedling and grain rot) was investigated. The N-octanoyl homoserine lactone (C8-HSL)-deficient mutant of *B. glumae* was aflagellate, and hence, it was unable to swim like the wild-type variant. The QsmR, an IcIR-type transcriptional regulator, was identified in the hierarchical gene regulation of *B. glumae* flagella. Although the nonmotile *qsmR*, *flhA*, and *flhDC* mutants produced toxoflavin, the pathogenicity was abolished in these mutants. Variations in the typical regulatory mechanisms of the flagellum genes were detected in *B. glumae*, which contained FlhDC that is generally present in peritrichous flagellum systems. The results showed that the functional flagella of *B. glumae* had a central role in pathogenicity (Kim et al. 2007). *Acidovorax citrulli* (cucurbit BFB) has saprophytic cycles in cucurbit seeds and enters into pathogenic cycle during seed-to-seedling transmission. Involvement of QS in the process of transformation of saprophytic to pathogenic cycle as well as in watermelon seed colonization and seed-to-seedling transmission was investigated. The mutants *aacl* (luxl homolog) and *aacR* (lux R homology) mutants of AAC00–1 colonized germinating watermelon seed at wild-type levels. But seed-to-seedling transmission of BFB was affected in a cell density-dependent manner. When seed inoculum was reduced, seed-to-seedling transmission of *A. citrulli* declined significantly (Johnson and Walcott 2013).

Bacteria-to-bacteria communication is associated with a QS process based on production of N-AHL. The AHL-mediated QS system in *Pseudomonas syringae* pv. *syringae* (*Pss*) required AHL synthase, AhlI, and the regulator AhIR, and it is regulated by AefR. The contribution of QS to the expression of a variety of traits likely to be involved in epiphytic fitness and virulence of *P. syringae* was investigated. An *aefR* (^ –) mutant and an *ahlI* (^ –) (double mutant, deficient in AHL production) exhibited impairment of alginate production, in addition to increased susceptibility to H_2O_2, compared with the wild-type strain. These mutants were hypermotile in culture. Further, they invaded the leaves at a faster rate resulting in higher levels of incidence of brown spot lesions in bean leaves after 48 h of incubation. One *aefR* (^ –) mutant showed motility and virulence to the maximum extent. AHL-deficient mutants, like the wild-type strain, induced water-soaked lesions on pods. In contrast, AHL-deficient mutants lost their ability for tissue maceration. In addition, significant reduction in disease incidence and in planta growth of *P. syringae* pv. *taba*ci was observed in transgenic tobacco plants that produce AHL compared to wild-type plants. QS in *P. syringae* regulated traits contributed to epiphytic fitness as well as to distinct stages of pathogenesis in susceptible plant (Quiñones et al. 2004). Two major plant virulence traits, namely production of EPS, amylovoran, and levan and tolerance to free oxygen radicals, were controlled in a bacterial cell density–dependent manner. When *Bacillus* spp. strain A24 acyl homoserine lactonase gene *aiiA* was expressed in *E. amylovora*, induction of AHL biosensors was entirely abolished, in addition to impairment of EPS production and tolerance to H_2O_2 and reduction in virulence on apple leaves (Molina et al. 2005). Variation in dynamics of sizes of population of *Xanthomonas campestris* pv. *vitians* (*Xcv*) in susceptible and resistant cultivars of lettuce was monitored. Population growth was generally higher in susceptible cultivars compared with resistant. Population levels reached in resistant cultivar were significantly lower than in susceptible cultivars (Bull et al. 2015).

5.2.3.3 Colonization of Plant Tissues by Bacterial Pathogens

5.2.3.3.1 Endophytic Interaction of Bacterial Pathogens with Host Plant Tissues

Bacterial pathogens do not produce any specialized structures to penetrate and grow through compact tissues of host plants. The production of macromolecules like enzymes is regulated by different major secretion pathways. Some exoenzymes and toxins are secreted through a type I-sec independent pathway. *Xanthomonas oryzae* pv. *oryzae* Philippine race 6 (PR6) strains could not infect rice lines expressing resistance gene *Xa21*. The *raxR* and *raxH* genes encoding a response regulator and a histidine protein kinase of a two component-regulatory system, respectively, were identified. Null mutants of the PR6 strain PX099 were defective in either *raxR* or *raxH* or both. These mutants induced significantly longer lesions and reached much higher population levels compared to the wild-type strain in Xa21 rice leaves. The mutants (*raxH* and *raxR*) were complemented to wild-type strain levels of AvrXa21 activity by introducing expression vectors carrying *raxR1* and *raxH*, respectively (Burdman et al. 2004). The role of Osmi 1861K inducible by *Xanthomonas oryzae* pv. *oryzae* (*Xoo*) in rice genotypes with varying levels of resistance to BLB disease was studied. Stem-loop real-time RT-PCR analysis revealed that OsmiR1861K was significantly upregulated in resistant genotype SH5 in comparison with less resistant genotype 8411 at 3 and 5 days after inoculation with *Xoo* strain Zhe173. Two target genes of OsmiR1861K were downregulated in SH5 at the same time after inoculation with Zhe173 strain. Pre-OsmiR1861K was significantly induced by Zhe173 in two rice genotypes. Higher constitutive levels of OsmiR1861K in resistant genotypes generally were observed. Induction of Osmi1861K by other *Xoo* strains was recorded in leaves of SH5 (Zhang et al. 2015). Molecular mechanisms of rice plant responses to two pathovars, *X. oryzae* pv. *oryzae* (*Xoo*) and *X. oryzae* pv. *oryzicola* (*Xoc*) were investigated. Differentially expressed genes (DEGs) were studied. The gene ontology (Go) analysis of DEGs showed that while stress-related genes were upregulated, the genes encoding photosynthesis components were downregulated in rice leaves infected by bacterial pathogens. Hormones such as methyl jasmonate (MeJA) and abscisic acid (ABA) as well as light in the control of gene expression had critical role in disease progress (Khojasteh et al. 2018).

The cell wall–degrading enzymes produced by *Xylella fastidiosa (Xf)* (grapevine PD) might be involved in cell wall degradation. Several ß-1,4-gluacanases, xylanases, and xylosidases and one polygalacturonase (PG)-encoding gene were identified in *Xf* genome. A functional PG was shown to be encoded by the *polyA*. A mutant defective in *pglA* became nonpathogenic and was compromised in its ability to colonize the tissues grapevine systemically. The results indicated that PG was required for successful infection of grapevine by *Xf* and its function was a critical virulence factor for *Xf* pathogenesis in grapevine (Roper et al. 2007). *X. fastidiosa* (CVC) and grapevine PD colonizes xylem vessels of host plants and the cuticular lining of the foregut of insects that function as obligate vectors, eventually forming persistent biofilm structures on these surfaces. Biofilm structures are important for efficient retention and transmission of the pathogen by vector insects. Bacterial pathogens employ surface adhesion strategies during establishment and systemic spread in their host cell surfaces such as pili, fimbriae, and afimbrial adhesions are involved in these processes. *X. fastidiosa* employed many of these structures for efficient colonization of its insect and plant hosts. Three afimbrial adhesions, XadA1, Hsf/XadA2, and XadA3 were predicted to be trimeric autotransporters with a C-terminal YadA – anchor membrane domain. In isogenic *Xf* mutants, cell-wall aggression and biofilm formation were severely impaired in the absence of *XadA3*. The mutants did not exhibit significant reduction in cell-surface attachment. Acquisition by insect vectors and transmission of mutants to grapevines was reduced on deletion of *XadA3*. The *XadA3* mutant was hypovirulent in grapevines. *XadA1* or *XadA2* deletion resulted in a greater reduction in virulence than the wild-type strain. The results indicated the importance of adhesion protein and their individual contributions to different aspects of *X. fastidiosa* interaction with grapevines (Feitosa-Junior et al. 2022).

Pantoea stewartii subsp. *stewartii (Psst)* produces stewartan exo/capsular polysaccharide (EPS) in a cell density-dependent manner governed by the EsaI/EsaR QS system. Stewartan EPS synthesis governed by the EsaI/EsaR QS system was required for effective adhesion of bacterial cells and development of spatially defined 3D biofilms. A nonvirulent mutant lacking the *esaI* gene strongly adhered to surfaces in addition to the development of densely packed less structurally defined biofilms in vitro. Exogenous application of AHL counteracted adhesion phenotype. QS mutants lacking EsaR repressor attached poorly to surfaces forming amorphous biofilms heavily enmeshed in excess EPS. The two QS mutant strains were found largely localized at the site of infection, in contrast to efficient dissemination of the WT strain within xylem, primarily a basipetal direction (Koutsoudis et al. 2006). Genetic investigations on the virulence of *Pseudomonas syringae* pv. *actinidiae* (*Psa*) (kiwifruit canker) were taken up to characterize genes involved in pathogenesis. A genetic screen was performed to identify transposon mutants altered in lipolytic activity, since mechanism of regulation, production, and secretion of enzymes play a crucial role in pathogen virulence. Altered lipolytic activity phenotypes were identified. The Tn5 mutants of *Psa* in functions of phosphomannomutase, AlgC, lipidA biosynthesis acyltransferase, glutamine-cysteine ligase, and type IV pilus protein, PilI were affected in in planta survival and/or growth in kiwifruit plants (Patel et al. 2017). Pathogenic *Pseudomonas syringae* pv. *phaseolicola* and nonpathogen strains of *P. syringae* could multiply to similar levels on bean leaves, when water was supplied to the apoplast during infection. In vivo imaging showed water-soaked regions where *P. syringae* pv. *tomato (Pst)* DC3000 aggressively proliferated. On the other hand, an avirulent strain of *Pst* DC3000 (*avrRpt2*), which activated effector-triggered immunity, failed to induce water-soaked lesions (Xin et al. 2016). The *Pst* DC3000 avrE– hopMI– mutant lacking water-soaking-inducing effectors could not induce an aqueous apoplast during infection (Xin et al. 2016).

Xanthomonas gardneri induces water-soaked symptoms in tomato. AvrIlah1, a transcription activator-like (TAL) effector (TALE) of *X. gardneri* targeted two basic helix-loop-helix (bHLH) transcription factors in plants. Activation of these transcription factors causes expression of two pectin-modifying genes, a pectate lyase and a pectinesterase. Ectopic expression of two bHLH transcriptions and the pectin lyase resulted in a water-soaked lead apoplast upon bacterial infection, using designer TALEs (dTALEs), using designer TALEs (dTALEs). Tobacco leaf infected by *X. gardneri* containing AvrHahe could draw externally added water from leaf surface into the apoplast (Schwartz et al. 2017). Induction of different symptoms in susceptible plants by bacterial pathogens was dependent on virulence factors in addition to *hrp*-controlled pathogenic factors secreted via T3SS. EPS produced by bacterial pathogens are involved in formation of water-soaking symptoms observed in many diseases. Large amounts of EPS are produced by bacterial pathogens both in vitro and in vivo. EPS produced by xanthomonads is called xanthan and it has multiple functions. The amounts of EPS produced appeared to be related to level of virulence of *Xanthomonas oryzae* pv. *oryzae* (*Xoo*). Compositional variations of EPS may influence virulence of *Xoo* (Singh et al. 2006). EPS formed by *X. axonopodis* pv. *manihotis* (cassava bacterial blight) was examined for its virulence capability. EPS-mutants were generated by targeted disruption of the biosynthetic gene *gumD*. Mutants had severely reduced virulence. Although two EPS– mutants could induce water-soaking and chlorosis, symptoms were restricted to inoculated spots. On the other hand, wild-type 156 strain rapidly spread and developed into flaccidity. The wild-type strain produced EPS

200-fold greater quantity than the EPS– mutants. Petioles injected with both EPS– mutants did not exhibit any visible symptoms, whereas EPS from the wild-type strain induced severe wilting in inoculated plants. Further, EPS conferred resistance to H_2O_2 and UV irradiation (Kemp et al. 2004).

The capsular EPS of *Erwinia amylovora* (*Ea*) (fire blight of pear and apple) is composed of complex EPS amylovoran. The gene clusters *cms* and *cps* controlling biosynthesis of levan and amylovoran were sequenced. Amylovoran negative mutants failed to induce any symptoms in inoculated pear seedlings (Bernhard et al. 1993). *E. amylovora* required pathogenicity factors for production of EPS amylovoran and to induce HR in nonhost plant species. Synthesis of virulence factors such as effector protein harpin and the synthesis of amylovoran are tightly regulated by environmental conditions. The global regulation of bacterial transcription is a general response to the environmental changes depending on many factors including histone-like proteins like H-NS (Dorman and Deighan 2003). H-NS is mainly a multifunctional gene regulator, predominantly negative, but H-NS may also activate genes by repressing a repressor (Dorman 2004). Attenuation of virulence of animal pathogens by small basic histone like protein H-NS is known. An *hns* homolog from *E. amylovora* was identified by complementing an *Escherichia coli hns* mutant with a cosmid library from *E. amylovora*. A difference in functions of the two *hns*-like genes in *E. amylovora* was recorded for production of EPS amylovoran. Levan produced was significantly increased by *hns* mutations. But synthesis of capsular EPS amylovoran and of levan was reduced, if *hns* from *E. amylovora* plasmid was overexpressed. Increase in amylovoran synthesis was due to mutation in chromosomal *hns* of *E. amylovora*. But mutations affected development of symptoms adversely on immature pear fruits (Hildebrand et al. 2006).

In *Xanthomonas campestris* pv. *campestris (Xcc)* production of EPS xanthomonadin pigment and signal molecule diffusible factor (DF) was governed by *pigB* gene. The *pig* mutants were able to produce EPS and xanthomonadin, following extracellular application of DF. Function *pigB* gene was required for epiphytic survival and natural host infection. The production of extracellular enzymes by *Xcc* was regulated by the products of *rpfABFCHG* gene cluster. The DSF, as cell-to-cell communication signal of *Xcc* played an important role in regulation of virulence factor production and biofilm dispersal (He et al. 2006). Cell-to-cell signaling in *Xcc* and *X. axonopodis* pv. *citri (Xac)* mediated by diffusible signal factor (DSF) was investigated. The production of pathogenesis factors such as synthesis of EPS xanthan was regulated by a cluster of genes (A–I) known as *rpf* (regulation of pathogenic factor) in *Xcc* (Slatter et al. 2000). Two of these genes *rpfH* and *rpfI* were not detectable in *Xac*. Two strains of *Xac*, one with mutation in *rpfF* gene resulting in inability to produce DSF and another mutant defective in *rpfC* with DSF overproducing ability were constructed. These mutants could produce lower concentrations of extracellular cyclic ß-1,2-glucans, endoglucanase and protease extracellular enzymes. These mutants were drastically compromised in their ability to induce canker symptoms in lemon leaves compared to the wild-type strain (Siciliano et al. 2006). Among the metalloproteases produced by bacterial pathogens, Prt proteases secreted by *Xanthomonas campestris* are well characterized. These enzymes play an important role in degradation of plant cell wall components and/or they may be involved in combating plant immune defenses. Alkaline protease AprA from *Pseudomonas syringae* pv. *tomato (Pst)* is a zinc-dependent metalloproteases which was required for full virulence on tomato and *Arabidopsis* plants. The *aprA* deletion mutant of *Pst* had reduced growth and induced fewer disease symptoms on tomato plants (Pel et al. 2014). Effector molecules of *P. syringae* pv. *tomato (Pst)* DC3000 were examined for their influence on miRNA biogenesis maturation and function. Effector protein AvrPtoB, when expressed transiently in *Arabidopsis* leaves, reduced accumulation of the miR393 precursors primary mcR303a and pri-miR393b. It appeared that AvrPtoB is unlikely to be a general repressor of miRNA transcription. AvrPto also might function as a suppressor of silencing (Navarro et al. 2006).

Host signaling pathways may be manipulated by secreted effector proteins. The AvrPphB protease from *Pseudomonas syringae* pv. *phaseolicola* could alter host signaling pathway. Transgenic AvrPphB *Arabidopsis* rps5–2 susceptible mutant showed a reduced PAMP response, including lower deposition of callose and reduced accumulation of H_2O_2. AvrPphB acted as a double-edged sword, promoting bacterial infection and also revealing percentage of the pathogen in the host plant equipped with specific R protein. The HopB1 effector from *P. syringae* pv. *tomato* (*Pst*) allowed blocking of the PTI response, without inducing ETI resistance. HopB1, a T8 family threonine protease specifically cuts certain SERK (somatic embryogenesis receptor-like kinase) kinases including the cell surface immune coreceptor BAK1. HopB1 is constitutively bound to FLS2 (Li et al. 2016). Elevated RpoS protein level led to decrease in production of virulent factor amylovoran in *Erwinia amylovora*. Virulence, biofilm formation, and survival of *E. amylovora* were significantly affected by amylovoran and its level was directly limited by ClpXP proteolytic complex. The *clpxP* mutants accumulated RpoS and this had direct impact in the expression of several virulence factors. The *rpoS* mutant of *E. amylovora* exhibited stimulated production of amylovoran (Lee and Zhao 2017). Involvement of Lon in the networks controlling virulence in *Erwinia amylovora* was investigated. Disabling *lon* gene led to increased expression of T3SS, and overproduction of amylovoran and nonmotile phenotype. Lon degraded HrpS, the key regulatory protein required for transcription of the T3SS genes. Further, Lon directly targeted a structural protein of T3SS pilus. HrpA. Lon protease was also implicated in the regulation of amylovoran production and motility (Lee et al. 2018).

Plant hormones play an important role in host defense responses (stress hormones) and they regulate plant growth and development (growth hormones). SA, JA, and ET are phytohormones with significant role in development of susceptibility/resistance of plants to infection by bacterial pathogens, which may suppress accumulation and/or block the signals produced by these hormones and proteolysis as an important part of this process. The XopJ protease from *Xanthomonas campestris* pv. *vesicatoria*, a T3SS effector protein from the YopJ cysteine protease family, interferes with the SA signaling by direct targeting of the proteosome. XopJ disrupts the turnover of NPR1 (nonexpressor of pathogenesis-related 1), the master regulator of SA responses. The protease degrades the RpT6 (regulatory particle AAA-ATPase E6) protein of the 19S of ubiquitinated proteins, including NPR1(Ma and Ma 2016). JA/ET pathways act antagonistically to the SA pathway and this situation is exploited by some pathogens, by activating JA signaling. JA-dependent responses are repressed by jasmonate zimdomain proteins (JA2s). JA2s bind to and repress several JA-responsive transcription factors. The HopX1 cysteine protease from *P. syringae* pv. *tabaci* (tobacco wild fire disease) destabilized JA2s in *Arabidopsis*, and induced JA pathway, leading to opening of stomata facilitating entry of bacteria into the plant. SA-dependent defenses were inactivated, favoring infection and disease development (Gimenez-Ibanez et al. 2014). The ET pathway could be manipulated by T3SS effector protease, XopD *from X. campestris* pv. *vesicatoria*. XopD directly targeted ET-repressive transcription factor SIERF4 of the tomato plant which was involved in ET synthesis. Reduction in ET accumulation led to higher level of susceptibility to *X. campestris* pv. *vesicatoria* (Kim et al. 2013).

Erwinia amylovora is primarily dispersed through ooze, which contains a mass of exopolysaccharides and bacterial cells, and it is exuded as droplets from infected host tissue. Ooze droplets were collected from field-inoculated apple trees. Ooze droplets harboring large cells populations were typically smaller in total volume and had darker color. The ooze was not exuded through natural openings, but it was present on erumpent mounds as small tears in tissues. Observations under SEM suggested that *E. amylovora*-induced wounds in tissue provided the exit holes for ooze extrusion from the host tissue. Gene expression analyses indicated that *E. amylovora* cells in stem sections located above ooze droplets and in ooze droplets were actively expressing critical pathogenicity genes such *hrpL*, *dspE*, and *amsK* (Slack et al. 2017). Molecular responses of leaves of apple cv. Free Redstar (resistant) and cv. Idared (susceptible) to fire blight disease were investigated. Expression pattern in the resistant cultivar, especially upregulated genes encoding proteins involved in HR and controlled cell death (BAX inhibitor and HIR protein) suggested that hypersensitive reaction might be the principal mechanism of resistance to fire blight in resistant cultivar. A potential role in the defense against *E. amylovora* might be also due to ß-1,3-glucanase (PR protein 2), which was upregulated only in the resistant cultivar (Markiewicz and Michalezuk 2015).

Plants develop initial defense reaction known as PTI following recognition of PAMPs, such as components of bacterial flagellum or the bacterial translational machinery. Bacteria inject several effector proteins into plant cell to counteract PTI via a specialized secretion system. The effector proteins change the transcriptome and proteome of host cells to reprogram the cells to become susceptible to the pathogen. The ETI is induced following recognition of the presence of effector molecules. Silencing of plant gene expression via host-encoded small RNAs was implicated in both PTI and ETI. The mechanism by which the bacterial pathogen could suppress the RNA-mediated silencing of endogenous gene expression facilitating induction of PTI has been identified (Katiyar-Agarwal et al. 2006; Navarro et al. 2006). Changes in the plant transcriptome induced by a bacterial effector molecule AvrRpt2 were analyzed by host cell induced convergent transcription at a locus in the plant genome. Overlapping and complementary transcripts could anneal to generate a ds-RNA. This ds-RNA was then processed into a type of small RNA known as natural antisense small interfering RNA (nat-siRNA), which are 21–24 nucleotides long. The small RNAs induced by AvrRpt2 silence protein-coding genes in the plant genome, resulting in enhancement of ETI. Likewise, recognition of a peptide from bacterial flagellin induces accumulation of another small RNA, the microRNAs (miRNA) miR393, which repressed production of proteins involved in auxin signaling in plant cell, leading to increase in level of PTI. Effector molecules of *P. syringae tomato* DC3000 were evaluated for their influence on miRNA biogenesis, maturation, and function. The effector protein AvrPtoB, when expressed transiently in *Arabidopsis* leaves, reduced accumulation of the miR393 precursors primary miR393a and pri-miR393b. It appeared that AvrPtoB is unlikely to be general repressor of miRNA transcription as a suppressor of silencing. In transiently expressed AvrPto-wild *Arabidopsis*, accumulation of miR393, miR171, and miR173 was reduced (Navarro et al. 2006).

5.2.3.3.2 Distribution of Bacterial Pathogens in Plants

The *hrp*-dependent factors and certain host-specific/nonspecific toxins may be involved in the suppression of host gene expression facilitating spread of the pathogen from initially infected cells to other tissues or organs which may express different symptoms characteristic of the pathogen. Plasmodesmata (PD), the symplasmic pores connect plant cells, facilitating intercellular trafficking of essential molecules such as proteins, sugars, hormones, and RNAs. Movement of these molecules is controlled by permeability of PD to the molecules. One of the regulatory components in PD permeability is callose, a polysaccharide in the form of ß-1,3-glucan which is localized in the neck region of PD (Wu et al. 2018). Callose deposition/degradation around PD necks and/cell walls was frequently observed during plant defense responses. Callose deposition is typically triggered by conserved pathogen-associated molecular patterns (PAMPs) such as bacterial flagellin22 and elongation factor EF-Tu. The flg22-induced callose response required FLS2, brassinosteroid insensitive 1-associated receptor kinase 1 (BAK1), and botrytis-induced kinase 1 (BIK1) (Lu et al. 2010). As a counteracting response, pathogens inject different effectors that manipulate host cell structure and function or defense responses. Many gram-negative bacterial pathogens employ T3SS to subvert and colonize their hosts. PD modification mediates cell-to-cell movement and disperses effectors to expand pathogen invasion to other tissues/organs. Several effectors of *Pseudomonas syringae* moved from one cell to another cell through PD. Many effectors were able to suppress PAMP-induced callose accumulation. The PD-dependent movement of bacterial effectors was investigated using effectors of *P. syringae* pv. *tomato* DC3000. These effectors moved in plant cells, but their movement was restricted when callose accumulated to high concentrations (Li et al. 2021). In addition to effector-targeted PD-associated proteins, another effector protein delivered by *P. syringae*, Hop01–1, localized at PD and modulated PD permeability (Aung et al. 2020).

The *eps* genes govern synthesis of EPS, and the ability to produce EPS is positively correlated with the virulence of strains of pathogens including *Erwinia amylovora, Xanthomonas campestris,* and *Clavibacter michiganensis* (Narayanasamy 2002, 2008). Chlorosis, the resultant of loss of chlorophyll, may be due to the response of susceptible plant to activity of toxins elaborated by the pathogen, ultimately resulting in tissue collapse and death of plants. *Xanthomonas campestris* pv. *vesicatoria (Xcv)* induced chlorosis as the early visible symptom in tomato and the infected tissues turn necrotic later. The locus present in *Xcv* responsible for induction of chlorosis in tomato and nonhosts was identified and characterized. The gene, chlorosis factor (*ecf*) encode a hydrophobic protein with similarity to four other proteins in plants, including HolPsyAE. Chlorosis-inducing activity was conferred by this fragment on *Xcv*89–1 strain. The product of *ecf* appeared to be transmitted to hosts and the associated phenotype was *hrp*-dependent (Morales et al. 2005). The cell death associated with HR was a genetically controlled and regulated process. PCD, observed frequently in a host exhibiting HR, represented high levels of resistance or immunity. The effector AvrPtoB from *Pseudomonas* acts inside the plant cell to inhibit PCD initiated by the Pto and Cf9 disease resistance proteins (Abramovitch et al. 2003). Inactivation of SA-mediated resistance is a critical step in pathogenesis. The CEL mutant of *P. syringae* pv. *tomato (Pst)* DC3000 reached higher population levels and caused more pronounced chlorosis and necrosis in NahG plants than in wild-type isolate CO1-O*Arabidopsis* plants. The NahG plants cannot accumulate SA, because of the degradation of SA by *nahG* encoded SA hydroxylase. The effector locus CEL of *Pst* DC3000 contains at least four characterized effector genes (*avrE, HopPtoM, hrpW,* and *hopPotnA*) that are conserved among various *P. syringae* pathovars (Deb Roy et al. 2004). Reduction in symptom severity was attributed to deletion of T3SS effector gene *hopPtoM* in *P. syringae* pv. *tomato* (*Pst*) DC3000 (Badel et al. 2003). In contrast, disease severity was increased without any effect on pathogen development due to mutation in another effector gene *hopPtoN* in *Pst* DC3000 (Lopez-Solanilla et al. 2004). The effectors AvrptoA and AvrPtoB enhanced necrosis associated with the disease caused by *Pst* DC3000 (Lin and Martin 2005).

Two *Xanthomonas campestris* pv. *campestris (Xcc)* mutant strains 8397 and 8396 defective in xanthan production were employed. Leaves of *Nicotiana benthamiana* and *Arabidopsis* preinfiltrated with xanthan exhibited symptoms in response to inoculation with strains 8397 and 8396. The bacterial populations in the infected leaves was about 30- or 25-fold higher than in control (water-treated) and infection reached levels similar to that of wild-type strain. The role of callose deposition and its suppression of host defense and susceptibility were determined by monitoring callose deposition in leaves inoculated with mutant strains. Leaves inoculated with either 8397 or 8396 strain showed higher level of callose deposition in the host cell walls than in leaves inoculated with wild-type strain 8004. In contrast, leaves pretreated with xanthan failed to accumulate callose in response to inoculation with mutant strains, indicating that xanthan might induce susceptibility by suppressing deposition of callose on plant cell walls. Direct correlation between absence of callose deposition and production of disease symptom by *X. campestris* pv. *campestris (Xcc)* in susceptible plants was observed (Yun et al. 2006). Processes contributing to development of symptoms/resistance to grapevine PD were investigated by assessing movement and multiplication of a *gfp*-expressing *Xylella fastidiosa (Xf)* strain. Stems and petioles of grapevine cultivars differing in levels of resistance to PD were inoculated with *Xf.* The pathogen reached greater population levels and colonized more number of xylem vessels in the stem of susceptible cultivars, compared to resistant cultivars (Baccari and Lindow 2011). *Acidovorax citrulli* (watermelon bacterial fruit blotch) reached the seed by penetrating ovary pericarp or by invasion of pistil. The effects of the infection pathways on localization of watermelon seeds by the pathogen were assessed. *A. citrulli* was detected in the testa. Percentage of *A. citrulli*-positive PE layer samples, was significantly greater for seeds from pistil-inoculated lots than for seeds from pericarp-inoculated plants. Mean percentage of pathogen-positive embryo samples was significantly greater for seeds from pistil-inoculated lots than for seeds from pericarp-inoculated seed lots. *A. citrulli* cells were present in the PE layers and cotyledons of pistil-inoculated seeds (Dutta et al. 2012).

One of the degenerative changes induced upon invasion of HLB pathogen '*Ca.* L. asiaticus' (Las) appeared to be swelling of middle lamella between cell walls surrounding sieve elements. Anatomical aberration was observed in samples from newly growing flushes in inoculated sweet orange and grapefruit trees at early presymptomatic stage of infection. As the symptoms became visible, presence of bacterial cells in infected tissues correlated with degree of progression. A large number of bacterial cells could be visualized in phloem sieve tubes in samples from presymptomatic young flushes. Highly symptomatic leaves, however, did not exhibit the presence of bacterial cells (Folimonova and Achor 2010). The presence of HLB pathogen '*Candidatus* Liberibacter asiaticus' DNA was detected by qPCR format in extracts of peduncles from Sanguenelli and Conners fruits. Further, pathogen DNA could be detected in extracts of seed coats. It was possible to detect pathogen DNA in seed coats peeled from seeds. But no Conners seedling which grew from these seeds on agar contained pathogen DNA. Though '*Ca.* L. asiaticus' could be detected in different seed tissues, none of the seedlings grown from seeds contained detectable amounts of pathogen DNA (Hilf 2011). Bacterial cells with morphology and physical dimensions appropriate for '*Ca.* L. asiaticus' were observed in the phloem sieve elements in the vascular bundles of grapefruit seed coats. Observations under TEM revealed round and elongated bacilliform-like shaped cells, characteristic of the pathogen in the phloem sieve elements of seed coat vascular bundle. Bacterial cells were detected in the vascular tissue of intact seed coats of grapefruit and pummelo and in fragmented vascular bundles excised from grapefruit seed coats (Hilf et al. 2013). Vertical transmission of *Ca.* L. asiaticus' from seeds to seedlings did not occur, in spite of the pathogen being present in different seed tissues (Bagio et al. 2020).

Various virulence factors employed by '*Ca.* L. asiaticus' (Las) for development, colonization, and transmission by ACP were studied. Las genome encoded T1SS, a complete general secretory pathway (Sec), and an autotransporter type V secretion system (T5SS) (Duan et al. 2009). The Las genome contained a complete Sec translocon (Prasad et al. 2016). Las could utilize a SA hydroxylase encoded by *sahA* to degrade SA resulting in abolition of plant defenses. The *sahA* gene was highly expressed in *Citrus* spp. Recombinant Las SahA purified from *E. coli* displayed strong enzymatic activity to degrade SA. In transgenic tobacco plants, SahA abolished SA accumulation and the hypersensitive response indued by incompatible pathogens (Li et al. 2017). Las proteins containing a Sec-dependent signal peptide were predicted in the psy62 strain. On the other hand, significant differences were detected in *Candidatus* species (CLas, CLam, and CLaf) that shared only 45 sec-dependent proteins. Among the genes encoding a functional signal peptide, 36 were more highly expressed in citrus than in ACP, suggesting that these genes might be crucial for Las survival and pathogenicity (Wang et al. 2017). Prophages were considered to contribute to virulence of Las. Four types of prophages were identified. Prophage gene *SC2-gp095* encoded a ROS-scavenging peroxidase whose expression was high in plant hosts. Transient expression of *SC2-gp095* in *Nicotiana benthamiana* detoxified H_2O_2 and resulted in downregulation of *rbohB* gene encoding the respiratory burst (Jain et al. 2015). Type IVC tight adherence (*tad*) pilus-encoding gene cluster is conserved in *Liberibacter* spp. Las genome contains seven pili genes (*flp 1–7*), which contain multiple conserved regions. Expression of Tad pilus encoding genes was higher in the insect host than in plant tissues. Las Tad pili might be involved in attachment to the cells and colonization of psyllid midgut (Andrade and Wang 2019).

Clavibacter michiganensis subsp. *michiganensis (Cmm)* (tomato canker) multiplied rapidly in cotyledon in the stem toward both root and shoot. Luminescent signals were observed in tomato seedling roots over time. High humidity increased disease severity. Light intensity was strongly correlated with bacterial populations in planta (Xu et al. 2012). *Cmm* colonized primarily xylem, resulting in systemic infection with development of leaf wilting and ultimate death of infected plants. Pathogenicity of *Cmm* was dependent on plasmid-borne virulence factors and serine protease located as the chromosome *chp/tomA* PAI. The GFP-labeled strain of *Cmm* extensively colonized the lumen of xylem vessels and preferentially attached to spiral secondary wall thickening of protoxylem. The wild-type strain of Cmm NCPPB382 moved acropetally leading to an extensive systemic colonization of the entire plant and the pathogen reached the apical tissues at 15 days after inoculation. In contrast, the strain Cmm 100 (lacking plasmids pCM1 and pCM2) or Cmm 27 (lacking *chp/tomA* PAI) remained confined to the area surrounding the inoculation site. Virulence factors located on the *chp/tomA* PAI or the plasmids might be required for spread of *Cmm* cells to other tissues/organs of infected plants (Chalupowicz et al. 2012). Paraffin scanning electron microscopy (PSEM) was employed to study the interaction of infection of *Arabidopsis thaliana (At)* and *Pseudomonas syringae* pv. *tomato (Pst)* DC 3000 (bacterial speck disease). *Pst* entered the leaves through wounds or stomata and multiplied to high population in intercellular spaces in leaves. Syringe-infiltrated *At* plants were asymptomatic. Using PSEM technique, presence of *Pst* cells were visualized in intercellular air spaces and on the leaf surface. Stomata aperture filled with bacteria between guard cells were also revealed by PSEM method (Caldwell and Iyer-Pascuzzi 2019).

Infection process of *Burkholderia glumae* (rice bacterial panicle blight) was tracked at different growth stages of rice plants. The pathogen preferentially colonized the growing point of rice plants. The uppermost leaf at the two-leaf stage had the most intensive colonization by *B. glumae*. The population of the pathogen then decreased significantly, and it recovered in the booting stage and broke out in panicles. The *gfp*-labeled *B. glumae* strain was detected first in the vascular bundles of lateral roots and tap roots, and later in root hairs, epidermal cells, and the main root cap. Following spraying bacterial suspension, *B. glumae* cells landed on leaf sheaths and leaves and colonized epidermal hairs and leaf hairs or invaded into the cells through stomata. Simultaneously the pathogen from vascular bundles of roots spread into vascular bundles of leaf sheaths and leaves (Li et al. 2016a). The *gfp*-labeled strain of *B. glumae* LMG2196 was initially distributed on the surface of glumes and colonized in glume hairs and cells of the edge of sterile lemma, palea, and lemma. The base of glume hairs was initially colonized and then penetrated into the inner surface of the palea and lemma. Then *B. glumae* spread on gynoecium and stamens through contact. The contact or friction among glumes or leaf sheaths provided avenues for further spread. The pathogen spread into the inner tissues was mainly through connective tissues of anther. Glume hairs had an important role in the initial colonization of panicle by *B. glumae* (Li et al. 2016b). Colonization processes of GFP-labeled *Xylella fastidiosa (Xf)* citrus strain (variegated chlorosis) showed that *Xf* multiplied at a faster rate and moved to other tissues in susceptible citrus plants than in resistant plants. The predominant pitted stem morphology in citrus made the bacterial movement difficult. In susceptible genotypes, the pathogen could move from primary to secondary xylem, whereas the pathogen was restricted to primary xylem in resistant plants. Further, induction of lignification occurred earlier in the resistant genotypes, in the presence of pathogen, resulting in a physical barrier possibly to impair bacterial colonization (Niza et al. 2015). The GFP-labeled strain of *Xanthomonas fragariae* (strawberry angular leaf spot) was constructed to visualize the distribution of the pathogen in infected strawberry tissues. *X. fragariae* colonized leaf mesophyll, the top portion of the petiole adjacent to the leaf blade, and occasionally colonized xylem vessels down to the middle of the petioles. Colonization of vascular bundles and limited systemic movement in planta appeared to be a passive process. *X. fragariae* was able to directly enter petioles and colonize the space under epidermis. The pathogen was detected occasionally by qPCR in these tissues (Wang et al. 2018).

'*Candidatus* Liberibacter asiaticus' (Las) was distributed in a patchy manner in infected leaves. Las was detected more frequently in the symptomatic than in asymptomatic plants. Las was detected more frequently and bacterial abundances were higher among horizontally growing roots just under the soil surface than among deeper vertically growing roots (Lowzada et al. 2016). *Diaphorina citri*, ACP, transmits '*Candidatus* Liberibacter asiaticus' (Las) in a persistent manner. Interaction of Las with ACP was investigated, using the fluorescence in situ hybridization (FISH) technique with three nucleic acid probe for monitoring localization of the pathogen in different tissues of ACP adults after feeding on infected plants. Las was detected in the filter chamber, midgut, Malpighian tubules, hemolymph, salivary glands, ovaries, and in muscle and fat tissues of ACP. Las was detected in the phloem of infected citrus leaves. Pleomorphic bodies or thin short rods were dispersed and present individually in citrus leaf phloem and in ACP hemolymph, but more densely aggregated in cells of salivary glands and other organs and tissues of ACP were detected (Ammar et al. 2011). Phloem-limited bacterium '*Candidatus* Liberibacter solancearum' (potato zebra chip, ZC) is transmitted by potato psyllid *Bactericera cockerelli* from infected plants to healthy plants. Patterns of pathogen translocation through phloem in potato and tomato plants were investigated. Leaves of plants inoculated using infective psyllids were collected from different positions in plants tested for the presence of the pathogen. '*Ca.* L. solanacearum' was detected at 2–3 weeks after inoculation, frequently in upper- and middle tier leaves. Pathogen population was frequently low in both tomato and potato, as revealed by qPCR format (Levy et al. 2011). The uneven distribution of '*Ca.* L. solanacearum' in inoculated potato leaves was observed. The influence of vasculature on the distribution of the pathogen was assessed. Following inoculation of a single potato leaf, '*Ca.* L. solanacearum' infected significantly a greater number of leaflets that had direct vascular connectivity with inoculated leaf than leaves with minimal connectivity (Cooper et al. 2015). Members of the serralysin family are considered virulence factors, and they are secreted by T1SS. Gene expression analysis showed that from 2 to 6 weeks, expression of '*Ca.* L. solanacearum' serralysin-like gene was at least two-fold higher than in first week, indicating that gene expression remained high as the disease progressed (Ravindran et al. 2018).

5.2.3.3.3 Influence of Infection by Bacterial Pathogens on Host Plant Physiology

5.2.3.3.3.1 Effects on Photosynthetic Activity and Carbohydrate Metabolism As the bacterial pathogen initiates the infection process, plants counteract the effects of pathogen-induced effectors. This leads to increased demands for assimilates in the host plant. Further, the pathogen manipulates plant carbohydrate metabolism to meet its requirements. Bacterial infections commonly result in chlorotic or necrotic spots and consequent reduction in photosynthetic activity. Fluorescence of chlorophyll a was used as a sensitive marker for efficacy of photosynthesis. Changes in chlorophyll fluorescence could be detected even prior to appearance of visible symptom (BonFigure et al. 2006). A decrease in photosynthesis occurred in incompatible interaction also. A decrease in photosynthesis was detectable earlier with avirulent strain than with virulent strain of *P. syringae* (Tao et al. 2003). Expression of sugar-regulated photosynthetic genes such as small subunit of ribulose-1,5-bisphosphate carboxylase (RbcS) and chlorophyll *a, b* binding protein (Cab) was analyzed. The results were similar to those of chlorophyll fluorescence analysis. However, in two incompatible interactions, repression of photosynthetic genes was not detected, even though photosynthetic activity decreased. During pathogenesis, it is possible that downregulation and simultaneous increased demand for assimilates may result in a transition of source tissue into a sink tissue. Enhanced expression and activity of cell wall invertases were observed in *P. syringae-Arabidopsis* interaction (BonFigure et al. 2006). Hexoses released by invertase activity may act as signaling molecules and repress photosynthetic genes. This downregulation of photosynthesis, in turn, decreases the net photosynthetic rate (Berger et al. 2007).

Clavibacter michiganensis subsp. *michiganensis (Cmm)* requires a long latent period. The physiological and morphological changes induced by *Cmm* in relation to water stress were studied under greenhouse conditions. *Cmm* reduced the growth and altered plant-water relations of tomato plants by reducing the stem hydraulic conductivity. NDW1 index (an indicator of water content of the leaves) showed potential for use to diagnose tomato bacterial canker early before symptom expression as noninvasive technique (Romero et al. 2018). A wide range of cellular processes may be regulated by extracellular ATP (eATP) functioning as a signaling molecule. A significant reduction was observed in PSII maximal photochemical efficiency (Fu/Fm), maximum quantum efficiency of PSII photochemistry at illumination, the PSII operating efficiency, rate of noncyclic electron transport through PSII (ETR), photochemical quenching (qP), and eATP level in the tissues of non-photochemical quenching (qN), and the quantum yield of regulated energy dissipation of PSII [Y(NPQ)] was significantly increased. Change of PSII photochemistry of pathogen infection could be mediated by eATP in *Phaseolus vulgaris* leaves infected with *Xanthomonas campestris* pv. *phaseoli (Xcp)* (Jiao et al. 2016). BLB infection by *X. oryzae* pv. *oryzae* (Xoo) greatly reduced the maximum photosynthetic rate (P_{max}), but increased the light compensation point (LCP) and dark respiratory rate (RD). Reduction of pigments and increased stomatl resistance (SR) induced by BLB infectin might be responsible for decline in photosynthesis in affected plants (Hu et al. 2018). Callose deposition in phloem of plants infected by '*Candidatus* Liberibacter' may reduce translocation of photosynthetic products, resulting in starch accumulation. Higher pathogen population and accumulation correlated with genes involved in phloem functionality and starch accumulation were linked to upregulation of genes involved in starch biosynthesis and repression of those related to starch breakdown (Bova et al. 2017).

5.2.3.3.3.2 Effects on Nitrogen Metabolism The importance of nitrogen (N) metabolism has been assessed as an outcome of the interaction between bacterial pathogens and their respective host plant species. The metabolic state of plants during the infection process may be in part controlled through transcriptional regulation of N metabolism enzymes. Most of the N metabolism genes are modulated in response to different pathogens. A smaller proportion of genes displayed a variable pattern of modulation, both induced and repressed by pathogens, with no obvious pattern linked to either pathogen lifestyle or compatibility of infection, as result of specific manipulation by pathogens. N fertilizers exert variable effects on the development of diseases induced by bacterial pathogens, depending on host-pathogen combinations. N fertilizers generally increase susceptibility of plants to biotrophs, whereas susceptibility of plants to necrotrophic pathogens may be reduced due by N fertilizers. The impact on growth under N-limiting conditions on the susceptibility of *Arabidopsis* Col–0 plants to necrotrophic bacterium *Erwinia amylovora* and to necrotrophic fungus *Botrytis cinerea* was assessed. N limitation reduced resistance of *Arabidopsis* to *E. amylovora*. On the other hand, N limitation reduced susceptibility of *Arabidopsis* to *B. cinerea* (Fagard et al. 2014). Contradictory effects of N supply on host plant defense have been observed. Activities of defense-related enzymes chitinase, chitosanase, and peroxidase in *Arabidopsis* were reduced under low N supply. Further, application of BION, a chemical elicitor of plant defense, reduced the level of induction of these enzymes under low N nutrition levels (Dietrich et al. 2004).

Variations in N supply may impact plant-pathogen interactions significantly through effect on virulence of the pathogen. Perception of N nutritional status by pathogens may influence signal controlling activation of virulence factors and metabolic adaptation. The NRT2 gene family belongs to the major facilitator superfamily (MFS) of transporters. NRT2.1 is the main high-affinity nitrate transporter in roots of *Arabidopsis* under low nitrate availability (Li et al. 2007). NRT2.1 represses responses to biotrophic pathogens probably favoring abiotic resistance by safeguarding energy. The reduced susceptibility of *nrt2.1* mutant to *Pseudomonas syringae* was considered to be due to a faster response of SA-dependent signaling pathway and reduced sensitivity to bacterial toxin coronatine. In a SA-deficient background, *nrt2.1* mutant was susceptible as the wild-type strain. Priming not only by SA but also by other secondary metabolites appeared to contribute to reduced susceptibility of *nrt2.1* mutants to *P. syringae*. In addition, several ribosomal proteins were overexpressed in the mutant, which might be another factor contributing to resistance to bacterial infection (Camanes et al. 2012). NRT2.6 participated in response to necrotrophic *Erwinia amylovora*. *Arabidopsis nrt2.6* mutant was more sensitive to *E. amylovora* than the wild-type strain. Hypersensitivity of the mutant was correlated with a reduced ROS accumulation, whereas no difference in transcriptional response to the pathogen or in other cellular responses such as callose synthesis and nitric oxide (NO) production was observed. Nitrate sensing or nitrate transport might have an impact on defense reactions. The results suggested that SA defense pathway, ROS production, and metabolite status may play important roles in the cross-talk between N availability and biotic stress (Dechorgnat et al. 2012). Different effects of bacterial pathogen infection on amino acid metabolism of infected plants have been reported. Whole amino acid pool increased in crucifers infected by compatible or incompatible strains of *Xanthomonas campestris* pv. *campestris* in crucifers. Disease intensity was greater in CaAS1-silenced pepper leaves infected by *Xanthomonas campestris* pv. *vesicatoria* than in controls in these areas and then allowed efficient defense-related HR, cell fortification, and downregulation of pathogen-induced senescence and other harmful effects associated with infection by *X. campestris* pv. *vesicatoria* (Hwang et al. 2011).

Effects of *Pseudomonas syringae* harpins on mitochondrial proteins of *A. thaliana* suspension culture cells were assessed by proteomics. Proteins might be generated from citric acid cycle and they were upregulated indicating involvement of mitochondria in the response to the harpin elicitor (Livaja et al. 2008). The differences between PTI and ETI in the *Arabidopsis thaliana (At)-P. syringae* pv.

tomato (Pst) pathosystem were investigated, using proteomics analysis. The components of basal defense [leaves inoculated with a mutant in the hypersensitive response and pathogenesis (*hrp*) cluster] from disease and resistance response following *Pst* DC3000 (wild-type) and DC3000 [expressing effector protein AvrRpm1-DC3000 (*avrRpm1*)] inoculations were resolved. The presence of several unique proteins in inoculated plants was recorded. The proteomic changes could be correlated with a transcriptomics analysis of the same response (Jones et al. 2006).

Bacterial pathogens establish first contact with plants via plasma membrane beyond the cell wall. Hence, the proteins localized in plasma membrane may have important roles in signal transduction. Investigation on plasma membrane proteins of rice (*Oryza sativa*) suspension cells with two strains of *Xanthomonas oryzae* pv. *oryzae (Xcc)* was taken up to determine variations in the proteomes of susceptible and resistant plants. Most of the differentially expressed pathogen-induced proteins showed similar patterns in response to both susceptible and resistant plants. Differences in protein abundance observed in the resistant plants pointed to specific roles of plasma membrane components of resistance responses (Chen et al. 2007). Bacterial pathogens coordinate population growth to colonize susceptible plant tissues and it is accomplished by secretion of N-AHL to commence the process of quorum-sensing. The interaction between *Xanthomonas axonopodis* pv. *citri* (*Xac*) (citrus canker) and citrus plants was investigated by proteomics. Sugar-related photosynthetic proteins such as Rubisco and Rubisco activase, as well as of adenosine triphosphate (ATP) synthase showed significant downregulation, whereas NADH hydrogenase was upregulated in infected leaves. This response is characteristic for a reduction in photosynthetic efficiency during disease development. In addition, an increase of proteins related to plant defense against pathogen was detected in *X. axonopodis* pv. *citri* treated leaves (Garavaglia et al. 2010). The role of lipoprotein VacJ on the physiology and virulence of *X. citri* subsp. *citri* was investigated. Mutation of *vacJ* led to decrease in biofilm forming ability in glass tubes and on leaf surfaces, swimming motility, bacterial growth in planta and virulence on mandarin orange plants compared with wild-strain. VacJ might play an important role in the virulence of *X. citri* subsp. *citri* that is associated with regulation of biofilm formation and swimming motility (Li and Yu 2020).

5.2.3.3.3.3 Effects of Micronutrients on Bacterial Disease Development The effects of micronutrient application on the development of *Xanthomonas euvesicatoria* (pepper bacterial leaf spot) were assessed. A model (ECHA1) consisting of concentrations of copper, manganese, potassium, and the iron/zinc ratio as independent variables was used for validation in three different commercial pepper fields. Relative gene expression of both *NPR1* and *PR1* genes was significantly higher in pepper grown in predicted low-risk sites compared with pepper from high risk sites. The level of risk at a particular location might be influenced by how macronutrient and micronutrient concentrations affect to plant disease resistance genes in the systemic acquired resistance (SAR) pathway (Dutta et al. 2017). The influence of micronutrients, zinc (Zn), manganese (Mn), and/or copper (Cu) supply and infection of citrus by '*Candidatus* Liberibacter asiaticus' (Las) on physiological and growth characteristics of sweet orange trees (*Citrus sinensis*) were assessed. Las affected adversely plant biomasss production, regardless of nutrient treatments, particularly root growth, and increased specific leaf dry weight as the disease developed progressively due to accumulation of starch in leaves. Further, individual treatment with Zn, Mn, or Cu could reduce deleterious effects of HLB on starch metabolism. A combination of Zn/Mn/Cu+ could reduce acquisition of Las from infected plants by its vector *Diaphorina citri* adults and nymphs, suggesting the possibility of restricting HLB disease spread by micronutrient treatment (da Silva et al. 2020).

5.2.3.3.3.4 Effects on Metabolism of Other Compounds The effects of induction of phenolic compounds in grapevine cv. Thompson Seedless on development of PD caused by *Xylella fastidiosa* were assessed. At 2 months after infection, catechin and astringin were detected at greater levels in xylem sap. Further, multiple catechins, procyanidins, and stilbenoids were present in greater concentrations in xylem cell walls. Most of the tissues involved in photosynthesis were abscised. The initial response of grapevine plants to infection by *X. fastidiosa* was greater production of phenolic compounds, but the diseased plant had no resource to support secondary metabolite production including defense-related phenolic compounds (Wallis and Chen 2012). Induction of apoplastic phenolics in response to inoculation of *Pseudomonas syringae* pv. *tabaci (Pst)* in leaves of bell pepper (*Capsicum annuum*) plants was studied. A large number of phenolics accumulated in leaf apoplast commencing from 24 h after inoculation. All fully developed leaves in a single plant had similar phenolic response based on the magnitude and composition of the phenolics. Significant differences in the magnitude and composition of phenolics were observed at 24 h after inoculation with other bacterial pathogens, *P. syringae* pv. *tomato* and *P. syringae* pv. *syringae*. Dynamics of induction of phenolics in the apoplast of pepper by *P. syringae* pv. *tabaci* were similar to those occurring in tobacco. Characterization of phenolics induced in pepper by *Pst* were structurally different from those induced in tobacco (Barker et al. 2020).

Among the three potato genotypes tested, Red La Sola tubers showed more severe symptoms of zebra chip (ZC) induced by '*Candidatus* Liberibacter solanacearum' (Lso) at 4 weeks after inoculation. Titers of Lso did not differ among cultivars. Tuber levels of amino acids, carbohydrates, and phenolics differed among cultivars (Wallis et al. 2014). Apparent tolerance of five potato cultivars/breeding clones which did not exhibit any or little symptoms of ZC disease in freshly cut or fried tuber slices was investigated.

Five apparently tolerant clones showed less pronounced shifts in host physiology following Lso infection than the susceptible cv. Atlantic. Lack of responses of tolerant clones to Lso infection might be a possible mechanism of resistance to ZC disease (Wallis et al. 2015). The usefulness of contents and types of phenolic compounds in apple leaves as potential markers of resistance to fire blight disease caused by *Erwinia amylovora* was assessed. The contents of phenolics and flavonoids in leaves were determined before and after inoculation with *E. amylovora* of highly resistant (Enterprise) and highly susceptible (Idared). Faster changes in basic levels of phenols occurred in resistant cultivar, although basic levels of phenols were similar in both susceptible and resistant plants before inoculation. The difference between cultivars was related to the content of the compounds and rate of release of free phenols from their glucosides. The levels of 15 phenolics were significantly higher in young leaves of Idared than in Enterprise. The amount of SA in Idared was lower than in Enterprise. Accumulation of salicylate in Idared was more pronounced than in Enterprise. Higher levels of naringenin glucosides, 4-hydroxybenzoic acid and gentisic acid were present in Enterprise. Activity of hydroxyquinone persisted for longer time in vitro. Hydroxyquinone was also effective in restricting spread of the disease in apple shoot also (Sklodowska et al. 2018).

'*Candidatus* Liberibacter solanacearum' (Lso) (potato zebra chip, ZC) vectored by psyllids, induces shifts in physiology of tubers leading to disease symptom development. Concentrations of asparagine, aspartic acid, glutamine, fructose, glucose, sucrose, a ferulic acid derivative, and quinic acid were lower in foliage of Lso-inoculated plants than in noninfected plants. However, levels of proline, serine, four phenolic compounds, and most terpenoids in foliage were greater in pathogen-inoculated plants than in noninfected plants. Levels of asparagine and aspartic acid in upper stems and levels of ellagitannins and most nonterpenoids in lower stems and sesquiterpenoid levels in lower stem were greater in Lso-inoculated plants than in noninfected plants. The results suggested that several defense-related terpenoid compounds might increase in plants inoculated with Lso by psyllid vectors (Wallis et al. 2015). The rootstock Cleopatra mandarin (*Citrus reticulata*) roots contained a higher population of HLB pathogen '*Candidatus* Liberibacter asiaticus' (Las) than the roots of Swingle citrumelo (*C. paradisi* x *Poncirus trifoliata*). Sucrose and hormone metabolism in Cleopatra mandarin and Swingle citrumelo were evaluated. In fibrous roots of Cleopatra mandarin higher expression of genes related to sucrose was consistent, compared to noninoculated seedlings and 5 weeks postinoculation (*wpt*). Sucrose content and gene expression related to sucrose cleavage in fibrous roots of Swingle citrumelo were less disrupted by Las infection than Cleopatra mandarin at 5 *wpt*. Genes associated with SA, ET, and ABA signaling, phospholipids D (PLD) and phospholipase A2 (PLA2) were activated by Las infection at 5 *wpt* in Cleopatra mandarin. Upregulation of PR1, lower response of sucrose metabolism genes and downregulation of biosynthesis of phytohormones indicated that Swingle citrumelo might activate a more effective defense against Las than Cleopatra mandarin (Wu et al. 2018).

5.2.3.3.3.5 Role of Environmental Conditions on Disease Progress Environmental conditions existing in a location has significant influence on the host plant and microbial plant pathogens. The effects of temperature on '*Ca.* L. solanacearum' (Lso) (potato zebra chip, ZC) were assessed in potato plants inoculated using infective psyllids at various temperatures under in vitro. Temperatures less than 17° C seemed to lessen the rate of development of Lso and symptoms of ZC disease. On the other hand, temperatures above 32° C affected pathogen development, indicating the temperature sensitivity of the ZC pathogen (Munyaneza et al. 2012). More severe symptoms were seen in freshly sliced tubers held at 3° C than at 6–9° C. Lso titers were significantly variable in both cultivars over time at these temperatures. Levels of phenolic compounds known to be associated with ZC symptom severity were higher in tubers held at 3° C (Wallis et al. 2017). The influence of light on the interaction of *Pseudomonas cichorii* JBC1 with tomato was investigated. A significant reduction in disease induced by *P. cichorii* on seedlings exposed to green or red light, compared with those grown under white light or darkness, was observed. Exposure to green and red light suppressed disease incidence into tomato leaves. Upregulation of defense-related genes phenylalanine-ammonia lyase (PAL) and PR-1a was significant due to exposure to green and red right in infected tomato seedlings, compared to that of white light or darkness. The results revealed inhibitory effects of green and red light against *P. chicorii* infecting tomatoes (Nagendran and Lee 2015).

5.2.3.3.4 Role of Toxins from Bacterial Pathogens in Symptom Development

Pseudomonas syringae pathovars produce different kinds of phytotoxins in their respective host plants. *P. syringae* pv. *tomato (Pst)* DC3000 produces coronatine in infected tomato. During infection of tomato leaves by *Pst*, change in the pattern of an isoform of glutamine was induced in response to *Pst* infection. Elevated levels of asparagine synthetase in infected leaves suggested that part of glutamine could be converted into asparagine for export to healthy plants to save nitrogen. The genes involved may be located in one or more clusters. Tissue necrosis is the symptom induced by phytotoxins. The *lemA* gene in *P. syringae* pv. *syringae* strain B728a is essential for production of extracellular protease and syringomycin and consequently for induction of lesion formation, whereas *lemA* is not required for pathogenicity (Narayanasamy 2008). *Pseudomonas syringae* pv. *tomato (Pst)* DC3000 produces COR in susceptible tomato plants capable of inducing necrotic lesions on treated leaves. COR has two distinct moieties functioning as intermediates in the biosynthetic pathway. Coronamic acid and coronafacic acid are synthesized by separate pathways and joined by an amide bond to form COR. Several pathovars of *P. syringae*, including *tomato, maulicola, glycinaeae,* and *atropurpurea,* are capable of producing COR, and it is known to function as a virulence factor promoting chlorosis in their respective hosts. Anthocyanin production, alkaloid accumulation, ethylene emission, tendril coiling, and root growth are modified by COR which acts as a virulence factor and contributes to disease development (Bender 2000). Reopening of stomata was accomplished by *Pst* DC3000 by producing COR, indicating its ability to act as a virulence factor. COR-defective mutants had reduced growth and symptom production in planta compared with wild-type strain. The mode of action of coronatine was studied at molecular level. Coronatine promoted opening of dark-closed stomata of broad bean and Italian rye grass. Effect of COR in the stomatal aperture was more pronounced in Italian rye grass. Further, COR activated membrane bound ATPase activity and induced stomatal opening (Mino et al. 1987). Plant protein RIN4 (RPM1 interacting protein 4) and activation of H^+-ATPase were necessary for COR to reopen stomata (Liu et al. 2009).

The mechanism underlying the strong induction of neutral leucine aminopeptidase (LAP-N) and acidic LAP (LAP-A) upon infection by *Pseudomonas syringae* pv. *tomato (Pst)* was investigated. Following infection of tomato by COR-producing and deficient-*Pst* strains, LAP expression was monitored. LAP RNA and activity were detected only with COR-producing *Pst* strains. Treatment of excised tomato shoots with COR induced increases in RNAs for JA regulated PR protein gene *PR-1.* The results indicated that COR mimicked the wound response (Pautot et al. 2001). Induction of JA-response genes in tomato-*Pst* DC3000 interaction depended on production of COR (Zhao et al. 2003). In tomato-*P. syringae* pv. *tomato (Pst)* interaction, intact COR molecule was required for both suppression of SA-mediated defense responses and full disease symptom expression in tomato (Uppalapati et al. 2008). *Pst* DC3000 produced exopolysaccharide alginate (AlgU), which regulated expression of genes such as *algD* related to alginate biosynthesis. The molecular mechanisms of AlgU in virulence of *Pst* DC3000 were studied. After inoculation of tomato with mutants Δ*algU*, but not Δ*alD*, the inoculated plants developed milder symptoms and bacterial populations were also less. Expression profiles of *Pst* DC3000 genes revealed that AlgU could regulate not only expression of genes encoding alginate biosynthesis but also expression of genes related to T3SS effectors and phytotoxin coronatine (COR). Further, Δ*algU* mutant exhibited full virulence in *Arabidopsis fls2 efr1* double mutant, which was compromised in the recognition of PAMPs. In addition, application of COR restored the phenotype of Δ*algU* mutant in the stomatal response. AlgU might have an important role in the virulence of *Pst* DC3000 by regulatory COR production (Ishiga et al. 2018).

The tetracycline repressor (TetR) family of bacterial regulatory proteins enable bacteria to sense and respond quickly to the environment. The TetR-like AefR (AHL) and **e**piphytic **f**itness **r**egulator) protein was demonstrated to regulate the synthesis of N-AHL and contribute to the epiphytic fitness of *P. syringae* pv. *syringae* (*Pss*) strain B728a on soybean leaf surfaces (Quiñones et al. 2004). A mutant strain DB4H2 of *P. syringae* pv. *tomato* (*Pst*) DC3000 contained a single Ta5 insertion in PSPT03576, an ORF that is predicted to encode protein belonging to TetR family of insertional regulators. The ORF PSPTo3576 was essential for the virulence of *Pst* DC3000 and the encoded protein TvrR (TetR-like **v**irulence **r**egulator) (Preiter et al. 2005). *P. syringae* pv. *syringae (Pss)* secreted a family of structurally closely related peptide derivatives known as syringolins. Syringolin A is the major variant among syringolins. Genes involved in biosynthesis of syringolin A encode proteins consisting of modules typical of nonribosomal peptide synthetases and type I polypeptide synthetases (Amrein et al. 2004). Expression of *syrB* gene that controls synthesis of syringomycin, a nonhost-specific phytotoxin produced by *Pss.* Expression of *syrB* was significantly less in plants receiving N fertilization compared with nonfertilized plants (Cao et al. 2005). *P. syringae* pv. *syringae (Pss)* contains *salA* gene, which is a key regulatory element for production of syringomycin and encodes a member of Lux A regulatory protein family. Further, *salA*, a member of GacS/GacA signal transduction system, is essential for bacterial virulence, syringomycin A production, and expression of *syrB1* synthetase gene. Expression of 16 genes was significantly higher in the wild-type strain for some toxin genes. Except for *sylD* synthetase gene for syringolin production, all ORFs controlled by SalA were located in the *syr-syp* genomic island (Lu et al. 2005). The production of syringomycin and expression of *syrB1*, a syringomycin synthetase gene in *P. syringae* pv. *syringae (Pss)* are induced by specific plant signal molecules. The *salA* mutant did not produce syringopeptin, indicating the involvement of SalA/SyrF regulon in its synthesis. The subgenomic analysis of transcriptional expression confirmed that *syr-syp* genes are induced following addition of arbutin and D-fructose to syringomycin minimal medium (SRM). Plant signal molecules are transmitted through the GacS, SalA/SyrF pathway to activate the coordinated transcriptional expression of *syr-syp* genes (Wang, Lu, Wang et al. 2006). The synthetase genes for syringomycin (*sypA, sypB,* and *sypC*) and four regulatory genes *(salA, syrF, syrG,* and *syrP*) and nine putative secretion genes dedicated to the production of two toxins were activated by the phenolic plant signal molecule arbutin (Lu et al. 2005). The syr-syp genomic island was organized into five polycistronic operons along with eight individual genes. Conserved promoter sequences of the *syr-syp* genes contributed to conserved promoter sequences of *syr-syp* genes and to coregulation of syringomycin and syringopeptin production (Wang, Lu, Yang et al. 2006).

Pseudomonas syringae pv. *phaseolicola (Psp)* (bean halo blight disease) produces phaseolotoxin, which induces characteristic haloes on infected leaves and pods. The chlorotic halo is due to the action of phaseolotoxin that inhibits ornithine carbamoyl transferase, a critical enzyme in the urea cycle. Phaseolotoxin, a nonhost-specific toxin, is produced by pathovars of *Pseudomonas syringae* pv. *phaseolicola*, pv. *actinidiae*, and pv. *syringae*. The production of phaseolotoxin is temperature-dependent being optimally produced between 18° C and 20° C, while production of toxin becomes undetectable at above 28° C (Nüske and Fritsche 1989). Phaseolotoxin effectively inhibits activity of ornithine carbamoyl transferase (OCTase) from the plant. *P. syringae* pv. *phaseolicola (Psp)* synthesizes phaseolotoxin-resistant OCTase (ROCT) to protect itself against its toxin, and it is the product of the gene *argK* and ensures a supply of arginine required for the growth of *Psp* (Aguilera et al. 2007). The regulatory mechanisms involved in phaseolotoxin biosynthesis were investigated. The global regulon integration host factor (IHF) which bound to the promoter region of *phtD* operon exerted a negative effect on the expression of this operon. This regulatory protein was identified in *P. syringae* pv. *phaseolicola (Psp)* as part of the phaseolotoxin synthesis system. The IHF protein was one of the elements involved in the regulation of phaseolotoxin synthesis by *Psp* (Arvizu-Gomez et al. 2011). Regulation of phaseolotoxin cluster is very complex. The global regulation GacA/GacS are required for the expression of all transcriptional units, except argK, which becomes constitutive in a *gacA*– mutant background. In *gacA*– background, downregulation of genes with operons *phtA, phtD, phtI*, and *phtM* whose expression was negligible at 18° C. Involvement of other genes included in the *Pht* cluster in regulation of gene *argK* was investigated. Genes *phtABC* located within *Pht* cluster in the transcriptional repression of genes *argK* at temperatures above 20° C was not permissive for phaseolotoxin biosynthesis. This repression was mediated by a protein present in both toxigenic and nontoxigenic strains of *P. syringae* and required the coordinated participation of *phtA, phtB*, and *phtC* products to carry out efficient argK repression (Aguilera et al. 2012). *Burkholderia glumae* (rice seedling and grain rot) produces the phytotoxin, toxoflavin, composed of two proteins, TRP-1 and TRP-2. Tox– mutants are nontoxigenic and also nonpathogenic to rice. Toxoflavin retarded the growth of leaves and roots of rice plants, in addition to induction of chlorosis in panicles in the grain rot phase. TRP-1 protein was encoded by the *toxA* gene. Mutants disrupted in the *tox* operon did not produce the toxoflavin and did not induce typical chlorotic lesions on infected rice panicles. The *tox* operon was indicated to be polycistronic by RT-PCR assay. The results suggested that toxoflavin might be synthesized in part through a biosynthetic pathway common to the synthesis of riboflavin (Suzuki et al. 2004). '*Candidatus* Liberibacter solanacearum' (Lso) (potato zebra chip, ZC) has a reduced genome, and its proteins were predicted to be prone to instability and aggregation. The soluble dihydrodipicolinate synthase was successfully overexpressed with GroEL/GroES, whereas soluble pyruvate kinase was successfully overexpressed with GroEL/GroES, dnaK/dnaJ/grpE, or a trigger factor. In addition, both '*Ca.* L. solanacearum' enzymes were active and had canconical tetrameric oligomeric structure in solution, consistent with other bacterial orthologs (Gilkes et al. 2019).

5.2.3.3.5 Expression of Pathogen Genes and Host Plant Genes

Bacterial pathogens possess genes that regulate different metabolic processes in in vitro and in planta environments to attain required population levels for their growth, reproduction, and infection of their respective host plant species and vectors. *Xylella fastidiosa* (grapevine PD) is transmitted by sharpshooter vectors. *X. fastidiosa* has to regulate traits crucial for both virulence on grapevine and colonization of vector tissues via production of a fatty acid signal molecule, DSF, whose production is dependent on *rpfF*. *X. fastidiosa rpf* mutants exhibited increased virulence to plants, but they could not be transmitted by insect vector from infected plant to healthy plant. Protein coding genes whose expression was significantly different between the wild-type and an *rpfF* mutant were identified. The number of protein-coding genes downregulated in the *rpfF* mutant was less, compared with the wild-type strain. The RpF function was required for regulation of 11 regulatory and δ factors, including *rpfE, yybA, PD1177, glnB, rpfG, PD0954, PDO199, PD2050, coir, rpoH*, and *rpoD*. RpF may generally be required for regulation of genes involved in attachment and biofilm formation, and suppressing most type IV pili and gum genes (Wang et al. 2012).

Xanthomonas oryzae pv. *oryzicola (Xoc)* (rice bacterial leaf streak, BLS) has virulence factors regulated by a DSF-dependent QS system. The pathogenicity-related gene *hshB* of *Xoc* encoded a hydrolase with a putative signal peptide and this gene was relatively conserved in the genus *Xanthomonas*. Mutation of *hshB* drastically impaired the extracellular protease activity, extracellular polysaccharide production, growth in minimal medium, and resistance to oxidative stress and bismerthiazol of *Xoc*. The *hshB* transcription was positively regulated by *clp* and DSF and induced by poor nutrition (Zhao et al. 2012). The presence of putative six hypothetical secreted proteins, absent in *Xoo*, were detected in *Xoc* strain BLS256. The *epv* was highly conserved in *Xanthomonas* spp. (except in *Xoo*). In-frame deletion of *epv* in *Xoc* strongly impaired virulence and EPS production. Expression of both *gumB* (a gene encoding EPS xanthan biosynthesis export protein) and another virulence-related gene, *pgK* (encoding phosphoglycerate kinase) were significantly reduced in the *epv*-deletion mutant, compared with the WT strain RS105. Further, *epv* was positively regulated by both diffusible signal factor and global regulator Clp in *X. oryzae* pv. *oryzicola* (Qian et al. 2012).

Transcriptional response of sweet orange cv. Valencia to infection of *Xanthomonas citri* subsp. *citri* wild-type (WT) and pthA4 mutant was studied. PthA4 bound to the effector, binding element (EBE) of CsLOB1 and activated gene expression of this susceptibility gene. The PthA4-mediated infection significantly upregulated expression of a variety of genes involved in cell wall degradation and modification, DNS packaging, G-protein, protein synthesis, sucrose metabolism, and cell division functions. On the other hand, the downregulated genes were mainly enriched in photosynthesis, transport, secondary metabolism, cytochrome P450, and various plant defense-associated mechanisms (Hu et al. 2016). Molecular interaction of *Xanthomonas translucens* with wheat was explored by studying global gene and protein expression changes in both leaves and roots of wheat (*Triticum aestivum*). Simultaneous analyses of above- and below-ground tissues were performed to identify common and specific changes in plant responses. Massive downregulation of photosynthesis, confirming the chloroplasts as critical signaling was inferred. Roots responded remarkably to foliar infection by *X. translucens* and their responses significantly differed from those locally triggered in infected leaves. Roots could be a site of energy production and synthesis of various secondary metabolites. Further, accumulation of JA, pipecolic acid, and/or the downstream mediator of H_2O_2 as long distal signals from infected leaves to roots paved the way for effective suppression of disease development in wheat (Garcia-Seco et al. 2017).

Expression analysis of selected defense-response genes and PTI marker genes in heirloom tomato challenged with virulent *Pseudomonas syringae* pv. *tomato* DC3000 strain (bacterial speck disease) was performed. Transcript levels of defense response genes, including *SlPR1a, SlPR-Q'b* (ß-1,3-glucanase), *SlGST* (glutathione-*S*-transferase), and peroxidase, were upregulated in the resistant cultivars. The PTI marker genes *SlPT15* (encoding pathogen-inducible ethylene response element-binding protein-like transcription factor) and *SlFLS2* (leucine rich-repeat receptor-like kinase) were strongly upregulated in the incompatible reaction with the resistant cultivar. In contrast, transcripts from all tested genes were downregulated in the compatible interaction in susceptible cultivars. Defense response and PTI marker gene were induced in the early infection process at 3 days postinoculation, and such induction was consistent with lower levels of disease severity in resistant cultivars than in susceptible cultivars (Veluchamy and Panthee 2015). Growth and resistance of tomato to *Pseudomonas syringae* pv. *tomato (Pst)* DC3000 were determined in 80 gene knockouts, identified earlier through genome-wide association mapping of global *Arabidopsis thaliana* populations. *Pst* DC3000 exhibited two distinct growth patterns in all gene knockouts, showing maximum growth after 4 or 7 days after inoculation (*dai*). The differential abundance of *Pst* DC3000 in the multiple gene knockouts indicated the presence of a number of new genes with potential role for defense against *Pst* DC3000 (Saleem et al. 2017). In another investigation, the virus-induced gene silencing (VIGS) approach was applied to identify genes involved in host basal defense response to *Clavibacter michiganensis* subsp. *michiganensis (Cmm)*, infection on *Nicotiana benthamiana*. A preliminary screen comprising 160 genes from tomato involved in defense-related signal identified genes involved in host-pathogen interactions. Expression of a set of 14 genes and three additional targets involved in defense-related signal was suppressed in the large-scale VIGS experiments. The effects of disease development and bacterial growth were determined using *Cmm-N. benthamiana* pathosystem. Disease susceptibility and bacterial populations in planta were enhanced by silencing genes encoding *N. benthamiana* homologs of ubiquitin-activating enzyme, snaking-2, extension-like protein, divinyl ethersynthase, 3-hydroxyl-3-methyl glutaryl-coenzyme A reductase-2, and Pto-like kinase (Balaji et al. 2011). Techniques for genetic manipulation of *Clavibacter michiganensis* subsp. *michiganensis* (*Cmm*) (tomato bacterial canker) were employed to investigate the host-pathogen interaction. Gene functional analysis of *C. michiganensis* knockout and complementation experiments on *celA, chpC*, and *chpG* was performed. Cell wall-degrading enzymes and serine proteases played important role in the development of wilt and canker symptoms (Hwang et al. 2019), whereas pbpC, an enzyme involved in peptidoglycan assembly was associated with stress resistance (Chen et al. 2021).

The clustered regularly interspersed short palindromic repeats (CRISPR/CRISPR-associated Cas) bacterial immune system capable of protecting against exogenous nucleic acids and bacteriophage invasion, was useful for genetic manipulation. The CRISPR/Cas-mediated defense system involves adaptation in which foreign DNA (protospacers) are acquired and incorporated into the CRISPR locus, a portion of the bacterial genome that acts as a repository to recognize the foreign DNA; expression of which involves transcription of the CRISPR RNA (crRNA) that acts as a guide to identify invading nucleic acids; and interference in which the Cas9-CrRNA complex interacts with foreign DNA, resulting in cleavage of the target DNA by Cas9 endonuclease (Marraffini 2015; Mohanraju et al. 2016). Gene functional analysis could be hastened by employing efficient gene editing technique and additional selectable markers. Highly efficient unmarked CRISPR/Cas9-mediated gene editing protocol for *C. michiganensis* was developed coupling expression of Cas9 and single-guide RNA (sgRNA) with homology-directed repair templates and negative selectable marker coda::upp within a single plasmid. CRISPR/Cas9-mediated transformation could be utilized for site-directed mutagenesis, in which an A to G point mutation was introduced at this 128th nucleotide of *C. michiganensis rpsL* gene to generate a streptomycin-resistant mutant. Complete gene knockout in which deletion of *Cm celA* or *katA* genes resulted in transformants that lacked cellulase and catalase activity respectively. The unmarked CRISPR/Cas9-mediated system was more useful and convenient, since it permitted reuse of template plasmid repeatedly to facilitate editing of multiple genes (Chen et al. 2022).

5.3 BIOLOGY OF PHYTOPLASMAL PATHOGENS

Pathogens causing yellows type of diseases, like aster yellows, were recognized as a distinct group of plant pathogens and called MLOs, following the investigations of Doi et al. (1967) and Ishiie et al. (1967). The trivial name

'Phytoplasma' was officially adopted in 1994 to replace the earlier designation 'mycoplasma-like organism' (MLO). Phytoplasmas associated with plant diseases are transmitted by leafhoppers, planthoppers, and psyllids. In plant host, the phytoplasmas are restricted to phloem tissues. On the other hand, they can colonize salivary glands, hemolymph, and other tissues of the vector insects. After acquisition, the phytoplasmas are circulated in the body of the vector and reach salivary gland before transmission to healthy plants.

5.3.1 Characteristics of Phytoplasmas

Despite intensive research efforts, none of the phytoplasmas could be grown in an artificial medium and, therefore, their cultural characteristics are not known. Phytoplasmas are restricted to phloem tissues of infected plants and exhibit wide variations in size and shape (pleomorphic) of phytoplasmal cells present in phloem tissues. Phytoplasmas are insensitive to penicillin, but sensitive to tetracyclines. In contrast, bacterial cells have distinct cell wall and are sensitive to penicillin. Phytoplasmas lack a cell wall and hence, they can pass through smaller pores, even though the diameter of a phytoplasmal cell may be larger (Musetti and Favalli 2004). *Spiroplasma citri* (citrus stubborn and corn stunt) could be grown in artificial medium and its pathogenicity was proved using vector insects (Davis and Lee 1982). Serological and nucleic acid properties have been frequently used for their detection and identification. PCR primers based on the sequences of the highly conserved 16S rRNA gene are employed for the identification of a wide range of phytoplasmas. Phytoplasmas cause certain characteristic symptoms of diagnostic value. Phytoplasmas induce general stunting or dwarfing of whole plants or plant parts. Chlorosis and smalling of leaves are associated with several phytoplasmal diseases. Antholysis of floral parts leading to virescence, phyllody and proliferation of different floral parts that are transformed into leaf-like structures is a characteristic effect of infection by phytoplasma. Partial or total sterility of infected plants may be seen frequently in rice, legume, vegetable, and oilseed crops. Proliferation of axillary buds and a large number of thin shoots may be commonly observed in rice and sugarcane. *Spiroplasma citri*, causing citrus stubborn disease, and phytoplasma causing AY, PD, tomato big bud (TBB) diseases, and clover club leaf (CCL), were visualized in the sieve elements of infected plants by treating the midribs with macerating enzymes (cellulase and macrozyme) and separating the sieve elements. The helical spiroplasma, pleomorphic phytoplasmas, and slender rod-shaped CCL agent could be visualized under dark-field microscope. Fluorescent DNA binding dye 4,6-diamidino-2-phenylindole-HCl (DAPI) was effective in revealing the presence of phytoplasmas in phloem tissues of infected plants (Kazmi et al. 2007). The nature of the cause of coffee crispness (Crespera) disease occurring in Colombia was explored using the DAPI fluorescence test. Petiole sections from diseased leaves were stained with DAPI. Characteristic fluorescence was observed in sections of diseased petioles, whereas no fluorescence was emitted from comparable healthy tissues. Further, ultrathin sections of glutaraldehyde-fixed petioles of diseased leaves revealed the presence of pleomorphic bodies in phloem cells, indicating the infection of coffee plants by phytoplasma (Galvis et al. 2007).

Variations in the strains of '*Candidatus* Phytoplasma mali' were observed when they were present in mixed infections in apple. Occurrence of multiple infections by strains of '*Ca.* P. mali' could be detected by single-strand conformation polymorphism (SSCP) analysis (Seemüller et al. 2010). Phylogenetic analysis revealed that '*Candidatus* Phytoplasma aurantifolia' phytoplasma isolates from symptomatic and asymptomatic acid lime trees had identical 16S rRNA sequences belonging to subgroup IIB. Quantitative PCR analysis indicated wide variations in the phytoplasma concentrations in leaves showing disease symptoms, whereas asymptomatic leaves contained very low pathogen DNA contents in the same tree. Acid lime trees from areas without disease symptoms contained medium level of pathogen population (Al-Ghaithi et al. 2017). Restriction fragment length polymorphism (RFLP) analyses of PCR-amplified 16S rDNA sequences using several restriction endonucleases such as *Alu*I, *Hae*III or *Rsa*I were useful for identification, differentiation and classification of phytoplasmas. Characteristic RFLP patterns for different strains could be resolved using agarose or acrylamide gel electrophoresis. Phytoplasma groups identified based on comprehensive RFLP analyses with different restriction enzymes have been considered with phylogenetic groups delineated based on sequence data. Based on extensive RFLP analyses of phytoplasma 16S rDNA sequences, phytoplasmas have been classified into different groups and subgroups (Lee et al. 2007). Other methods, including heteroduplex mobility assay (HMA), based on the 16S-23S ISR (intergenic spacer region) have also been employed to differentiate 16SrI subgroups (Wang and Hiruki 2005). The terminal restriction fragment length polymorphism (T-RFLP) technique based on the 23 S rRNA gene sequences, which distinguish the various 16SrI groups, was applied for identification of phytoplasmas (Hodgetts et al. 2007). Further, less conserved sequences such as ribosomal protein gene clusters or the 16S-23S rRNA intergenic spacer region have also been employed for identification of subgroups. They were consistent with genomic subclusters delineated earlier based on DNA sequence homology investigations with probe hybridization and later based on in silico RFLP analysis. Phytoplasma lineages are referred to as '*Candidatus* Phytoplasma species' (Davis and Sinclair 1998; Lee et al. 2007).

5.3.2 Dissemination of Phytoplasmas

5.3.2.1 Spread through Planting Materials

Phytoplasmas primarily spread through infected planting materials such as tubers, corms, suckers, setts, and stem pieces obtained from infected mother plants or asymptomatic transplants. Indexing of mother plants by different method of detection is advocated to restrict dissemination of phytoplasmas through planting materials. Certification programs are in operation in several countries to ensure availability of pathogen-free planting materials (Narayanasamy 2001, 2011). In a few cases, involvement of parasitic dodder in transmission of the phytoplasma under experimental conditions has been demonstrated, as in the case of Italian clover phyllody phytoplasma (ICPh). The phytoplasma was transmitted to dodder *Cuscuta campestris* from its original host *Chrysanthemum leucanthium* and then to periwinkle (*Catharanthus roseus*) and then to *Chrysanthemum carnatum*. PCR assays confirmed the presence of ICPh in these additional hosts (Webb et al. 1999). Transmission of poinsettia branch-inducing phytoplasma (PoiBIp) and its association phenotype during cutting propagation was reported. Economic values of all commercial poinsettias are due to their small bushy shape induced by grafting a scion infected poinsettia branch-inducing strain of '*Ca*. Phytoplasma pruni' (Neriya et al. 2011). The densities of phytoplasmas were consistently lower in tissues located at the top of plants, regardless of tissue types. Commercially raised stock plants contained significantly different levels of PoiBIp load. Stock plants with low infection densities tended to produce smaller proportions of plants exhibiting higher degrees of branching both before and after pinching. The results suggested that uneven distribution of PoiBip within and among stock plants might result in production of cuttings with variable phytoplasma densities affecting phenotypic uniformity (Lee et al. 2020).

5.3.2.2 Dissemination of Phytoplasmas through Biological Vectors

Phytoplasmas depend on their transmission from infected plants to healthy plants largely through insects and this appears to be the only natural mode of transmission of phytoplasmas. The insects acquire the phytoplasmas after feeding on infected plants. The presence of phytoplasmas in different organs of the vector may be detected by various immunoassays and nucleic acid-based techniques. Grapevine Flavescence dorée (FD) phytoplasma was detected in the salivary glands of leafhopper vector *Euscelidius variegatus* (Lherminier et al. 1989). Nucleic acid–based methods have been applied to detect phytoplasmas in their vectors. The presence of Italian clover phyllody phytoplasma (ICPh) was detected in the salivary glands and midgut of the leafhopper *E. variegatus* by in situ hybridization procedure (Webb et al. 1999). Papaya bunchy top disease (BTD) phytoplasma could be detected in the leafhopper *Empoasca papayae* obtained from BTD-affected papaya fields in Cuba, indicating the availability of inoculum for new infections through the infective insects (Arocha et al. 2008). '*Candidatus* Phytoplasma mali' (*Cpm*) (AP) is transmitted by two phloem-feeding psyllids, *Cacopsylla picta* and *C. melanoneura*, in Italy (Mayer et al. 2009). The pathogen multiplies in salivary glands and is translocated into another healthy plant via saliva while feeding (Hogenhout et al. 2008). Vertical transmission or transovarial transmission of *Cpm* in the psyllid vectors was examined. The pathogen was detected by SYBR Green real-time PCR format in egg batches with each developmental stage and reached the peak concentration in F_1 adults. A significant increase was detected between the titer of fourth and fifth instar nymphs and between fifth instar nymphs and F_1 adults. No detectable differences in the titers of '*Ca*. P. mali' in parental females and males were detected (Mittelberger et al. 2017).

Detection of the phytoplasma in the vector may not provide evidence for its vectoring ability to transmit the phytoplasma to healthy plants. Hence, as an alternative to the biological transmission assays was developed. PCR detection of phytoplasmas in the insect feeding medium (sucrose) after allowing the insect to feed on the medium was adopted. Transmissibility of FD phytoplasma by *Euscelidius variegatus* was positively correlated with its detection of FD in the insect feeding medium by PCR assay. PCR assay detected PD phytoplasma in the feeding medium after allowing the field-collected *Orosius albicinctus* and *Anaceratagallia laevis* to feed (Tanne et al. 2001). Sugar beet virescence phytoplasma was detected and quantified using real-time TaqMan PCR assay in DNA extracts of beet leafhopper, *Circulifer tenellus*. The pathogen could be detected in the extracts of single or groups of five beet leafhoppers and the concentration of phytoplasma was variable in individual leafhopper (Crosslin et al. 2006). Leafhoppers and treehoppers captured in carrot fields were tested by nested PCR assay for the presence of AY phytoplasma. The leafhoppers *Macrosteles quadripunctulatus, M. sexnotatus*, and *M. laevis* carried the phytoplasma belonging to the subgroups 16SrI-B and 16SrXII-A subgroups. *M. quadripunctulatus* and *M. sexnotatus* were positive for AY phytoplasma, whereas *M. laevis* carried stolbur phytoplasma (Duduk et al. 2008). Corn stunt disease phytoplasma *Spiroplasma kunkelii* is transmitted by phloem-feeding leafhopper *Dalbulus maidis*. A field-deployable real-time PCR procedure detected *S. kunkelii* in leafhopper tissues as well as in corn plant tissues infected by corn stunt pathogen (Wei et al. 2006). PCR and RT-PCR assays showed the presence of ROLD phytoplasma using pathogen-specific primers for amplification of DNAs from infected plants. *Nephotettix cincticeps* was the primary vector transmitting the phytoplasma. *Inazuma dorsalis* reported as vector earlier was rare in the affected fields (Li et al. 2015). The presence of the phytoplasma in leafhopper and positive transmission to broccoli seedlings were confirmed based on the presence of fragments corresponding to 16S rDNA of the phytoplasma (Kreyci et al. 2018). '*Candidatus* Phytoplasma pyri' was associated with PYLR disease. The infective potential of the leafhopper *Cacopsylla pyri* was assessed, using an artificial feeding medium for insects and healthy peach trees for inoculation with phytoplasma by infective insects. Insect species captured inside peach plots carrying '*Ca*. P. pyri' was *Cacopsylla pyri*. High percentages of peach trees exposed to *C. pyri* were infected with the phytoplasma (Sabaté et al. 2018).

5.3.3 Sources of Inoculum

'*Candidatus* Phytoplasma solani' (grapevine Bois noir, BN phytoplasma) and its principal vector *Hyalesthes obsoletus* appeared to originate from the Mediterranean region. Multilocus sequence typing of *tuf*, *vmp1*, and *stump* genes of isolates of '*Ca*. P. solani' was applied to track from a tentative reservoir plant (inoculum source) through an associated vector population to the infected grapevine. The molecular characterization and phylogenetic analyses of methionine aminopeptidase (*map*) gene sequences of Flavescence dorée phytoplasma (FDP) affecting grapevine and related strains revealed a high prevalence of the *map*-type FD2 in grapevines and an *Orientus ishidae* pool. The ecological cycle of FDP might not be dependent exclusively on the grapevine-specific feeding diet of *S. titanus* but also on other insect vector(s) and/or plant host(s) (Casati et al. 2017).

The molecular basis of the interaction between *Spiroplasma citri* (citrus stubborn disease) and the leafhopper vector species was studied. The propagative nature of the relationship between *S. citri* and *Circulifer haematoceps* involves multiplication of the pathogen in the leafhopper cells. In *S. citri*, an adhesion-related protein named P89 was directly involved in the spiroplasma–vector cell interaction (Yu et al. 2000). Proteins of *C. haematoceps* were screened by putative *S. citri*-binding molecules. Spiroplasma proteins showed affinity for leafhopper proteins. The insect proteins were identified as high mannose N-glycoprotein, which facilitated interaction of the glycoprotein-lectin type with *S. citri* proteins. Lectin detection in *S. citri* showed the presence of only one protein of 24-kDa. One spiroplasma protein with a similar molecular mass of 24-kDa had an insect protein-binding capacity. This protein, spiralin, was the most abundant membrane protein in *S. citri*. Experiments with purified spiralin and insect glycoprotein confirmed the binding of spiralin to insect glycoproteins. The results indicated a key role for spiralin in the transmission of *S. citri* by mediating spiroplasma adherence to epithelial cells of the insect vector gut or salivary gland (Killiny et al. 2005). The molecular mechanism underlying the specificity of transmission of '*Candidatus* Phytoplasma asteris' OY strain line W(OY) by vector was investigated. Line W(OY) contained a cell surface membrane protein, antigenic membrane protein (Amp), abundantly. Localization of phytoplasma to the microfilaments of the visceral smooth muscle surrounding intestinal tract of vectors was visualized. Amp forms a complex with three insect proteins, actin, myosin heavy chain, and myosin light chain. The presence of Amp-microfilament complexes was detected only in all OY-transmitting leafhopper species. The results suggested that formation of the Amp-microfilament complex might be a critical factor in influencing the transmission of phytoplasmas by insects (Suzuki et al. 2006).

5.3.4 Process of Infection by Phytoplasmas

All phytoplasmas have to be introduced into the susceptible host plant cells by their biological vector(s) as a first step in the process of infection. An adhesion-like gene was identified in *Spiroplasma kunkelii* (corn stunt disease), which could be cultured in artificial medium (Wei et al. 2006). Phytoplasmas have a unique and complex cycle that involves colonization of the plant phloem and various tissues of organs of vectors which pose different conditions. Further, many phytoplasmas have wide host ranges, including plants grown in different ecosystems. The dynamic architecture of phytoplasma genomes may account for variations in their genome sizes due to the presence of repetitive elements of various types and adaptation of phytoplasmas to diverse environments existing in the plant and insect hosts. The number of genes with a role in phytoplasma-host interactions has been identified. Genes encoding surface membrane proteins and effector proteins have important roles in the successful establishment of infection by phytoplasmas. They have great ability to dramatically alter their gene expression upon switching between plant and insect hosts (Marcone 2014).

5.3.4.1 Pathogenicity of Phytoplasmas

Peanut witches' broom (PnWB) phytoplasma infected periwinkle (*Catharanthus roseus*) for the purpose of studying the biology of phytoplasmas. A differential display (DD) strategy was applied to isolate periwinkle cDNAs, which were differentially expressed following infection by PnWB phytoplasma. Seven DD transcripts were identified as truly differentially expressed following PnWB phytoplasma infection. One (homologous with *phi.1* gene) of them was upregulated, while others were downregulated. Four downregulated genes shared homology with genes encoding the *psaDa* gene, ML domain gene, eukaryotic translation initiation factor SUS1 gene, and plastidic aldolase NAPLDP1 gene (Chen and Lin 2011). The role of AAA+ proteins of apple proliferation phytoplasma '*Candidatus* Phytoplasma mali' in virulence and suppression of virulence was investigated. Full-length deduced protein sequences of AAA+ ATPase AP460 of single- and multiple-strain accessions were compared. Two regions could be distinguished in this protein. Region 1, located in the N-terminal part, is characterized by 14 highly conserved substitutions associated with suppression of virulence. In region 2, more diverse proteins located in the C-terminal part, highly conserved substitutions associated with two types of virulence, were identified. Two different mechanisms of suppression existed, of which the supposed effect on ATP function appeared to be more important than the suppression-associated substitutions in the region (Seemüller et al. 2018).

5.3.4.2 Colonization of Plant Tissues

The molecular mechanisms of movement of phytoplasmas from inoculated site to other plant organs were investigated. Distribution of lethal yellowing (LY) phytoplasma in different parts of coconut palm was studied by employing PCR assay for detection of phytoplasma DNA. The stem, young leaves, inflorescences, and stem apex contained high pathogen DNA contents, whereas intermediate leaves and roots without apex had low pathogen DNA concentrations. LY phytoplasma DNA was not detected in mature leaves. All parts of asymptomatic plants except the stem had lower contents of LY phytoplasma DNA, compared to symptomatic plants. Stems of asymptomatic plants showed accumulation of phytoplasma consistently. The phytoplasma might move from photosynthetic source tissues to sink tissues via phloem mass flow mass process. Accumulation of phytoplasma in the (lower) stem prior to symptom expression suggested that the pathogen might move first from leaves which are fed by the infective vectors and after sufficient increase in population might move to other plant parts. Nested PCR analysis indicated the presence of LY phytoplasma in 28% embryos tested. Phytoplasma DNA was detected in the embryos from fruit of any fruiting branches, regardless of age. Seedlings growing from infected and asymptomatic plants did not express disease symptoms (Oropeza et al. 2011).

The sequence and nature of events that establish the relationship of phytoplasmas with their respective host plants and vector insects have been studied in some pathosystems. Ultrastructural reorganization of sieve elements (SEs) in tomato plants infected by stolbur disease caused by '*Candidatus* Phytoplasma solani' was investigated. The presence of membranous connections between phytoplasma and SE plasma membrane and a close association of pathogen cells with sieve element reticulum (SER) and SE cytoskeleton were observed (Buxa et al. 2015). Following introduction of phytoplasmas directly inside by SEs by stylets of vector insect, they might float freely in the SE lumen as round or elongated cells. The phytoplasmas in the lumen exhibited actin labeling. The presence of host cell actin filaments on the phytoplasmas to adhere to SE structures and then another to the SE plasma membrane was detected. Free-floating, dividing phytoplasmas were observed in the lumen. The adhesion structure that anchored the phytoplasma body to SE plasma membrane was tubular and oriented perpendicularly to the SE plasma membrane (Musetti et al. 2016). A putative membrane protein containing a Mollicutes-adhesion motif (MAM) was identified in onion yellows phytoplasma (Tran-Nguyen et al. 2008). The presence of actin on the SE plasma membrane in the proximity of the adhesion structure indicated an important role in anchoring (Hafke et al. 2013). Actin was localized at one side of a large portion of the phytoplasma bodies floating in SE lumen. Actin was present simultaneously on phytoplasma membrane and SE plasma membrane. Actin-binding ability seemed to have multiple functions in phytoplasma spread with the susceptible plant hosts (Buxa et al. 2015).

Floral transition in plants is a critical event that triggers the switch from vegetative growth to reproductive development. Meristem switching genes involved in pathogen-induced alterations of growth pattern (PIAGP) have been identified in the tomato-multiple top (PPT) phytoplasma pathosystem. Single gene mutants and transgenic tomato lines that exhibit phenotypes similar to the disease symptoms were selected, and comparative analysis of expression profiles of selected genes in healthy and PPT phytoplasma-infected plants were performed through qRT-PCR assay. Each type of symptom induced by phytoplasma was associated with shoot apical meristem modification. The phases include premature floral meristem termination, suppressed floral meristem initiation, delayed conversion of vegetative meristems to inflorescence meristem, and repetitive initiation and outgrowth of lateral vegetative meristem. The pathogen-induced changes in growth patterns were correlated with transcriptional reprogramming of key meristem switching genes (Wei et al. 2013). The pattern of colonization and spread of witches' broom disease of Mexican lime (WBDL) caused '*Candidatus* P. aurantifola' within the lime trees was investigated. After gaining entry into the susceptible lime plants, '*Ca.* P. aurantifolia' moved downward into the roots as well as upward into the apical leaves. Initially the concentration of the pathogen was high in apical leaves, but a decrease in concentration occurred before establishment of infection and subsequent expression of symptoms. Severe symptoms of WBDL appeared in the grafted branches at 10 months after inoculation (Mazraie et al. 2019).

5.3.4.3 Pathophysiology of Plants Infected by Phytoplasmas

5.3.4.3.1 Effects on Photosynthesis

Photosynthesis in plants is a vital function involved in the production of carbohydrates that are required for the growth of both host plants and phytoplasmas. Diseases induced by phytoplasmas exhibit yellowing or chlorosis to differing degrees, indicating the adverse effects of the pathogen on photosynthesis. Reduction in photosynthesis in grapevines plants infected by FD phytoplasma was recorded. Deregulation in some stomata control of leaf gas exchange during FD infection was indicated. In FD-infected plants net photosynthesis and transpiration gradually decreased during the season, the effect being less in the absence of water stress. FD-infected leaves had higher nonphotochemical quenching than healthy ones. Metabolic, but not stomatal, leaf gas exchange limitation in FD-infected and recovered grapevine were the plant responses to FD disease (Vitali et al. 2013). Physiological and biochemical changes in sweet orange (*Citrus sinensis*) infected by phytoplasma were assessed. Phytoplasma induced yellowing of leaves, stunting and rolled leaves, and fruits, stunted roots of plants and witches' broom symptoms. Primary metabolites, including chlorophyll *a, b* and total chlorophyll contents, were significantly reduced. On the other hand, carotenoid, proline and soluble protein were increased, whereas sugar, hydrogen peroxide, and malondialdehyde were decreased in phytoplasma-infected plants. Secondary metabolites such as phenolics, glycine, betaine, and anthocyanin were increased whereas ascorbic acid content registered a reduction in infected plants. Activities of antioxidative enzymes, ascorbate peroxidase, superoxide dismutase, catalase, and peroxidase activities increased in phytoplasma-infected plants (Rasool et al. 2020).

Effects of maize bushy stunt phytoplasma (MBSP) on host selection behavior of the leafhopper vector *Dalbulus maidis* were assessed. Choice assays of contrasting leaves of healthy (mock-inoculated) vs infected maize were performed during asymptomatic and symptomatic phases of plant infection, with leafhopper males or females previously exposed to infected plants or not. During asymptomatic phase of maize infection, the bacteriliferous insects preferred leaves of healthy plants. The symptomatic phase of infection triggered a pull-push effect on non-bacteriliferous females that were first attracted to symptomatic leaves, but hours later moved to healthy leaves. Non-bacteriliferous males initially preferred symptomatic leaves and later became equally distributed between treatments. Oviposition of leafhoppers reduced drastically on symptomatic leaves. Changes in vector behavior induced by MBSP favored primary spread of the phytoplasma, since bacteriliferous females preferred healthy leaves at early (asymptomatic) stages of maize plants. At later stages, secondary spread might be favored, because non-bacteriliferous females were initially attracted to symptomatic leaves, permitting acquisition and subsequent transmission of the pathogen when they moved to healthy plants (Ramos et al. 2020).

5.3.4.3.2 Effects on Hormone Metabolism

Phytoplasmas induce general stunting of plants, smalling and bunching of leaves, and modification of floral parts into leafy structures in many plant hosts, indicating hormonal imbalances in the metabolism of infected plants. Endogenous levels of gibberellic acid (GA) were significantly reduced in infected plants. Reduction in GA3 content in diseased plants was correlated with downregulation of genes responsible for biosynthesis of bioactive GAs (GA2ox1 – GA3ox1) and genes involved in formation of GA precursors. GA content of infected plants could be restored by external application of GA to levels comparable to those in healthy plants and to attenuate the intensity of big bud symptom induced by potato purple top (PPT) phytoplasma. Plants infected by phytoplasma had prolonged low GA2ox1 and GA3ox1 under GA deficiency conditions (Ding et al. 2013). Recovery of plants infected by phytoplasma is observed to a different extent. Spontaneous complete and stable remission of symptoms of FD in grapevine occurred under some conditions. The molecular basis of this phenomenon of recovery from FD-infection of grapevine was investigated. Several genes coding for enzymes in H_2O_2 metabolism in the ascorbate-glutathione cycle, defense responses, and biosynthesis of hormones were identified. Large amounts of H_2O_2 were detected in tissues of 2-years-recovered R plants, essentially linked to an upregulation of genes involved in the production of H_2O_2 (germin-like protein and glycolate oxidase), whereas overexpression of some scavenging genes reduced the quantity of H_2O_2. FD-infected plants reacted to phytoplasma infection, with SA-mediated signaling, although this response was not effective against FD pathogen infecting grapevine (Gambino et al. 2013). '*Candidatus* Phytoplasma solani' infected tomato and several weed species. Phytoplasmas colonize the sieve elements of susceptible plants causing drastic changes in phloem function and impairment of assimilate translocation. Infected plants exhibited yellowing of shoots, leaf curling and general stunting of plants. Infection of citrus by '*Ca.* P. solani' and Fe starvation resulted in development of chlorotic leaves and altered thylakoid organization in infected plants. Fe starvation was responsible for additive effects in chlorophyll content, and leaf chlorosis RNA sequence analysis showed that both infection by phytoplasma and Fe deficiency resulted in altered expression of a large subset of common genes involved in photosynthetic light reaction, porphyrin/chlorophyll metabolism, and flowering control. In Fe-deficient plants, phytoplasma infection perturbed Fe deficiency response in roots (Buoso et al. 2019).

Phytoplasmas are restricted to SE of infected plants involved in the translocation of nutrients and a broad spectrum of signals. In *Arabidopsis*, two sieve element occlusion-related (SEOR) genes, AtSEOR1 (At3go1680) and AtSEOR2 (At3go1670), were essential for formation of E protein filaments through a heteromeric assemblage of two SEOR proteins. *Atseor1ko* mutant line of *Arabidopsis thaliana* supported considerably lower population of phytoplasma, though the phloem flow was not affected. The effects of chrysanthemum yellows phytoplasma infection on phytohormones showed that increased salicylic acid (SA) and abscisic acid (ABA) levels and decreased contents of indole-3-acetic acid (IAA) were present in plants infected by phytoplasmas. Phytohormone levels were monitored in midribs and laminar tissues of wild-type *Atseor1ko* and *Atseor2ko Arabidopsis* lines at the commencement and completion of phytoplasma infection. Proteins AtSEOR1 and AtSEOR2 occurred as conjugates in the form of filaments in sieve elements of *A. thaliana*. A reduction in the phytoplasma population was present in plants infected by defective-mutant *Atseor1ko*. The impact of SEOR2 was revealed in *Atseor1ko* plants by a lack of detectable phytoplasma population at an early stage of infection (symptomless plants) and a lower phytoplasma titer at a later stage (after symptom expression). Time-correlated HPLC-MS/MS-based phytohormone analysis revealed increased jasmonate levels in midribs of *Atseor1ko* plants at an early stage of infection and appreciably enhanced levels of IAA and ABA at early and late stages (Bernardini et al. 2020).

REFERENCES

Abrahamian P, Timilsina S, Minsavage GV, Sushmita KC, Goss EM et al. (2018) The type III effector AvrBsT enhances *Xanthomonas perforans* fitness in field-grown tomato. Phytopathology 108: 1355–1362.

Abramovitch RB, Kim YJ, Chen S, Dickman MB, Martin GB (2003) *Pseudomonas* type III effector AvrPtoB induces plant disease susceptibility by inhibition of host programmed cell death. EMBO J 22: 60–69.

Abramovitch RB, Martin GB (2004) Strategies used by bacterial pathogens to suppress plant defenses. Curr Opin Plant Biol 7: 356–364.

Aguilera S, Lopez-Lopez K, Nieto Y, Garciduenas-Pina R, Hernandez-Guzman G et al. (2007) Functional characterization of the gene cluster from *Pseudomonas syringae* pv. *phaseolicola* NPA3121 involved in synthesis of phaseolotoxin. J Bacteriol 189: 2834–2843.

Aguilera S, Torre-Zavala SD, Hernandez-Flores JL, Murillo J, Bravo J et al. (2012) Expression of the gene for resistance to phaseolotoxin (*argK*) depends on the activity of genes *phtABC* in *Pseudomonas syringae* pv. *phaseolicola*. PLoS ONE 7: e46815. https://doi.org/10.1371/journal pone.0046815

Akhavan A, Bahar M, Saeid G, Lak MR (2009) Factors affecting epiphytic populations of *Xanthomonas axonopodis* pv. *phaseoli* in epidemiology of bean common blight. J Sci Technol Agric Naur Resour 13: 265–277.

Aldridge P, Hughes T (2002) Regulation of flagellar assembly. Curr Opin Microbiol 5: 160–165.

Al-Ghaithi AG, Al-Sadi AM, Al-Hammadi MS, Al-Shariqi RM, Al Yahyai RA et al. (2017) Expression of phytoplasma-induced witches' broom disease symptoms in acid lime (*Citrus aurantifolia*) trees is affected by climatic conditions. Plant Pathol 66: 1380–1388.

Amino A (2010) Bacterial adhesions to host components in periodontitis. Periodontology 2000(52): 2–37.

Ammar E-D, Shatters RG Jr, Hall DG (2011) Localization of '*Candidatus* Liberibacter asiaticus' associated with citrus huanglongbing disease, in its psyllid vector using fluorescence in situ hybridization. J Phytopathol 159: 726–734.

Amrein H, Makart S, Granado J, Shakya R, Schneider-Pokorny J et al. (2004) Functional analysis of genes involved in the synthesis of syringolin A by *Pseudomonas syringae* pv. *syringae*. Molec Plant-Micro Interact 17: 90–97.

Anderson JC, Pascuzzi PE, Xiao F, Sessa G, Martin GB (2006) Host mediated phosphorylation of type III effector AvrPto promoter *Pseudomonas* virulence and avirulence in tomato. Plant Cell 18: 502–514.

Andrade M, Wang N (2019) The Tad pilus apparatus of '*Candidatus* Liberibacter asiaticus' and its regulation by VisNR. Molec Plant-Micro Interact 32: 1175–1187.

Antolinez CA, Fereres A, Moreno A (2017) Sex-specific probing behavior of the carrot psyllid *Bactericera trigonica* and its implication in the transmission of '*Candidatus* Liberibacter solanacearum. Eur J Plant Pathol 147: 627–637.

Araki H, Tian D, Goss EM, Jakob K, Halldorsdottir SS et al. (2006) Presence/absence of polymorphism or alternative pathogenicity islands in *Pseudomonas viridiflava*, a pathogen of *Arabidopsis*. Proc Natl Acad Sci USA 103: 5887–5892.

Arnold DL, Lovell HC, Jackson RW, Mansfield JW (2011) *Pseudomonas syringae* pv. *phaseolicola*: from 'has been' to supermodel. Molec Plant Pathol 12: 617–627.

Arocha Y, Almeida R, Jones P (2008) Using ribosome probe for detection of phytoplasma associated with bunchy top symptom of papaya (BTS) in plant and insect hosts. Rev Prot Veg 23: 16–20.

Arora SK, Ritchings BH, Almira EC, Long S, Ramphal R (1997) A transcriptional activator, FleQ regulates mucin adhesion and flagellar gene expression in *Pseudomonas aeruginosa* in a cascade manner. J Bacteriol 179: 5574–5581.

Arthur JC (1885) Proof that the disease of trees known as pear blight is directly due to bacteria. NY Agric Exp Sta Bull 2 n.s.: 1–4.

Arvizu-Gomez J, Hernandez-Morales A, Pastor-Palacios G, Brieba LG, Alvarez-Morales A (2011) Integration of host factor (IHF) binds to the promoter region of the *phtD* operon involved in phaseolotoxin synthesis is in *Pseudomonas syringae* pv. *phaseolicola* NPS 3121. BMC Microbiol 11: 90.

Asselin JE, Bonasera JM, Kim JF, Oh C-S, Beer SV (2011) *Eop1* from a Rubus strain of *Erwinia amylovora* functions as a host range-limiting factor. Phytopathology 101: 935–944.

Athinuwat D, Brooks S, Burr TJ, Prathuangwong S (2018) Flagella and pili of *Xanthomonas axonopodis* pv. *glycines* are associated with motility, biofilm formation and virulence on soybean. J Phytopathol 166: 590–600.

Aung K, Kim P, Li Z, Joe A, Kvitko B et al. (2020) Pathogenic bacteria target plant plasmodesmat to colonize and invade surrounding tissues. Plant Cell 32: 595–611.

Baccari C, Killiny N, Ionescu M, Almeida RPP, Lindow SE (2014) Diffusible signal factor-repressed extracellular traits enable attachment of *Xylella fastidiosa* to insect vectors and transmission. Phytopathology 104: 27–33.

Baccari C, Lindow SE (2011) Assessment of the process of movement of *Xylella fastidiosa* within susceptible and resistant grape cultivars. Phytopathology 101: 77–84.

Backus EA, Morgan DJ (2011) Spatiotemporal colonization of *Xylella fastidiosa* in its vector supports the role of egestion in the inoculation mechanism of foregut-borne plant pathogens. Phytopatholgy 101: 912–922.

Backus EA, Shugart HJ, Rogers EE, Morgan RJ, Shatters S (2015) Direct evidence of egestion and salivation of *Xylella fastidiosa* suggests sharp shooters can be 'flying syringes'. Phytopathology 105: 608–620.

Badel JL, Nomura K, Bandyopadhyay SRS, Collmer A, He SY (2003) *Pseudomonas syringae* pv. *tomato* DC3000 HopPtom (CEL ORF3) is important for lesion formation but not growth in tomato and is secreted and translated by the Hrp type III secretion system in chaperone dependent manner. Molec Microbiol 49: 1239–1251.

Badel JL, Shimizu R, Oh HS, Collmer A (2006) *Pseudomonas syringae* pv. *tomato avrE1/hopM1* mutant is severely reduced in growth and lesion formation in tomato. Molec Plant-Micro Interact 19: 99–111.

Bagio TZ, Canteri MG, Júnior RPL (2020) Revisiting seed transmission of '*Candidatus* Liberibacter asiaticus' in different citrus species and hybrids. Trop Plant Pathol 45: 334–341.

Balaji V, Sessa G, Smart CD (2011) Silencing of host basal defense response-related gene expression increases susceptibility of *Nicotiana benathamiana* to *Clavibacter michiganensis* subsp. *michiganensis*. Phytopathology 101: 349–357.

Barker CJ, Smith JM, Yarberry AJ, Rice C (2020) Induction of apoplast phenolics in pepper (*Capsicum annuum*) leaves in response to pathogenic bacteria. Physiol Molec Plant Pathol 109: 101453.

Barny MA, Gaudriault S, Brisset MN, Paulin JP (1999) Hrp-secreted proteins: their role in pathogenicity and HR elicitation. Acta Hortic 489: 353–358.

Beanland L, Hoy CW, Miller SA, Nault LR (2000) Influence of aster yellows phytoplasma on the fitness of aster leafhopper (Homoptera: Cicadellidae). Ann Entomol Soc Amer 93: 271–276.

Bender CL (2000) Regulatory interactions between the HRP type III protein secretion system and coronatine biosynthesis *in Pseudomonas syringae* pv. *tomato* DC3000. Micrbiology (Reading) 146: 2447–2456.

Berger S, Benediktyova Z, Matous K, BonFigure KB, Mueller MJ et al. (2007) Visualization of dynamics of plant pathogen interaction by novel combination of chlorophyll fluorescence imaging and statistical analysis: differential effects of virulent and avirulent strains of *Pseudomonas syringae* and of oxylipins on *Arabidopsis thaliana*. J Exp Bot 58: 797–806.

Bernardini C, Pagliari L, De Rosa V, Almeida-Trapp M, Santi S et al. (2020) Pre-symptomatic modified phytohormone profiles associated with lower phytoplasma titers in an *Arabidopsis seor1ko* line. Sci Rep 10:14770. https://doi.org/110.1038/s41598-020-71660-0

Bernhard F, Coplin DL, Geider K (1993) A gene cluster for amylovoran synthesis in *Erwinia amylovora* characterization and relationship to *cps* genes in *Erwinia stewartii*. Molec Gen Genet 239: 158–168.

BonFigure KB, Schreiber U, Gabbler A, Roitsch T, Berger S (2006) Infection with virulent and avirulent *Pseudomonas syringae* strains differentially affects photosynthesis and sink metabolism in *Arabidopsis* leaves. Planta 225: 1–12.

Bova LP, Cristofani-Yaly M, Machado MA (2017) Physiological, anatomic, and gene expression changes in *Citrus Sunki, Poncirus trifoliata* and their hybrid after '*Candidatus* Liberibacter asiaticus' functions. Phytopathology 107: 590–599.

Bové JM (2006) Huanglongbing: a destructive newly emerging century-old disease of citrus. Plant Pathol 88: 7–37.

Bronstein PA, Marichi M, Cartinhour S, Schneider DJ, De Lisa MP (2005) Identification of a twin-arginine translocation system in *Pseudomonas syringae* pv. *tomato* DC3000 and its contribution to pathogenicity and fitness. J Bacteriol 187: 8350–8361.

Budde IP, Ullrich MS (2000) Interactions of *Pseudomonas syringae* pv. *glycinea* with host and nonhost plants in relation to temperature and phytotoxin synthesis. Molec Plant-Micro Interact 13: 951–961.

Bull CT, Gebben SJ, Goldman PH, Trent M, Hayes RJ (2015) Host genotype and hypersensitive reaction influence population levels of *Xanthomonas campestris* pv. *vitians* in lettuce. Phytopathology 105: 316–324.

Buoso S, Pagliari L, Musetti R, Marhni M, Marroni F et al. (2019) '*Candidatus* Phytoplasma solani' interferes with distribution and uptake of iron in tomato. BMC Genom 20: 703. https://doi.org/10.1186/s12864.019-6062-x

Burbank LP, Stenger DC (2016) A temperature-independent cold-shock protein homolog acts as a virulence factor in *Xylella fastidiosa*. Molec Plant-Micro Interact 29: 335–344.

Burdman S, Shen Y, Lee SW, Xuc Q, Ronald P (2004) RaxH/RaxR: a two-component regulatory system in *Xanthomonas oryzae* pv. *oryzae* required for AvrXa21 activity. Molec Plant-Micro Interact 17: 602–612.

Bütner D, He SY (2009) Type III protein secretion in plant pathogenic bacteria. Plan Physiol 150: 1656–1664.

Buxa SV, Degola F, Folizzotto R, De Marco F, Loschi A et al. (2015) Phytoplasma infection in tomatoes associated with reorganization of plasma membrane ER stacks and actin filaments in sieve elements. Front Plant Sci 6: 650. https://doi.org/10.1111/j.1364-3703.200800472.x

Caldwell D, Iyer-Pascuzzi AS (2019) A scanning electron microscopy technique for viewing plant-microbe interactions of tissue and cell-type resolution. Phytopathology 109: 1302–1311.

Camanes G, Pastor V, Cerezo M, Garcia-Agustin P, Herrero FV (2012) A deletion in the high-affinity transporter NRT2.1 alters metabolomic and transcriptomic responses to *Pseudomonas syringae*. Plant Signal Behav 7: 619–622.

Canale MC, Komada KMA, Lopes JRS (2020) Latency and incubation of '*Candidatus* Liberibacter asiaticus' in citrus after vector inoculation. Trop Plant Pathol 45: 320–376.

Canale MC, Tomaseto AF, Haddad ML, Colleta-Filho HD, Lopes JRS (2017) Latency and persistence of '*Candidatus* Liberibacter asiaticus' in its psyllid vector *Diaphorina citri* (Hemiptera: Liviidae). Phytopathology 107: 264–272.

Cao T, Duncan RA, Mc Kenry MV, Shackel KA, Delong TM et al. (2005) Interaction between nitrogen-fertilized peach trees and expression of *syrB*, a gene involved in syringomycin production in *Pseudomonas syringae* pv. *syringae*. Phytopathology 95: 581–586.

Carr EA, Zaid AM, Bonasera JM, Lorbeer JW, Beer SV (2013) Infection of onion leaves by *Pantoea ananatis* leads to bulb infection. Plant Dis 97: 1524–1528.

Casati P, Jermini M, Quaglino F, Corbani G, Schaerer S et al. (2017) New insights on Flavescence dorée phytoplasma ecology in the vineyard agro-ecosystem in southern Switzerland. Ann Appl Biol 171: 37–51.

Chalupowicz L, Cohen-Kandli M, Dror O, Eichenlaub R, Gartemann K-H et al. (2010) Sequential expression of bacterial virulence and plant defense genes during infection of tomato with *Clavibacter michiganensis* subsp. *michiganensis*. Phytopathology 100: 252–261.

Chalupowicz L, Zellermann E-M, Fluegel M, Dror O, Eichenlaub R, Gartemann K-H, Savidor A, Sessa G, Iraki N, Barash I, Manulis-Sasson S (2012) Colonization and movement of GFP-labeled *Clavibacter michiganensis* subsp. *michiganensis* during tomato infection. Phytopathology 102: 23–31.

Chang JH, Desveaux D, Creason AL (2014) The ABCs and 123s of bacterial secretion systems in plant pathogenesis. Annu Rev Phytopathol 52: 317–345.

Chen F, Yuan Y, Li Q, He Z (2007) Proteomic analysis of rice plasma membrane reveals proteins involved in early defense response to bacterial blight. Proteomics 7: 1529–139.

Chen WY, Lin C-P (2011) Characterization of *Catharanthus roseus* genes regulated differentially by peanut witches' broom phytoplasma infection. J Phytopathol 159: 505–510.

Chen X, Bai K, Lyu Q, Jiang N, Li J, Luo L (2021) Role of penicillin-binding proteins in the viability, morphology, stress tolerance, and pathogencity of *Clavibacter michiganensis*. Phytopathology 111: 1301–1312.

Chen X, Tan Q, Lyu C, Yu C, Jiang N, Li J, Luo L (2022) Unmarked gene editing in *Clavibacter michiganensis* using CRISPR/Cas9 in *Clavibacter michiganensis* using CRISPR/Cas9 and 5-fluorocytosine counterselection. Molec Plant-Micro Interact 35: 4–14.

Chiang Y-H, Coaker G (2015) Effector-triggered immunity: NLR immune perception and downstream defense repressor. TAB 13: e0813. https://doi.org/10.1199/tab.0183

Cicero JM, Fisher TW, Brown JK (2016) Localization of '*Candidatus* Liberibacter solancearum' and evidence for surface appendages in potato psyllid vector. Phytopathology 106: 142–154.

Cicero JM, Fisher TW, Qureshi JA, Stansly PA, Brown JK (2017) Colonization and intrusive invasion of potato psyllid by '*Candidatus* Liberibacter solanacearum'. Phytopathology 107: 36–49.

Cohn J, Martin GB (2003) Pathogen recognition and signal transduction in plant immunity. In: Ezekowitch RABH, Hoffmann JA (eds.), *Infection, Disease, Innate Immunity*. Humana Press Inc., Totowa, NJ, USA, pp. 3–27.

Cohn M, Bart RS, Shybut M, Dahlbeck D, Gomez M (2014) *Xanthomonas axonopodis* virulence is promoted by a transcription activator-like effector mediated induction of a SWEET sugar transporter in cassava. Molec Micro-Plant Interact 27: 1186–1198.

Constantin EC, Haegeman A, Van Vaerenbergh J, Baeyen S, Van Maldergehem C et al. (2017) Pathogenicity and virulence gene content of *Xanthomonas* strains infecting Araceae, formerly known as *Xanthomonas axonopodis* pv. *diffenbachiae*. Plant Pathol 66: 1539–1554.

Cooper WR, Alcala PE, Barcenas NM (2015) Relationship between plant vascular architecture and within-plant distribution of '*Candidatus* Liberibacter solanacearum' in potato. American Pot Res 92: 91–99.

Cox KL, Meng F, Wilkins KE, Li F, Wang P et al. (2017) TAL effector-driven induction of a SWEET gene confer susceptibility to bacterial blight of cotton. Nature Commun 8: 15588–15601.

Crosslin JM, Vandemark GJ, Munyaneza JE (2006) Development of a real-time quantitative PCR for detection of the Columbian Basin purple top phytoplasma in plants and leafhoppers. Plant Dis 90: 663–667.

Crossman L, Dow JM (2004) Biofilm formation and dispersal in *Xanthomonas campestris*. Microbes Infect 6: 623–629.

Da Silva JR, de Alvarenga FV, Boaretto RM, Lopes JRS, Quaggio JA, Filho HDC, Mattos D Jr (2020) Following effects of micronutrient supply in HLB-infected trees: plant responses and '*Candidatus* Liberibacter asiaticus' by Asian citrus psyllid. Trop Plant Pathol 45: 597–610.

Dahan J, Wenninger EJ, Thompson B, Eid S, Olsen N et al. (2017) Relative abundance of potato psyllid haplotypes in southern Idaho potato fields during 2012 to 2015, and incidence of '*Candidatus* Liberibacter solanacearum' causing zebra chip disease. Plant Dis 101: 822–829.

Damsteegt VD, Postinkova EN, Stone AL, Kuhlmann M, Sechler A et al. (2010) *Murraya paniculata* and related species as potential hosts and inoculum reservoirs of '*Candidatus* Liberibacter asiaticus', causal agent of huanglongbing. Plant Dis 94: 528–533.

Daniels R, Vanderleyden J, Michels J (2004) Quorum sensing and swimming migration in bacteria. FEMS Microbiol Rev 28: 261–289.

Darrasse A, Darsonval A, Boreau T, Brisset MN, Durand K et al. (2010) Transmission of plant pathogenic bacteria by non-host seeds without induction of an associated defense reaction at emergence. Appl Environ Microbiol 76: 6787–6796.

Davis RE, Lee I-M (1982) Pathogenicity of spiroplasma, mycoplasma-like organisms and vascular-limited fastidious walled bacteria. In: Mount GS, Lacy GH (eds.), *Phytopathogenic Prokaryotes*, volume 1. Academic Press, New York, pp. 492–513.

Davis RE, Sinclair WA (1998) Phytoplasma identity and disease etiology. Phytopathology 88: 1372–1376.

Deb Roy S, Thilmony R, Kwack YB, Nomura K, He SY (2004) A family of conserved bacterial effectors inhibits salicylic acid-mediated basal immunity and promotes disease necrosis. Proc Natl Acad Sci USA 101: 9927–9932.

Dechorgnat J, Patrit O, Krapp A, Fagard M, Daniel-Vedele F (2012) Characterization of the *Nrt 2.6* gene in *Arabidopsis thaliana*: a link with plant responses to biotic and abiotic stress. PLoS ONE 7: e42491.

Dietrich R, Plob K, Heit M (2004) Constitutive and induced resistance to pathogens in *Arabidopsis thaliana* depends on nitrogen supply. Plant Cell and Environ 27: 896–906.

Ding Y, Wu W, Wei W, Davis RE, Lee I-M et al. (2013) Potato purple top phytoplasma-induced disruption of gibberellin homeostasis in tomato plants. Ann Appl Biol 162: 131–139.

Doi Y, Ternaka M, Yora K, Asuyama H (1967) Mycoplasma and PLT group-like microorganisms with mulberry dwarf, potato witches' broom, aster yellows and paulownia witches' broom. Ann Phytopathol Soc Jpn 33: 259–266.

Dorman CJ (2004) H-NS: a universal regulator for a dynamic genome. Nat Rev Microbiol 2: 391–400.

Dorman CJ, Deighan P (2003) Regulation of genes expression by histone-like proteins in bacteria. Curr Opin Genet Dev 13: 179–184.

Duan Y, Zhou L, Hall DG, Li W, Doddapaneni H, Lin H, Liu L, Vehling CM, Gabriel DW, Williams KP, Dickerman A, Sun Y, Gottwald T (2009) Complete genome sequence of citrus huanglongbing bacterium, '*Candidatus* Liberibacter asiaticus' obtained through metagenomics. Molec Plant-Micro Interact 22: 1011–1020.

Dubrow ZE, Carpenter SCD, Carter ME, Grinage A, Gris C et al. (2022) Cruciferous weed isolates of *Xanthomonas campestris* yield insight into pathovar genomic relationships and genetic determinants of host and tissue specificity. Molec Plant-Micro Interact 35: 791–802.

Duduk B, Calarai A, Paltrinieri A, Duduck N, Bertacccini A (2008) Multigenic analysis for differentiation of aster yellows phytoplasmas infecting carrots in Serbia. Ann Appl Biol. https://doi.org/10.1111/j.1774-7348.2008.00285.x

Dunger G, Arabolaze AL, Gottig N, Orellano EG, Ottado J (2005) Participation of *Xanthomonas axonopodis* pv. *citri hrp* cluster in citrus canker and nonhost plant responses. Plant Pathol 54: 781–788.

Dutta B, Avei U, Hahn MG, Walcott RR (2012) Location of *Acidovorax citrulli* in infested watermelon seeds is influenced by the pathway of bacterial invasion. Phytopathology 102: 461–468.

Dutta B, Langston B, Luo X, Carlson S, Kichler J, Gitaitis RD (2017) A risk assessment model for bacterial leaf spot of pepper (*Capsicum annuum*) caused by *Xanthomonas euvesicatoria* based on concentrations of macronutrients, micronutrients, and micronutrient ratios. Phytopathology 107: 1331–1338.

Dutta B, Sanders H, Langston DB, Booth C, Smith S, Gitaitis RD (2014) Long-term survival of *Acidovorax citrulli* in Citron melon (*Citrullus lanatus* var. *citroides*) seeds. Plant Pathol 63: 1130–1137.

Dutta B, Schneider RW, Robertson CL, Walcott RR (2016) Embryo localization enhances the survival of *Acidovorax citrulli* in watermelon seeds. Phytopathology 196: 330–338.

Esteves MB, Kleina HT, de Melo Sales T, Oliveira TP, de Lara AR et al. (2019) Transmission efficiency of *Xylella fastidiosa* subsp. *pauca* sequence types by sharpshooter vectors after in vitro acquisition. Phytopathology 109: 286–293.

Fagard M, Launay A, Clement G, Courtial J, Dellagi A et al. (2014) Nitrogen metabolism meets phytopathology. J Exp Bot 66: 5643–5656.

Fatmi M, Schaad NW (2002) Survival of *Clavibacter michiganensis* subsp. *michiganensis* in infected stems under natural field conditions in California, Ohio and Morocco. Plant Pathol 51: 149–154.

Fayette J, Jones JB, Pernezny K, Roberts PD, Raid R (2018) Survival of *Xanthomonas campestris* pv. *vitians* on lettuce in crop debris, irrigation water, and weeds in south Florida. Eur J Plant Pathol 151: 341–353.

Feitosa-Junior OR, Souza APS, Zaini PA, Baccari C, Ionescu M et al. (2022) The XadA trimeric autotransporter adhesions in *Xylella fastidiosa* differentially contribute to cell aggregation, biofilm formation, insect transmission and virulence to plants. Molec Plant-Micro Interact 35: 857–866.

Feitosa-Junior OR, Stefanello E, Zaini PA, Nascimento R, Pierry PM et al. (2019) Proteomic and metabolomic analyses of *Xylella fastidiosa* OMV-enriched fractions reveal association with virulence factors and signaling molecules of the DSF family. Phytopathology 109: 1344–1353.

Felipe V, Romero AM, Montecchia MS, Vojnou AA, Bianco MI et al. (2018) *Xanthomonas vesicatoria* virulence factors involved in early stages of bacterial spot development in tomato. Plant Pathol 67: 1936–1943.

Ferrante P, Scortichini M (2018) *Xanthomonas arboricola* pv. *fragariae*: a confirmation of the pathogenicity of the pathotype strain. Eur J Plant Pathol 150: 825–829.

Filloux A, Hachani A, Bleves S (2008) The bacterial type VI secretion machine: yet another player for protein transport across membranes. Microbiology 154: 1570–1583.

Fleigmann J, Felix G (2016) Immunity: flagellin seed from all sides. Nat Plants 2: 16136.

Folimonova SY, Achor DS (2010) Early events of citrus greening (huanglongbing) disease development at ultrastructural level. Phytopathology 100: 949–958.

Fory PA, Triplett L, Ballen C, Abello JF, Duitama I et al. (2014) Comparative analysis of two emerging rice seed bacterial pathogens. Phytopathology 104: 436–444.

Frenkel O, Bornestein M, Sulhani R, Sharabani G, Sofer M et al. (2016) Secondary spread of *Clavibacter michiganensis* subsp. *michiganensis* in nurseries and the conditions leading to infection of tomato seedlings. Eur J Plant Pathol 144: 569–579.

Ftayeh RM, von Tiedemann A, Rudolph KWE (2011) A new selective medium for isolation of *Clavibacter michiganensis* subsp. *michiganensis* from tomato plants and seed. Phytopathology 101: 1355–1364.

Galvis CA, Leguizamon JE, Gaitan AL, Mejia JF, Alvarez E et al. (2007) Detection and identification of a group 16SrIII-related phytoplasma associated with coffee crispness disease in Colombia. Plant Dis 91: 248–252.

Gambino G, Boccacci P, Margaria P, Palmano S, Gribaudo I (2013) Hydrogen peroxide accumulation and transcriptional changes in grapevines recovered from flavescence dorée disease. Phytopathology 103: 776–738.

Garavaglia BS, Thomas L, Gottig N, Dunger G, Garofalo CG et al. (2010) A eukaryotic-acquired gene by a biotrophic phytopathogen allows prolonged survival on the host by counteracting the shut-down of plant photosynthesis. PLoS ONE 5: e8950–2010.

Garcia-Seco D, Chiapello M, Bracale M, Pesce C, Bagnaresi P, Dubois E, Moulin L, Vannini C, Koebnik R (2017) Transcriptional and proteome analysis reveal new insight into proximal and distal responses of wheat to foliar infection by *Xanthomonas translucens*. Sci Rept 7: 10157. https://doi.org/10.1038/s41598-017-10568-8

Gasparich GE (2010) Spiroplasmas and phytoplasmas: microbes associated with plant hosts. Biologicals 38: 195–203.

Gilkes JM, Frampton RA, Smith GR, Dobson RCJ (2018) Potential pathogenicity determinants in the genome of '*Candidatus* Liberibacter solanacearum' the causal agent of zebra chip disease of potato. Austr Plant Pathol 47: 119–134.

Gilkes JM, Sheen CR, Frampton RA, Smith GR, Dobson RCJ (2019) The first purification of functional proteins from unculturable, genome-reduced, bottlenecked α-proteobacterium '*Candidatus* Liberibacter solanacearum'. Phytopathology 109: 1141–1148.

Gimenez-Ibanez S, Boter M, Fernández-Barbero G, Chini K, Rathjen JP et al. (2014) The bacterial effector Hopx1 targets JAZ transcriptional repressors to activate jasmonate signaling and promote infection in *Arabidopsis*. PLoS Biol 12: e1001792.

Gochez AM, Minsavage GC, Potnis N, Canteros BI, Stall RE et al. (2015) A functional XopAG homologue in *Xanthomonas fuscans* pv. *aurantifolii* strain limits host range. Plant Pathol 64: 1207–1214.

Goel AK, Lundberg D, Torres MA, Matthews R, Akimoto-Tomiyama C et al. (2008) The *Pseudomonas syringae* type III effector HopAM1 enhances virulence on water-stressed plants. Molec Plant-Micro Interact 21: 361–370.

Gotting N, Garavaglia BS, Daurelio LD, Valentine A, Gehring C et al. (2008) *Xanthomonas axonopodis* pv. *citri* uses a plant natriuretic peptide-like protein to modify host homeostasis. Proc Natl Acad Sci USA 105: 18631–18636.

Granato LM, Oliveira TS, Boscariol-Camargo RL, Galdeano DM, Silva NV et al. (2020) '*Candidatus* Liberibacter asiaticus' putative effectors: in silico analysis and gene expression in citrus leaves displaying distinct huanglongbing symptoms. Trop Plant Pathol 45: 646–657.

Gu K, Yang B, Tian D, Wu L, Wang D et al. (2005) *R* gene expression induced by a type III effector triggers disease resistance in rice. Nature 435: 1122–1125.

Guan W, Wang T, Huang Q, Tian E, Liu B, Yang Y, Zhao Y (2020) LuxR-type regulator, AcrR regulates flagellar assembly and contributes to virulence, motility, biofilm formation and growth ability of *Acidovorax citrulli*. Molec Plant Pathol 21: 489–501.

Guerra ML, Malafaia CB, Macedo AJ, Silva MV, Mariano RLR et al. (2020) Biofilm formation by *Xanthomonas campestris* pv. *viticola* affected by abiotic surfaces and culture media. Trop Plant Pathol 43: 146–151.

Guo W, Gao J, Chen Q, Ma B, Fang Y, Liu X, Chen G, Liu I (2019) Crp-like protein plays both positive and negative roles in regulating the pathogenicity of bacterial pustule pathogen, *X. axonopodis* pv. *glycines*. Phytopathology 109: 1171–1183.

Hadas R, Kritzman G, Kleitman F, Gefen T, Manulis S (2005) Comparison of extraction producers and determination of detection threshold for *Clavibacter michiganensis* subsp. *michiganensis* in tomato seeds. Plant Pathol 54: 643–649.

Hafke JB, Ehlers K, Foller J, Holl SR, Beckers, van Bel AJE (2013) Involvement of the sieve-element cytoskeleton in electrical responses to cold shocks. Plant Physiol 162: 707–719.

Hajri A, Loiseau M, Cousseau-Suhard P, Renaudin I, Gentit P (2017) Genetic characterization of '*Candidatus* Liberibacter solanacearum' haplotypes associated with apiaceous crops in France. Plant Dis 101: 1383–1390.

Hartung JS, Halbert SE, Pelz-Stelinski K, Brlansky RH, Chen C et al. (2010) Lack of evidence for transmission of '*Candidatus* Liberibacter asiaticus' through seed taken from affected fruit. Plant Dis 94: 1200–1205.

Hastings JW, Greenberg EP (1999) Quorum sensing: the explanation of a curious phenomenon reveals a common characteristic of bacteria. J Bacteriol 181: 2667–2668.

He YW, Xu M, Lin K, Ng YJA, Wen CM et al. (2006) Genome scale analysis of diffusible signal factor regulon in *Xanthomonas campestris* pv. *campestris*: identification novel cell-to-cell communication of dependent genes and functions. Molec Microbiol 59: 610–622.

Henry E, Toruño TY, Jauñeau A, Deslandes L, Coaker G (2017) Direct and indirect visualization of bacterial effector delivery into diverse plant cell types during infection. Plant Cell 29: 1555–1570.

Hildebrand M, Aldridge P, Geider K (2006) Characterization of *hns* genes from *Erwinia amylovora*. Molec Gen Genom 275: 310–319.

Hilf ME (2011) Colonization of citrus seed coats by '*Candidatus* Liberibacter asiaticus': implications for seed transmission of the bacterium. Phytopathology 101: 1242–1250.

Hilf ME, Sims KR, Folimonova SY, Achor DS (2013) Visualization of '*Candidatus* Liberibacter asiaticus' cells in the vascular bundle of citrus seed coats with fluorescence in situ hybridization and transmission electron microscopy. Phytopathology 103: 545–554.

Hodgetts J, Ball T, Boonham N, Mumford R, Dickinson M (2007) Use of terminal restriction fragment length polymorphism (T-FRLP) for identification of phytoplasmas in plants. Plant Pathol 56: 357–365.

Hogenhout SA, Oshima K, Ammar E, Kakizawa S, Kingdom HN, Namba S (2008) Phytoplasmas: bacteria that manipulate plants and insects. Molec Plant Pathol 9: 403–423.

Hu H, Shen L, Zhagn G, Gu Q, Zheng K (2018) Influence of bacterial leaf blight on the photosynthetic characteristics of resistant and susceptible rice. J Phyotpathol 166: 547–554.

Hu Y, Duan S, Zhang Y, Shantharaj D, Jones JB et al. (2016) Temporal transcription profiling of sweet orange in response to PthA4-mediated *Xanthomonas citri* subsp. *citri* infection. Phytopathology 106: 442–451.

Hurley B, Lee D, Mott A, Wilton M, Liu J et al. (2014) The *Pseudomonas syringae* type III effector HopF2 suppresses *Arabidopsis* stomatal immunity. PLoS ONE 9: e114921.

Hwang IS, An SH, Hwang BK (2011) Pepper asparagine synthesis 1 (CaAS1) is required for plant nitrogen assimilation and defense responses to microbial pathogens. Plant J 67: 749–752.

Hwang IS, Oh EJ, Lee HB, Oh CS (2019) Functional characterization of two cellulase genes in the Gram-positive pathogenic bacterium *Clavibacter michiganensis* for wilting in tomato. Molec Plant-Micro Interact 32: 491–501.

Ikawa Y, Ohnishi S, Shoji A, Furutani A, Tsuge S (2018) Concomitant regulation by a LacI-type transcriptional repressor XylR on genes involved in xylan and xylose metabolism and the type III secretion system in rice pathogen *Xanthomonas oryzae* pv. *oryzae*. Molec Plant-Micro Interact 31: 605–613.

Ionescu M, Zaini PA, Baccari C, Tian S, da Silva AM et al. (2014) *Xylella fastidiosa* outer membrane vehicles modulate plant colonization by blocking attachment surfaces. Proc Natl Acad Sci USA 111: E3910–E3918.

Ishiga T, Ishiga Y, Betsuyaku S, Nomura N (2018) AlgU contributes to the virulence of *Pseudomonas syringae* pv. *tomato* DC3000 by regulating production of the phytotoxin. J Gen Plant Pathol 84: 189–201.

Ishiie T, Doi Y, Yora K, Asuyamaa H (1967) Suppressive effects of an antibiotic of tetracycline group on symptom development of mulberry dwarf disease. Ann Phytopathol Soc Jpn 33: 267–275.

Jacques MA, Josi K, Darrasse A, Samson R (2005) *Xanthomonas axonopodis* pv. *phaseoli* var *fuscans* is aggregated in stable biofilm population size in the phyllosphere of field-grown beans. Appl Environ Microbiol 71: 2008–2015.

Jain M, Fleitas LA, Gabriel DW (2015) Prophage-encoded peroxidase in '*Candidatus* Liberibacter asiaticus' is a secreted effector that suppresses plant defenses. Molec Plant-Micro Interact 28: 1330–1337.

Jiao Q-S, Feng H-Q, Tian W-Y, Bai J-Y, Sun K et al. (2016) Extracellular ATP functions in alleviating the decrease of PSII photochemistry caused by the infection of *Xanthomoans campestris* pv. *phaseoli*. Plant Pathol 65: 819–825.

Jimenez-Guerrero I, Perez-Montano F, Da Silva GM, Wagner N, Shkedy D, Zhao M, Pizarro L, Bar M, Walcott R, Sessa G, Pupko T, Burdman S (2020) Show me your secret(ed) weapons: a multifaceted approach reveals a wide arsenal of type III-secreted effectors in the cucurbit pathogenic bacterium *Acidovorax citrulli* and novel effectors in the Acidovorax genus. Molec Plant Pathol 21: 17–37.

Jin Q, Hu W, Brown I, McGhee G, Hart P et al. (2001) Visualization of secreted Hrp and Avr proteins along the Hrp pilus during type III secretion in *Erwinia amylovora* and *Pseudomonas syringae*. Molec Micobiol 40: 1129–1139.

Johnson KL, Walcott RR (2013) Quorum-sensing contributes to seed-to-seedling transmission of *Acidovorax citrulli* on watermelon. J Phytopathol 161: 561–573.

Jones AM, Thomas V, Bennett MH, Mansfield J, Grant M (2006) Modifications to the *Arabiopsis* defense proteome occur prior to significant transcriptional change in response to inoculation with *Pseudomonas syringae*. Plant Physiol 142: 1603–1620.

Kaneshiro WS, Alvarez AM (2003) Variability and survival of *Clavibacter michiganensis* subsp. *michiganensis* (*Cmm*) in seed. Phytopathology 93: S43.

Karavina C, Tigere TA, Chihiya J (2008) The contribution of soil and crop debris inocula in the outbreak of bacterial common blight in field beans (*Phaseolus vulgaris* L.) under Zimbabwean conditions. J Sustain Dev Afr 10: 221–233.

Katiyar-Agarwal S, Morgan RA, Dahlbeck D, Borsani O, Villegas A Jr, Zhu J, Staskawicz BJ, Jim H (2006) A pathogen-inducible endogenous sRNA in plant immunity. Proc Natl Acad Sci USA 103: 18002–18007.

Kawaguchi A, Tanina K, Inoue K (2010) Molecular typing and spread of *Clavibacter michiganensis* subsp. *michiganensis* in greenhouse in Japan. Plant Pathol 59: 76–83.

Kazmi MR, Farooq SI, Fateh FS, Hameed S, Fahmeed F et al. (2007) Identification of phytoplasma from mango trees showing sudden death phenomenon through molecular technique. Pak J Bot 39: 2153–2157.

Kemp BP, Horne J, Bryant A, Cooper RM (2004) *Xanthomonas axonopodis* pv. *manihotis gumD* gene is essential for EPS production and pathogenicity and enhances epiphytic survival on cassava (*Manihot esculenta*). Physiol Molec Plant Pathol 64: 209–218.

Khojasteh M, Khahani B, Taghavi M, Tavakol E (2018) Identification and characterization of responsive genes in rice during compatible interactions with pathogenic pathovars of *Xanthomonas oryzae*. Eur J Plant Pathol 151: 141–153.

Killiny N, Catroviejo M, Saillard C (2005) *Spiroplasma citri* spiralin acts in vitro as a lectin-binding to glycoproteins from its insect vector *Circulifer haemotoceps*. Phytopathology 95: 541–548.

Killiny N, Rashed A, Almeida RP (2012) Disrupting the transmission of a vector-borne plant pathogen. Appl Environ Microbiol 78: 638–643.

Kim J, Kang Y, Choi O, Jeong Y, Jeong JE et al. (2007) Regulation of polar flagellum genes in mediated by quorum sensing and FlhDC in *Burkholderia glumae*. Molec Microbiol 64: 165–174.

Kim J-G, Li X, Roden JA, Taylor KW, Aakrre CD et al. (2009) *Xanthomonas* 73S and interacts with a tomato atypical receptor-like kinase and TFT1. Plant Cell 21: 1305–1323.

Kim J-G, Stork W, Mudgett MB (2013) *Xanthomonas* type III effector XopD desunoylates tomato transcription factor SIERF4 to suppress ethylene responses and promote pathogen growth. Cell Host Microbe 13: 143–154.

Koizumi M (1985) Citrus canker: the world situation. In: Timmer IW (ed.), *Citrus Canker an International Perspective*. University of Florida, Institute of Food and Agricultural Sciences, Gainesville, USA, pp. 2–7.

Komatsu T, Kondo N (2015) Winter habitat of *Xylophilus ampelinus*, the cause of bacterial blight of grapevine, Japan. J Gen Plant Pathol 81: 237–242.

Koutsoudis MD, Tsaltas D, Minogue TD, Von Bodman SB (2006) Quorum sensing regulation governs bacterial adhesion, biofilm development and host colonization in *Pantoea stewartii* subsp. *stewartii*. Proc Natl Acad Sci USA 103: 5983–5988.

Kranthausen HJ, Horner G, Zimmermann S, Voegele RT, Brandle F (2018) Competence of *Xanthomonas campestris* from cruciferous weeds and wallflower (*Erysimum cheiri*) to induce black rot in cabbage. Eur J Plant Pathol 151: 275–289.

Kraus CM, Mazo-Molina C, Smart CD, Martin GB (2017) *Pseudomonas syringae* pv. *tomato* strains from New York exhibit virulence attributes intermediate between typical race 0 and race 1 strains. Plant Dis 101: 1442–1448.

Kreyci PF, Eckstein B, Lopes JRS, Ferreira J, Bedendo IP (2018) Transmission of '*Candidatus* Phytoplasma pruin'-related strain associated with broccoli stunt by four species of leafhoppers. J Phytopthol 166: 502–505.

Krugner R, Sisterson MS, Chen J, Stenger DC, Johnson MW (2014) Evaluation of olive as a host of *Xylella fastidiosa* and associated sharpshooter vectors. Plant Dis 98: 1186–1193.

Kutschera A, Schombel U, Wrobel M, Gisch N (2019) Loss of *wbpL* disrupts O-polysaccharide synthesis and impairs virulence of plant-associated *Pseudomonas* strains. Molec Plant Pathol 20: 1535–1549.

Laia M, Moreira L, Dezajacomo J, Brigati J, Ferreira C et al. (2000) New genes of *Xanthomonas citri* subsp. *citri* involved in pathogenesis and adaptation revealed by transposon-based mutant library. BMC Microbiol 9: 12.

Lamichhane JR, Messean A, Morris CE (2015) Insights into epidemiology and control of diseases of annual plants caused by *Pseudomonas syringae*. J Gen Plant Pathol. https://doi.org/10.1007/10327-015-0605

Leach JE, Roberts PD, Guo A, Barton-Willis P (1989) Multiplication of *Xanthomonas campestris* pv. *oryzae* in leaves. Proceedings of the International Workshop Bacterial Blight of Rice, The International Rice Research Institute, Philippines, pp. 43–53.

Lee I-M, Zhao Y, Davis RE, Wei W, Martini M (2007) Prospects of DNA based systems for differentiation and classification of phytoplasmas. Bull Insectol 60: 239–244.

Lee JH, Ancona V, Zhao Y (2018) Lon protease modulates virulence traits in *Erwinia amylovora* by direct monitoring of major regulators and indirectly through the Rcs and Gac-Csr regulatory systems. Molec Plant-Micro Interact 19: 827–840.

Lee JH, Zhao Y (2017) ClpXP-dependent RpoS degradation encodes full activation of type III secretion system, amylovoran production and motility in *Erwinia amylovora*. Phytopathology 107: 1346–1352.

Lee S, Chu C-Y, Chu C-C (2020) Variability of phytoplasma infection density in poinsettia and evaluation of its association with the level of branching host plants. Plant Dis. https://doi.org/10.1094/PDIS-10-20-2304-RE

Levy J, Ravindran A, Gross D, Tamborindeguy C, Pierson E (2011) Translocation of '*Candidatus* Liberibacter solanacearum', the zebra chip pathogen in potato and tomato. Phytopathology 101: 1285–1291.

Lherminier J, van Terwisscha ST, Boudon-Padicu E, Caudwell A (1989) Rapid immunofluorescence detection of the grapevine flavescence dorée mycoplasma-like organism in the salivary glands of the leafhopper Euscelidius variegatus Kbm. J Phytopathol 125: 353–360.

Li J, Pang Z, Trivedi P, Zhou X, Yang X, Ying X, Jia H, Wang N (2017) '*Candidatus* Liberibacter asiaticus' encodes a functional salicylic acid (SA) hydroxylase that degrades SA to suppress plant defenses. Molec Plant-Micro Interact 30: 620–630.

Li J, Wang N (2014) Foliar application of biofilm formation-inhibiting compounds enhances control of citrus canker caused by *Xanthomonas citri* subsp. *citri*. Phytopathology 104: 134–142.

Li L, Kim P, Yu L, Cai G, Chen S et al. (2016a) Activation dependent destruction of a coreceptor by a *Pseudomonas syringae* effector dampens plant immunity. Cell Host Micro 20: 504–514.

Li L, Wang L, Liu L-M, Hou Y-X, Li Q-Q et al. (2016b) Infection process of *Burkholderida glumae* before booting stage of rice. J Phytopathol 164: 825–832.

Li S, Hao W, Lu G, Huang J, Liu C, Zhou G (2015) Occurrence and identification of a new vector of rice orange leaf phytoplasma in South China. Plant Dis 99: 1483–1487.

Li Y, Wang Y, Okamoto M, Crawford NM, Siddiqi MY, Glass ADM (2007) D-cs section o ANR2.1 tnR2.1:tNRT2.2 include high-affinity nitrate transporter gene cluster. Plant Physiol 143: 425–433.

Li Y, Yu Q (2020) Role of the lipoprotein VacJ/MlaA in the physiology and virulence of the plant pathogen *Xanthomonas citri* subsp. *citri*. Physiol Molec Plant Pathol 112: 101546.

Li Z, Variz H, Chen Y, Liu S-L, Aung K (2021) Plasmodesmata-dependent intercellular movement of bacterial effectors. Front Plant Sci 12: 640277.

Lim MTS, Kunkel BN (2005) The *Pseudomonas syringae avrRpt2* gene contributes to virulence on tomato. Molec Plant Micro-Interact 18: 625–633.

Lima WC, Paquola ACM, Varani AM, van Sluys MA, Menck CFN (2008) Laterally transferred genomic islands in Xanthomonadales related to pathogenicity and primary metabolism. FEMS Microbiol Lett 281: 87–97.

Lin H, Gudmestad NC (2013) Aspects of pathogen genomes, diversity, epidemiology, vector dynamics, and disease management for a newly emerged disease of potato: zebra chip. Phytopathology 103: 524–537.

Lin NC, Martin GB (2005) An *avrPtoA/avrPtoB* mutant of *Pseudomonas syringae* pv. *tomato* DC3000 does not elicit Pto-mediated resistance and is less virulent on tomato. Molec Plant-Micro Interact 18: 43–51.

Lin Y, Liao Y-Y, Huang R-X, Li A-Z, An S-Q, Tang J-L, Tang D-J (2021) Extracellular amylase is required for full virulence and regulated by the global posttranscriptional regulator RsmA in *Xanthomonas campestris* pv. *campestris*. Phytopathology 111: 1104–1113.

Liu J, Elmore JM, Fuglsang AT, Palmgren MG, Staskawicz BJ, Coaker G (2009) RIN4 functions with plasma membrane H+-ATPases to regulate stomatal apertures during pathogen attack. PLoS ONE 7: e1000139.

Livaja M, Palmieri MC, von Rad U, Durner J (2008) The effect of the bacterial effector protein harpin on transcriptional profile and mitochondrial proteins of *Arabidopsis thaliana*. J Proteomics 71: 148–159.

Lopez-Solanilla E, Brostein Ph, Schneider AR, Collmer AC (2004) HopPtoN is a *Pseudomonas syringae* TTSS effector and cysteine protease that suppresses pathogen-induced necrosis associated with both compatible and incompatible plant interactions. Molec Microbiol 54: 353–365.

Lowzada ES, Ed Vazquez O, Braswell WE, Yanev G, Devanaboina M et al. (2016) Distribution of '*Candidatus* Liberibacter asiaticus, above and below ground in Texas citrus. Phytopathology 106: 702–709.

Lozano-Duran R, Bourdais G, He SY, Robatzek S (2014) The bacterial effector HopM1 suppresses PAMP-triggered oxidative burst and stomatal immunity. New Phytol 202: 259–269.

Lu D, Wu S, Gao X, Zhang Y, Shan L, He P (2010) A receptor-like cytoplasmic kinase, BIK1 associates with a flagellin receptor complex to initiate plant innate immunity. Proc Natl Acad Sci USA 107: 496–501.

Lu SE, Wang N, Wang J, Chen ZJ, Gross DC (2005) Oligonucleotide microarray analysis of the Sal A regulon controlling phytotoxin production by *Pseudomonas syringae* pv. *syringae*. Molec Plant-Micro Interact 18: 324–333.

Lu Y, Ishimaru CA, Glazebrook J, Samac DA (2018) Comparative genome analysis of *Clavibacter michiganensis* subsp. *insidiosus* and pathogenicity on *Medicago truncatula*. Phytopathology 108: 172–185.

Ma K-W, Ma W (2016) Phytohormone pathways as targets to facilitate infection. Plant Molec Biol 91: 713–725.

Ma W, Dong FT, Stavrinides J, Guttman DS (2006) Type III effector diversification via both pathoadaptation and horizontal transfer in response to a coevolutionary arms race. PLoS Genet 2: e209.

Makino S, Sugio A, White F, Bogdanove AJ (2006) Inhibition of resistance gene-mediated defense in rice by *Xanthomonas oryzae* pv. *oryzicola*. Molec Plant-Micro Interact 19: 240–249.

Malvino ML, Bott AJ, Green CE, Majumdar T, Hind SR (2022) Influence of flagellin polymorphism gene regulation and responsive memory on the mobility of *Xanthomonas* species that cause bacterial spot disease of solanaceous plants. Molec Plant-Micro Interact 35: 157–169.

Mann RA, Smith TH, Buhlmann A, Blom J, Goesmann A et al. (2013) Comparative genomics of 12 strains of *Erwinia amylovora* identifies a pangenome with a large conserved core. PLoS ONE 8: e55644.

Marraffini LA (2015) CRISPR-Cas immunity in prokaryotes. Nature 526: 55–61.

Marcone C (2014) Molecular biology and pathogenicity of phytoplasmas. Ann Appl Biol 165: 199–221.

Markiewicz M, Michalezuk L (2015) Molecular response of resistant and susceptible apple genotypes to *Erwinia amylovora* infection. Eur J Plant Pathol 143: 515–526.

Mattinen L, Nissinen R, Rupi T, Kalkkinen N, Pirhonen M (2007) Host extract-induced changes in the secretome of the plant pathogenic bacterium *Pectobacterium atrosepticum*. Proteomics 7: 3527–3537.

Mayer CJ, Jarausch B, Jarausch W, Jelkmann W, Vilcinskas A et al. (2009) *Cacopsylla melanoneura* has no relevance as vector of apple proliferation in Germany. Phytopathology 99: 729–738.

Mazraie MA, Izadpanah K, Hamzehzarghani H, Salehi M, Faghihi MM (2019) Spread and colonization of '*Candidatus* Phytoplasma aurantifolia' in lime plants (*Citrus aurantifolia* [(Christm.) Swingle] as revealed by real-time PCR assay. J Plant Pathol 101: 629–637.

McGraw SL, Park DH, Jones R, Bentley MA, Rico A et al. (2016) GABA(γ-aminobutyric acid) uptakes via the GABA permease GabP represses virulence gene expression in *Pseudomonas syringae* pv. *tomato* DC3000. Molec Plant-Micro Interact 29: 938–949.

Mello AFS, Yokomi RK, Chen, Fletcher J (2010) Citrus stubborn severity is associated with *Spiroplasma citri* titer, but not with bacterial genotype. Plant Dis 94: 75–82.

Melotto M, Underwood W, Koczan J, Nomura K, He SY (2006) Plant stomata function in innate immunity against bacterial invasion. Cell 126: 969–980.

Melotto M, Zhang L, Oblessuc PR, He SY (2017) Stomatal defense a decade later. Plant Physiol 174: 561–571.

Mendoza-Herrera A, Harrison K, Yao J, Ibanez F, Tamborindeguy C (2018) Infection by '*Candidatus* Liberibacter solanacearum' haplotype A and B in *Solanum lycopersicum* 'Moneymaker'. Plant Dis 102: 2009–2015.

Meng X, Bonasera JM, Kim JF, Nissinen RM, Beer SV (2006) Apple proteins that interact with DspA/E, a pathogenicity effector of *Erwinia amylovora*, the fire blight pathogen. Molec Plant-Micro Interact 19: 53–61.

Mino Y, Matsushita Y, Sakai R (1987) The effect of coronatine on stomatal opening in leaves of broad bean and Italian rye grass. Ann Phytopathol Soc Jpn 53: 53–55.

Mittelberger C, Obkircher L, Oettl S, Oppedisano T, Pedrazzoll F et al. (2017) The insect vector *Cacopsylla picta* vertically transmits the bacterium '*Candidatus* Liberibacter mali' to its progeny. Plant Pathol 66: 1015–1021.

Mohanraju P, Makarova KS, Zetsche B, Zhang F, Koonin ZV, van der Oost J (2016) Diverse evolutionary roots and mechanistic variations of the CRISPR-Cas system. Science 353: aad15147.

Molina L, Rezzonico F, Defago G, Duffy B (2005) Autoinduction in *Erwinia amylovora* evidence of an acylhomoserine lactone signal in the fire blight pathogen. J Bacteriol 187: 3206–3213.

Moloto VM, Goszczynska T, du Toit LJ, Coutinho TA (2017) A new pathovar of *Pseudomonas syringae* pathovar *alli* isolated from onion plants exhibiting symptoms of blight. Eur J Plant Pathol 147: 591–603.

Mondal KK, Verma G, Manju, Junaid A, Mani C (2016) Rice pathogen *Xanthomonas oryzae* pv. *oryzae* employs inducible *hrp*-dependent XopF type III effector protein for its growth, pathogenicity, and for suppression of PTI response to induce blight disease. Eur J Plant Pathol 144: 311–323.

Monier JM, Lindow SE (2003) Differential survival of solitary and aggregated bacterial cells promotes aggregate formation on leaf surfaces. Proc Natl Acad Sci USA 100: 15977–15982.

Morales CQ, Posada J, Macneale E, Franklin D, Rivas I et al. (2005) Functional analysis of the early chlorosis factor gene. Molec Plant-Micro Interact 18: 477–486.

Mougous JD, Cuff ME, Raunser S, Shen A, Zhou M et al. (2006) A virulence locus of *Pseudomonas aeruginosa* encodes protein secretion apparatus. Science 312: 1526–1530.

Munyaneza JE, Sengoda VG, Buchman JL, Fisher TW (2012) Effects of temperature on '*Candidatus* Liberibacter solanacearum' and zebra chip potato disease symptom development. Plant Dis 96: 18–23.

Musetti R, Favalli MA (2004) Microscopy techniques applied to the study of phytoplasma diseases: traditional and innovative methods. www.Formatex.org/microscopy2/papers/72–80.pdf

Musetti R, Pagilari L, Buxa SV, Degola F, De Marco F et al. (2016) OHMS: phytoplasmas dictate changes in sieve-element ultrastructure to accommodate their requirements for nutrition, multiplication and translocation. Plant Signal Behav 11: e1138191. https://doi.org/10.1080/15592324.2016.1138891

Mutaqin K, Comer JL, Wayadande AC, Melcher U, Fletcher J (2011) Selection and characterization of *Spiroplasma citri* mutants by random transposon mutagenesis. Canad J Microbiol 57: 525–532.

Nachappa P, Shapiro AA, Tamborindeguy C (2012) Effect of '*Candidatus* Liberibacter solanacearum' on fitness of its insect vector, *Bactericera cockerelli* (Hemiptera: Tiozidae) on tomato. Phytopathology 102: 41–46.

Nagendran R, Lee YH (2015) Green and red light reduces the disease severity by *Pseudomonas cichorii* JBC 1 in tomato plants via upregulation of defense-related gene expression. Phytopathology 105: 412–418.

Nagendran R, Lee YH (2019) SrfC of Pseudomonas cichorii JBC 1 affects its attachment to the host surface and host tissue infection. Plant Pathol 68: 1099–1108.

Narayanasamy P (2001) *Plant Pathogen Detection and Disease Diagnosis*, second edition. Marcel Dekker, Inc, New York.

Narayanasamy P (2002) *Microbial Plant Pathogens and Crop Disease Management*. Science Publsihers, Enfield, USA.

Narayanasamy P (2008) *Molecular Biology in Plant Pathogenesis and Disease Management*, volume 2. Springer Science + Business Media B.V., Heidelberg, Germany.

Narayanasamy P (2011) *Microbial Plant Pathogens: Detection and Disease Diagnosis, Volume 2, Bacterial and Phytoplasmal Pathogens*. Springer Science, Heidelberg, Germany.

Narayanasamy P (2017) *Microbial Plant Pathogens – Detection and Management in Seeds and Propagules*, volume 1. Wiley-Blackwell, Chichester, UK.

Nascimento R, Gouran H, Chakraborty S, Gillespie HW, Almeida-Souza HO et al. (2016) The type II secreted lipase/esterase LesA is a key virulence factor required for *Xylella fastidiosa* pathogenesis in grapevines. Sci Rep 6: 18598.

Navarro L, Dunoyer P, Jay F, Arnold B, Dharmasiri N et al. (2006) A plant miRNA contributes to antibacterial resistance by representing auxin signaling. Science 312: 436–439.

Neriya Y, Sugawara K, Maejima K, Hashimoto M, Momatsu K et al. (2011) Cloning expression analysis and sequence diversity of genes encoding two different immunodominant membrane proteins in poinsettia branch-inducing phytoplasma (PoiBI). FEMS Microbiol Lett 361: 115–122.

Nissinen RM, Ytterberg AJ, Boyadanove AJ, van Wijk KJ, Beer SV (2007) Analysis of the secretomes of *Erwinia amylovora* and selected *hrp* mutants reveal novel type III secreted proteins and an effect of HrpJ on extracellular harpin levels. Molec Plant Pathol 8: 55–67.

Niza B, Coletta-Filho HD, Merefa MV, Takita MA, de Souza AA (2015) Differential colonization patterns of *Xylella fastidiosa* infecting citrus genotypes. Plant Pathol 64: 1259–1269.

Nüske J, Fritsche W (1989) Phaseolotoxin production by *Pseudomonas syringae* pv. *phaseolicola*: the influence of temperature. J Basic Microbiol 29: 441–447.

Ochiai H, Inoue Y, Takeya M, Sasaki A, Kaku H (2005) Genome sequence of *Xanthomonas oryzae* pv. *oryzae* suggests contribution of large number of effector genes and insertion sequences to its race diversity. J Agric Res Quart 39: 275–287.

Okamoto M, Tanaka Y, Abrams SR, Kamiya Y, Seki M, Nambara E (2009) High humidity induces abscisic acid 8′-hydroxylase in stomatal and vasculature to regulate local and systemic abscisic acid responses in *Arabidopsis*. Plant Physiol 149: 825–834.

Okechukwu RU, Ekpo EJA (2008) Survival of *Xanthomonas campestris* pv. *vignicola* in infested soil, cowpea seed and cowpea debris. Trop Agric Res Extn 11: 45–48.

Orlovskis Z, Canale MC, Thole V, Pecher P, Lopes JAS et al. (2015) Insect-borne plant pathogenic bacteria getting a ride goes beyond physical contact. Curr Opin Insect Sci 9: 16–23.

Oropeza C, Cordova I, Chumba A, Narvaez M, Saenz LR et al. (2011) Phytoplasma distribution in coconut palms affected by lethal yellowing disease. Ann Appl Biol 159: 109–117.

Panchal S, Roy D, Chitrakar R, Price L, Breitbach ZS et al. (2016) Coronatine facilitates *Pseudomonas syringae* infection of *Arabidopsis* leaves at night. Front Plant Sci 7: 880.

Park D, Mirabella R, Bronstein PA, Preston GM, Haring MA et al. (2010) Mutations in γ-aminobutyric acid (GABA) transaminase genes in plants or *Pseudomonas syringae* reduce bacterial virulence. Plant J 64: 318–330.

Parker JK, Witsotsky SR, Johnson IG, Hijaz FM, Killiny N et al. (2014) Viability of '*Candidatus* Liberibacter asiaticus' prolonged by addition of citrus juice to culture medium. Phytopathology 104: 15–26.

Patel HK, Ferranta P, Xianfa M, Javvadi G, Subramoni S et al. (2017) Identification of loci of *Pseudomonas syringae* pv. *actinidiae* involved in lipase activity and their role in colonization of kiwifruit leaves. Phytopathology 107: 645–653.

Pautot V, Holzer FM, Chaufaux J, Walling LL (2001) The induction of tomato leucine aminopeptidase gene (*LapA*) after *Pseudomonas syringae* pv. *tomato* infection is primarily a wound response triggered by coronatine. Molec Plant-Micro Interact 14: 214–224.

Pel MJC, van Dijken AJH, Bardoel BW, Seidl MF, van der Ent S et al. (2014) *Pseudomonas syringae* evades host immunity by degrading flagellin monomers with alkaline protease AprA. Molec Plant-Micro Interact 27: 603–610.

Perino C, Gaudriault S, Vian B, Barny MA (1999) Visualization of harpin secretion in planta during infection of apple seedlings by *Erwinia amylovora*. Cell Microbiol 1: 131–141.

Pernezny K, Jones JB (2002) Common bacterial blight of snap bean in Florida. Univ Florida Coop Extn Ser Plant Pathol Fact Sheet No. 62.

Picchi SC, Takita MA, Coletta-Filho HD, Machado MA, de Souze AA (2016) N-acetylcysteine interferes with biofilm formation, motility and epiphytic behavior of *Xanthomonas citri* subsp. *citri*. Plant Pathol 65: 561–569.

Potnis N, Krasileva K, Chow V, Almeida NF, Patil PB et al. (2011) Comparison of genomics reveals diversity among xanthomonads infecting tomato and pepper. BMC Genom 12: 14.

Prasad S, Xu J, Zhang Y, Wang N (2016) SEC-translocon-dependent extracytoplasmic proteins of '*Candidatus* Liberibacter asiaticus'. Front Microbiol 7: 1989.

Preiter K, Brooks DM, Penaloza-Vazquez A, Sreedharan A, Bender CL, Kunkel BN (2005) Novel virulence gene of *Pseudomonas syringae* pv. *tomato* strain DC3000. J Bacteriol 187: 7805–7814.

Qian G, Zhang Y, Zhou Y, Liu C, Zhao Y et al. (2012) epv, encoding a hypothetical protein is regulated by DSF-mediating quorum-sensing as well as global regulator Clp and is required for optimal virulence in *Xanthomonas oryzae* pv. *oryzicola*. Phytopathology 102: 841–847.

Quiñones BC, Pujol CJ, Lindow SE (2004) Regulation of AHL production and its contributions to epiphytic fitness in *Pseudomonas syringae*. Molec Plant-Micro Interact 17: 521–531.

Raiol-Junior LL, Baia ADB, Luiz FQBF, Fassini CG, Marques VV et al. (2017) Improvement in the excised citrus leaf assay to investigate inoculation of '*Candiatus* Liberibacter asiaticus' by the Asian citrus psyllid *Diaphorina citri*. Plant Dis 101: 409–413.

Ramos A, Esteves MB, Cortes MTB, Lopes IRS (2020) Maize bushy stunt phytoplasma favors its spread by changing host preference of the insect vector. Insects 11: 600. https://doi.org/10.3390/insects11090600

Rapicavoli JN, Kinsinger N, Perring TM, Backus EA, Shugart HJ et al. (2015) O-antigen modulates insect vector acquisition of the bacterial plant pathogen *Xylella fastidiosa*. Appl Environ Microbiol 78: 638–643.

Rashed A, Nash TD, Paetzold L, Workneh F, Rush CM (2012) Transmission efficiency of '*Candidatus* Liberibater solanacearum' and potato zebra chip disease progress in relation to pathogen titer, vector numbers and feeding sites. Phytopathology 102: 1079–1085.

Rasool A, Jahan MS, Shazad U, Tariq A, Calica PN (2020) Effect of plasma infection on primary and secondary metabolites and antioxidative enzyme activities of sweet orange (*Citrus sinensis* L.). J Plant Pathol Microbiol 11: 519.

Ravindran A, Saenkham P, Levy J, Tamborindeguy C, Lin H, Gross DC, Pierson E (2018) Characterization of the serralysin-like gene of '*Candidatus* Liberibacter solanacearum' associated with potato zebra disease. Phytopathology 108: 327–335.

Roberts R, Liu AE, Wan L, Geiger AM, Hind SR, Rosli HG, Martin GB (2020) Molecular characterization of differences between tomato immune receptors flagellin sensing 3 and flagellin sensing 2. Plant Physiol 183: 1825–1837.

Roberts SJ, Brough J, Everett B, Redstone S (2004) Extraction methods for *Xanthomonas campestris* pv. *campestris* from brassica seed. Seed Sci Technol 32: 439–453.

Rojas ES, Gleason ML (2012) Epiphytic survival of *Erwinia tracheiphyla* on muskmelon (*Cucumis melo* L.). Plant Dis 96: 62–66.

Romero AM, Vega D, Pizzorno R, Cordon G, Correa OS (2018) Hydraulic and leaf reflectance attractions induced by *Clavibacter michiganensis* subsp. *michiganensis* in tomato plants. Eur J Plant Pathol 152: 567–572.

Roper MC, Greve LC, Warren JG, Labavitch JM, Kirpatrick BC (2007) *Xylella fastidiosa* requires polygalacturonase for colonization and pathogenicity in *Vitis vinifera* grapevines. Molec Plant-Micro Interact 20: 411–419.

Roy D, Panchal S, Rosa BA, Melotto M (2013) *Escherichia coli* 0157:H7 induces strong plant immunity than *Salmonella enterica* Typhimurium SL1344. Phytopathology 103: 326–332.

Ryan RP, Vorhofter FJ, Potnis N, Jones JB, Van Sluys MA et al. (2011) Pathogenomics of *Xanthomonas*: understanding bacterium-plant interactions. Nat Rev Microbiol 9: 344–355.

Sabaté J, Rodon J, Artigues M, Lavika A, Battle A (2018) Transmission of '*Candidatus* Phytoplasma pyri' by naturally infected *Cacopsylla pyri* to peach, an approach to the epidemiology of peach yellow leaf roll (PLR) in Spain. Plant Pathol 67: 978–986.

Saleem M, Ji H, Amirullah A, Traw MB (2017) *Pseudomonas syringae* pv. *tomato* DC3000 in multiple gene knockouts predicts interactions among hormonal, biotic and abiotic stress responses. Eur J Plant Pathol 149: 779–786.

Santos CL, Correia-Neves M, Moradas-Ferreira P, Mendes MV (2012) A walk into the LuxR regulators of Actinobacteria: phylogenomic distribution and functional diversity. PLoS ONE 7: e46758.

Santos LVS, Melo EA, Silva AMF, Felix KCS, Quezado-Duval AM, Albuquerque GMR, Gama MAS, Souza ES (2020) Weeds as alternate hosts of *Xanthomonas euvesicatoria* pv. *euvesicatoria* and *X. campestris* pv. *campestris* in vegetable-growing fields in the state of Pernambuco, Brazil. Trop Plant Pathol 45: 484–492.

Schellenberg B, Ramel C, Dudler R (2010) *Pseudomonas syringae* virulence factor syringolin A counteracts stomatal immunity by proteosome inhibition. Molec Plant-Micro Interact 23: 1287–1293.

Schornack S, Meyer A, Romer P, Jordan T, Lahaye T (2006) Gene-for-gene mediated recognition of nuclear-tagged AvrBs3-like bacterial effector proteins. J Plant Physiol 163: 256–272.

Schwartz AR, Morbitzer R, Lahaye T, Staskawicz BI (2017) TALE-induced bHLH transcription factors that activate a pectate lyase contribute to water-soaking in bacterial spot of tomato. Proc Natl Acad Sci USA 114: E897–E903.

Seemüller E, Kiss E, Sule S, Schneider B (2010) Multiple infection of apple trees by distinct strains of '*Candidatus* Liberibacter mali' and its pathological relevance. Phytopathology 100: 863–870.

Seemüller E, Zikeli K, Furch ACU, Winsing A, Jelkmann W (2018) Virulence of '*Candidatus* Phytoplasma mali' strains is closely linked to conserved substitution in AAA+ ATPase AP460 and their supposed effect on enzyme function. Eur J Plant Pathol 150: 701–711.

Sena-Vélez M, Graham JH, Giron JA, Redondo C, Cubero J (2020) Characterization of the extracellular matrix of biofilms formed by *Xanthomonas citri* subsp. *citri* strains with different host ranges. Trop Plant Pathol 45: 306–319.

Sena-Vélez M, Redondo C, Gell I, Ferragud E, Johnson E et al. (2015) Biofilm formation and motility of *Xanthomonas* strains with different citrus host range. Plant Pathol 64: 767–775.

Sharabani G, Manulis-Sasson S, Bornstein M, Shulhani R, Lofthouse ML, Chalupowicz L, Shtienberg DC (2013) The significance of gutation in the secondary spread of *Clavibacter michiganensis* subsp. *michiganensis* in tomato greenhouses. Plant Pathol 62: 578–586.

Shi X, Lin H (2018) The chemotaxis regulator *pilG* of *Xylella fastidiosa* is required for virulence in *Vitis vinifera* grapevines. Eur J Plant Pathol 150: 351–362.

Shiotani H, Uematsu T, Tsukamoto Y, Shimizu Y, Ueda K et al. (2009) Survival and dispersal of *Xanthomonas citri* pv. *citri* from infected Satsuma mandarin fruit. Crop Protect 28: 19–23.

Siciliano F, Torres P, Sendin L, Bermejo C, Fillippone P, Vellice G, Ramallo J (2006) Analysis of the molecular basis of *Xanthomonas axonopodis* pv. *citri* pathogenesis in *Citrus limon*. Electronic Biotechnol. https://doi.org/10.225/vol19-issue 3-full text-20

Singer AU, Schulze S, Skarina T, Xu X, Cui H et al. (2013) A pathogen type III effector with a novel E3 ubiquitin ligase architecture. PLoS Pathog 9. https://doi.org/10.1371/journal.ppat.1003121

Singh VB, Kumar A, Kirubakaran SI, Ayyadurai N, Samish Kumar R et al. (2006) Comparison of exopolysaccharides produced by *Xanthomonas oryzae* pv. *oryzae* strains BX01 and BX08 that vary in degrees of virulence in rice. Phytopathol Z 154: 410–413.

Sklodowska M, Mickicinski A, Wielanek M, Kuzniak E, Schiczewski P (2018) Phenolic profiles in apple leaves and the efficacy of selected phenol against fire blight (*Erwinia amylovora*). Eur J Plant Pathol 151: 213–228.

Slack SM, Zeng Q, Outwater CA, Sundon GW (2017) Microbiological examination of *Erwinia amylovora* exopolysaccharide ooze. Phytopathology 107: 403–411.

Slatter H, Alvarez-Morales A, Barber CE, Daniels MJ, Dow JM (2000) A two-component system involving an HD-GYP domain protein links cell-cell signaling to pathogenicity gene expression in *Xanthomonas campestris*. Molec Microbiol 38: 986–1003.

Smith EF, Townsend CO (1907) A plant tumour of bacterial origin. Science 25: 671–673.

Sobiczewski P, Iakimova ET, Mokincinski A, Wegrzynowicz-Lesiak E, Dyki B (2017) Necrotrophic behavior of *Erwinia amylovora* in apple and tobacco leaf tissue. Plant Pathol 66: 842–855.

Solé M, Scheibner F, Hoffmeister A-K, Hartman N, Hause G et al. (2015) *Xanthomonas campestris* pv. *vesicatoria* secretes proteases and xylanases via the Xps-type III secretion system and outer membrane vesicles. J Bacteriol 197: 2879–2893.

Song Z, Zhao Y, Zhou X, Wu G, Zhang Y et al. (2015) Identification and characterization of two novel DSF-controlled virulence-associated genes within *nodB-rhgB* locus of *Xanthomonas oryzae* pv. *oryzicola*. Phytopathology 105: 588–596.

Sperandio V, Torres AG, Jarvis B, Nataro JP, Kaper JB (2003) Bacteria-host communication: the language of hormone. Proc Natl Acad Sci USA 100: 8951–8956.

Su P, Song Z, Wu G, Zhao Y, Zhang Y et al. (2018) Insights into the roles of two genes of the histidine biosynthesis operon in pathogenicity of *Xanthomonas oryzae* pv. *oryzicola*. Phytopathology 108: 542–551.

Su Q, Wang X-X, Leng M, Qi Y-H, Pang F-Y, Tang J-L, Lu G-T (2021) A HU-like protein is required for full virulence in *Xanthomonas campestris* pv. *campestris*. Molec Plant Pathol 22: 1574–1586.

Suzuki F, Sawada H, Azegami K, Tsuchiya K (2004) Molecular characterization of the *tox* operon involved in toxoflavin biosynthesis of *Burkholderia glumae*. J Gen Plant Pathol 70: 97–107.

Suzuki S, Oshima K, Kakizawa S, Arshida R, Jung HY et al. (2006) Interaction between the membrane of a pathogen and insect microfilament complex determines insect vector specificity. Proc Natl Acad Sci USA 103: 4252–4257.

Szczesny R, Buttner D, Escolar L, Schulze S, Seiferth A et al. (2010) Suppression of the AvrBs1-specific hypersensitive response by YopJ effector homolog AvrBs1 from *Xanthomonas* depends on a SNF1-related kinase. New Phytol 187: 1058–1074.

Tamir-Ariel D, Rosenberg T, Navon N, Burdan S (2012) A secreted lipolytic enzyme from *Xanthomonsa campestris* pv. *vesicatoria* is expressed in planta and contributes to its virulence. Molec Plant Pathol 13: 556–567.

Tancos KA, Borejsza-Wysocka E, Kuchne S, Breth D, Cox KD (2017) Fire blight symptomatic shoots and the presence of *Erwinia amylovora* in asymptomatic apple budwood. Plant Dis 101: 186–191.

Tanne E, Boudon-Padieu E, Clair D, Davidovich M, Melamed S (2001) Detection of phytoplasma by polymerase chain reaction of insect feeding medium and its use in determining vectoring ability. Phytopathology 91: 741–746.

Tao Y, Xie Z, Chan W, Glazebrook J, Chang HS et al. (2003) Quantitative nature of *Arabidopsis* responses during compatible and incompatible interactions with bacterial pathogens *Pseudomonas syringae*. Plant Cell 15: 317–330.

Thinakaran J, Pierson E, Kunta M, Munyaneza JE, Rush CM et al. (2015) Silverleaf night shade (*Solanum elaeginifolium*), reservoir host for '*Candidatus* Liberibacter solanacearum', the putative causal agent of zebra chip disease of potato. Plant Dis 99: 910–915.

Tian Y, Zhao Y, Shi L, Cui Z, Hu B et al. (2017) Type VI secretion systems of *Erwinia amylovora* contribute to bacterial competition, virulence and exopolysaccharide production. Phytopathology 107: 654–661.

Tian Y, Zhao Y, Wu X, Liu F, Hu B et al. (2015) The type VI protein secretion system contributes to biofilm formation and seed-to-seedling transmission of *Acidovorax citrulli* on melon. Molec Plant Pathol 16: 38–47.

Toruño TY, Stergiopoulos I, Coaker G (2016) Plant pathogen effectors: cellular probes interfering with plant defenses in spatial and temporal manners. Annu Rev Phytopathol 54: 419–441.

Toussaint V, Benoit DL, Carise O (2012) Potential of weed species to serve as a reservoir for *Xanthomonas campestris* pv *vitians*, the causal agent of bacterial leaf spot of lettuce. Crop Protect 41: 64–70.

Tran-Nguyen LTT, Kube M, Schneider B, Reinhardt R, Gibb KS (2008) Comparative genome analysis of '*Candidatus* Phytoplasma australiense' (subgroup tuf-Australia1; rp-A) and 'Ca. Phytoplasma austeris' strains OY-M and AY-WB. Bacteriology 190: 3979–3991.

Trivedi P, Wang V (2016) Host immune responses accelerate pathogen evolution. ISME J 8: 727–731.

Uppalapati SR, Ishiya Y, Wangdi T, Urbanczyk-Wochniak F, Ishiya T et al. (2008) Pathogenicity of *Pseudomonas syringae* pv. *tomato* seedlings; phenotypic and gene expression analyses of the virulence function of coronatine. Molec Plant-Micro Interact 21: 383–395.

Ura H, Furyya N, Iiyama K, Hidaka M, Tsuchiya K et al. (2006) *Burkholderia gladioli* associated with symptoms of bacterial grain rot and leaf sheath browning or rice plants. J Gen Plant Pathol 72: 98–103.

van der Wolf JM, Evenhuis A, Kastelein P, Krijger MC, Funke VZ et al. (2018) Risks for infection of strawberry plants with an aerosolized inoculum of *Xanthomonas fragariae*. Eur J Plant Pathol 152: 711–722.

van der Wolf JM, van der Zouwen PS (2010) Colonization of cauliflower blossom (*Brassica oleracea*) by *Xanthomonas campestris* pv. *campestris* via flies (*Calliphora vomitoria*) can result in seed infestation. J Phytopathol 158: 726–732.

Veluchamy S, Panthee DR (2015) Differential expression analysis of a select list of genes in susceptible and resistant heirloom tomatoes with respect to *Pseudomonas syringae* pv. *tomato*. Eur J Plant Pathol 142: 653–663.

Vidaver AK, Lambrecht PA (2004) Bacteria as plant pathogens. Plant Health Instr. https://doi.org/10.1094/PHJ-1-2004-0809-01

Vitali M, Chitaara W, Galetto L, Bosco D, Marzachi C et al. (2013) Flavescence dorée phytoplasma deregulates stomatal control of photosynthesis in *Vitis vinifera*. Ann Appl Biol 162: 335–346.

Walcott RR, Giataitis RD, Castro AC (2003) Role of blossoms in watermelon seed infestation by *Acidovorax avenae* subsp. *citrulli*. Phytopathology 93: 528–534.

Wallis CM, Chen J (2012) Grapevine Phenolic compounds in xylem sap and tissues are significantly altered during infection by *Xylella fastidiosa*. Phytopathology 102: 816–826.

Wallis CM, Munyaneza JE, Chen R, Novy R, Bester G et al. (2015) '*Candidatus* Liberibacter solanacearum' titers in and infection effects on potato tuber chemistry of promising germplasm exhibiting tolerance to zebra chip disease. Phytopathology 105: 1573–1584.

Wallis CM, Rashed A, Wallingford AK, Paetzold L, Workneh F, Rush CM (2014) Similarities and differences in physiological responses to '*Candidatus* Liberibacter solanacearum' infection among different potato cultivars. Phytopathology 104: 126–133.

Wallis CM, Rashed A, Workneh F, Paetz L, Rush CM (2017) Effects of holding temperature on the development of zebra chip symptoms, '*Candidatus* Liberibacter solanacearum' titers, and phenolic levels in 'Red La Soda' and 'Russet Norkotah' tubers. Amer J Pot Res 94: 334–341.

Wang D, Korban SS, Pusey PL, Zhao Y (2011) Characterization of the RcsC sensor kinase from *Erwinia amylovora* and other enterobacterial. Phytopathology 101: 710–717.

Wang D, Korban SS, Zhao Y (2010) Molecular signature of differential virulence in natural isolates of *Erwinia amylovora*. Phytopathology 100: 192–198.

Wang H, Mc Tavish C, Turechek WW (2018) Colonization and movement of *Xanthomonas fragariae* in strawberry tissues. Phytopathology 108: 681–690.

Wang K, Hiruki C (2005) Distinctions between phytoplasmas at subgroup level detected by heteroduplex mobility assay. Plant Pathol 54: 625–633.

Wang N, Li J-L, Lindow SE (2012) RpfF-dependent regulon of *Xylella fastidiosa*. Phytopathology 102: 1045–1053.

Wang N, Lu SE, Wang J, Chen ZF, Gross DC (2006) The expression of genes encoding lipodepsipeptide phytotoxins by *Pseudomonas syringae* pv. *syringae* is coordinated in response to plant signal molecules. Molec Plan-Micro Interact 19: 257–269.

Wang N, Lu SE, Yang Q, Sze SH, Gross DC (2006) Identification of the *sry-syp* box in the promoter regions of genes dedicated to syringomycin and syringopeptin production by *Pseudomonas syringae* pv. *syringae* E301D. J Bacteriol 188: 160–168.

Wang N, Pierson EA, Seutbal JC, Xu J, Levy JG et al. (2017) '*Candidatus* Liberibacter'-host interface: insights into pathogenesis and disease control. Annu Rev Phytopathol 55: 451–482.

Webb DR, Bonfiglioli RG, Carraro L, Osler R, Symmons RH (1999) Oligonucleotides as hybridization probes to localize phytoplasmas in host plants and insect vectors. Phytopathology 89: 894–901.

Wei W, Davis RE, Nuss DL, Zhao Y (2013) Phytoplasmal infection derails genetically preprogrammed meristem and alters plant architecture. Proc Natl Acad Sci USA 110: 19149–19154.

Wei W, Opgenorth DC, Davis RE, Chang CJ, Summers CG et al. (2006) Characterization of a novel adhesion-like gene and design of a real-time PCR rapid, sensitive and specific detection of *Spiroplasma kunkelii*. Plant Dis 90: 1233–1238.

Weißhaupt S, Kohl L, Kunz S, Hinze M, Ernst M, Schimid A, Voegele RT (2016) Alternative inoculum sources for fire blight: the potential role of fruit mummies and nonhost plants. Plant Pathol 65: 470–483.

Wichmann G, Ritchie D, Kousik CS, Bergelson J (2005) Reduced genetic variation occurs among genes of the highly clonal plant pathogen *Xanthomonas axonopodis* pv. *vesicatoria*, including the effector gene avrBs2. Appl Environ Microbiol 71: 2418–2432.

Wöhner TW, Richter K, Sundin GW, Zhao Y, Stockwell VO et al. (2018) Inoculation of *Malus* genotypes with a set of *Erwinia amylovora* strains indicates a gene-for-gene relationship between the effector gene *eop1* and both *Malus floribunda* 821 and *Malus* 'Evereste'. Plant Pathol 67: 938–947.

Wright CA, Beattie GA (2004) *Pesudomonas syringae* pv. *tomato* cells encounter inhibitory levels of water stress during the hypersensitive response of *Arabidopsis thaliana*. Proc Natl Acad Sci USA 101: 3269–3274.

Wu CF, Lin JS, Shaw GC, Lai EM (2012) Acid-induced type VI secretion system is regulated by ExoR-ChvG/ChvT signaling cascade in *Agrobacterium tumefaciens*. PLoS Pathog 8: e1002938.

Wu G, Su P, Wang B, Zhang Y, Qian G, Liu F (2017) Novel insights into Tat pathway in *Xanthomonas oryzae* pv. *oryzae* stress adaption and virulence identification and characterization of Tat-dependent translocation proteins. Phytopathology 107: 1011–1021.

Wu J, Alferez FM, Johnson EG, Graham JHC (2018) Upregulation of PR1 and less disruption of hormone and sucrose metabolism in roots is associated with lower susceptibility to '*Candidatus* Liberibacter asiaticus'. Plant Pathol 67: 1426–1435.

Xin XF, Nomura K, Aung K, Velasquez AC, Yao J, Boutrot F, Chang JA, Zipfel C, He SY (2016) Bacteria establish an aqueous living space in plants crucial for virulence. Nature 539: 524–529.

Xu X, Rajashekara G, Paul PA, Miller SA (2012) Colonization of tomato seedlings by bioluminescent *Clavibacter michiganensis* subsp. *michiganensis* under different humidity regimes. Phytopathology 102: 177–184.

Yan Q, Rogan CJ, Anderson JC (2019) Development of a *Pseudomonas syringae-Arabidopsis* suspension cell infection system for investigating host metabolite-dependent regulation of type III secretion and pattern-triggered immunity. Molec Plant-Micro Interact 32: 527–539.

Yang B, Sugio A, White FF (2006) *Os8N3* is a host disease susceptibility gene for bacterial blight of rice. Proc Natl Acad Sci USA 1003: 10503–10508.

Yang B, White FF (2004) Diverse members of AvrBs3/PthA family of type III effectors are major virulence determinants in bacterial blight disease of rice. Molec Plant-Micro Interact 17: 1192–1200.

Yang B, Zhu W, Johnson LB, White FF (2000) The virulence factor AvrXa7 of *Xanthomonas oryzae* pv. *oryzae* is a type III secretion pathway-dependent nuclear-localized double-stranded DNA-binding protein. Proc Natl Acad Sci USA 97: 9807–9812.

Yang Q, Li X, Tu H, Pan SQ (2017) Agrobacterium-delivered virulence protein VirE2 is trafficked inside host cells via a myosin XI-K-powered ER/actin network. Proc Natl Acad Sci USA 114: 2982–2987.

Yu J, Wyadande AC, Fletcher J (2000) *Spiroplasma citri* surface protein P89 implicated in the adhesion to cells of the vector *Circulifer tenallus*. Phytopathology 90: 716–722.

Yun MH, Torres PS, Oridi ME, Rigano LA, Gonzalez-Lamothe R et al. (2006) Xanthan induces plant susceptibility by suppressing callose deposition. Plant Physiol 141: 178–187.

Zeng W, He SY (2010) A prominent role of the flagellin receptor FLAGELLIN-SENSING2 in mediating stomatal response to *Pseudomonas syringae* pv. *tomato* DC3000 in *Arabidopsis*. Plant Physiol 153: 1188–1198.

Zhang S, Mei J, Wang T, Wang C, Zhang W et al. (2015) Identification and expression analysis of Osmi R1861K in rice leaves in response to *Xanthomonas oryzae* pv. *oryzae*. J Gen Plant Pathol 81: 108–117.

Zhang X, Babadoost M (2018) Characteristics of *Xanthomonas cucurbitae* isolates from pumpkin and survival of the bacterium in pumpkin seeds. Plant Dis 102: 1779–1784.

Zhang Y, Zhang F, Li B, Yang YZ, Ibrahim M et al. (2017) Characterization and functional analysis of *clpB* gene from *Acidovorax avenae* susbp *avenae* RS-1. Plant Pathol 66: 1369–1379.

Zhao Y, Qian G, Fan J, Yin F, Zhou Y et al. (2012) Identification and characterization of a novel genes, *hshB* in *Xanthomonas oryzae* pv. *oryzicola*, coregulated by quorum-sensing and *clp*. Phytopathology 102: 252–259.

Zhao Y, Thilmony R, Bender CL, Schaller A, He SY et al. (2003) Virulence systems of *Pseudomonas syringae* pv. *tomato* promote bacterial speck disease in tomato by targeting the jasmonate pathway. Plant J 36: 485–499.

Zheng X-Y, Zhou M, Yoo H, Pruneda-Paz JL, Spivey NW et al. (2015) Spatial and temporal regulation of biosynthesis of the plant immune signal salicylic acid. Proc Natl Acad Sci USA 112: 9166–9173.

Zhou Z, Wu Y, Yang Y, Du M, Zhang X et al. (2015) An Arabidopsis plasma membrane proton ATPas modulates JA signaling and is explated by the *Pseudomonas syringae* effector protein AvrB for stomatal invasion. Plant Cell 27: 2032–2041.

Zipfel C, Robatzek S, Navarro L, Oakeley EJ, Jones JDG et al. (2004) Bacterial disease resistance in *Arabidopsis* through flagellin perception. Nature 428: 764–767.

Additional References

Altmann F (1996) *N*-glycosylation in insects revisited. Trend Glycosci Glycotechnol 8: 101–114.

Ammar E-D, Fulton D, Bai X, Meulia T, Hogenhout SA (2004) An attachment tip and pili-like structure in insect- and plant pathogenic spiroplasma of the class mollicutes. Arch Microbiol 181: 97–105.

Angeles-Ramos R, Vidaver AK, Flynn P (1991) Characterization of epiphytic *Xanthomonas campestris* pv. *phaseoli* and pectolytic xanthomonads recovered from symptomless weeds in the Dominican Republic. Phytopathology 81: 677–681.

Azarabadi S, Abdollahi H, Torabi M, Salehi Z, Nasiri J (2017) ROS generation, oxidative burst and dynamic expression profiles of ROS-scavenging enzymes of superoxide dismutase (SOD), catalase (CAT), ascorbate peroxidase (APX) in response to *Erwinia amylovora* in pear (*Pyrus communis*). Eur J Plant Pathol 147: 279–294.

Barbara DJ, Morton A, Clark MF, Davies DL (2003) Immunodominant membrane proteins from two phytoplasmas in the aster yellows clade (chloroante aster yellows and clover phyllody) are highly divergent in the major hydrophilic region. Microbiology 148: 157–167.

Bartoli C, Roux F, Lamichhane JR (2016) Molecular mechanisms underlying the emergence of bacterial pathogens: an ecological perspective. Molec Plant Pathol 17: 303–310.

Beger S, Sinha AK, Roitsch T (2007) Plant physiology meets phytopathology: plant primary metabolism and plant-pathogen interactions. J Exp Bot 58: 4019–4026.

Bender CL, Alarcon-Chaidez F, Gross DC (1999) *Pseudomonas syringae* phytotoxin: mode of action, regulation and biosynthesis by peptide and polyketide synthetases. Microbiol Molec Biol Rev 63: 266–272.

Block A, Alfano JR (2011) Plant targets for *Pseudomonas syringae* type III effectors: virulence targets or guarded decoys? Curr Opin Micro 14L: 39–46.

Block CC, Shepherd LM, Mbofung-Curtis GC, Sernett JM, Robertson AE (2019) Re-evaluation of seed transmission of *Clavibacter michiganensis* subsp. *nebraskensis* in *Zea mays*. Plant Dis 102: 110–116.

Bonas U, Schulte R, Fenselau S, Minsavage GV, Staskawicz BJ et al. (1991) Isolation of a gene cluster from *Xanthomonas campestris* pv. *vesicatoria* that determined pathogenicity and the hypersensitive response in pepper and tomato. Molec Plant-Micro Interact 4: 81–88.

Boonrod K, Munteanu B, Jarausch B, Jarausch W, Krczal G (2012) An immunodominant membrane protein (Imp) of '*Candidatus* Phytoplasma mali' binds to plant actin. Molec Plant-Micro Interact 25: 889–895.

Boudon-Padieu E, Larrue J, Clair D, Hourdel J, Jeanneau A et al. (2004) Detection and prophylaxis of elm yellows phytoplasmas in France. Invest Agrar Sist Recur For 13: 71–80.

Burbank LP, Stenger DC (2017) The DinJ/RelE toxin-antitoxin system suppresses bacterial proliferation and virulence of *Xylella fastidiosa* in grapevine. Phytopathology 107: 388–394.

Burrill JJ (1878) Pear blight. Trans Illinois State Hortic Soc 114–116.

Collmer A, Badel JL, Charkowski AO, Deng Wl, Fouts DW et al. (2000) *Pseudomonas syringae* pv. *syringae* Hrp type III secretion system and effector proteins. Proc Natl Acad Sci USA 97: 8770–8777.

Cornara D, Sicard A, Zeilinger AR, Porcelli F, Purcell AH, Almeida RPP (2016) Transmission of *Xylella fastidiosa* to grapevine by the meadow spittle bug. Phytopathology 106: 1285–1290.

Cui Y, Zou L, Zou H, Li Y, Zakria M et al. (2013) HrpE3 is a type III effector protein required for full virulence of *Xanthomonas oryzae* pv. *oryzicola* in rice. Molec Plant Pathol 14: 678–692.

Danet J-L, Foissac X, Zreck L, Salar F, Verdin E (2003) '*Candidatus* phlomobacter fragariae' is the prevalent agent of marginal chlorosis of strawberry in French production fields and is transmitted by the planthopper *Cixius wagneri* (China). Phytopathology 93: 644–649.

Daugrois JH, Boisne-Noc R, Rott P (2014) Leaf surface colonization of sugarcane by *Xanthomonas albilineans* and subsequent disease progress vary according to the host cultivar. Plant Dis 98: 191–96.

Denance N, Legendre B, Briand M, Olivier V, de Boisseson C et al. (2017) Several subspecies and sequence types are associated with the emergence of *Xylella fastidiosa* in natural settings in France. Plant Pathol 66: 1054–1064.

Devash Y, Okon Y, Hemis Y (1980) Survival of *Pseudomonas* from tomato in soil and seeds. Phytopathol Z 99: 175–185.

Dow JM, Clarke BR, Milligan DE, Tang JL, Daniels MJ (1990) Extracellular proteases from *Xanthomonas campestris* pv. *campestris*, the black rot pathogen. Appl Environ Micro 56: 2994–2998.

Dutta B, Barman AK, Srinivasan R, Avei U, Ullman DE et al. (2014) Transmission of *Pantoea ananatis* and *P. agglomerans*, causal agents of center rot of onion (*Allium cepa*), by onion thrips (*Thrips tabaci*) through feces. Phytopathology 104: 812–819.

Dutta B, Gitaitis R, Barman A, Avci U, Marasigan K et al. (2016) Interaction between *Frankliniella fusca* and *Pantoea ananatis* in the center rot epidemic of onion (*Allium cepa*). Phytopathology 106: 956–962.

Dutta B, Gitaitis R, Sanders H, Booth C, Smith S, Langston DB Jr (2014) Role of blossom colonization in pepper seed infestation by *Xanthomonas vesicatoria*. Phytopathology 104: 232–239.

Dutta B, Gitaitis R, Smith S, Langston Jr D (2013) Interactions of seedborne bacterial pathogens with host and nonhost plants in relation to seed infestation and seedling transmission. PLoS ONE 9: e99215.

Dutta B, Ha Y, Lessi JT, Avci U, Sparks AC et al. (2015) Pathways of bacterial invasion and watermelon seed infection by *Acidovorax citrulli*. Plant Pathol 64: 537–544.

Dutta B, Schern H, Gitaitis RD, Walcott RR (2012) *Acidovorax citrulli* seed inoculum load affects seedling transmission and spread of bacterial fruit blotch of watermelon under greenhouse conditions. Plant Dis 96: 705–711.

Eckshtain-Levi N, Munitz T, Živanović M, Traore SM, Sproer C et al. (2014) Comparative analysis of type III secreted effector genes reflects divergence of *Acidovorax citrulli* strains into three distinct lineages. Phytopathology 104: 1152–1162.

Ember I, Bodor P, Zsofi-Palfi Z, Ladanyi M, Pasti G et al. (2018) Bois noir affects the yield and wine quality of *Vitis vinifera* L. cv. 'Chardonnay'. Eur J Plant Pathol 152: 185–197.

Englehardt S, Lee J, Gabbler Y, Kemmerling B, Haapalainen ML et al. (2009) Separable roles of the *Pseudomonas syringae* pv. *phaseolicola* accessor protein HrpZ1 in ion-conducting pore formation and activation of plant immunity. Plant J 57: 706–717.

Ference CM, Gochez AM, Behalu F, Wang N, Graham JH, Jones JB (2018) Recent advances in the understanding of *Xanthomonas citri* subsp. *citri* pathogenesis and citrus canker disease management. Molec Plan Pathol 19: 1302–1318.

Francis MI, Kostenyuk I, Orbovic V, Lorkutov A, Zolotukhin M, Graham JH (2011) Automated needle-free injection method for delivery of bacterial suspension into citrus leaf tissues. J Phytopathol 159: 347–351.

Galan JE, Collmer A (1999) Type III secretion machines: bacterial devices for protein delivery into host cells. Science 284: 1322–1328.

Galetto L, Bosco D, Balestrini R, Genre A, Fletcher J, Marzachi C (2011) The major antigenic membrane protein of 'Candidatus Phytoplasma asteris' selectively interacts with ATP synthase and actin of leafhopper vector. PLoS ONE 6: e22571. https://doi.org/10.137/Journal.pone.0022571

Garcia AL, Torres SCZ, Heredia M, Lopes SA (2012) Citrus responses to *Xylella fastidiosa* infection. Plant Dis 96: 1245–1249.

Gelan JE, Collmer A (1999) Type III secretion machines: bacterial devices for protein delivery into host cells. Science 284: 1322–1329.

Goto M, Ohata K (1956) New bacterial diseases of rice brown stripe and grain rot. Ann Phytopathol Soc Jpn 21: 46–47.

Gottwald TR (2010) Current epidemiological understanding of citrus huanglongbing. Annu Rev Phytopathol 48: 119–139.

Gottwald TR, Graham JH, Schubert TS (1997) An epidemiological analyses of the spread of citrus canker in urban Miami, Florida and synergistic interaction with Asian citrus leaf miner. Fruits 52: 371–378.

Hilf ME, Luo W (2018) Dynamics of '*Candidatus* Liberibacter asiaticus' colonization of new growth of citrus. Phytopathology 108: 1165–1171.

Himeno M, Kitazawa Y, Yoshida T, Maejima K, Yamaji Y et al. (2014) Purple top symptoms are associated with reduction of leaf cell death in phytoplasma-infected plants. Sci Rep 4: 4111.

Hopkins DL, Thompson CM (2002) Seed transmission of *Acidovorax avenae* subsp. *citrulli* in cucurbits. HortScience 37: 924–926.

Hoshi A, Ishii Y, Kakizawa S, Oshima K, Namba S (2007) Host-parasite interaction of phytoplasmas from a molecular biological perspective. Bull Insectol 60: 105–107.

IRPCM (2004) '*Canidatus* Phytoplasma', a taxon for wall-less, nonhelical prokaryotes that colonize plant phloem and insects. Internatl J Syst Evol Microbiol 54: 1243–1255.

Jackson RW, Athanassopoulos E, Tscamis G, Mansfield JW, Sesma A et al. (1999) Identification of a pathogenicity island which contains gene for virulence and avirulence on a large native plasmid in bean pathogen *Pseudomonas syringae* pathovar *phaseolicola*. Proc Natl Acad Sci USA 96: 10875–10880.

Jalai A, Mehdi S, Sangtarash MH (2017) Comparative genomic analysis of wide and narrow host range strains of *Xanthomonas citri* subsp. *citri*, showing differences in the genetic content of their pathogenicity and virulence factors. Austr Plant Pathol 46: 49–61.

Jones D, Dangl JL (2006) The plant immune system. Nature 444: 323–329.

Kakizawa S, Oshima K, Ishii Y, Hoshi A, Maejima K et al. (2009) Cloning of immunodominant membrane protein genes of phytoplasmas and their in planta expression. FEMS Microbiol Lett 203: 92–101.

Kang Y, Choi O, Jeong Y, Jeong JE, Jim JY et al. (2007) Regulation of polar flagellum genes is mediated by quorum sensing and FlhDC in *Burkholderia glumae*. Molec Microbiol 64: 165–179.

Kim J-G, Taylor KW, Mudgett MB (2011) Comparative analyses of the XopD type III secretion (T3S) effector family in plant pathogenic bacteria. Molec Plant Pathol 12: 715–730.

Kloek AP, Brooks DM, Kunkel BN (2000) A *dsbA* mutant of *Pseudomonas syringae* exhibits reduced virulence an partial impairment of type III secretion. Molec Plant Pathol 1: 139–150.

Koinuma H, Miyazaki A, Wakaki R, Fujimoto Y, Iwabuchi N et al. (2018) First report of '*Candidatus* Phytoplasma pruni' infecting cassava in Japan. J Gen Plant Pathol 84: 300–304.

Kosovac A, Radonjic S, Hrncic S, Krstić O, Toševski I, Jović J (2016) Molecular tracing of the transmission routes of bois noir in Mediterranean vineyards of Montenegro and experimental evidence for the epidemiological role of *Vitex agnus-castus* (Lamiaceae) and associated *Hyalesth obsoletus* (Cixiidae). Plant Pathol 65: 285–298.

Krugner R, Ledbetter CA (2016) Rootstock effects on almond leaf scorch disease incidence and severity. Plant Dis 100: 1617–1621.

Kunta M, da Graca JV, Malik NSA, Luzada ES, Setamou M (2014) Quantitative distribution of '*Candidatus* Liberibacter asiaticus' in the aerial parts of the huanglongbing-infected citrus trees in Texas. HortScience 49: 65–68.

Lee D, Bourdais G, Yu G, Robatzek S, Coaker G (2015) Phosphorylation of the plant immune regulator RPM1-INTERACTING PROTEIN 4 enhances plant plasma membrane H+-ATPase activity and inhibits flagellin-triggered immune responses in *Arabidopsis*. Plant Cell 27: 2042–2056.

Lee I-M, Davis RE (1983) Phloem-limited prokaryotes in sieve elements isolated by enzyme-treatment of diseased plant tissue. Phytopathology 73: 1540–1543.

Lee I-M, Hammond RW, Davis RE, Gunderson DE (1993) Universal amplification and analysis of pathogen 16S rDNA for classification and identification of mycoplasma-like organisms. Phytopathology 83: 834–842.

Lee JH, Zhai YF (2016) Integration of host factor is required for RpoN-dependent *hrpL* gene expression and controls motility by positively regulating *rsmB* sRNA in *Erwinia amylovora*. Phytopathology 106: 29–36.

Li Y, Hao G, Galvani CD, Meng Y, De La Fuente L, Hoch HC, Burr TJ (2007) Type I and type IV pili of *Xylella fastidiosa* affect twitching motility, biofilm formation and cell-to-cell aggregation. Microbiol 153: 719–726.

Lopes SA, Luiz FQBF, Oliveira HT, Cifuentes-Arenas JC, Raiol-Junior LL (2017) Seasonal variation of '*Candidatus* Liberibacter asiaticus' titers in new shoots of citrus in distinct environments. Plant Dis 101: 583–590.

Mahmood T, Jan A, Kakrishima M, Komatsu S (2006) Proteomic analysis of bacterial blight defense-responsible proteins in rice leaf blades. Plant Cell 6: 2265–2276.

Mann RS, Ali JG, Hermann SL, Tiwari S, Pelz-Stelinski KS et al. (2012) Induced release of a plant-defense volatile deceptively attracts insect vectors to plants infected with a bacterial pathogen. PLoS Pathog 8: e100.2610.

Martini X, Hoffmann, M, Coy MR, Stelinski LL, Pelz-Stelinski KS (2015) Infection of an insect vector with a bacterial plant pathogen increases in propensity for dispersal. PLoS ONE 10: e0129373.

Mattick JS (2002) Type IV pili and twitching mobility. Annu Rev Microbiol 56: 298–314.

Minato N, Himeno M, Hoshi A, Maejima K, Komatus K et al. (2014) The phytoplasmal virulence factor TENGU causes plant sterility by downregulating of the jasmonic acid and auxin pathways. Sci Rep 4: 7399.

Monger WA, Jeffries CJ (2018) A survey of '*Candidatus* Liberibater solanacearum' in historical seed from collections of carrot and related Apiaceae species. Eur J Plant Pathol 150: 803–815.

Moragrega C, Matias J, Aleta N, Montesinos E, Rovira M (2011) Apical necrosis and premature drop of Persian (English) walnut fruit caused by *Xanthomonas arboricola* pv. *juglandis*. Plant Dis 95: 1565–1570.

Morton A, Davies DL, Blomquist CL, Barbara DJ (2003) Characterization of homologues of the apple proliferation immunodominant membrane protein genes from three related phytoplasmas. Molec Plan Pathol 4: 109–114.

Nadarsah G, Stavrinides J (2011) Insects as alternative hosts for phytopathogenic bacteria. FEMS Microbiol Rev 35: 555–575.

Namba S (2019) Molecular and biological properties of phytoplasmas. Proc Jpn Acad Ser B 95: 401–418.

Navarro L, Jay F, Nomura K, He SY, Voinnet O (2005) Suppression of the microRNAs pathway by bacterial effector proteins. Science 321: 964–967.

Nomura K, DebRoy S, Lee YH, Pumplin N, Jones J et al. (2006) A bacterial virulence protein suppresses host innate immunity to cause disease. Science 313: 220–223.

Oshima K, Kakizawa S, Arashida R, Ishii Y, Hoshi A et al. (2007) Presence of two glycolytic gene clusters in a severe pathogenic line of '*Candidatus* Phytoplasma asteris'. Molec Plant Pathol 8: 481–489.

Popa O, Hazjani-Cove E, Landan G, Martin W, Dagan T (2011) Directed networks reveal genomic barriers and DNA repair bypass to lateral gene transfer among prokaryotes. Genome Res 21: 599–609.

Quecine MC, Silva TM, Carvalho G, Saito S, Mondin M et al. (2016) A stable *Leifsonia xyli* subsp. *xyli* GFP-tagged strain reveals a new colonization niche in sugarcane tissues. Plant Pathol 65: 154–162.

Rashed A, Workneh F, Paetzold L, Gray J, Rush CM (2014) Zebra chip disease development in relations to plant age and time of '*Candidatus* Liberibacter solanacearum' infection. Plant Dis 98: 15–26.

Rashed A, Workneh F, Paetzold L, Rush CM (2015) Emergence of '*Candidatus* Liberibacter solanacearum' infected seed potato in relation to the time of infection. Plant Dis 99: 274–280.

Rauf S, Gisch N, Schaffer M, Illig T, Westphal L et al. (2015) A lectin S-domain receptor kinase mediates lipopolysaccharide sensing in *Arabidopsis thaliana*. Nat Immunol 16: 426–433.

Ruggiero A, Punz P, Landi S, Costa A, van Oosten MM, Grillo S (2017) Improving plant water use efficiency through molecular genetics. Horticulture 3: 31. https://doi.org/10.3390/horticulture 3020031

Saettler AW (1989) Assessment of yield loss caused by common blight on beans in Uganda. Annu Rep Bean Imp Coop 35: 113–114.

Schneider B, Ahrens U, Kirkpatrick BC, Seemüller E (1993) Classification of plant pathogenic mycoplasma-like organisms using restriction-site analysis of PCR amplified 16S rDNA. J Gen Microbiol 139: 519–527.

Schumann GL (1991) *Plant Diseases: Their Biology and Social Impact*. The American Phytopathological Society, St Paul, MN, USA.

Smith ARW, Hodson A (2011) Effects of *Pseudomonas syringae* pv. *morsprunorum* strain C28, outer membrane and LPS preparations on cherry callus. J Phytopathol 159: 358–367.

Sugawara K, Homma Y, Komatsu K, Himeno M, Oshima K et al. (2013) The alteration of plant morphology by small peptides released from the proteolytic processing of bacterial peptide TENGU. Plant Physiol 162: 2005–2014.

Sugio A, Kingdom HN, MacLean AM, Grieve VM, Hogenhout SA (2011) Phytoplasma protein effector SAP11 enhanced insect vector reproduction by manipulating plant development and defense hormonr biosynthesis. Proc Natl Acad Sci USA 108: E1254–E1263.

Supuran CT, Scozzafava A, Mastrolorenzo A (2001) Bacterial proteases: current therapeutic use and future prospects for development of new antibiotics. Expert Opin Therap Pat 11: 221–259.

Thebaud G, Yvon M, Alary R, Sauvion N, Labonnne G (2009) Efficient transmission of '*Candidatus* Phytoplasma prunorum' is delayed by eight months due to a long latency in its host-alternating vector. Phytopathology 99: 265–273.

Thomma BPHJ, Penninck JA, Brockaert WF, Cammune BP (2001) The complexity of disease signaling in *Arabidopsis*. Curr Opin Immunol 13: 63–68.

van der Wolf JM, van der Zouwen PS, van der Heijden L (2013) Flower infection of *Brassica oleracea* with *Xanthomonas campestris* pv. *campestris* results in high levels of seed infection. Eur J Plant Pathol 136: 103–111.

Vega D, Romero AM (2016) Survival of *Clavibacter michiganensis* subsp. *michiganensis* in tomato debris under greenhouse conditions. Plant Pathol 65: 545–550.

Walcott RR (2005) Bacterial fruit blotch of cucurbits. Plant Health Instr. https://doi.org/10.1094/PHI-I-2005-1025-02

Wallis CM, Rashed A, Chen J, Paetzold L, Workneh F, Rush CM (2015) Effects of potato-psyllid vectored '*Candidatus* Liberibacter solanacearum' infection on potato leaf and stem physiology. Phytopathology 106: 189–198.

Wang S, Sun J, Fan F, Tan Z, Zou Y (2016) A *Xanthomonas oryzae* pv. *oryzae* effector XopR associates with receptor-like cytoplasmic kinases and suppresses PAMP-triggered stomatal closure. Sci China Life Sci 59: 897–905.

Wei W, Davis RE, Lee I-M, Zhao Y (2007) Computer simulated RFLP analysis of 16S rDNA genes identification of ten new phytoplasma groups. Internatl J Syst Evol Microbiol 57: 1855–1867.

Wei W, Kakizawa S, Suzuki S, Jung H-Y, Nishigawa H et al. (2004) In planta dynamic analysis of onion yellows phytoplasma using localized inoculation by insect transmission. Phytopathology 94: 244–250.

Workneh F, Henne DC, Childers AC, Paetzold L, Rush CM (2012) Assessments of the edge effect in intensity of potato zebra chip disease. Plant Dis 96: 943–947.

Wu S-W, Kumar R, Iswanto ABB, Kim J-Y (2018) Callose balancing at plasmodesmata. J Exptl Bot 69: 5325–5339.

Xue C, Liu Z, Dai L, Bu J, Liu M et al. (2018) Changing host photosynthetic, carbohydrate and energy metabolisms play important roles in phytoplasma infection. Phytopathology 108: 1067–1077.

Zhang J, Li W, Xiang T, Liu Z, Laluk K et al. (2010) Receptor-like cytoplasmic kinases integrate signaling from multiple plant immune receptors and are targeted by a *Pseudomonas syringae* effector. Cell Host Micro 7: 290–301.

Zhao Y, He SY, Sundin GW (2006) The *Erwinia amylovora avrRpt2(EA)* gene contributes to virulence on pear and AvrRpt2(EA) is recognized by *Arabidopsis* RPS2 when expressed in *Pseudomonas syringae*. Molec Plant-Micro Interact 19: 644–654.

Zhou X, Teper D, Andrade MO, Zhang T, Chen S et al. (2018) A phosphorylation switch on Lon protease regulates bacterial type III secretion system in host. MBIO 9: 10–1128.

6 Detection and Identification of Phyllosphere Viruses and Viroids

6.1 INTRODUCTION

Phyllosphere viruses and viroids are ultramicroscopic and obligate intracellular parasites of plants and induce various kinds of symptoms or remain asymptomatic in susceptible plants. External symptoms may be useful only to group them as mosaic, vein clearing, dwarfing, and phyllody, and it is not possible to detect and identify them precisely. Morphological, biochemical, immunological, and genetic characteristics form the basis for development of various techniques for early detection and precise identification, differentiation, and quantification of viruses in infected plants, seeds/propagules, and their biological vectors, which have an important role in the dissemination of viruses from infected plants to healthy plants. Various immunoassays, polymerase chain reaction (PCR)–based assays and loop-mediated isothermal amplification (LAMP) assays, are applied for detection and vary in their sensitivity, reliability, and rapidity. Nanotechnology and the remote sensing method have the potential for wide application for detection and identification of plant viruses in plant materials. Cost-effectiveness, field applicability, sensitivity, and reliability are the attributes considered for selecting technique(s) suitable for field survey, quarantine operations, and production of virus-free plants, seeds, and propagules that are important and a basic requirement for development of management systems for various crops. Viroids have the simplest structure among the microbial pathogens, as they exist only as naked RNA molecules with all power for inducing diseases as the viruses and other pathogens. Molecular techniques are employed for their detection, identification, and differentiation in infected plants and planting materials.

Viruses can infect all organisms from highly evolved human beings to primitive prokaryotes. However, they were proved to be the cause of plant diseases only in the late 19th century. Although significant advancements have been made in understanding different aspects of host plant–virus interactions, management of plant virus diseases is still restricted largely to preventive methods such as elimination of infected sources of inoculum, use of virus-free seeds and planting materials, and reducing the vector population involved in the dispersal of viruses from infected plants to healthy plants. The development of techniques for rapid, sensitive, specific, and reliable detection and identification of viruses and their strains with varying levels of virulence in plants, seeds, and planting materials has been useful for reducing incidence and spread of virus diseases by applying different tactics. Plant viruses are ultramicroscopic, intracellular, obligate pathogens possessing limited morphological characteristics useful for their precise identification. Viruses are simple in constitution, primarily containing a protein coat that encloses the viral genomic nucleic acid—ribonucleic acid (RNA) or deoxyribonucleic acid (DNA). Viruses do not have any metabolic activity, facilitating infection of susceptible host plants. Hence, they need to be introduced into cells of susceptible host plants by their respective biological vector(s) or by propagative methods (grafting/budding) practiced for horticultural crops and under experimental conditions. Most of the plant viruses depend on different kinds of insects, mites, mealy bugs, nematodes, and fungi for transmission from infected plants to healthy plants. Transmission of certain viruses may occur through parasitic dodder under natural conditions. Mechanical inoculation of plants with virus inoculum is performed under experimental conditions to study different aspects of viruses. Various techniques based on the biological, immunological, and genetic characteristics of viruses have been developed for their rapid, reliable and sensitive detection, identification, differentiation, and quantification. As these techniques differ in the levels of sensitivity, reliability, and rapidity, they have to be selected depending on the cost-effectiveness and field applicability so that crop growers can take appropriate disease management decisions. The usefulness and limitations of the different techniques are discussed hereunder.

6.2 DETECTION AND IDENTIFICATION OF PLANT VIRUSES

6.2.1 Plant Virus Taxonomy

The system of naming plant viruses was initially based on the type of major symptoms induced and host plant species infected (e.g., *Tobacco mosaic virus*). As the number of viruses reported on plants increased rapidly, this system was found to be inadequate. The need for a comprehensive system of nomenclature and classification for viruses based on various characteristics of the purified viruses became essential. The International Committee on Nomenclature of Viruses (ICNV) was established in 1966 to develop a universal taxonomic system for viruses. The name of the committee was later changed to the International Committee on Taxonomy of Viruses (ICTV) to develop a system for naming and classification of viruses. With increasing knowledge on the characteristics of plant viruses, they have been classified into 20 families and 71 genera, with another 17 genera to be assigned later to appropriate families (Mayo and Brunt 2007). The ninth report of the ICTV on Virus Taxonomy

DOI: 10.1201/9781003456827-6

was presented in 2011. This report included 2,284 virus and viroid species distributed among 349 genera, 19 subfamilies, 87families, and 6 orders. Each genus contains a type species and other species. The type species is a characterized virus representative of the species as a whole. Most species are members of a genus (King et al. 2011). Plant viruses are placed in two groups based on the nature of viral genome as RNA viruses and DNA viruses. The RNA viruses are further grouped into single-stranded (ss) + sense RNA viruses, double-stranded (ds) + sense RNA viruses, and ss – sense RNA viruses. On the other hand, DNA viruses, being numerically fewer than RNA viruses, are divided into ss-DNA viruses and reverse transcribing ds-DNA viruses. Particle morphology and presence of membranous envelope, immunological properties, nucleotide sequences of genomic nucleic acid and amino acid sequences of the viral coat protein, and presence of virus-associated proteins have also been used as parameters for classification of viruses and their strains (Narayanasamy 2011).

6.2.2 Biological Methods

6.2.2.1 Symptomatology

Phyllosphere viruses may infect several plant species which express either localized symptoms on inoculated or infected leaves as local lesions, vein clearing or vein banding or systemic symptoms in organs/tissues away from site of inoculation. Systemic symptoms include general stunting or dwarfing of whole plants or specific organs, malformation/discoloration of leaves and floral tissues. Symptoms expressed visibly may reflect corresponding changes in internal tissues and/or derangements in the physiological functions in infected plants. The formation of local lesions, either chlorotic or necrotic, on some plant species may be useful, not only for detection of the virus but also to estimate the concentration of the virus present in the inoculum. Plant species responding with local lesions following virus infection such as *Nicotiana glutinosa, Gomphrena globosa, Phaseolus vulgaris, Vigna unguiculata, Datura metel, D. ferox*, and *Chenopodium amaranticolor* are known as diagnostic/indicator hosts (Narayanasamy and Duraiswamy 2003). *Pepper vein mottle virus* (PMMoV), earlier known as *Tobacco mosaic virus*-pepper strain, was identified by inoculating extracts of infected seeds onto to leaves of *N. glutinosa*. Countable necrotic local lesions developed after inoculation was useful for quantitative estimation of PMMoV (Toyoda et al. 2004). *Zucchini yellow mosaic virus* (ZYMV) isolates infecting cucurbits could be detected and differentiated based on type of symptoms induced by them on nine host plant species which reacted with local lesions or systemic symptom (Bananej et al. 2002). *Cucumber mosaic virus* (CMV) isolates associated with banana mosaic disease (BMD) were grouped into four pathotypes based on symptoms induced in banana, *Nicotiana glutinosa*, and cowpea (Chou et al. 2009). Lisianthus (*Eustoma russellianum*) was used as the indicator host plant to study distribution of *Iris yellow spot virus* (IYSV) in onion and transmission of IYSV by *Thrips tabaci* (Srinivasan et al. 2011). *Potato yellow blotch virus* (PYBV) infection was observed in a potato breeding line in Scotland. PYBV was comparable to *Potato virus A* in infecting solanaceous and nonsolanaceous indicator plant species with the exception of *Solanum deimssum*. But different symptoms were induced by PYBV and PVA in *Nicotiana benthamiana* and *N. occidentalis* P1 (Nisbet et al. 2019).

As the viruses moved from infected/inoculated sites, they become systemic, inducing different secondary symptoms in tissues away from infection sites. The secondary symptoms may be expressed as color changes, teratological symptoms, necrosis, and abnormal growth forms. Mosaic patterns, chlorosis, leaf roll, leaf curl, enations, galls, and tumors may be observed in different aerial plant parts as well as general stunting of plants and reduction in internodes and stems. Some of the external symptoms may indicate the identity of the viruses involved. Viruses cause certain internal symptoms due to anatomical and cytological changes in infected tissues. The production of intracellular virus-specific inclusion bodies has diagnostic value in the case of some viruses. Members of 20 virus groups induce inclusions which form one of the main characteristics of the respective group. These inclusions are useful for diagnostic and taxonomic purposes and for identification and characterization of virus-specific, noncapsid proteins and possible sites of virus synthesis. Various types of inclusions and their usefulness for virus detection and identification are described in detail in an earlier publication (Narayanasamy 2001). Epidermal strips from leaves and stems of squash and watermelon plants infected by *Squash vein yellowing virus* (SqVYV) were stained with Azure A (AA) and Luxol Brilliant-Green (LBG). The presence of cylindrical inclusions were visualized in SqVYV-infected watermelon and squash along the plasma membrane of the cell wall of epidermal tissues. However, additional inclusions that stained with LGB and light pink with AA were observed in vascular tissue, indicating the presence of protein and RNA respectively (Adkins et al. 2007). Symptomatic leaf tissues infected by *Citrus leprosis virus*, on examination under transmission electron microscope (TEM) revealed the presence of viroplasms in the nuclei of infected parenchyma cells and rod-shaped particles resembling those typically observed during infection by dichorhaviruses. The causal virus was tentatively named *Citrus chlorotic spot virus* (CiCSV) (Chabi-Jesus et al. 2018).

6.2.2.2 Modes of Transmission of Viruses

Plant viruses may be transmitted by different methods under natural and experimental conditions. Plant viruses may transmitted by artificial inoculation (mechanical/grafting/budding) or naturally by biological agencies (vectors/dodder). In a few cases, natural root grafts may result in infection of healthy plants. The most common natural method of transmission is through biological vectors belonging to different kinds of insects, mites, mealybugs, nematodes and fungi. Grapevine viruses can be detected by graft-inoculating *Vitis rupestris* St. George. *V. vinifera* cv. Cabernet Frane could be employed to detect *Grapevine leafroll- associated viruses* (Rowhani et al. 2005). Members of the genera *Ilarvirus* and *Trichovirus* infecting *Prunus* spp. could be detected by sap inoculation to *Chenopodium quinoa, Nicotiana occidentalis, N. benthmiana* and *Cucumis sativus* (Rowhani et al. 2005).

6.2.2.3 Cross-Protection between Related Viruses/Strains

The relationship between strains of a virus or closely related viruses may be established by demonstrating cross-protection between them in infected plants, which is adopted as a management strategy for viruses like *Citrus tristeza virus* (CTV). The efficacy of three mild strains of CTV in protecting grapefruit cv. Ruby Red on sour orange rootstock was assessed by visual observations and ELISA tests at 16 days after inoculation with mild strains. The mild strain DD102 provided better protection against severe strain of CTV to grapefruit trees (Powell et al. 1999). Cross-protection test was performed for specific diagnosis of *Citrus psorosis virus* (CPsV). *Psorosis B* was used as challenge virus to demonstrate cross-protection against *Psorosis A* in the protected citrus plants. Healthy sweet orange seedlings inoculated with *Psorosis B* exhibited characteristic symptoms within 6 months. In contrast, plants protected with *Psorosis A* did not show symptoms of infection due to cross-protection offered by *Psorosis A*. Cross-protection against *Psorosis B* in citrus plants developed in citrus plants, as no observable symptoms of the virus could be noted (Martin et al. 2002).

6.2.2.4 Physical and Biochemical Methods

Plant viruses are purified from extracts obtained from infected plant tissues. In methods based on the principles of electrophoresis, use of pH gradients or paper curtains have been employed to purify *Southern bean mosaic virus* and for separation of *Tobacco mosaic virus* strains from plant extracts. In the case of unstable viruses like *Citrus infectious variegation virus*, the electrophoretic technique is especially valuable (Narayanasamy and Duraiswamy 2003). Infection by viruses could be detected by differences in the isozyme patterns of specific enzymes in healthy and infected leaves by sodium dodecyl sulfate (SDS)-polyacrylamide gel electrophoresis (PAGE) technique. Several proteins, both structural and nonstructural proteins of viruses are synthesized in virus-infected plants, in addition to the proteins of the host plants. The presence of a unique protein band (32 kDa) only in leaves infected by *Wheat yellow head virus* (WYHV) was detected by employing the SDS-PAGE technique. The amino acid sequences of the virus-induced protein was most closely related to the nucleoprotein of *Rice hoja blanca virus* (Seifers et al. 2005). The SDS-PAGE technique is useful in determining the physical properties of purified viruses that may indicate the identity of the virus and its strains. The relative molecular mass of the CP of *Apricot latent virus* was determined by SDS-PAGE procedure. The dissociated CP migrated as a single band (Ghanem-Sabanadzovic et al. 2005). Likewise, the molecular mass of the dissociated nucleoproteins (NPs) of *Calla lily chlorotic spot virus* (CCSV), *Tomato spotted wilt virus* (TSWV), *Impatiens necrotic spot virus* (INSV), and *Watermelon silver mottle virus* (WSMoV) were determined by the SDS-PAGE procedure (Lin et al. 2005). In another investigation, molecular masses of dissociated nucleoproteins (NPs) of *Chickpea chlorotic stunt virus* (CpCSV-FB) was purified using sucrose density gradient in which virus separated as a single component. SDS-PAGE analysis of purified virus produced a strong and a faint protein band with molecular masses of 20-kDa corresponding to the major CP (Abraham et al. 2006). The SDS-PAGE procedure was employed to purify the recombinant CP protein expressed in *Escherichia coli* and the purified protein was used for generating antibodies specific to the target virus. Tricine-SDS-PAGE system was employed for the purification of the CP of *Grapevine leafroll-associated virus 8* (GLRaV-8) which was serologically distinct from the CP of GLRaV-1 by Western blot analysis (Monis 2000). Bacterially expressed CP of *Faba bean necrotic yellows virus* (FBNYV) moved as a distinct band (Kumari et al. 2001). The CP gene of *Citrus psorosis* virus (CPsV) was expressed in *E. coli* BL21 as a fusion protein. The recombinant protein was purified using affinity chromatography. The fractions were checked for the presence of recombinant protein by SDS-PAGE technique (Loconsole et al. 2006). The purified recombinant fusion N protein of *Tomato spotted wilt virus* (TSWV) was analyzed by SDS-PAGE procedure, followed by Western blot analysis (Wu et al. 2009).

The extent of the incidence of wheat streak mosaic disease caused by *Wheat streak mosaic virus* (WSMV) was assessed using Landsat 5 Thematic Mapper (TM) images in two counties of Texas. WSMV was widely distributed in both counties and diseased plants could be separated from healthy plants based on images. The TM images showed promise for detection and quantification of incidence of wheat streak mosaic disease (Mirik et al. 2011). A novel method involving construction of cell-based sensors for virus detection, based on membrane (antibody) engineered bacteria was developed for detection of *Tobacco mosaic virus* (TMV) and *Cherry leaf roll virus* (CLRV). Sensors containing modified *Escherichia coli* XL-1 Blue MRF Bacteria were used. Detection principle was based on the measurement of changes in bacterial membrane potential as a result of virus-antibody-binding (Gramberg et al. 2012). The effectiveness of near-infrared spectrometry (NIRS) for early detection of cassava frogskin disease CFSD was assessed by testing cassava accessions. Six classification models were compared for their efficacy. The models, support vector machine with a linear kernel (SVM), Bayesian generalized linear model (BGLM), parallel random forest (PRANDF), and partial least squares (PLS) models showed high efficiency in differentiating diseased and healthy accessions with an overall accuracy >80%. The SVM and BGLM models were highly accurate (99.07% and 98.92%, respectively) and had a reproducibility value of 0.98 in differentiating accessions according to their CFSD status. NIRS was found to be a viable alternative for detection of CFSD with advantages of early and precise detection to other diagnostic methods (Freitas et al. 2020).

Virus particle morphological characteristics have been useful to group them as rod-shaped, flexuous rod-shaped, icosahedra-shaped, or bullet shaped viruses. Serological methods did not detect *Citrus psorosis virus* (CPsV) and *Indian citrus ringspot virus* (ICRSV) in Citrus Ringspot-affected plants. But the pathogens were graft-transmissible. However, observations under electron microscope revealed the presence of a rhabdo-like virus which was graft-transmitted to several citrus species and mechanically to herbaceous hosts. Virus particles were bacilliform and resembled rhabdovirus nucleocapsids deprived of lipoprotein envelope. Partial sequences of the viral nucleoprotein and RNA polymerase genes showed a distinct genetic relatedness with cytorhabdoviruses. The virus was called *Iranian citrus ringspot-associated virus* (IrCRSaV) (Sadeghi et al. 2016).

6.2.2.5 Immunoassays

Various immunoassays have been applied for detection, identification, and quantification of plant viruses and their strains in aerial plant organs. Antisera specific to several viruses could be prepared more easily, because of their structural simplicity. When a purified virus preparation is injected into a test animal system polyclonal antibodies (PAbs) are generated and they may recognize different binding sites (epitopes) on the same virus protein molecules. Hybridomas are somatic cell hybrids formed by fusion of ß-lymphocytes (antibody-producing cells) with myeloma cells (capable of multiplying indefinitely). The hybridoma has the ability to produce specific antibody from lymphocyte and grow on synthetic medium in vitro indefinitely from myeloma cells. Monoclonal antibodies (MAbs) produced by a hybridoma are identical and specifically react only with an epitope present on the immunogen. The methods of production of PAbs and MAbs are described in detail in earlier publications (Narayanasamy 2005, 2011, 2017). Binding of an antibody specifically to its homologous antigen indicates positive reactions which may be visualized or recognized clearly. The development of labeled antibody techniques was a significant advancement in application of immunoassays for virus detection in plants and other environmental samples. Labels or markers such as enzymes, fluorescent dyes or radioactive materials are attached to the antigen or antibody. These methods have higher levels of specificity, sensitivity, rapidity and reliability in detection and identification of viruses in plants particularly asymptomatic plants/tissues. Among the labeled antibody methods, enzyme-linked immunosorbent assay (ELISA) and its variants have been employed predominantly for detection of viruses infecting various crops. The possibility of automation for handling large numbers of samples is an added advantage for certification and quarantine programs. ELISA tests have been very useful to detect viruses in seed and vegetatively propagated plant materials (propagules) which are generally asymptomatic with low virus titer.

6.2.2.5.1 Enzyme-Linked Immunosorbent Assay

ELISA is the preferred immunoassay for detection, identification, and quantification of various viruses and their strains, because of its greater sensitivity and specificity of results. Three variants of ELISA, double antibody sandwich (DAS)-ELISA, triple antibody sandwich (TAS)-ELISA, and plate-trapped antigen (PTA) ELISA are frequently used for virus detection. In order to reduce the cost of ELISA tests, horseradish peroxidase (HRP), ß-glucosidase isolated from *Escherichia coli*, and penicillinase were tested for their effectiveness as markers. Alkaline phosphatase amplification (AMP) and penicillinase (PNC) systems could detect *Maize streak virus* (MSV) in both maize plants and also in insect vectors. The color change from blue to yellow (positive) was pronounced in PNC system and ELISA plate reader was not required to record the results (Afolabi and Thottappilly 2008). Antisera against nucleocapsids proteins (NPs) of *Melon yellow spot virus* (MYSV) and *Watermelon silver mottle virus* (WSMoV) were employed for diagnosis of diseases caused by these viruses. Samples from symptomatic plants were tested by ELISA test using MAbs specific to the NP of WSMoV or MYSV. Reverse-transcription polymerase chain reaction (RT-PCR) assay was applied to confirm the results of ELISA tests. Of the watermelon samples assayed, 22.4% and 9.2% samples were singly infected by WSMoV and MYSV respectively, whereas 0.17% samples showed mixed infections by both viruses. Aphid-borne viruses, *Zucchini yellow mosaic virus (ZYMV), Papaya ringspot virus*-watermelon type (PRSV-W) and *Cucumber mosaic virus* (CMV) were also detected in cucurbits. The results indicated the existence of host preference of watermelon by WSMoV and melon by MYSV (Peng et al. 2011). In order to detect new isolates of *Wheat streak mosaic virus* (WSMV) or new viruses, wheat plants exhibiting virus-like symptoms were assayed. WSMV, *Triticum mosaic virus* (TriMV), *High Plains wheat mosaic virus* (HPWMoV) and *Foxtail mosaic virus* (FoMV) were detected by ELISA (Kumssa et al. 2019).

ELISA tests have been extensively employed for detection of a large number of viruses affecting different horticultural crops. The presence of *Grapevine leafroll-associated virus 3* (GLRaV-3) was detected in all infected plants tested by the ELISA technique (Rowhani et al. 1997). Five hybridoma cell lines immunoreactive with *Grapevine leafroll-associated viruses* were cloned and characterized. The DAS-ELISA format using 15F1 MAb and R102 PAb was effective in detecting GLRaV-5 and GLRaV-8 viruses in crude extracts of infected grapevine tissues. The MAbs enhanced specificity and sensitivity of detection of GLRaV-5 and -8 viruses in crude extracts of infected grapevine tissues. The MAbs enhanced specificity and sensitivity of detection of GLRaV-5 and -8 viruses by reducing background due to presence of host-specific antibodies in polyclonal antiserum (Monis 2000). Infected grapevine planting materials are the main mode of long distance dissemination of *Grapevine leafroll-associated virus* 3 (GLRaV-3). A single- chain antibody fragment (scFv) LR3cp-1 specific to coat protein (CP) of GLRa-V-3 was selected from phage display library. The antibody bound specifically to the entire length of GLRaV-3 particles and exhibited a high binding affinity (Orecchia et al. 2008). In a later investigation, the scFv antibody (scFvLR3) targeting the CP of GLRaV-3 was expressed in *E. coli*. The recombinant protein was fused to the light chain constant domain of human immunoglobulin to create the bivalent reagent. The indirect ELISA format showed that the bivalent reagent bound efficiently to IgGLR3 and captured the target virus particles from the sap of grapevine infected by GLRaV-3 (Cogotzi et al. 2009).

Grapevine samples were collected during the dormant period from three different locations. DAS-ELISA was the most specific method providing the highest degree of confidence in positive results for detection of *Grapevine leafroll-associated virus 3* (GLRaV-3) (Olmos et al. 2016). An MAb was employed for rapid detection of *Grapevine leafroll-associated virus 4* (GLRaV-4) strain Ob (GLRaV-4 Ob) by ELISA. Based on the relatively small size of the CP, reduced amino acid identity and distinct serological properties, GLRaV-4 Ob was identified as a divergent strain of GLRaV-4 (Reynard et al. 2015). The frequencies of detection of *Grapevine virus A* (GVA), *Grapevine leafroll-associated virus 3* (GLRaV-3), *Grapevine fan leaf virus* (GFLV), *Grapevine fleck virus* (GFkV), GLRaV-1, GLRaV-2, and *Grapevine virus B* (GVB) were determined (Leo et al. 2015). Infection of apricot trees by a recombinant isolate of *Plum pox virus* (PPV) was detected by DAS-ELISA format using four strain-specific MAbs in Italy (Myrta et al. 2005). ELISA test was effective in detecting infection of geranium (*Pelargonium* spp.) by *Prunus necrotic ringspot virus* (PNRSV). Positive reaction with antibodies generated against ilarviruses indicated the identity of *Prunus necrotic ringspot virus* (PNRSV) infecting rose geraniums (Kulshreshtha et al. 2005). The ELISA test was employed for detection of PNRSV in cherry orchard of *Prunus cerasus* or Monmorency and *P. avium* cv. Hedelfingen in New York. Sour cherry showed higher level of infection by PNRSV than sweet cherry. RT-PCR assay also confirmed the infection of cherry trees by PNRSV (Oliver et al. 2009). The MAbs generated against PNRSV reacted strongly with the virus but not with other viruses infecting cherry (Shang et al. 2013).

Serological detection of *Citrus psorosis virus* (CPsV) was achieved by performing TAS-ELISA format. CPsV was detected in extracts of citrus plants infected with any of the three isolates (PL3, PL4, or PL5) (Martin et al. 2000). Coat protein (CP) gene of CPsV was cloned and sequenced. The recombinant protein was employed to produce polyclonal antiserum against an Italian isolate of CPsV. The PAb As-Ps-Rc1 was effective in detecting CPsV in citrus leaves in DAS-ELISA and indirect ELISA formats (Loconsole et al. 2006). Commercial citrus groves in Cyprus were examined for the infection by *Citrus tristeza virus* and *Citrus variegation virus* (CVV) by biological indexing and ELISA tests. Mother plants free of infection by these viruses were identified, although some of them were infected by *Citrus psorosis virus* (CPsV). Field-deployable lateral flow immunoassay (LFIA) was developed to detect *Citrus tristeza virus* (CTV). PAbs were generated in rabbit against purified virions from *Nicotiana benathamiana* inoculated with an infectious recombinant cDNA clone of CTV (rCTV). LFIA detected CTV within 10 min in extracts from infected plants. The results of LFIA were in agreement of ELISA and RT-quantitative PCR (qRT-PCR) LFIA was cost-effective, rapid, and field-deployable for detection of CTV (Maheshwari et al. 2017).

DAS-ELISA format detected the presence of seven mechanically transmissible *Potato viruses A, Y, V, M, S, X* and aphid-transmitted *Potato leafroll virus* (PLRV) in in vitro generated potato plantlets. Virus-specific antisera were employed in the tests (Khan et al. 2003). A recombinant antibody fusion protein V3HCL showing specific reactivity for PLRV was labeled with biotin. This procedure was effective in detecting PLRV. The scFv reagents derived from synthetic phage display platforms could provide effective alternatives to assays employing immunoreagents (Al-Mrabeh et al. 2009). Variants of plant viruses may arise due to selection or mutation with different infection potential (virulence). Such variants have to be identified rapidly to make necessary changes, if required, in disease management strategy. A mutant strain of *Potato virus M*, called *Potato virus M-ID* (PVM-ID), differing in coat protein sequences in the amino-terminus region was identified using ELISA (Cavileer et al. 1998). ELISA tests were effective in detecting isolates of *Potato virus Y* necrotic strain (PVY^N) causing tobacco veinal necrosis disease using PAbs and MAb4C3 antibodies (Crosslin et al. 2005). The CP gene of the ordinary strain of *Potato virus Y* (PVY^O) was cloned into the expression vector, pET-28a (+). The insert was sequenced and analysis showed that CP gene was in frame with intact N-terminal 6X histidine tags. The approximately 35-kDa recombinant fusion protein was observed in inclusion bodies of induced *E. coli* BL21 cells. This fusion protein was purified and used as antigen to raise PAbs in rabbits. In Western blot and dot blot immunobinding assay (DIBA) both PVY^O-CP IgG and PVY^O IgG strongly reacted with the recombinant CP. The PVY^O-CP IgG could detect PVY^O in infected samples. A PVY^O-CP ELISA was developed and compared with conventional ELISA kit based on purified virus particles (PVY^O-ELISA) of field samples and was effective as PVY^O ELISA kit (Jeevalatha et al. 2013). The genes encoding CP and triple gene block protein (TGBp1) of *Potato virus M* (PVM) were cloned into expression vector pET-45b (+) (N-terminal 6x Histag) and expressed in *E. coli* Rosetta gami-2 (DE3). The purified recombinant antigens were used for generating PAbs. PVM was detected by using PAbs in Western blot analysis, plate-trapped ELISA, and DAS-ELISA as a coating in infected potato leaf extracts. The antibodies against recombinant nonstructural protein detected the TGBp1 only in Western blot analysis (Cerovska et al. 2012).

PAbs generated against nucleocapsids were employed to detect viruses employing in ELISA formats. *Tomato yellow ring virus* (TYRV), occurring in Iran was detected using PAbs targeting nucleoprotein (Hassani-Mehraban et al. 2005). The virus inducing chlorotic spots in leaves of *Phalaenopsis* orchids was established in *Chenopodium quinoa* and *Nicotiana benthamiana*. This isolate reacted positively with POTY monoclonal antibody used for detecting *Potato virus Y* in indirect ELISA format, indicating that the virus infecting *Phalaenopsis* might be a potyvirus. The isolate 7–2 was identified as a member of *Potyvirus*, based on results of electron microscopy, immunoblotting and sequence analysis and called *Phalaenopsis chlorotic spot virus* (PhCSV) (Zheng, Chen, Chen et al. 2008). Another virus causing chlorotic ringspot in leaves of orchid *Phalaenopsis* was identified as an isolate of *Capsicum chlorosis virus* (CaCV), based on reactions of antiserum generated against the nucleocapsid protein (NP) of the virus isolate. The antiserum cross-reacted with viruses belonging to the genus *Tospovirus* in indirect ELISA format (Zheng, Chen, Yang et al. 2008). A nonstructural NSs protein encoded by viral strand of *Watermelon silver mottle virus* (WSMoV) SRNA formed filamentous inclusion bodies in the cytoplasm of plant cells. The NSs protein of WSMoV was expressed by *Zucchini yellow mosaic virus* (ZYMV) vector in squash. The PAbs reacted strongly with NSs crude antigen of WSMoV and weakly with *Capsicum chlorosis virus* (CaCV), but not with *Calla lily chlorotic spot virus* (CCSV) belonging to the same serogroups. On the other hand, the MAbs reacted with NSs proteins of members of WSMoV serogroups in the indirect ELISA format (Chen et al. 2006). Nucleocapsid protein gene of *Tomato spotted wilt virus* (TSWV) was expressed in *Escherichia coli* and the recombinant protein was used to generate PAbs and MAbs capable of reacting with TSWV (Wu et al. 2009). Use of skim milk powder applied as a blocking step, improved the reliability of the assay. IYSV was detected in leek and onion plantings (Smith et al. 2006). The cause of chickpea chlorotic stunt disease was established by employing DAS-ELISA and TAS-ELISA formats. The PAb to *Chickpea chlorotic stunt virus* –FB was used as capture antibody and the MAbs were employed as detecting antibodies in the TAS-ELISA format, allowing more sensitive detection of CpCSV antigen (Abraham et al. 2006).

PAbs were produced in rabbit using bacterially expressed recombinant coat proteins (rCPs) of two serologically distinct *Alfalfa mosaic virus* AMV strains. The antisera against rCPs had antibody contents comparable to antisera prepared using soybean sap containing homologous viruses, as revealed by indirect ELISA format. PAbs against rCPs were specific, sensitive and detected all AMV isolates that originated from soybean fields from geographically different regions of the United States. Phylogenetic analysis also revealed that all isolates were clustered together along with reference AMV strains belonging to group 1. PAbs against rCP could be employed as effective probes for detection of the virus and diagnosis of the disease induced by the virus (Khatabi et al. 2012). PAbs generated against recombinant *E. coli*-expressed nucleocapsid (NP) of the virus reacted specifically with *Soybean vein necrosis-associated virus* (SVNaV). Phylogenetic analysis utilizing NP sequences from several SVNaV isolates showed closer relationships, but distinct from all established serogroups or distinct serotypes within *Tospovirus*. SVNaV might represent a new and distinct soybean-infecting tospovirus serotype (Khatabi et al. 2013). Coat protein (CP) genes of *Bean leaf roll virus* (BLRV) and *Pea enation mosaic virus* (PEMV) were bacterially expressed as fusion proteins and used for production of antisera in rabbits. No cross-reaction between BLRV and PEMV was observed in antigen-coated plate (ACP)-ELISA format, and it could be employed to assess incidence of BLRV and PEMV in alfalfa and pea and vetch (Vemulapati et al. 2014). Virus-associated protein (VAP) coded by ORF II of *Banana streak MY virus* (BSMYV) (banana leaf streak disease) was expressed in *E. coli* and PAbs were generated against purified recombinant VAP (rVAP) fusion protein in rabbits. Immunocapture (IC)-PCR assay using VAP antibody was highly specific and it could differentiate episomal viral infection from integrated endogenous BSV (eBSV) sequences. The rVAP of BSMYV could be employed for large-scale application via in quarantine, certification and germplasm exchange programs (Selvarajan et al. 2016). DAC-ELISA detected *Tobacco streak virus* (TSV) in symptomatic cotton plants and *Chenopodium amranticolor* and *C. quinoa* could be used as indicator plants for detection TSV in cotton (Vinodkumar et al. 2017). *Cymbidium mosaic virus* (CymMV) was detected using dipstick assay. CyMV, *Odontoglossum ringspot virus* (ORSV), and *Orchid fleck virus* (OFV) were present in mixed infections in orchids. In addition, PAbs were generated against *Escherichia coli*-expressed OFV phosphoprotein (OFV-P) and employed in plate-trapped antigen (PTA)-enzyme-linked immunosorbent assay (ELISA) format. Orchids (*Phalaenopsis* hybrid) infection by OFV could be detected by PTA-ELISA efficiently (Bratsch et al. 2021).

6.2.2.5.2 Dot Immunobinding Assay

The dot immunobinding assay (DIBA) is similar to the ELISA format. Microtiter plates are replaced by nitrocellulose or nylon-based membrane on which antigen is immobilized. *Grapevine rupestris stem-pitting-associated virus* (GRSPaV) was detected in leaf petioles and cortical scrapings from dormant canes (Minafra et al. 2000). The DIBA test could be performed using a chemiluminescent compound NBT/BCIP as a substrate to detect viruses. *Soybean mosaic virus* (SbMV) was detected by DIBA test (Ahangaran et al. 2009). The DIBA test was applied to identify banana plants infected by *Banana bunchy top virus* (BBTV) and *Cucumber mosaic virus* (CMV) facilitating eradication of infected plants as a disease management strategy (El-Dougdoug et al. 2006). A rapid diagnostic DIBA procedure was developed to detect CMV in the laboratory and in field. Although the DIBA test was less sensitive than TAS-ELISA format, it offered advantages of simplicity, field applicability during surveys and storability of blots for long periods without loss of reactivity (Zein and Miyatke 2009).

6.2.2.5.3 Tissue Blot Immunoassay

Tissue blot immunoassay (TBIA) is performed by transferring viral antigens from freshly cut plant surface to nitrocellulose membrane. The non-requirement of any sample preparation step and lack of reaction with host plant proteins are important advantages of TBIA over ELISA formats. This method is simple, sensitive, and rapid in terms of providing the results. The imprinted membranes could be stored for long periods and transported to the laboratory for further analysis. *Barley mild mosaic virus* (BaMMV) and *Barley yellow mosaic virus* (BaYMV) could be detected in leaf and stem samples from greenhouse-inoculated and field-grown plants by direct stem blot immunoassay (DSBIA). Stem tissue was more suitable for direct blotting on nitrocellulose membranes. Five of seven fields examined showed infection by one or both viruses. Direct stem blot immunoassay DSBIA was effective in routine indexing of barley genotypes for BaMMV or BaYMV under field conditions (Johnson et al. 2007). Viruses spreading via propagation practices (grafting/budding) have to be detected rapidly and reliably. *Citrus psorosis virus* (CPsV) was detected readily in young shoots and leaves by direct tissue blot immunoassay (DTBIA). CPsV was consistently detected in tissue prints of young tender leaves or shoots. DTBIA had the advantage of being simpler, cheaper, and faster than ELISA and has potential for application when suitable plant tissues could be used for testing (Martin et al. 2002). CPsV was readily detected by TBIA procedure in samples from tender shoots, flowers, and young leaves collected during spring, but not in hardened old leaves (Loconsole et al. 2006).

An in situ immunoassay (ISIA) was developed for detection of *Citrus tristeza virus* (CTV). The fixed tissues were incubated with PAbs or MAbs, followed by application of enzyme-conjugated with secondary antibodies and exposure to a substrate mixture consisting of nitroblue tetrazolium (NBT) and 6-bormo-4-chloro-3-indolyl phosphate (BCIP). The development of purple color indicating positive reaction could be observed under a light microscope. By employing isolate-specific MAbs, mild and severe strains of CTV could be differentiated in ISIA test (Lin et al. 2002). Tissue blots were incubated with pre-reaction solutions of CTV-specific antibodies and labeled secondary antibodies. This procedure allowed detection of both CTV + decline-inducing isolate T-36 by employing CTV-specific PAb and MAbs (Lin et al. 2006). Bacterially expressed coat protein (BE-CP) of *Faba bean necrotic yellows virus* (FBNYV) was employed to generate PAbs. FBNYV was reliably detected, as the PAbs showed strong specific reaction in TBIA test (Kumari et al. 2001). TBIA test was effective for tracking the distribution of *Turnip mosaic virus* (TuMV) and its chimeric viruses in cotyledons and leaves of cabbage and radish (Suchiro et al. 2004). Necrotic strain of *Potato virus Y* (PVY) (isolate T01) induced necrotic symptoms in older leaves of systemically infected tobacco plants doubly infected with *Cucumber mosaic virus* (CMV). PVY was detected at high concentrations in almost all sections including apical tissues in doubly infected tobacco plants. In contrast, PVY isolate was not detectable in the shoot apex of tobacco plants infected only by PVY. The results indicated that protein of CMV, facilitated its spread to apical tissues also. In the absence of CMV, PVY was restricted to lower leaves of tobacco (Ryang et al. 2004). A diagnostic kit based on the principle of TBIA was recommended by the International Center for Agricultural Research on the Dry Areas for detection of 19 viruses infecting legumes. *Soybean mosaic virus* (SbMV) could be detected by tissue printing immunoassay. Tissue prints of infected soybean tissues on nitrocellulose membranes were treated with antibody-enzyme conjugate, followed by incubation with substrate solution containing NBT/BCIP or Fastred. Color development clearly indicated the presence of SBMV antigen in the samples (Ahangaran et al. 2009).

6.2.2.5.4 Immunoblot Assay

The immunoblot assay involves transfer of electrophoresed viral proteins from gel matrix onto a membrane and allowing access subsequently for specific immunoprobes. Native p25 coat proteins (CP) of five isolates of *Citrus tristeza virus* (CTV) were compared based on serological analysis of peptide maps using both MAbs and PAbs. Serological analysis of peptide maps allowed differentiation of CTV isolates (Albiach-Marti et al. 2000). CP gene sequence of Italian isolate of *Citrus psorosis virus* (CPsV) was expressed in *E. coli* BL21 as a fusion protein and antibodies to the recombinant protein was generated. A protein band of expected size for CPsV was identified in Western blotting performed with total protein extract of CPsV-infected *Chenopodium quinoa*. (Loconsole et al. 2006). A potyvirus infecting chickpea was identified as *Chickpea yellow mosaic virus* (CYMV). The total protein sample preparations of CYMV were probed with the *Potyvirus* MAb on immunoblots. A single band of ca. 32-kDa was identified in virus protein preparation purified from chickpea (Larsen et al. 2003). Potyvirus-like long flexuous filamentous particles were visualized in *Phalaenopsis* orchid plants under electron microscope in addition to pin-wheel inclusion. Leaf extracts of mock-inoculated and virus-infected *Chenopodium quinoa* plants were transferred to membranes. The virus infecting *Phalaenopsis* orchid was detected and identified as *Phalaenopsis chlorotic spot virus* (PhCSV) by the SDS-PAGE technique (Zheng, Chen, Chen et al. 2008). Antiserum generated against *Angelonia flower break virus* (AnFBV) reacted strongly and specifically with polypeptide corresponding to the CP of the virus (Adkins et al. 2006).

Bacterially expressed proteins were used to produce specific antisera against recombinant RdRp and CP of *Grapevine leafroll-associated virus 2* (GLRaV2). In Western blots, antisera against RdRp (1: 5,000) or recombinant CP reacted strongly with homologous proteins expressed in *E. coli*. CP antiserum was able to detect infection by GLRaV-2 in all samples (Xu et al. 2006). Virus-specific antiserum was generated against recombinant CP protein of GLRaV-Pr isolate and employed in Western blotting. A band of ca. 37-kDa recognized by the antiserum was consistently detected in all extracts from GLRaV-Pr-infected plants. The pattern expressed from the CP coding ORF 5 had an electrophoretic mobility indistinguishable from the GLRaV-Pr CP expressed from infected tissues. Based on the sequences of the entire genome of GLRaV-Pr, the virus was considered a new species within the genus *Ampelovirus* (Maliogka et al. 2009). The PAbs specific to orchid virus inducing chlorotic ring spot on leaves and PAbs specific to *Watermelon silver mottle virus* (WSMoV) reacted strongly with crude sap of infected leaves of *Phalaenopsis* orchid plants. A protein band measuring about 30-kDa was recognized in Western blotting using the antisera of WSMoV or the orchid isolate. The virus was identified as an isolate of *Capsicum chlorosis virus* (CaCV), based on the sequence analysis of viral N gene determined from cDNA clone (Zheng, Chen, Chen et al. 2008). The viral NSs protein of *Tomato spotted wilt virus* (TSWV) expressed in *E. coli* was transferred to a nitrocellulose membrane. Anti-TSWV MAb1 or anti-HisTagMAb reacted strongly with TSWV N protein (29-kDa) and recombinant TSWV N protein (48-kDa) (Wu et al. 2009).

6.2.2.5.5 Immunochromatographic Assay

A simple immunochromatographic assay (ICA) was developed based on the interaction between the target virus and antibodies and their conjugates with colored colloidal particles. The test strip is dipped into extracts of samples. Then the sample liquid flows through membranes and triggers immunochemical interactions resulting in visible coloration in test and reference lines. Plant viruses such as rod-shaped *Tobacco mosaic virus* (TMV) and filamentous *Potato virus X* (PVX) and *Potato virus Y* (PVY) could be detected by ICA. The ICA could be used for rapid diagnosis of new virus diseases under field conditions (Boyzova et al. 2009).

6.2.2.5.6 Microscopy-Based Immunoassays

An immunochemistry approach was applied to detect and monitor the distribution of *Potato virus Y* (PVY) in tobacco leaves infected singly and doubly with *Cucumber mosaic virus* (CMV) also. PVY antigen was confined to external phloem cells in singly infected tobacco plants. On the other hand, PVY was uniformly distributed in whole tissues including the external phloem, xylem parenchyma and internal phloem cells in doubly infected tobacco plants. The results indicated that presence of CMV facilitated accumulation of PVY in stem tissues which could be associated with enhanced spread of PVY into young tissues of tobacco plants (Ryang et al. 2004). *Citrus psorosis virus* (CPsV) was consistently detected by ISEM technique (Martin et al. 2004). Involvement of an *Ampelovirus* species, *Pineapple mealybug wilt-associated virus 2* (PMWaV-2) was investigated to establish the identity of the etiology of the disease. The relationship between an undescribed *Ampelovirus* sharing highest homology with PMWaV-1 and a putative deletion mutant sharing highest homology with PMWaV-2 was established by ISEM and reverse transcription (RT)-PCR assay (Sether et al. 2005). ISEM procedure effectively detected PMWaV-1, -2, and -3 in germplasm collections (Gambley et al. 2008). Decoration technique was applied to detect and identify *Chickpea chlorotic stunt virus* (CpCSV). The antiserum specific to the virus strongly decorated purified CpCSV-FB isolate (Abraham et al. 2006). The bivalent reagent CL-LR3 produced by fusing the bacterially expressed antibody-specific to *Grapevine leaf roll-associated virus 3* (GLRaV-3) detected GLRaV-3 effectively (Cogotzi et al. 2009). Distribution of the coat protein (CP) of *Tomato yellow leaf curl Sardinia virus* (TYLCSV) was monitored in tomato and *Solanum nigrum* plants using the immunogold labeling technique. By using gold-labeled African cassava mosaic virus (ACMV)-CP antiserum, the label was detected mostly inside mature and immature sieve elements and the nuclei of companion and phloem parenchyma cells of TYLCSV-infected leaves. Pattern of immunogold (IGL) distribution was similar in both TYLCSV-infected tomato and *S. nigrum* and ACMV-infected *Nicotiana benthamiana* plants. In companion cells, the label was absent in palisade and spongy parenchyma cells, indicating TYLCSV was phloem-limited (Medina et al. 2006).

6.2.2.6 Nucleic Acid–Based Techniques

Closely related organisms share greater nucleotide sequences similarity compared to distantly related ones. A highly specific nucleotide sequence present only in the target virus may be used for constructing primers and probes for detection and identification of the virus and its strains. Generally nucleic acid–based methods are more sensitive and specific in detecting the viruses and providing results rapidly compared to immunoassays and biological methods. The diagnostic methods based on assessment of similarities between nucleic acids of viruses and their strains are useful for identification of new or unusual viruses or strains (Narayanasamy 2011, 2017).

6.2.2.6.1 Nucleic Acid Hybridization Techniques

Bamboo mosaic virus (BaMV) and its associated satellite RNA could be detected by employing either ^{32}P or DIG-labeled probes synthesized from cDNA clones of BaMV genomic (L probe) and sat-BaMV (S probe) RNA. ^{32}P labeled L and S probes were more sensitive (25-fold) than DIG-labeled probes in detecting the virus in infected leaf extracts (Hsu et al. 2000). ^{32}P labeled cDNA probe prepared from N gene of *Calla lily chlorotic spot virus* (CCSV) was employed to hybridize with ds-RNA of CCSV, *Watermelon silver mottle virus* (WsMoV) and *Tomato spotted wilt virus* (TSWV) (Lin et al. 2005). Probes labeled with nonradioactive chemicals are preferred, because of the hazards associated with radioactive labels. The sensitivity of cRNA probes labeled with DIG was assessed using dot blot, Northern blot and microplate hybridization methods. Specific riboprobes from 3′ ends of various isolates of *Apricot latent virus* were employed for detection of the virus. In dot blot assays, the riboprobe pApr-47 specifically hybridized with total nucleic acid (TNA) extracts from infected apricot and *Nicotiana occidentalis* (used as indicator plant), but not with healthy controls (Ghanem-Sabanadzovic et al. 2005). The northern hybridization procedure efficiently detected infection of rose geraniums (*Pelargonium* spp.) by *Plum necrotic ringspot virus* (PNRSV), as the ELISA format (Kulshreshtha et al. 2005). Virus from *Angelonia angustifolia* was detected by northern blot technique in *N. benthamiana* plants. The largest of four bands comigrated with equivalent dsRNA of *Carnation mosaic virus* (CarMV), and this band was considered to be *Angelonia flower break virus* (AnFBV) genomic RNA (Adkins et al. 2006).

DIG-labeled probes specific to *Cucumber mosaic virus* (CMV), *Papaya ringspot virus*-watermelon strain (PRSV-W), *Squash mosaic virus* (SqMV), and *Zucchini yellow mosaic virus* (ZYMV) were employed to detect the viruses in crude extracts from infected leaves by dot blot hybridization. DIG-labeled probes were specific and sensitive in identifying the viruses infecting cucurbits rapidly and precisely (Meng et al. 2007). Fluorescent in situ hybridization may be employed to localize plant virus transcripts. *Tomato yellow leafcurl virus* (TYLCV) and *Potato leafroll virus* (PLRV) were detected using this format. Transcripts of both viruses were localized specifically to the phloem sieve elements of infected plants. TYLCV and PLRV could also be detected in their respective biological vectors *Bemisia tabaci* and *Myzus persicae* (Ghanim et al. 2007). Dot blot hybridization was applied for detecting *Pepino mosaic virus* (PepMV) in tomato crops, using a DIG-RNA probe complementary to a fragment of RNA-dependent RNA polymerase (RdRp) of PepMV. Sensitivity of ELISA and dot blot hybridization was in equivalence, whereas RT-PCR assay was more sensitive (Alfaro-Fernández et al. 2009). The dot blot hybridization assay detected *Grapevine virus A* (GVA) using in vitro transcribed nonradioactive RNA probe for routine detection of GVA from phloem scrapings of dormant wood samples (Komínek et al. 2008). *Southern tomato virus* (STV), a ds-RNA virus belonging to the genus *Amalgavirus* was detected by dot blot hybridization using DIG-labeled RNA probe based on the nucleotide sequence of STV putative CP protein. The presence of STV was detected in extracts of leaves, fruits, roots and seed by the assay (Puchades et al. 2017).

Individual probe for specific virus and a polyprobe capable of detecting all viruses in coinfections were developed. Sensitivity of detection of viruses was on par with tests for single virus detection and ribopoly-6 hybridization for *Cucumber mosaic virus* (CMV), *Pepino mosaic virus* (PepMoV), and *Potato virus* Y (PVY). Ribopoly-6 could detect RNA transcripts corresponding to the CP of any of the six viruses tested without compromising sensitivity of detection. Both ELISA and ribopoly-6 probe assays were equally effective, indicating poly-6 tissue printing technique could be employed for large-scale field testing (Aparicio et al. 2009). A polyprobe containing sequences of eight viruses and two viroids fused in tandem was employed for detection of viruses and viroids infected fruit trees, using nonradioactive molecular hybridization method. Poly-10 probe could detect *Apple mosaic virus (ApMV)*, *Prunus necrotic ringspot virus* (PNRSV), *Prune dwarf virus* (PDV), *American plum line pattern virus* (APLPV), *Plum pox virus* (PPV), *Apple chlorotic leaf spot virus* (ACLSV), *Apricot latent virus* (ApLV), *Plum bark necrosis* and *Stem-pitting associated virus* (PBNSPaV), *Hop stunt viroid* (HSVd), and *Peach latent mosaic viroid* (PLMVd). The efficiency of detection of viruses and viroids was comparable to that of individual riboprobe allowing detection of up to 2.5 pg of viral or viroid RNA. The results indicated the potential of poly10 for routine detection of fruit tree viruses and viroids (Peiro et al. 2012). A unique polyprobe, carrying partial sequences of different viruses or viroids fused in tandem could allow simultaneous (polyvalent) detection of up to ten different pathogens using a nonradioactive hybridization protocol. A polyprobe carrying seven different conserved regions on *Nib* gene could detect 32 potyviruses assayed with no signal in respective healthy tissues. The results indicated the potential of polyprobe to detect several potyviruses simultaneously (Sánchez-Navarro et al. 2018). Global artichoke is infected by more than 25 viruses. Digoxigenin-labeled polyprobe was assembled from synthetic 50-nt ss-DNA sequences derived from genomes of the most commonly occurring viruses infecting global artichoke. The polyprobe could detect *Artichoke Italian latent virus, Artichoke latent virus, Artichoke mottled crinkle virus,* and other viruses infecting artichoke. These viruses could detected simultaneously by this inexpensive, reliable, and sensitive technique (Minutillo et al. 2021).

6.2.2.6.2 Polymerase Chain Reaction–Based Techniques

PCR and its variants have been the cornerstone for rapid advancements made in all branches of biological sciences, including plant pathology, particularly in the field of pathogen detection and disease diagnosis. PCR-based assays offer several advantages, of which increasing the concentration of the pathogen-related sequences present in the naturally infected plants at titers below the levels of detection is the major advantage. An initially very low number of DNA molecules can be amplified to reach a concentration that becomes detectable. Both genomic ds-DNA and ss-DNA of plant viruses can be directly amplified by PCR formats. Majority of plant viruses have RNA genomes and they need a reverse transcription step to generate complementary DNA (cDNA) prior to PCR amplification. The PCR assay provides distinct advantages of a high level of sensitivity, speed, specificity, and reliability of results. Various purification procedures have been developed to obtain plant viruses unassociated with inhibitory compounds such as polysaccharides and polyphenolic compounds that may inhibit PCR amplification. Increasing availability of nucleotide sequence data of plant viruses facilitates extensive application of PCR-based assays in viral disease diagnosis. The efficiency of DNA extraction, quality, and stability of viral genome could be improved by addition of supplements like sodium sulfite (antioxidant) (Narayanasamy 2011).

The RT-PCR assay was more sensitive than ELISA test for detection of *Prunus necrotic ringspot virus* (PNRSV) and *Prune dwarf virus* (PDV). These viruses could be detected at any time throughout the growing season (Mekuria et al. 2003). The location of PDV in young leaves and flower buds could be determined by in situ RT-PCR format. The CP gene of PDV was used as a target to produce cDNA copy that was amplified by PCR and visualized using a direct detection method using digoxigenin-labeled probes (Silva et al. 2003). *Prunus necrotic ringspot virus* (PNRSV), *Potato virus Y*, and *Apple scar skin viroid* (ASSVd) could be reliably detected by RT-PCR assay, whereas nested PCR format was more effective in detecting *Plum bark necrosis stem-pitting associated virus* (PPBNSPaV) (Sipahioglu and Ocak 2006). The presence of PNRSV in cherry orchard of *Prunus cerasus* cv. Montomorency and *Prunus avium* cv. Hedelfungen leaves in New York was confirmed by RT-PCR assay and amplification of viral CP gene in total RNA extracted from infected leaves. In addition, latent infection of a majority of trees by PNRSV was also revealed by RT-PCR assay (Oliver et al. 2009).

Use of virus-free plants and planting materials is the basic disease management strategy for reducing incidence and spread of viruses affecting asexually propagated crops. *Grapevine A virus* (GAV), *Grapevine B virus* (GBV) and *Grapevine leafroll-associated virus 3* (GLRaV-3) were detected in infected plants using spot-PCR format. Specific amplification of genomic fragments of the viruses was achieved by RT-PCR procedure (La Notte et al. 1997). Molecular variants of *Grapevine A virus* (GAV) could be detected reliably and specifically by inoculating *Nicotiana benthamiana* and employing RT-PCR assay (Goszczynski and Josste 2002). RT-PCR assay employing CP-specific primers detected *Rupestris stem-pitting-associated virus* (RSPaV) in all grapevine samples. In contrast, the primer pair targeting replicase gene sequencer could detect RSPaV only in symptomatic plants (Habili et al. 2006). The effectiveness of elimination of *Grapevine leafroll-associated virus 1* (GLRaV-1) and *Grapevine rupestris stem-pitting-associated virus* (GRSPaV) from *Vitis vinifera* cv. Agiorgitiko following combined application of in vitro thermotherapy and tissue culture methods was assessed using nested-PCR format. Meristem or shoot tip cultures coupled with thermotherapy were effective in eliminating both GLRaV-1 and GRSPaV to varying degrees. Relatively high percentages of plantlets were virus-free (Skiada et al. 2009). *Peach mosaic virus* (PcMV) could be detected by RT-PCR assay, employing PcMV-derived primers PM16AF and PM16AFR. Semi-nested RT-PCR procedure was developed, involving use of cDNA generated by PM16AFR. PcMV-specific fragments were amplified making this protocol to be very specific for detection of PcMV and its strains (James et al. 2006). Polyvalent degenerate oligonucleotide (PDO)-RT-CR assay was applied for detection of *Banana mild mosaic virus* (BanMMV) and *Banana virus X* (BVX) infecting *Musa* spp. PDO inosine-containing primers were well suited to detect only BanMMV. Species-specific primers were designed for detection of BVX. A polyclonal anti-BanMMV antiserum was employed for immunocapture (IC) of BanMMV virions from leaf extracts, leading to development of PDO-IC-RT-PCR assay (Teycheney et al. 2007).

Potato virus Y (PVY) isolates (PVY^N) causing veinal necrosis in tobacco were detected rapidly by RT-PCR format coupled with one-tube RT-PCR (cRT-PCR) in a duplex reaction (Crosslin et al. 2005). RT-PCR assay was effective in detecting *Potato mop top virus* (PMTV) by employing primers targeting CP gene in RNA3 of PMTV. The procedure was very effective in detecting the virus in seed potato tubers and ware potatoes (Xu et al. 2004). RT-PCR with novel primers could detect virus isolates from all major groups within the genus. The extent of specificity of new primers Nib2F and Nib3R was compared with specificity of two routinely applied primer pairs. Reactions with NIb2and NIb3R produced amplicons of *Potyvirus* and its reliability indicated its potential for use as a routine diagnostic assay (Zheng et al. 2010). Two sets of primers employed for diagnosing *Andean potato mottle virus* (APMoV) in seed tubers, amplified two amplicons of 315-b and 732-bp products, indicating its potential for routine testing in quarantines (Lee and Rho 2015). A rapid viral nucleic acid release (NAR) buffer suitable for RT-PCR-based diagnostic methods for detection of potato viruses was developed. Using NAR buffer viral nucleic acid could be isolated efficiently from potato leaves, thrips, and aphids for PCR and RT-PCR-based detection of RNA and DNA genomes of plant viruses (Raigond et al. 2018).

Simultaneous detection of multiple viruses infecting citrus was essential to reduce time, labor and costs. *Sweet potato chlorotic fleck virus* (SPCFV) was often found in mixed infection along with *Sweet potato feathery mottle virus* (SPFMV) and *Sweet potato virus 2* (SPV2). *Ipomoea setosa* was top-grafted onto symptomatic sweet potato plants. The presence of SPCFV, SPFMV-O and SPV2 in *I. setosa* scion shoots could be detected by employing virus-specific primers (Jones and Dwyer 2007). The multiplex RT-PCR assay combined with magnetic nanobeads (MNB-RT-PCR) was developed for efficient detection of *Sweet potato virus G* (SPVG), *Sweet potato virus C* (SPVC), and *Sweet potato chlorotic fleck virus* (SPCFV). MNB-RT-PCR assay was more rapid and sensitive than conventional RT-PCR format in detecting these viruses (Deng et al. 2014). Viruses associated with different cassava brown streak disease (CBSD) symptom phenotypes, including the asymptomatic type, were investigated by RT-PCR using virus-specific primers. The two virus species were detected in samples as single (*Cassava brown streak virus* (CBSV) or *Ugandan cassava brown streak virus* (UCBSV) or mixed infections (CBSV + UCBSV) (Munganyinka et al. 2018). The multiplex (m)-RT-PCR format, using primers designed from their conserved regions was applied for detection of *Arabis mosaic virus* (ArMV), *Grapevine fan leaf virus* (GFLV), *Grapevine fleck virus* (GFV), and *Grapevine leafroll-associated virus 1, 2, 3* (GLRaV-1, -2, -3). Fragments specific for viruses were simultaneously amplified from infected samples and they identified by their molecular sizes in agarose gel electrophoresis (Gambino and Gribaudo 2006). Infectious clones of putative *Grapevine red blotch virus* (GRBV) were engineered from the genome. Systemic GRBV infection was detected by multiplex PCR assay in *Vitis vinifera* exhibiting foliar symptoms. Spliced GRBV transcripts of replicase-associated protein coding region accumulated in leaves of agroinfected vines (Yepes et al. 2018).

Primers and TaqMan probes were employed for detection of *Citrus psorosis virus* (CPsV) and *Citrus variegation virus* (CVV) by simplex real-time (q) RT-PCR assay. The multiplex (m)RT-PCR assay detected CPsV, CVV, and *Citrus tristeza virus* (CTV) simultaneously in a single reaction (Loconsole et al. 2010). A multiplex RT-PCR assay was applied for simultaneous detection of strawberry viruses, *Strawberry crinkle virus* (SCV), *Strawberry mild yellow edge virus* (SMYEV), *Strawberry mottle virus* (SMoV) and *Strawberry vein banding virus* (SVBV) which occurred in different combinations of mixed infections. The upper limit of detection of these viruses in tissue extracts was 1/200. When concentrations of the templates were different, sensitivity of the assay was low (Thompson et al. 2003). Two multiplex RT-PCR formats with coamplification of an internal control from a plant mRNA was developed for reliable sensitive detection of *Apple stem-pitting virus* (ASPV), *Apple chlorotic spot virus* (ACLSV), and *Apple stem grooving virus* (ASGV). Virus-specific primers were designed to allow two multiplex RT-PCR amplifications accompanied by one pair of primers of the mitochondrial *nad5* gene. The mRNA representing *nad5* gene was present even in dormant trees (Ma et al. 2008). Duplex RT-PCR protocol was evaluated by nine European laboratories, using crude plant extracts as template for virus detection. The primer pair PNRSV-10F/PNRSV-10R amplified a 348-bp fragment of *Prunus necrotic ringspot virus* (PNRSV) CP, whereas the primer pair PDV-17F/PDV-12R amplified a 303-bp fragment of *Prunus dwarf virus* (PDV)-CP. Sensitivity and specificity values of simultaneous detection using crude extracts were equal or higher than that of simplex assays, indicating the superiority of the duplex assay system (Massart et al. 2008).

Multiplex RT-PCR assay was applied to detect *Sugarcane mosaic virus* (SCMV), *Sugarcane streak mosaic virus* (SCSMV), and *Sugarcane streak yellow leaf virus* (SCYLV). Primers designed from CP genes of respective viruses amplified specific products for these viruses. The multiplex RT-PCR assay was equally efficient as the uniplex format for the detection these viruses (Viswanathan et al. 2010). A multiplex RT-PCR assay was developed to detect and discriminate *Maize stripe virus* (MSpV-Sorg), *Sugarcane streak mosaic virus* (SStMV-Sorg), and a rhabdovirus, *Indian sorghum mosaic virus* (SMV), infecting sorghum crops. Three sets of primers were capable of reacting in both uniplex and multiplex formats. The specificities of multiplex primers were confirmed by sequencing m-RT-PCR products. The multiplex RT-PCR assay efficiently detected the viruses present in field-collected sorghum samples (Srinivas et al. 2013). PCR assays were applied to detect *Sugarcane white streak virus* (SWSV) and *Sugarcane striate virus* (SStrV) in *Saccharum* species and interspecific hybrids in the Miami World Collection of sugarcane and related grasses. SStrV was detected in different varieties, whereas SWSV infection was not detected in any variety. Six phylogenetic groups were identified in SStrV isolates (Boukari et al. 2020).

Potato virus Y (PVY) and its variants were detected using multiplex RT-PCR format targeting P1 cistron of PVY. Isolates of PVY were classified into two major groups: isolates producing leaf necrosis in tobacco and isolates that did not produce leaf necrosis in tobacco (PVY^{O}) (Nie and Singh 2002). Later multiplex assay was employed in conjunction with P1 assay to detect and identify PVY isolates. The recombinants, PVY^{NTN} and $PVY^{N:O}$ were identified based on amplification of three products or one product around respective recombination junctions (Nie and Singh 2003). Another multiplex RT-PCR format was developed based on *Hinc*II restriction digests of P1 amplicons to differentiate PVY^{O} and PVY^{N}, PVY^{NTN} NA-PVY^{N} or NA-PVY^{NTN} (Crosslin et al. 2005). A single multiplex RT-PCR assay capable of assigning PVY strain types and detect mixed infections with respect to major strain types was developed. This format identified samples containing strain mixtures. It was possible to distinguish 16 mixed infections that could not be detected earlier (Lorenzen et al. 2006). Simultaneous detection of *Potato virus A* (PVA), *Potato virus S* (PVS), PVX, and PLRV was accomplished by applying multiplex RT-PCR designed from sequences of 18S RNA for amplification of fragments unique to the target viruses. The assay was effective in detecting the target viruses individually, as well as in different combinations, and it had greater sensitivity for detection of PVX than the commercial DAS-ELISA format. The multiplex RT-PCR assay had greater sensitivity and reliability for detection of potato viruses in mixed infections (Du et al. 2006).

The multiplex RT-PCR assay was applied to detect *Potato virus X, Potato virus Y* (PVY), *Potato leaf roll virus* (PLRV), *Potato virus M* (PVM) and *Potato virus A* (PVA). Five pairs of highly specific primers were selected for use in the quintuplex RT-PCR assay, and amplification conditions were optimized. This system was amenable for customization for simplex, duplex, triplex, tetraplex, and quintuplex RT-PCR formats by changing only the primer pairs to detect any combination of PVX, PVY, PLRV, PVM, and PVA. The RT-PCR procedure was faster, cheaper, and more sensitive and reliable than DAS-ELISA for detection of potato viruses in mixed infections (Zhang et al. 2017). A multiplex RT-PCR assay with high-fidelity PCR procedure was employed to detect six important potato viruses simultaneously. The results of multiplex RT-PCR assay were consistent as in conventional PCR format. This multiplex RT-PCR assay showed potential for application in seed certification and regulatory programs to diagnose viruses that might be present in tubers with surface or internal necrotic symptoms (Cating et al. 2015). TaqMan R qPCR-based one-step procedure called direct reverse transcription quantitative PCR (DiRT-qPCR) format had greater sensitivity for detection of PVY, PLRV, and PVS depending in virus species (Stammler et al. 2018). Isolates from potato leaves and tubers were characterized using RT-PCR assay and whole genome sequencing. All analyzed PVY-positive samples contained the same recombinant PVY strain: PVY^{NTN}. Phylogenetic analysis showed that the three South Korean isolates were most closely related to PVY^{NTNa} isolates from Japan and Vietnam (Chikh-Ali et al. 2019).

The multiplex RT-PCR format was applied for detection of *Dasheen mosaic virus* (DsMV), *Turnip mosaic virus* (TuMV), *Konjac mosaic virus* (KoMV), and *Zandtedeschia mild mosaic* virus (ZMMV) infecting Calla lily. Specific primers for each virus were designed based on the sequence of 3′ terminal region of respective viruses. Fully expanded leaves of Calla lily were tested by m-RT-PCR procedure. The presence of virus in single or mixed infections was detected in Calla lily samples. Application of m-RT-PCR assay provided more reliable detection rapidly in addition to being cost-effective and showed potential for large-scale use for production of virus-free stocks (Hu et al. 2010). Simultaneous detection of viruses frequently infecting tobacco, *Cucumber mosaic virus, Tobacco mosaic virus, Tobacco etch virus*, and *Potato virus Y* was possible by applying a one-step multiplex RT-PCR procedure. Four sets of virus-specific primers were employed to function in the same reaction agents and cycling conditions. The one-step RT-PCR protocol was high throughput, specific and sensitive for simultaneous detection of target viruses in tobacco plants (Zhang et al. 2013). A duplex RT-PCR protocol was applied for simultaneous detection *of Onion yellow dwarf virus* (OYDV) and *Shallot latent virus* (SLV). Virus-specific primers efficiently amplified products specific to OYDV and SLV. Duplex RT-PCR format detected OYDV and SLV in garlic selections (Majumder et al. 2008). A multiplex RT-PCR format was applied for detection of *Garlic virus B* (GarV-B) and *Garlic virus D* (GarV-D) using both degenerate genus-specific and species-specific primer sets (Taglienti et al. 2018). RT-PCR assay was developed for detection of *Squasn vein yellowing virus* (SqVYV) infecting squash plant (*Cucurbita pepo*) and a product of expected size 957-bp was amplified. Mixed infection of cucurbits by SqVYV and PRSV-W could be detected by this RT-PCR protocol (Adkins et al. 2007). The multiplex PCR coupled with the LiquiChip assay was applied for identification of *Zucchini yellow mosaic virus, Cucumber green mottle mosaic virus,* and *Cucumber mosaic virus* by coamplification with plant mRNA as an internal control. The assay was highly specific for distinguishing individual viruses from mixed population and it was more sensitive than multiplex RT-PCR assay (Kuan et al. 2018).

Specific primers were designed from sequences of conserved region of RdRp gene and 3′ UTR region of viral RNA of *Onion yellow dwarf virus* (OYDV) isolates. RT-PCR was performed on total RNA extracted from leaves and bulbs of infected plants. PCR products unique to OYDV were detected, establishing the identity of OYDV in infected onion plants (Arya et al. 2006). Partial CP nucleotide sequences of *Bean yellow mosaic virus* (BYMV) from 8 isolates from black pod syndrome (BPS)-symptomatic *Lupinus. angustifolius* plants, showed that all belonged to the BYMV general phylogenetic group (Kehoe et al. 2014). RT-PCR assay detected a viral replicative strand and short interfering RNA (siRNA) matching the new *Ilavirus* in symptomless tomato plants (Saqib et al. 2015). Degenerate primers were employed for detection of *Lily symptomless virus* (LSV) using primers designed based on amino acid sequences of CP in four LSV strains. RT-PCR assay amplified a 483-bp fragment from LSV (Nesi et al. 2009). Based on phylogenetic analysis of the individual tospoviral proteins indicated that *Calla chlorotic leaf spot virus* (CCSV) might be closely related to *Tomato zonate spot virus* (isolated from tomato) infecting tomato in China. The degenerate primer pairs gL2740/gL3920c and gm410/gm870c primer pairs could detect also *Melon yellow spot virus* (Chen et al. 2012).

Following emergence of *Groundnut ringspot virus* (GRSV) and *Tomato chlorotic spot virus* (TCSV) which induced indistinguishable symptoms, identification of the causal agent became difficult. When used in conjunction with local lesion-passaged virus isolates, the genome segment-specific RT-PCR assays facilitated high-throughput screening of plants or thrips (vectors) samples for interspecies reassortments. These tospoviruses commonly present in mixed infections under field conditions could be reliably identified by RT-PCR assays (Tantiwanich et al. 2018). Two-step multiplex RT-PCR assay was developed to detect *Bean yellow mosaic virus* (BYMV) and *Canna yellow streak virus* (CaYSV) and *Canna yellow mottle virus* (CaYSV) in infected plants. Immunosorbent electron microscopy revealed the presence of filamentous potyvirus particles in rhizome and leaf tissues (Chauhan et al. 2015). Different diagnostic techniques, including end-point RT-PCR assay and RT-qPCR assay, were employed to target single and multiple gene targets on *Rose rosette virus* (RRV) (Babu et al. 2018). Nested RT-PCR assay was applied for generic detection of subgroups of genetically related viruses *Grapevine leafroll-associated virus 4, 5, 6, 9,* (GLRaV-4, -5, -6, -9) and two isolates (GLRaV-De and GLRAV-Pr) that represented new species belonging to this lineage. This format involves a one-step RT-PCR for generic detection of *Vitivirus* and *Foveavirus (Closteroviridae)* using degenerate primers that target heat shock protein HSP70h gene followed by a nested

PCR which could detect all virus-members of the lineage and differentiate them from other grapevine closteroviruses. All isolates of GLRaV-4, -5, -6,-9, -De and -Pr either single or in multiple infections were successfully amplified when the degenerate primers LR5-ClusdoL and HSP-nest2 were employed. No amplicon was obtained for the isolates of GLRaV-1, -2, and -3 or healthy control (Maliogka et al. 2008). Detection of grapevine RNA viruses commenced with cDNA synthesis of purified viral RNAs, followed by performance of PCR assay. The cDNAs have to be validated by an internal control. Primer sets were designed based on sequences of *Vitis vinifera* phosphophenol pyruvate carboxylase gene. The results indicated that these primers might be useful reference markers in RT-PCR-based virus detection assays both for validation of cDNA synthesis and detection of trace amounts of genomic DNA (Olah et al. 2019).

Degenerate primers containing inosine were used in a polyvalent RT-PCR assay for detection of the genera *Trichovirus, Capilovirus* and *Foveavirus*. A product of 362-bp was amplified from nucleic acid extracts from *Prunus* and *Malus* leaf samples. All viruses tested could be detected by nested RT-PCR procedure (Foissac et al. 2005). *Peach mosaic virus* (PcMV) was detected using RT-PCR procedure employing PcMV-derived primers PM16AFF and PM16AFR. Semi-nested RT-PCR was employed using cDNA generated by PM17AFR for amplification with PM-AF1 and PM-AFR primers. The nested RT-PCR format enhanced specificity of detection of *Peach mosaic virus* (James et al. 2006). Grapevine cv. Cabernet Franc (CF) with unusual symptoms of red blotches, marginal reddening and red vines was examined for viral origin. Metagenomic analysis of cDNA libraries obtained from ds-RNA enriched nucleic acid (NA) preparations from bark scrapings on an Illumina platform revealed sequences having a distant relationship with members of *Geminiviridae*. The virus was, identified as *Grapevine red blotch-associated virus* (GRBaV) (Al-Rwahnih et al. 2013). Full-length genome of *Citrus yellow vein-clearing virus* (CYVCV) isolate AY (citrus yellow vein-clearing disease, CYVCD) was amplified by long-distance RT-PCR format. Four of genomic cDNA clones were inoculated into Jincheng (*C. sinensis*) seedlings through *Agrobacterium tumefaciens* (At)-mediated vacuum infiltration and the inoculated plants were infected as confirmed by RT-PCR and by direct tissue blot immunoassay. CYVCV was established as the primary cause of CYVCD (Cui et al. 2018). Genomic fragments of *Watermelon green mottle virus* (WGMMV) were amplified by RT-PCR using degenerate primers of cucurbit-infecting tobamovirus. WGMMV was identified as a new tobamovirus infecting watermelon (Cheng et al. 2019).

Plant viruses with DNA genome could be detected using PCR assays in different crops. Two specific primers were designed based on the sequences of *Banana bunchy top virus* (BBTV)-Eg-DNA2.2 (component 2). BBTV was detected in all banana samples tested including asymptomatic plants (Soweha 2005). Primers based on the CP gene sequences were effective in detecting BBTV in infected banana leaves. Major DNA fragment of 500-bp was amplified from extracts of leaves of infected banana plants (El-Dougdoug et al. 2006). A simple procedure for extracting DNA from banana plants was developed for extracting DNA from BBTV-infected banana leaves and aphid vectors. Sodium sulfite and polyvinyl pyrrolidone were included with extraction buffer. BBTV could be detected in CaCl2-dried leaf samples from BBTV-infected plants (Selvarajan et al. 2015). PCR assay was applied to determine contents DNA of *Tomato leafcurl virus* (ToLCV) strains inducing severe and mild symptoms in tomato plant. An isolate of bipartite *Tomato leafcurl virus New Delhi virus*-Severe (ToLCVNDV-Svr) induced severe symptoms in tomato. On the other hand, a monopartite virus (*Tomato leafcurl Joydepur virus*-Mild) (ToLCVJV-Mild) was detected in tomato plants showing mild symptoms (Maruthi et al. 2005). Transmission potential of whitefly vector *Bemisia tabaci* isolates from tomato to pepper and from pepper to tomato for *Tomato yellow leafcurl Thailand virus* (TYLCTHV) was assessed investigated. TYLCTHV isolate (SPN-Tom1A) was detected in viruliferous whiteflies. *B. tabaci* Asia I was highly efficient in transmitting TYLCTHV from tomato to tomato. The presence of TYLCTHV in bodies of viruliferous whiteflies fed on tomato and pepper was confirmed by PCR assay. Pepper plants infected by TYLCTHV might not function as sources of virus inoculum for tomato crops (Yule et al. 2019). Betasatellites associated with monopartite begomoviruses plays a key role in expression of severe form of the disease in dicotyledonous plants. An improved set of universal primers, CLB36F an CLB37R with potential was employed to detect a vast range of begomoviruses across the crops. The improved universal primer pair detected the betasatellites of all the samples (Kumar et al. 2018). *Rice black-streaked dwarf virus* (RBSDV), *Rice stripe virus* (RSV) and *Southern rice black-streaked dwarf virus* (SRBSDV) were designed and multiple-one step RT-PCR procedure was applied using the virus-specific primers. The method was simple, rapid and sensitive, and effective for detecting the viruses in samples of rice, maize, small brown planthoppers, and white-backed planthoppers collected from different regions in China (Zhou et al. 2018). *Potato virus Y* strains are transmitted by aphid vector, *Myzus persicae*. Transmissible virus fraction is only retained in the aphid vector's mouthparts. Quantitative PCR assay was developed to detect and quantify PVY present in the stylets of the vector. A specific method based on TaqMan chemistry with higher sensitivity than conventional RT-PCR was employed. A significant amount of PVY virions was detected within the dissected stylets of *M. persicae*. The assay estimated the portion of the virus partially attached to the stylets and the number of virus targets could be determined uniformly. The protocol could be used for rapid and sensitive detection of PVY strains (Khelifa 2019).

6.2.2.6.3 Reverse Transcription Loop-Mediated Isothermal Amplification Assay

Eiken Chemical Co. Tokyo, Japan, developed a novel nucleic acid amplification procedure called LAMP, which amplified DNA under constant temperature with high specificity, sensitivity, rapidity and efficiency. The reverse transcription (RT)-LAMP technique is based on the principle of standard displacement reaction and stem-loop structure that amplifies the target under isothermal conditions. Amplification efficiency of LAMP method is high due to continuous amplification under isothermal conditions, producing a large amount of precipitate of magnesium pyrophosphate as a byproduct, leading to turbidity. Using four specially designed primers that include six short sequences matching target DNA fragment, the DNA is amplified under isothermal conditions within a short period of time. Positive reactions are indicated by detecting the turbidity, using an inexpensive turbidimeter. Further, the higher efficiency of the LAMP procedure facilitates visual observations of amplification via naked eye by using a UV-lamp in the presence of an intercalary dye such as SYBR Green I or ethidium bromide. The reaction performed in single-tube by mixing the buffers, primers, reverse transcriptase, and DNA polymerase and incubating the mixture at 63° C for 60 min. The increase in turbidity of the reaction mixture correlates with the amount of DNA synthesized. Measurement of turbidity is very simple and no expensive equipment is required for LAMP procedure. The time required for the RT-LAMP assay is much less compared to RT-PCR–based assays (Notomi et al. 2000; Parida et al. 2004).

RT-LAMP amplification of viral DNA provided efficient detection of *Japanese yam mosaic virus* (JYMV) in infected plants. JYMV was efficiently detected in leaves, propagules, and roots of Japanese yam plants. Virus titer was positively correlated with turbidity of reaction mixture determined by turbidimeter easily (Fukuta et al. 2003 *Tomato spotted wilt virus* (TSWV) was detected by immunocapture (IC) RT-PCR technique. Genomic RNA of TSWV was reverse transcribed, followed by amplification under isothermal (65° C) conditions for 60 min. Intensity of turbidity of the reaction mixture was proportional to the concentrations of amplicons. The RT-LAMP procedure was more sensitive and rapid in detecting TSWV in infected chrysanthemum plants than IC-RT-PCR format (Fukuta et al. 2004). A set of four primers matching a total of six sequences of CP gene of *Potato virus Y* was designed in such a way that a loop could be formed and elongated during DNA amplification. Assessment of turbidity visually and spectrometrically could save time and reduce the cost of data analysis significantly (Nie 2005). A closed-tube real-time RT-LAMP format was applied for fluorescent detection of strains of PVY in potato. A set of primers for RT-LAMP assay was designed by aligning 40 sequences of CP genes of different isolates of PVY to amplify the conserved region of CP genes of different isolates of PVY to amplify conserved region of CP gene. The RT-LAMP assay was more sensitive than RT-PCR format and provided the results rapidly (Przewodowska et al. 2015). The RT-LAMP technique was applied to visually detect *Potato leaf roll virus* (PLRV) infecting potato. RNA was extracted from PLRV-infected leaves and a set of primers (F3, B3, FIP, BIP, LF, and LB) designed for RT-LAMP reaction of the CP gene of PLRV was used for amplification of virus-specific product (Ahmadi et al. 2013).

A one-step accelerated RT-LAMP protocol was developed for detection and identification of *Plum pox virus* (PPV). Six primers based on the sequences of a conserved region in the C-terminus of CP coding region of PPV were employed in the assay. Isolates of strains D, M, EAC, and W of PPV were detected efficiently and identified precisely. PPV could be consistently detected in both herbaceous and woody host plant species infected by PPV (Varga and James 2006). In a later investigation, a modified one-step RT-LAMP procedure was applied for detection of PPV in virus suspensions from plant extracts. In place of gel electrophoresis, a homogenous color test was performed on amplified product. As amplification and visualization occurred homogenously in unopened tubes, the risk of contamination of subsequent samples by former amplification products was significantly minimized. This LAMP format provided reliable detection of PPV rapidly for both single samples and for large-scale surveys (Hadersdorfer et al. 2011). *Banana streak virus* (BSV) could be reliably and sensitively detected by employing RT-LAMP assay. Four primers matching a total of six sequences of the ORF III polyprotein gene concerned were employed. RT-LAMP assay could be applied extensively, because of its simplicity for routine screening of tissue culture-derived materials for freedom from BSV for producing virus-free stocks (Peng et al. 2012). One-step RT-LAMP technique was developed for detection of *Indian citrus ringspot virus* (ICRSV). Four primers were designed to target coat protein (CP) gene of ICRSV. Amplification product was visualized after adding SYBR Green I dye. The standardized RT-LAMP assay was highly specific and efficiently detected ICRSV in all samples tested. One-step RT-LAMP format was cost-effective, simple, sensitive, and specific providing the results rapidly (Kokane et al. 2021).

A LAMP assay was developed for detection of *Potato yellow mosaic Panama virus* (PYMPV), *Tomato leafcurl Sinoha virus* (ToCSiV), and *Tomato yellow mottle virus* (TYMoV), using a set of three primer pairs specific to each of these viruses. Products of amplification specific to PYMPV, ToLCSiV, and TYMoV were detected in total DNA extracts obtained from different plant tissues such as leaves, stems, flowers, fruits, and roots of infected tomato plants collected from different production regions. The LAMP technique was more rapid and less expensive than PCR. Further, the LAMP assay was successful in detecting all three viruses in sap extracts from infected tomato leaves. In contrast, the PCR assay failed to detect these viruses under the same experimental conditions (Herrera-Vasquez et al. 2018). Whitefly-transmitted begomovirus, *Tomato leafcurl New Delhi virus*-Potato was a significant obstacle for obtaining virus-free potato stocks. LAMP assay was developed for specific detection of ToLCNDV-Potato, using six primers that could recognize CP gene sequence of the target virus. The results were determined based on visual observation of turbidity, color change using SYBR Green dye and also by gel electrophoresis. ToLCNDV-Potato could be detected in infected plants collected from potato fields. The presence of the virus in infected potato tuber and asymptomatic plants could be detected by LAMP format. Print-capture LAMP assay developed in this investigation could reduce the cost and time of assay in large-scale application as in tuber certification programs (Jeevalatha et al. 2018). One-step RT-LAMP assay was developed to detect *Tomato chlorosis virus* (ToCV) either in total RNA or crude RNA extracted from infected plants. RT-LAMP assay was highly efficient and more sensitive than conventional RT-PCR assay for detection of ToCV (Karwitha et al. 2016). The LAMP assay was developed for efficient and reliable detection of *Tomato brown rugose fruit virus* (ToBRFV). Amplification of virus-specific product occurred at isothermal conditions of 65° C for 35 min. The LAMP assay was as sensitive as qPCR format and more sensitive in detecting ToBRFV than the RT-PCR format (Sarkes et al. 2020).

Zucchini yellow mosaic virus (ZYMV) in squash and melon plants was detected using RT-LAMP assay by employing a set of four primers targeting ZYMV. The assay was more sensitive than RT-PCR assay for detecting of ZYMV in infected squash melon tissues. Reactions of RT-LAMP assay could be visualized by staining the amplified products with SYBR Green I dye. RT-LAMP assay was superior to RT-PCR assay, because of its simplicity, rapidity, and sensitivity in detecting ZYMV infection in cucurbitaceous crops (Kuan et al. 2014). One-step accelerated RT-LAMP protocol was applied for detection of *Melon yellow spot virus* (MYSV). Four primers were designed to anneal onto a conserved region in MYSV N gene. A ladder-like pattern of amplified products was observed after electrophoresis. The RT-LAMP assay was specific, rapid and more sensitive than the RT-PCR format (Zeng et al. 2016). Three viruses infecting watermelon, *Watermelon mosaic virus* (WMV), *Squash mosaic virus* (SqMV), and *Zucchini yellow mosaic virus* (ZYMV) were detected using one-step RT-LAMP format. Coat protein (CP) gene sequences of WMV and SqMV and cylindrical inclusion protein sequence of ZYMV were used for designing virus-specific primers. Sensitivity of the RT-LAMP assay was greater than that of conventional RT-PCR format (Zhao et al. 2016). The RT-LAMP assay was applied for detection of *Barley yellow mosaic virus* (BaYMV) and *Barley mild mosaic virus* (BaMMV), employing two pairs of oligonucleotide primers (inner and outer primers) to amplify CP gene of BaYMV. The assay detected only BaYMV, but not eight other viruses tested, indicating the specificity of assay. All samples tested were positive for BaYMV indicating the effectiveness of the assay (Chen et al. 2021). The LAMP technique was employed for early and precise detection of *Wheat dwarf virus* (WDV) in wheat plants. WDV was detected rapidly and had high specificity for WDV amplification with accuracy in equivalence to qPCR format (Hao et al. 2021).

6.2.2.6.4 Combination of Diagnostic PCR with Other Techniques

The sensitivity and specificity of detection of plant viruses in plants and their vectors may be enhanced by combining two techniques that are based on different modes of action. *Onion yellow dwarf virus* (OYDV) was detected in infected garlic plants by applying immunocapture (IC)-RT-PCR assay (Mahmoud et al. 2007). *Soybean mosaic virus* (SMV) was detected by performing IC-RT-PCR procedure using a pair of primers designed to amplify a fragment with coding region of SMV coat protein (CP) gene sequences. Infection of soybean samples by SMV was confirmed by detecting virus-specific amplicon in the agarose gel after staining with ethidium bromide (Ahangaran et al. 2009). The IC-RT-PCR procedure was sensitive and minimized problems with extraction of total RNA from infected plant tissues (Lima et al. 2014). *Plum pox virus* (PPV), causing Sharka disease in *Prunus domestica* cv. Cacanska was detected by IC-RT-PCR assay in infected shoots (grown under tissue culture conditions) treated with ribavirin (1-ß-ribofuraanosyl-1,2,4-triazole-3-carboxamide) with antiviral activity. The IC-RT-PCR assay revealed that about 15% plants treated with ribavirin were free of PPV infection (Paunovic et al. 2007). IC-RT-PCR assay was employed to detect *Citrus tristeza virus* (CTV), using primers corresponding to both ends of the CP gene. The presence of CTV in all samples was confirmed by IC-RT-PCR assay. Regardless of the absence of visible symptoms in field trees, severe CTV variants appeared to be widespread and present for long time in the Adriatic region, making it a reservoir of severe CTV strains (Cerni et al. 2009). *Tomato spotted wilt virus* was detected efficiently by IC-RT-PCR assay (Wu et al. 2009). The IC-RT-PCR protocol could be applied to typing PVY isolates in potato leaf tissues collected from fields (Chikh-Ali et al. 2013).

A single-tube multiplex IC-RT-PCR format was applied for simultaneous detection of tobamoviruses, *Pepper mild mottle virus* (PMMoV) and *Tobacco mild green mosaic virus* (TMGMV). The assay was efficient in detecting both viruses simultaneously. Two RT-PCR products corresponding to PMMoV (317-bp) and TMGMV (510-bp) were detected directly in crude plant extracts. Specificity of primer pairs PM317-F/-R and CPTMG-F/-R specific for PMMoV and TMGMV respectively ensured reliable detection and differentiation of these viruses in single and duplex IC-RT-PCR assays (Kim et al. 2006). A dual IC-RT-PCR protocol was applied for detection of *Lily symptomless virus* (LSV) and *Arabis mosaicvirus* (ArMV) infecting lily by coating PCR tubes with antibodies specific to them, (Zheng et al. 2013). Single-tube multiplex RT-PCR format was employed to detect virus with RNA/DNA genome. Two PCR amplicons, 650-bp for *Cucumber mosaic virus* (CMV) and 450-bp for *Piper yellow mottle virus* (PYMoV) were identified in the gel. The protocol was useful for detection of viruses with RNA and DNA genomes present in mixed infections in black pepper (Bhat and Siju 2007). A cocktail of ELISA-nested RT-PCR procedure was developed for detection of *Bean pod mottle virus* (BPMV) in soybean seeds. This protocol was more sensitive than IC-RT-PCR format (Jin et al. 2016). The IC-RT-PCR format was performed using different combinations of three virus extraction buffers and two virus nucleic acid releasing buffers along with one virus RNA releasing condition for detection of *Grapevine leaf roll-associated virus 3* (GLRaV-3). The virus could be detected reliably in the vines of five infected grapevine cultivars, facilitating application of the protocol in certification programs (Kumar et al. 2015). Quantitative real-time PCR assay showed variation in expression of *Apple stem grooving virus* (ASGV) and *Apple stem pitting virus* (ASPV) in different parts during different seasons. ASPV and ASGV were detected in bark using immuno-molecular diagnostic methods. Testing leaves and bark during fall, appeared to be optimum time for diagnosis of ASGV and ASPV in apple trees (Nabi et al. 2018).

6.2.2.6.5 Real-Time PCR Assay

The real-time PCR assay overcomes the limitation of the conventional PCR assay. PCR-amplified products are detected by real-time PCR as the reaction occurs progressively without the requirement of post-PCR steps. Principles and application of real-time PCR assay are discussed in earlier publications (Narayanasamy 2011, 2017). Real-time RT-PCR format detected *Tomato spotted wilt virus* (TSWV) in all infected samples collected from tomato, potato, pepper, lettuce, and aster crops. Real-time RT-PCR effectively detected TSWV in all ELISA-positive samples of leaf tissues of plant species (Dietzgen et al. 2005). *Cucumber vein yellowing virus* (CVYV) could be detected efficiently by applying real-time RT-PCR format using specific primers designed from a nucleotide sequence of the RNA polymerase gene (*NIb*) conserved among all CVYV strains (Picó et al. 2005). The real-time PCR assay was effective in detecting *Beet necrotic yellow vein virus* (BNYVV). Real-time RT-PCR assays were more sensitive in detecting the target virus compared to conventional RT-PCR format (Harju et al. 2005). Fluorescent TaqMan-based real-time RT-PCR assays specific to *Potato virus Y* strains PVY^N and PVY^O were applied to detect single nucleotide polymorphism (SNP) (A/G_{2213}) that could be employed as a molecular determinant. PVY^{NI} isolates inducing tobacco necrosis were detected and differentiated by targeting the molecular determinant (Balme-Sinibaldi et al. 2006). The current season spread of *Potato virus Y* (PVY) in potato fields was monitored using ELISA and real-time RT-PCR assays. Real-time RT-PCR assay detected presence of PVY in more samples of leaves than ELISA test. PVY infection determinant in growing tubers showed a significant positive correlation with PVY incidence in tubers at harvest (Fageria et al. 2013). *Sweet potato feathery mottle virus* (SPFMV), *Ipomoea vein mosaic virus* (IVMV), and *Sweet potato virus G* (SPVG) infecting sweet potato could be detected by simplex reactions directly from infected leaves of sweet potato plants. Real-time PCR assay proved to be a faster and safer method, in addition to being more sensitive for detection and quantification of sweet potato viruses. However, the requirement of expensive instruments limited the application of real-time PCR assay by most countries (Kokkinos and Clark 2006). A rapid, sensitive, and reliable one-tube real-time RT-PCR assay was employed for detection and quantification of *Little cherry virus* (LChV) in *Prunus* spp. Primers and TaqMan probe based on conserved regions of capsid protein gene were employed (Katsiani et al. 2018).

Primers and probes targeting specific regions of *Grapevine leafroll-associated virus 1, 2, 3, 4, 5* and *-9* (GLRaV-1, -2, -3, -4, -5 and -9) with 100% sequence identity were employed in the TaqMan PCR assay. The TaqMan assay was more sensitive than conventional one-step RT-PCR for testing different isolates of GLRaV-1, -2, -3, -4, -5, and -9 in both purified RNA or crude extracts from infected grapevine plants (Osman et al. 2007). Real-time RT-PCR assay was applied for detection of Flavescence dorée (FD), while real-time nested PCR assay following RT-PCR was used for detection of Bois noir (BN), GLRaV-3 and GVA using the same crude extract as template (Margaria et al. 2009). A real-time multiplex PCR assay was applied to detect and differentiate PPV strains. SYBR Green I format was inexpensive and provided more reliable results rapidly, since use of electrophoresis was not required (Varga and James 2006). The highest sensitivity of PPV detection in winter was provided by conventional real-time RT-PCR format which detected PPV in 97.5% of trees that developed sharka symptoms later. The results indicated that immobilized targets of PPV and spot real-time RT-PCR could be adopted for large-scale testing to assist certification programs to supply PPV-free *Prunus* plants (Capote et al. 2009). Spread of *Citrus tristeza virus* (CTV) to most citrus-growing countries occurred primarily through exchange of infected budwood and secondary spread of CTV was through aphid vectors. The TaqMan real-time RT-PCR assay was developed using purified RNA target or coupled to tissue-print and squash procedure for detection and quantification of CTV. Tissue-print ELISA test revealed differences in the number and intensity of immunoprecipitate localized in the vascular area of imprints. Using real-time RT-PCR, CTV isolates from different hosts and origin could be detected and quantified consistently. Tissue-print and squash real-time RT-PCR formats exhibited high sensitivity, specificity, and reliability for detection and quantification of CTV in immobilized plant materials (Bertolini et al. 2008). Populations of *Citrus tristeza virus* may consist of severe isolates causing seedling yellows (SY) and/or stem pitting (SP) in sweet orange and grapefruit. Quantitative real-time RT-PCR format was applied for reliable detection and precise identification of CTV variants, using a general set and three TaqMan-locked nucleic acids (LNAs) probes targeting sequences characteristic of severe, mild (non-SY, non-SP) and T36-like isolates. Specific amplification occurred in fresh or silica-desiccated CTV-infected samples. Fresh tissues were more suitable than desiccated tissues for efficient detection of CTV. Analysis of pathogenicity of distinct CTV isolates from 20 countries showed that mild isolates reacted with mild probe, whereas severe SP and SY isolates reacted with severe SP or T36-like and often with the second probes. This protocol including three separate real-time RT-PCR reactions was more rapid, sensitive, and specific than previously used methods (Ruiz-Ruiz et al. 2009). The efficiency of tissue-print (TP) real-time RT-PCR assay was compared with validated TP-ELISA test, using specific monoclonal antibodies 3DF1 and 3CA5 for detection of CTV isolates. Comparative estimates indicated that TP-real-time RT-PCR was the most sensitive procedure and it might be employed to monitor the sanitary status of commercial citrus trees (Vidal et al. 2012). In a later investigation, reverse-transcription droplet digital (RT-dd) PCR assay and RT-qPCR format exhibited high degree of linearity and quantitative correlation. But RT-ddPCR assay was more sensitive than RT-qPCR format (Wang et al. 2020). *Citrus tatter leaf virus* (CTLV) RT-qPCR assay had high amplification efficiency and showed 100% CTLV detection rate, when they were tested on known CTLV-positive trees infected with different isolates of CTLV (Park et al. 2018).

The presence of *Sugarcane yellow leaf virus* (SCYLV) could be detected using RT-PCR and qRT-PCR formats in cultivars considered immune. Semi-quantitative titers in these cultivars were significantly lower than in susceptible cultivars (Zhu et al. 2010). SYBR Green I based one-step real-time RT-PCR assay was applied for quantifying *Southern rice black-streaked dwarf virus* (SRBSDV) in rice rapidly and precisely. Primers for the assay were designed based on conserved sequence in S9 RNA of SRBSDV isolates and they were specific to targets. The assay was more sensitive than the conventional RT-PCR for detection of SRBSDV (Zhou et al. 2014). The qRT-PCR assay that could react with all *Squash mosaic virus* (SqMV) isolates of three genotypes was developed. A conserved sequence segment in the 3′ untranslated region of RNA2 was identified for primer and probe design and employed. An internal amplification control with endogenous gene sequence (18S ribosomal RNA) was incorporated in the duplex qRT-PCR system to enhance reliability and interpretation of results (Li et al. 2016). Simultaneous detection and identification of viruses causing tomato yellow leafcurl disease (TYLCD) and their vectors were achieved by applying a panel of quantitative multiplexed real-time PCR assays. The target viruses and their vectors were *Tomato yellow leafcurl virus*-Israel (TYLCV-IL), *Tomato leafcurl virus* (ToLCV), *Bemisia tabaci* Middle East Asia Minor 1 species (MEAM 1, B biotype) and *B. tabaci* Mediterranean species (MED, Q biotype (van Brunschot et al. 2013). A new strain, *Tomato leafcurl New Delhi virus* (ToLCNDV)-ES, infecting commercial zucchini melon and tomato crops in Spain was detected by tissue-print hybridization test. TaqMan qPCR assay was more sensitive than tissue-print hybridization to detect viral genomic DNA-A and DNA-B. ToLCNDV-ES DNA titers were significantly lower in tomato than in zucchini plants (Simón et al. 2018).

6.2.2.6.6 DNA Array Technology

DNA array technology is a highly efficient method for detection of multiple pathogens simultaneously. DNA array technology miniaturizes traditional bioanalytical detection system, so that hundreds of biomolecules with unique identity could be detected simultaneously in one single experiment. In the microarray platform (glass slide) thousands of diagnostic probes may be spotted and hybridized with labeled targets like nucleic acids extracted from the specimen and chemically modified with a fluorescent or other reporter (compound/molecule). Identification/detection of DNA sequences (genes or gene mutations) and determination of expression levels or abundance of genes may be accomplished by applying DNA microarray technique. The presence of several viruses may be detected in a single generic assay. A single glass microscope slide forming the microarray, can accommodate up to 30,000 DNA probes. DNA arrays identify gene sequences from each virus present in the mixture in a single assay. Microarray exposed to fluorescently labeled cDNA from the sample, followed by scanning with microarray scanner is employed for detecting the targets in the sample concerned. Microarrays are named differently as biochip, DNA chip, gene chip, and gene array (Shi et al. 2003). Microarray technique was efficient in detecting potato viruses, *Potato virus X* (PVX), *Potato virus Y* (PVY), *Potato virus A* (PVA), and *Potato virus* S (PVS) in plants infected either individually or in mixture of these viruses. Closely related viruses or strains of these viruses could be detected and discriminated. The assay was able to detect sequence variants with greater than 90% sequence identity (Boonham et al. 2002). In a later study, short synthetic single-stranded oligomers (40-nt) were employed as capture probes instead of PCR products. Potato viruses, PVA, PVS, PVX, PVY, *Potato virus M* (PVM), and *Potato leaf roll virus* (PLRV) could be effectively detected in both single and multiple infections by using a microchip. Predominantly occurring strains of PVY and PVS were detected and differentiated by employing oligonucleotide-based microarray procedure (Bystricka et al. 2005). Conventional PCR and reverse transcription-ligase detection reaction-PCR (MRLP)-based universal microarray system was applied to detect potato virus infections in potato leaves and tubers. The MRLP amplicons were analyzed by zip-code microarray hybridization and subsequent scanning. Further, positive disease incidence detected by MRLP microarray was higher than by PCR for corresponding viruses in field samples (Wang et al. 2012). A 25-mer oligonucleotide microarray was developed targeting four distinct potyviruses that included 85 probes designed from conserved and variable sequence regions of nuclear inclusion b (NIb) gene, RNA-dependent RNA polymerase (RdRp) gene, coat protein (CP) gene and the 3′ untranslated region (UTR), specific to four targeted potyviruses at both species and strain levels (Wei et al. 2009).

Cucumber mosaic virus (CMV) serogroups and subgroups could be detected and differentiated by the membrane-based macroarray detection system. Coat protein (CP) genes of 14 isolates were amplified using cy3-labeled generic-and species-specific primers (Deyong et al. 2005). A nonradioactive macroarray system with oligonucleotide probes immobilized on nylon membranes was applied for detection of *Cucumber mosaic virus* (CMV), *Potato virus Y* (PVY), and *Potato leaf roll virus* (PLRV). All target viruses could be detected readily in the macroarray with strong signals observed relative to the background on the membranes. The labeled target DNA hybridization occurred specifically to the infecting virus and no-cross hybridization to probes for heterologous viruses was seen. Conventional RT-PCR and real-time RT-PCR assays were more sensitive than the macroarray system (Agindotan and Perry 2007). As diagnostic methods used earlier were time-consuming, a 70-mer oligonucleotide microarray capable of detecting simultaneously a broad spectrum of known viruses as well as new viruses infecting grapevine by cross-hybridization to highly conserved probes was developed. The array contained and species-specific regions of 44 plant viral genomes. Accuracy of detection of 10 grapevine viruses by microarray was compared with that of RT-PCR. Specific probe hybridization patterns obtained for each of *Grapevine leafroll-associated virus 1, 2, 3.4, 7, -9, Grapevine fan leaf virus, Grapevine rupestris stem-pitting-associated virus, Grapevine virus A*, and *Grapevine virus B* revealed the powerful nature of this approach for high-throughput plant certification purposes (Engel et al. 2010). Detection of 12 potato viruses including PVY, PLRV and PVX simultaneously was possible using a complementary DNA (cDNA) macroarray system. A suitable region in the genome of the selected viruses and their respective cDNAs were cloned into plant vectors. PCR reactions were carried out using one primer pair for each of viruses. The macroarray system was more sensitive than ELISA and RT-PCR assays (Maoka et al. 2010). A crop-specific macroarray was developed to detect 38 of the most prevalent or emergent viruses infecting grapevines. The reusable macroarray consisted of 1,578 virus-specific 60- to 70-mer oligonucleotide probes and 19 plant and internal control probes spotted onto a nylon membrane (Thompson et al. 2014).

6.2.2.6.7 Metagenomics Next-Generation Sequencing Technique

Metagenomics is an approach for studying microbial populations in a sample by analyzing the nucleotide sequence content. RNA viruses, viroids and the RNA stages of actively replicating DNA viruses can be directly detected. The nature of causal agent of a new disease affecting cotton (*Gossypium hirsutum*) was investigated. The complete nucleotide sequences of two independent virus genomes isolated from cotton blue disease (CBD)–resistant and –susceptible cultivars were determined. The virus genome was closely related to *Cotton leaf roll dwarf virus. Aphis gossypii* could transmit the virus efficiently. Further, a constructed full-length infectious cDNA clone from the virus induced typical symptoms of the disease systemically in CBD-resistant cotton plants. The new disease was caused by an atypical *Cotton leaf roll dwarf virus* (Agrofoglio et al. 2017). The presence of *Tomato chlorosis virus* (ToCV) was confirmed by specific RT-PCR either in the sampled plants or in the whiteflies collected from the leaves of infected plants. Phylogenetic analysis suggested that the Chinese ToCV-SDSG isolate was closely related to ToCV-Florida isolate (Zhao et al. 2014). A novel geminivirus infecting *Prunus* spp. was characterized and identified using molecular methods. The virus provisionally named *Prunus Geminivirus A* (PrGVA) was identified using Illumina sequencing from an asymptomatic plum tree. PrGVA identity was confirmed by rolling cycle amplification, cloning and Sanger sequencing of its complex genome. Apart from the non-nucleotide motif present in its virion strand origin of replication, other conserved motifs of PrGVA suggested its geminiviral origin (Rwahnih et al. 2018).

Identification of novel disease-inducing virus was one of the first aspects of plant health impacted by application of next-generation sequencing (NGS), also known as high-throughput sequencing (HTS), resulting in the discovery of more than 100 new plant viruses, new virus variants, or new plant hosts for known viruses (Barba et al. 2014). The development of new techniques like NGS, capable of detecting multiple viruses present in infected plants is essential when disease symptoms are absent, unspecific, or triggered only when plants are infected by multiple viruses. As the majority of plant viruses have RNA as their genome and those that have DNA genomes produce RNA transcripts, analysis of RNA sequences from plant samples is an effective method for virus detection. Sequencing of total small RNAs (sRNAs) has proved to be an effective method of plant virus detection (Wu et al. 2010). NGS may in a single experiment determine the sequence of hundreds of thousands to millions of virus RNAs, which can be reassembled to obtain the genomic sequence of the target virus genomes. NGS has significantly increased the number of novel plant viruses discovered and characterized, both in host plants and their biological vectors. NGS procedure may facilitate the design of new detection primers of broader specificity for improving sensitivity of conventional detection assays (Marais et al. 2014). Although there are several advantages of NGS procedure, the high cost may be a limiting factor for wider application by researchers in all countries (Hadidi et al. 2016).

The genetic structure and diversity of *Cucumber mosaic virus* (CMV) were assessed. Pepper (*Capsicum annuum*) samples collected during field survey were subjected to RT-PCR analysis. CMV isolates were analyzed by full-genome sequencing. Genetic diversity analyses indicated that the *2b* gene of CMV was under weaker purifying selection than other viral genes. Phylogenetic analyses of RNA1 showed that CMV isolates from pepper could be divided into three clusters in subgroup 1 (Kim et al. 2014). Three tomato-infecting tobamoviruses were characterized using biological, immunological and genetic characteristics. Complete genome sequences of *Tobacco mosaic virus* (TMV), *Tomato mosaic virus* (ToMV) and *Tomato mottle mosaic virus* (ToMMV) were obtained and their genetic relationships to other tobamoviruses were evaluated. The multiplex RT-PCR was employed for species-specific detection and identification of TMV, ToMV, and ToMMV. Further, a resistance-breaking strain of ToMMV on selected tomato cultivars that were resistant to ToMV was identified, suggesting the need for monitoring incidence of tobamoviruses to recognize changes in the virulence patterns of tobamoviruses infecting tomato (Sui et al. 2017). *Tomato necrotic spot virus* (ToNSV) was characterized using genomic, biological, and immunological properties. The complete genomic sequence identified ToNSV as a new ilarvirus distinct from tomato-infecting ilarviruses. Coat protein coding region of the isolates, occurring in the United States, showed 94–98% identity, while the same isolates had 99% amino acid sequence identity. ToNSV was identified as a new subgroup causing necrotic spotting disease of tomato (Bratsch et al. 2019).

The complete genome sequence and genome organization of *Eggplant mild leaf mottle virus* (EMLMV) were studied. Extracted viral RNA was subjected to SOLiD next-generation sequence analysis, followed by ds-cDNA synthesis or PCR amplification. The viral genome included a putative single, large open reading frame (ORF) that encoded a polyprotein. The deduced amino acid sequence of the EMLMV polyprotein was relatively similar to that of *Sweet potato mild mottle virus* (SPMMV), with 37% shared sequence identity. Among the four ipomoviruses, only SPMMV and the putative genus member EMLMV contained a helper component proteinase (HC-Pro) gene (Dombrovsky et al. 2012). Sweet potato viruses *Sweet potato feathery mottle virus* (SPFMV) strain RC and *Sweet potato virus* strain WA (*Crinivirus* sp.), *Sweet potato leaf-curl Georgia virus* (*Begomovirus* sp.) and *Sweet potato pakakuy virus* strain BC (*Sweet potato badnavirus B*) were detected by employing small RNA deep-sequencing (SRDS) procedure. Identity of the viruses established by the SRDS procedure was confirmed by PCR assay and sequencing of amplicons. The SRDS procedure had the potential for wide applicability for simultaneous detection of several viruses affecting different crops (Kashif et al. 2012). Sweet potato samples were analyzed by small RNA deep-sequencing and PCR/RT-PCR assays. Eleven virus species involved in diseases of sweet potatoes were detected. Further, coinfection by two or more virus species was also recorded in some infected plants (Ma et al. 2019).

Deep sequencing of sRNAs enriched from symptomatic and asymptomatic grapevine was performed. A large number of viral small RNAs that shared identical nucleotide sequences with genomes of 27 plant viruses was profiled. Two cDNA libraries were constructed from symptomatic and an asymptomatic grapevine plants. Deep sequencing of the two cDNA libraries showed that the most abundant viral small RNAs aligned with the genomes of viruses included in *Badnavirus* (genus), *Caulimoviridae* (family). The virus was tentatively named *Grapevine vein clearing virus* (GVCV) (Zhang et al. 2011). NGS provides an alternative approach for identification of viral pathogens in grapevine. Unknown viruses causing asymptomatic infection are identified by the homologies they share with genomic sequences of the known cognate species. NGS procedure cataloged the genomic sequences of viruses and other pathogens extracted as DNA or RNA from infected grapevines (Al-Rwahnih et al. 2013). Several grapevine transcriptomes were screened to identify viruses and viroids that infect grapevines. Five viruses infecting grapevine were identified. The results showed that 2.16% of the transcriptome was comprised of viral RNAs with more than half of the viral reads derived from *Grapevine rupestris stem pitting-associated virus* (GRSPaV). Further, single nucleotide polymorphisms (SNPs) were identified for each virus. Plant transcriptome data using bioinformatic analyses could be effectively applied for identification of five viruses in the grapevine cv. Cabernet Sauvignon (Jo et al. 2018). Complete genome sequence of *Plum bark necrosis stem pitting-associated virus* (PBNSPaV) isolates from four *Prunus* sources was determined by pyrosequencing and they had same organization as the reference isolate. The least divergent isolate showed overall 91.8% of nucleotide identity with type isolate. Complete sequencing of divergent isolates resulted in development of a more broad spectrum detection test targeting a conserved region in the *P61*gene. The results indicated a much broader diversity of PBNSPaV isolates than recognized earlier (Marais et al. 2014). Infection of cherry trees in Italy and plum trees in Azerbaijan by a new virus was observed. Double-stranded (ds) RNAs purified from infected trees were analyzed by deep sequencing contigs showing weak, but significant, identity with various members of the family *Betaflexiviridae* were constructed. Comparison of nucleotide sequences indicated that virus isolates present in *Prunus* plants belonged to the same virus species. Genomic organization of the isolates was similar to that of some members of *Betaflexiviridae*. Phylogenetic analysis of the deduced encoded proteins showed a clustering with *Potato virus* T (PVT). The new member of the genus *Tepovirus* was identified using RT-PCR assay (Marais et al. 2015).

NGS was effectively employed to identify multiple viruses within a metagenomic sample. Simultaneous detection of 11 recognized citrus viruses was achieved by employing electronic probes (e-probes), which were designed and screened against raw sequencing data to minimize bioinformatic processing time. The efficiency of e-probes-based approach was validated with NGS datasets generated from different RNA preparations: ds-RNA from Mexican lime generated with different *Citrus tristeza virus* (CTV) genotypes, ds-RNA from field samples, and small RNA and total RNA from grapefruit infected with CTV T3 genotypes (Jooste et al. 2017). A naturally occurring strain of *Citrus tristeza virus* (CTV) from lemon trees of more than 100 years old was characterized. Virus-derived small interfering RNAs, induced by Dicer processing of ds-RNAs formed during RNA virus replication were used as targets for sequencing analysis showed that 94.05% of the 432,632 sequences corresponded to L192GR isolate of CTV. Comparative analysis of complete genomes revealed molecular homology with CTV-VT isolate FS2–2 from Florida with 98.2% nucleotide sequences identity. Most isolates of CTV-VT group could induce seedling yellows syndrome and react positively with MAb MCA13, which was typically associated with severe CTV isolates. But MCA 13-positive L192GR induced very mild or even no symptoms in the seedling indicator plants. The results indicated the need for combining molecular data with biological characteristics of CTV strains (Varveri et al. 2015). Transmission electron microscopy revealed the presence of rod-like particles of *Citrus leprosis* (CL) in the nucleus and cytoplasm of infected cells in leprosis disease lesions. NGS of total RNA extracts from three symptomatic samples revealed the genome to be distinct although highly related (>92% nucleotide identity) viruses whose genetic organization was similar to that of dichorhaviruses. Genome sequences of these viruses exhibit 62% nucleotide sequence identity with those of *Orchid fleck virus* and *Coffee ringspot virus*. The mites *Brevipalpus phoenicis* transmitted the virus to *Arabidopsis*. The new CL-associated dichorhavirus was named *Citrus leprosies N dichorahvirus* (Ramos-Gonzalez et al. 2017). Identification of novel viruses generated through reassortments of virus genomic fragments is possible by the NGS technique (also known as HTS). Reassortments of genomic RNA fragments created a novel strain of *Orchid fleck virus* (OFV) (*Dichorhavirus*) infecting citrus in Mexico. A typical strain of OFV Cit1 was associated with citrus leprosies virus disease complex. RT-PCR assay could not detect OFV Cit1 from field collected diseased citrus samples. In addition, high-throughput RNA sequencing of these isolates revealed a new virus strain, OFV Cit2, which had a highly variable RNA1, but an RNA2 identical to OFV Cit1 (Roy et al. 2020).

The NGS procedure was employed to identify and characterize several related novel viruses from infected apple trees. Viruses associated with apple rubbery wood disease (ARWD) were called *Apple rubbery wood virus* (ARWV) 1 and 2. Additional samples with ARWD symptoms were positive to PCR assay employing primers based on sequences of ARWV1 and 2 genomic RNA segments. These two viruses were most closely related to members of the order *Bunyavirales*. Three RNA segments (large, medium, and small) were characterized and the viruses might represent a new genus under family *Phenuiviridae* and the virus was provisionally named *Rubodvirus* (Rubbery wood virus) (Rott et al. 2018). Full genome sequence by next-generation sequencing of small interfering RNAs procedure was applied to establish the identity of the virus causing an emerging yellows disease of pepper (PYD) in Greece. Recovery of full genome sequence of a *Polerovirus* closely related to *Pepper vein yellows virus* (PeVYV) was accomplished. The virus was tentatively called *Pepper yellow vein virus* (PeYV). Sequence similarity and phylogenetic analysis highlighted the common ancestry of the three poleroviruses (PeVYV, PeYV and *Pepper yellow leafcurl virus* (PYLCV), associated with PYD. The results indicated that PYD was induced by at least three species closely related pepper-infecting viruses (Lotos et al. 2017). *Tomato spotted wilt virus* (TSWV) is a negative-stranded RNA virus with broad host range. Changes in *Arabidopsis thaliana* transcriptomic profile in response to systemic infection by TSWV were determined using the NGS procedure. Plant immune defense and protein degradation genes were induced by TSWV infection, whereas photosynthesis pathway was adversely affected. Transcriptomic analysis revealed the dynamic nature of TSWV infection in *A. thaliana* at whole transcriptome level (Xu et al. 2020).

HTS platforms could be employed for greater capability of surveillance and screening plants material for presence of novel viruses in mixed infections. Appropriate sampling of potentially infected plants is essentially required for effective surveillance when managed honeybee (*Apis mellifera*) hives were placed in flowering crops to deliver essential pollination services. They could provide valuable surveillance by effectively sampling entire crops during collection of pollen and nectar. The ability of honey bees to create a representative sample from a large number of plants was combined with sensitivity of HTS virus detection in honeybees to function as a powerful system of early detection of plant viruses. National virus survey of honeybees was performed, using HTS screening of pooled honeybee samples across Australia. HTS data identified simultaneously multiple viruses in these samples. *Cucumber green mottle virus* was detected in several states of Australia, before detection of virus presence in diseased plant material (Roberts et al. 2018). HTS procedure was applied to examine the presence of novel virus/viroid in 60 pistachio trees. A virus tentatively named *Pistachio Ampelovirus* A (PAVA) was identified in the National Clonal Germplasm Repository (NCGR) and orchards in California. A putative viroid called *Citrus bark-cracking viroid*-pistachio (CBCVd-pis, *Cocadvirioid, Pospiviroidae*) was also detected in the samples (Al-Rwahnih et al. 2018).

Massively parallel sequencing (MPS) or NGS can detect pathogens that are not anticipated to be in the test sample. The effectiveness of the portable MPS device as a surveillance tool for detection of viral and bacterial plant pathogens in both infected plant and insect vector tissues was tested. The Oxford Nanopore Technology MinION sequencer is a portable, single molecule genome-sequencing instrument that works by taking frequent electrical measurements, while a single strand of DNA passes through a protein nanopore. *Plum pox virus* (PPV) transmitted by aphid vector *Myzus persicae* and '*Candidatus* Liberibacter asiaticus' (Las) (citrus huanglongbing/greening disease) transmitted by psyllid *Diaphorina citri* were tested using MinIon device coupled with whole transcriptome amplification (WTA) to sequence the metatranscriptome of plant and insect tissues infected by PPV and Las. The percentage of generated reads assigned to PPV was 95% from infected tissue and 3% from viruliferous *M. persicae*. Sequencing of

D. citri resulted in 22% of the reads mapping as Las. PPV and Las were detected in both plum and insect tissue samples (Badial et al. 2018). The cDNA from infected tissues was sequenced with a single strand by the Oxford Nanopore sequencing technique (ONT). This procedure confirmed the presence of *Wheat streak mosaic virus* (WSMV) in samples. In addition, *Triticum mosaic virus* (TrMV) was present in all four samples tested, while BYDV was detected in three samples. Deep coverage sequencing of full-length, single-strand WSMV revealed variation compared with WSMV Sidney-81 reference strain and might represent new variants that could overcome resistance gene *Wsm2* (Fellers et al. 2019).

The NGS approach was applied to detect and identify *Sweet potato feathery mottle virus* (SPFMV) and *Sweet potato chlorotic stunt virus*. Small RNAs (sRNA, 20–24 nt) in total RNA extracts from coinfected plants were targeted. NGS analysis of total RNA extracted from a single apple plant, an infection complex which included *Apple chlorotic leaf spot virus* (ACLSV) and *Apple stem pitting virus* (ASPV) was performed. Five isolates/variants of ASPV were identified present in the same sample (James et al. 2017). A full-length cDNA clone of *Pepper mild mottle virus* (PMMV) isolate HNu infecting Chinese Capsicum cv Yellow Lantern was constructed. Virus progeny derived from the cDNA clone replicated and moved systemically in inoculated pepper, inducing symptoms similar to those observed on plants infected by PMMV-Hnu in Yellow Lantern under field conditions. The results showed that PMMV-Hnu was the causal agent of symptoms induced in Yellow Lantern plants (Hu et al. 2020). Single infection by *Pepino mosaic virus* (PepMV) Chilean 2 (CH2) strain was observed. Phylogenetic analysis of CH2 isolates randomly selected from both areas revealed a remarkable divergence between CH2 isolates with increased nucleotide variability in the locations where both strains were present (Alcaide et al. 2020).

6.2.2.6.8 CRISPR/Cas Detection System

Plant virus diagnostic approaches based on natural viral proteins and genomes have been applied. In order to improve specificity of testing, systems based on clustered regularly interspaced short palindromic repeats (CRISPR) and CRISPR-associated (Cas) proteins, are more effective for detection of microbial pathogens, including viruses. The CRISPR-Cas system is based on RNA-guided endonucleases and derived from an effective immune system protecting bacteria against invading nucleic acids (Makarova et al. 2020). CRISPR-Cas12a system was applied for specific detection of *Tobacco mosaic virus* (TMV), *Tobacco etch virus* (TEV), and *Potato virus X* (PVX) with + sense RNA strand in infected *Nicotiana benthamiana* plants. Different sets of oligonucleotides were used to amplify a specific genomic region (coat protein) of the virus to be detected. Specific amplification of different virus-derived dsDNAs by RT-PCR was confirmed by gel electrophoresis. The multiplexed detection of all three viruses in mixed infections was also possible with CRISPR-Cas12a system. CRISPR-Cas13a/d system was used to directly detect RNA viruses from infected plant tissues, bypassing the need for viral genome amplification step and obtaining a readout that determined viral concentration in the plant (Marqués et al. 2022).

6.2.2.6.9 Heteroduplex Mobility Analysis

Heteroduplex mobility analysis (HMA) depends on differences in the rate of migration between a DNA heteroduplex and a DNA homoduplex that can be useful to identify mismatches or deletion in DNA sequences. The presence of begomoviruses in cassava plants was detected by employing the HMA procedure. A highly conserved core region of coat protein (CP) gene of field isolates was amplified, followed by denaturing annealing with a representative of reference strains. HMA profiles of the isolates were able to differentiate four different virus species and 11 different virus strains. In addition, the HMA profiles showed good correlations with sequencing results and phylogenetic comparisons with other sequenced cassava viruses (Berry and Rey 2001). Serological and molecular methods were applied to detect *Grapevine leafroll-associated virus 2* (GLRaV-2) isolates from several grapevine accessions. HMAs were performed to assess genetic variability among GLRaV-2 isolates in ORF coding CP gene of 17 isolates and six strains used as reference. Eight diverse GLRaV-2 variants were identified among the infected grapevines tested. The most common strain was indistinguishable from reference strains from Semillon and Pinot noir 95 accessions. Three other GLRaV-2 variants from Southern and Central Italy were different from all reference strains. Two diverse GLRaV-2 variants were detected in a grapevine accession from Italy (Angelini et al. 2004).

6.3 DETECTION OF VIROIDS

Among the phyllosphere plant pathogens, viroids are the smallest and simplest in structure containing only an RNA molecule, but capable of inducing various kinds of plant diseases in susceptible plants as other microbial pathogens. They possess small, single-stranded, covalently closed circular RNAs with a highly base-paired, rather rod-like, stiff conformation. Viroid genomic RNAs do not code for any viroid-specific protein. They are entirely dependent on the enzyme machinery of their host for replication of the genomic RNAs (Tabler and Tsagris 2004). Viroids can replicate independently, following introduction into cells/tissues of susceptible plants, reaching titers sufficient enough to induce characteristic symptoms or remain latent in certain plant species.

6.3.1 Classification of Viroids

Viroids have a GC-rich genome consisting of 246–401 nucleotides. They have been classified into two families, based on their biological, biochemical, and structural properties. Viroids with a central conserved region (CCR) and without hammerhead ribozymes are included in the family *Pospiviroidae*, whereas viroids lacking CCR capable of self-cleaving through hammerhead ribozymes are placed in the family *Avsunviroidae*. Members of *Pospiviroidae* replicate in the nucleus, whereas members of *Avsunviroidae* can replicate in the chloroplasts (Flores et al. 2005). Viroids do not exhibit distinct variations in their structure. It may be difficult to differentiate them based on the variability of structural characteristics.

6.3.2 Biological Methods

Viroids may induce different symptoms, which are sometimes similar to those caused by viruses in some plant species. *Potato spindle tuber viroid* (PSTVd) was found to be a simple single-stranded RNA molecule with no protein coat. PSTVd induced abnormally elongated tubers with spindle or cylindrical shape and prominent eyes characteristic of PSTVd infection in potato plants (Diener 1979). PSTVd dahlia isolate could be transmitted experimentally to tomato and potato. Mild chlorosis and stunting were seen in tomato whereas no symptoms could be observed in aerial plant parts or tubers in potato (Verhoeven et al. 2016). *Apple scar skin viroid* (ASSVd) causes restricted symptoms on fruits only and the symptom type may differ depending on apple cultivar (Koganezawa et al. 2003).

Citrus spp. show infection by *Citrus exocortis viroid*, *Citrus bent leaf viroid*, *Hop stunt viroid*, *Citrus viroid III* and *Citrus viroid IV (Pospiviroidae)* under natural conditions. Biological indexing by grafting infected scions onto Arizona 861-S1Etrog citron plants was performed for detecting *Citrus exocortis viroid* (CEVd), *Hop stunt viroid* (HSVd), and *Citrus viroid III* (Malfitano et al. 2005). The diagnostic/assay/bioindicator plants expressing clear symptoms rapidly were employed for detection and differentiation of viroids. *Citrus aurantifolia, Chrysanthemum morifolium* cv. Mistletoe, *Cucumis sativa* cv. Sporu, and *C. sativus* were used as bioindicator plants for detection of viroids. Ability to be transmitted through seeds is one of the biological characteristics of viroids. PSTVd could be transmitted via pollen, true seeds of potato and also through seeds of tomato. *Peach latent mosaic viroid* (PLMVd) was transmitted through pollen to healthy peach cultivars. Likewise, *Citrus exocortis viroid* (CEVd) could be transmitted through seeds of *Impatiens walleriana* and *Verbena* x hybrid, while *Pepper chat fruit viroid* (PCFVd) and *Tomato chlorotic dwarf viroid* were transmissible through true seeds (Narayanasamy 2011, 2017).

6.3.3 Physicochemical Methods

The PAGE technique is effective in separating nucleic acid molecule based on their differential mobility in an electrical field. Citrus viroids could be detected by employing the sequential polyacrylamide gel electrophoresis (sPAGE) procedure in the extracts of young bark tissues from infected plants. *Hop stunt viroid, Citrus exocortis viroid, Citrus viroid III, Citrus viroid IV*, and *Citrus bent leaf viroid* were detected by sPAGE technique (Malfitano et al. 2005). The return (R)-PAGE procedure was applied as an adjunct method to RT-PCR assay for detection of viroids. RT-PCR assay was employed for detection of PSTVd and *Chrysanthemum stunt viroid* (CSVd) in potato tubers and chrysanthemum samples, respectively. The amplified product of 199-bp, as viroid-specific band, could be visualized by R-PAGE procedure. Bands representing PSTVd and CSVd were located adjacently to each other (Bostan et al. 2010).

6.3.4 Nucleic Acid–Based Techniques

As the viroids lack a protein component, immunoassays are practically inapplicable for detection of viroids in plants and propagative materials. Nucleic acid–based techniques are employed for sensitive, rapid, reliable, and precise detection and identification of viroid(s) in infected plants, since biological methods are time-consuming, labor-intensive, and yield inconsistent results.

6.3.4.1 Nucleic Acid Hybridization Techniques

The ^{32}P-lebeled cDNA probes were allowed to hybridize with the immobilized *Potato spindle tuber viroid* (PSTVd) RNA in solid matrix nitrocellulose membrane. As the hazardous nature of radioactive probes was realized, non-radioactive probes such as digoxigenin and biotin were increasingly utilized later for preparing probes. Digoxigenin (DIG)-labeled cDNA probes were prepared for full-length cDNA clone of *Chrysanthemum stunt viroid* (CSVd). The pathogen was detected in dot blot hybridization, which was more sensitive than return PAGE procedure (Li et al. 1997). The dot blot method employing DIG-labeled probes were useful for effective routine detection of CSVd (Mumford et al. 2000). Citrus-infecting viroids, *Citrus exocortis viroid* (CEVd), *Citrus bent leaf viroid* (CBLVd), *Hop stunt viroid* (HSVd), *Citrus viroid III* (CVd-III), and *Citrus viroid IV* (CVd-IV) could be detected in inoculated Etrog citron using the northern blot hybridization method and viroid-specific probes (Barbosa et al. 2005). Citrus cultivars of Washington Navel and Naveline oranges were infected by *Citrus tristeza virus* (CTV), whereas Clementine trees were infected by *Citrus infectious variegation virus* (CIVV). All these virus-infected trees were coinfected by HSVd and 50% of them were infected also by CEVd (Barbosa et al. 2007). Northern and dot blot procedures employing viroid-specific DIG-labeled probes were useful for reliable detection of CEVd, CBLVd, Citrus viroid (CVd-I), HSVd, CVd-III and CVd-IV (Malfitano et al. 2005). The dot blot hybridization method employing DIG-labeled viroid-specific probes was effective in detecting *Peach latent mosaic viroid* (PLMVd), as well as *Hop stunt viroid* (HSVd) infection in peach trees (Hassan and Ryšánek 2004).

A polyprobe was developed for simultaneous detection of four viroids infecting grapevine by cloning tandem full-length sequences of *Hop stunt viroid* (HSVd), *Australian*

Grapevine viroid (AGVd), and *Grapevine yellow speckle viroid-1* and *-2* into a single vector. The cRNA probe detected all four viroids with sensitivity similar to that obtained using individual probes (Zhang et al. 2012). Dot blot hybridization assay using a DIG-labeled cRNA polyprobe was developed for simultaneous detection of *Apple scare skin viroid, Apple dimple fruit viroid, Apple fruit crinkle viroid, Pear blister canker viroid, Hop stunt viroid*, and *Peach latent mosaic viroid* infecting pome and stone fruit trees. Detection limits of the polyprobe were similar to those of single probes, indicating that the poly-6 probe was sensitive and reliable for detection of homologous viroids in the dot blot hybridization assay (Lin et al. 2011). Simultaneous detection of four viroids was accomplished by using a polyprobe and nonisotopic dot-blot hybridization procedure. A polyprobe was constructed by cloning tandem-full-length sequences of *Hop stunt viroid* (HSVd), *Australian grapevine viroid,* and *Grapevine yellow speckle viroid 1* and *2* into single vector. The cRNA probe detected all four viroids with sensitivity similar to that obtained using individual probes (Zhang et al. 2012). The polyprobe technique was applied for simultaneous detection of viruses and viroids infecting tomato crops. Three polyprobes were developed covering 4 viroids (poly 4), 12 viruses (poly 12), or 4 viroids plus 12 viruses (poly 16). Detection limits of three polyprobes were comparable to the individual probes. Tests on field samples showed that all positive samples detected with individual probes were also detected with the corresponding poly 12 (98%) or poly 16 (100%) probes. Tomato seed testing showed that both single and polyprobes could be employed to detect an infected seed present in the tomato seedlots (Sánchez-Navarro et al. 2019).

The tissue-printing hybridization (TPH) procedure was efficient in detecting *Citrus exocortis viroid* (CEVd), *Chrysanthemum stunt viroid* (CSVd), *Hop stunt viroid* (HSVd), and *Avocado sun blotch viroid* (ASBVd) in their respective natural host plant species. The TPH technique could be employed to determine the sites of viroid accumulation, in addition to providing reliable detection of viroids rapidly (Romero-Durban et al. 1995). *Potato spindle tuber viroid* (PSTVd) was detected in stem rachis of two tomato cultivars by the TPH method earlier in susceptible tomato cv. Rutgers. Distribution and localization of PSTVd in different tissues were visualized using the TPH technique (Stack-Lorenzen et al. 1997). The print-PCR format simplified sample preparation, in addition to providing the opportunity of testing a large number of potato plants for freedom of viroid infection. Dot-PCR based on dotting plant sap on filter paper could be employed for testing in vitro potato plantlets and tuber tissues for indexing. Samples containing one infected plant and nine healthy plants could be tested reliably for PSTVd infection (Weidemann and Buchta 1998). Sensitivity of detection of citrus viroids could be enhanced by first inoculating *Citrus medica* (Etrog citron) and employing DIG-labeled RNA or DNA probes (Palacio-Bielsa et al. 1999). TPH or imprint hybridization detected *Peach latent mosaic viroid* (PMVd) and *Hop stunt viroid* (HSVd) in citrus plants. RT-PCR assay confirmed the results of TPH procedure (Michelutti et al. 2005). In a later investigation, incidence of viroid diseases infecting *Prunus* spp. was observed in high proportions. Peaches and apricots contained PLMVd infection, whereas HSVd infection was detected in apricot, peach or plum. Mixed infection by PLMVd and HSVd was also detected in two samples (Gümüs et al. 2007). Detection of *Apple scar skin viroid* (ASSVd), *Apple dimple fruit viroid* (ADFVd) and *Pear blister canker viroid* (PBCVd) in field-grown trees was achieved by applying TPH and dot blot hybridization assay, using nonradioactive labels. The TPH assay showed potential for specific detection and discrimination of viroids infecting pome fruits (Lolic et al. 2007). The TPH assay was employed to detect *Apple scar skin viroid, Apple dimple fruit viroid, and Pear blister canker viroid* in commercial orchards and nurseries. The TPH technique was found to be a valuable tool for selecting viroid-free mother trees for obtaining healthy stocks of pome tree fruit crop (Di Serio et al. 2010).

6.3.4.2 Reverse Transcription–Polymerase Chain Reaction Assays

The RT-PCR assay has been applied extensively for detection of a large number of viroids affecting various crops. RT-PCR assays with appropriate primers are sensitive, specific, rapid, and reliable for detection, identification, and differentiation of *Peach latent mosaic viroid* (PLMVd) and *Hop stunt viroid* (HSVd) (El-Dougdoug 1998), PLMVd and HSVd (Hassan et al. 2004), mixed infections of pear with *Pear blister canker viroid* (PBCVd + HSVd) (Hassen et al. 2006), and HSVd and HLVd (Eastwell and Nelson 2007) in pome fruit trees and hop crops. Dot blot hybridization and TaqMan assays efficiently and specifically detected *Chrysanthemum stunt viroid* (CSVd) and CSVd/*Potato spindle tuber virioud* (PSTVd). When a CSVd-primer-probe set and PSTVd primer-probe set were employed, TaqMan amplification of viroid genomes resulted in two distinct patterns, based on which respective viroids could be detected in the RNA extracts from infected plants (Mumford et al. 2000). *Potato spindle tuber viroid* (PSTVd) was detected using an exogenous internal standard, in vitro (Kox et al. 2005). In a later investigation, RT-PCR assay was applied to detect PSTVd, using the primer pair PSTVd-321/PSTVd-323. PCR analyses showed that all samples testing positive by dot blot hybridization generated specific RT-PCR amplificon, indicating the infection of samples by PSTVd (Di Serio 2007). A duplex RT-PCR format in combination with a plant RNA as internal control was developed for precise and reliable detection of PSTVd (Hataya 2009).

Multiplex RT-PCR format was developed for simultaneous detection of PSTVd and *Tomato chlorotic viroid* (TCDVd). Amplified cDNAs generated a 271-bp PCR product that was used as the template for PSTVd and 191-bp for TCDVd. The multiplex RT-PCR procedure detected both PSTVd and TCDVd precisely in field-infected tomato plants and petunia (*Petunia* x hybrid) plants (Matsushita et al. 2010). Simultaneous detection of multiple viroids in infected stone fruit trees was achieved by employing RT-PCR assay including specific primers to viroids as well as for host mRNA and internal control. *Peach latent mosaic viroid* (PLMVd) and *Hop stunt viroid* (HSVd)-infected samples from different cultivars and orchards were tested. All targets were amplified under the same RT-PCR conditions, and products specific to specific to these viroids were amplified by the assay (Hassan et al. 2009). The RT-PCR assay was employed to assess the sanitary status of citrus varieties and/or clones with respect to the extent of viroid infection to select elite mother plants of citrus for vegetative propagation. All mother plants tested were infected by *Citrus exocortis viroid* (CEVd) and *Citrus bent leaf viroid* (CBLVd), although they were free of *Citrus tristeza virus* (CTV) and *Citrus variegation virus* (CVV). The viroids could be eliminated from lemon, mandarin, and Jaffa orange plants by adopting micrografting technique (Kaparia-Isaia et al. 2007). The multiplex RT-PCR assay was employed for simultaneous detection of *Chrysanthemum stunt viroid* (CSVd) and *Chrysanthemum chlorotic mottle viroid* (CChMVd). The multiplex RT-PCR assay could detect both viroids from doubly infected plants without any nonspecific amplification. RT-PCR format provided the advantage of saving time, cost and labor over conventional RT-PCR assay without compromising the sensitivity and reliability of results (Hosakawa et al. 2007).

One-step multiplex RT-PCR format was applied for simultaneous detection and differentiation of *Citrus exocortis viroid, Citrus bent leaf viroid, Citrus viroid III*, and *Citrus viroid IV* in infected citrus plants. The ubiquitin gene, with greater expression and stability than other housekeeping genes, was used as the internal control. Several citrus plants were infected by more than one viroid and a few plants were infected with up to four viroids. The sensitivity of uniplex and multiplex RT-PCR formats was at the same level. One-step multiplex RT-PCR assay detected the viroids efficiently in samples from sweet orange, lemon, mandarin, and mandarin hybrid (Wang et al. 2009). Citrus budlings from citrus cultivars originating from different countries were tested by the multiplex RT-PCR assay for detection of *Citrus exocortis viroid* (CEVd), *Citrus bent leaf viroid* (CBLVd), *Hop stunt viroid* (HSVd), *Citrus dwarfing viroid* (CDVd), and *Citrus bark cracking viroid* (CBCVd). The multiplex RT-PCR assay showed positive reactions for CEVd, CBLVd, HSVd, CDVd, and CBCVd. Citrus viroids could spread through budlings to the countries importing them (Al-Harthi et al. 2013). Several viroids can infect hop (*Humulus lupulus*) crops and some of them being highly virulent causing dieback symptoms. Citrus cuttings collected from field-grown plants were forced to sprout in a growth chamber in sterilized agriperlit. One step-RT-PCR format was employed to detect presence of viroids in leaves from forced cuttings. *Hop stunt viroid* (HSVd) and *Citrus viroid III* (CVd-III) were readily detected in leaves (Guardo et al. 2005). Simultaneous detection of three hop viroids belonging to three different genera infecting hop crops viz., *Hop latent viroid* (HLVd), *Hop stunt viroid* (HSVd), and *Apple fruit crinkle viroid* (AFCVd) was achieved by applying a multiplex RT-PCR format. All samples were positive for HLVd. A low level of infection by AFCVd in hop crops was recorded in China (Lu et al. 2012). One-step multiplex RT-PCR (mRT-PCR) format was applied for simultaneous detection of *Hop latent viroid* (HLVd), *Hop stunt viroid* (HSVd), *Apple fruit crinkle viroid* (AFGVd), and *Citrus bark cracking viroid* (CBCVd) infecting hop plants. Specificities of single and mRT-PCR assays for all four viroids were comparable. Sensitivity of mRT-PCR was compared with that of dot blot hybridization and single RT-PCR assay on biolistically infected hop plants. The mRT-PCR format was robust, simple, and rapid in providing results (Guček et al. 2019).

The quantitative real-time RT-PCR (qRT-PCR) format was developed to enhance the sensitivity and reliability of detection of *Peach latent mosaic viroid* (PLMVd), using a specific set of primers and probes and an efficient extraction method for nucleic acids from infected peach plants. The accuracy and efficiency of the assay were assessed using different PLMVd peach isolates as templates, including three Calico variants. Strong signal was detected from all variants tested (Luigi and Faggioli 2011). The diagnostic method consisting of reverse transcription, followed by qRT-PCR assay was applied for testing planting materials for the presence of PLMVd in plants handled by certification personnel. The qRT-PCR assay was more sensitive than conventional RT-PCR format. The qRT-PCR assay could detect and quantify *Plum latent mosaic viroid* (PLMVd) from 3 months after postinoculation, whereas the reference RT-PCR did not detect the pathogen until 8 months postinoculation (Parisi et al. 2011). The multiplex RT-PCR assay was developed for simultaneous detection of *Hop stunt viroid* (HSVd), *Grapevine yellow speckle viroid 1* (GYSVd-1), *Grapevine yellow speckle viroid 2* (GYSVd-2), *Australian grapevine viroid* (AGVd), and *Citrus exocortis viroid* (CEVd) from commercial vineyards and germplasm collections in Italy. HSVd and GYSVd-1 were detected in field-grown grapevines, whereas GYSVd-2 and AGVd were identified in two grapevine table cultivars and germplasm collections. The presence of GYSVd and AGVd were detected in vines grown in southern Italy (Gambino et al. 2014).

A detection and identification system was developed for comprehensive diagnosis of diseases caused by eight pospiviroids, based on real-time RT-PCR format. Two real-time RT-PCR formats based on SYBR Green and TaqMan technologies were combined. Tomato seed samples were first screened for all pospiviroids by SYBR Green procedure. Samples positively reacting with universal primer set were subjected to TaqMan real-time PCR method, with specific primer and probe sets. The two-step detection system was effective in samples containing one infected seed out of 400 seeds, placing an efficient safety umbrella to prevent introduction of pospiviroids through tomato seeds (Yanagisawa et al. 2017). *Potato spindle tuber viroid* (PSTVd) could be detected using SYBR Green-based real-time RT-PCR format coupled with melting curve analysis along with or without internal control from potato. Real-time RT-PCR format was more sensitive than conventional RT-PCR. Amplification of both PSTVd RNA and internal control RNA was indicated by two melting peaks in the duplex real-time RT-PCR format (Jeevalatha et al. 2015). An easy-to-use method was developed for detection of *Potato spindle tuber viroid* (PSTVd) based on RNA binding to positively charged anion-exchange resin beads that provided up to 100-fold enhancement of sensitivity of detection. The RT-LAMP method was applied for rapid, specific, sensitive, and visual detection of PSTVd in potato plants. The results of the assay could be visualized as ladder-like bands in an agarose gel and using SYBR Green I dye. Target specificity for RT-LAMP primers was assessed by including a restriction digestion step, after RT-LAMP assay. Analytical sensitivity of the assay was greater than those of real-time PCR and RT-PCR assays (Verma et al. 2020).

Pepino mosaic virus (PepMV) could be detected in tomato leaf tissue and water samples, whereas PSTVd was detected in seed and leaf extracts from tomato. The RT-LAMP assay was optimized for detection of PepMV and PSTVd in vitro and on-site conditions in crude leaf homogenates within 30 minutes (Mehle et al. 2017). The multiplex real-time quantitative RT-PCR assay was performed to detect diversified strains of *Pepper chat fruit viroid* (PCFVd) and *Columnea latent viroid* (CLVd) in a single tube. The assay was efficient in detecting strains with sequence diversity of PCFVd and CLVd (Sombat et al. 2018). Universal primers were employed in one-step RT-PCR and RT-LAMP techniques to detect six viroids in infected plants. The results showed that a pair of degenerate primers could be used in one-step RT-PCR to amplify *Columnea latent viroid, Chat fruit viroid, Potato spindle tuber viroid, Tomato chlorotic dwarf viroid,* and *Tomato planta macho viroid* from Solanaceae seed and plants. Further, five primers were used in RT-LAMP procedure to amplify these viroids. Generally the RT-LAMP method was more sensitive than the RT-PCR assay (Tseng et al. 2021).

6.4 VARIABILITY IN VIRUSES

Variability in different characteristics of plant viruses may be due to changes in environment, differences in the levels of susceptibility/resistance of host plant species/cultivars, and exposure to chemicals and adverse environments resulting in mutations in virus and host plant genes. As the internal cellular conditions in various plant hosts and vectors are widely different, viruses have to adapt themselves to establish infection successfully. Viruses do not possess any physiological functions such as respiration. Direct interaction between viral genome and products of host genes may result in required alterations in the viral genomes, leading to variations in their virulence. Existence of variations in virulence of viruses was exploited to develop methods of utilizing mild strains to cross-protect susceptible plants against severe strains of the same virus or related viruses. Various techniques have been applied to assess variability in viruses and their strains (Narayanasamy 2011, 2017).

6.4.1 Biological Characteristics of Plant Viruses

Plant viruses may induce different kinds of symptoms when they are introduced into susceptible plants by their respective natural biological vector(s). Variations in the type and intensity of symptoms induced by strains of a plant virus may be observed. Cultivars possess different levels of resistance/susceptibility to viral pathogens, resulting in development of different grades of symptom intensity or absence of symptom expression. In some host-virus combinations, as in the case of *Tobacco mosaic virus* (TMV) and tobacco cv. Xanthi nc, systemic spread of the virus is arrested and only necrotic local lesions are formed on inoculated leaves. Such hypersensitive response (HR) was practically utilized for development of tobacco varieties showing field resistance, and the virus did not spread due to lack of natural vector for TMV (Narayanasamy 2001). Variations in virulence of a virus may be due to selection of a strain from a mixture of strains or mutations in a virus population present in a location. Introduction of a resistance gene from wild or immune host into a cultivar or exposure to adverse environmental conditions, such as high temperature may lead to mutation in viral gene. The production of new mutants may be possible by application of mutagens like nitrous acid. Mild strains produced artificially or selected from naturally occurring mixture of strains may be utilized to protect susceptible cultivars against severe strains of the same virus. Different strains of a virus may be isolated by single-local lesion inoculation (analogous to single-spore isolation of fungal pathogens), since each local lesion is considered to be initiated by a single virus particle. The nature of relationships between viruses and their vectors may show distinct differences in acquisition access and inoculation access periods, presence, and length of latent periods required for the vectors to become viruliferous and duration for which the virus is retained by the vector species (Narayanasamy and Doraiswamy 2003; Narayanasamy 2008)).

Angelonia flower break virus (AnFBV), causing flower break and mild foliar symptoms, was biologically characterized. *Nicotiana benthamiana* showed local and systemic symptoms, whereas *N. rustica* and *N. sylvestris* developed local lesions when infected by AnFBV (Adkins et al. 2006). Variations in biological characteristics of *Zucchini yellow mosaic virus* (ZYMV) were studied by inoculating different plant species and by sequencing of genomic portion of the C-terminal part of Nib and N-terminal part of the CP gene. All isolates induced necrotic local lesions on *Chenopodium amaranticolor*, while only four isolates could produce necrotic lesions on *Gomphrena globosa* (Bananej et al. 2002). *Cucumber mosaic virus* (CMV) isolates causing banana mosaic disease were inoculated onto Cavendish banana plants using *Aphis gossypii*. Isolates of CMV could be differentiated into four pathotypes based on symptoms induced on banana, *Nicotiana glutinosa* and *Vigna unguiculata* (cowpea). Three different symptom types (A, B, and C) could be recognized using *N. glutinosa* as a differential host plant (Chou et al. 2009). Diversity of biological characteristics of *Wheat yellow mosaic virus* (WYMV) isolates occurring in Japan was assessed based on reactions of wheat cultivars to 14 isolates of WYMV collected from various locations. The isolates of WYMV were classified into pathotypes I, II, or III (Ohki et al. 2014). *Citrus tristeza virus* (CTV) isolate (L192GR) occurring naturally in lemon trees of more than 100 years old in Greece was characterized. This isolate showed molecular homology with CTV-VT isolate FS2–2 from Florida. Most isolates of CTV-VT group induced seedling yellows syndrome and reacted positively with MCA13, which is typically associated with severe CTV isolates (Varveri et al. 2015).

Isolates of *Cucumber green mottle mosaic virus* from Spain and other countries grouped into two major clusters (I and II) based on partial genome sequences. Representative isolates of cluster II, but not cluster I produced local lesions on *Chenopodium amaranticolor* (Crespo et al. 2017). *Tomato leafcurl New Delhi virus* (ToLCNDV) DNA-A and DNA-B, when agroinoculated, induced the typical symptoms in tomato and zucchini squash plants. Betasatellite DNA was not detected in infected tomato and zucchini squash plants, revealing that betasatellite DNA was not required for infection by the virus. Cucumber, melon, watermelon, zucchini squash, tomato, eggplant, and pepper were readily infected by ToLCNDV through transmission by whiteflies. (Ruiz et al. 2017). *Potato yellow blotch virus* (PYBV), causing isolated yellow patches on leaves of potato breeding lines was recovered. Sequence analysis showed that PYBV was closely related to *Potato virus A* (PVA) with an overall identity of 72% at a nucleotide level for the whole genome. Differential symptoms induced by PYBV and PVA on *Nicotiana benthamiana, N. hesperist*, and *N. occidentalis* P1 might be useful to differentiate these viruses (Nisbet et al. 2019).

6.4.2 Variability in Physicochemical Characteristics of Viruses

The stability of viruses in both extracted plant sap and purified preparations was determined based on assessment of longevity in vitro (LIV), dilution end point (DEP), and thermal inactivation point (TIP). *Tobacco mosaic virus* and *Tomato spotted wilt virus* (TSWV) were found to be highly stable and highly unstable, respectively, based on these parameters. Properties of purified preparations such as electrophoretic properties, molecular weight of virus particles, coat protein, and nucleic acid species present in monopartite/multipartite viruses, sedimentation coefficient of viral components, strandedness (single/double), electrophoretic mobility, viral genome organization, amino acid sequence of viral coat protein, and virus-associated proteins and nucleotide sequence of viral genome, and presence of satellite virus and satellite RNA have been useful for classification of viruses into different groups and genera (Narayanasamy 2001).

6.4.3 Variability in Immunological Characteristics

6.4.3.1 Immunoblot Analysis

Immunoblot analysis is useful to determine relationship between viruses and their strains. *Angelonia flower break virus* (AnFBV) infecting Angelonia plants contained a 38-kDa protein which appeared as a major polypeptide band in the gel after performing SDS-PAGE analysis of disrupted preparations. Immunoblot analysis with AnFBV antiserum showed a clear strong reaction with the 38-kDa band. This protein was identified as the AnFBV coat protein, in the reciprocal tests with antisera against *Pelargonium flower break virus* (PFBV), *Carnation mosaic virus* (CarMV), and *Saguaro cactus virus*, positive reactions were observed only with homologous viruses alone, indicating the capacity of immunoblotting technique to differentiate AnFBV from other viruses (Adkins et al. 2006).

6.4.3.2 Enzyme-Linked Immunosorbent Assay

Potato virus Y (PVY) produces, in addition to coat protein (CP), a helper component protein (HC-Pro) that helps in aphid transmission of PVY. MAbs and PAbs generated against HC-Pro could be used to differentiate strains of PVY. *Potato virus Y* (PVY) strain causing cracking of tubers was observed in Western Washington. Strain typing revealed infections primarily by PVY^{O}, PVY^{NTN}, and $PVY^{N\text{-}wi}$. Mixed infections of $PVY^{O/NTN}$ and $PVY^{O/N\text{-}wi}$ were also detected using ELISA. These PVY strain types were recovered from plants originating from cracked seed tubers. $PVY^{N\text{-}wi}$ was an important constituent of PVY strain composition in Washington (Inglis et al. 2019). Differentiation of strains of *Rice yellow mottle virus* (RYMV) was achieved by employing the antigen-coated plate (ACP)–ELISA format using PAbs. RYMV isolates from different locations were classified into three main serogroups: Sg1, Sg2, and Sg3. Existence and levels of serodiversity of RYMV isolates were reliably determined by applying SDI values (Sere et al. 2007). *West African Rice yellow mottle virus* (RYMV) isolates could be grouped into three distinct serogroups by employing a double antibody sandwich (DAS)-ELISA format using a combination of the universal Mab5B, and MAbs specific to serotypes of RYMV. These serogroups were correlated to two RYMV pathotypes that were differentiated based on their reaction on a set of differential rice cultivars. Similarly a panel of 16 MAbs raised against *Bean yellow mosaic virus* (BYMV) effectively differentiated strains of BYMV by performing triple antigen sandwich (TAS)-ELISA tests. Variations in serological characteristics exhibited relationship with four pathotypes (I–IV) recognized based on pathogenic potential (virulence) of the isolates. Two serotypes of *Sweet potato feathery mottle virus* (SPFMV) occurring in Uganda could be differentiated by a specific MAb generated against CP of SPFMV. Incidence and distribution of serotypes of *Barley yellow dwarf virus* (BYDV) were monitored using specific Mabs. The extent of incidence of PAV serotype transmitted by *Rhopalosiphum padiavenae* and MAV serotype transmitted by *Macrosiphuum avenae* was directly correlated to their respective aphid vector populations (Quillec et al. 2000; Kareyeija et al. 2000; Narayanasamy 2011).

Based on serological relationships and phylogenetical analysis of nucleocapsid proteins (NPs), 16 official and tentative species have been assigned to the genus *Tospovirus,* and they are clustered into three major serogroups and four monospecific serotypes (Jan et al. 2003). *Calla lily chlorotic spot virus* (CCSV) isolates from Taiwan were identified as a tospovirus serologically, but distantly, related to Watermelon silver mottle virus (WSMoV) based on the serological relationship established with reactions of CCSV nucleocapsid protein (NP) in indirect ELISA format and low-intensity banding in immunoblotting. The MAbs generated against CCSV NP or WSMoV NP reacted specifically with homologous antigens, but not with heterological antigens in both ELISA and immunoblot analysis (Lin et al. 2005). Isolates of *Citrus psorosis virus* (CPsV) from 53 field sources was analyzed for variability of CP gene of the virus. Nine serogroups and at least ten different epitopes were recognized. Reaction patterns in TAS-ELISA format grouped the field sources into nine patterns called A–I, the most frequent pattern being A with 39.6% of the sources (Alioto et al. 2003). *Prunus necrotic ringspot virus* (PNRSV) infects *Prunus* trees and rose plants. Reactivity of PNRSV isolates to MAbs generated against PNRSV strain infecting rose was assessed using an ELISA test. The most common PNRSV serotype in rose was different from the most prevalent serotype in *Prunus* spp. Pathogenicity tests showed that all rose isolates could infect *Prunus persica* seedlings, but only three of the four isolates from *Prunus* were weakly virulent to *Rosa indica* (Moury et al. 2001). Variability in the isolates of *Plum pox virus* (PPV) was assessed by employing the TAS-ELISA format, using PPV-D and PPV-M specific MAbs. All isolates showed positive reaction to at least one or the other MAb as well as to PPV-D and/or PPV-M-specific primers in PCR assays. Using TAS-ELISA procedure, isolates were identified as PPV-D, PPN-M and as a mixture of D and M strains (Zagrai et al. 2008). Isolates and strains of *Cucumber mosaic virus* (CMV) were differentiated into two subgroups I and II using PAbs and MAbs in different ELISA formats and immunodiffusion test. The results of PCR assay and RFLP analysis confirmed those of immunoassays (Hristova et al. 2002). *Tomato mosaic virus* (ToMV) and *Tobacco mosaic virus* (TMV) commonly infected tomato as single or mixed infection. MAb19.H1 generated against ToMV was effective in detecting and differentiating ToMV from TMV, when applied in plate-trapped antigen (PTA)-ELISA format (Duarte et al. 2002). MAbs generated against coat protein of *Grapevine leafroll-associated virus* (GLRaV) showed varying levels of specificity of reaction with different isolates of GLRaV-1, as revealed by DAS-ELISA. MAb IG10 reacted with both GLRaV-1 and GLRaV-3 coat proteins, indicating that these two viruses might share some antigenic determinants (epitopes) (Seddas et al. 2006).

6.4.4 Variability in Genetic Characteristics

6.4.4.1 Double-Stranded RNA Analysis

Double-stranded (ds) RNAs, the replicative form (RF) of single-stranded (ss) RNA virus or as the infecting viral genomes of ds RNA viruses (*Phytoreovirus* genus), may be detected in infected plants. The ds-RNA profiles generated through Northern blot hybridization may be useful for identifying and classifying an unknown virus or strain. The relationship between the new virus *Angelonia flower break virus* (AnFBV) and *Carnation mottle virus* (CarMV) was established by performing ds RNA analysis. Four major bands were detected in *Nicotiana benthamiana* plants infected by AnFBV. The largest band (about 4-kb) comigrated with putative ds RNA form of AnFBV genomic RNA (Adkins et al. 2006). The nature of the causal agent of cassava frogskin disease (CFSD) occurring in Colombia was investigated. Virus-like particles, inclusion-like bodies and ds RNA viral genomic segments were detected in CFSD-associated cassava plants. The virus inducing CFSD was identified as a member of *Reoviridae* and most closely related to *Rice ragged stunt virus* (RRSV), as revealed by phylogenetic analysis. Isolates in cassava samples collected from different locations were identified as *Cassava frog skin virus* (Calvert et al. 2008).

6.4.4.2 Hybridization Assays

Hybridization with cDNA or cRNA probes is an easy and powerful procedure for determining the differences located in any region of the viral genome. DIG-labeled probes were employed for detection of *Citrus tristeza virus* (CTV). Hybridization of DIG-labeled probes with purified ds RNA or concentrated total RNA extracts spotted on nylon membranes allowed detection of CTV nucleic acid in infected tissues (Narváez et al. 2000). *Cucumber mosaic virus* (CMV) is highly heterologous in its coat protein. The CP genes of 14 different isolates were amplified using cy3-generic primers. The probes targeted regions optimal for differentiation between subgroups 1 and 2 or between subgroups 1a and 1b. The microarray method allowed a clear differentiation of 14 different CMV isolates into serogroups 1 and 2 (Deyong et al. 2005).

6.4.4.3 Polymerase Chain Reaction–Based Techniques

Plant-infecting RNA viruses possess greater potential compared to DNA viruses for rapid genetic changes due to point mutations and genome rearrangement/recombination that are considered to be the principal mechanism of evolution of RNA viruses. The most conserved region is the one encoding viral polymerase enzyme. The CP gene is generally variable and virus-specific. The epitopic characteristics determine serotype and species of the virus and it is used frequently in phylogenetic analyses. Citrus decline-inducing strains of Citrus tristeza virus (CTV) could be differentiated from non-decline-inducing strains by employing the RT-PCR assay. Two internal (T30) and CP3 specific for decline-inducing strains of CTV and two terminal primers HCP1 and HCP2 for the ends of the capsid protein gene were employed. The protocol was able to differentiate decline-inducing and non-decline-inducing strains (CTV) based on amplification of a 320-bp fragment only in decline-inducing CTV strains (Huang et al. 2004). Severe isolates of CTV induced seedling yellows (SY) and/stem pitting (SP) symptoms in sweet orange and grapefruit. These isolates were identified and differentiated by performing real-time RT-PCR assay by employing a general primer set and three TaqMan locked nucleic acid (LNA) probes targeting characteristic of severe, mild (non-SY, non-SP) and T36-like isolates. Pathogenically distinct CTV isolates from different countries were analyzed by real-time RT-PCR protocol. The mild isolates reacted only with the mild probe, whereas severe SP and SY isolates reacted with the severe-S or the T36-like probes, respectively (Ruiz-Ruiz et al. 2009).

The multiplex TaqMan-based real-time RT-PCR format was applied to identify severe strains of *Citrus tristeza virus* (CTV) and separate genotypes that react with MAb13. Three strain-specific probes were developed using intergenic sequences between the major and minor CP genes (CPi) in a multiplex reaction. Probe CPi-VT3 was designed for VT and T3 genotypes. Probe CPi-T36 and T36 genotypes and probe CPi-T36-NS to identified isolates in an outgroup clade of T36-like genotypes in California. The CTV isolates from California, Florida and a large panel of CTV isolates from an international collection could be differentiated (Yokomi et al. 2017). The CTV isolates induced stem pitting syndrome. CTV commonly occurred as a population of different strains, masking the phenotypic effect of individual strains. Single-variant CTV isolates were used to determine phenotypic expression of CTV strains in commercial grapefruit cultivars as well as CTV indicator plants. Under field conditions, two grapefruit varieties inoculated with RB isolate caused mild stem pitting which did not negatively affect horticultural attributes during eight years of experimentation (Cook et al. 2016). Selected CTV isolates were subjected to deep sequencing of small RNAs. Full-length sequences were assembled, annotated and trifoliate orange resistance-breaking (RB) isolates of CTV were identified. The California RB isolates were further distinguished into two subclades. Group I included CA-RB-115 and CA-RB-AT25 with 99% nucleotide sequence identity with RB type strain NZ RB-G90 and group II included CA-RB-AT35 with 99% and 96% sequence identity with Taiwan Pumelo/SP/T1 and HA 18–9 respectively. California RB isolates induced mild symptoms, compared with severe isolates. The presence of RB genotype of CTV was considered to be important, because *P. trifoliata* and its hybrids were used commonly as rootstocks in California (Yokomi et al. 2017). Complete genome sequences of *Citrus yellow vein clearing virus* (CYVCV) isolates (19) were sequenced and characterized. Sequence identity of the isolates ranged from 97% to 99.8%, as revealed by comparison of whole genome sequences, indicating a very low level of sequence homogeneity among CYVCV isolates. Phylogenetic analysis of these 23 genomic sequences suggested that all CYVCV isolates were clustered into the same clade (Zhou et al. 2017).

Assessment of genetic diversity in *Plum pox virus* (PPV) isolates causing Sharka was performed. Two major serologically distinct strains PPV-D and PPV-M were recognized using strain-specific MAbs. RT-PCR assay was performed using primers that amplified a 243-bp fragment showed that restriction sites of *Alu*I and *Rsa*I were present in plum and apricot samples. An amplified 836-bp cDNA fragment derived from the P3–6K, coding region of both isolates had restriction profiles typical of strain PPV-D. Nucleotide sequence analysis of the entire CP coding region of the plum and apricot isolates resulted in the identification of a unique deletion of six nucleotides in the N-terminal region in the plum isolate. Several nucleotide substitutions in the CP-coding region were commonly present in plum and apricot (Spiegel et al. 2004). *Plum pox virus* (PPV) isolates collected from a small focus of infection in apricot in Italy were characterized. Strain PPV-BR was a recombinant of PPV-D strain, as revealed by reactivity with MAbs and RFLP analyses of *P3–6K1* gene. PPV isolates could be grouped into two main clusters, one comprising all PPV-D isolates and another all PPV-M isolates. PPV-M cluster split into two clusters comprising typical PPV-M isolates or all recombinant isolates (PPV-Rec (Myrta et al. 2005). Mutations, indel events and interstrain combination events may contribute to genetic diversity of PPV. Isolates of PPV strain Winona (W) were analyzed including nine new strain W isolates with complete genome sequence. Other strains of PPV with more than one isolate with known complete genome sequence were also included in this investigation. Five interstrain recombination events were detected among PPV isolates (PPV-C and PPV-M strains). Among PPV-W isolates, four were recombinants, one of which P2–1 was a mosaic, with three identified recombination events. Further, a new interstrain recombination event was identified between isolates of strain M isolate and strain Rec, a known recombinant. This investigation provided evidence for intrastrain recombination in PPV isolates contributing to genetic diversity of PPV (James et al. 2016).

Peach mosaic virus (PcMV) and *Cherry mottle leaf virus* (CMLV), though serologically related, induced distinct diseases in peach and cherry, respectively. Complete nucleotide sequence data showed that PcMV shared 73% identity with CMLV. The CP amino acid sequence identity between isolates of PcMV ranged from 97% to 99% versus 83% identity when compared with the CP of CMLV. In addition, phylogenetic analysis supported placement of

PcMV and CMLV as members of the genus *Trichovirus* (James et al. 2006). CMLV isolates shared 78.4–82.7% of identities at the nucleotide sequence level between each other. The results of phylogenetic analysis also indicated the high genome sequence divergence among CMLV isolates (Su et al. 2016). Genetic diversity of *Grapevine red blotch-associated virus* (GRBaV) was studied. Phylogenetic analyses of viral replication-associated protein (Rep) gene fragment from 42 isolates of GRBaV differentiated two distinct clades 1 and 2 of the virus. Comparison of GRBaV genomes provided evidence of recombination underlying some of the variation existing in isolates within clade 1. Phylogenetic analyses of coat and replicase-associated protein sequences among single-stranded DNA viruses showed GRBaV to form a group within *Geminiviridae* (Krenz et al. 2014). HTS of RNA from representative source trees for *Apricot ringpox disease* (ARPD) did not reveal additional unique virus sequences that might be associated with respective diseases (Villamor et al. 2016). The diagnostic method involving the use of primer pair targeting movement protein gene was developed to determine the genetic diversity of *Prune dwarf virus* (PDV) isolates. A high degree of nucleotide sequence diversity in the isolates of PDV was observed. Phylogenetic grouping based on host or geographical origin of isolates provided reliable discrimination of PDV isolates. Highly supported recombination signals were detected in five PDV isolates with breakpoints located at different positions in the central part of RNA3 within or close to the intergenic region (Predajna et al. 2017).

Genetic variability of isolates of *Grapevine leafroll-associated virus 1* (GLRaV-1) was investigated. RT-PCR assay was employed to amplify segments of heat-shock protein homolog (*HSP70h*) gene, coat protein (*CP*), CP duplicate 2 (*CPd2*) gene, and PRF9 (*p24*), and the amplicons were cloned and sequenced. Pairwise comparison of nucleotide sequences revealed intra- and inter-isolate sequence diversity, with *CPd2* and *HSP70h* being the most and divergent respectively among the four regions studied. Global phylogenetic analyses of sequences from the four genomic regions revealed segregation of GLRaV-1 isolates into three major clades. Putative recombination events were revealed among *HSP70h*, *CP*, and *p24* sequences (Alabi et al. 2011). In another investigation, complete genome sequences of two isolates of GLRaV-1 WA-CH and WA-PN differed in the number of nucleotides present. The genomes of both isolates encoded nine putative ORFs and the arrangements of these ORFs were similar to that of Australian and Canadian isolates. The RT-PCR assay and RFLP procedure were employed to differentiate variants of GLRaV-1 present in vineyards. Northern analysis of total RNAs from GLRaV-1 infected grapevine samples revealed three subgenomic RNAs (sgRNAs) corresponding tentatively to CP, p21, and p24 (Donda et al. 2017). Genetic diversity of field isolates of *Grapevine leafroll-associated virus 2* (GLRaV-2) in Pacific Northwest (PNW) vineyards was determined by sequencing the entire CP and a portion of the heat-shock protein 70 homolog (*HSP70h*) genes. Phylogenetic analysis of CP and HSP70h nucleotide sequences revealed segregation of GLRaV-2 isolates into six lineages with virus isolates from PNW distributed in PN, H4 and RG lineages. Multiple alignments of CP amino acid sequences showed lineage specific differences (Jarugula et al. 2010). Plants of two grapevine cultivars were infected by a novel genetic variant of *Grapevine leafroll-associated virus 3* (GLRaV-3) called ID45 which exhibited ≤80% nucleotide sequence identity level to known GLRaV-3 isolates in its most conserved *HSP70h* gene. The complete genome sequence of GLRaV-3 ID45 was determined using a combination of high-throughput and conventional Sanger sequencing. ID45 variant represented a new phylogenetic group IX of GLRaV-3. RT-PCR format was developed to distinguish ID45 variant from the prevalent GLRaV-3 phylogroup I present in single or mixed infections in Idaho State (Thompson et al. 2019).

Isolates of *Grapevine virus A* (GVA) implicated with Kober stem grooving disorder of the rugose wood disease complex, were analyzed for their genetic diversity. Analysis of a portion of RNA-dependent RNA polymerase (RdRp) and complete CP sequences showed inter- and intra-isolate sequence diversity. Global phylogenetic analysis of CP sequences revealed segregation of virus isolates into four major clades with no geographic clustering (Alabi et al. 2014). *Grapevine vein clearing virus* (GVCV) inducing vein clearing and vine decline has a circular, ds-DNA genome. Sequence variants of GVCV could not be phylogenetically grouped into clades according to geographical location and grapevine cultivar. Seven nucleotides were changed in the sequenced RT and ZF regions of GVCV from a grafted Traminette vine and one in the sequenced region of GVCV from grafted Cayuga White (Guo et al. 2014). Genetic diversity of *Sugarcane bacilliform virus* (SCBV) isolates was assessed. PCR assay with degenerate primer targets, 720-base pair (bp) fragment of the reverse transcriptase/ribonuclease H (RT/RNase H) genomic region was applied. PCR-amplified fragments from 84 SCBV positive samples were cloned, sequenced and analyzed for their genetic diversity. Considerable haplotype diversity was seen within individual SCBV isolates (Wu et al. 2016).

Diagnostic efficiency of two primers RSP13-RSP14 and RSP2-RSP21 was compared using RT-PCR assay for detection of *Grapevine rupestris stem pitting-associated virus* (GRSPPaV) in plant extracts from infected grapevines. The CP gene was also amplified by RT-PCR and the amplicons were examined followed by analyzing amplicons by single-stand conformation polymorphism (SSCP) and sequencing. The results revealed the existence of wide variability within this virus, independent of cultivar or geographic origin. Sequence analysis indicated the differences in four groups of sequence variants. GRSPaV is composed of a wide range of sequence variants, and five sequence variants were sequenced completely. The efficiency of RT-PCR varied widely depending on the spectrum of primers. Pair of degenerate primer designed based on the consensus consequence of the genomic region of primers was comparable to the broad spectrum primers RSP13 and RSP14 in detecting multiple GRSPaV variants for identification of new and distinct sequence variants (Terlizzi et al. 2010). NGS technology was applied to characterize a novel virus infecting grapevine. Phylogenetic analysis based on the predicted RP and CP proteins firmly placed the virus within *Vitivirus* genus. The presence of the virus in the berries at diverse development stages was detected and sequences of the new virus was detected in grapevine samples sharing 74.4% and 97.8% nucleotide identity. The new virus was called *Grapevine virus L* (GVL) (Debat et al. 2019). Microarray or RT-PCR assay revealed the presence of *Grapevine virus E* (GVE) in the vineyards. All GVE-infected vines were coinfected by *Grapevine leafroll-associated virus 3* in New York. All GVE infected vines were also infected by GLRaV and other vitiviruses in California also. Genetic diversity among isolates of GVE, was assessed by employing RT-PCR to amplify a viral cDNA and a 675-nucleotide region. All 20 GVE isolates sequenced were closely related with >98% nucleotide identity to the SA94 isolate from South Africa (Vargas-Asencio et al. 2016).

Complete nucleotide sequences of six isolates of *Apple necrotic mosaic virus* (ApNMV) ApNMV, causing different types of mosaic symptoms in apple, were prepared. Phylogenetic analysis indicated that the Chinese ApNMV isolates were clustered in subgroup 3 together with *Prunus necrotic ringspot virus* (PNRSV) and *Apple mosaic virus* (ApMV). The six isolates investigated, showed substantial sequence diversity (Xing et al. 2018). A novel *Tepovirus* isolate infecting *Prunus* species was characterized through analysis with the deep sequencing procedure. Sequence comparisons indicated that the virus isolates from *Prunus* plants belonged to the same virus species. Genome organization of the isolates was similar to that some members of *Betaflexiviridae*. Phylogenetic analysis of the deduced encoded proteins showed a clustering with the sole member of the genus *Tepovirus*, *Potato virus T* (PVT). Hence, the virus was named *Prunus virus T* (PrVT) (Marais et al. 2015). *Plum pox virus* (PPV) is a complex of nine strains adapted to different *Prunus* hosts. Two unusual PPV isolates Tat-2 and Tat-4 were detected in sour cherry in Russia. Two complete genomes of Tat-2 and Tat-4 isolates were sequenced using the Illumina HiSeq 2500 platform. Both isolates possessed 9792-nt, excluding the poly (A) tail with organization typical of PPV and had 99.4% and 99.7% identity between each other at nucleotide and amino acid levels. Phylogenetic analysis placed Tat-2 and Tat-4 in a separate clade, distinct from C and CR strains. Tat-2, Tat-4 and related isolates represented a new cherry-adapted PPV strain and they were called PPV-CV (Cherry Volga) (Chirkov et al. 2018). Isolates of *Papaya leaf distortion mosaic virus* (PLDMV) were grouped based on host range into two biotypes – P biotype capable of infecting papaya and C biotype incapable of infecting papaya. Like the *Papaya ringspot virus* biotype W, the non-papaya infecting papaya biotype of PLDMV was able to infect cucurbitaceous crops. CP sequences of three strains of PLDMV-P shared high identities of 95–97%, while they shared identities of 88–89% with that of PLDMV-C biotype. Phylogenetic analysis based on CP amino acid sequences grouped PLDMV strains in one cluster, distinct from other potyvirus clusters. In PLDMV cluster P biotype strains (LDM, YM, of M) showed close relationship among them (Maoka and Hataya 2005). was widespread in Cuba, causing considerable yield losses in quality and quantity. Partial coat protein gene of all Cuban *Papaya ringspot virus* (PRSV-P) isolates (107 sequences) from Americas and the Caribbean Islands were subjected to phylogenetic analyses which indicated the most recent ancestor in 1942 had highest posterior density (HPD) of 95% (Mederos et al. 2019).

Genetic diversity of isolates of *Strawberry mild yellow edge virus* (SMYEV) was assessed using phylogenetic analysis. The CP gene of SMYEV isolates was sequenced. All isolates contained 729 nt, encoding a protein of 242 amino acids (aa), and they shared 81.5–99.6% nucleotide identity. Four possible subgroups of the isolates were indicated by phylogenetic analysis. Recombination analysis showed that some isolates were recombinants formed as a result of recombination events (Torrico et al. 2016). Genetic diversity and evolutionary mechanisms influencing *Figure mosaic virus* (FMV) populations were studied. Nucleotide sequences from four genomic regions (each within RNAs 1, 2, 3, and 4) from FMV isolates from different countries were analyzed. FMV genetic variation was low and some geographically distant FMV isolates clustered together as indicated by phylogenetic analysis. Genetic drift appeared to have an important role in shaping FMV populations (Walia et al. 2014). Genetic diversity of isolates of *Banana bract mosaic virus* (BBrMV) was assessed, based on viral genome-linked protein (*VPg*) gene. Phylogenetic analysis based on nucleotide sequence showed that except for two isolates (TN1 and TN2), all Indian isolates clustered together. Functional motifs of *VPg* gene were conserved. Gene flow between Indian populations of BBrMV from banana was relatively low (Anuradha and Selvarajan 2018).

Induction of veinal necrosis in indicator plants forms the basis of biological differentiation of *Potato virus Y* (PVY) PVY isolates. One-step fluorescent RT-PCR assay on a single nucleotide polymorphism (SNP) was applied to examine the relationship between biological and molecular characteristics. PVY isolates were assigned to their respective group based on SNPs. Further, mixed infection involving PVY^N and PVY^O strains were detected and differentiated by this assay (Jacquot et al. 2005). The strain PVY^{NTN} induces potato tuber necrotic ringspot disease (PTNRD) and it could be differentiated by employing strain-specific endonucleases that cleave their respective PCR products. Differentiation of these two PVY strains based on the restriction profiles was reliable and reproducible (Lorenzen et al. 2006). *Potato virus Y* (PVY) strain PVY^O induces mild mosaic in tobacco and necrosis and severe stunting in potato cultivars with *Ny* gene. Whole-genome sequencing and phylogenetic analysis of PVY^O and PVY^O-05 strains were performed. PVY^O-05 isolates formed a separate lineage within PVY^O genome cluster in whole genome phylogenetic tree. On the other hand, PVY^{O} sequences separated into at least two distinct lineages on the whole-genome phylogenetic tree. Phylogenetic analysis of recombinants showed that $PVY^{N:O}$ and $PVY^{N\text{-}Wi}$ recombinants acquired their PVY^{O} segments from two separate PVY^{O} lineages, whereas PVY^{NTN} recombinants acquired its PVY^{O} segment from the same lineage as $PVY^{N:O}$ (Karasev et al. 2011). Light Cycler technique was developed for detection and differentiation of PVY strains in potatoes. Specific characteristics of melting curves allowed detection and differentiation of each of the PVY strains, O-FL, ORB, N, and NTN in a single reaction. Melting curves of extracts of tubers from market revealed the presence of O-FL, O-RB, and NTN strains of PVY (Rotem et al. 2016). Phylogenetic analyses showed that CRM2 belonged to phylogenetic subgroup Y^{0} and eight strains to Y^{05}, three to Y^{02} and one to Y^{C1}. Biologically differentiated strains of PVY^{O} and PVY^{Z} were within phylogenetic groups Y^{0} or Y^{05}. Likewise, other strains were placed in different phylogenetic subgroups (Kehoe and Jones 2016). *Potato virus A* may infect potato cultivars lacking hypersensitivity genes *Na* and *NaKE*. Three strain groups (= pathotypes) unrelated to phylogenetic groupings, were differentiated among five isolates of PVA inoculated into a potato differential cultivar. The results suggested that PVA and PVY populations apparently might have arisen in the Andes about 18 centuries ago and possibly taken to Europe (Fuentes et al. 2021).

Population structure, genetic diversity, and sources of genetic variation in field isolates of *Tomato spotted wilt virus* (TSWV) from peanut in the United States, were investigated. Selected regions of nucleocapsid, movement protein, and RNA-dependent RNA polymerase (RdRp) genes were amplified and sequenced to identify haplotypes and inter-genetic relationships between isolates of TSWV. Evidence for positive selection on single amino acid in each gene on a background predominant purifying selection acting upon each locus was obtained. Estimates of mutation rate in TSWV population suggested that mutation exerted significant effect on shaping pathogen population (Kaye et al. 2011). Emergence of resistance breaking strains of TSWV was triggered by cultivation of resistant peppers (*Capsicum* spp.) carrying the *Tsw* locus. Incidence of natural TSWV RB variants in China, was observed. Variations in the TSWV nonstructural (NSs) proteins formed the key determinants associated with *Tsw* resistance-breaking isolates. Sequence analysis of the NSs revealed that RB isolates did not have amino acid mutation sites common to other RB isolates (Jiang et al. 2017).

The population structure of *Pepino mosaic virus* (PepMV) was analyzed by comparing the nucleotide sequences of RNA-dependent RNA polymerase (RdRp), triple gene block (TGB) and CP genes of isolates. Spanish isolates were mostly composed of highly similar isolates belonging to the European strain (Pagán et al. 2006). Genetic diversity within *Pepper veinal mottle virus* (PVMV) and its relationship with other potyviruses was investigated. CP gene sequence diversity revealed three clades that corresponded to the geographic locations. PVMV isolates were differentiated into two species by employing RT-PCR for amplification of targeted fragments, based on large indel in CP gene (Moury et al. 2005). Strain-specific primers for severe (Svr) and California/Logan/CA were employed in PCR assay and also for Worlan (Wor)-like strains of *Beet curly top virus* (BCTV). Whole genome sequencing with overlapping primers were found that Wor-like strains included Wor, Colardo, and previously undescribed strain called Kimberlyl. The results confirmed a shift from Svr dominant BCTV strains becoming undetectable later (Strausbaugh et al. 2017). Genetic diversity of *Zucchini yellow mosaic virus* (ZYMV) isolates infecting pumpkin, melon, cucumber and zucchini was assessed. Leaves from symptomatic plants were sampled and subjected to high-throughput sequencing and 15 complete new ZYMV genomic sequences were generated. Isolates from different geographical locations were placed in a major phylogroup B, based on comparison of their nucleotide sequences. Analysis of 63 complete genomes revealed recombination events in isolates from Australia. The results indicated possible entry of virulent crop pathogen strains not only into northern Australia but also spreading between different agricultural regions of Australia (Maina, Coutts, Edwards, de Almeida, Ximenes et al. 2017). Bell pepper leaf samples from plants showing leafroll and interveinal yellowing and weeds present in pepper fields were subjected to polerovirus generic amplicon. High similarity (876–981 nt) with poleroviruses was observed. NGS was used to acquire a larger part of the genome which resulted in the reconstruction of isolate 105D's partial genome (5,496 nt). Sequence similarity analysis revealed the presence of a divergent polerovirus isolate belonging to a new species, tentatively named *Pepper leafroll chlorosis virus* (PeLRCV) (Kamran et al. 2018).

Melon samples with virus-like symptoms were evaluated for virus infection using RT-PCR assay. Infection of samples by one or multiple viruses was detected. Genetic diversity analysis of *Watermelon mosaic virus* (WMV) suggested that all isolates examined, clustered into group A and purifying selection could restrict variability in population of viruses infecting water melon (Wang, Wu et al. 2017). Three distinct subgroups IA, 2B, and II of *Cucumber mosaic virus* (CMV) isolates have been recognized. Further, some isolates associated with satellite RNAs (sat RNAs), a type of non-encoding transcript that may alter the symptoms of CMV infections. Genetic diversity of CMV and sat RNA isolates obtained from legumes and tomato was investigated. Phylogenetic analysis of CMV coat protein sequence of legume and tomato isolates were classified within subgroups IA and IB with limited genetic diversity. CMV sat RNAs collected from nine field legumes exhibiting mild symptoms and one tomato with a necrotic syndrome contained functional necrogenic motif and they were grouped within necrogenic group of CMV-sat RNAs (Giakountis et al. 2018). Genetic diversity of *Sweet potato feathery mottle virus* (SPFMV) and *Sweet potato virus C* (SPVC) was investigated for their genetic diversity. Phylogenetic analysis revealed that all 17 new SPFMV genomes fitted with within either major phylogenetic A, minor phylogroup II, or major phylogroup B. Within phylogroup A's minor phylogroup II. High frequency of interspecies and inter-phylogroup combination events reflect widespread occurrence of mixed SPVC and SPFMV infections within sweet potato plants (Maina et al. 2018). Several viruses, including eight viruses belonging to *Alexivirus (Alfaflexivirdae)*, infected garlic crops in many countries. Garlic bulbs imported from China and Spain tested by ELISA and RT-PCR techniques revealed the presence of garlic viruses A, B, D, and X. The most genetically distinct populations contained GarV-X, while isolates of GarV-B and GarV-D appeared to be genetically more uniform (Bereda et al. 2017). Whole-genome sequences of two *Dasheen mosaic virus* (DsMV) strains, Hawaiian strain I (KY242358) and Hawaiian strain II (KY242359) were determined. Hawaiian strains I and II were 77% and 85% identical, respectively, with other completely sequenced DsMV isolates. Hawaii strains were most closely related to *Vanilla mosaic virus* (VanMV) (KY505964.1), a strain of DsMV infecting vanilla. Analyses of DsMV genome sequences detected three recombinants from China and India (Wang, Wu et al. 2017). Genetic variability of new isolates of *Turnip mosaic virus* (TuMV) occurring in Korea was investigated. Infectious clones generated from these isolates were subjected to phylogenetic analysis. All new isolates and existing isolates belonged to the basal BR group, indicating that the pathotype could infect both *Brassica* spp. and *Raphanus* spp. (Gong et al. 2019). New isolates (14) of *Pea seedborne mosaic virus* (PSbMV) obtained from semi-leafless field pea (*Pisum sativum*) were sequenced and their partial CP nucleotide sequences were analyzed. On inoculation of plants of cv. Green Feast and PBA Gunyah pea (both carrying resistance gene *sbm2*) with PSbMV pathotype P-2 isolate W1, resistance was overcome in a small proportion of plants of each cultivar, indicating the presence of resistance-breaking variants (Congdon et al. 2016).

Sequence analysis of CP genes of *Sugarcane mosaic virus* (SCMV) isolates (33) and *Sorghum mosaic virus* (SrMV) isolates (ten) was performed to assess their genetic diversity. Near full-length CP gene sequenced were analyzed along with the sequence data of all available isolates in GenBank. Isolates of SCMV could be classified into five groups which include three groups already identified. The sugarcane (SCE), maize (NZ), and Thailand groups and two nearly identified noble sugarcane (NSCE) and Brazil groups were differentiated (Xu et al. 2008). Nucleotide sequences of CP genes of *Sugarcane streak mosaic virus* (SCSMV) isolates and *Sorghum mosaic virus* (SrMV) isolates were sequenced and compared with sequences in GenBank to perform phylogenetic analysis. Variations in sequences due to geographic locations were also seen. All isolates of SCSMV clustered into a single Chinese group (Wang, Li, Huang et al. 2017). Isolates of *Sugarcane yellow leaf virus* (SCYLV) from 3 Hawaiian cultivars (one resistant and two susceptible) were fully sequenced and the sequences were aligned to full and partial sequences. YLS-segment coding for CP exhibited the least genetic diversity, whereas the other sequence fragments A – D, representing the ORFs 0–5, expressed two-fold higher diversity as revealed by phylograms. Hawaiian SCYLV strains clustered together next to the Peru strains, apart from the BRA-strains and well apart from REU-strains (ElSayed et al. 2011).

Resistance spectrum analyses were taken up by testing *rymv1–4* and *rymv1–5* alleles that were present in rice accessions Tog5438 and Tog5674 respectively, against isolates of *Rice yellow mottle virus* (RYMV). A hypervirulent pathotype T, capable of overcoming all known sources of high resistance was spatially localized in West-Central Africa. Pathotype T appeared to be more abundant. Deep-sequencing analyses revealed that resistance breakdown did not coincide always with fixed mutations. On the other hand, virulence mutations in a small percentage of virus population might be sufficient for resistance breakdown (Hébrard et al. 2018). *Rice black-streaked dwarf virus* (RBSDV) and *Southern rice black-streaked dwarf virus* (SRBSDV) are involved in maize dwarf disease (MRDD) in China. RBSDV segment 8 ((S8) contained the only mutation in the genomes of these viruses included in the genus *Fijivirus*. Analysis of the S8 was analyzed in virus isolates different locations in China along with 14 S8 sequences from GenBank revealed that pathogen associated with MRDD was RBSDV, whereas pathogen associated with the disease in other locations was SRBSDV. Six recombination events were detected in S8 in RBSDV and two recombination events were detected in S8 of SRBSDV which was more highly genetically differentiated (Zhou et al. 2018).

Begomovirus, *Curtovirus*, *Mastrevirus*, and *Topocuvirus* genera are placed in the family *Geminiviridae*, based on biological properties and viral genome organization. They have mostly bipartite genome divided into two components named DNA- and DNA-B. Monopartite begomoviruses lack DNA-B. Major kind of variation among the genomes of begomoviruses is the absence of DNA-B in viruses like *Tomato yellow leafcurl virus* (TYLCV) from Israel (Is) and from Sardinia (Sa). Sequence identity of 89% nucleotide sequences was suggested as threshold for DNA as the basis of differentiation of begomoviruses by International Committee on Taxonomy of Viruses (ICTV) (Fauquet et al. 2005). Status of infection in the germplasm of the sweet potato collection of the US Department of Agriculture (USDA) by *Sweet potato leaf curl virus* (SPLCV) and other *Begomovirus* spp. was studied. All accessions of sweet potato were tested by RT-PCR assay for the presence of SPLCV. Nucleotide polymorphisms among isolates of SPLCV were evaluated by sequencing the *AV1* coat protein gene from 24 SPLCV-infected accessions. Most of the *AV1* genes derived from previously classified SPLCV species clustered together, some of which formed well-defined monophyletic clades (Barkley et al. 2011). *Tomato yellow leafcurl virus* (TYLCV) isolates from infected tomato, pepper, cucurbits, and beans from five countries were analyzed using combinational analysis of PCR and restriction enzyme (PCR-RE) analysis. Sequences of six isolates were analyzed to differentiate them by PCR-RE combination analysis. Oligonucleotide primers for six TYLCV isolates were designed to amplify approximately 740-bp, including the intergenic region (IR) and parts of V1 and V2 ORF. Unique restriction enzyme sites were analyzed to identify isolate-specific restriction enzyme sites on the PCR products of each isolate tested. Three enzymes, *Dde*I, *Fau*I, and *Bss*SI, were selected by in silico analysis (Park et al. 2014). *Tomato leafcurl New Delhi virus* (ToLVNDV) introduced into Spain did not have betasatellite DNA in infected tomato and full-length genomic sequences from zucchini squash and tomato showed high genetic homogeneity in ToLCNDV populations in Spain, indicating a single introduction event (Ruiz et al. 2017).

The complete genome sequence determination showed that *Bhendi yellow vein mosaic virus* (BYVMV) contained DNA-A comprising 2,757-nt and association of a betasatellite suggested that the virus isolate was monopartite. Genome organization was characteristic of begomoviruses encoding seven sense strands and five ORFs (AC1–AC5) in complementary strand. Phylogenetic analysis of the isolates of BYVMV clustered on a well-defined branch separately, and the virus appeared to be a recombinant (Venkataravanappa et al. 2013). Isolates of BYVMV in samples of okra (*Hibiscus esculentus*) infected by yellow vein mosaic disease (YVMD) from several locations in India were examined. A high degree of genetic variability and subpopulation structure within begomovirus species could be inferred (Venkataravanappa et al. 2015). Virus infection inducing severe mosaic and mild leaf curl symptoms in tomato was observed in Madhya Pradesh, India. Bipartite begomovirus was associated with the disease as revealed by sequence analyses of their cloned full-length genome. The highest nucleotide identity and close phylogenetic relationship with *Mungbean yellow mosaic India virus* (MYMIV) were observed. Hence, the virus was identified as an isolate of MYMIV infecting tomato in India (Agnihotri et al. 2019).

HTS of the small RNA (sRNA) technique has been effectively employed to differentiate viruses and their strains in single and multiple infections of various crops. *Papaya ringspot virus* biotype (TRSV-W) from different cucurbitaceous crops was investigated for possible genetic connectivity between crop viruses. Samples from cucumber, pumpkin, rockmelon, honeydew melon (*Cucumis melo*) and watermelon (*Citrullus lanatus*) were subjected to HTS analysis. Recombination analysis revealed 13 recombination events among the 49 complete genomes. Two Timorese isolates were recombinants, while all 12 Australian isolates were non-recombinants. Following inoculation with an Australian PRSV isolate, zucchini (*Cucurbita pepo*), watermelon, rockmelon and honeydew melon developed systemic foliage symptoms characteristic of PRSV-W (Maina, Coutts, Edwards, de Almeida, Ximenes et al. 2017). The Illumina NextSeq 500 platform was employed to determine papaya plant viromes. The presence of PRSV, *Lettuce chlorosis virus* (LCV), and *Tomato yellow leafcurl virus*-IL (TYLCV-IL) was detected by RT-PCR assay of leaves (Alabi et al. 2017). Grapevine plants showing red blotch symptoms were examined for viral origin in North America. Based on the molecular characteristics, *Grapevine red blotch-associated virus* (GRBaV was identified as the causal agent. Phylogenetic analysis of the predicted that amino acid sequence CP of GRBaV was similar to those of the CP of 23 geminiviruses belonging to *Geminiviridae* (Bahder et al. 2016). Grapevine cv. Riesling plants showing decline and leafroll symptoms were examined using HTS analysis of viromes of infected plants. Total RNAs as templates and Illumina HiSeq 2500 platform were employed for the analysis. The presence of five viruses and three viroids, *Grapevine leafroll-associated virus 1* (GLRaV-1) and GLRaV-3 (genus *Ampelovirus, family Closteroviridae), Grapevine rupestris stem pitting-associated virus* (GRSPaV) was detected by HTS analysis in the infected vines. Complete genomes of two novel and highly divergent isolates of GLRaV-3 were determined. Two independent rapid amplifications of cDNA ends (RACE) strategies were applied to obtain sequences at both 5′ and 3′ termini of the viral genomes. These two isolates were widely divergent from all other isolates of GLRaV-3. High degrees of variation in both length and sequence in the terminal untranslated regions (UTRs) of GLRaV-3 variants were recorded (Xiao et al. 2019).

6.4.4.4 Single-Strand Conformation Polymorphism Analysis

Single-strand conformation polymorphism (SSCP) analysis is a powerful alternative to assess genetic diversity in genomes of isolates/strains of plant viruses. *Grapevine virus A* (GVA, (*Vitivirus*) associated with Shiraz disease was extremely heterogenic. Divergent strains of GVA clustered into three molecular groups I, II, and III that could be differentiated based on nucleotide similarity. Comparative molecular analysis of the movement protein (MP) gene (ORF3) of single GVA variants of the three groups of variants transmitted from various grapevines to *Nicotiana benthamiana* was performed. SSCP analysis of a 234-nt sequence showed that group II variants could be differentiated from group I and group III. Further, SSCP analysis of 234-bp RT-PCR products revealed the presence of slowly migrating bands with diffused appearance indicative of GVA variants of molecular group II associated with majority of grapevines affected by Shiraz disease. SSCP analysis might be useful to rapidly identify variants of group II currently associated with various grapevines with different stages of infection (Goszczynski 2002). Distribution and genetic diversity of *Potato yellow mosaic Panama virus* (PYMPV) were studied. The DNA of these isolates from tomato plants were subjected to rolling circle amplification (RCA) analysis for detecting the causal begomovirus. RCA-positive samples were tested by PCR assay with specific primer pair to identify PYMPV. Four distinct patterns were differentiated by SSCP procedure (Davino et al. 2018).

Genetic variation in two groups of *Citrus tristeza virus* (CTV) isolates from Spanish source obtained by successive host passages, and successive aphid transmission were studied. SSCP analysis of genes p18 and p20 was performed. Estimation of genetic diversion within and between isolates and evaluation of genetic differentiation between populations was performed. Genetic variation of different genes, repeated inoculation in the field, homologous RNA recombination between sequence variations, and the presence of different defective RNAs are some of the factors that may contribute to biological variability of CTV isolates (Ayllón et al. 2006). In another investigation, polymorphism in the CP gene of CTV by cleavase fragment length polymorphism (CFLP) and SSCP analyses were employed to differentiate 12 CTV isolates with different biological properties from different geographical locations. Biologically characterized sources of CTV were analyzed by IC-RT-PCR followed by CFLP analysis targeting CP gene. Most of the CTV isolates were grouped precisely based on the CP sequences of variants constituting the isolates. In plants inoculated with a mixture of mild and severe isolates, the patterns were more closely related to the severe isolates in plants with mixed infections. The CFLP procedure showed potential for identification and differentiation of CTV isolates reliably and precisely (Marqués et al. 2006). The CTV CP gene of the isolates from different countries was amplified by RT-PCR, cloned, and characterized by SSCP analysis. Phylogenetic analysis indicated the existence of seven well-defined clusters. The asymmetric PCR-ELISA typing (PET) assay was developed in the frame of this clustering pattern, using a set of eight hybridization probes. Visual based application of this assay provided results comparable to ELISA test (Nolasco et al. 2009). *Citrus psorosis virus* (CPsV) contains kinked filaments composed of three negative, strand RNA molecules, and a 48-kDa coat protein. Two different intensities of symptoms in affected trees could be differentiated: psorosis A (PsA) and psorosis B (PsB). Isolates of PsA and PsB induced similar symptoms following graft inoculation. PsA and PsB-inducing subisolates from the same tree showed identical SSCP profiles in homologous segments of RNA1 and RNA3, whereas segments of RNA2 generated profiles that could be used as basis for differentiation of PsA- and PsB-associated sequence variants. SSCP analysis of RNA2 population present in different tissues of psorosis-affected plants showed that PsA-inducing isolates contained PsB-associated sequence variant predominantly present in blistered twigs and gummy pustules affecting old leaves. Characteristic of PsB isolates and PsB-associated sequence variant accumulated preferentially in bark lesions of the trunk and limbs (Velázquez et al. 2012).

6.5 VARIABILITY IN VIROIDS

6.5.1 Variability in Nucleic Acid Characteristics

6.5.1.1 Hybridization-Based Techniques

Dot blot hybridization was effective in detecting and differentiating viroids infecting *Citrus* spp. Riboprobes specific to *Citrus exocortis viroid* (CEVd) and *Hop stunt viroid* (HSVd) were transcribed using plasmids. All citrus plants tested, exhibited coinfection by *Citrus tristeza virus* (CTV) or *Citrus psorosis virus* (CPsV) or *Citrus variegation virus* (CVV) and HSVd or CEVd (Barbarossa et al. 2007). Infection of *S. jasminoides* and *S. rantonnetti* by *Potato spindle tuber viroid* (PSTVd) was observed in addition to potato in Italy. Dot blot hybridization in conjunction with PCR amplification was employed to assess molecular variability of isolates of PSTVd from different host plant species. Primers PSTVd-32 and PSTVd-33 were employed to amplify cDNA of about 360-bp corresponding to full-length PSTVd RNA from infected samples. All samples tested positive by dot blot hybridization and generated an RT-PCR amplicon of the same fragment. Isolates of PSTVd from *S. jasminoides* and *S. rantonnetti* showed limited variability (Di Serio 2007).

6.5.1.2 Polymerase Chain Reaction–Based Techniques

Apple dimple fruit viroid (ADFVd) and *Apple scar skin viroid* (ASSVd) induced similar symptoms. The RT-PCR assay employed differently labeled primers in conjunction with universal primer for detection and differentiation of two field isolates of ADFVd. Sequencing of 18 full-length cDNA clones revealed five new sequence variants. Sequence comparison showed nine polymorphic positions distributed in different regions of the ADFVd molecule. Sequence conservation existing between certain regions of the ADFVd and ASSVd, as well as their genome sequence divergence in other regions, were used to design viroid-specific primers, each labeled with a different fluorescent dye for rapid detection and discrimination of the two viroids by RT-PCR amplification. A cDNA of the expected size and fluorescence (254-bp red and 350-bp green) was amplified from tissues infected by ADFVd and ASSVd, respectively. Differentiation of these two viroids present together in infected samples could be detected and differentiated by RT-PCR assay (Di Serio et al. 2002). In a later investigation, genetic diversity of isolates of ASSVd determined by RT-PCR assay showed that variability was more common in the pathogenic domain of viroid genome (Walia et al. 2009).

Peach latent mosaic viroid (PLMVd) isolates inducing peach calico (PC) have an additional sequence that contains PC pathogenicity determinant (Malfitano et al. 2005). Sequencing of the PCR products showed that nine new PLMVd variants were present in the samples from infected peach trees showing symptoms. The isolates showed variability in the RNA length (335–338 nt). Sequences of all isolates clustered together in group III. Isolates of *Hop stunt viroid* (HSVd) are primarily divided into three groups based on overall homology as plum-type, hop-type and citrus type. Two HSVd isolates from apricot cv. Saturn were not clustered in any of the groups proposed earlier (Mandic et al. 2007). *Hop stunt viroid* (HSVd) can infect several woody plant species in *Prunus* spp., *Citrus* spp. and *Vitis* spp. In addition, HSVd in combination with *Citrus exocortis viroid* (CEVd) might infect grapevine also. Primers specific for HSVd, CEVd, and *Grapevine speckle viroid 1* (GYSVd-1), GYSVd-2, and *Australian grapevine viroid* (AGVd) were employed. Based on the sequences of PCR amplicons, the presence of both HSVd and CEVd in infected grapevine plants could be detected. Phylogenetic analysis showed that Brazilian grapevine HSVd variants clustered with other citrus and grapevine variants (Eiras et al. 2006). HSVd isolates from naturally infected apricot, plum, and peach were characterized. Phylogenetic analyses indicated that one apricot isolate clustered with the recombinant group, whereas other isolates from apricot, plum, and peach, clustered with the hop group, revealing the genetic diversity of HSVd isolates (Gazel et al. 2008).

Genetic diversity of *Citrus exocortis viroid* (CEVd) from the citrus cv. Clementinier (*Citrus reticulata*) was assessed by sequencing cloned cDNA copies of natural isolates. Nucleotide sequence analysis showed that differences in the isolates could be recognized in four sequences at 14 positions of 371-nt of CEVd. Sequence variation might occur naturally without passage in an indicator plant species. Selection pressures induced by different host plant species might intensify further by environmental factors, resulting in emergence of a new master sequence. The presence in citrus that has a long life span might provide conditions for accumulation of several sequence variants in each infected tree (Elleuch et al. 2006). Specific variants of *Hop stunt viroid* (HSVd) causing citrus cachexia from six locations in China were examined for their genetic diversity. Neither any sequence variant-cultivar relationship, nor distinct regional specificity could be detected among HSVd populations examined. Three clusters consisting of three HSVd variants were identified by phylogenetic analysis. Most Chinese isolates might contain mixtures of cachexia and non-cachexia variants. Interactions between *Citrus cachexia viroid* (CCaVd) and non-CCaVd variants might play an important role in suppressing cachexia symptom expression (Wang et al. 2010). Phylogenetic analysis revealed that the genotypes of *Potato spindle tuber viroid* (PSTVd) were grouped in a cluster of genotypes from tomato or *Physalis peruviana*. This isolate exhibited a mutation ($U_{257} \rightarrow A$) that was linked to lethal symptom expression in tomato (van Brunschot et al. 2014).

6.5.1.3 Heteroduplex Mobility Analysis

Hop stunt viroid (HSVd) exhibits high variability, plasticity, and wide sequence distribution, comparing well with *Peach latent mosaic viroid* (PLMVd) with extreme variability. Further, HSVd has a wide host range which includes cucumber, grapevine, citrus, plum, peach, pear, apricot, and almond. Hence, determinination of the extent of variability and monitoring occurrence of HSVd was taken up. Isolates of HSVd in grapevines were examined in locations where hop crops were also grown. As *Hop latent viroid* (HLVd) also occurred along with HSVd, specific primers were employed to differentiate them in mixed infections. The cDNAs of HSVd from individual samples were analyzed by temperature gradient gel electrophoresis (TGGE) and heteroduplex mobility analysis. HSVd was present as dominant populations in all samples tested. As the high incidence of HSVd infection in grapevines was observed, the potential danger of growing hop adjacent to vineyards should be considered seriously (Matoušek et al. 2003).

REFERENCES

Abraham AD, Menzel W, Lessemann DE, Varrelmann M, Vetten HJ (2006) *Chickpea chlorotic stunt virus*: a new *Polerovirus* infecting cool-season food legumes in Ethiopia. Phytopathology 96: 437–446.

Adkins S, Hammond J, Gera A, Maroon-Lango CJ, Sobolev I et al. (2006) Biological and molecular characterization of a novel carmovirus isolated from *Angelonia*. Phytopathology 96: 460–467.

Adkins S, Webb SE, Achor D, Roberts PD, Baker CA (2007) Identification and characterization of a novel whitefly-transmitted member of the family *Potyviridae* isolated from cucurbits in Florida. Phytopathology 97: 145–154.

Afolabi AS, Thottappilly G (2008) Comparative studies on alkaline phosphatase (ALP), alkaline phosphatase amplification (AMP), horseradish peroxidase (HRP), penicillinase (PNC) and avidin-biotin penicillinase amplification of ELISA in detecting *Maize streak virus* (MSV) in maize plants and *Cicadulina mbila* (leafhopper) insect vector. Sci Res Essay 3: 524–530.

Agindotan B, Perry KL (2007) Macroarray detection of plant RNA viruses randomly primed and amplified complementary DNAs from infected plants. Phytopathology 97: 119–127.

Agnihotri AK, Mishra SP, Ansar M, Tripathi RC, Singh R et al. (2019) Molecular characterization of *Mungbean yellow mosaic India virus* infecting tomato (*Solanum lycopersicum* L.). Austr Plant Pathol 48: 159–165.

Agrofoglio Y, Delfosse VC, Casse MF, Hopp HE, Kresic IB et al. (2017) Identification of a new cotton disease caused by an atypical *Cotton leafroll dwarf virus* in Argentina. Phytopathology 107: 369–376.

Ahangaran A, Mohammadi GM, Habibi MK, Khezri S, Shahraeen N (2009) Use of a rapid serological and nucleic acid-based methods for detecting *Soybean mosaic virus*. J Agric Sci Technol 11: 91–97.

Ahmadi S, Almasi MA, Fatehi F, Struik PC, Moradi A (2013) Visual detection of *Potato leaf roll virus* by one-step reverse transcription loop-mediated isothermal amplification of DNA with hydroxynaphthol blue dye. J Phytopathol 161: 120–124.

Alabi OJ, Al-Rwahnih M, Jiffon JL, Setamou M, Brown JK et al. (2017) A mixed infection of *Lettuce chlorosis virus, Papaya ringspot virus* and *Tomato yellow leafcurl virus*-IL detection in a Texas papaya orchard affected by a virus-like disease outbreak. Plant Dis 101: 1094–1102.

Albiach-Marti MR, Guerri J, Cambra M, Garnsey SM, Moreno SM (2000) Differentiation of *Citrus tristeza virus* isolates by serological analysis of p25 coat protein of *Faba bean necrotic yellows virus*. J Phytopathl 149: 543–550.

Alcaide C, Rabadan MP, Juarez M, Gomez P (2020) Long term circulation of two strains of *Pepino mosaic virus* in tomato crops and its effect on population and genetic variability. Phytopathology 110: 49–57.

Alfaro-Fernández A, Sánchez-Navarro JA, Cebrian MC, Cordoba-Selles MC, Pallas V, Jorda C (2009) Simultaneous detection and identification of *Pepino mosaic virus* (PepMV) isolates by multiplex one-step RT-PCR. Eur J Plant Pathol 125: 143–158.

Al-Harthi SA, Al-Sadi AM, Al-Saady AA (2013) Potential of citrus budlings originating in the Middle East as sources of citrus viroids. Crop Protect 48: 13–15.

Alabi OJ, Rwahnih MA, Karthikeyan G, Poojari S, Fuchs M, Rowhani A, Naidu RA (2011) *Grapevine leafroll-associated virus* occurs as genetically diverse populations. Phytopathology 101: 1446–1456.

Alibi OJ, Rwahnih MA, Mekuria TA, Naidu RA (2014) Genetic diversity of *Grapevine virus A* in Washington and California vineyards. Phytopathology 104: 548–560.

Alioto D, Malfitano M, Troisi A, Peluso A, Martin S et al. (2003) Variability in the coat protein gene of *Citrus psorosis virus* in Campania southern Italy. Arch Virol 148: 2155–2166.

Al-Mrabeh A, Ziegler A, Cowan G, Torrance L (2009) A fully recombinant ELISA using in vivo biotinylated antibody fragments for the detection of *Potato leafroll virus*. J Virol Meth 159: 200–205.

Al-Rwahnih M, Dave A, Anderson MM, Rowhani A, Ugemoto JK et al. (2013) Association of a DNA virus with grapevines affected by red blotch disease in California. Phytopathology 103: 1069–1076.

Al-Rwahnih M, Rowhani A, Westrick N, Diaz-Lara A, Trouillas FP et al. (2018) Discovery of viruses and virus-like pathogens in pistachio using high throughput sequencing. Plant Dis 102: 1419–1425.

Angelini E, Bertazzon N, Borgo M (2004) Diversity among *Grapevine leaf roll-associated virus 2* isolates detected by heteroduplex mobility assays. J Virol Meth 92: 151–163.

Anuradha S, Selvarajan R (2018) Molecular characterization of *Banana bract mosaic virus* from India reveals recombination and positive selection in the VPg gene. J Plant Pathol 100: 523–531.

Aparicio F, Solar S, Aramburu J, Galipienso L, Nuez, Pallas V, Lopez C (2009) Simultaneous detection of six RNA plant viruses affecting tomato crops using a single digoxigenin-labeled polyprobe. Eur J Plant Pathol 123: 117–123.

Arya M, Baranwal VK, Ahlawat YS, Singh I (2006) RT-PCR detection and molecular characterization of *Onion yellow dwarf virus* associated with garlic and onion. Curr Sci 9: 1220–1234.

Ayllón MA, Rubio L, Sentandreu V, Moya A, Guerri J, Morenos P (2006) Variation in two gene sequences of *Citrus tristeza virus* after host passages. Virus Genes 32: 119–128.

Babu B, Knox G, Paraet ML, Ochia-Corona FM (2018) Rose rosette disease: recent advances on molecular diagnostic tools. HortScience 53: 596–600.

Badial AB, Sherman D, Stone A, Gopakumar A, Wilson V, Schneider W, King J (2018) Nanopore sequencing as a surveillance tool for plant pathogens in plant and insect tissues. Plant Dis 102: 1648–1652.

Bahder BW, Zalo FG, Jayanth M, Sudhansharma MR (2016) Phylogeny of geminivirus coat protein sequences and digital PCR aid in identifying *Spissistilus festinus* as a vector of *Grapevine redblotch-associated virus*. Phytopathology 106L: 1223–1230.

Balme-Sinibaldi V, Tribodet M, Croizat F, Lefeuvre P, Kerlan C et al. (2006) Improvement of *Potato virus Y* (PVY) detection and quantification using PVY^N- and PVY^O-specific real-time PCR assays. J Virol Meth 134: 261–266.

Bananej K, Keshavarz T, Vahdat A, Salekdeh GH, Glasa M (2002) Biological and molecular variability of *Zucchini yellow mosaic virus* in Iran. J Phytopathol 156: 654–659.

Barba M, Czosnek KH, Hadidi A (2014) Historical perspective, development and applications of next-generation sequencing in plant virology. Viruses 6: 106–136.

Barbarossa L, Loconsole G, Vovlas C (2007) Virus and virus-like diseases of citrus in Epinis. J Plant Pathol 89: 273–276.

Barbosa CJ, Pina JA, Perez-Panadés J, Bernard L, Serra P et al. (2005) Mechanical transmission of citrus viroids. Plant Dis 89: 749–754.

Barbosa M, Loconsole G, Vovlas C (2007) Virus and virus-like diseases of citrus in Epirus. J Plant Pathol 89: 273–276.

Barkley NA, Pinnow DL, Wang ML, Ling KS, Jarret RL (2011) Detection and classification of SPLCV isolates in the US Sweet Potato germplasm collection via a real-time PCR assay and phylogenetic analysis. Plant Dis 95: 1385–1391.

Bereda M, Paduch-Cichal E, Dabrowska E (2017) Occurrence and phylogenetic analysis of allexiviruses identified on garlic from China, Spain, and Poland commercially available on the Polish retail market. Eur J Plant Pathol 149: 227–237.

Berry S, Rey MEC (2001) Differentiation of cassava-infecting begomoviruses using heteroduplex mobility assays. J Virol Meth 92: 151–163.

Bertolini E, Moreno A, Capote N, Olmos A, de Luis A et al. (2008) Quantitative detection of *Citrus tristeza virus* in plant tissues and single aphids by real-time RT-PCR. Eur J Plant Pathol 120: 177–188.

Bhat AI, Siju S (2007) Development of a single-tube multiplex RT-PCR for the simultaneous detection of *Cucumber mosaic virus* and *Pepper yellow mottle virus* associated with stunt disease of black pepper. Curr Sci 93: 973–976.

Boonham N, Walsh K, Preston S, North J, Smith P, Barker J (2002) The detection of tuber necrotic isolates of *Potato virus Y* and the accurate discrimination of PVY^O, PVY^N, and PVY^C strains by RT-PCR. J Virol Meth 102: 103–112.

Bostan H, Gazel M, Elibuyuk IO, Caglayan K (2010) Occurrence of Pospiviroid in potato, tomato, and some ornamental plants in Turkey. Afr J Biotechnol 9: 2613–2617.

Boukari W, Filloux D, Daugrois JH, Fernandez E, Mollov D, Kaye C, Hincapie M, Wang ASJ, Roumagnac P, Rott P (2020) Molecular detection of *Sugarcane strate virus* and *Sugarcane white streak virus* and their prevalence in the Miami World Collection of sugarcane and related grasses. Plant Pathol 69: 1060–1069.

Boyzova NA, Safenkova IV, Chirkov SN, Zherdev AV, Blintsov AN et al. (2009) Development of immunochromatographic test systems for express detection of plant viruses. Appl Biochem Microbiol 45: 204–209.

Bratsch SA, Grinstead S, Creswell TC, Ruhl GE, Mollov D (2019) Characterization of *Tomato necrotic spot virus*, a subgroup 1 *Ilarvirus* causing necrotic foliar, stem and fruit symptoms in tomatoes in the United States. Plant Dis 103: 1391–1396.

Bratsch SA, Olszewski N, Lockhart B (2021) Incidence of cymbidium mosaic, odonotoglossm ringspot, and orchid fleck virus in orchids in Minnesota andproduction of antibodies for use in ELISA to detect orchid fleck virus. Eur J Plant Pathol 159: 543–554.

Bystricka D, Lenz O, Marz I, Piherova L, Kmoch S et al. (2005) Oligonucleotide-based microarray: a new improvement in microassay detection of plant viruses. J Virol Meth 128: 151–155.

Calvert LA, Cuvero M, Lozano I, Villareal N, Arroyave J (2008) Identification of the three strains of a virus associated with cassava plants affected by frogskin disease. J Phytopathol. https://doi.org/10.1111/j.1439-0434.2008.0141.x

Capote N, Bertolini E, Olmos A, Vidal E, Martinez MC et al. (2009) Direct sample preparation methods for the detection of *Plum pox virus* by real-time RT-PCR. Internatl Microbiol 12: 1–6.

Cating RA, Funke CN, Kaur N, Hamm PB, Frost KE (2015) A multiplex reverse transcription (RT) high fidelity PCR protocol for detection of six viruses that cause potato tuber necrosis. Amer J Potato Res 92: 536–540.

Cavileer TD, Clarke RC, Corsini DL, Berger PH (1998) A new strain of potato carlavirus. Plant Dis 81: 98–102.

Cerni S, Skoric D, Ruscic J, Krajacic M, Papic T et al. (2009) East Adriatic- a reservoir of severe *Citrus tristeza virus* strains. Eur J Plant Pathol 124: 701–706.

Cerovska N, Moravec T, Plchova H, Hoffmeisterova H, Dedic P (2012) Production of polyclonal antibodies to the recombinant *Potato virus M* (PVM) nonstructural triple gene block protein 1 and coat protein. J Phytopathol 160: 251–254.

Chabi-Jesus C, Tassi AD, Guerrra-Peraza O, Kitajima EW, Salaroli RB et al. (2018) Identification and characterization of *Citrus chlorotic spot virus*, a new *Dichorhavirus* associated with citrus leprosis-like symptoms. Plant Dis 102: 1588–1598.

Chauhan RP, Wijayasekara D, Webb MA, Verchot J (2015) A reliable and rapid multiplex RT-PCR assay for detection of two potyviruses and a pararetrovirus infecting canna plants. Plant Dis 99: 1695–1703.

Chen T-C, Huang C-W, Kuo Y-W, Liu F-L, Hsuan C-H et al. (2006) Identification of common epitopes on a conserved region of NSs proteins among tospoviruses of *Watermelon silver mottle virus* serogroups. Phytopathology 96: 1296–1304.

Chen T-C, Li J-T, Lin Y-P, Yeh Y-C, Kang Y-C et al. (2012) Genomic characterization of *Calla lily chlorotic spot virus* and design of broad-spectrum primers for detection of tospoviruses. Plant Pathol 61: 183–194.

Chen Z, Mao S, Zhang W, Fan X, Wu W, Liu C, Zhao K, Lu R (2021) Rapid visual detection method for *Barley yellow mosaic virus* using reverse transcription loop-mediated isothermal amplification (RT-LAMP). Plant Dis 105: 2658–2663.

Cheng Y-H, Huang C-H, Jan F-J (2019) Identification and characterization of *Watermelon green mottle mosaic virus* as a new cucurbit-infecting tobamovirus. Ann Appl Biol 174: 31–39.

Chikh-Ali M, Gray SM, Karasev AV (2013) An improved multiplex IC-RT-PCR assay distinguishes nine strains of *Potato virus Y*. Plant Dis 97: 1370–1374.

Chikh-Ali M, Rodriguez-Rodriguez M, Green KJ, Kim D-J, Chung S-M et al. (2019) Identification and molecular characterization of recombinant *Potato virus Y* (PVY) in potato from South Korea, PVY^{NTN} strain. Plant Dis 103: 137–142.

Chirkov S, Sheveleva A, Ivanov P, Zakubanskiy A (2018) Analysis of genetic diversity of Russian sour cherry *Plum pox virus* isolates provides evidence of a new strain. Plant Dis 102: 569–575.

Chou C-N, Chen C-E, Wu M-L, Su H-J, Yeh H-H (2009) Biological and molecular characterization of Taiwanese isolates of *Cucumber mosaic virus* associated with banana mosaic disease. J Phytopathol 157: 85–93.

Cogotzi L, Giampetruzzi A, Nölke G, Orecchia M, Elico V et al. (2009) An assay for the detection of *Grapevine leafroll-associated virus 3* using a single-chain fragment variable antibody. Arch Virol 154: 19–26.

Congdon BS, Coutts BA, Renton M, Banovic M, Jones RAC (2016) *Pea seedborne mosaic virus* in field pea: widespread infection, genetic diversity and resistance gene effectiveness. Plant Dis 100: 2475–2482.

Cook G, van Vuren SP, Breytenbach JHJ, Steyn C, Burger JT, Maree HJ (2016) Characterization of *Citrus tristeza virus* single variant sources in grapefruit in greenhouse and field trials. Plant Dis 100: 2251–2256.

Crespo O, Jansen D, Garcia C, Ruiz L (2017) Biological and molecular diversity of *Cucumber green mottle mosaic virus* in Spain. Plant Dis 101: 977–984.

Crosslin JM, Hamm PB, Shiel PJ, Hane DC, Brown CR, Berger PH (2005) Serological and molecular detection of tobacco veinal necrosis isolates of *Potato virus Y* (PVY^N) from potatoes grown in the western United States. Amer J Potato Res 62: 263–269.

Cui T, Bin Y, Yan J, Mei P, Li Z et al. (2018) Development of infections cDNA clones of *Citrus yellow vein clearing virus* using a novel and rapid strategy. Phytopathology 108: 1212–1218.

Davino S, Panno S, Caruso AG, Davino M, Vasquez JAH (2018) High genetic stability of *Potato yellow mosaic Panama virus* infecting tomato in Panama. J Phytopathol 100: 59–65.

Debat H, Zavallo D, Brisbane RS, Voncina D, Almeida RPP et al. (2019) *Grapevine virus L*: a novel vitivirus in grapevine. Eur J Plant Pathol 155: 319–328.

Deng X-G, Zhu F, Chen Y-J, Liu J, Zhu T et al. (2014) A more sensitive and rapid multiplex RT-PCR assay combining with magnetic nanobeads for simultaneous detection of viruses in sweet potato. Eur J Plant Pathol 140: 111–117.

Deyong Z, Willingmann P, Heinze C, Adam G, Pfunder M et al. (2005) Differentiation of *Cucumber mosaic virus* isolates by hybridization to oligonucleotides in a macroarray format. J Virol Meth 133: 101–105.

Di Serio F (2007) Identification and characterization of *Potato spindle tuber viroid* infecting *Solanum jasminoides* and *S. rantonnetti* in Italy. J Plant Pathol 89: 297–300.

Di Serio F, Afechtal M, Attard D, Choueiri E, Gumus M et al. (2010) Detection by tissue printing hybridization of pome fruit viroids in Mediterranean basin. 21st international conference viruses and other graft-transmissible diseases of fruit crops. Julius-Kühn Arch 427: 357–360.

Di Serio F, Malfitano M, Alioto, Ragozzino A, Flores R (2002) *Apple dimple fruit viroid*: sequence variability and its specific detection by multiplex fluorescent RT-PCR in the presence of *Apple scar skin viroid*. J Plant Pathol 84: 27–34.

Diener TO (1979) *Viroids and Virioid Diseases*. Wiley, New York.

Dietzgen RG, Twin J, Tally J, Selladurain S, Carroll ML et al. (2005) Genetic variability of *Tomato spotted wilt virus* in Australia and validation of real-time for its detection in single and bulked samples. Ann Appl Biol 146: 517–530.

Dombrovsky A, Sapkota R, Lachman O, Antignus Y (2012) *Eggplant mild leaf mottle virus* (EMLMV), a new putative member of the genus *Ipomovirus* that harbors an HC-Pro gene. Virus Genes 44: 329–337.

Donda BP, Jargula S, Naidu RA (2017) An analysis of the complete genome sequence and subgenomic RNAs reveals unique features of the Ampelovirus, *Grapevine leafroll-associated virus 1*. Phytopathology 107: 1069–1079.

Du ZY, Chen JS, Hiruk C (2006) Optimization and application of a multiplex RT-PCR system for simultaneous detection of five potato viruses using 18S RNA as an internal control. Plant Dis 90: 185–189.

Duarte KMR, Gomes LA, Andrino PG, Leal Junior GA, da Silva FHB et al. (2002) Identification of tomato mosaic tobamovirus (ToMV) using monoclonal antibodies. Sci Agric 59: 109–112.

Eastwell KC, Nelson ME (2007) Occurrence of viroids in commercial hop (*Humulus lupulus* L.) production areas in Washington State. Plant Health Prog: 1–8.

Eiras M, Targon MLPN, Fajardo TVM, Flores R, Kitajima EW (2006) *Citrus exocortis viroid* and *Hop stunt viroid* doubly infecting grapevines in Brazil. Fitopatol Bras 31: 440–446.

El-Dougdoug KA (1998) Occurrence of *Peach latent mosaic viroid* in apple (*Malus domestica*). Ann Agric Sci (Cairo) 43: 21–30.

El-Dougdoug KA, Hazaa MM, Gomaa HA, El-Maaty SA (2006) Eradication of banana viruses from naturally infected banana plants. 1. Biological and molecular methods of detection of *Cucumber mosaic virus* and *Banana bunchy top virus* in naturally infected banana plants. J Appl Sci Res 2: 1156–1163.

Elleuch A, Khouaja FD, Fajardo TVM, Flores R, Kitajima EW (2006) *Citrus exocortis viroid* and *Hop stunt viroid* doubly infecting grapevines in Brazil. Fitopatol Bras 31: 440–446.

ElSayed AI, Weig AR, Komor E (2011) Molecular characterization of *Hawaiian Sugarcane yellow leaf virus* genotypes and their phylogenetic relationship to strains from other sugarcane- growing countries. Eur J Plant Pathol 129: 399–412.

Engel EA, Escobar PF, Rojas LA, Rivera PA, Fiore N et al. (2010) A diagnostic oligonucleotide microarray for simultaneous detection of grapevine viruses. J Virol Meth 163: 445–451.

Fageria MS, Singh M, Nanayakkara U, Pelletier KY, Nie X et al. (2013) Monitoring current season spread of *Potato virus Y* in potato fields using ELISA and Real-time RT-PCR. Plant Dis 97: 641–644.

Fauquet CM, Mayo MA, Maniloff J, Desselberger U, Ball LA (2005) *Virus Taxonomy. VIII Report of the ICTV*. Elsevier Academic Press, London, pp. 751–756.

Fellers JP, Webb C, Fellers MC, Rupp JS, De Wolf E (2019) Wheat virus identification within infected tissue using nanopore sequencing technology. Plant Dis 103: 2199–2203.

Flores R, Randles JW, Owens RA, Bar-Joseph M, Diener TO (2005) Subviral agents: Viroids. In: Fauquet CM, Mayo MA, Maniloff J, Desselberger U, Ball LA (eds.), *Virus Taxonomy VIII*. Academic Press, New York, pp. 1147–1161.

Foissac X, Svanella-Dumas L, Gentit P, Dulucq M, Marias M et al. (2005) Polyvalent degenerate oligonucleotides reverse transcription polymerase chain reaction: a polyvalent detection and characterization tool for Trichoviruses, Capilloviruses, and Fovea viruses. Phytopathology 96: 137–144.

Freitas EL, Brito AC, de Oliveira SAS, de Oliveira EJ (2020) Early diagnosis of cassava frogskin disease in powdered tissue samples using near-infrared spectroscopy. Eur J Plant Pathol 156: 547–558.

Fuentes S, Gibbs AJ, Adams IP, Wilson C, Botermans M et al. (2021) *Potato virus A* isolates from three continents: their biological properties, phylogenetics and prehistory. Phytopathology 111: 217–226.

Fukuta S, Iida T, Mizukami Y, Ishida A, Ueda T et al. (2003) Detection of *Japanese yam mosaic virus* by RT-LAMP. Arch Virol 148: 1713–1720.

Fukuta S, Oshishi K, Yoshida K, Mizukami Y, Ishida A et al. (2004) Development of immunocapture reverse transcription loop-mediated isothermal amplification for the detection of *Tomato spotted wilt virus* from chrysanthemum. J Virol Meth 121: 49–55.

Gambino G, Gribaudo I (2006) Simultaneous detection of new grapevine viruses by multiplex reverse transcription-polymerase chain reaction with coamplification of a plant RNA as an internal control. Phytopathology 96: 1123–1229.

Gambino G, Navarro B, Torchetti EM, La Notte P, Schneider A, Mannini F, Di Serio FC (2014) Survey on viroids infecting grapevine in Italy: identification and characterization *of Australian grapevine viroid* and *Grapevine yellow speckle viroid 2*. Eur J Plant Pathol 140: 199–205.

Gambley CF, Steele V, Geeting ADW, Thomas JE (2008) The genetic diversity of ampeloviruses in Australian pineapples and their association with mealybug wilt disease. Austr Plant Pathol 37: 95–107.

Gazel M, Serce CV, Caglayan K, Faggioli F (2008) Sequence variability of *Hop stunt viroid* isolates from stone fruits in Turkey. J Plant Pathol 90: 23–28.

Ghanem-Sabanadzovic NA, Abbadi H, Rawanih M, Castellano MA, Myrta A (2005) Identification and partial characterization of an isolate of *Apricot latent virus* from Palestine. J Plant Pathol 87: 37–41.

Ghanim M, Brumin M, Popovski S (2007) A simple, rapid and inexpensive method for localization of *Tomato yellow leaf-curl virus* and *Potato leafroll virus* in plant and insect vectors. J Virol Meth 159: 311–314.

Giakountis A, Tsarmpopoulos I, Chatzivassilou EK (2018) *Cucumber mosaic virus* isolates from Greek legumes are associated with satellite RNAs that are necrogenic for tomato. Plant Dis 102: 2268–2276.

Gong J, Ju H-K, Kim I-H, Seo E-Y, Cho I-S, Hu W-X, Han J-Y, Kim J-K, Choi R, Lim Y-P, Hammond J, Lim H-Y (2019) Sequence variations among 17 new radish isolates of *Turnip mosaic virus* showing differential pathogenicity and infectivity in *Nicotiana benthamiana*, *Brassica rapa*, and *Raphanus sativus*. Phytopathology 109: 904–912.

Goszczynski DE, Josste AEC (2002) The application of single-strand conformation polymorphism (SSCP) technique for the major analysis of molecular heterogeneity of *Grapevine virus A*. Vitis 41: 77–82.

Gramberg B, Kintzios S, Schmidt U, Mewis I, Ulrichs C (2012) A basic approach towards the development of bioelectric bacterial biosensors. J Phytopathol 160: 106–111.

Guardo M, Marletta T, Boninelli G, Leonardi A, Sorrentino G et al. (2005) Detection of viroids in forced citrus cuttings. J Plant Pathol 87: 141–144.

Guček T, Jakse J, Matoušek J, Radišek S (2019) One-step multiplex RT-PCR for simultaneous detection of four viroids from hop (*Humulus lupulus*). Eur J Plant Pathol 154: 273–286.

Gümüs M, Paylan IC, Matic S, Myrta A, Sipahioglu HM et al. (2007) Occurrence and distribution of stone fruit viruses and viroids in commercial plantings of *Prunus* species in western Anatolia, Turkey. J Plant Pathol 89: 265–268.

Guo Q, Honesty S, Xu ML, Zhang Y, Schoelz J et al. (2014) Genetic diversity and tissue and host specificity of *Grapevine vein clearing virus*. Phytopathology 104: 539–547.

Habili N, Farrokhi N, Lima MF, Nocholas P, Randles JW (2006) Distribution of *Rupestris stem pitting-associated virus* variants in two Australian vineyards showing different symptoms. Ann Appl Biol 148: 91–96.

Hadersdorfer J, Neumuller M, Treutter D, Fischer TC (2011) Fast and reliable detection of *Plum pox virus* in woody host plants using the Blue LAMP protocol. Ann Appl Biol 159: 456–466.

Hadidi A, Flores R, Candresse T, Barba M (2016) Next-generation sequencing and genome editing in plant virology. Front Microbiol 7: 1325. https://doi.org/10.3389/fmicb.2016.01325

Hao X, Wang L, Zhang X, Zhong Q, Hajano J, Xu L, Wu Y (2021) A real-time loop-mediated isothermal amplification for detection of *Wheat dwarf virus* in wheat and the insect vector *Psammnotettix alienus*. Plant Dis 105: 413–420.

Harju V, Skelton A, Clover GRG, Ratti C, Boonham N et al. (2005) The use of real-time RT-PCR (TaqMan) and post ELISA virus release for the detection of *Beet necrotic yellow vein virus* types containing RNA5 and its comparison with conventional RT-PCR. J Virol Meth 123: 73–80.

Hassan M, Gomez G, Pallas V, Myrta A, Rysanek P (2009) Simultaneous detection and genetic variability of stone fruit viroids in Czech Republic. Eur J Plant Pathol 124: 363–368.

Hassan M, Ryšánek P (2004) Occurrence of mixed infection of *Peach latent mosaic viroid* and *Hop stunt viroid* in Czech Republic. Acta Fytotech Zootech 7: 81–82.

Hassan M, Zouhan M, Ryšánek P (2004) Development of a PCR method for *Peach latent mosaic viroid* and *Hop stunt viroid* detection for certification of plant material. Acta Fytotech Zootech 7: 83–85.

Hassani-Mehraban A, Saaijer J, Peters D, Goldbach R, Kormelink R (2005) A new tomato-infecting tospovirus from Iran. Phytopathology 95: 852–858.

Hassen IF, Roussel S, Kummert J, Fakhfakh H, Marrakchi M et al. (2006) Development of a rapid RT-PCR test for the detection of *Peach latent mosaic viroid, Pear blister canker viroid, Hop stunt viroid*, and *Apple scar skin viroid* in fruit trees in Tunisia. J Phytopathol 154: 217–223.

Hataya T (2009) Duplex reverse transcription-polymerase chain reaction system to detect *Potato spindle tuber viroid* using an internal control mRNA and non-infectious positive control RNA. J Gen Plant Pathol 75: 167–172.

Hébrard E, Pinel-Galzi A, Oudare A, Poulicard N, Aribi J et al. (2018) Identification of a hypervirulent pathotype of *Rice yellow mottle virus*: a threat to genetic resistance deployment in west-central Africa. Phytopathology 108: 299–307.

Herrera-Vasquez JA, Puchades AV, Elvira-Gonzalez L, Jaen-Sanjur JN, Carpino C et al. (2018) Fast detection by loop-mediated isothermal amplification (LAMP) of the three begomovirus species infecting tomato in Panama. Eur J Plant Pathol 151: 243–250.

Hosakawa M, Shiba H, Kawabe T, Nakashima A, Yazawa S (2007) A simple and simultaneous detection method for two different viroids infecting chrysanthemum by multiple direct RT-PCR. J Jap Soc Hortic Sci 76: 60–65.

Hristova D, Yordanova A, Mavrodieva V (2002) Differentiation of Bulgarian isolates of *Cucumber mosaic virus* by serological methods. J Phytopathol 150: 334–339.

Hsu YH, Annamalai P, Lin CS, Chen YY, Chang WC et al. (2000) A sensitive method for detecting *Bamboo mosaic virus* (Ba MV) and establishment of BaMV-free meristem-tip cultures. Plant Pathol 49: 101–107.

Hu W, Qin L, Yan H, Miao W, Cui H, Liu W (2020) Use of an infectious cDNA clone of *Pepper vein mottle virus* to confirm the etiology of a disease in *Capsicum chinense*. Phytopathology 110: 80–84.

Hu WC, Huang CH, Lee SC, Wu CI, Chang YC (2010) Detection of four calla potyviruses by multiplex RT-PCR using nad5 mRNA as internal control. Eur J Plant Pathol 126: 43–52.

Huang Z, Rundell PA, Guan X, Powell CA (2004) Detection and isolates differentiation of *Citrus tristeza virus* in infected field trees based on reverse transcription-polymerase chain reaction. Plant Dis 88: 625–629.

Inglis DA, Gundersen B, Beisssinger A, Benedict C, Karasev AV (2019) *Potato virus Y* in seed potatoes sold at Garden Stores in Western Washington: prevalence and strain composition. Amer J Potato Res 96: 225–243.

Jacquot E, Tribodet M, Croizat F, Balme-Sindibald V, Kerlan C (2005) A single nucleotide polymorphism based technique for specific characterization of Y^O and Y^N isolates of *Potato virus Y* (PVY). J Virol Meth 125: 83–93.

James D, Phelan J, Jesperson G (2017) The Geneva Complex- an interesting disease perspective based on the results of NGS analysis. Proceedings of the 24th International Conference on Virus and Other Graft Transmissible Diseases of Fruit Crops (ICVF), Thessalonki, p 43.

James D, Sanderson D, Varga A, Sheveleva A, Chirkov S (2016) Genome sequence analysis of new isolates of the Winona strain of *Plum pox virus* and the first definitive evidence of intrastrain recombination events. Phytopathology 106: 407–416.

James D, Varga A, Croft H, Rast H, Thompson D, Hayes S (2006) Molecular characterization, phylogenetic relationships and specific detection of *Peach mosaic virus*. Phytopathology 96: 137–144.

Jan F-S, Chen T-C, Yeh S-D (2003) Occurrence, importance, taxonomy, and control of thrips-borne potyviruses. In: Huang HC, Acharya SN (eds.), *Advances in Plan Disease Management*. Kerala Agriclutural University, Trivandrum, India, pp. 391–411.

Jarugula S, Alabi OJ, Martin RR, Naidu RA (2010) Genetic variability of natural populations of *Grapevine leafroll-associated virus 2* in Pacific Northwest vineyards. Phytopathology 100: 698–707.

Jeevalatha A, Kaundal P, Kumar R, Raigond B, Kumar R et al. (2018) Optimized loop-mediated isothermal amplification assay for *Tomato leafcurl New Delhi virus* –(potato) detection in potato leaves and tubers. Eur J Plant Pathol 150: 565–573.

Jeevalatha A, Kaundal P, Sharma MN, Thakur P, Chakrabarti SK et al. (2013) Expression of coat protein of an ordinary strain of *Potato virus Y* on *Escherichia coli* and production of polyclonal antibodies for diagnosis. J Phytopathol 161: 671–674.

Jeevalatha A, Kumar R, Raigond B, Sundaresh S, Sharma S et al. (2015) Duplex real-time RT-PCR assay for the detection of *Potato spindle tuber viroid* (PSTVd) along with *ef1*-α gene of potato. Phytoparasitica 43: 317–325.

Jiang L, Huang Y, Sun L, Wang B, Zhu M et al. (2017) Occurrence and diversity of *Tomato spotted wilt virus* isolates breaking the *Tsw* resistance gene of *Capsicum chinense* in Yunnan, southwest China. Plant Pathol 66: 980–989.

Jin J, Shen J, Cai W, Liao F, Gao X et al. (2016) A cocktail ELISA semi-nested RT-PCR assay to detect *Bean pod mottle virus* in soybean seeds. J Phytopathol 164: 904–912.

Jo Y, Song MK, Choi H, Park J-S, Lee J-W et al. (2018) In silico identification of viruses and viroids infecting grapevine cultivar Cabernet Sauvignon using a grapevine transcriptome. J Plant Pathol 100: 91–96.

Johnson G, Park J-C, Kim Y-K, Kim M-J, Lee M-J et al. (2007) Direct stem blot immunoassay (DSBIA): a rapid, reliable, and economical detection technique suitable for testing large number of barley materials for first monitoring and resistance screening to *Barley mild mosaic virus* and *Barley yellow mosaic virus*. Plant Pathol J 23: 260–265.

Jones RAC, Dwyer GI (2007) Detection of *Sweet potato chlorotic fleck virus* and *Sweet potato feathery mottle virus*-strain O in Australia. Austr Plant Pathol 36: 591–594.

Jooste TL, Visser M, Cook G, Burger JT, Maree HJ (2017) In silico probe-based detection of citrus viruses in NGS data. Phytopathology 107: 988–993.

Kamran A, Lotos L, Amer MA, Al-Saleh MA, Alshahwan IM et al. (2018) Characterization of *Pepper leafroll chlorosis virus*, a new *Polerovirus* causing yellowing disease of bell pepper in Saudi Arabia. Plant Dis 102: 318–326.

Kaparia-Isaia TH, Kyriakou A, Ioannides J, Papayiannis L, Minas GR (2007) Sanitation of local citrus varieties and/or clones in Cyprus. Acta Hortic 741: 301–306.

Karasev AV, Hu X, Brown CJ, Kerlan C, Nikolaeva O et al. (2011) Genetic diversity of the ordinary strain *Potato virus Y* (PVY) and origin of recombinant PVY strains. Phytopathol 101: 1052–1060.

Kareyeija RF, Kreuze FF, Gibson RW, Valkonen JPT (2000) Two serotypes of *Sweet potato feathery mottle virus* in Uganda and their interaction with resistant sweet potato cultivars. Phytopathology 90: 1250–1255.

Karwitha M, Feng Z-K, Shen Y, Essendi W, Zhang W-N et al. (2016) Rapid detection of *Tomato chlorosis virus* from infected plant and whitefly by one-step reverse transcription loop-mediated isothermal amplification. J Phytopathol 164: 255–263.

Kashif M, Pietilä S, Artola K, Jones RAC, Tugume AK et al. (2012) Detection of viruses in sweet potato from Honduras and Gautemala augmented by deep sequencing of small RNAs. Plant Dis 96: 1430–1437.

Katsiani AT, Pappi P, Olmos A, Efthimiou KE, Maliogka VI et al. (2018) Development of a real-time RT-PCR for the universal detection of LChV1 and study of the seasonal fluctuation of the viral titer in sweet cherry cultivars. Plant Dis 102: 899–904.

Kaye AC, Moyer JW, Parks EJ, Carbone I, Cubeta MA (2011) Population genetic analysis of *Tomato spotted wilt virus* in peanut in North Carolina and Virginia. Phytopathology 101: 147–153.

Kehoe MA, Buirchell BJ, Coutts BA, Jones RAC (2014) Black pod syndrome of *Lupinus angustifolius* is caused by late infection with *Bean yellow mosaic virus*. Plant Dis 98: 739–745.

Kehoe MA, Jones RAC (2016) Improving *Potato virus Y* strains nomenclature: lessons from comparing isolates obtained over a 73-year period. Plant Pathol 65: 322–333.

Khan SA, Hoque MI, Sarker RH, Muehlbach HP (2003) Detection of important plant viruses in in vitro regenerated plants by double antibody sandwich method of ELISA. Plant Tissue Cult 13: 21–29.

Khatabi B, He B, Hajimorad MR (2012) Diagnostic potential of polyclonal antibodies against bacterially expressed recombinant coat protein of *Alfalfa mosaic virus*. Plant Dis 96: 1352–1357.

Khatabi B, Wen R-H, Hershman DE, Kennedy BS, Newman MA et al. (2013) Generation of polyclonal antibodies and serological analyses of nucleocapsid protein of soybean necrosis-associated tospovirus serotype. Eur J Plant Pathol 133: 783–790.

Khelifa M (2019) Detection and quantification of *Potato virus Y* genomes in single aphid species. Plant Dis 103: 2315–2321.

Kim J-H, Choi G-S, Kim J-S, Lee S-H, Choi J-K, Ryu K-H (2006) Development of single-tube multiplex immunocapture RT-PCR assay for simultaneous detection of two pepper tobamoviruses. Plant Pathol 22: 164–167.

Kim MK, Seo JK, Kwak HR, Kim JS, Kim HK et al. (2014) Molecular genetic analysis of *Cucumber mosaic virus* populations infecting pepper suggests unique patterns of evolution in Korea. Phytopathology 104: 993–1000.

King AMQ, Adama MJ, Carstens EB, Lefkowitz EJ (2011) *Virus Taxonomy – Classification and Nomenclature of Viruses. Ninth Report of the International Committee on Taxonomy of Viruses*. Elsevier/Academic Press, Amsterdam, pp. 1–1272.

Koganezawa H, Yang X, Zhu SF, Hashimoto J, Hadidi A (2003) Apple scar skin viroid in apple. In: Hadidi A, Flores R, Randles JW, Semancik JS (eds.), *Viroids*. CSIRO Publishing, Collingwood, Australia, pp. 137–141.

Kokane AD, Kokane SB, Warghane AJ, Gubayad MG, Sharma AK, Reddy MK, Ghosh DK (2021) A rapid and sensitive reverse transcription-loop-mediated isothermal amplification (RT-LAMP) assay for the detection of *Indian citrus ringspot virus*. Plant Dis 105: 1346–1355.

Kokkinos CD, Clark CA (2006) Real-time PCR assays for detection and quantification of sweet potato viruses. Plant Dis 90: 783–788.

Komínek P, Komínková-Bryxiová M, Jandurová OM, Holleinova V (2008) An RNA probe for the detection of *Grapevine virus A*. J Phytopathol 156: 762–764.

Kox LF, Boxman ILA, Jansen CCC, Roenhorst JW (2005) Reliability of nucleic acid amplification techniques: modified target RNA as exogenous internal standard for a real-time RT-PCR for *Potato spindle tuber viroid*. Eur J Plant Pathol 35: 117–124.

Krenz B, Thompson JR, Mc Lane HL, Fuchs M, Perry KL (2014) *Grapevine red blotch-associated virus* is widespread in the United States. Phytopathology 104: 1232–1240.

Kuan C-P, Chang W-S, Yang T-C (2018) Use of fluorescent microsphere-based assay for detection of three cucurbit infecting viruses. Plant Dis 102: 2324–2329.

Kuan C-P, Deng T-C, Huang H-C, Chi H-H, Lu Y-L (2014) Use of reverse transcription loop-mediated isothermal amplification for the detection of *Zucchini yellow mosaic virus*. J Phytopathol 162: 238–244.

Kulshreshtha S, Verma N, Hallan V, Raikhy G, Singh KM et al. (2005) Detection and identification of *Prunus necrotic ringspot virus* in *Pelargonium*. Austr Plant Pathol 34: 599–601.

Kumar R, Palicherl SR, Mandal B, Kadiri S (2018) PCR-based detection of betasatellite associated the begomoviruses using improved universal primers. Austr Plant Pathol 47: 115–118.

Kumar S, Rai R, Baranwal VK (2015) Development of an immunocapture-reverse transcription- polymerase chain reaction (IC-RT-PCR) using modified viral RNA release protocol for the detection of *Grapevine leafroll-associated virus 3* (GLRaV-3). Phytoparasitica 43: 311–316.

Kumari SG, Makkouk KM, Katul L, Vetten HJ (2001) Polyclonal antibodies to the bacterially expressed coat protein of *Faba bean necrotic yellows virus*. J Phytopathol 149: 543–550.

Kumssa TT, Rupp JS, Fellers MC, Fellers JP, Zhang G (2019) An isolate of Wheat streak mosaic virus from foxtail overcomes Wsm2 resistance in wheat. Plant Pathol 68: 783–789.

La Notte P, Minafra A, Saldarelli P (1997) A spot-PCR technique for phloem-limited grapevine viruses. J Virol Meth 66: 103–108.

Larsen RC, Kaiser WJ, Wyatt SD, Buxton-Druffel KL, Berger PH (2003) Characterization of a new potyvirus naturally infecting chickpea. Plant Dis 87: 1366–1371.

Lee S, Rho J-Y (2015) Development of a OCR diagnostic system for detecting *Andian Potato mottle virus* associated with quarantine in Korea. Amer J Potato Res 92: 546–550.

Leo G, Luison D, Grande SB, Alabanese G, Faggioli F (2015) Grapevine viruses' detection and sanitary selection in grapevine germplasm of Calabria (Southern Italy). J Phytopathol 163: 690–695.

Li R, Berendsen S, Ling KS (2016) A duplex real-time RT-PCR system with an internal control offers sensitive and reliable broad spectrum detection of *Squash mosaic virus* variants. Plant Dis 100: 625–629.

Li SF, Hataya T, Furuya K, Horita H, Sano T et al. (1997) Occurrence of chrysanthemum stunt disease in Kokkaido and detection of *Chrysanthemum stunt viroid* by electrophoresis and hybridization. Annu Rept Soc Plant Protect North Japan 48: 113–117.

Lima JAA, Nascimento AKQ, Radaelli P, Silva AKF, Siva FR (2014) A technique containing immunoprecipitation and RT-PCR for RNA plant virus detection. J Phytopathol 162: 426–433.

Lin L, Li R, Mock R, Kinard G (2011) Development of a polyprobe to detect six viroids of pome and stone fruit trees. J Virol Meth 171: 91–97.

Lin YH, Chen TC, Hsu HT, Liu FL, Chy FH et al. (2005) Serological comparison and molecular characterization for verification of *Calla lily chlorotic spot virus* as a new *Tospovirus* species belonging to *Watermelon silver mottle virus* serogroups. Phytopathology 95: 1482–1488.

Lin YJ, Rundell PA, Powell CA (2002) In situ immunoassay (ISIA) of field grapefruits inoculated with mild isolates of *Citrus tristeza virus* indicates mixed infection with severe isolates. Plant Dis 86: 458–461.

Lin YJ, Rundell PA, Xie L, Powell CA (2006) Pre-reaction of *Citrus tristeza virus* (CTV)- specific antibodies increases speed of direct tissue immunoassay for CTV. Plant Dis 90: 675–679.

Loconsole G, Castellano MA, Dell'Orco M, Bosica D, Savino V (2006) Serological detection of *Citrus psorosis virus* using a polyclonal antiserum to recombinant virus coat protein. J Plant Pathol 88: 171–177.

Loconsole G, Saponari M, Savino V (2010) Development of real-time PCR-based assays for simultaneous and improved detection of citrus viruses. Eur J Plant Pathol 128: 251–259.

Lolic B, Aflechtal M, Matic S, Myrt A, Di Serio F (2007) Detection by tissue-printing of pome fruit viroids and characterization of *Pear blister canker viroid* in Bosnia and Herzegovina. J Plant Pathol 89: 369–373.

Lorenzen JH, Piche LM, Gudmestad NC, Meacham T, Shiel P (2006) A multiplex PCR assay to characterize *Potato virus Y* isolates and identify strain mixtures. Plant Dis 90: 935–940.

Lotos L, Olmos A, Orfanidou C, Efthmiou K, Avgelis A et al. (2017) Insights into the etiology of *Polerovirus*-induced pepper yellow disease. Phytopathology 107: 1567–1576.

Lu W, Zhang Z, Xu P, Liu S, Wang H et al. (2012) Simultaneous detection of three viroid species infecting hops in China by multiplex RT-PCR. J Phytopathol 160: 308–310.

Luigi M, Faggioli F (2011) Development of quantitative real-time RT-PCR for the detection and quantification of *Peach latent mosaic viroid*. Eur J Plant Pathol 130: 109–116.

Ma BG, Niu J-X, Morley-Bunker M, Pan L-Z, Zheng H-P (2008) Detection of three pear viruses by multiplex RT-PCR assays with coamplification of an internal control. Austr Plant Pathol 37: 117–132.

Ma S, Zheng Q, Ye J, Feng W, Zhou G, Zhang T (2019) Identification of viruses infecting sweet potato in southern China by small RNA. J Gen Plant Pathol 85: 122–127.

Maheshwari Y, Selvaraj V, Hajeri S, Ramadugu C, Keremane M et al. (2017) On-site detection of *Citrus tristeza virus* (CTV) by lateral flow immunoassay using polyclonal antisera derived from virions produced by a recombinant CTV. Phytoparasitica 45: 333–340.

Mahmoud SYM, Abo-El Meaty SA, El-Borollosy AM, Abdel-Ghaffar MH (2007) Identification of onion yellow dwarf potyvirus as one of the major viruses infecting garlic in Egypt. Amer Eur J Agric Environ Sci 2: 746–755.

Maina S, Barbetti MJ, Martin DP, Edwards UR, Jones RAC (2018) New isolates of *Sweet potato feathery mottle virus* and *Sweet potato virus C*: biological and molecular properties and recombination analyses based on complete genomes. Plant Dis 102: 1899–1914.

Maina S, Coutts BA, Edwards OR, de Almeida L, Ximenes A et al. (2017) *Papaya ringspot virus* populations from East Timorese and northern Australian cucurbit crops: biological and molecular properties and absence of genetic connectivity. Plant Dis 101: 985–993.

Majumder S, Baranwal VK, Joshi S (2008) Simultaneous detection of *Onion yellow dwarf virus* and *Shallot latent virus* in infected leaves and cloves of garlic by duplex RT-PCR. J Plant Pathol 90: 371–374.

Makarova KS, Wolf YL, Iranzo J, Shamkov AS, Alkhnbashi OS et al. (2020) Evolutionary classification of CRISPR-Cas system: a burst of Class 2 and derived variants. Nat Rev Microbiol 18: 67–83.

Malfitano M, Barone M, Duran-Vila N, Alioto D (2005) Indexing of viroids in citrus orchard of Campania, Southern Italy. J Plant Pathol 87: 115–121.

Maliogka VI, Dovas CI, Katis NI (2008) Generic and species-specific detection of viruses belonging to an evolutionarily distinct lineages within the *Ampelovirus* genus. J Virol Meth 154: 41–47.

Maliogka VI, Dovas CI, Lotos L, Efthiniou K, Katis NI (2009) Complete genome analysis and immunodetection of a member of a novel virus species belonging to the genus *Ampelovirus*. Arch Virol 154: 209–218.

Mandic B, Al Rwahnih M, Myrta A, Gomez G, Pallas V (2007) Incidence and genetic diversity of *Peach latent mosaic viroid* and *Hop stunt viroid* in stone fruits in Serbia. Eur J Plant Pathol. https://doi.org/10.1007/s.10658-007-9205.x

Maoka T, Hataya T (2005) The complete nucleotide sequence and biotype variability of *Papaya leaf distortion mosaic virus*. Phytopathology 95: 128–135.

Maoka T, Sugiyama S, Maruta Y, Hataya T (2010) Application of cDNA macroarray for simultaneous detection of 12 potato viruses. Plant Dis 94: 1248–1254.

Marais A, Faure C, Couture C, Bergey B, Gentit P et al. (2014) Characterization by deep sequencing of divergent *Plum bark necrosis stem pitting-associated virus* (PBNSPaV) isolates and development of a broad-spectrum PBNSPaV detection assay. Phytopathology 104: 660–666.

Marais A, Faure C, Mustafaye E, Barone M, Alioto D et al. (2015) Characterization by deep sequencing of *Prunus virus T*, a novel *Tepovirus* infecting *Prunus* species. Phytopathology 105: 135–140.

Margaria P, Turina M, Pamano S (2009) Detection of flavescence dorée and bois noir phytoplasmas, *Grapevine leafroll-associated virus 1* and *3* and *Grapevine virus A* from the same crude extract by reverse transcription real-time TaqMan assays. Plant Pathol 58: 838–845.

Marqués M-C, Sánchez-Vicente J, Ruiz R, Montagul-Martinez R, Márquez-Costa R, Gomez G, Carbonell A, Darós J-A, Rodrigo G (2022) Diagnostics of infections produced by the plant viruses TMV, TEV, and PVX with CRISPR-Cas12 and CRISPR-Cas13. ACS Synthetic Biol 11: 2384–2393.

Marqués NT, Bailey AM, Niblett CL, Nolasco G (2006) Differentiation of *Citrus tristeza virus* (CTV) isolates by cleavase fragment length polymorphism (CFLP) analysis of the major coat protein gene. Phytopathol Mediterr 45: 99–109.

Martin RR, James D, Lévesque CA (2000) Impacts of molecular diagnostic technologies on plant disease management. Annu Rev Phytopathol 38: 207–239.

Martin S, Alioto D, Milne RG, Ganrsey SM, Garcia ML et al. (2004) Detection of *Citrus psorosis virus* by ELISA, molecular hybridization, RT-PCR and immunosorbent electron microscopy and its association with citrus psorosis disease. Eur J Plant Pathol 110: 747–757.

Martin S, Milne RG, Alioto D, Guerri J, Moreno P (2002) Detection of *Citrus psorosis virus* in field trees by direct tissue blot immunoassay in comparison with ELISA, symptomatology, biological indexing and cross-protection tests. Plant Pathol 51: 134–141.

Maruthi MN, Alan SN, Kader KA, Rekha AR, Cork A et al. (2005) Nucleotide sequencing, whitefly transmission and screening tomato for resistance against two newly described begomoviruses in Bangladesh. Phytopathology 95: 1472–1481.

Massart S, Brostaux Y, Barbarossa L, Cesar V, Cieslinska M et al. (2008) Interlaboratory evaluation of a duplex RT-PCR method using crude extracts for the simultaneous detection of *Prune dwarf virus* and *Prunus necrotic ringspot virus*. Eur J Plant Pathol 122: 539–547.

Matoušek J, Orctova L, Patzak J, Svoboda P, Ludvikova I (2003) Molecular sampling of *Hop stunt viroid* (HSVd) from grapevines on hop production areas in Czech Republic and hop production. Plant Soil Environ 49: 168–175.

Matsushita Y, Usugi T, Tsuda S (2010) Development of a multiplex RT-PCR detection and identification system for *Potato spindle tuber viroid* and *Tomato chlorotic dwarf viroid*. Eur J Plant Pathol: 165–170.

Mayo MA, Brunt AA (2007) Plant virus taxonomy. In: Khan JA, Dijkstra J (eds.), *Handbook of Plant Virology*. The Haworth Press, New York, pp. 11–20.

Mederos DC, Giolitti F, Torres C, Portal O (2019) Distribution and phylodynamics of *Papaya ringspot virus* in *Carica papaya* in Cuba. Plant Pathol 68: 239–250.

Medina V, Pinner MS, Bedford ID, Achon MA, Gemeno C (2006) Immunolocalization of *Tomato leafcurl Sardinia virus* in natural host plants and its vector or *Bemisia tabaci*. J Plant Pathol 88: 299–308.

Mehle N, Kogovsek P, Racki, Jakomin T, Gutierrez-Aguire I (2017) Filling the gaps in diagnosis of *Pepino mosaic virus* and *Potato spindle tuber viroid* in water and tomato, seeds and leaves. Plant Pathol 66: 1191–1201.

Mekuria G, Ramesh SA, Alberts C, Bertozzi T, Wirthensohn M (2003) Comparison of ELISA and RT-PCR for detection of *Prunus necrotic ringspot virus* and *Prune dwarf virus* in almond. J Virol Meth 114: 65–69.

Meng J, Gu Q-S, Lin S-M, Peng B, Liu L-F et al. (2007) Dot blot hybridization for detection of five cucurbit viruses by digoxigenin-labeled cDNA probes. Agric Sci China 6: 1450–1455.

Michelutti R, Myrta A, Pallás V (2005) A preliminary account on the sanitary status of stone fruits at the clonal Genebank in Harrow, Canada. Phytopathol Mediterr 44: 71–74.

Minafra A, Casati P, Elico V, Rowhani A, Saldarelli P, Savino V, Martelli GP (2000) Serological detection of *Grapevine rupestris stem pitting-associated virus* (GRSPaV) by a polyclonal antiserum to recombinant coat protein. VITIS 39: 115–118.

Minutillo SA, Spano R, Gallitelli D, Mascia T (2021) Simultaneous detection of 10 viruses in globe artichoke by a synthetic oligonucleotide-based DNA polyprobe. Eur J Plant Pathol 160: 991–997.

Mirik M, Jones DC, Price JA, Workneh F, Ansley RJ, Rush CM (2011) Satellite remote sensing of wheat infected by *Wheat streak mosaic virus*. Plant Dis 95: 4–12.

Monis J (2000) Development of monoclonal antibodies reactive to a new grapevine leaf roll-associated closterovirus. Plant Dis 84: 858–862.

Moury B, Cardin L, Onesto JP, Candresse T, Poupet A (2001) Survey of *Prunus necrotic ringspot virus* and its variability in rose and *Prunus* spp. Phytopathology 91: 84–91.

Moury B, Pallox A, Caranta C, Gognalons P, Souche S et al. (2005) Serological, molecular, and pathotype diversity of *Pepper veinal mottle virus* and *Chilli veinal mottle virus*. Phytopathology 95: 227–232.

Mumford RA, Walsh K, Boonham N (2000) A comparison of molecular methods for routine detection of viroids. Bull EPPO/OEPP 30: 431–435.

Munganyinka K, Ateka EM, Kihuram AN, Kanyange MC, Tairo F et al. (2018) Cassava brown streak disease in Rwanda, the associated viruses and disease phenotypes. Plant Pathol 67: 377–387.

Myrta A, Al-Rowhani M, Savino V (2005) Presence of a recombinant isolate of *Plum pox virus* in Apulia. J Plant Pathol 87: 127–130.

Nabi SU, Mir JI, Sharma OC, Singh DB, Zaffer S et al. (2018) Optimization of tissue and time for rapid serological, and molecular detection of *Apple stem pitting virus* and *Apple stem grooving virus* in apple. Phytoparasitca 46: 705–713.

Narayanasamy P (2001) *Plant Pathogen Detection and Disease Diagnosis*, second edition. Marcel Dekker Inc., New York, USA.

Narayanasamy P (2005) *Immunology in Plant Health and Its Impact on Food Safety*. The Haworth Press Inc., New York.

Narayanasamy P (2008) *Molecular Biology in Plant Pathogenesis and Disease Development*, volumes 1–3. Springer, Heidelberg, Germany.

Narayanasamy P (2011) *Microbial Plant Pathogens – Detection and Disease Diagnosis, Volume 3, Viral and Viroid Pathogens*. Springer Science + Business Media B.V., Heidelberg, Germany.

Narayanasamy P (2017) *Microbial Plant Pathogens Detection and Management in Seeds and Propagules*, volume 1. Wiley-Blackwell, Chichester, Oxford, United Kingdom.

Narayanasamy P, Duraiswamy S (2003) *Plant Viruses and Viral Diseases*. New Century Book House, Chennai, India.

Narváez G, Skander BS, Ayllon MA, Rubino L, Guerri J et al. (2000) A new procedure to differentiate *Citrus tristeza virus* in isolates by hybridization with digoxigenin-labeled cDNA probes. J Virol Meth 85: 83–92.

Nesi B, Trinchello D, Lazzereschi S, Grassotti A (2009) Production of *Lily symptomless virus*-free plants by shoot meristem tip culture and in vitro thermotherapy. HortScience 44: 217–219.

Nie X (2005) Reverse transcription loop-mediated isothermal amplification of DNA for the detection of *Potato virus Y*. Plant Dis 89: 605–610.

Nie X, Singh RP (2002) A new approach for simultaneous differentiation of biological and geographical strains of *Potato virus Y* by uniplex and multiplex RT-PCR. J Virol Meth 104: 41–54.

Nie X, Singh RP (2003) Specific differentiation of recombinant $PVY^{N:O}$ and PVY^{NTN} isolates by multiplex RT-PCR. J Virol Meth 113: 69–77.

Nisbet C, Monger WA, Ross S, Holmes RF, Nova Y et al. (2019) Biological and molecular characterization of *Potato yellow blotch virus*, a new species of the genus *Potyvirus*. Plant Pathol 68: 251–260.

Nolasco G, Santos C, Silva G, Ponseca F (2009) Development of an asymmetric PCR-ELISA typing method for *Citrus tristeza virus* based on the coat protein gene. J Virol Meth 155: 97–108.

Notomi T, Okayama H, Matsubuchi H, Yonekawa T, Watanabe K et al. (2000) Loop-mediated isothermal amplification of DNA. Nucleic Acids Res 28: e63.

Ohki T, Netsu O, Kojima H, Onuki M, Maoka T et al. (2014) Biological and genetic diversity of *Wheat yellow mosaic virus* (Genus *Bymovirus*). Phytopathology 104: 313–319.

Olah R, Deak T, Turesan M, Szenasi M, Borde A et al. (2019) Use of an introns containing grapevine gene as internal control for validation of cDNA synthesis in virus detection by RT-PCR. Eur J Plant Pathol 149: 765–770.

Oliver JE, Freer J, Andersen RL, Cox KD, Robinson TL, Fuchs M (2009) Genetic diversity of *Prunus necrotic ringspot virus* isolates within a cherry orchard in New York. Plant Dis 93: 599–606.

Olmos A, Bertolini E, Ruiz-Garcia AB, Martinez C, Peiro R, Vidal E (2016) Modeling the accuracy of three detection methods of *Grapevine leafroll-associated virus 3* during the dominant period using a Bayesian approach. Phytopathology 106: 510–518.

Orecchia M, Nölke G, Saldarelli P, Del'Orco M, Uhde-Holzem K et al. (2008) Generation and characterization of a recombinant antibody fragment that binds to the coat protein of *Grapevine leafroll-associated virus 3*. Arch Virol. https://doi.org/10.1007/s00705-008-0.100-3

Osman F, Leutenegger C, Golino D, Rowhani A (2007) Real-time RT-PCR (TaqMan®) assays for the detection of *Grapevine leafroll-associated virus 1*, *5*, and *9*. J Virol Meth 141: 22–29.

Pagán I, Cordaba-Selles MC, Martinez-Priego L, Fraile A, Malpica JM et al. (2006) Genetic structure of the population of *Pepino mosaic virus* infecting tomato crops in Spain. Phytopathology 96: 274–279.

Palacio-Bielsa A, Foissac X, Duran-Villa N (1999) Indexing citrus viroids by imprint hybridization. Eur J Plant Pathol 105: 897–903.

Parida M, Posada G, Inoue S, Hasebe F, Morita K (2004) Real-time reverse-transcription loop-mediated isothermal amplification for rapid detection of West Nile virus. J Clinic Microbiol 42: 252–263.

Parisi O, Lepoivre P, Jijakli MH (2011) Development of a quick quantitative real-time PCR for the in vitro detection and quantification of *Peach latent mosaic viroid*. Plant Dis 95: 137–142.

Park J, Kil E-J, Kim J, Shin Y-G, Heo N-Y, Lee S (2014) Rapid detection and identification of six *Tomato yellow leafcurl virus* isolates from different regions using polymerase chain reaction and restriction enzyme analysis. J Phytopathol 162: 209–217.

Park J-W, Kunta M, McCollum G, Gonzalez M, Vedasharan P et al. (2018) Development of a sensitive real-time PCR detection method for *Citrus tatter leaf virus*. J Plant Pathol 100: 67–73.

Paunovic S, Ruizic D, Vujovic T, Milenkovic S, Jerenovic D (2007) In vitro production of *Plum pox virus*-free plum by chemotherapy with ribavirin. Biotechnol Biotechnol Equip 21: 417–421.

Peiro A, Pallás V, Sanchez-Nivarro JA (2012) Simultaneous detection of eight viruses and two viroids affecting stone fruit trees by using a unique polyprobe. Eur J Plant Pathol 132: 469–475.

Peng J, Fan Z, Huang J (2012) Rapid detection of *Banana streak virus* by loop-mediated isothermal amplification assay in South China. J Phytopathol 160: 248–250.

Peng J-C, Yeh S-D, Huang L-H, LI J-T, Cheng Y-E et al. (2011) Emerging threat of thrips-borne *Melon yellow spot virus* on melon and watermelon in Taiwan. Eur J Plant Pathol 130: 205–214.

Picó B, Sifres A, Nuez F (2005) Quantitative detection of *Cucumber yellow vein virus* in susceptible and resistant plants using real-time PCR. J Virol Metho 128: 14–20.

Powell CA, Pelosi RR, Rundell PA, Stover E, Cohen M (1999) Cross-protection of grapefruit from decline inducing isolates of *Citrus tristeza virus*. Plant Dis 83: 989–991.

Predajna L, Sihelska N, Benedikova D, Soltys K, Candresse T et al. (2017) Molecular characterization of *Prune dwarf virus* cherry isolates from Slovakia shows their substantial variability and reveals recombinant events in PDV RNA3. Eur J Plant Pathol 147: 877–885.

Przewodowska A, Zacharzewska B, Choluj J, Treader K (2015) A one-step, real-time reverse transcription loop-mediated isothermal amplification assay to detect *Potato virus Y*. Amer J Potato Res 92: 303–311.

Puchades AV, Carpino C, Alfaro-Feranandez A, Font-San-Ambrosio MI, Davino S et al. (2017) Detection of *Southern tomato virus* by molecular hybridization. Ann Appl Bio 171: 172–178.

Quillec FL, Plantegenest M, Riault G, Gedryver CA (2000) Analyzing and modeling temporal disease progress of *Barley yellow dwarf virus* serotypes in barley fields. Phytopathology 90: 860–866.

Raigond B, Verma A, Kochhar T, Roach S, Sharma S et al. (2018) Development of simplified and rapid nucleic acid release protocol for PCR-based detection of potato viruses. Phytoparasitica 46: 255–263.

Ramos-Gonzalez PL, Chabi-Jesus C, Guerra-Peraza O, Tassi AD, Kijajima ES et al. (2017) *Citrus leprosis virus N*: a new dichorhavirus causing citrus leprosis disease. Phytopathology 107: 963–973.

Reynard J-S, Schneeberger PHH, Frey JF, Schaerer S (2015) Biological, serological and molecular characterization of a highly divergent strain of *Grapevine leafroll-associated virus 4* causing grapevine leaf roll disease. Phytopathology 105: 1262–1269.

Roberts JMK, Ireland KB, Tay WT, Paini D (2018) Honey bee-assisted surveillance for early plant virus detection. Ann Appl Biol 173: 283–293.

Romero-Durban J, Cambra M, Duran-Vila N (1995) A simple imprint-hybridization method for detection of viroids. J Virol Meth 55: 37–47.

Rotem N, Shtein O, Rosner A, Levy D, Rabinowitch HD (2016) Detection and differentiation of *Potato virus Y* strains by melting analysis of an oligonucleotide virus probe. Amer J Potato Res 93: 620–625.

Rott ME, Kesanakurti P, Berwarth C, Rast H, Boyes I et al. (2018) Discovery of negative-sense RNA viruses in trees infected with Apple rubbery wood disease by next-generation sequencing. Plant Dis 102: 1254–1263.

Rowhani A, Uyemoto JK, Golino DA (1997) A comparison between serological and biological assays in detecting *Grapevines leafroll-associated viruses*. Plant Dis 81: 779–801.

Rowhani A, Uyemoto JK, Golino DA, Martelli GP (2005) Pathogen testing and certification of *Vitis* and *Prunus* species. Annu Rev Phytopathol 43: 261–278.

Roy A, Stone AL, Otero-Colina G, Wei G, Brlansk RH et al. (2020) Reassortment of genome segments creates stable linkages among strains of *Orchid fleck virus* infecting citrus in Mexico. Phytopathology 110: 106–120.

Ruiz L, Simon A, Velasco L, Jansen D (2017) Biological characterization of *Tomato leafcurl New Delhi virus* from Spain. Plant Pathlogy 66: 376–382.

Ruiz-Ruiz S, Moreno P, Guerri J, Ambros S (2009) Discrimination between mild and severe *Citrus tristeza virus* isolates with a rapid and highly-specific real-time reverse transcription-polymerase chain reaction method using TaqMan LNA probes. Phytopathology 99: 307–315.

Rwahnih MA, Alabi OJ, Westrick NM, Golino D (2018) *Prunus geminivirus A*: a novel *Grablovirus* infecting *Prunus* spp. Plant Dis 102: 1246–1253.

Ryang BS, Kobaori T, Matsumoto T, Kosaka Y, Ohki ST (2004) *Cucumber mosaic virus* 2b protein compensates for restricted systemic spread of *Potato virus Y* in doubly infected tobacco. J Gen Virol 85: 3405–3414.

Sadeghi MS, Afsharifar A, Izadpanah K, Loconsole G, De Stradis A et al. (2016) Isolation and partial characterization of a novel *Cytorhabdovirus* from citrus trees. Plant Dis 100: 66–71.

Sánchez-Navarro JA, Cooper CN, Pallás V (2018) Polyvalent detection of members of the genus *Potyvirus* by molecular hybridization using a genus probe. Phytopathology 108: 1522–1529.

Sánchez-Navarro JA, Corachan L, Font I, Alfaro-Fernandez A, Pallás V (2019) Polyvalent detection of twelve viruses and four viroids affecting tomato by using a unique polyprobe. Eur J Plant Pathol 155: 361–368.

Saqib M, Wylie SJ, Jones MGK (2015) Serendipitous identification of a new *Ilfavirus*-like infecting tomato and its subsequent characterization. Plant Pathol 64: 519–527.

Sarkes A, Fu H, Feindel D, Harding M, Feng J (2020) Development and evaluation of a loop-mediated isothermal amplification (LAMP) assay for detection of *Tomato brown rugose fruit virus*. PLoS ONE 15: 0230403. https://doi.org/10.137/journal.pone.0230403

Seddas A, Haidar MM, Grief C, Jacquet C, Cloquemin G et al. (2006) Establishment of a relationship between *Grapevine leafroll closteroviruses 1* and *3* by use of monoclonal antibodies. Plant Pathol 49: 80–85.

Seifers DL, Harvey TL, Martin TJ, Haber S, She YM et al. (2005) Association of a virus with wheat displaying yellow head disease symptoms in the Great Plains. Plant Dis 89: 888–895.

Selvarajan R, Balasubramanian V, Gayathrie T (2016) Highly efficient immunodiagnosis of episomal *Banana streak virus* using polyclonal antibodies raised against recombinant virus-associated protein. J Phytopathol 164: 497–508.

Selvarajan R, Balasubramanian V, Sasireka T (2015) A simple, rapid, and solvent free nucleic acid extraction protocol for detection of *Banana bunchy top virus* by polymerase chain reaction and loop-mediated isothermal amplification. Eur J Plant Pathol 142: 389–396.

Sere Y, Onasanya A, Akator K, Afalobi A, Abo ME (2007) Serological differentiation indices and phylogenetic analysis of *Rice yellow mottle virus* isolates in Cot d'Ivoire. J Biol Sci 7: 1147–1154.

Sether DM, Melzer MJ, Busto J, Zee F, Hu JS (2005) Diversity and mealybug transmissibility of ampeloviruses in pineapple. Plant Dis 89: 450–456.

Shang H-W, Zhu L-J, Zhao W-C, Xu D, Wang Y-F (2013) Preparation and characterization of a monoclonal antibody to *Prunus necrotic ringspot virus*. J Phytopathol 161: 371–375.

Shi L, Hu W, Su Z, Lu X, Tong W (2003) Microarray technologies and applications. In: *Applied Mycology and Biotechnology, Volume 3, Fungal Genetics*. Elsevier, Amsterdam, The Netherlands, pp. 271–293.

Silva C, Tersco S, Nolasco G, Oliveira MM (2003) Cellular locations of *Prune dwarf virus* in almond sections by in situ reverse transcription-polymerase chain reaction. Phytopathology 93: 278–285.

Simón A, Ruiz L, Velasco L, Janssen D (2018) Absolute quantification of *Tomato leafcurl New Delhi virus* Spain strain, ToLCNDV-ES: virus accumulation in a host-specific manner. Plant Dis 102: 165–171.

Sipahioglu HM, Ocak MO (2006) Use of dried high-phenolic laden host leaves for virus and viroid preservation and detection by PCR methods. J Virol Meth 137: 120–124.

Skiada FG, Grigoriadour K, Maliogka VI, Katis NI, Eleftheeriou EP (2009) Elimination of *Grapevine leafroll-associated virus 1* and *Grapevine rupestris stem pitting-associated virus* from grapevine cv. Agiogitko and a micropropagation protocol for mass production of virus-free plantlets. J Plant Pathol 91: 177–184.

Smith TN, Wylie SJ, Coutts BA, Jones RA (2006) Localized distribution of *Iris yellow spot virus* within leeks and reliable large scale detection. Plant Dis 90: 729–733.

Sombat S, Reanwarakorn K, Ling K-S (2018) Developing a multiplex real-time RT-PCR for simultaneous detection of *Pepper chat fruit viroid* and *Columnea latent viroid*. Austr Plant Pathol 47: 615–621.

Soweha HE (2005) Detection of banana bunchy top nanavirus using polymerase chain reaction in different Egyptian cultivars. Internatl J Agric Bot 7: 698–700.

Spiegel S, Kovalenko EM, Varga A, James D (2004) Detection and partial molecular characterization of two *Plum pox virus* isolates from plum and wild apricot in southwest Kazakhstan. Plant Dis 88: 973–979.

Srinivas KP, Reddy PS, Hema M, Sreenivasulu P (2013) Multiplex RT-PCR in parallel detection of three RNA viruses associated with stripe and mosaic diseases in sorghum in India. Eur J Plant Pathol 137: 211–215.

Srinivasan R, Diffie S, Sundaraj S, Mullis SW, Riley D et al. (2011) Evaluation of *Lisianthus* as an indicator host for Iris yellow spot virus. Plant Dis 95: 1520–1527.

Stack-Lorenzen P, Guilton MC, Werner R, Muhlbach HP (1997) Detection and tissue distribution of *Potato spindle tuber viroid* in infected tomato plants by tissue imprint hybridization. Arch Virol 142: 1289–1296.

Stammler J, Oberneder A, Kellermann A, Hadersdorfer J (2018) Detecting potato viruses using direct reverse transcription quantitative PCR (DiRT-qPCR) without RNA purification: an alternative to DAS-ELISA. Eur J Plant Pathol 152: 237–248.

Strausbaugh CA, Eujayal IA, Wintermantel WM (2017) *Beet curly top virus* strains associated with sugar beet in Idaho, Oregon, and a Western U. S. Collection. Plant Dis 101: 1373–1382.

Su L, Bhagwat B, Bernardy M, Wiersma P, Cheng Z et al. (2016) Molecular analysis of three new *Cherry mottle leaf virus* isolates reveals intra-species genetic diversity and interspecies gene transfer. Phytoparasitica 44: 571–574.

Suchiro N, Natsuaki, Watanabe T, Okuda S (2004) An important determinant of the ability of *Turnip mosaic virus* to infect *Brassica* spp. and/or *Raphanus sativus* is in its P3 protein. J Gen Virol 85: 2087–2098.

Sui X, Zheng Y, Li R, Padmanabhan C, Tian T et al. (2017) Molecular and biological characterization of *Tomato mottle mosaic virus* and development of RT-PCR detection. Plant Dis 101: 704–711.

Tabler M, Tsagris M (2004) Viroids: petite RNA pathogens with distinguished talents. Trends Plant Sci 9: 339–348.

Taglienti A, Tiberini A, Manglli A, Rea R, Paoletti S et al. (2018) Molecular characterization of allexiviruses in a complex mixture of garlic viruses in Latium (central Italy). Eur J Plant Pathol 150: 797–780.

Tantiwanich Y, Chiemsombat P, Naidu RA, Adkins S (2018) Integrating local lesion assays with conventional RT-PCR for detection of interspecies tospovirus reassortments and mixed tospovirus infections. Plant Dis 102: 715–719.

Terlizzi F, Ratti C, Filippini G, Pisi A, Credi R (2010) Detection and molecular characterization of Italian *Grapevine rupestris stem pitting-associated virus* using primers targeting the polymerase domain and partial genome sequencing of novel variant. Ann Appl Biol 159: 478–490.

Teycheney P-Y, Acina I, Lockhart BEL, Candresse T (2007) Detection of *Banana mild mosaic virus* and *Banana virus X* by polyvalent degenerate oligonucleotide RT-PCR (PDO-RT-PCR). J Virol Meth 142: 41–49.

Thompson BD, Dahan J, Lee J, Martin RR, Karasev AV (2019) A novel genetic variant of *Grapevine leafroll-associated virus 3* (GLRaV-3) from Idaho grapevines. Plant Dis 103: 509–518.

Thompson JK, Fuchs M, McLane H, Celebi-Toprak F, Fischer KF et al. (2014) Profiling viral infections in grapevine using a randomly primed reverse transcription-polymerase chain reaction/macroarray multiplex platform. Phytopathology 104: 211–219.

Thompson JR, Wetzel S, Klerks MM, Vakovac D, Schoen CD et al. (2003) Multiplex RT-PCR detection of four aphid-borne strawberry viruses in *Fragaria* spp. in combination with a plant mRNA-specific internal control. J Virol Meth 111: 85–93.

Torrico AK, Celli MG, Cafrune EE, Krishbaum DS, Conci VC (2016) Genetic variability and recombination analysis of the coat protein of *Strawberry mild yellow edge virus*. Austr Plant Pathol 45: 401–409.

Toyoda K, Hikichi Y, Takeuchi S, Okumura A, Nasu Y et al. (2004) Efficient inactivation of *Pepper mild mottle virus* (PMMoV) in harvested seeds of green pepper (*Capsicum annuum* L.) assessed by a reverse transcription polymerase chain reaction (RT-PCR) based amplification. Sci Rep Fac Agric Okayama Univ 93: 29–32.

Tseng Y-W, Wu C-F, Lee C-J, Chen Y-K, Jan R-J (2021) Universal primers for rapid detection of six pospiviroids in Solanaceae plants using one-step reverse transcription PCR and reverse transcription-loop-mediated isothermal amplification. Plant Dis 105: 2867–2872.

Van Brunschot SL, Gambley CF, De Barro PJ, Grams R, Thomas JE et al. (2013) Panel of real-time PCRs for the multiplexed detection of two tomato-infecting begomoviruses and their cognate whitefly vectors. Plant Pathol 62: 1132–1146.

Van Brunschot SL, Verhoeven JTJ, Persley DM, Geering ADW, Drenth A et al. (2014) An outbreak of *Potato spindle tuber viroid* in tomato is linked to imported seeds. Eur J Plant Pathol 139: 1–7.

Varga A, James D (2006) Use of reverse transcription loop-mediated isothermal amplification for the detection of *Plum pox virus*. J Virol Meth 128: 184–190.

Vargas-Asencio J, Al-Rwahnih M, Rowhani A, Celebi-Toprak F, Thompson JR et al. (2016) Limited genetic variability among American isolates of *Grapevine virus E* from *Vitis* spp. Plant Dis 100: 159–163.

Varveri C, Olmos A, Pina JA, Marroquin C, Cambra M (2015) Biological and molecular characterization of a distinct *Citrus tristeza virus* originating form a lemon tree in Greece. Plant Pathol 64: 792–798.

Velázquez K, Piña JA, Navarro L, Moreno P, Guerri J (2012) Association of *Citrus psorosis B* symptoms with sequence variant of the *Citrus psorosis virus* RNA2. Plant Pathol 61: 448–456.

Vemulapati B, Druffel KL, Husebye E, Eigenbrode SD, Pappu HR (2014) Development and application of ELISA assays for the detection of two members of the family *Luteoviridae* infecting legumes: *pea enation mosaic virus* (genus *Enamovirus*) and *Bean leafroll virus* (genus *Luteovirus*). Ann Appl Biol 165: 130–136.

Venkataravanappa V, Prasanna HC, Reddy CNL, Reddy MK (2015) Evidence for two predominant viral lineages, recombination and subpopulation structure in begomoviruses associated with yellow vein mosaic disease of okra in India. Plant Pathol 64: 508–518.

Venkataravanappa V, Reddy CNL, Jalali S, Reddy MK (2013) Molecular characterization of a new species of begomovirus associated with yellow vein mosaic of bhendi (okra) in Bhubhaneswar, India. Eur J Plant Pathol 136: 811–822.

Verhoeven, JTJ, Westernberg M, van Ede EPM, Visser K, Roenhorst JW (2016) Identification and eradication of *Potato spindle tuber viroid* in dahlia in the Netherlands. Eur J Plant Pathol 146: 444–447.

Verma G, Raigon B, Pathania S, Koccjar T, Naga K (2020) Development and comparison of reverse transcription-loop mediated isothermal amplification assay (RT-LAMP), RT-PCR, and real-time PCR for detection of *Potato spindle tuber viroid* in potato. Eur J Plant Pathol 158: 951–964.

Vidal E, Yokomi RK, Moreno A, Bertolini E, Cambra E (2012) Calculation of diagnostic parameters of advanced serological and molecular tissue-print methods for detection of *Citrus tristeza virus:* a model for other plant pathogens. Phytopathology 102: 114–121.

Villamor DEV, Mekuria TA, Pillai SS, Eastwell K (2016) High-throughput sequencing identifies novel viruses in nectarine: insights to the etiology of stem-pitting disease. Phytopathology 106: 519–527.

Vinodkumar S, Nakkeeran S, Malathi VG, Karthikeyan G, Amala Balu P (2017) *Tobacco streak virus* an emerging threat to cotton cultivation in India. Phytoparasitica 45: 729–743.

Viswanathan R, Karuppaiah R, Balamuralikrishnan M (2010) Detection of three major RNA viruses infecting sugarcane by multiplex reverse transcription-polymerase chain reaction (multiplex RR-PCR). Austr Plant Pathol 39: 79–84.

Walia JJ, Willemsen A, Elci J, Caglayan K, Falk BW et al. (2014) Genetic variation and possible mechanisms driving the evolution of worldwide *Figure mosaic virus* isolates. Phytopathology 104: 108–114.

Walia Y, Kumar Y, Rana T, Bhardwaj P, Ram R et al. (2009) Molecular characterization and variability analysis of *Apple scar skin viroid* in India. J Gen Plant Pathol 75: 307–311.

Wang P, Guo P, Chen J-H, Dong Q-F, Ding X-F et al. (2012) Application of multiplex reverse transcription-ligase detection reaction-polymerase chain reaction (MRLP)-mediated universal DNA microarray for detecting potato viruses in field samples. Eur J Plant Pathol 132: 217–227.

Wang X, Zhou C, Thang K, Zhu Y, Li Z (2009) A rapid one-step multiplex RT-PCR assay for the simultaneous detection of five citrus viroids in China. Eur J Plant Pathol 124: 175–180.

Wang X, Zhou Y, Li Z, Tang K, Liu Y et al. (2010) Molecular, biological and phylogenetic analysis of *Hop stunt viroid* associated with citrus cachexia disease. J Phytopathol 158: 372–377.

Wang X-Y, Li W-F, Huang Y-K, Zhang R-Y, Shan H-L et al. (2017) Molecular detection and phylogenetic analysis of viruses causing mosaic symptoms in new sugarcane varieties in China. Eur J Plant Pathol 148: 931–940.

Wang Y, Wu B, Borth WB, Hamin I, Green JC et al. (2017) Molecular characterization and distribution of two strains of *Dasheen mosaic virus* on Taro in Hawaii. Plant Dis 101: 1980–1989.

Wang Y, Yang Z, Zhao J, Li R, Wang Q, Li J, Li Z, Zhou Y (2020) Development of a sensitive and reliable reverse-transcription-droplet digital polymerase chain reaction (RT-ddPCR) assay for the detection of *Citrus tristeza virus*. Eur J Plant Pathol 156: 1175–1180.

Wei T, Pearson MN, Blohm D, Nolte M, Armstrong K (2009) Development of shoot oligonucleotide microarray for the detection and identification of multiple potyviruses. J Virol Meth 162: 109–118.

Weidemann HL, Buchta U (1998) A simple and rapid method for detection of *Potato spindle tuber viroid* (PSTVd) by RT-PCR. Potato Res 41: 1–8.

Wu J, Yu C, Yang C, Zhou X (2009) Monoclonal antibodies against the recombinant nucleocapsid protein of *Tomato spotted wilt virus* and its application in virus detection. J Phytopathol 157: 344–349.

Wu Q, Luo Y, Lu R, Lan N, Lai EC et al. (2010) Virus discovery by deep sequencing and assembly of virus-derived small silencing RNAs. Proc Natl Acad Sci USA 107: 1606–1611.

Wu X-B, Alabi OJ, Damaj MB, Sun S-R, Mirkov TE et al. (2016) Prevalence and RT/RNase H, genealogy of *Sugarcane bacilliform virus* isolates from China. J Phytopathol 164: 595–607.

Xiao H, Li C, Al Rwahnih M, Dolija V, Meng B (2019) Metagenomic analysis of Riesling grapevine reveals a complex viromes including two new and divergent variants of *Grapevine leafroll-associated virus 3*. Plant Dis 103: 1275–1285.

Xing F, Robe BL, Zhang Z, Wang H, Li S (2018) Genomic analysis, sequence diversity and occurrence of *Apple necrotic mosaic virus*, a novel virus associated with mosaic disease of apple trees in China. Plant Dis 102: 1841–1847.

Xu D-L, Park J-W, Mirkov TE, Zhou G-H (2008) Viruses causing mosaic disease in sugarcane and their genetic diversity in southern China. Arch Virol. https://doi.org/10.1007/s00705-008-0072-3

Xu H, De Haan TL, De Boer SH (2004) Detection and confirmation of *Potato mop top virus* in potatoes produced in the United States and Canada. Plant Dis 88: 363–367.

Xu M, Chen J, Huang Y, Shen D, Sun P et al. (2020) Dynamic transcriptional profiles of *Arabidopsis thaliana* infected by *Tomato spotted wilt virus*. Phytopathology 110: 153–163.

Xu ZY, Hong N, Xing B, Wang GP (2006) Molecular characterization of a Chinese isolate of *Grapevines leafroll-associated virus 2* and production of an antisera to recombinant virus proteins. J Plant Pathol 88: 89–94.

Yanagisawa H, Shiki Y, Matsushita Y, Ooishi M, Takane N, Tsuda S (2017) Development of a comprehensive detection and identification on molecular based system for eight pospiviroids. Eur J Plant Pathol 149: 11–23.

Yepes LM, Cieniewicz E, Krenz B, McLane H, Thompson JR et al. (2018) Causative role of *Grapevine red blotch virus* in red blotch disease. Phytopathology 108: 902–909.

Yokomi RK, Selvaraj V, Maheshwari Y, Saponari M, Giampetruzzi A et al. (2017) Identification and characterization of *Citrus tristeza virus* isolates breaking resistance in trifoliate orange in California. Phytopathology 107: 901–908.

Yule S, Chiemsombat P, Srinivasan R (2019) Detection of *Tomato yellow leaf curl Thailand virus* transmitted by *Bemisia tabaci* Asia 1 in tomato and pepper. Phytoparasitica 47: 143–153.

Zagrai L, Zagrai I, Ferencz B, Gagborannu I, Kovacs K, Petricele I, Popescu O, Pamfil D, Capote N (2008) Serological and molecular typing of *Plum pox virus* isolates in the North of Romania. J Plant Pathol 90: S1.41–S1.46.

Zein HS, Miyatke K (2009) Development of rapid, specific, and sensitive detection of *Cucumber mosaic virus*. Afr J Biotechnol 8: 751–769.

Zeng R, Xu L-H, Gao S-G, Ni X-H, Chen C-L et al. (2016) One-step reverse transcription loop-mediated isothermal amplification assay for rapid detection of *Melon yellow spot virus*. Eur J Plant Pathol 145: 119–124.

Zhang J, Wang R, Song J, Luo Z, Yang J et al. (2013) One-step multiplex RT-PCR for simultaneous detection of four viruses in tobacco. J Phytopathol 161: 92–97.

Zhang W, Zhang Z, Fan G, Gao Y, Wen J et al. (2017) Development and application of a universal and amplified multiplex RT-PCR assay to detect five potato viruses. J Gen Plant Pathol 83: 33–45.

Zhang Y, Singh K, Kaur R, Qiu W (2011) Association of a novel DNA virus with grapevine vein-clearing and vine decline syndrome. Phytopathology 101: 1081–1090.

Zhang Z, Peng S, Jiang D, Pan S, Wang H et al. (2012) Development of a polyprobe for the simultaneous detection of four grapevine viroids in grapevine plants. Eur J Plant Pathol 132: 9–16.

Zhao L-M, Li G, Gao Y, Liu Y-L, Sun G-Z et al. (2014) Molecular detection and complex genome sequences of *Tomato chlorosis virus* isolates from infectious outbreaks in China. J Phytopathol 162: 634–637.

Zhao L-M, Liu Y, Wu Y, Hao X (2016) Rapid detection of watermelon viruses by reverse transcription loop-mediated isothermal amplification. J Phytopathol 164: 330–336.

Zheng L, Rodoni BC, Gibbs MJ, Gibbs AJ (2010) A novel pair of universal primers for the detection of potyviruses. Plant Pathol 59: 211–220.

Zheng Y, Zhang W, Lu X, Zhang G, Wang H, Zhang X, Feng J, Long H (2013) Specific and rapid detection of *Lily symptomless virus* and *Arabis mosaic virus* in lily by dual IC-RT-PCR. J Phytopathol 161: 823–827.

Zheng Y-X, Chen C-C, Chen Y-K, Jan F-J (2008) Identification and characterization of a potyvirus causing chlorotic spots on *Phalaenopsis* orchids. Eur J Plant Pathol 121: 87–95.

Zheng Y-X, Chen C-C, Yang C-J, Yeh S-D, Jan F-J (2008) Identification and characterization of a tospovirus causing chlorotic ringspot in *Phalaenopsis* orchids. Eur J Plant Pathol 120: 199–209.

Zhou T, Du L, Lan Y, Sun F, Fan Y, Zhou Y (2014) Development of SYBR Green I-based one-step real-time RT-PCR assay for quantifying *Southern rice black-streaked dwarf virus* in rice. J Phytopathol 162: 26–32.

Zhou Y, Chen HM, Cao MJ, Wang XF, Jin X et al. (2017) Occurrence, distribution and molecular characterization of *Citrus yellow vein clearing virus* in China. Plant Dis 101: 137–143.

Zhou Y, Zhang X, Wang D, Weng, Di H, Zhang L, Dong L, Zhang H, Zu H, Li X, Wang Z (2018) Differences in molecular characteristics of segment 8 in *Rice black-streaked dwarf virus* and *Southern rice black-streaked dwarf virus*. Plant Dis 102: 1115–1123.

Zhu YJ, Lim STS, Schenik S, Aranas A, Komor E (2010) RT-PCR and quantitative RT-PCR deduction of *Sugarcane yellow leaf virus* (SCYLV) in symptomatic and asymptomatic plants of Hawaiian sugarcane cultivars and correlation of SCYLV titre to yield. Plant Pathol 127: 263–273.

Additional References

Abraham AD, Varrelmann M, Vettan HF (2012) Three distinct nanoviruses, one of which represents a new species, infect faba bean in Ethiopia. Plant Dis 96: 1045–1053.

Abrokowah F, Dzahini-Obiatey H, Galyuon I, Osae-Awuku F, Muller E (2016) Geographical distribution of *Cacao swollen shoot virus* molecular variability in Ghana. Plant Dis 100: 2011–2017.

Adams LA, Abidrabo P, Miano DW, Alicai T, Kinyua ZM et al. (2013) High throughput real-time RT-PCR assays for

specific detection of cassava brown streak disease- causing viruses and their application to testing of planting material. Plant Pathol 62: 233–242.
Almasi MA, Almasi G (2018) Colorimetric immunocapture loop-mediated isothermal amplification assay for detection of *Impatiens necrotic spot virus* (INSV) by GeneFinder™ dye. Eur J Plant Pathol 150: 533–538.
Arabatova J, Lehtok K, Pehu E, Pehu T (1998) Localization of the P1 protein of potato associated with (1) cytoplasmic inclusion bodies and (2) cytoplasm of infected cells. J Gen Virol 79: 2319–2323.
Barzon I, Lavezzo F, Militeldo V, Topp O, Palú G (2011) Applications of next generation sequencing technologies to diagnostic virology. Internatl J Molec Sci 12: 7861–7884.
Belgrader P, Bennett W, Hadley D, Richards J, Stratton P et al. (1999) PCR detection of bacteria in seven minutes. Science 284: 449–450.
Cabauatan PG, Arboleda M, Azzam O (1998) Differentiation of *Rice tungro bacilliform virus* strains by restriction analyses and DNA hybridization. J Virol Meth 76: 121–126.
Candresse T, Cambra M, Dallot S, Lanneau M, Asensio M et al. (1998) Comparison of monoclonal antibodies and polymerase chain reaction assays for typing of isolates belonging to the D and M serotypes of plum pox potyvirus. Phytopathology 88: 198–204.
Chang Y-H, Huang C-H, Chang C-J, Jan F-J (2019) Identification and characterization of *Watermelon green mottle virus* as a new cucurbit-infecting tobamovirus. Ann Appl Biol 174: 31–39.
Chauhan RP, Hamon HF, Rajakaruna P, Webb MA, Payton M et al. (2015) Reliable detection for *Bean yellow mosaic virus, Canna yellow streak virus* and *Canna yellow mottle virus* in Canna varieties with red foliage. Plant Dis 99: 188–194.
Chingandu N, Dongo L, Gutierrez OA, Brown JK (2019) The previously unidentified, divergent *Badnavirus* species *Cacao red vein-banding virus* is associated with cacao swollen shoot disease in Nigeria. Plant Dis 103: 1302–1308.
Crosslin JM, Hamm PB, Hane DC, Jaeger J, Brown CR et al. (2006) The occurrence of PVY^O, PVY^N, and $PVY^{N:O}$ strains of *Potato virus Y* in certified potato seed lot trials in Washington and Oregon. Plant Dis 90: 1102–1105.
da Silva RM, de Souto ER, Pedroso JC, Arakava R, Almeida AMR et al. (2008) Detection and identification of TMV infecting tomato under protected cultivation in Parana State. Braz Arch Biol Technol 51: 903–909.
Diener TO (1971) *Potato spindle tuber viroid* 'virus' IV. A replicating, low-molecular weight RNA. Virology 45: 411–428.
Duran-Vila N, Semancik JS, Broadbent P (2000) Viroid diseases cachexia and exocortis. In: Timmer LW, Garnsey SM, Graham JH (eds.), *Compendium of Citrus Diseases*. The American Phytopathological Society, St Paul, MN, USA.
Faria JC, Maxwell DP (1999) Variability in geminivirus isolates associated with *Phaseolus* spp. in Brazil. Phytopathology 89: 262–268.
Fernandez RLAE, Filloux D, Martin DP, Rott P, Roumagnac P (2016) Prevalence of *Sugarcane yellow leaf virus* in sugarcane-producing regions in Kenya revealed by reverse transcription loop-mediated isothermal amplification method. Plant Dis 100: 260–268.
Figueira AR, Geraldino-Duarte PS, Nunez AMP, Lent JV, Galvino-Costa SBF et al. (2019) Characterization of *Soybean yellow shoot virus*, a new member of the family *Potyviridae* infecting soybean plants in Brazil. Plant Dis 103: 1172–1180.
Galvino-Costa SBF, Figueira AR, Rabelo-Filho FAC, Moraes FHR, Nikolaeva OV et al. (2012) Molecular and serological typing of *Potato virus Y* isolates from Brazil reveals a diverse set of recombinant strains. Plant Dis 96: 1451–1458.
Gil-Salas FM, Peters J, Boonham N, Cuadrado IM, Jansen DC (2012) Coinfection with *Cucumber vein yellowing virus* and *Cucurbit yellow stunting disorder virus* leading to synergism in cucumber. Plant Pathol 61: 468–478.
Glasa M, Malinowski T, Predajna L, Pupola N, Dekena D et al. (2011) Sequence variability, recombination analysis and specific detection of the W strain of *Plum pox virus*. Phytopathology 101: 980–985.
Green KJ, Chikh-Ali M, Hamasake RT, Melzer J, Karasev AV (2017) *Potato virus Y* isolates from *Physalis peruviana* are unable to systemically infect potato within the PVY^C strain group. Phytopathology 107: 1433–1439.
Harper K, Creamer R (1995) Hybridization detection of insect-transmitted plant viruses with digoxigenin-labeled probes. Plant Dis 79: 563–567.
Harrison BD, Robinson DJ (1999) Natural genomic and antigenic variation in whitefly-transmitted geminiviruses (Begomoviruses). Annu Rev Phytopathol 37: 369–398.
Hassani-Mehraban A, Cheewachaiwit S, Relevante C, Kormelink R, Peters D (2011) *Tomato necrotic ring virus* (TNSV), a recently described tospovirus species infecting tomato and pepper in Thailand. Eur J Plant Pathol 130: 449–456.
Herranz MC, Sánchez-Navarro JA, Aparicio F, Pallás V (2005) Simultaneous detection of six stone fruit viruses by non-isotopic molecular hybridization using unique riboprobe or polyprobe. J Virol Meth 124: 49–55.
Hilf ME, Karasev AV, Albiach-Marti MR, Dawson WO, Garnsey SM (1999) Two paths of sequence divergence in *Citrus tristeza virus* by real-time reverse transcription-polymerase chain reaction assays. Phytopathology 100: 319–327.
Hsu HT, Barzuna L, Hsu YH, Bliss P-W, Perry KL (2000) Identification and subgrouping of *Cucumber mosaic virus* with mouse monoclonal antibodies. Phytopathology 90: 615–620.
Hull WC, Huang C-H, Lee S-C, Wu C-J, Chang Y-C (2010) Detection of four calla potyviruses by multiplex RT-PCR using nad5 mRNA as an internal control. Eur J Plant Pathol 126: 43–52.
James AP, Geijskes RJ, Dale JL, Harding RM (2011) Development of a novel rolling-circle amplification technique to detect *Banana streak virus* that also discriminates between integrated and epidermal virus sequences. Plant Dis 95: 57–62.
Jones S, Baizan-Edge A, Macfarlane S, Torrance L (2017) Viral diagnosis in plants using next-generation sequencing: computational analysis in practice. Front Plant Sci 8: 1770. https://doi.org/10.3389/fpls.2017.01770
Just K, Leke WN, Sattar MN, Luik A, Krarnhedew A (2014) Detection of *Tomato yellow leafcurl virus* in imported tomato fruit on northern Europe. Plant Pathol 63: 1454–1460.
Köklii G, Rojas A, Kvarnheden A (2006) Molecular identification and the complete nucleotide sequence of a *Tomato yellow leafcurl virus* isolate from Turkey. J Plant Pathol 188: 61–66.
Konate G, Traore O, Coulibaly MM (1997) Characterization of *Rice yellow mottle virus* isolates in Sudan-Sahelian areas. Arch Virol 142: 1117–1124.
Krauthausen J-J, Leinhos GME, Muller J, Radtke PC, Jehle JA (2012) Identification and incidence of *Iris yellow spot virus* in *Allium* field crops in southwest Germany. Eur J Plant Pathol 134: 345–356.

Kreuze JF, Perez A, Unitveros M, Quispe D, Fuentes S et al. (2009) Complete viral genome sequence and discovery of novel viruses by deep sequencing of viruses. Virology 388: 1–7.

Lamprecht RL, Kasdorf GGF, Stiller M, Staples SM, Nel LH et al. (2010) *Soybean blochy mosaic virus*, a new *Cytorhabdovirus* found in South Africa. Plant Dis 94: 1348–1354.

Letham LJ, Jones RAC (1997) Occurrence of tomato spotted wilt tospovirus in native flora, weeds and horticultural crops. Austr J Agric Res 48: 359–369.

Li R, Salih S, Hurtt S (2004) Detection of geminiviruses in sweet potato by polymerase chain reaction. Plant Dis 88: 1347–1351.

Lin S-S, Hou RF, Yeh S-D (2000) Heteroduplex mobility and sequence analyses for assessment of variability of *Zucchini yellow mosaic virus*. Phytopathology 90: 228–235.

Ling KS, Wechter WP, Walcott RR, Keinath AP (2011) Development of a real-time RT-PCR assay for *Squash mosaic virus* useful for broad spectrum detection of various serotypes and its incorporation into a multiplex seed health assay. J Phytopathol 159: 649–656.

Luigi M, Faggioli F (2013) Development of a quantitative real-time RT-PCR (qRT-PCR) for detection of *Hop stunt viroid*. Eur J Plant Pathol 137: 231–235.

Luigi M, Faggioli F, Barba M, Giunchedi L (2020) The molecular characterization of ASVd isolates associated with dapple fruit and fruit rugosity in plum seedlings suggests as possible role of breeding in viroid dissemination. The 21st international congress virus and other graft-transmissible diseases of fruit crops. Julius-Kuhn Arch 427: 101–104.

Luuden S, Meng B, Avery J Jr, Qiu W (2010) Association of Grapevine fan leaf virus, Tomato ringspot virus and Grapevine rupestris stem pitting-associated virus with a grapevine vein-clearing complex in var. Chardonnay. Eur J Plat Pathol 126: 135.

Maina S, Barbetti MJ, Edwards OR, de Almeida L, Ximenes A et al. (2018) *Sweet potato feathery mottle virus* and *Sweet potato virus C* from east Timorese and Australian sweet potato: biological and molecular properties and biosecurity implications. Plant Dis 102: 589–599.

Maina S, Coutts BA, Edwards OR, de Almeida L, Kehoe MA et al. (2017) *Zucchini yellow mosaic virus* populations from east Timorese and northern Australian cucurbit crops: molecular properties, genetic connectivity and biosecurity implications. Plant Dis 101: 1236–1245.

Maliogka VI, Dovas CI, Efthimiou K, Katis NI (2004) Detection and differentiation of *Comoviridae* species using a semi-nested RT-PCR and a phylogenetic analysis based on the polymerase protein. J Phytopathol 152: 404–409.

Maliogka VI, Dovas CI, Lessemann DE, Winter S, Katis NI (2006) Molecular identification, reverse transcription-polymerase chain reaction detection, host reactions and specific cytopathology of *Artichoke yellow ringspot virus* infecting onion crops. Phytopathology 96: 622–629.

Marquez-Martin B, Aragón-Caballero L, Fiallo-Olivé E, Navas-Castillo I, Moriones E (2011) *Tomato leaf deformation virus*, a novel begomovirus associated with a severe disease of tomato in Peru. Eur J Plant Pathol 129: 1–7.

Matoušek J, Orctova L, Steger G, Riesner D (2004) Biolistic inoculation of plants with viroid nucleic acids. J Virol Meth 122: 153–164.

Minutillo SA, Marais A, Mascia T, Faure C, Svanella-Dumas L, Theil S, Payet A, Perennec S, Schoen L, Gallitelli D, Candresse T (2015) Complete nucleotide sequence of *Artichoke latent virus* shows it to be a member of the genus *Macluravirus* in the family *Potyviridae*. Phytopathology 105: 1155–1160.

Misbeth S, Al-Musa A, Anfoka G, Salem N (2007) Molecular characterization *in Grapevine virus A* isolates from Jordan. Phytopathol Mediterr 46: 195–200.

Moon JS, Allen RG, Domier LL, Hewings AD (2000) Molecular and biological characterization of a trackable Ilinois isolate of *Barley yellow dwarf virus*-PAV. Plant Dis 84: 483–486.

Moura MF, Mituti T, Marubayashi JM, Gioria R, Kobori RF et al. (2011) A classification of *Pepper yellow mosaic virus* isolates into pathotypes. Eur J Plant Pathol 131: 549.

Mulenga RM, Leg JP, Ndunguru J, Miano DW, Mutitu EW et al. (2016) Survey molecular detection and characterization of geminiviruses associated with cassava mosaic disease in Zambia. Plant Dis 100: 1379–1387.

Myrta A, Terlizzi di B, Boscia D, Caglayan K, Gabriel I et al. (1998) Detection and serotyping of Mediterranean *Plum pox virus* isolates by means of specific monoclonal antibodies. Acta Virol 42: 251–253.

Narayanasamy P (2003) *Plant Viruses and Viral Diseases*. New Century Book House, Chennai, India.

Nava AR, Patte CP, Hiebert E, Poston JE (2004) Detection and variability of begomoviruses in tomato from Andean states of Venezuela. Plant Dis 90: 61–66.

Olmos A, Cambra M, Dasi MA, Candresse T, Esteban O et al. (1997) Simultaneous detection and typing of plum pox potyvirus (PPV) and PCR-ELISA. J Virol Meth 68: 127–137.

Owens RA, Diener TO (1981) Sensitive and rapid diagnosis of *Potato spindle tuber viroid* disease by nucleic acid hybridization. Science 213: 670–672.

Papyiannis LC, Kyriakou A, Kaparia-Isaia T (2007) Typing of *Plum pox virus* (PPV) strains in Cyprus. Austr Plant Dis Notes 2: 29–30.

Perry KL, McLane H, Hyder MZ, Dangl GS, Thompson JR, Fuchs MF (2016) *Grapevine red blotch-associated virus* is present in free-living *Vitis* spp. proximal to cultivated grapevines. Phytopathology 106: 663–670.

Prefarres F, Hoareau M, Chiroleu F, Reynaud B, Dintinger J et al. (2011) A novel synthetic quantification standadrd including virus and internal report targets: application for detection and quantification of emerging begomomviruses in tomato. J Virol 8: 389. www. Virologyj.xom/xontent/8/1/389

Rosner A, Maslenin L (1999) Differentiation of PVY^{NTN} by unique single-restriction cleavage of PCR products. Potato Res 42: 215–221.

Saldarelli P, Giampetruzzi A, Morelli M, Malossini U, Pirolo C et al. (2015) Genetic variability of *Grapevine Pinot gris virus* and its association with grapevine leaf mottling and deformation. Phytopathology 105: 555–563.

Sasaya T, Nou Y, Koganezawa H (198) Biological and serological comparisons of *Bean yellow mosaic virus* (BYMV) isolates in Japan. Ann Phytopathol Soc Jpn 64: 24–33.

Shamloul AM, Hadidi A, Zhu SF, Singh RP, Sagredo BC (1997) Sensitive detection of *Potato spindle tuber viroid* using RT-PCR and identification of a viroid variant naturally infecting pepino plants. Canad J Plant Pathol 19: 89–91.

Song Z, Liu K, Li Z, Zhou C (2016) RT-PCR-RFLP or genetic diversity analysis of the *Citrus tatter leaf virus* strain of *Apple stem grooving virus*. Eur J Plant Pathol 144: 687–692.

Srinivas KP, Subba Reddy Ch V, Ramesh P, Lava Kumar P, Sreenivasulu P (2010) Identification of a virus naturally infecting sorghum in India as *Sugarcane streak mosaic virus*. Eur J Plant Pathol 127: 13–19.

Stobbe AH, Schneider WL, Hoyt PR, Melcher U (2014) Screening metagenomic data for viruses using the E-probe diagnostic nucleic acid assay. Phytopathology 104: 1125–1229.

Suehiro N, Matsuda K, Okuda S, Natsuaki T (2005) A simplified method for obtaining plant viral RNA for RT-PCR. J Virol Meth 125: 67–73.

Takeshita M, Uchiba T, Takanami Y (1999) Sensitive detection of cucumber mosaic cucumovirus RNA with digoxigenin-labeled cRNA probe. J Fac Agric Kyushy Univ 43: 349–354.

Thompson KG, Dietzgen RG, Thomas JE, Teakle DS (1996) Detection of *Pineapple bacilliform virus* using polymerase chain reaction. Ann Appl Biol 129: 57–64.

Thottappilly G, Dahal G, Lockhart BEL (1998) Studies on a Nigerian isolate of banana streak badnavirus. I. Purification and enzyme-linked immunosorbent assay. Ann Appl Biol 132: 253–261.

Trucco V, de Breuil S, Bejerman N, Lenardon S, Giolitti F (2016) *Bean leafroll virus* (BLRV) in Argentina molecular characterization and detection in alfalfa fields. Eur J Plant Pathol 146: 207–212.

Vaira AM, Veechiati M, Masenga V, Accotto GP (1996) A polyclonal antiserum against a recombinant viral protein combines specificity with versatility. J Virol Meth 48: 209–212.

Varveri C (2000) Potato Y potyvirus detection by immunological and molecular techniques in plants and aphids. Phytoparasitica 28: 141–148.

Walkey DGA, Lyons NF, Taylor JD (1992) An evaluation of a virobacterial agglutination test for detection of plant viruses. Plant Pathol 41: 462–471.

Wang D, Li G, Du SS (2017) Occurrence of viruses infecting melon in Xingjiang of China and molecular characterization of *Watermelon mosaic virus* isolates. Eur J Plant Pathol 147: 919–931.

Webster CG, Wylie SJ, Jones MGK (2004) Diagnosis of plant viral pathogens. Curr Sci 86: 1604–1607.

Wu W-Q, Guo X-G, Zhang H-M, Yang J, Lv M-F et al. (2013) Simultaneous detection and survey fo three rice varieties in China. Plant Dis 97: 1181–1186.

Xie WS, Hu JS (1995) Molecular cloning, sequence analysis and detection of *Banana bunchy top virus* in Hawaii. Phytopathology 85: 339–347.

Yambao MLM, Cabauatan PG, Azzam O (1999) Differentiation of *Rice tungro spherical virus* variants by RT-PCR and RFLP. Internatl Rice Res Notes 23 (2): 22–24.

Zanini AA, Cuellar WJ, Celli MG, Luque AV, Medina RD et al. (2018) Distinct strains of the re-emergent *Cassava common mosaic virus* (genus: *Potexvirus*) infecting cassava in Argentina. Plant Pathol 67: 1814–1820.

Zhang L, Jelkmann W (2017) Construction of full-length infectious cDNA clones of *Apple chlorotic leaf spot virus* and their agroinoculation to woody plants by a novel method of vacuum infiltration. Plant Dis 101: 2110–2115.

Zheng Y-X, Chen CC, Kreuze JF, Gibson RW, Valkonen JPT (2000) Two serotypes of *Sweet potato feathery mottle virus* in Uganda and their interaction with resistant sweet potato cultivars. Phytopathology 90: 1250–1255.

7 Biology of Phyllosphere Viral and Viroid Pathogens

7.1 INTRODUCTION

Phyllosphere viruses and viroids cause local and systemic infections that are expressed as different symptoms such as necrosis, mosaic, leafcurl, enations, ringspots, stunting, and progressive decline in growth and yield of affected plants. Some of the viruses spread both vertically and horizontally through seeds and propagules to subsequent generations and long or short distances. Viruses depend largely on their biological vectors for their transmission from infected plants to healthy plants, and they determine the type of relationships with their vectors which remain inoculative for different periods of time, depending on a nonpersistent, persistent, or propagative type of relationship. Some of the propagative viruses are transmitted through eggs of the female vectors to many subsequent generations. Viruses may have a wide or restricted host range of alternative plant species, which help in their survival in the absence of crop plants. Viruses contain either RNA or DNA, as their genomes are enclosed in a protein coat and are devoid of any enzymatic activity. They are dependent entirely on the synthetic machinery of the host plant for replication to reach concentrations sufficient enough to induce different kinds of symptoms characteristic of the virus or latent infections. Replication of viruses commences with release of viral nucleic acid from the protein coat, synthesis of progeny virions, and cell-to-cell and systemic movement to other tissues of infected plants. Virus-encoded silencing suppressors counteract RNA-based silencing host defense mechanisms. Viroids, lacking a protein component, utilize synthetic machinery of the host plants for their replication and become systemic in susceptible plants which exhibit various symptoms. Viroids exist in the form of different strains virulent on various crop plants causing different magnitudes of disease severity.

Phyllosphere plant viruses and viroids infecting aerial plan parts cause localized or systemic infections, inducing symptoms characteristic of the associated viruses. When viruses or viroids infect susceptible cultivars, injuries caused to different tissues involved in photosynthesis, absorption and transport of water and nutrients, and translocation of photosynthates to storage organs result in reduction in quantity and quality of the produce. Phytopathometry is the measurement of disease and constitutes the basis for reliable assessment of crop losses, since the magnitude of loss is directly related to disease severity. Quantitative measurement of disease incidence/severity refers to the number of diseased sampling units, expressed as the percentage of the total number of sampling units examined, as in the case of most virus and viroid diseases (Narayanasamy 2002). Symptom severity induced by different strains of viruses has been assessed in some crops. *Citrus tristeza virus* (CTV) exists in the form of different strains varying in their virulence and consequently in the severity of disease symptoms induced. CTV strains inducing stem pitting symptoms, cause ultimate death of the infected plants. In contrast, mild strains may or may not induce any visible symptoms (Nikolaeva et al. 1998). The real-time polymerase chain reaction (RT-PCR) format was applied to differentiate severe strains of CTV from mild strains, paving the way for development of methods of eradicating citrus trees infected by severe strains, which might otherwise serve as sources of infection for further spread of the disease induced by severe strains in citrus orchards (Ruiz-Ruiz et al. 2009).

7.2 IMPACT OF VIRUS INFECTION ON PLANT DEVELOPMENT

7.2.1 Biological Characteristics of Viruses

7.2.1.1 Types of Symptoms

Phyllosphere viruses cause external and internal symptoms in infected plants. Formation of mosaic, leafcurl, ringspots, partial or total sterility of infected plants, and malformation of leaves or floral parts may be observed. *Cucumber mosaic virus* isolates causing banana mosaic disease (BMD) were classified into four pathotypes, based on symptoms induced on banana, *Nicotiana glutinosa*, and cowpea (*Vigna unguiculata*), following inoculation of test plants using *Aphis gossypii*. Three symptom types, A, B, and C, were distinguished using *N. glutinosa* as a differential host plant (Chou et al. 2009). Gentian (*Gentiana* spp.) plants were infected severely by *Impatiens necrotic spot virus* (INSV), resulting in systemic symptoms restricted to a few upper leaves. Asymptomatic shoot cuttings from infected stock plants grown asexually in greenhouse developed systemic symptoms. When gentian plants were infected with INSV, the virus was preferentially transported from infected leaves to the root and winter buds (Nekoduka and Sano 2018). *Citrus leprosis virus* (CiLV-C) infecting citrus plants in Mexico is transmitted by the tenuipalpid mite *Brevipalpus phoenics*. Symptom expression in citrus plants required several weeks or months when inoculated by the vector. Common bean (*Phaseolus vulgaris*) developed local necrotic lesions within 5 days. Black bean IAC Una was useful as assay host by developing local lesions in

DOI: 10.1201/9781003456827-7

2 days at 28–30° C (Garita et al. 2013). Strains of *Papaya ringspot virus* (PRSV) occurring in Cuba were molecularly characterized. Diversity of PRSV strains increased during 1985–1990, and viral dispersion occurred from Mexico and other unknown locations (Mederos et al. 2019).

Rice black-streaked dwarf virus (SRBSDV) spread to maize crops after the harvest of rice crops, causing characteristic symptoms on both crops (Huang et al. 2011). *Wheat streak mosaic virus* (WSMV) and *Triticum mosaic virus* (TriMV) transmitted by wheat curl mite *Aceria tosichella* occurred together. Incidence of TriMV and frequency of coinfection with WSMV and High Plains virus (HPV) was assessed. WSMV was most prevalent, whereas HPV was detected in a few counties. TriMV-positive samples were coinfected with WSMV, whereas WSMV and HPV were mainly detected in single infections (Byamukama et al. 2013). Strains of *Potato virus Y* (PVY) differ in their virulence and rate of spread in various countries around the world. A shift to a higher proportion of $PVY^{N\text{-}Wi}$ in 2013, compared to 2012, was observed. PVY also affected tuber quality. A clear association between PVY plant infections and suberized canoe-shaped cracks in progeny tubers was recorded. Nonsymptomatic plants also showed such cracks in fewer tubers (Benedict et al. 2015). Strain-specific PVY spread might be due to variation with potato cultivar and selective PVY^{O} resistance to Gold Rush, mostly expressed at the plant-to-plant transmission level (MacKenzie et al. 2018). Isolates of *Potato virus S* (PVS) were obtained from three cultivated potato species (*Solanum* spp.) from Andean countries and Scotland. Systemic infection in *Chenopodium quinoa* was induced by nine isolates, whereas three isolates were restricted to inoculated leaves, indicating that they belonged to biologically defined strains PVS^{A} and PVS^{O}. Nine wild species were inoculated with PVS; eight species reacted with symptomless systemic infection, while *Solanum megistacrolobum* developed systemic hypersensitive resistance (SHR) with one PVS^{O} and two PVS^{A} isolates. Andean potato cultivars developed mostly asymptomatic primary infection, but predominantly symptomatic secondary infection. PVS^{A} and PVS^{O} isolates induced similar foliage symptoms (Santillan et al. 2018). The impact of time of infection by *Bean pod mottle virus* (BPMV) on soybean yield, yield components, and grain quality components were assessed. The results indicated possible risk of BPMV infection resulting in reduction in both quantity and quality of soybean yield (Byamukama et al. 2015). The effects of infection of faba bean by *Pea necrotic yellow dwarf virus* and *Pea enation mosaic virus* were assessed by immunoassays and low altitude true-color imaging data in Germany (Saucke et al. 2019).

Viruses may induce varying intensities of stunting, leafcurl, yellowing, mosaic mottling, necrosis, and shoe-string symptoms in tomato crops. *Tomato bud necrosis virus* and *Tomato leafcurl virus* were predominantly detected using the immunoassay and PCR-based technique (Nagendran et al. 2018). The effects of rate and timing of infection of sugarcane by *Sugarcane yellow leaf virus* (SCYLV) on yield were investigated in field trials. No significant effect of SYLCV on yield was noted in plant crop, regardless of crop cultivar and soil type. However, yield reduction in ratoon crops showed variations from nonsignificance to 27% infection, depending on cultivar and soil type (Boukari et al. 2019). *Cassava common mosaic virus* (CsCMV) (cassava common mosaic disease, CCMD) was widespread, and it was detected using DAS-ELISA and RT-PCR assays (Zanini et al. 2018). The presence of *Cassava brown streak virus* (CBSV) and *Ugandan Cassava brown streak virus* (UCBSV) in various cassava cultivars was detected and quantified by RT-PCR and qRT-PCR assays. CBSV was present in all samples with symptoms. Tolerant cassava genotypes were infected only by CBSV and accumulated higher titer in roots than in aerial organs. Susceptible genotypes were coinfected with CBSV and UCBSV with varying titers of these viruses (Ogwok et al. 2015). The differential distribution of whitefly-transmitted *Squash yellowing virus* (SqVYV) and *Cucurbit leaf crumple virus* (CuLCrV) and aphid-transmitted *Papaya ringspot virus* type W (PRSV-W) which severely impacted watermelon production was determined by tissue-blot nucleic acid hybridization assays *Cucurbit yellow stunting disorder virus* (CYSDV) was also detected in different host plant tissues. SqVYV and PRSV-W were distributed differently compared with CuLCrV in watermelon plants (Turechek et al. 2010). Cucurbit viruses infect cucurbit and several weed species and they were detected using dot-blot binding immunoassay (DIBA), employing antisera against cucurbit-infecting viruses. The highest incidence of *Papaya ringspot virus*-watermelon strain (PRSV-W) followed by *Watermelon mosaic virus 2* (WMV-2) and *Zucchini yellow mosaic virus (ZYMV)* was detected. Mixed virus infections were common involving two or three viruses in various combinations. PRSV-W infection in some of the samples of weed species was also observed (Ali et al. 2012). *Tobacco mosaic virus* (TMV) infecting tomato is transmissible through tomato seeds. TMV strain occurring in Sapopena, Brazil, could not infect *Petunia hybrida*, a known host for TMV and *Tomato mosaic virus* (ToMV). The TMV-Sapopena strain could be differentiated, based on the differences in the type of symptoms induced on tomato cultivars and by its inability to infect *P. hybrida* (da Silva et al. 2008).

Plant viruses induce external macroscopic symptoms. In addition, characteristic internal microscopic symptoms are also induced by them in infected plants. Cytological effects, such as starch accumulation, inhibition of plastid development, chloroplastic destruction, and phloem degeneration, are some of the changes caused by viruses in different crops. A virus-specific effect is the formation of intracellular inclusions in some host plants following virus infection. These inclusions have characteristics useful for identification of viruses. Inclusions may be formed in either cytoplasm or nucleus of infected cells, and they may be either amorphous or crystalline in nature. Among the characteristics considered for classification of viruses, characteristics of inclusion bodies and their intracellular location are also included. Members of 20 virus groups induce inclusions which have diagnostic value. These inclusions are useful for characterization of virus-specific, noncapsid proteins and possible sites of virus synthesis. *Cucumber mosaic virus* (CMV), *Cauliflower mosaic virus* (CaMV), *Peanut chlorotic streak virus*, *Tobacco mosaic virus*, *Potato virus X* and *Potato virus Y* produce characteristic inclusions in infected plants. The presence of viruses in aggregates has been visualized using transmission electron microscopic techniques (Narayanasamy 2001).

7.2.1.2 Coinfection of Plants by Viruses

Multiple infections or coinfection of a plant species by two or more viruses are observed frequently resulting in synergism or antagonism under natural conditions viruses. Mechanisms and viral products involved in viral synergistic interactions remain unclear in most cases. *Wheat streak mosaic virus* (WSMV) and *Maize chlorotic mottle virus* (MCMV) were present in corn plants together and concentrations of both viruses increased indicating a synergistic relationship between them. *Tobacco etch virus* genome encodes a broad range of pathogenicity enhancer that transactivates replication of heterologous viruses, such as *Cucumber mosaic virus* (CMV) and *Tobacco mosaic virus* (TMV). The viral RNA silencing suppressor protein P1/HC-Pro appeared to be involved in the synergistic effect possibly by decreasing the targeting of other virus by RNA silencing machinery (Vance 1999) or complementation of a movement deficiency by one of the viruses (Ryang et al. 2004). *Tomato spotted wilt virus* (TSWV), transmitted by thrips, and *Tomato chlorosis virus* (ToCV), transmitted by whiteflies, occurred together as mixed infections in tomato crops. The effects of synergistic interaction between TSWV and ToCV were investigated using tomato cv. Anastasia, which was susceptible to ToCV but resistant to TSWV under normal conditions, and also in tomato cv. Moneymaker, which was susceptible to both viruses. A dramatic increase in ToCV was observed in leaf tissues of cv. Moneymaker coinfected with TSWV. By using in situ RNA hybridization, confinement of ToCV to phloem cells of plants either singly or doubly infected by the viruses was indicated. No difference in distribution of TSWV in plant tissues could be detected between singly and doubly infected plants. All cv. Anastasia plants inoculated with ToCV exhibited typical symptoms of virus infection, whereas only 20% plants inoculated with TSWV showed TSWV symptoms, when ToCV preceded TSWV inoculation. Resistance to TSWV was compromised in cv. Anastasia plants. Similar breakdown of resistance to TSWV was not observed in plants inoculated with ToCV and TSWV simultaneously. However, breakdown of TSWV resistance was seen in other commercial cultivars. All three resistant cultivars possessed TSWV resistance gene *Sw5*. DAS-ELISA tests confirmed virus infections of test plants. Preinoculation of tomato plants with ToCV, enhanced susceptibility to TSWV, whereas coinoculation did not affect resistance of tomato plants (Garcia-Cano et al. 2006). Tomato crops were severely affected by *Tomato severe rugose virus* (ToSRV) and *Tomato chlorosis virus* (ToCV) in Brazil. Both ToSRV and ToCV are transmitted by *Bemisia tabaci* in a persistent, circulative, and semipersistent manner. The kinetics of systemic invasion by ToSRV and ToCV in tomato plants in single and coinfections were evaluated. Both viruses moved systemically in tomato plants as early as one day after inoculation. The results indicated that field-infected tomato plants might function as sources of inoculum soon after infection and before onset of symptoms (Favara et al. 2019).

Interactions between *Potato virus X* (PVX) and *Potato virus* S (PVS) in potato plants were investigated. Mixed infections with PVX and PVS increased the titer of PVS and enhanced expression of foliar symptoms in primarily and secondarily infected plants of cv. Royal Blue (Nyalugwe et al. 2012). The effects of coinfection on titer and accumulation of genetically different isolates of *Potato virus Y* from single and coinfections in potato, tobacco, as well as efficiency of transmission of isolates of PVY by *Myzus persicae* were assessed. PVY isolates representing three strain groups, PVY^{O} (pathotype and serotype O), PVY^{NW} (pathotype N and serotype O), and PVY^{NTN} (pathotype and serotype N) were included. DAS-ELISA, multiplex RT-PCR format, aphid transmission tests, and bioassays were employed to detect these PVY strains in different samples. Concentrations of some isolates of PVY were significantly reduced during coinfection both in potato and tobacco plants. The effects of coinfection might differ depending on virus isolates and host plant. Rates of virus transmission attained the peak when single aphids were used for transmission of PVY^{NTN} isolates and the lowest was with PVY^{O} isolates. Individual aphids (*M. persicae*) could simultaneously transmit two PVY isolates (Syller and Grupa 2014). *Potato virus Y* (PVY), due to continuous mutation and genome recombination, forms novel strains with differing virulence on various crop hosts. Responses of tomato to different strains of PVY, particularly emerging strains such as PVY^{NTN} and $PVY^{N:O}$ and in combination with *Potato virus X* were determined. Plants infected with PVY^{O}, $PVY^{N:O}$, PVYN or PVY^{NTN} strains developed symptoms of mosaic, leaf deformation, whereas PVX infection resulted in mild local lesion and varying intensity of mosaic and leaf deformation symptoms. Combined infections of PVX + PVY induced more severe symptoms. Depending on PVY strains, level of synergism might vary, with PVY^{NTN} exerting the greatest effect. Combined infection by PVX and PVY enhanced PVX titer and reduced PVY concentration. Comparison of transcriptional analyses of 46 key genes involved in signal molecule synthesis and signal transduction revealed diverse gene expression profiles on both single and combined infections. Generally, coinfections resulted in greater number of genes being differentially regulated than single infections. Changes in expression of genes involved in oxidative stress and salicylic acid (SA) pathways were distinctive and clearly recognizable in coinfections of PVX and PVY^{NTN}, leading to further enhancement of reactive oxygen species (ROS) and SA-producing capabilities and a reduction of ROS scavenging ability over single infections (Liang et al. 2016).

Squash leaf curl virus (SLCV)–infected melon plants exhibited mild symptoms that faded with lapses of time after inoculation. *Watermelon chlorotic stunt virus* (WmCSV)–infected plants produced symptoms which were intensified with time. Dual inoculation with SLCV and WmCSV led to more severe symptoms than WmCSV alone. Dual infection did not affect WmCSV DNA level, but SLCV DNA level was increased by several-fold due to the presence of WmCSV (Sufrin-Ringwald and Lapidot 2011). The interactive effects of another combination of *Cucurbit yellow stunting disorder virus* (CYSDV), and *Cucurbit vein yellowing virus* (CVYV) infecting zucchini squash and transmitted by *Bemisia tabaci* were investigated. Infection of zucchini plants by CYSDV did not result in formation of any recognizable symptom with virus titer decreasing during 15–45 days postinoculation (dpi) in vitro inoculation conditions. After this period, CYDSV could not be detected. On the other hand, CVYV induced conspicuous symptoms restricted to vein clearing in some apical leaves. *Zucchini yellow mosaic virus* in a single inoculation accumulated similarly, but plants showed severe symptoms of stunting, leaf deformation, and mosaic yellowing. In mixed infections with CYSDV and CVYV, visible symptoms were induced by ZYMV. Regarding virus titers, no consistent change in CVYV accumulation occurred in the presence of ZYMV. In contrast, dramatic increase in CYSDV titer was observed. Synergistic effect on the titer of crinivirus was not recorded in coinfection with ZYMV (Gill-Salas et al. 2011).

Pepper plants (*Capsicum annuum*) with *Tsw* resistance genes exhibited unusually severe symptoms of *Tomato spotted wilt virus* (TSWV) infection. Biological clones of TSWV isolated from individual local lesions formed on *Nicotiana glutinosa*, produced two different types of symptoms after inoculation onto *Tsw* resistant pepper plants: (i) typical symptoms induced by resistance-breaking (RB) isolates, characterized by chlorosis and stunting, and (ii) severe symptoms as recorded in the pepper plants under natural conditions, after simultaneous inoculation of RB and non-resistance-breaking (NRB) isolates. Synergistic interaction between RB and NRB isolates resulted in enhancement of the ability of NRB isolate to infect resistant pepper plants (Aramburu et al. 2015). Two genotypes of *Iris yellow spot virus* (IYSV) infecting onion and lisianthus (*Eustoma grandiflorum*) plants, ISYVNL and IYSLVBR occurred together in Japan. These genotypes were isolated from single necrotic lesions in infected plants. BR genotype was more widespread than NL genotype. The presence of these genotypes depended on the invasion of viruliferous *Thrips tabaci* that might carry either one or both genotypes (Fuji et al. 2015). Interactive relationship between vitiviruses, *Grapevine virus A* and *B* (GVA and GVB), *Grapevine leafroll-associated viruses 1, 2,* or *3* (GLRaV-1, -2 or -3) was investigated. GVB was preferentially associated with GLRaV-2 and GVA was associated with preferentially with GLRaV-1 or GLRaV-3 (Rowhani et al. 2018).

Interactive effects of *Triticum mosaic virus* (TriMV) and *Wheat streak mosaic virus* (WSMV) were determined. Plants of winter wheat Millennium susceptible to WSMV and Mace resistant to WSMV were mechanically inoculated with TriMV, WSMV, and TriMV + WSMV. Mace was also resistant to TriMV. Double infection of winter wheat by TriMV and WSMV aggravated disease severity and loss of biomass was significant in susceptible cultivars (Byamukama et al. 2012). The efficiency of TriMV transmission by three genetically distinct wheat curl mite (WCM) populations, Nebraska (NE), Montana (MT), and South Dakota (SD), was evaluated. TriMV potentially exerted adverse effects on WCM survival (McMechan et al. 2014). The impact of coinfection of corn (maize) by two strains of *Maize chlorotic dwarf virus*-severe and mild (MCDV-S and MCDV-M) and coinfection of MCDV and *Maize dwarf mosaic virus* (MDMV) in sweet corn hybrid was investigated under greenhouse conditions. Virus disease symptom severity was enhanced, and plants did not recover over time due to coinfection by MDMV-OH and MCDV-S (Morales et al. 2014). Vector-borne viruses may modify plant-vector interactions in favor of viral survival and dispersal due to the ability of these vectors to promote several changes in plant phenotype. Ecological interactions between *Barley yellow dwarf virus* (BYDV) and its vector *Rhopalosiphum padi* were investigated. Olfactory response and performance of *R. padi* to wheat singly and doubly infected by fungal pathogen *Gibberella zeae* and BYDV were assessed. Non-viruliferous aphids preferred odors of BYDV-infected plants during coinfection. Population of aphids increased on BYDV-infected wheat plants. Coinfected wheat plants exhibited synergistic effect, making them more attractive to aphids which increased in population significantly. The results indicated that coinfection might be helpful for virus spread (Santos et al. 2016).

7.2.1.3 Host Range of Viruses

Plant viruses may have a wide or restricted host range, which may function as alternative hosts and influence the spread of the viruses. *Alfalfa mosaic virus* (AMV) has a wide host range including 150 species in 22 families. Likewise, *Potato virus Y* (PVY) can infect more than 30 plant species (Narayanasamy 2001). Other viruses with wide host ranges include *Tomato yellow leafcurl virus* (Papayiannis et al. 2011), *Tomato chlorosis virus* (Kil et al. 2015), and *Tomato spotted wilt virus* (Macharia et al. 2016). Infection of wild relatives of rice by *Rice yellow mottle virus* (Allarangaye et al. 2006) was observed. The alternative host plant species may also function as overwintering hosts for some viruses and their biological vectors. *Stellaria aquatica* could be an overwintering host for *Tomato yellow leafcurl virus* (TYLCV) in the absence of tomato in greenhouses (Kil et al. 2015). Host range of *Tomato chlorosis virus* (ToCV)

Brazilian isolate (25 plant species) and the preference of *B. tabaci* for oviposition on the host plant species identified as susceptible to the virus were revealed by the investigation of Fariña et al. (2019). Root samples from 61 weed species included in 24 families were tested by RT-PCR assay for the presence of ToCV. The presence of ToCV in 17 weed species including *Stellaria media, Chenopodium album, Ipomoea hederacea, Phytolacca Americana, Solanum americanum* and *S. nigrum* was detected by amplification of virus-specific sequences. Tomato was the best source of ToCV inoculum, whereas sweet pepper was the least susceptible to ToCV and less preferred by whiteflies among the solanaceous host plant species tested (Mituti et al. 2018). The capacity of the virus to accumulate, the relationship between the virus titer in the source plants and transmission by whiteflies were determined in the selected set of weeds and crops. Cucurbit hosts allowed *Cucurbit yellow stunting disorder virus* (CYSDV) to retain the highest titers and were efficient sources for virus acquisition by whiteflies. A positive relationship was observed between virus titer in source plants and transmission rate in the cucumber plants. Noncucurbit hosts had significantly low virus titers, and they differed in their ability to serve as sources of inoculum (Wintermantel et al. 2016).

Weeds present in and adjacent to cucurbit crops might be sources of inoculum of *Cucumber green mottle virus* (CGMMV) facilitating persistence of the disease in a location. Of the 11 weed species tested, six ELISA-positive for CGMMV remained asymptomatic. RT-PCR assays and bioassays on melon, cucumber, and *Nicotiana benthamiana* confirmed the infection of the previous weed species by CGMMV. The results revealed the critical role of weed species in providing inoculum to watermelon in addition to being bridge hosts for CGMMV (Shargil et al. 2017). Papaya and watermelon were cultivated together frequently in close proximity or overlapping cycles providing chances of watermelon functioning as sources of inoculum of *Papaya meleira virus* (PMeV-Mx) for papaya in Mexico. Watermelon seedlings (*Citrullus lanatus*) inoculated with latex from papaya were infected by PMeV-Mx, as confirmed by RT-PCR and qRT-PCR formats. The results showed that watermelon could play an important role in the spread of PMeV-Mx in papaya orchards in Mexico (García-Cámara et al. 2018).

Poleroviruses infecting sugar beet, spinach, radish, and brassica were identified as *Beet western yellows virus* (BWYV)-JP, *Beet leaf yellowing virus* (BLYV), *Brassica yellows virus* (BrYV) and BrYV-R (radish strain) based on host range and molecular analysis. Aphid vector *Myzus persicae* was employed to inoculate 100 different plants. BLYV shared a similar host range to *Beet mild yellowing virus* (BMYV). Host range of BrYV was similar to that of *Turnip yellows virus* (TuYV), BLYV, and BWYV and they were more closely related to each other and to *Cucurbit aphid-borne yellows virus* (AABYV) than to BMYV (Yoshida and Tamada 2019). Tospoviruses *Groundnut ringspot virus* (GRSV) and *Tomato chlorotic spot virus* (TCSV) infected solanaceous crops and weeds in Florida. These viruses occurred sympatrically with *Tomato spotted wilt virus* (TSWV) in tomato and pepper. Impatiens and lettuce were identified as nonsolanaceous hosts for GRSV and TCSV under greenhouse conditions (Webster et al. 2015). *Grapevine vein clearing virus* (GVCV) induced vein-clearing and vine-decline disease. *Ampelopsis cordata*, wild relatives of grapevine, exhibited infection by GVCV. The grape aphid *Aphis illinoisensis* was tested for its ability to transmit GVCV. *A. illinoisensis* could transit GVCV under greenhouse conditions. The results indicated *A. cordata* could be a host for GVCV and *A. illinoisensis* might be the vector involved in the dispersal of GVCV and vine-decline disease spread (Petersen et al. 2019).

7.2.1.4 Modes of Transmission of Viruses

Plant viruses have to be introduced into the cells of susceptible plants by different agencies viz., biological vectors or propagation methods or humans. Transmissibility of viruses through pollen, seed, and propagules occurs under natural conditions. The most common mode of transmission of phyllosphere viruses is through vectors, insects, and mites. Such natural transmission of plant viruses is discussed in detail in an earlier publication (Narayanasamy 2017, 2020).

7.2.1.4.1 Abiotic Transmission of Viruses

Manual or commercial engraving tools and laboratory-fabricated needle arrays have been used for vascular puncture inoculation (VPI) to place the plant viruses in the desired location in plant tissues. Transfer of *Maize rayado fino virus* (MRFV) from sap of an infected plant using VPI resulted in an average infection rate in maize of 70%. Further, maize, oat, and triticale were also infected by inoculating *Oat blue dwarf virus* (OBDV) following adoption of this method (Weiland and Edwards 2011). Transmission of viruses via seeds and propagules was studied intensively, because of its impact on the seed industry and epidemiological significance. The nature of transmission of *Tobacco mosaic virus* (TMV) in flue-cured tobacco seeds was investigated using RT-qPCR assay. Using the cross-pollination method, it was possible to detect maternal, but not paternal, sources of TMV resulting in seed transmission. TMV was transmitted through embryos in tobacco seeds (Ellis et al. 2020). Seed transmission of *Tomato yellow leafcurl virus* (TYLCV) reported earlier in Korea was reevaluated. The seedborne nature of TYLCV was confirmed with detection of viral genomic DNA in seeds and reproductive floral organs of tomato and *Nicotiana benthamiana* plants infected by Mediterranean isolates of TYLCV. However, virus contents of seeds were reduced following surface disinfection of seeds, suggesting most of the virus might be located in externally on seed surface. Seed transmission of TYLCV was not a general property, although it might be possible under certain conditions (Pérez-Padilla et al. 2020).

Transmission of plant viruses may occur through leaf contact as in *Tobacco mosaic virus* (TMV) and *Potato virus* X (PVX), which are highly stable and infectious under greenhouse conditions. Potato viruses are known to be transmitted through cutting knives and other handling equipment. *Zucchini yellow mosaic virus* (ZYMV) was transmitted from infected to healthy zucchini (*Cucurbita pepo*) plants under greenhouse conditions by direct contact between leaves, crushed or trampled and to a lesser extent on ZYMV-contaminated blades. Hygienic practices to restrict spread of ZYMV have to be applied in greenhouse-grown crops (Coutts et al. 2013). Potato virus Y^O strain (PVY^O) could be transmitted from infected potato plants to healthy plants, when leaves rubbed against each other under greenhouse conditions. PVY^O was transmitted also when cut surfaces of infected tubers were rubbed onto potato plant leaves and to a limited extent, when blades contaminated with infective sap were used to cut healthy potato tubers. However, tuber-to-tuber transmission did not occur when blades were used to cut healthy tubers, after cutting infected tubers (Coutts and Jones 2015). The role of tuber-cutting and plant wounding in transmission of PVY strains, PVY^O, $PVY^{N:O}$, and PVY^{NTN} was assessed. Seed tuber cutting with the same knife did not result in PVY transmission to healthy tubers (Fageria et al. 2015). Transmission of *Citrus vein clearing virus* (CYVCV) via contaminated knife blade inoculation from infected sour orange and rough lemon plants to healthy plants in China was observed (Zhang et al. 2019).

7.2.1.4.2 Biotic Transmission of Viruses

7.2.1.4.2.1 Seed Transmission Biotic transmission of plant viruses requires involvement of biotic agent(s) as carrier(s) of the virus(es) until they reach specific susceptible host plant cells/tissues. Seeds and propagules carry several viruses, and they form the primary source of inoculum even in locations/countries far away from the production fields. Hence, use of virus-free seeds is emphasized as a disease management tactic as in the case of *Barley stripe mosaic virus, Lettuce mosaic virus, Bean common mosaic virus, Cowpea mosaic virus,* and *Soybean mosaic virus*. Seedborne viruses have been detected in seeds by applying various techniques. Phyllosphere viruses transmitted through seeds belong to the genera *Alfamovirus, Ampelovirus, Comovirus Cucumovirus, Enamovirus, Mandarivirus, Ophiovirus, Potyvirus, Trichovirus*, and viruses belonging to unassigned genera – *Babuvirus, Fujivirus, Hordeivirus, Pecluvirus, Potexvirus, Sobemovirus*, and *Tobamovirus*. The grow-out test, ELISA test, and PCR-based assays are performed to determine the extent of seed infection by viruses. Perennial fruit crops like grapevines are likely to be infected by several viruses because of long exposure, resulting in disease complexes. In annual crops that are asexually propagated like potato, multiplex RT-PCR system could be applied for simultaneous detection of potato viruses, *Potato virus Y, Potato virus X, Potato virus A, Potato virus S*, and *Potato leafroll virus*. The multiplex RT-PCR (mRT-PCR) and other diagnostic assays like IC-RT-PCR procedures are applied to enhance the sensitivity and reliability of detection of viruses. Mixed infections of grapevines by viruses and phytoplasmas causing grapevine yellows disease (GYs) could be resolved by applying the real-time RT-PCR format (Narayanasamy 2017).

Pollen from infected plants may carry the viruses as they germinate on floral organs. Transmission through infected pollen of *Common bean mosaic virus* (Nelson and Down 1933; Medina and Grogan 1961), *Alfalfa mosaic virus* (Frosheiser 1964), *Barley stripe mosaic virus* (Carroll 1980), and *Apple latent spherical virus* (Nakamura et al. 2011) was demonstrated earlier. *Asparagus virus 2* (AV-2) sequence analysis indicated presence of mixed infection by two distinct AV-2-infected tobacco plants showed that AV-2 was localized in the meristem and associated with pollen grains (Kawamura et al. 2014). *Asparagus virus 2* (AV-2) was detected in 75% of the seedlings when both female and male parents were infected. Infection of parents by AV-2 resulted in reduction in successful pollination. A high percentage of AV-2-free female plants contracted virus infection, following pollination with infected pollen. Vertical transmission of *Asparagus virus-2* via seed and horizontal transmission via pollen were observed (Jaspars et al. 2015). Transmission of *Cucumber green mottle mosaic virus* (CGMMV) via pollen to healthy plants and onto subsequent generation of seedlings was investigated. The rate of CGMMV transmission to fruits in plants fertilized with pollen from infected plants was less than in mechanically inoculated plants. CGMMV might be transmitted both horizontally via cucumber pollen and vertically to subsequent generation in infected seeds (Liu et al. 2014). Pollen transmission of members of *Potyviridae* family occurs commonly. Transmission of *Zucchini yellow mosaic virus* (ZYMV) via seed occurred in *Cucurbita pepo* subsp. *texana*, as determined by RT-PCR assay. ZYMV infection of *C. pepo* seeds resulted in virtually symptomless infection which could not be detected by immunoassay (Simmons et al. 2011). Seed transmission of *Zucchini yellow mosaic virus* (ZYMV) in different cucurbits, including wild *Cucurbita* species was observed. Seed infection by ZYMV via pollen occurred. Seed transmission of ZYMV in virus-resistant transgenic plants was also observed (Harth et al. 2017). *Pelargonium zonate spot virus* (PZSV) was detected in both seed and pollen grains of infected tomato plants. Seed-to-seedling transmission of PZSV to mother plants occurred, but they remained asymptomatic. The results indicated that seed mother plants pollinated with PZSV-infected pollen did not express visible symptoms (Lapidot et al. 2010). Cytopathological effects of colonization of onion tissues by *Artichoke yellow ringspot virus* (AYRSV) were assessed by electron microscopy and decoration techniques. Large cytoplasmic inclusions were formed in mesophyll cells of infected plants. Cytoplasmic inclusions contained voluminous accumulations of entangled membranous elements, more or less developed and specifically arranged particle aggregates and conspicuous globules possibly of lipid nature. A high proportion of infected cells contained AYRSV particles (Maliogka et al. 2006).

Infected pollen may play a key role in both seed and plant-to-plant transmission of the members of *Ilarvirus* genus. The route of infection and the pattern of distribution of *Prunus necrotic ringspot virus* (PNRSV) in apricot pollen were investigated. PNRSV was detected inside as well as on the outside of infected apricot pollen grains by dot blot hybridization technique. Flower buds with anthers contained pollen mother cells (PMCs) and PNRSV was detected in all flower bud tissues, including sepals, petals, and anthers. Different degrees of infection of anthers were revealed by in situ hybridization assay. Viral RNA was localized only in the vascular tissue in some, whereas other anthers showed complex invasion of PNRSV with PMCs also being infected. PNRSV was not detected in the nucleus of any of the pollen grains. PNRSV was detected in the cytoplasm of vegetative cell and the virus was sometimes localized around and inside generative cell. In the germinated infected pollen grains, viral RNA was evenly distributed inside pollen grains, as well as in the pollen tube with a higher concentration at the tip or growth zone of germinating pollen tube. PNRSV was able to move from vascular system to infect anther connective tissue, using high number of cytoplasmic connection bridges characteristic of this tissue. PNRSV invaded the tapetum cells and then pollen mother cells (PMCs). Pollen tubes on germination, could reach the apricot embryo sacs. PNRSV could be detected in the ovules and nucellus in the megaspore stage. The virus was localized in later developmental stages in the funiculus, as well as in the corresponding ovule, and the distribution of the virus was uneven in the ovary at this stage. Later PNRSV invaded all parts of the ovule including inner and outer integuments surrounding the nucleus, which enclosed the embryo sac. PNRSV was detected in the embryos in immature seeds at globular stage, using dot blot hybridization technique. PNRSV RNA was detected in embryos in the globular developmental stage. Viral infection of seed integuments, nucellus, and testa in immature seeds could be detected using in situ hybridization assay. Infection of testa mainly affected areas in and around its vascular tissue. The presence of viral RNA was detected in the periphery of fruits, whereas crosses between infected pistils and infected pollen resulted in total invasion of entire fruits, indicating the vertical transmission of PNRSV from gametes to apricot seedlings (Amari et al. 2007, 2009). The route of infection of cucumber by *Cucumber green mottle virus* (CGMMV) was investigated by using scanning electron microscopy (SEM). The presence of CGMMV in pollen from infected plants was detected by SEM observations. Transmission of CGMMV to seeds and subsequently to the seedlings growing from infected seeds was revealed by RT-PCR assay (Liu et al. 2014).

Mechanisms of virus transmission through seed may vary depending on the host-virus combination. Mechanism of seed transmission of *Pepper mild mottle virus* (PMMoV) in pepper (*Capsicum annuum*) was investigated. The virus immunolocalized in pepper seed was visualized using fluorescence microscopy techniques. Two distinct pathways of virus distribution of PMMoV in the epidermis and parenchyma was seen, but not in the endosperm or embryo, whereas in other seeds the virus was restricted to the surface of the epidermis and parenchyma (Genda et al. 2005). In a later investigation, the location of PMMoV within seeds, as they developed in inoculated seedlings of pepper was followed by detecting the virus coat protein using immunofluorescence microscopy procedure. The presence of PMMoV was inferred by the fluorescence around immature seeds and placenta in the ovaries on the fourth branching node at 20 days postanthesis (*dpa*). The area with fluorescence progressively expanded from the placenta into the integument and the parenchyma and finally reached the tip of immature seeds by 34 *dpa*. The embryo and endosperm beyond the endothelium did not fluoresce (Genda et al. 2011). Vegetative propagation is the principal mode of *Sweet potato leafcurl virus* (SPLCV) dispersal through planting materials. Sweet potato seeds were planted in whitefly-free greenhouse and production of disease symptoms was observed in some plants. The presence of SPLCV in all leaves and floral tissues showing symptoms was detected by PCR assay. SPLCV was detected in dissected endosperm and embryos. SPLCV-specific ss- and ds- DNAs were detected in seedlings growing from SPLCV-infected seeds and the virus replicated in developing seedlings (Kim et al. 2015).

Transmission of *Turnip yellow mosaic virus* (TYMV) via seeds of three commercial Chinese cabbage (*Brassica pekinensis*) seed batches was assessed. Seedlings growing from the seeds were assayed at cotyledon stage and adult plants using DAS-ELISA and RT-PCR assay (Alfaro-Fernández et al. 2016). *Tomato yellow leafcurl virus* (TYLCV) was seed transmissible in tomato seeds. Four susceptible sweet pepper cultivars were agro-inoculated with TYLCV infectious clone. Super Yellow was highly susceptible, whereas other cultivars were less susceptible. Replication of TYLCV in infected seeds was confirmed by strand-specific amplification using virus-sense and complementary sense-specific primer sets. Asymptomatic sweet pepper could function as a source of inoculum of TYLCV to tomatoes and other susceptible crops (Kil et al. 2018). The role of *Soybean mosaic virus* (SMV)-encoded proteins in seed- and aphid- transmission were investigated. Mutants and chimeric recombinants between SMV413 (efficient aphid- and seed-transmissible strain) and an isolate of SMVG2 (lacking aphid- and seed transmissibility) were constructed. Seed transmission of SMV was influenced by P1, HC-Pro, and CP genes. Replacement of the P1 coding region of SMV 413 with that of SMVG2 significantly enhanced seed transmissibility of SMV413. Substitution in SMV413 of the two amino acids that varied in the CPs of the two isolates with those from SMVG2, G to D, in the DAG motif and Q to P near the carboxyl terminus significantly reduced seed transmission. The Q to P substitution in SMV413 also abolished virus-induced seed coat-mottling in plant introduction PI 68671. The results revealed the association of P 1, CP, and DAG motif with seed transmission of a potyvirus and that HC-Pro in interactions with CP were important functions in the virus infection cycle (Jossey et al. 2013). Highly contaminated watermelon seeds collected from *Cucumber green mottle mosaic virus* (CGMMV)-infected experimental plants were used. A high level of transmission of CGMMV was observed in bioassays performed using melon plants following mechanical inoculation of seed extract prepared from CGMMV contaminated seeds (Sui et al. 2019).

7.2.1.4.3 Transmission of Viruses by Vectors

Among the different natural modes of viral dissemination, transmission of phyllosphere viruses from infected plants to healthy plants by insects and mites is more common and important. Plant viruses are transmitted by their vectors with varying efficiency and have a biological relationship indicating the existence of specificity of transmission by a species or related species of the vector. The levels of efficiency and nature of biological relationship are entirely dependent upon the virus concerned, but not by the vector. Plant viruses have the potential to survive in two different environments existing in plant and vector and they take control of the interaction to direct the plant and vector to perform essential functions for the multiplication and survival in the absence of plant hosts (Narayanasamy 2011).

Transmission characteristics of viruses by their vectors are determined by allowing the test insects to feed on source (infected) plants with typical symptoms induced by the virus for predetermined periods which may vary depending on the virus-vector combination. The insects may be able to transmit the virus to the healthy plants, soon after termination of the acquisition access period (AAP), and such viruses are known as nonpersistent viruses/stylet-borne viruses. Semipersistent viruses/circulative viruses require a longer AAP and transmit them after completion of a latent period. Persistent viruses or propagative viruses are acquired after a long AAP on infected plants and may be transmitted after a long latent period, and the vector remains viruliferous for several weeks or until death. Some of them are capable of replication in the vector and may be transmitted vertically to the next generation through the eggs of female insects. This type of transmission is called congenital transmission or transovarial transmission. The terms 'incubation period' and 'latent period' are used to represent, respectively, the time between inoculation of susceptible plant and development of visible symptoms and time between inoculation and infectiousness of the host plant. The vector efficiency may be influenced by the virus, vector species, vector population, movement, and the environment to which the vectors are exposed. The nature of the virus-vector relationship and vector specificity may be useful for putative identification of the virus concerned. Plant viruses may be grouped based on the taxonomic group of vectors as aphid-, leafhopper-, mealybug-, and mite-transmitted viruses. The insects included in Hemiptera have piercing-sucking mouthparts comprising a needle-like stylet bundle consisting of two mandibular and two maxillary stylets. The interlocked maxillar stylets form two canals between them. The narrower salivary canal delivers saliva into the feeding punctures in plant tissues and the wider food canals take up sap into the cibarium (sucking pump), the esophagus, and the rest of the alimentary canal. These insects may be specialized as phloem, xylem, or mesophyll feeders, whereas others can feed on all these tissues. Likewise, viruses transmitted by them are phloem-limited, whereas others are not tissue-specific and infect all plant tissues (Narayanasamy 2011).

Potato virus Y (PVY) disease management system depends of pathogen (disease) pressure and vector pressure under field conditions. Vector pressure is dependent of the relative efficiency factors (REFs) of aphid species. Potential transmission risk is calculated and expressed as relative efficiency factor (REF). REFs reflect transmission efficiency of *Myzus persicae*, the most efficient among aphid species involved in PVY transmission. Relative transmission efficiency factors (REFs) for isolates of PVY strains PVY^{N}, PVY^{NTN}, and $PVY^{N\text{-}Wi}$ were determined. Biotype Mp2 exhibited comparable average transmission efficiencies of *Myzus persicae* for isolates of PVY^{N}, PVY^{NTN}, and $PVY^{N\text{-}Wi}$ (Verbeek et al. 2010). Seed potatoes are affected commonly by *Potato virus Y* (PVY) and *Potato virus A* (PVA). The efficiency of transmission of UK strains by *M. persicae* was nominally set to a REF of 1 and REFs for other species were calculated relative to this. For PVA transmission, *Macrosiphum euphorbiae* (REF 0.91) was almost as efficient as *M. persicae* (Fox et al. 2017). Difference in the vector efficiency might be responsible for increase in prevalence of a strain of PVY in a location. Transmission efficiency of a clone of *M. persicae* for five isolates, each of PVY^{O} and $PVY^{N\text{-}Wi}$, was determined. No apparent difference in the transmission by *M. persicae* was observed. Both PVY PVY^{O} and $PVY^{N\text{-}Wi}$ isolates had similar transmission phenotype in being transmitted efficiently by *M. persicae*. Differential aphid transmissibility is not likely to be responsible for enhancement of prevalence of $PVY^{N\text{-}Wi}$ (Mello et al. 2011). Transmission efficiencies of *M. persicae* of PVY^{NTN}, $PVY^{N:O}$, and PVY^{O} differed depending on potato cultivar serving as virus source. PVY^{O} was transmitted most efficiently from mixed-strain infection sources (Mondal, Lin et al. 2017). *Potato leafroll virus* (PLRV) is transmitted by green peach aphid *Myzus persicae*, and potato aphid *Macrosiphum euphorbiae*. PLRV incidence was positively correlated with aphid abundance, as well as proportion of PLRV-positive aphids. In Russet Burbank, aphid abundance only was positively correlated with PLRV infection. Aphid density could be considered a consistent indicator of risk of infection by PLRV across potato cultivation (Mondal, Wenninger et al. 2017).

Sweet potato feathery mottle virus (SPFMV) is transmitted more efficiently by *Myzus persicae* and *Aphis gossypii* than other species of aphids. The effects of virus source plants, and single or mixed infections on transmission of SPFMV by aphid vectors were determined. The rate of SPFMV transmission from *Ipomoea hederacea* with a natural mixed infection by *A. gossypii* was significantly greater than on other host-virus combinations. Positive relationship between virus titer and transmission by both aphid species was observed. Plants with mixed infections were better sources of inoculum suitable for transmission by aphid vectors than plants infected singly. *A. gossypii* was a more efficient transmitter of SPFMV than *M. persicae* (Wosula and Davis 2012). The effects of aphid abundance, aphid species diversity and virus titers in infected plants on spread of these viruses in Louisiana and Mississippi fields were determined. *Aphis gossypii, Myzus persicae, Rhopalosiphum padi,* and *Therioaphis trifolii* were the most abundant aphid species vectoring sweet potato viruses. SPFMV was more commonly detected than the other viruses (Wosula et al. 2013). *Broad bean wilt virus* (BBWV-1), a nonpersistent virus, has several aphid species as vectors with varying efficiency. Transmission rates of two genetically divergent BBWV-1 isolates PV-132 (USA) and Ben (Spain) were assessed by using isofemale lines of nine aphid species from eight different genera collected in Spain. The results showed that virus concentration in source plant was a key factor in BBWV-1 transmissibility. The Spanish isolate Ben was transmitted more efficiently than the American isolate PV-132 by most aphid species tested, due to higher virus titer attained in Ben plants. Transmission rates of different aphid species showed significant variations (Ferriol et al. 2013). Influence of *Cucurbit aphid-borne yellows virus* (CABYV) with a persistent mode of transmission by *Aphis gossypii* was investigated. Direct changes induced by CABYV on aphid feeding behavior were minor, when viruliferous aphids were fed on infected plants. But nonviruliferous aphids remained for a longer time on infected plants feeding from phloem in CABYV-infected plants compared to noninfected plants. It is possible that CABYV might indirectly manipulate aphid feeding behavior through its shared host plant to favor acquisition of the virus. Alterations in vector behavior of showing preference to infected plants prior to acquisition of CABYV and to healthy plants after acquisition of the virus, creating suitable conditions for rapid spread of CABYV (Carmo-Sousa et al. 2016). *Pentalonia nigronervosa* is involved in *Banana bunchy top virus* (BBTV) transmission in a persistent manner. The aphids could be identified through direct sequencing of PCR products. About 78% of *P. nigronervosa* collected from symptomatic plants in infected plantations in KwaZulu-Natal tested positive for BBTV (Robbertse et al. 2019).

Bemisia tabaci transmits begomoviruses exclusively in a persistent-circulative mode. *B. tabaci* is a complex of cryptic species comprising several species formerly known as biotypes, which are differentiated primarily by DNA markers. Transmission efficiencies of different populations of *B. tabaci* vary significantly within the same species. Persistence of *Tomato yellow leafcurl virus* (TYLCV) and *Tomato yellow leafcurl Sardenia virus* (TYLCSV) in *B. tabaci* as infective entities for longer than the latent period, sometimes until their entire life indicated the possibility of virus replication within the vector insects. Immunolocalization of coat protein (CP) revealed that TYLCV circulated throughout the insect body as intact virions. Viral DNA could be amplified from ovaries of *B. tabaci* after TYLCV acquisition (Ghanim et al. 1998). Transcript of selected viral genes of TYLCV and *Tomato mottle virus* (TMoV) were quantified in whiteflies after AAP on source plants. Contents of TmoV transcripts decreased rapidly after transfer from tomato (virus host) to cotton (virus nonhost). In contrast, TYLCV transcripts increased after the transfer and were readily detected after 7 days (Ghanim et al. 2009). Response of whitefly to *Tomato yellow leafcurl China virus* (TYLCCV) was determined by employing next-generation sequencing (NGS). *B. tabaci* genes were involved in 157 biochemical pathways that were differentially expressed in response to acquisition and retention of TYLCCV. *Begomovirus* could alter cell cycle and primary metabolism and activate whitefly immune responses. Further, TYLCCV invaded ovaries and fat body tissues and induced autophagy-related processes (Luan et al. 2011). The primary salivary glands (PSGs) function as the major organ of *B. tabaci* in transmission pathways of begomoviruses. Transcriptome of PSGS was sequenced. Several highly expressed genes had functions that might be involved in virus transmission or passage (Su et al. 2012).

Primary vector of *Sweet potato chlorotic stunt virus* (SPCSV) was identified as *Bemisia tabaci* biotype B. Later, *Trialeurodes abutiloneus* was also found to be vector exhibiting semipersistent relationship with SPCSV. Mixed infection of SPCSV with aphid-transmitted *Sweet potato feathery mottle virus* (SPFMV) in sweet potato was frequently observed in Peru. *Bemisia afer*, in high populations was also recorded, when population of *B. tabaci* was low. Transmission dynamics of SPCSV in single and double infection with SPFMV by *B. afer* and *B. tabaci* were investigated. SPCSV was transmitted by both whitefly species with similar transmission efficiency. The results revealed that *B. afer* was an additional vector capable of spreading the sweet potato virus disease (SPVD) induced by both SPCSV and SPFMV (Gamarra et al. 2010). Cassava brown streak disease (CBSD) remaining a potential threat to cassava cultivation in African countries, is caused by two RNA viruses, *Cassava brown streak virus* (CBSV) and *Ugandan Cassava brown streak virus* (UCBSV). *Bemisia tabaci* acquired CBSD within an AAP and IAP of 5–10 min each and retained infectivity for up to 48 h. CBSV and UVBSV showed a semipersistent relationship with whitefly vectors (Maruthi et al. 2017).

Female adults of *Bemisia tabaci* were fed on TYLCV-infected and noninfected eggplants. The presence of TYLCV in whitefly following AAP of 72 h was confirmed by testing plants with PCR assay. *B. tabaci* fed more often from phloem sieve elements and made larger number phloem contacts in eggplants than nonviruliferous whiteflies. Further, duration of salivation phase in phloem sieve elements (El) preceding sustained sap ingestion was longer in viruliferous than in nonviruliferous whiteflies ($P < 0.05$). This type of probing behavior significantly enhanced inoculation efficiency of TYLCV by *B. tabaci*. TYLCV directly manipulated settling, probing and feeding behavior of *B. tabaci* in a way that enhanced vector efficiency and spread of the virus. Further, TYLCV-*B. tabaci* interactions are mutually beneficial to both virus and its vector, because *B. tabaci* fed more efficiently after acquisition of TYLCV (Moreno-Delafuente et al. 2013). Transmission efficiency of ipomoviruses by *B. tabaci* varied and for some species appeared to be relatively low. Only a low rate of transmission of *Sweet potato mild mottle virus* was observed. Use of large numbers of viruliferous whiteflies resulted in high rates of transmission of *Cucumber vein yellowing virus* (CVYV) and *Squash vein yellowing virus* (SqVYV). Feeding behavior indicated that whiteflies rarely punctured cells until they reached phloem tissue elements (Dombrovsky et al. 2014). Surveys to assess whitefly populations of *B. tabaci* Mediterranean (MED) and Middle East-Asia Minor 1 (MEAM 1) and incidence of *Tomato leaf curl New Delhi* virus showed that the virus incidence and the single and mixed populations of MED and MEAM 1 were related (Berkin et al. 2018). Transmission efficiency of *Cotton leafcurl Multan virus* (CLCuMuV), one of the major begomoviruses associated with cotton leafcurl disease by four whitefly species, were compared. The indigenous Asia II species could transmit the virus. By sequencing the virus and using immunofluorescence technique differential transmission of the virus was recorded. Varying efficiencies of CLCuMuV to cross the midgut of different whitefly species was observed. Most begomoviruses associated with leafcurl disease shared similar transmission characteristics (Pan et al. 2018).

Crinivirus-whitefly interaction is in semipersistent mode of transmission. *Lettuce infectious chlorosis virus* (LIYV) specifically associated with the ibarium or anterior foregut of whiteflies capable of transmitting the virus, but not in nonvector whiteflies (Chen et al. 2011). Global transcriptional changes occurring in *B. tabaci* MEAM 1 (biotype B) during its interaction with semipersistent *Tomato chlorosis virus* (ToCV) were examined using high-throughput Illumina paired-end RNA-Seq technology. Gene expression was temporarily influenced in whiteflies, due to virus acquisition from ToCV-infected source plants and greater number of genes exhibited differences in expression within the first 24 h of feeding, compared with whiteflies fed on virus-free controls (Kaur et al. 2016). NGS technologies provide a powerful tool box for genetic analysis that determines whether an insect transmits plant viruses, activity factors that influence transmission efficiency, and an understanding of the biology that drives transmission of diverse plant viruses. NGS technologies have been applied efficiently for rapid detection and identification of several known and unknown viruses and viroids infecting plants. Entire viral genome sequences could be rapidly determined or even confirm the presence of multiple viruses in mixed infections. Metagenomic analysis on whiteflies was able to identify typical begomoviruses, using sequencing of the virus-enriched nucleic acid extracts of infected plants. A previously unknown whitefly-transmitted *Carlavirus*, closely related to *Cowpea mild mottle virus*, was identified in whitefly via metagenomic analysis (Rosario et al. 2014).

Leafhoppers and planthoppers are involved in transmission of several viruses and also phytoplasmas. Rhabdoviruses are transmitted in a persistent-propagative manner, and some are transovarially transmitted to the next generation through eggs of leafhopper vectors. Acquisition thresholds range from less than 1 min for rhabdoviruses infecting mesophyll and phloem and up to 15 min for phloem-limited viruses. *Maize stripe virus* (MStV) and *Rice stripe virus* (RSV) are transmitted in a persistent-circulative-propagative manner. The efficiency of transmission declines with increase in age of insects. Most tenuiviruses are transovarially transmitted through high proportions of the female insects. Members of the genus *Reovirus* are also transovarially transmitted through female insects. Members of *Marafivirus* multiply in their vectors, but not transmitted transovarially (Hogenhout et al. 2008). The extent of beet leafhopper *Circulifer tenellus*, carrying *Beet curly top virus* (BCTV) was determined. Of the 800 insects tested, 18.9% were positive for BCTV and the vector could transmit the virus to several crops including bean, tomato, and pepper. Complete virus genomes from positive insects collected were determined. The genome showed 94–98% identities with BCTV-Worland sequence (AY2134867) (Rondon et al. 2016). *Rice gall dwarf virus* (RGDV), a reovirus, is transmitted by the leafhopper *Recilia dorsalis* in a persistent-propagative manner. The effects of propagative transmission by leafhopper population were assessed from eggs laid by viruliferous females. Viral infection directly prolonged the nymphal stages of insect vector. However, the viruliferous population had lower survival rates compared to control. In addition, adult emergence rate of viruliferous populations was lower than that of control. The mean longevity of viruliferous populations was shorter than that of nonviruliferous population (Figure 7.1). The qRT-PCR assay and Western blot assays revealed differential expression of vitellogenin (Vg) in viruliferous versus nonviruliferous adult females. It was inferred that RGDV reduced levels of Vg transcript and protein product resulting in the lower fecundity of its vector (Figure 7.2). Different adverse effects of RGDV on *R. dorsalis* appeared to restrict expansion of viruliferous vector populations (Chen et al. 2016).

Characteristics of transmission of *Wheat yellow striate virus* (WYSV) by *Psammotettix alienus* were determined. Virus titer increased with increase in time after acquisition of the virus, suggesting WYSV might replicate in *P. alienus*. Female insects could transovarially transmit WYSV to the next generations at relatively high efficiency. Bacilliform particle aggregates were present in the cytoplasm of salivary gland and midgut tissues. The results showed that WYSV had persistent, circulative and propagative relationship with its vector *P. alienus* (Du et al. 2020).

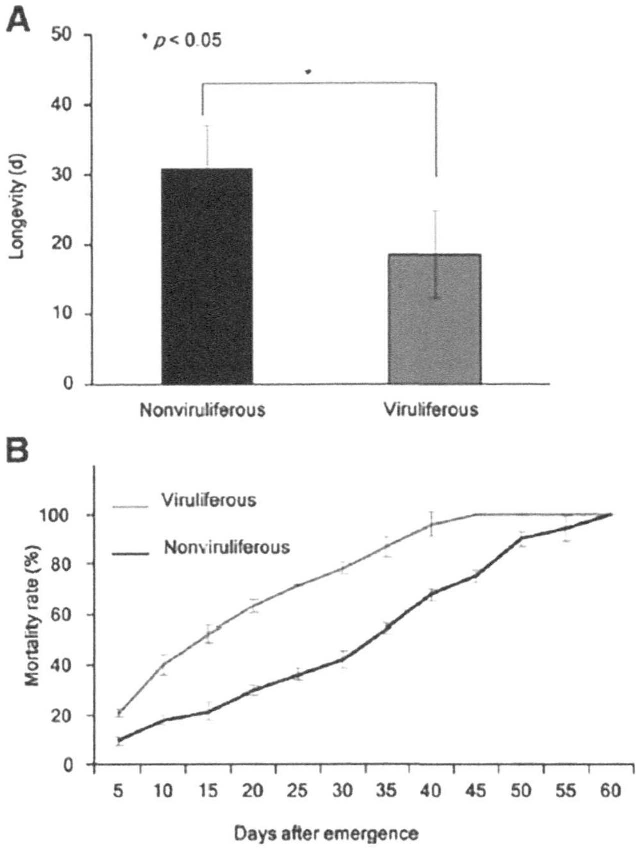

FIGURE 7.1 Effects of *Rice gall dwarf virus* (RGDV) on longevity of viruliferous *Recilia dorsalis*. (A) Reduction in longevity of viruliferous *R. dorsalis*. (B) Enhanced mortality (%) of viruliferous *R. dorsalis*.

Source: (Courtesy of Backus et al. 2015 and with kind permission of the American Phytopathological Society, MN, USA)

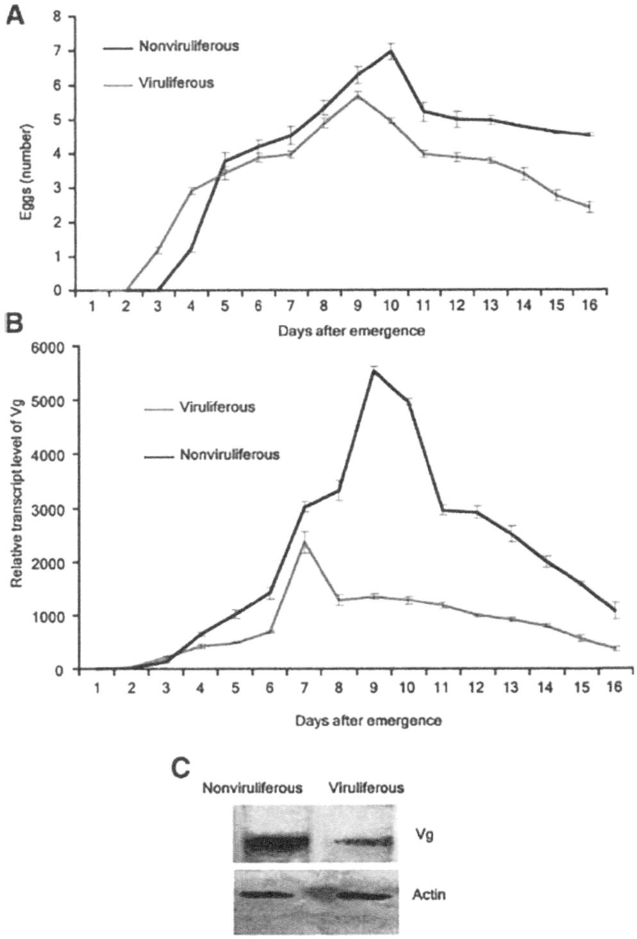

FIGURE 7.2 Effects of *Rice gall dwarf virus* (RGDV) on fecundity and expression of vitellogenin (Vg) in viruliferous *R. dorsalis*.

Source: (Courtesy of Chen et al. 2016 and with kind permission of the American Phytopathological Society, MN, USA) (Orcid.org/0002–4926–9364)

All plant-infecting viruses are placed in the genus *Tospovirus*, family *Bunyaviridae*. *Tomato spotted wilt virus* (TSWV) is the type member of the genus *Tospovirus* enclosing 14 recognized virus species. *Tospovirus* has unique relationship with thrips vector, since the adults can transmit, only if the virus is acquired at the larval stage. *Frankliniella occidentalis* could transmit 5 *Tospovirus* species (Hogenhout et al. 2008). Differential retention of *Tomato spotted wilt virus* (TSWV) in thrips vector *F. occidentalis* for different periods was examined. About half of the transmitting thrips maintained the capacity to infect leaf discs, indicating persistent propagative nature of transmission of TSWV. However, a considerable proportion of the thrips could transmit the virus in their early phases of adult life, but did not retain their infectivity in the remainder of the lifetime (Mautino et al. 2012). Appearance of TSWV isolates capable of overcoming resistance genes *Sw-5* and *Tsw*, respectively, in tomato and pepper cultivars was observed. The risk of spread of the resistance-breaking isolates via the principal vector *F. occidentalis* was assessed. Transmission rate of TSWV isolates was related to virus titers accumulated in thrips, but not virus accumulation in source plants. There was no correlation between transmission efficiency by thrips and the genotype or between transmission efficiency and the ability of overcoming resistance in pepper or tomato (Debreczeni et al. 2013). Ability of the vector *Frankliniella occidentalis* to transmit non-resistance breaking TSWV isolates when confined to blossoms on plants with or without *Sw-5* resistance allele was assessed. *Sw-5*$^{+}$ plants inoculated by adult thrips feeding on blossom clusters or small fruit developed infections in reproductive tissues (Houle and Kennedy 2017). *Groundnut bud necrosis virus* (GBNV)–infected and nonviruliferous thrips showed significant differences in the developmental period. Preadult longevity and fecundity of the viruliferous thrips were affected due to the presence of GBNV. *T. palmi* preferred GBNV-infected cowpea plants to healthy plants (Daimei et al. 2017).

The extent of modification of vector competence of wheat curl mite (WCM) *Aceria tosichella* genotypes following feeding on wheat plants coinfected by *Wheat streak mosaic virus* (WSMV) and *Triticum mosaic virus* (TriMV) was assessed. Coinfection reduced WSMV transmission by type 2 WCM from 50% to 35.6%. In contrast, coinfection increased transmission of TriMV. Reduction in WSMV transmission might result in reduced vector survival in the presence of TriMV in wheat plants. In another experiment, Type 1 mites did not transmit TriMV from plants coinfected with WSMV. A high relative incidence of TriMV-positive plants might be due to coinfection with WSMV (Oliveira-Hofman et al. 2015). The mite *Aceria tulipae* was transmitted after molting of the vector, and nymphs could acquire the virus. The presence of the virus in the midgut, body cavity, and salivary gland was detected indicating a circulative persistent relationship between WSMV and *A. tulipae. Wheat spot mosaic virus* was also transmitted by *A. tulipae* (Paliwal 1980). Wheat curl mite (*Aceria tosichella*) could transmit *Wheat streak mosaic virus* (WSMV), *Triticum mosaic virus* (TriMV), and *Wheat mosaic virus* (WMoV) to wheat (*Triticum aestivum*) in the Great Plains region of the United States. These viruses were present in single, double, or triple combinations in infected plants (Byamukama et al. 2016). Wheat curl mite (WCM) *Aceria tosichella* (Eriophyidae) could transmit *Wheat streak mosaic virus* (WSMV), and the virus-vector relationship showed that *A. tosichella* could also transmit *High Plains Wheat mosaic virus* (HPWMoV) and *Triticum mosaic virus* (TriMV) and the transmission of HPWMoV and TrMoV and TrMoV occurred across the Great Plains, and coinfection of wheat by two or all three viruses was common (Tatineni and Hein 2018).

An eriophyid-borne virus is transmitted neither by other mites nor by more than one eriophyid species. Seeds of wheat, corn or grasses might also spread the viruses to virus-free areas infested by *A. tosichella*. Fecundity of *A. tosichella* was reduced on plants infected with TriMV. Negative differential off-host survival of *A. tosichella* from TriMV-infected plants, compared with those tested from noninfected and WSMV-infected plants, was observed (de Lillo et al. 2018). Three mite populations transmitted Sydney 81 at higher rates compared with Type WSMV. Two mite populations, Nebraska and Type 2 transmitted Sydney 81 at higher rates, compared with other three populations (Wosula et al. 2016). Wheat curl mite (WCM), *Aceria tosichella*, efficiently transmitted *Wheat streak mosaic virus* (WSMV). WSMV is not transmitted transovarially, and only immature WCMs could acquire the virus. All stages in the life cycle of the mite could transmit the virus, but immature mites were more efficient in transmitting the virus. WSMV persisted in the vector for at least 7 days in adults kept on virus-immune plants, but transmission efficiency decreased with time (Tatineni and Hein 2018). *Rose rosette virus* (RRV) was transmitted by *Phyllocoptes fructiphilus*. RRV belongs to the genus *Emaravirus*, order *Bunyavirales. P. fructiphilus* could establish, lay eggs and develop nymphs and adults on all rose genotypes. Mites required a long acquisition/latent period, but a short inoculation period for successful transmission of RRV (de Lillo et al. 2018). The mites do not have wings. They might be transported by insects during pollination, and dispersed by wind also. (Vazquez-Iglesias et al. 2020).

Grapevine leafroll disease is induced by a complex of phylogenetically related closteroviruses, some of which are transmitted by mealybugs. Two viruses, *Grapevine leafroll-associated virus 4* and *-9* (GLRaV-4 and 9), were mealybug-borne. No evidence of vector specificity was inferred, since different vector species could transmit GLRaV-3 and *Planococcus ficus* could transmit five GLRAVs (Tsai et al. 2010). *P. aceris* transmitted GLRaV-1, -3, -4, -5, -6, and -9 from infected grapevine plant to healthy plants. Further, *P. aceris* was able to transmit *Grapevine virus A* (GVA), and *Grapevine virus B* (GVB). First instar nymphs were the most efficient in transmitting GLRaV-1, -3, and GVA (Le Maguet et al. 2012). Under field conditions, about half of all inoculated plants became infected with *Grapevine leafroll-associated virus 3* (GLRaV-3). Vineyard infected plants functioned as effective virus sources 1 year after inoculations (Blaisdell et al. 2016). *Grapevine virus A* (GVA) and GLRaV-1 and -3 often caused coinfections. These viruses could coexist in grapevine plants as well as in their insect vectors. After feeding on GVA, GLRaV-1 and GLRaV-3 coinfected grapevine plants and nymphs of *P. ficus* and *P. citri* exhibited similar virus acquisition rates and retained low quantities of viruses until the third post-acquisition day. However, the vector species differed in transmission efficiency. *P. ficus* was more efficient in transmitting GVA and GLRaV-3, whereas *P. citri* transmitted GLRaV-1 more efficiently (Pacifico et al. 2016).

7.2.1.4.3.1 Mechanism of Virus Transmission by Aphids

Identification of viral molecules that control specific interactions with aphid vectors has been possible by employing mutagenesis and reverse genetics approaches. Viral determinants include viral capsid protein and additional nonstructural virus-encoded proteins. The mouthparts of aphids are well adapted to their feeding habits, and the flexible stylet bundle passes through intercellularly without damaging plant tissues. The aphid introduces the stylets deeper under the epidermis until they reach the sieve elements. During feeding, aphids produce different saliva with different compositions and functions. Viruses from different plant tissues are acquired and transmitted to susceptible plants during the process of feeding. Based on virus uptake during sap ingestion and virus inoculativity during the putative regurgitation, it was suggested that in non-circulative (nonpersistent) transmission, aphids may act as 'flying syringes' (Harris 1977). *Cauliflower mosaic virus* (CaMV), the noncirculative (semipersistent) virus species, was precisely located within the vector and it was present exclusively at the tip of the stylet. The CP of viruses belonging to *Cucumovirus*, *Alfamovirus*, *Carlavirus*, and *Crinivirus* is the only viral component determining virus retention within their respective aphid vector. An additional virus-induced helper component (HC) was required for aphid transmission. The mode of action, as per the bridge hypothesis, was that two independent interacting domains might link the CP on one side and the putative attachments sites within the insect on the other side (Pirone and Blanc 1996).

Viruses belonging to *Potyvirus* and *Caulimovirus* produce HCs which are well characterized and they are involved in transmissible complexes that are retained within the vectors. The N-terminal part of *Potyvirus*-encoded polyprotein contains the functional product that helps the aphid-mediated transmission of virus particles. Helper component-protease (HC-Pro) is a multifunctional protein involved in many steps of the *Potyvirus* infection process. HC-Pro might be involved in synergistic effects with other viruses and virus movement through plasmodesmata and in access to plant vasculature (Schaad and Carrington 1996; Rojas et al. 1997). In addition, role of HC-Pro as a suppressor of gene silencing and host necrosis response also indicated as its functions (Hu et al. 2009). The HC of *Cauliflower mosaic virus* (CaMV) is required specifically and solely for aphid transmission. Gene II encodes a protein of 18-kDa called P2 that is totally dispensable for CaMV infectivity. Protein P2 is not involved in other potyviruses. Aphid-transmitted viruses require coat proteins as determinants of transmission as well as a helper component-proteinase (HC-Pro) for mediating transmission. Potyvirus HC-Pro is a multifunctional protein is required for aphid transmission, and also for viral polyprotein processing, long-distance systemic movement within infected plants, maintenance of viral RNA genome amplification, synergistic interactions with unrelated viruses in mixed infections, and suppression of posttranscriptional gene silencing (PTGS) (Stenger et al. 2005). An additional factor was identified as viral protein P3 which was very intimately associated with the CP in mature virus particles and mediated P2 attachment to virions (Leh et al. 2001; Plisson et al. 2005). A two-step pattern for acquisition of *Cauliflower mosaic virus* (CaMV) by its aphid vector appears to be controlled by an inclusion body specifically and solely involved in vector-mediated transmission known as the transmission body (TB). When the vector punctures an infected cell, the TB is disrupted and the released P2 attaches to its receptor within the aphid stylets, whereas P3 is solubilized in a free conformation that is not retained together with P2. Subsequently, the P2-loaded aphid acquired P3-virion complexes from viral populations or from the phloem, due to the very high affinity between P2 and virion-bound P3 (Drucker et al. 2009). *Citrus tristeza virus* (CTV) is efficiently transmitted by *Toxoptera citricida* in a semipersistent manner. The efficiency of transmission is influenced by AAP for CTV. Maximum CTV titer reached was at AAP 6 h for isolates CT11A and CT16–2 and at 4 h for isolate CTLJ. The results suggested that the transmission efficiency was positively correlated with viral titer in the insect (Liu et al. 2019).

Circulative-propagative transmission of bullet-shaped viruses (*Cytorhabdovirus* and *Nucleorhabdovirus*) by aphids, leafhoppers, and planthoppers occurs under natural conditions. Aphid-transmitted rhabdoviruses exhibit a high degree of vector specificity. Rhabdoviruses very rarely induce disease in their vectors in which they replicate at a low rate, in contrast to the serious effects induced in planthopper vectors in which they replicate to reach high levels. Entry of enveloped viruses into host cells involves a virus recognition step, followed by membrane fusion. Rhabdoviruses enter the cells via the endocytotic pathway and fusion occurs with endosomal compartment. Trimeric rhabdovirus glycoprotein (G protein) protrudes from virion's lipid envelope and plays a major role during viral entry. It is responsible for binding the virus and endosomal membranes. Fusion with the endosome membrane is triggered by a conformational change of the G protein induced by low pH of the cellular compartment. The function of G protein and the aphid gut endocytosis process may be based on gene function conservation among members of *Rhabdoviridae*. These viruses are phloem-limited, and the phloem-feeding aphids are suitable for their transmission. Nanoviruses and luteoviruses, in spite of having similar mode of transmission, rely on different viral determinants for virus internalization in the vector. Capsid protein of luteoviruses is required for efficient virus transport. On the other hand, nanoviruses need possibly additional HC for their transmission (Franz 1999).

Transmission electron microscopy (TEM) was applied to monitor route of luteovirus movement in the vector. Internal endocytosis was revealed, commencing with clathrin-mediated endocytosis, followed by transport of virions enclosed in vesicles from the apical to the basal pole of the cell and ending with fusion of virus-containing vesicles to the basal plasmalemma. Transcriptome analysis of the expression of intestinal genes after luteovirus acquisition revealed that only 1.9% of genes were differentially expressed in the presence of the virus (Brault et al. 2010). As most of the virus could be internalized both in vector and nonvector species, the gut epithelium is not very specific for uptake of luteoviruses and the basal lamina and basal plasmalemma surrounding anterior silk gland cells were able to function as selective barriers to transmission. After internalization in the aphid body, persistence of virions in the haemolymph relied on interactions with a protein synthesized by a bacterium hosted as an endosymbiont located in sieve tubes. This protein could bind to purified luteoviruses and stimulate virus transmission by aphid when added to artificial diet, suggesting that some phloem proteins might promote virus acquisition by aphids (Brault et al. 2010). Persistent viruses (propagative/nonpropagative) can be transmitted to plants by injecting the virus into the insect hemolymph. Often injected viruses are transmitted at higher rates than orally acquired viruses, because of movement of the virus across the insect gut is a barrier to virus passage to hemolymph. In persistent-circulative transmission, luteoviruses may enter the epithelial cells of the gut of the vector by endocytosis. Virus particles may move rapidly through the intestinal cell layer. The virus could be observed inside these cells at 4 h after acquisition. Virus particles accumulated at high concentrations in epithelial cells when aphids fed on a virus source. But when aphids were removed from virus source, virus concentration in these cells began to decrease rapidly after the second day. Transport of virus through salivary gland cells was achieved via endocytosis and exocytosis (Garret et al. 1996; Gray and Gildow 2003). Four aphid proteins associated with transmission of *Cereal yellow dwarf virus*-RPV by *Schizophis graminum* were identified. These proteins bound purified virus and two of them were identified as luciferase- and cyclophilin-like proteins with possible function related to endocytosis pathway. Aphid transmission of luteoviruses is highly specific, and each virus species is generally transmitted efficiently by one or two aphid species. However, one aphid species can transmit several luteoviruses, as in the case of *Myzus persicae* (Reinbold et al. 2003; Burrows et al. 2007).

Persistent and nonpersistent viruses change host plants to enhance transmission by insect vector by applying different strategies consistent with the mode of transmission. Changes in small RNAs (sRNAs) in *Myzus persicae*, after acquisition feeding, were assessed by small RNA sequencing (sRNA-seq) in nonpersistent (represented by *Potato virus Y*) and persistent (represented by *Potato leafroll virus*). Viruses replicating in host tissues induce production of virus-derived small interfering RNA (viRNA). Small RNA (sRNA) sequencing was employed to visualize the sRNA response of *M. persicae* to *Potato virus Y* (PVY) and *Potato leafroll virus* (PLRV). Aphids exposed to PLRV produced significantly fewer 22-mers aligned to the aphid genome and an abundance of 26–27-mers. In addition, expression of *Buchnera aphidicola* tRNA-derived sRNAs was influenced by PLRV and to a lesser extent by PVY, suggesting that plant viruses might alter the aphid endosymbiont relationship. Expression of the PLRV silencing suppressor peroxide in plants assisted the increase in MpDNV titer in the absence of PLRV infection. The results showed that based on the mode of aphid transmission, viruses might induce distinct effects in their vectors in respect of post-transcriptional gene regulation, symbiosis with *Buchera* spp. (Pinheiro et al. 2019).

7.2.1.4.3.2 Mechanism of Transmission of Viruses by Whiteflies

Hamiltonella encoded a 63-kDa GroEL (earlier known as a symbionin present in whiteflies) interacted with TYLCV virions on their way to salivary glands and protected virions from the insect immune system, ensuring efficient transmission of the virus to healthy plants. *Hamiltonella* GroEL-TYLCV interaction was highly specific in *B. tabaci*, since GroEL proteins from other symbiotic bacteria harbored in *B. tabaci* did not react with TYLCV CP (Sanchez-Campos et al. 1999). A later investigation revealed that *Rickettsia*-infected *B. tabaci* acquired more TYLCV from infected plants, retained the virus for longer period and showed about twofold greater efficiency, compared to an uninfected strain of the insects (Kliot et al. 2014). *Tomato yellow leafcurl virus* (TYLCV) apparently replicated in its vector and showed reduced fitness, whereas *Tomato mottle virus* (TMoV) did not appear to replicate in whiteflies and did not affect vector fitness. TYLCV DNA was transmitted transovarially to embryos/eggs of female whiteflies and could be sexually transmitted between them (Ghanim et al. 1998; Ghanim and Czosnek 2000). Transmission, retention, and circulation of begomovirus in *B. tabaci* are facilitated by interaction with several proteins of whitefly vector and its endosymbionts. Yeast two-hybrid assay between *B. tabaci* cDNA expression library and the coat protein (CP) of *Tomato leafcurl New Delhi virus* (ToLCNDV) and *Cotton leafcurl Rajasthan virus* (CLCuV), revealed that collagen was the common protein interacting with both viruses. Collagen protein was localized in gut layers of *B. tabaci*. Pull-down and blot assays confirmed the association of endogenous collagen with ToLCNDV CP. Immunolocalization analysis also showed colocalization of ToLCNDV particles and collagen within insect gut (Rana et al. 2019).

Tomato yellow leafcurl virus (TYLCV) was retained by the whitefly vectors for their entire life time. The Israeli isolate, TYLCV-IL, can be transferred from the male whitefly to females and vice versa during mating. TYLCV-IL could be transferred transovarially from viruliferous female to their next generation through reproductive system. Luteoviruses mainly cross the aphid hind gut to the haemolymph, while begomoviruses cross the filter chamber a specialized organ within the midgut. Luteoviruses enter accessory salivary glands of the aphid vector, whereas begomoviruses enter the primary salivary glands of whitefly vectors (Ghanim et al. 1998, Ghanim et al. 2009). Involvement of heat shock proteins present in whitefly in binding with coat protein (CP) of TYLCV was reported (Gotz et al. 2012). Further, whitefly bacterial symbiont-derived GroEL protein interacted with TYLCV virions and also to protect the virus during circulation in the haemolymph of the vector (Gottlieb et al. 2010). A global transcriptome analysis of whiteflies, *Bemisia tabaci* MEAM1 fed on *Tomato yellow leafcurl China virus* (TYLCCNV)–infected tobacco, and in viruliferous whiteflies, 1,606 genes were differentially expressed and 298 of those genes were mapped to 157 pathways. Regulation of several cell-cycle regulatory genes as well as the downregulation of genes involved in lipid and amino acid metabolism, suggested that TYLCCNV appeared to behave like a pathogen in whiteflies (Luan et al. 2011). Upregulation of several genes involved in cellular and humoral immunity, including those that participate in autophagy, lysosome function, antimicrobial peptide production, coagulation, and melanization. Numerous genes involved in signal transduction during antiviral immune responses, including toll, TGF-ß, and mitogen-activated protein kinase pathways, were downregulated in whiteflies exposed to TYLCCNV and genes that function in apoptotic pathway were also downregulated. Insect genes predicted to be involved in fatty acid metabolism were downregulated at 24 h, while genes involved in carbohydrate metabolism were downregulated at 72 h (International Whitefly Genome Initiative 2015).

Mechanism of transovarial transmission of *Tomato yellow leafcurl virus* (TYLCV) and *Papaya leafcurl China virus* (PaLCuCNV) to the next generation in whiteflies was investigated. TYLCV was transmitted effectively via ovary of whitefly females of *Bemisia tabaci* older adults, but not by whitefly adults of very young age. Transovarial transmission of TYLCV to offspring increased with whitefly adult age. The vitellogenin (Vg) protein precursor of major yolk protein was involved in transovarial transmission of TYLCV. Specific interaction between virus CP and whitefly vitellogenin was essentially required for virus entry into whitefly ovary. Knockdown of the expression of Vg, inhibited entry of TYLCV and transovarial transmission efficiency decreased. Binding between TYLCV CP and Vg was important for TYLCV to overcome transovarial transmission barriers. TYLCV CP and Vg might bind together in the haemolymph and then transported into the oocytes via the existing transportation system. In contrast, another begomovirus, *Papaya leafcurl China virus* (PaLCuCNV) CP did not interact with whitefly Vg, resulting in failure of transovarial transmission of the virus. When partial CP sequence of PaLCuCNV was replaced with TYLCV, mutated PaLCuCNV became transovarially transmissible and mutated TYLCV became nontransmissible. This part of the CP sequence determined transovarial transmissibility. However, it had 33 different amino acids in the exchanged fragment. Both MEAM1 and MED whitefly could transovarialy transmit TYLCV to their progeny efficiently and the virus could be maintained at least for two generations by the whitefly in the absence of susceptible host plant. Further, adult progenies developed in TYLCV nonhost plants could infect healthy plants in the field, resulting in new infections. Transovarial transmission may contribute significantly to the global spread of TYLCV, as well as outbreak of TYLCV) in the field (Wei et al. 2017).

Some whitefly-transmitted viruses exhibit a noncirculative, semipersistent transmission relationship. In *Bemisia tabaci*-transmitted *Lettuce infectious yellows virus* (LIYV), LIYV-encoded protein determining its transmission and location of virus retention by *B. tabaci* were identified. Vector and nonvector whiteflies *B. tabaci* biotypes A and B were fed sequentially on artificial diets containing virions. LIYV capsid protein produced by expression in *E. coli*, followed by solutions containing different antibodies were used to bind LIYV virus or capsid protein in whiteflies. Examination of these whiteflies by wide-field fluorescence microscopy and confocal-laser scanning microscopy showed that fluorescence was localized only in a precise region of foregut, the anterior foregut region (cibarium) of the A biotype *B. tabaci*. On the other hand, no binding could be observed in nonvector B biotype fed on the same preparations. The results suggested that among the LIYV capsid proteins, only CPm was found within anterior foregut of only in A biotype, but not in B biotype. A recombinant LIYV CPm mutant was not transmissible by A biotype *B. tabaci* and it did not bind, but when the mutation was restored to give wild-type CPm, LIYV transmission by the vector was restored following specific binding in the anterior foregut (Chen et al. 2011). Viral CP of *Geminiviridae* is implicated as the sole determinant of whitefly transmissibility. Evidence indicated that CP served as viral attachment protein (VAP) to insect vector gut, a first step in acquisition of *African cassava mosaic virus* (ACMV). A recombinant CP of *Tomato yellow leafcurl virus* (TYLCV) fed to whiteflies bound to the midgut, and in competition assays with wild-type virus reduced the amount of virus in whiteflies (Wang et al. 2014a). Interactive effects of viruses in plants may be either synergistic or antagonistic in nature, influencing development of a virus by another present in mixed infections in plants. Influence of mixed infections of melon plants by two unrelated viruses, *Cucurbit yellow stunting disorder virus* (CYSDV) and *Watermelon mosaic virus* (WMV) on transmission, respectively, by whitefly and aphid was investigated. Transmission of CYSDV (*Crinivirus*) and WSMV (*Potyvirus*) from coinfected plants was compared with singly infected plants in respect of virus titer. Synergistic interactions in mixed infections resulted in significant increase in whitefly transmission of CYSDV. On the other hand, aphid transmission of WMV remained unaltered, although virus titer was reduced (Domingo-Calap et al. 2020).

7.2.1.4.4 Mechanism of Transmission of Viruses by Leafhoppers

During the process of persistent-propagative transmission by leafhoppers, virions accumulated in the midgut lumen to disperse into its haemolymph or nervous tissues, then into salivary gland from where the viruses might be injected into the host plant during feeding. Further, some of the circulative-propagative viruses like plant rhabdoviruses, tenuiviruses, and reoviruses can spread into the ovary from where virions might be transmitted to the progeny for several generations. *Rice stripe virus* (RSV), a *Tenuivirus*, could be transovarially transmitted to the progenies for the next 40 generations. *Rice dwarf virus* (RDV), a *Phytoreovirus*, was transmitted transovarially in *Nephotettix cincticeps* and maintained through generations of progeny for 6 years (Honda et al. 2007). Maternal transmission of endoparasitic microbes in an insect host is probably related to vitellogenin (Vg), a female-specific protein synthesized primarily by the fat body and secreted into hemolymph, from where it is absorbed by growing oocytes via receptor-mediated endocytosis (Tufail and Takeda 2009). Transovarially transmitted yeast-like symbionts (TLs) in brown planthopper (*Laodelphax striatellus*) were wrapped in Vg outside the ovary. The mechanism of transovarial transmission of *Rice stripe virus* (RSV) and the central role of vitellogenin (Vg) in *L. striatellus* was elucidated. Vg could bind to PC3 in vivo and in vitro and colocalized in the germarium. RSV filamentous ribonucleoprotein particles (RNPs) only accumulated in the terminal filaments and pedicel areas prior to

Vg expression and was not present in the germarium until Vg was expressed, where RSV RNPs and Vg had localized. These two proteins localized in the nurse cells. Knockdown of Vg expression due to RNA interference resulted in inhibition of the invasion of ovarioles by RSV. RSV RNPs might enter the nurse cells of the germarium via endocytosis through binding with Vg. The virus then entered finally the oocytes through nutritive cords, using the same route as for Vg transport (Huo et al. 2014).

Maize fine streak virus (MFSV) is transmitted by *Graminella nigrifrons* in a circulative-persistent manner. Transmission characteristics for MFSV and temporal increases in virus titer in infected leafhoppers were consistent with persistent propagative transmission of MFSV as in the case of other rhabdoviruses. The proportion of infected and inoculative insects increased with increasing post-first access to disease period (PADP), and no transmission occurred with a PADP shorter than 4 weeks, the time when MFSV was first detected in salivary glands. Immunolabeling and examination using immunofluorescence confocal laser-scanning microscopy (iCLSM) were applied to monitor accumulation of MFSV in dissected organs and hemocytes of leafhoppers fed on infected source plants. MFSV was detected in cells of midgut and filter chamber, neural tissues, visceral muscles and hemocytes in a few *G. nigrifrons*. After a 3- to 4-week PADP, MFSV infection was detected further in muscle tissues, tracheal cells, hindgut, and male accessory glands and accumulation around nuclei of cells. In salivary glands, at a 4-week PADP, MFSV was present in both nuclear and cytoplasmic locations, the periphery of secretory lobes (acini) and neural tissues. The percentage of MFSV-positive midguts increased by the 6-week PADP. Inoculative insects were detected at the same time that MFSV was detected in the salivary glands at 4-week PADP (Todd et al. 2010). Localization and distribution of *Wheat dwarf virus* (WDV, ss-DNA virus) in its leafhopper vector *Psammotettix alienus* were studied. The results suggested that WDV might move not only to the salivary glands from the anterior and middle gut via the hemocoel but also might pass directly through the sheath of filter chamber and might be readily transmitted to healthy wheat plants within 5 min of access period on infected plants. When bacterially expressed recombinant capsid protein (CP) was incubated with the internal organs of leafhoppers, presence of CP-immunoreactive antigens was detected at the anterior and middle gut. In addition, when leafhoppers fed with an antiserum raised against CP, accumulation of WDV in the gut cells, hemocoel, and salivary glands were significantly reduced. The results indicated that transmission of WDV was determined by a CP-mediated virion-vector retention mechanism (Wang et al. 2014b).

7.2.1.4.5 Mechanism of Transmission of Viruses by Eriophyid Mites

The role of HC-Pro homologues encoded by non-aphid transmitted viruses of the family *Potyviridae* in mediating transmission by their vectors was investigated. *Wheat streak mosaic virus* (WSMV) shares a common genome organization with aphid-transmitted species of *Potyvirus* genus. Although both tritimoviruses and potyviruses encode HC-Pro homologues, sequence conservation is low. Hence, the role of HC-Pro in semipersistent transmission of WSMV by wheat curl mite (WCM), *Aceria tosichella* was investigated. Transmissibility of WSMV HC-Pro was due to homologues of an aphid-transmitted potyvirus (*Turnip mosaic virus*). In contrast, both WSMV Sidney81 and a chimeric WSMV genome bearing HC-Pro of a divergent strain (WSMV-El Batan) were efficiently transmitted by *A. tosichella*. Replacing portions of WSMV-Sidney81 HC-Pro with corresponding regions from *Oat necrotic mottle virus* (ONMV) showed that determinants of wheat curl mite transmission mapped to the 5′ proximal half of HC-Pro. WSMV genomes containing HC-Pro of heterologous species could form virions, indicating that loss of vector transmissibility did not affect encapsidation. HC-Pro was required for virus transmission by a vector other than aphids also (Stenger et al. 2005).

7.2.1.4.6 Mechanism of Transmission of Viruses by Thrips

Tospoviruses are circulative-propagative in their thrips vector. *Tomato spotted wilt virus* (TSWV) replicates in western flower thrips *Frankliniella occidentalis*. Changes in gene expression between virus-free thrips reared in healthy plants and viruliferous thrips raised on TSWV-infected tomato plants were assessed. Infection of thrips by TSWV led to downregulation of genes related to ribosomes, amino acid metabolism, carbohydrate metabolism, compared with virus-free thrips. TSWV infection resulted in decreased protein synthesis and amino acid and carbohydrate metabolisms in infected thrips and also changed cellular and immune responses in *Frankliniella occidentalis*. High levels of mucin expression might contribute to infection, replication and diffusion of TSWV within thrips. Genes associated with inhibition of virus replication were upregulated (Zhang et al. 2013). *Iris yellow spot virus* (IYSV) is transmitted by *Thrips tabaci*. The rate of IYSV replication in the larva, pupa, and adult stages of *T. tabaci* and the other non-vector thrips species and color forms such as *Frankliniella occidentalis, F. schultzei* (dark) and *F. schultzei* (pale) was evaluated. Nucleocapsid (N) protein levels increased in all three stages of IYSV-fed *T. tabaci*, indicating replication of IYSV. The N protein levels did not increase at pupal and adult stages. Nonstructural (NS) protein was not detected in first instar of either vector or non-vector thrips species. After a 4-h post-acquisition period, a significant increase in NSs proteins was only observed in IYSV-fed *T. tabaci*, clearly differentiating vectors and nonvectors of IYSV (Birithia et al. 2013).

7.2.1.5 Infection Process of Plant Viruses

Plant viruses constitute a group of obligate, intracellular parasites that are entirely dependent on host plant tissues for replication. Their strategies for successful infection and induction of symptoms (pathogenesis) are, therefore, different and unique. Plant viruses do not possess any known physiological functions and enzymatic activities, and hence their entry into host plants depends on the presence of wounds created by mechanical damage on plant surface or feeding injuries made by vectors that assist viruses to reach susceptible host cells. Viruses have to be deposited directly into susceptible host cells, because of their inability to enter through intact cells, to initiate the process of infection. Viral infection of susceptible host plants is a complex process. Four principal stages in the process of infection resulting in development of disease symptoms have been recognized: (i) entry of virus into host plant cells, (ii) replication of virus in the primarily infected cells, (iii) movement of virus from infected cells to adjacent cells through plasmodesmata (PD), and (iv) long-distance movement through vascular system of host plant. Plant cells have cell walls forming formidable barrier to the entry of viruses. There is no receptor-mediated mechanism operating for effecting initial transfer of the virus into a plant cell. The cell wall barrier has to be breached, and the virus particle needs to be placed inside the cells through wounds or injured by feeding vectors (Llave 2010).

After entry of the virus particle into the susceptible cell, the first event is uncoating of viral coat protein and the viral genome is released into the cell cytoplasm. The infecting viral genome is translated directly, if it is a + sense RNA. Single-stranded (ss) RNA or mRNAs are formed and translated to produce early products such as virus replicase and perhaps other virus-specific proteins. Virus replicase or replicase-associated protein(s) required for synthesis of subgenomic mRNAs are formed. The viral genomes are synthesized by the viral replicase or replication-associated proteins. Proteins required relatively late in the virus replication cycle, such as CP and movement protein (MP) essential for cell-to-cell movement of the viruses, are synthesized. When they reach sufficient concentrations, CP subunits and viral genomes are assembled together to form new progeny virions, usually in the cytoplasm. Mature virions move from initially infected cell to adjacent cells in systemic infections. After overcoming the first line of defense of the host plant, the virus has to face nonspecific defensive reactions with which the plant responds to some molecular patterns that are common to different kinds of pathogens. Plants can recognize the virulence factors of viruses and create a new specific resistance layer that is induced only, when challenged by invading viruses expressing the virulence factors (Jones and Dangl 2006). Viral RNA induces specific plant defense responses in which a large number of plant proteins participate. Dicer-type ds-RNases, ss-RNases belonging to the Argonaute-type protein family, assemble in RNA-induced silencing complex (RISC), RNA polymerases, and RNA helicases are formed. This antiviral complex set of cellular proteases known as RNA silencing is apparently universal and the virus needs to escape the defenses that are host-mediated barriers in its life cycle. Initiation of this cycle depends on the nature of genetic material of the virus concerned (Dunoyer and Voinnet 2005).

Functioning of silencing suppression activity was investigated using different methods. The virus-encoded gene of interest is cloned into an appropriate shuttle vector that is transferred to a strain of *Agrobacterium tumefaciens* (At). *A. tumefaciens* suspensions designed to deliver the candidate gene and a reporter gene green fluorescent protein (*gfp*) or ß-glucuronidase (GUS) into plant cells and mixed and then pressure-infiltrated into leaves of *Nicotiana benthamiana* plants. Potential silencing suppressor is identified by the ability of the transiently expressed viral gene to enhance and/or sustain visibly higher levels of expression of the reporter gene (Johansen and Carrington 2001). Another approach depends on examining whether expression of a silenced transgene can be restored upon infection of a particular virus (silencing reversal). *Potato virus X* (PVX) has been commonly employed as virus vector due to its apparent inability to reverse RNA silencing (Voinnet et al. 2000). Another procedure examined the silencing suppression function of a viral protein using transgenic plants expressing the test protein and they are genetically crossed with plants carrying a reporter gene that is post-transcriptionally silenced. Progenies are tested for expression of the reporter gene is a consequence of being released from silencing. This approach was used to confirm silencing suppression ability of P1/HC-Pro protein of *Tobacco etch virus* (TEV) independently. Another procedure to examine the silencing suppression by viruses is the use of grafting of plant parts to evaluate the ability of virus-encoded silencing suppressor to obstruct movement of systemic RNA silencing signals. Grafting experiments were very useful in identifying suppressor activity associated with p25 of PVX and it was shown to interfere with spread of silencing signal. The P1/HC-Pro of TEV did not affect systemic signaling of silencing (Mallory et al. 2001).

RNA silencing has two distinct stages. At the initiation stage, host cells sense the presence of ds-RNA as abnormal and employ a ds-RNA-specific RNase as small interfering RNAS (siRNAs). Viral RNA may also be converted into a ds-RNA target by one of the RNA-dependent RNA polymerases (RdRps) encoded by the host plant (Yu et al. 2003; Molnar et al. 2005). The siRNA produced by the action of DCLs is then recruited into an RISC. At the maintenance stage, silencing of homologous RNA persists in the absence of the ds-RNA trigger. This is accomplished through a siRNA amplification process in which host RdRp synthesizes new ds-RNA, using siRNA as a primer and the homologous cellular RNA as template. Local induction of the generation of sequence-specific signals that spread systemically throughout the plant is the unique characteristic of RNA silencing in plants (Voinnet 2005). Geminiviruses possess C1, a multifunctional pathogenicity factor, which could suppress responses to jasmonates in the host plant. Sulfur, essential for plant life, is considered to have a role in plant defense in a phenomenon known as sulfur-induced resistance (SIR) or sulfur-enhanced defense (SED). *Geminivirus* C2 protein was shown to repress expression of genes involved in sulfur assimilation pathway in *Arabidopsis*. This adverse effect of geminivirus could be nullified by exogenous supply of jasmonate. The results suggested that infection of plants by geminiviruses might be counteracted through manipulation of jasmonate signaling pathway (Lozano-Duran et al. 2012).

The plant virus genome encodes suppressors of RNA silencing to specifically counteract the RNA-silencing-based defense mechanisms to ensure successful systemic infection of the host plant. In potyviruses, the protein P1/HC-Pro has multiple functions such as influencing aphid transmission polyprotein processing, genome amplification, and long-distance virus movement. P1/HC-Pro can strongly suppress RNA silencing most likely by acting on a maintenance step affecting the assembly and/or targeting of RISC complex (Chapman et al. 2004). The role of P1/HC-Pro in genome amplification and long-distance movement was highly correlated with its silencing suppressor activity. Suppression of RNA silencing-based host defense was critical for vigorous vital RNA replication followed by efficient systemic spread of the virus (Kassachau and Carrington 2001). Plant virus mutants, following deletion of genes encoding silencing suppressor proteins, may induce systemic or localized RNA silencing against themselves and other RNA molecules containing homologous sequences. *Cucumber mosaic virus* (CMV) mutant of the subgroup 1A strain Fny (Fny-CMV Δ2b), which could not express the 2b silencing suppressor protein, cross-protected tobacco and *Nicotiana benathamiana* plants against disease induction by wild-type Fny-CMV strain (Ziebell et al. 2004). The 2b protein of CMV has dual functions, as an RNA silencing suppressor protein and also as an inducer of disease symptoms. Hybrid virus FRad532b-CMV (renamed CMV-FRad2b-Pro) generated by replacement of the *2b* gene of the strain Fny-CMV with that from Rad35-CMV, displaced markedly low pathogenicity than Fny-CMV on *Nicotiana* species. However, replacement of proline with leucine at position 55 of the 2b protein of CMV-FRad2b-Pro (protein Rad2b-Pro) created a virus (CMV-FRad2b-Leu) that induced severe symptoms. Transgenic expression of Rad2b-Leu, but not of Rad2b-Pro in *Arabidopsis* induced symptom-like phenotypes and rescued the accumulation of the 2b deletion mutant FnyCMVd2b (Xu et al. 2013).

In *Grapevine leafroll-associated virus 3* (GLRaV-3) presence of a RNA silence suppressor in this genus *Closterovirus* was identified. RNA silencing suppressors are present in the 3′ end of the viral genome. Only p19.7 exhibited RNA silencing suppressor activity in diverse silencing inducing systems. This suppressor protein could overcome strong silencing inducers. The presence of silencing suppressor in *Ampelovirus* was detected in grapevine plants (Gouveia et al. 2012). *Turnip crinkle virus* (TCV) is highly virulent to *Arabidopsis* plants and silencing suppressor-defective mutants *TCVΔCP* could induce highly localized RNA silencing. *TCVΔCP* mutant provided some protection against wild-type TCV, but lacked complete protection. Cross-protection provided by *TCVΔCP* required the challenge virus to have similar sequence, which is characteristic of RNA silencing. Use of mutant plants involving key silencing pathway genes of RNA machinery including *dcl2, dcl4* and triple *dcl2/dcl3/dcl4* mutants was associated with highly localized RNA silencing. Cross-protection provided by *TCVΔCP* was dependent on host RNA silencing (Chen et al. 2014). Transcripts of open reading frames (ORFs) of *Sugarcane yellow leaf virus* (SCYLV), representing different parts of the virus determined in the sink leaves, source leaves, and mature internodes of two (SCYLV)-susceptible and one SCYLV-resistant Hawaiian sugarcane cultivars. Sink leaves contained higher transcript levels of ORF0–1, which coded for a suppressor protein than source leaves and internodes. Transcript levels for ORF3–4 which coded for CP and movement protein were significantly lower in source leaves in two of three cultivars than in sink leaves and nodes. Transcription of viral ORFs, particularly of the gene for the suppressor protein might be modulated in the sugarcane plant (ElSayed and Komor 2013). Transgenic tobacco plants expressing *Cauliflower mosaic virus* multifunctional protein, Tav (transactivator viroplasmin) under control of an artificially inducible promoter were developed. Following induction of Tav expression, pathogenesis-related (PR) protein *1a* gene expression was upregulated in the transgenic lines which exhibited visible chlorosis. Expression of Tav also induced salicylic acid (SA)- and ethylene-responsive PR genes. In a deletion analysis, chlorosis and PR1a induction function in transgenic tobacco were mapped to a region in Tav that had a role in pathogenesis in a susceptible host, elicitation of the hypersensitive response in a resistant host, suppression of RNA silencing and suppression of *Tomato bushy stunt virus* P19-mediated cell death in tobacco (Waliullah et al. 2015).

Virus disease symptoms may be due to toxic effect(s) of some components of viruses. Induction of plant 'suicide' defenses related to hypersensitive response (HR) has been studied. In general, HR is associated with a defense response perceived by receptors known as *R* genes, which confine the virus to the inoculated area resulting in inhibition of potential virus replication, leading to systemic infection of other plant tissues. Several virus specific *R* genes have been identified in plants, which not only bring about HR in response to particular virus but also prevent viral replication. HR and resistance are related, yet they are independent phenomena. For example, interaction of the product of *R* gene Rx-1 from potato with capsid protein (CP) of *Potato virus X* (PVX) could cause cell death. This gene was able to block virus replication before the concentration of viral CP required to trigger the necrotic response is reached (Bendahmane et al. 1999). In contrast, production of HR necrotic lesions may not hinder the virus from propagating throughout the plant, giving way to systemic infection. Resistance of *Arabidopsis* to *Cucumber mosaic virus* (CMV), derived from induction of *RCY1* gene by viral CP, was suppressed by sporadic mutations in resistance gene which exerts diverse effects on the development of local necrotic lesions (Sekine et al. 2006). Viruses replicate to increase in population in initially infected cells in different cycles to facilitate their spread to adjacent and other tissues, inducing symptoms characteristic of the virus. *Rice black-streaked dwarf virus* (RBSDV) (maize rough dwarf disease) was inoculated in eight maize cultivars using viruliferous planthoppers *Laodelphax striatellus* in the greenhouse. Among the susceptible varieties, the time points with maximum replication rates of the highly susceptible Ye478 and XH6 were significantly earlier than those of other susceptible varieties. Virus titers correlated negatively with levels of resistance of maize varieties (Xiaohua et al. 2019).

Different viral gene products, in addition to coat protein (CP), may be able to induce HR. Contribution of host resistance gene to reduce viral pathogenicity is variable and complex as revealed by the soybean-*Soybean mosaic virus* (SMV) pathosystem. Mechanical inoculation of soybean with SMV *Rsv1*gene elicited resistance to the majority of SMV isolates with no apparent symptoms. However, following graft inoculation, necrotic lesions were formed in stem, petioles, and leaf veins with cytological and histological characteristics of HR (Hajimorad and Hill 2001). On the other hand, isolate SMV-G7 caused lethal systemic HR following mechanical inoculation, whereas SMV-G7d, a variant derived from SMV-G7, induced systemic mosaic symptoms. Analysis of recombinant viruses revealed that ability to induce systemic HR resided in protein P3 (Hajimorad et al. 2005). However, wild-type SMV strain with P3 sequence from either SMV-G7 or SMV-G7d was unable to infect plants with *Rsv1* gene, whereas SMV-G7 and SMV-G7d with wild-type P3 were no longer pathogenic to these plants (Hajimorad et al. 2006). Concurrent modifications in proteins P3 and helper component HC-Pro were required to confer ability to overcome *Rsv1*-derived resistance on SMV wild-type isolate. However, single mutations in P3 were able to confer virulence on other *Rsv1* cultivars (Eggenberger et al. 2008; Hajimorad et al. 2011). *Cucumber mosaic virus* Fast New York strain (CMV-Fny) containing a mutated 2b protein (CMV-Fny Δb) was evaluated for its ability to infect bell pepper or Calwonder plants in comparison to wild-type CMV-Fny. Plants inoculated with mutant did not develop local or systemic symptoms, whereas CMV-Fny infected plants developed systemic chlorosis. CMV-FnyΔ2b accumulated in inoculated Calwonder leaves and inconsistently infected some noninoculated leaves at a low titer. Accumulation of CMV-FnyΔ2b in protoplasts at lower concentration compared to CMV-Fny was recorded. CMV2b protein might be required for systemic infection of Calwonder pepper plants and accumulation of the virus in inoculated protoplasts (Masiri et al. 2011). Local lesion response of tobacco to *Tobacco mosaic virus* (TMV) was associated with inhibitor of virus replication (IVR) gene (Elad et al. 2012).

Pepino mosaic virus (PepMV)–induced diseases manifest with a wide range of symptoms including systemic necrosis. Substitution of lysine for glutamic acid at position 67 in triple gene block 3 (TGB3) protein resulted in enhanced virus accumulation. Overexpression of both TGB3 and polymerase domain (POL) of the RNA-depended RNA polymerase (RdRp) led to necrosis, although only local expression of POL triggered HR-like symptoms. The results indicated that necrosis-inducing activity of POL was present in its highly conserved 'palm' domain and necrosis was jasmonic acid–dependent (Sempere et al. 2016). *Brassica carinata* (Ethiopian mustard) TZ-SMN-44–6 plants were sap inoculated with *Turnip mosaic virus* (TuMV) to study programmed cell death (PCD) pathways, using light microscopy and electrolyte leakage. Leaf responses to avirulent (TuMV-avir) and virulent (TuMV-vir) isolates and mock inoculation were compared. Generally TuMV-vir caused significantly greater cell deformation than TuMV-avir. Isolate TuMV-avir induced significantly greater electrolyte leakage than isolate TuMV-vir or mock inoculation. Both isolates might trigger morphological changes consistent with apoptotic-like PCD and necrosis-like PCD that depended on isolate virulence and stage of infection respectively (Nyalugwe et al. 2016). *Potato virus Y^{NTN}* (PVY^{NTN}) strain became systemic in susceptible plants. Healthy and PVY^{NTN}-infected potato tuber tissues were profiled for host gene expression. Tuber necrosis was characterized by high virus concentration and pronounced expression of defense-associated genes. Common defense responses associated with cell wall fortification, oxidative and biotic stress were induced. Responses in necrotic tissues of susceptible tubers resembled hypersensitive response of resistant primarily infected plants, but hormonal signaling was an important factor differentiating it from resistant plant tissues (Sedlar et al. 2018). HR to PVY is conferred by the *Ny-1* gene in potato cultivars. Tetraploid potato breeding clone pB07–37 possessed the allele *Ny-1* in a duplex state. Upon inoculation with PVY, size of necrotic lesions in leaves of PB07–37 was reduced, compared to cv. Rywal plants displaying *Ny-1* in a simplex dosage. Reproducible PVY induced proteins (60) involved in qualitative changes were differentially expressed in inoculated leaves (Szajko et al. 2019).

7.2.1.5.1 Movement of Viruses in Plants

7.2.1.5.1.1 Cell-to-Cell Movement of Viruses

Plant viruses replicating in the initially infected cells and then moving out of these cells to adjacent cells is known as cell-to-cell movement. Interaction with plant tissues of different kinds, results in interference with cellular transport process, which is a potentially effective form of changing physiological activities of host plant. Virus multiplication and movement are essential for disease symptom production. Infective virus cycle in susceptible plant generally commences through epidermal cells into which the vectors deposit the virions. In both types of movements, particularly in cell-to-cell movement, involvement of virus-encoded MPs is required. MPs could act by forming ribonucleoprotein complexes with viral genome or tubular structures that hold viruses to allow them to cross plasmodesmata (Lucas 2006; Sanchez-Navarro et al. 2006). Plasmodesmata are modified to a lesser or greater extent to allow cell-to-cell movement of viruses and then to distant plant organs including flowers and seeds through vascular system. Plasmodesmata (PD) function as cytoplasmic bridges and establish membrane (endoplasmic reticulum [ER] and plasma membrane), as well as cytoplasmic continuity between adjacent plant cells. Viral genomes are far larger than the size exclusion limit (SEL) of PD. Viruses move first from replication sites to PD at the cell periphery and then traverse the intracellular channels to enter neighboring cells (Nelson and Citovsky 2005).

Many virus-coded MPs may be delivered to PD via ER, whereas actin and myosin filaments may regulate flow of proteins in ER membrane (Boevink and Oparka 2005). Many viruses form replication centers enriched in endoplasmic reticulum and many viral proteins are translocated to ER. The role of ER translocation and plasmodesmatal targeting was investigated using calreticulin, a cellular protein that localized to PD (Chen et al. 2005). CP has a role in virus movement, in addition to providing protection to viral genome against inactivation. Requirements of *Cucumber mosaic virus* (CMV) for cell-to-cell movement appeared to be different. Both MP and CP were essential for cell-to-cell movement. CMV variants lacking CP could not move from infected cell to adjacent cells. Virion assembly was not a prerequisite for CMV movement, since assembly-defective CMV variants efficiently spread and induced local lesions. Disease development could be arrested by blocking cell-to-cell and long-distance movement of plant viruses which have direct effect on symptom expression (Narayanasamy 2008). As the severity of symptoms is not necessarily correlated with virus titer, disease development might be the result of specific interaction between virus and host components. Movement proteins of viruses interact with a large variety of host proteins promoting virus movement. The movement protein (TGBp2) of *Potato virus X* (PVX) interacted with cytoplasmic protein Tip-1 which in turn interacted with ß-1,3-glucanase (Fridborg et al. 2003), an enzyme that participates in regulation of callose levels playing an important role in the regulation of the SEL of PD (Iglesias and Meins 2000). Constitutive or tissue-specific expression of viral MPs might trigger typical viral infection symptoms, such as an abnormal accumulation of sugars, reduced photosynthesis, chlorosis and dwarfism. Transgenically expressed MPs preferentially locate to the plasmodesmata. Symptoms of virus diseases might be due to functional interference of these proteins with cytoplasmic communication channels. One of the effects of overexpression of MPs might be enhancement of SEL of PD (Olesinski et al. 1996). Changes in carbohydrate distribution, originated in mesophyll tissue following overexpression of MP were recorded in plant infected by *Tobacco mosaic virus* and *Cauliflower mosaic virus* (Love et al. 2005). Tobacco plants responded to constitutive expression of the MP from *Tomato spotted wilt virus* (TSWV) (NSm) by plasmodesmata sealing. Symptom development correlated with obstruction of PD by callose deposits. Symptoms of transgenic plants expressing NSm might be caused by a basal defense response that might counteract prolonged MP interference by altering plasmodesmatal function (Rinne et al. 2005).

7.2.1.5.1.2 Long-Distance Movement of Viruses Intracellular and intercellular movement of virus replication complexes (VRCs) of *Tobacco mosaic virus* (TMV) was monitored in intact leaf tissue. VRCs in primarily infected cells were associated with ER. High intracellular mobility by VRCs was seen later. VRCs traversed the PD and then process of formation and movement of VRCs in adjacent cells was detected. Application of inhibitors of filamentous actin and myosin blocked rapid movement of VRCs and the spread to adjacent cells (Kawakami et al. 2004). Tobamoviruses like *Tomato mosaic virus* (ToMV) have viral replication complexes containing a viral 126-kDa replicase component (126-k) which are generated transitorily in association with a viral movement protein. Transient expression of the *Tobamovirus* 126-k protein resulted in formation of cytoplasmic bodies with shapes similar to those of the replication complexes. ToMV MP proteins were transiently and simultaneously expressed in *Nicotiana benthamiana*. The 126-k protein was evenly distributed in the cytoplasm with occasional formation of small 126-k bodies, when MP was not expressed. These small 126-k bodies moved in the cytoplasmic stream and along the ER network. Small mobile 126-k bodies were captured at the terminal part of the MP-derived filamentous structures, resulting in the formation of 126-k-MP complexes and their translocation to another cellular site (Sasaki et al. 2014). Viruses included in *Potyvirus* genus have a genome with about 1,000 nucleotides encoding a polyprotein that is processed by three proteinases P1, HC-Pro, and NIa to synthesize ten mature viral proteins, P1, HC-Pro, P3, K1, C1, 6K2, VPg, NIa-Pro, Nib and CP. All viral proteins are involved different stages of viral cycle. The N-terminal part of HC-Pro controls aphid transmission, virulence (symptom severity), genome amplification, and virus accumulation. The central region controls both long-distance movement and replication maintenance functions. The C-terminal region of HC-Pro is a cysteine type proteinase and has a role in virus cell-to-cell movement. HC-Pro is also a pathogenicity determinant (Redondo et al. 2001). Involvement of HC-Pro in different steps of virus cycle such as transmission, replication, systemic movement and as a suppressor of posttranscriptional gene silencing (PTGS), in addition to cell-to-cell movement has been indicated (Pilsson et al. 2003). The movement complexes, comprising many proteins, function in a coordinated manner with different roles for protein components of the complex. Gating appears to be a basic requirement for cell-to-cell movement of nontube-forming viruses. Actin is involved in depolymerization leading to an increase in stomatal exclusion limit (SEL). In contrast, deposition of callose in response to defense wound responses, may lead to closure of plasmodesmata (Roberts et al. 2005). Interaction between *Potato virus* X (PVX) TGB2 and proteins that interact with ß-1,3-glucanase, a callose-degrading enzyme was suggested to be a strategy of PVX employed to gate PD to accelerate callose degradation (Fridborg et al. 2003). Interaction of *Tobacco mosaic virus* (TMV) MP with pectin methylesterase (PME) may regulate the activity of PME, loosening cell wall around PD and allowing PD to open more easily. Alternatively MP might recruit additional PME to PD to assist gating rather than relying on PME for targeting (Chen et al. 2005). Transport of viral genome is facilitated by viral MPs targeting cell periphery to mediate passage throughout PD. PME is an essential host factor enabling MPs in PD gating. Direct interaction of PME with *Tobacco mosaic virus* (TMV) MP is required for viral movement and in turn for virus viability. Level and timing of protein accumulation with regard to virus inoculation and MP expression might be crucial for functional outcome of PME-MP interaction and in turn for successful virus infection. Beneficial effect of downregulation of PME gene expression on development of the TMV-induced gene silencing (VIGS) approach was adopted to downregulate PME gene expression which resulted in reduction of TMV symptom severity and correspondingly to reduction of TMV-RNA accumulation in systemically infected leaves. The results indicated that *Nicotiana benthamiana* plants with hindered expression of PME might survive TMV infection which killed nonsilenced plants within a week (Bubici et al. 2015).

The majority of plant viruses depend on the phloem of host plants for long-distance movement, whereas some viruses are able to move through xylem. Different mechanisms may operate singly or in combination depending on the virus concerned. Host defense system silencing signal spreads between cells and also long distances from the lower to upper leaves. Movement of viruses to other plant organs depends on an RNA-dependent RNA polymerase (RdRp), whereas the signal for cell-to-cell movement does not depend on RdRp. Barriers formed as constitutive or postinfectional responses may block the systemic spread of the virus at different stages. Long-distance movement of *Potato leafroll virus* (PLRV) was studied using potato clones showing resistance to virus multiplication and accumulation, when inoculated through grafting or aphid transmission. Clones M62759 and PS1706 were highly resistant and susceptible to PLRV. Both clones expressed high level of resistance to virus multiplication. PLRV could be detected only in a few or none of the progeny plants of clone 62579, indicating strong inhibition of virus transport to tubers (Syller 2003). *Potato virus Y* (PVY) infects potato plants systemically inducing varying levels of severity of symptoms. Distribution of PVY in highly susceptible potato cultivar was monitored in different plant organs and tissues and location of viral RNA within plant tissues was traced. PVY accumulated in leaves, stem, shoot tips, roots, and tubers. Level of virus accumulation was specific for each organ or tissue. Symptomatic leaves and stem contained highest amounts of viral RNA and viral particles. Viral cytoplasmic inclusion bodies were localized in close vicinity to the epidermis and in trichomes. Viral RNA, viral particles, and cytoplasmic inclusion bodies colocalizes within the same type of cells or in close vicinity (Kogovšek et al. 2011).

Movement of *Cucumber mosaic virus* strain M (CMV-M) in infected zucchini-squash (*Cucurbita pepo*) was investigated. Reduction in cell-to-cell movement of CMV-M in inoculated leaves resulted in inhibition of systemic infection of zucchini squash. However, coinfection of zucchini plants along with *Zucchini yellow mosaic virus* (ZYMV) strain A facilitated CMV-M to move systemically. In upper leaves of plants doubly infected with CMV-M and ZYMV-A CMV-M CP gene-specific production was absent in plants inoculated with both viruses. ZYMV-encoded HC-Pro or other proteins of ZYMV might be involved in systemic movement of CMV-M into uninoculated top leaves away from the site of inoculation (Choi et al. 2002). Involvement of *Plum pox virus* (PPV) HC-Pro in long-distance movement in tobacco plants was observed. PPV could systemically infect several *Nicotiana* spp. However, it replicates in tobacco leaves, but could not spread systemically in this host. Failure of PPV to move long-distance could be complemented in transgenic tobacco plants, expressing the 5′ terminal region of the *Tobacco etch virus* (TEV) genome that naturally infected tobacco systemically. The TEV transgene with a mutation in the HC-Pro coding sequence could not facilitate systemic movement of PPV in tobacco. Activity of HC-Pro in the systemic movement of potyvirus of HC-Pro might be highly host-specific (Sáenz et al. 2002). Interaction between *Cucumber green mottle mosaic virus* (CGMMV) proteins and host factors was investigated. CGMMV helicase (HEL) domain interacted with a *Citrullus lanatus* small heat shock protein (sHSP). The sHSP was regulated on initial CGMMV infection in both *Nicotiana benthamiana* and *C. lanatus* plants, but not in systemically infected leaves. Subcellular localization of sHSP was changed after CGMMV infection. The sHSP inhibited viral RNA accumulation and retarded disease development in *N. benthamiana* transgenic plants. However, sHSP silencing did not have any significant effect on CGMMV infection (Liu et al. 2019). The tripartite genome of *Cucumber mosaic virus* (CMV) encodes replicase function in RNA1 and RNA2, with RNA3 encoding MP and CP. Both MP and CP are required for local cell-to-cell movement of CMV in squash. The role of CP in systemic movement was host-specific. CMV can infect cucumber systemically, whereas *Tomato aspermy virus* (TAV) was not able to do so. Systemic movement of *Cucumber mosaic virus* (CMV) in cucumber plants was studied by analyzing the structure that is translocated and its putative interaction with phloem components in the phloem exudates (PEs) samples. Viral particles interacted with a PE protein p48 which was a cucumber homolog of phloem protein 1 (PP1) from pumpkin. PP1 from pumpkin processes PD gating property and ability to translocate in sieve tubes. Interaction between encapsidated

CMV particles and P48 protein appeared to make viral RNA more resistant to RNase. The results indicated that the interaction with p48 protein modified the CMV particle structure and suggested that the virus particles might interact with cucumber homolog of PP1 during translocation in sieve tubes of cucumber (Requena et al. 2006).

Long-distance movement of *Wheat streak mosaic virus* (WSMV) was investigated through mutations within HC-Pro coding region of WSMV. Mutants bearing the GUS reporter gene were evaluated for their ability to establish primary infection foci. In vitro assays showed that HC-Pro proteinase activity was present in the mutants that could not establish foci of infection. The results showed that a complete HC-Pro deletion mutant could infect plants systemically and that certain substitutions in this dispensable gene probably prevented systemic infection by WSMV via interference (Stenger et al. 2006). *Triticum mosaic virus* (TriMV) was studied to follow the process of infection. In vitro RNA transcripts generated from full-length cDNA clones of TriMV were infectious on wheat. Wheat seedlings inoculated with in vitro transcripts induced mosaic and mottling symptoms similar to wild-type virus and the progeny virus was efficiently transmitted by wheat curl mites, indicating that cloned virus retained pathogenicity, movement and natural transmissibility by the vector mites. GFP-tagged TriMV could be detected in leaves, stems, and crown (Tatineni et al. 2015). *Wheat yellow mosaic virus* (WYMV) has a bipartite genome consisting of RNA1 and RNA2. The pathotypes I and II were discriminated by their pathogenicity to wheat cv. Fukuho. Pathotype I could infect Fukuho, but pathotype II was not pathogenic to this cultivar. A chimeric pathotype II virus harboring a cylindrical inclusion (CI) cistron from pathotype I facilitated systemic infection of Fukuho, indicating that CI protein was involved in pathogenicity. The differential pathogenicity between pathotypes I and II might depend on the ability for cell-to-cell movement or long-distance viral movement in which CI protein was involved. Involvement of CI protein in pathogenicity was clearly demonstrated (Ohki et al. 2019). *Maize chlorotic mottle virus* (MCMV), a positive sense ss-RNA virus, was investigated to determine subcellular localization of MCMV CP by tagging with fluorescent proteins and its role in infection. The CP fusion proteins could be imported into the nucleus of host cells and the N-terminal region rich in basic amino acids was essential for its nuclear localization. Bimolecular fluorescence complementation assays in maize protoplasts revealed that MCMV CP interacted with nuclear import factor importin-α1a (ZmIMPα1a) and importin-α1b (ZmIMPα1ß), suggesting that nuclear import of MCMV was mediated by importing α/ß pathway in plants (Zhan et al. 2016).

7.2.1.5.1.3 Colonization of Plant Tissues by Viruses Various plant tissues of plants provide differing degrees of support for viral replication. However, in some plant species, viruses replicate to reach fairly high titers without perceptibly inconveniencing the host and remain asymptomatic without producing any visible symptoms. Virus concentration in various tissues may be detected and quantified at different intervals after infection or inoculation, by employing suitable molecular techniques. Locations of *Citrus tristeza virus* (CTV) antigens could be visualized directly and precisely by in situ immunoassay (ISIA) in plant tissues. ISIA technique provided a unique advantage of determining the 3-dimensional distribution of CTV within the host (Lin et al. 2000). *Tomato spotted wilt virus* (TSWV) infected *Ranunculus asiaticus* grown for tuber production, potted plants and cut flowers. Distribution of TSWV in tuber and root tissues of *R. asiaticus* could be traced using tissue-blot immunoassay. TSWV was not uniformly distributed in *Ranunculus* tubers and the percentages of infected rootlets per tuber differed. TSWV could replicate in different organs of *R. asiaticus* plants (Whitefield et al. 2003). The presence of TSWV in different parts of potato tubers was monitored using ELISA test. Rates of TSWV transport/movement from infected foliage to tubers and from infected tubers to germinating sprouts/plants were assessed. Cultivars of potato had significant effect of TSWV movement from plant to tuber and vice versa. TSWV was unevenly distributed in infected tubers (Bulajić et al. 2014). Electron microscopy and decoration methods were applied to assess cytopathological effects of colonization of onion tissues by *Artichoke yellow ring spot virus* (AYRSV). Large cytoplasmic inductions were present only in mesophyll cells of infected plants. Cytoplasmic inclusions contained large accumulations of entangled membranous elements, more or less developed and specifically arranged particle aggregates and conspicuous globules of a presumed lipid nature. The cytopathologic structures primarily consisted of complexes of two cisternae of rough ER in a parallel arrangement. A high percentage of infected cells contained AYRSV particles (Maliogka et al. 2006). *Lilium* spp. are frequently infected by *Lily mottle virus* (LMoV) and *Cucumber mosaic virus* (CMV) simultaneously, causing severe symptoms. Western blotting showed that CPs of both LMoV and CMV accumulated in lily chloroplasts. Immunogold labeling further revealed that CPs of both viruses were localized in the stroma and thylakoid membranes of chloroplasts. In addition, CPs of both viruses were rapidly transported into isolated, intact chloroplasts. The CPs of both viruses exhibited species selection in their transmembrane transport into chloroplasts. Accumulation of high levels of CPs of LMoV and CMV inside chloroplast seemed to contribute to a synergistic interaction inducing severe mosaic symptoms (Zhang et al. 2019).

Distribution of *Grapevine leafroll-associated virus 3* (GLRaV-3), *Grapevine virus A* (GVA) and *Grapevine fan leaf virus* (GFLV) in embryonic callus of grapevine was monitored using digoxigenin-labeled oligonucleotide probe in in situ hybridization. Four months after culture initiation, a mosaic of infected and uninfected cells was present in callus originated by GFLV-infected plants. In callus originated by GVA- and GLRaV-3-infected explants the viruses were concentrated in a few cells surrounded by areas of virus-free cells. These two viruses were generally localized in different clusters of cells inside the callus and levels of infection were lower than those observed in GFLV-infected callus (Gambino et al. 2010). *Tomato necrotic streak virus* (TomNSV) moved downward to the roots from the initial point of infection and then upward from the root to tissues of active growth, such as flowers, fruits, and young leaves, which expressed symptoms of TomNSV infection (Badillo-Vargas et al. 2016).

Infection process of *Tomato yellow leafcurl virus* (TYLCV) operating in tomato fruit was studied. Real-time quantitative RT-PCR assay showed that TYLCV accumulated with increasing titers in early developing tomato fruit tissues from anthesis until 21 days post-anthesis. TYLCV DNA and transcripts of the CP gene localized specifically to the phloem tissue of young fruit, as well as sepals and petals. Embryos of developing seeds were infected. Expression of a host histone H4 gene was used as a marker for S-phase and the gene was expressed in phloem cells of young tomato fruit. TYLCV was transported to developing tomato fruit, where virus titer gradually increased because of movement of virus to other tissues and probably due to virus replication in fruit tissues (Just et al. 2017). Replicase (P1 protein) of *Prune dwarf virus* (PDV) is an important viral protein essentially required for infection of PDV. Replicase anchors viral RNA and builds replication complex along with RNA-dependent RNA polymerase (RdRp) in tobacco. PDV replication was strongly connected with tonoplasts. Further, PDV replicase and CP were also firmly associated with membranes of the ER. Potential involvement of these membrane structures in the process related to viral infection could be inferred (Koziel et al. 2017). Colonization of seed tissues by *Pepper mild mottle virus* (PMMoV) was monitored using fluorescence microscopy. Two distinct patterns of virus distribution were recognized. The presence of PMMoV in the epidermis and parenchyma was detected, but not in the endosperm and embryo, whereas in other seeds virus presence was restricted to the surface of epidermis and parenchyma (Genda et al. 2005). Infection cycle of PMMoV was studied through vertical transmission of the virus. Location of PMMoV within seeds, as they developed on inoculated seedlings of pepper was followed. The presence of PMMoV was revealed by the fluorescence around immature seeds and placenta in the ovaries. Area with fluorescence increased progressively expanded from the placenta into integument and the parenchyma and finally reached the top of immature seeds. The embryos or endosperm beyond the endothelium did not fluoresce (Genda et al. 2011).

The route of infection and pattern of distribution of *Prunus necrotic ringspot virus* (PNRSV) in apricot pollen were monitored. PNRSV was present inside as well as on the outside of infected apricot pollen grains. PNRSV was detected in all flower buds, including sepals, petals, and anthers. Viral RNA was localized only in the vascular tissue in some, while other anthers showed complex invasion of PNRSV with pollen mother cells (PMCs) also being infected. The virus could be detected in all tissues of flower buds, including anthers containing unicellular pollens. PNRSV infection was uniform in the early stages of maturation of pollen grains, cytoplasm exhibiting high levels of invasion by the virus. PNRSV was not present in the nucleus of the pollen grains examined. In the bicellular pollen grains, PNRSV was detected in the cytoplasm of vegetative cells and the virus was sometimes localized around and inside the generative cell. The presence of PNRSV inside the generative cell was also observed. In the germinated infected pollen grains, viral RNA was evenly distributed inside the pollen grains, carrying the virus to the embryo sacs followed by infection of developing apricot seed (Amari et al. 2007, 2009). *Pea enation mosaic virus* (PEMV) contains two genomes: PEMV-1 and PEMV-2 and a satellite RNA3 associated with PEMV-2. PEMV genomes accumulated to high levels in stipules, whole pea, and seed coat, but at much lower levels in pea embryo tissues. PEMV-1and PEMV-2 genomic RNAs accumulated to high titers in all pod wall and seed coat samples tested. PEMV-1 and PEMV-2 genomes could often be detected in embryo tissues, but at much lower concentrations. However, no evidence for seed transmission of PEMV was obtained (Timmerman-Vaughan et al. 2009).

7.2.1.5.1.4 Expression of Disease Symptoms Plants may respond to virus infection by producing hypersensitive response (HR), resulting in formation of local necrotic lesions on the inoculated leaves or systemic symptoms in other organs. However, in some plant species, the virus becomes systemic and no visible symptoms are induced. Recovery from severe phase of disease (cross-protection) confers significant level of resistance to infection by the same virus or closely related virus/strain. The link between the ability of plant viruses to reprogram host gene expression and their pathogenicity has been studied. Interference of viruses with cellular transport process leads to alteration of the physiology of susceptible plants. Infective viral cycle in a susceptible plant generally commences through epidermal cells following introduction of the virus by its vector or abrasion of plant surface. After completion of viral genome replication, the progeny virions move from infected cells and reach vascular system to invade distal plant organs. MPs can act by forming ribonucleoprotein complexes with viral genomes or tubular complexes that hold the virions and allow them to cross plasmodesmata (Narayanasamy 2008).

Development of disease symptoms may be due to specific interaction between virus and host components. Plant virus MPs have been shown to interact with a range of host proteins to facilitate movement of the virus in susceptible plants. Host proteins with which viral MPs interact may be localized in the nucleus, cytoplasm, endoplasmic reticulum, microtubules, and plasma membrane. Most of these interactions may be essential for the intra- and intercellular movement of viruses and these interactions have significant effects on symptom expression. TGBp2, *Potato virus*

X MP, interacted with cytoplasmic protein Tip-1 which in turn interacted with ß-1,3-glucanase (Fridborg et al. 2003). This enzyme participates in the regulation of callose levels which have an important role in regulation of the SEL of the plasmodesmata (Iglesias and Meins 2000). Protein P3 from potyviruses is an inducer of the responses mediated by several dominant resistance genes (Hajimorad et al. 2006; Chowda Reddy et al. 2011). Protein P3, along with the small 6-K1 peptide, was bound to *Soybean mosaic virus* polyprotein, and they condition the symptomatology and host range of the virus (Desbiez et al. 2003; Salvador et al. 2008). Replacement of P1-coding sequence of *Plum pox virus* (PPV) by that of another potyvirus *Tobacco vein mottle virus* (TVMV) had no effect on the virus infectivity in plants that were hosts of both PPV and TVMV, but inhibited infection by PPV on *Prunus persica* (Salvador et al. 2008). The relationship between virus pathogenicity determinant domain and induction of mosaic symptoms by *Pepper mild mottle virus* (PMMV) was investigated. Mutants were generated by a base substitution in the 130-K replication protein gene. Mutant with amino acid residue substitution at the 349 position alone did not induce any visible symptom, although the mutant reached relatively high concentration. Activity to suppress posttranscriptional gene silencing (PTGS) was impaired in the mutant that failed to induce symptom. The development of symptoms might be controlled by interaction between host PTGS and its suppression by the 130 K replication protein rather than virus accumulation (Tsuda et al. 2007).

Cauliflower mosaic virus (CaMV) with ds-DNA genome causes chlorosis in cruciferous crops for which gene VI protein (P6) was responsible. Primary role for P6 in the CaMV infection cycle is to modify the host translation machinery, termed translocation transactivator (TAV), facilitating translation of the polycistronic CaMV 35S RNA. A variant of P6 derived from CaMV D4 strain did not induce chlorosis upon transformation into *A. thaliana*. Two variants of P6 induced strong chlorosis and stunting in *Arabidopsis*. Complementation assay proved that P6 expressed in the D4–2 line was functional, since it was able to support the systemic infection of a CM1841 mutant (Yu et al. 2003). Geminiviruses do not have DNA polymerase activity and they have to depend on host factors for amplification of their genomes. As a consequence, an early step in the geminivirus infection process, is the induction of the required replication machinery. *Geminivirus* replication enhancer (REn) proteins (also called C3, Ac3, L3, or AL3) could enhance viral DNA accumulation and enhance infectivity and symptom expression, as in *Tomato leafcurl virus* (TLCV) (Selth et al. 2005). Some viral genes in combination with host plant factors may function as symptom determinants. Hybrids of *Cymbidium ringspot virus* (CymRSV) and *Carnation Italian ringspot virus* (CIRSV) belonging to *Tobamovirus* were examined to identify viral symptom determinants responsible for the generalized necrosis in infected plants. Necrotic response of *Nicotiana benthamiana* was due to the products of ORF1 (p35 and p19) of the hybrids. Symptoms in *N. bethamiana* infected by hybrids were entirely different. Hybrids containing chimeric ORF1 did not induce lethal necrosis, even when the level of viral replication was not reduced significantly. The production of necrotic phenotype required the presence of Wt p33 in addition to the p19 (Burgyán et al. 2000).

Multiple infections of plants by viruses occurring in synergism or antagonism mode are observed under natural conditions. Potyvirus-associated synergistic interactions are seen more frequently. Mechanism of synergistic interaction operating between CMV and *Potato virus Y* was suggested to be due to complementation of a movement deficiency by one of the viruses (Ryang et al. 2004). The effects of synergistic interaction between *Tomato spotted wilt virus* (TSWV) and *Tomato chlorosis virus* (ToCV) occurring in mixed infection were assessed. Tomato cv. Anastasia susceptible to ToCV, but resistant to TSWV and cv. Moneymaker susceptible to both viruses were used. A dramatic increase in ToCV was observed in leaf tissues of Moneymaker tomato plants coinfected by TSWV. RNA hybridization analysis showed confinement of ToCV in phloem cells of tomato plants either singly or doubly infected by the viruses and no difference in distribution of TSWV could be seen in both types of infections. When ToCV preceded TSWV inoculation, resistance to TSWV was compromised in Anastasia tomato plants. However, breakdown of TSWV resistance did not occur following simultaneous inoculation with ToCV and TSWV. But loss of TSWV resistance in commercial tomato cultivars containing resistance *Sw -5* gene was observed due to the presence of ToCV along with TSWV. Preinoculation with ToCV led to increased susceptibility to TSWV, whereas coinoculation did not alter resistance levels of tomato plants (Garcia-Canoe et al. 2006).

MicroRNAs (miRNAs), the posttranscriptional regulators, have a vital role in nutrient and water uptake and transfer in plants. Expression alterations of 11 miRNAs and their target mRNAs in tomato root and stem tissues were quantified following infection with *Cucumber mosaic virus* (CMV)-Fny strain, CMV-FnyΔ2b (a CMV-Fny 2b deletion mutant), CMV-Fny-sat T1 [CMV-Fny plus an aggressive satellite (sat) RNA variant at T1] and *Tomato aspermy virus* (TAV-Bj). Stem tissues became hollow in CMV-Fny-sat T1infected tomatoes, and a fibrosin phenotype at 21 days postinoculation onwards in the stems of TAV-Bj infected plants was detected (Feng et al. 2013). The *Citrus tristeza virus* (CTV) locus provided broad-spectrum resistance to CTV in *Poncirus trifoliata*. The difference in plant 24-nt sRNAs distribution corresponding to the *CTV-20* gene was detected earlier in susceptible species, following CTV infection. The presence of differences in methylation status of a specific region of *CTV-20* in two susceptible species, sour and sweet orange, following CTV infection was observed. On the other hand, no significant differences were observed in the tolerant Citrange Carrizo following infection. Further, hyper-methylation of the whole *CTV-20* gene was observed in Citrange Carrizo. Methylation in regulation of CTV-20 expression may have a role in CTV susceptibility (Fanelli et al. 2018).

7.2.1.5.1.5 Pathophysiology of Virus-Infected Plants
Several plant defense-related functions may be activated or suppressed by virus infection, depending on the compatibility/incompatibility of virus-host interactions. The development of symptoms in different host plant species following infection by viruses with different levels of virulence indicates the extent of derangement in essential physiological process required for normal growth and reproduction of plants. Photosynthesis is one of the basic physiological functions required for synthesis of carbohydrates and other compounds carried out in the presence of light. Disease development in tobacco after inoculation with *Cucumber mosaic virus* (CMV) was accelerated when plants were grown in high light (HL) intensity compared with those grown in low light (LL) intensity. Characteristic defense-related reactions such as release of ROS, accumulation of SA, expression of SA-mediated pathogenesis-related (PR) proteins, and the systemic development of disease were light-dependent. Plants exposed to HL or medium light had serious oxidative damage caused by CMV infection compared with LL-plants. Further, the levels of SA and mRNA transcripts for nonexpressor of pathogenesis-related gene 1 (NPR1) in CMV-inoculated plants were the highest under LL with decreasing photo flux density. In HL exposed plants, the expression of *PR1* and *PR2* were repressed, but the expression of *PR5* was not significantly altered, revealing a complex cross talk between light intensity and the SA-signaling pathway in regulating the response to tobacco infection by CMV. The results showed that light intensity reduced level of resistance to CMV in tobacco and reduction in resistance was correlated to repression of the SA-dependent signaling pathway and a change in cellular redox strains in plant cells (Chen et al. 2015).

Wheat streak mosaic virus (WSMV) infection affected photosynthetic parameters such as chlorophyll content, net photosynthetic rate, stomatal conductance, intercellular CO_2 concentration, and transpiration rate, which were reduced in WSMV-infected wheat plants. Early infection reduced photosynthetic parameters to much lower levels, compared to late infection. Photosynthetic parameters showed a negative linear relationship with grain yield and shoot biomass (Pradhan et al. 2015). The influence of *Citrus tristeza virus* (CTV) infection on photosynthetic activity and antioxidant metabolism was determined in highly susceptible citrus genotype Mexican lime (*Citrus aurantifolia*) using severe strain T318 and mild strain T385. Marked reduction in photosynthetic activity in infected plants was observed. This adverse effect could be attributed to a reduction in the carboxylative efficiency, whereas limitation of CO_2 diffusion through stomata had a reduced impact. CTV infection did not damage the antennae and did not reduce the efficiency of light harvesting complexes. CTV infection also accelerated activity of ascorbate peroxidase activity in new shoots formed in infected plants (Pérez-Clemente et al. 2015).

Responses of two potato cultivars Igor and Nadine to two isolates of *Potato virus Y* (PVY), the aggressive PVY^{NTN} and the mild PVY^{N} in different stages of infection were evaluated. Microarray and quantitative real-time PCR analyses were performed to identify differentially expressed genes after inoculation with each virus isolate. PVY^{NTN} accumulated in higher concentration in both potato cultivars. Microarray and qRT-PCR results implied an earlier accumulation of sugars and lower photosynthesis in leaves inoculated with virulent isolate compared with those inoculated with PVY^{N} isolate. PVY^{NTN} isolate did not activate early differential expression of the Fe-superoxide dismutase and pectin methylesterase (PME I) genes, indicating delay in plant response relative to that following PVY^{N} inoculation. Differences in the expression of ß-glucanase-1 gene were also observed in early plant responses to inoculation with each virus isolate (Kogovšek et al. 2010). Pretreatment of tobacco leaves with a low concentration of H_2O_2 suppressed HR-type necrosis associated with resistance to *Tobacco mosaic virus* (TMV). H_2O_2-mediated induced resistance to disease symptom correlated with enhanced activities of catalase (CAT), ascorbate peroxidase (APX), and guaiacol peroxide (POX) after viral infection. Induction of genes encoding antioxidants superoxide dismutase (SOD), CAT, and APX was also increased at early stage of TMV infection. Application of SOD and CAT suppressed necrosis induced by viral infection. Replication of TMV was not significantly affected due to H_2O_2 pretreatment of plants. SA, a primary component of plant defense, did not appear to function in H_2O_2-meidated symptom resistance. H_2O_2-mediated induced resistance to necrotic symptoms in tobacco appeared to depend on enhanced antioxidant potential of treated plants (Hafez et al. 2012). Transgenic lines of tobacco cv. Samsun NN variably silenced for RNA-dependent RNA polymerase (RDR-1) gene were assayed for peroxidase (POX) activity, following systemic infection with *Potato virus Y* strain O (PVY^{O}). Guaiacol-POX specific activity increased more than twofold in uninoculated upper leaves of susceptible lines silenced for RDR, compared to wild-type. The transgenic line R5–1 susceptible to PVY^{O} infection, with a much lower accumulation of RDR-2 transcript, showed two new POX activity bands compared to the R-14–1 line that was moderately suppressed for RDR-1. Transgenic lines with greater degree of susceptibility to the virus infection showed lower levels of transcript accumulation for both RDR-1 and MYB1 genes. The results suggested that downregulation of RDR-1 transport could lead to a change in pattern of expression of defense-related genes such as POX and MYB1 and enhanced accumulation of PVY^{O} in tobacco plants (Rakshandehroo et al. 2012). Begomoviruses and their vectors, whiteflies, can influence activities of plant defense proteins. The effects of *Tomato yellow leafcurl virus* (TYLCV) infection on POD activity of tomato plants were assessed. Leaves of different ages exhibiting differential POD activity in response to TYLCV infection and infestation with *Bemisia tabaci* B biotype were examined. Tomato leaf tissues exhibited increased

POD activity in the presence of TYLCV, mature leaves showing greater activity than young leaves. Generally, both infected and healthy leaves showed higher POD activity during whitefly infestation. In the infected young leaves, POD activity was much reduced in healthy leaves and increased gradually with period of exposure to *B. tabaci* B infestation (Dieng et al. 2011).

Physiological performance of grapevine infected by *Grapevine leafroll-associated virus 1* and *3* (GLRaV-1 and -3) was evaluated. Gas exchange was negatively affected in infected plants. At ripening stage, reduction in stomatal conductance was higher than in net CO_2 assimilation rate (A), resulting in higher intrinsic water use efficiency (A/g) in infected leaves. Virus infection had a negative effect on light saturated rate of CO_2 fixation at high irradiances and carboxylation efficiency, but in contrast, apparent quantum yield of CO_2 fixation was significantly higher. Infection by GLRaV-1 and -3 led to significant decrease in photosynthetic pigments, soluble sugars and soluble protein contents, while starch and anthocyanins were significantly reduced, but K, Mg, B, Cu, Zn, and Mn were not affected by these viruses (Moutinho-Pereira et al. 2012). Infection of grapevine plants by GLRaV-3 induced reddish purple discoloration of leaf blades and downward rolling of leaf margins. These symptoms were considered to be due to expression of genes encoding enzymes involved in anthocyanin synthesis and accumulation of flavonoids in infected leaves. Enhancement of total anthocyanin accumulation was related to development of GLD symptoms in the infected plantlets under PEG stress. Upregulated gene expression attained the highest level in GLRaV-3 infection was the key factor in accelerated accumulation of anthocyanin as indicated by expression of two genes *MYBA1* and *UFGT* involved in anthocyanin biosynthesis (Cui et al. 2017). Plants remaining asymptomatic allowed virus replication. The effects of GLRaV-3 infection in asymptomatic plants of grapevine cv. Malvasia de Banyalbufar (MdB) and Cabernet Sauvignon (CS) were assessed. Net CO_2 assimilation (AN) and electron transport rate (Jflux) were the parameters primarily affected by GLRaV-3. Reduction in AN in infected plants was attributed to restrictions in CO_2 diffusion caused by anatomical leaf changes and a reduction of Rubisco activity. The cultivar MdB expressed these effects more prominently. Reduction of AN led to a decrease in total O_2 uptake rate by the cytochrome oxidase pathway, producing slight differences in plant growth. Although plants remained asymptomatic, vital physiological functions were affected in these plants also (Montero et al. 2017).

7.3 IMPACT OF VIROID INFECTION IN PLANT DEVELOPMENT

7.3.1 Characteristics of Viroids

Viroids are obligate, subviral plant pathogens comprising small unencapsidated, covalently closed, circular, and non-encoding single-stranded (ss) RNA molecules (246–401 nucleotides). They share structural similarities with some satellite RNAs associated with certain plant viruses. Whereas viroids can replicate independently, satellite RNAs depend on the helper virus for their replication. Viroids are RNA parasites making efficient use of synthetic machinery of the host plant for their replication. Infection of fruit crops such as citrus, grapevine, and apple by one or more viroids simultaneously may be observed often. Precise sensitive and reliable detection techniques have to be employed for identification and differentiation and quantification of viroids, since they induce overlapping symptoms resulting in complexes.

7.3.1.1 Biological Properties of Viroids

Although there may be similarities in the symptoms induced by viroids and viruses, some of the viroids such as *Potato spindle tuber viroid* (PSTVd) cause characteristic symptom of abnormally elongated potato tubers which assume a spindle or cylindrical shape with prominent eyes. In citrus following graft inoculation of sensitive selection of Etrog citron, viroids induce distinguishable symptoms (Roistacher 1991). Likewise, symptoms of infection of apple by *Apple scar skin viroid* (ASSVd) are restricted to apple fruits (Koganezawa et al. 2003). *Peach latent mosaic viroid*, (PLMVd) induces irregularly shaped discolored and deformed fruits with cracked structures and enlarged pits (Malfitano et al. 2003). *Potato spindle tuber viroid* (PSTVd) produced local necrotic lesions in *Scoplia sinensis* following mechanical inoculation (Singh 1971). Cucumber (*Cucumis sativus*) was used as assay host for *Hop stunt viroid* (HSVd) by maintaining the plants at 30° C (Sasaki et al. 1981). *Gynura aurantica* plants were effective as a bioindicator for detection and identification of *Citrus exocortis viroid* (CEVd) infecting grapevines (Flores et al. 1985). Bioindexing using Arizona 861-S1 Etrog citron plants was widely employed for detection and identification of CEVd, HSVd, and *Citrus viroid III* (CV-III) (Malfitano et al. 2005). Grapevine viroids, *Grapevine yellow speckle viroid 1* and *2* (GYSVd-1 and -2), *Hop stunt viroid* (HSVd), *Citrus exocortis viroid* (CEVd), and *Potato spindle tuber viroid* (PSTVd) could be detected and differentiated by inoculating cucumber cv. Alpha, tomato cv. Castle Rock, and *Gynura aurantica* as bioindicators (Nasr-Eldin et al. 2012). PSTVd appears to have the widest host range, infecting several crops and wild relatives of *Solanum* spp. (Lenarčic et al. 2012). *Tomato apical stunt viroid* (TASVd) could infect several ornamental species in addition to tomato and potato (Anonymous 2011).

Transmission of viroids may occur from infected plants through seeds, pollen, propagules, and contaminated tools used for propagation. The duration of time for which PSTVd could remain infective in extracted tomato leaf sap was determined after application to cotton, wood, rubber tires, leather, metal, plastic, human skin, and string. PSTVd was infective for 24 h in all surfaces except human skin. PSTVd survived the longest period on leather, plastic, and string. The results indicated the importance of decontamination of clothing, tools, and machinery to prevent PSTVd infection and spread to other plants (Mackie et al. 2015). Transmission of PSTVd occurred via true seeds and pollen of potato (Fernow et al. 1970) and through seeds of tomato (Benson and Singh 1964). Transmission of *Grapevine yellow speckle viroid 1* (GYSVd-1) and *Hop stunt viroid* (HSVd) occurring through seeds of eight grapevine cultivars was reported (Wah and Symons 1999). Pollen transmission of *Peach latent mosaic viroid* (PLMVd) via pollen from peach trees of three cultivars infected with an Italian isolate of PLMVd was recorded (Barba et al. 2007). Likewise, the route of transmission of *Hop latent viroid* (HLVd) from infected plant to seeds was determined (Matoušek et al. 2008). Transmission of *Tomato chlorotic dwarf viroid* (TCDVd) via eggplant seed varied widely. TCDVd was not transmitted through pollen, but vertical transmission was highly efficient when healthy eggplants were pollinated with infected pollen (Gramazio et al. 2019). *Columnea latent viroid* (CLVd) was localized in the seed coat and in the seedlings. RT-PCR assay showed the presence of CLVd in both seed coat and the seedlings. CLVd-infected pollen carried the risk of infection to eggplants via pollen during cross-pollination (Bhuvitarkorn and Reanwarakorn 2019).

Potato spindle tuber viroid (PSTVd) could infect large number of plant species, including many solanaceous crops and ornamental plant species. *S. jasminoides* was a better source of inoculum for PSTVd than *B. suveolens*. The results indicated the possibility of PSTVd transmission between crops while handling (Verhoeven et al. 2010a, 2010b). None of the seven infected cultivars of dahlias exhibited symptoms of infection by PSTVd in early stages of plant development but growth suppression could be seen during later stages. The new PSTVd isolates might spread to tomato and potato (Tsushima et al. 2019). PSTVd, *Tomato chlorotic dwarf viroid*, *Tomato apical stunt viroid*, and *Columnea latent virioid* infected many plant species belonging to Compositae and Solanaceae families. However, only *Calendula officinalis*, *Petumax* hybrid and *Solanum melongena* expressed symptoms of infection (Matsushita and Tsuda 2015). Infection of volunteer plants in tomato, pepper, and chili crops by PSTVd was commonly observed in Western Australia. *Solanum nigrum*, *Datura leichhadtii* (thorn apple), *Nicandra physalodes*, *Conzya bonariensis* and native species, *Atriplex semilunaris*, *Rhagodia eremaea*, and *Streptoglossa* spp. were infected by PSTVd to varying extent (Mackie et al. 2016). *Tomato planta macho viroid* (TPMVd) and *Pepper chat fruit viroid* (PCFVd) may infect horticultural crops commonly. Symptoms of infection were observed in tomato, potato, green pepper, and *Solanum muricatum*. Both TPMVd and PCFVd were seed transmissible. The rate of seed transmission of TPMVd in tomato and *Petunia* x *hybrida* was high, whereas seed transmission of PCFVd was at lower level in these hosts. Pollen transmission of both viroids occurred at high rates. The results showed availability of spectrum of alternative hosts with potential to spread through the infected propagules of crops and infected seeds (Yanagisawa and Matsushita 2017).

7.3.1.2 Infection Process of Viroids

Plant pathogenic viroids are classified into two families: *Pospiviroidae* and *Avsunviroidae*. After mechanical inoculation, development of systemic infection by viroids depends on successful completion of different processes: (i) import of viroid molecules into a subcellular organelle, chloroplast, or nucleus of the cell for *Avsunviroidae* or *Pospiviroidae*, respectively; (ii) replication inside the respective organelle; (iii) export out of the chloroplast or nucleus after replication; (iv) trafficking into neighboring cells; (v) entry into phloem; (vi) long-distance trafficking within phloem to upper leaves; and (vii) exit from phloem to invade nonvascular cells and repeat replication and trafficking cycle. As the viroids do not have protein component as the viruses, the viroid RNA has to interact with host-encoded enzymes that may degrade RNAs (Ding 2009).

After entry into susceptible cells, viroids recruit host proteins for intracellular cell-to-cell and long-distance trafficking. *Potato spindle tuber viroid* (PSTVd) is localized in the nucleus for predicted replication through activities of DNA-dependent RNA polymerase II (RdRpII) via rolling circle mechanism (Schindler and Muhlbach 1992). Tomato Vir P1 protein was identified based on its property of interacting with PSTVd replicating in the host cell nucleus. In addition to PSTVd RNA, Vir P1 bound to *Hop stunt viroid* (HSVd) also. The amino acid sequence of Vir P1 did not contain any known RNA-binding motifs; however, a bromo domain and two molecular localization signals were identified (Martinez de Alba 2000). A specific and saturable receptor mediated the cytoskeleton-independent import of PSTVd plus strands into the nucleus, where processing to the final circular forms might occur. PSTVd could move from cell-to-cell, if injected into mesophyll cells. Two PSTVd strains could enter the phloem for long-distance transport, but they were unable to exit phloem, suggesting regulated cell-to-cell transport (Zhu et al. 2002). PSTVd and *Avocado sun blotch viroid* (ASBVd) were translocated through phloem, forming complexes with phloem proteins, and this mode of trafficking might be sustained by viroid replication and governed by viroid RNA motif and plant development. This motif was not required for trafficking in the reverse direction, but it was dependent on leaf development stages (Qi et al. 2004). PSTVd might move from the photoassimilate-source leaf into sink organs as observed in phloem transport of other compounds. In situ localization of PSTVd provided evidence for the mode of phloem transport of viroids. PSTVd did not invade the shoot apical meristem (SAM). However, selective transport of PSTVd RNA into sepals, but not into other floral organs, of *Nicotiana benthamiana* was recorded (Zhu et al. 2001).

Potato spindle tuber viroid (PSTVd) RNA movement within sieve tubes did not simply mass flow from source to sink organs, but was transported into selective sink organs. Further, two PSTVd mutants could enter phloem to spread systemically, but could not exit phloem in leaves of tobacco (Zhu et al. 2002). Movement of PSTVd in sections of infected tomato plants was monitored using in situ hybridization assay. Viroid signals were observed in vascular tissues as well as in other cells in subapical region, but not in the SAM (Forster et al. 2002). In contrast, *Peach latent mosaic viroid* (PLMVd) could invade SAM of infected peach tree (Zhu et al. 2001, 2002). PLMVd could replicate in the albino leaf sectors and bypass the RNA surveillance system, resulting in invasion of SAM and changes in protoplastids (Rodio et al. 2006). PLMVd was well distributed in shoots, leaves, bark, roots, and fruits. Young developing fruits appeared to be the best source of PLMVd (Delgado et al. 2005). In order to understand the viroid infection, identification of structural elements that may interact with host factors or serve as reaction substrates necessary in establishing infection is important. Several viroid-interacting proteins of potential functional significance have been identified. The chloroplast protein PARPB33 interacted with *Avocado sun blotch viroid* (ASBVd), while VirP1 interacted with PSTVd. On the other hand phloem proteins interacted with several viroids (Darós and Flores 2002; Kalantidis et al. 2007). VirP1, a protein from tomato was an efficient host for PSTVd and was isolated by its capacity for its specific binding to PSTVd RNA (Maniataki et al. 2003). Six proteins capable of interacting with *Peach latent mosaic viroid* (PLMVd) were identified. The presence of these proteins was detected in peach leaf extract partially purified by affinity column chromatography followed by mass spectrometry (Ding 2009). Viroids may modify host gene expression interfering with normal developmental pathways. Mature RNA or its replicative intermediates are likely to be the primary pathogenic effectors interacting with host protein or RNA (Diener 2001). Specific viroid pathogenicity determinants in both nuclear and chloroplastic viroids are responsible for cellular changes associated with viroid disease symptoms. A U257A substitution in PSTVd was recovered as a spontaneous mutation in transgenic tobacco plants expressing the viroid cDNA (Zhu et al. 2002). Peach calico symptoms, characterized by extreme chlorosis of infected tissues, was shown to be associated with the insertion of an extra 12–13 nt sequence, which folded into a hairpin to the left terminal loop of *Peach latent mosaic viroid* (PLMVd) (Malfitano et al. 2003; Rodio et al. 2006).

The pattern of seed transmission of *Potato spindle tuber viroid* (PSTVd) was investigated using in situ hybridization to histochemically analyze distribution of PSTVd in each developmental stage of petunia. PSTVd was detected in the reproductive tissues of infected female x infected male and infected female and healthy male. After pollination, PSTVd was detected in the developed embryo and endosperm in all three crosses (Matsushita and Tsuda 2014). PSTVd-D accumulated slowly and induced very mild symptoms in tomato cv. Rutgers plants. In contrast, PSTVd-I replicated faster and induced severe symptoms. The number of PSTVd-sRNA reads extensively decreased in PSTVd-D infected leaf and stem tissues, particularly those derived from the regions containing the nucleotides 119, 127, and 202. Relative expression levels (number of reads by deep-sequencing) of various host microRNAs changed after infection with PSTVd-I and PSTVd-D in comparison to healthy controls. In PSTVd-I and PSTVd-D infected stem tissues, miR319 reached high concentrations in stem tissues and later decreased below that in healthy leaves (Tsushima et al. 2015).

Knowledge on viroid invasion of SAM and its distribution in plant reproductive tissues, particularly the embryo, is vital to determine the extent of seed transmission. Distribution of *Chrysanthemum chlorotic mottle viroid* (CChMVd) was analyzed histochemically in the shoot meristems and flower buds of CChMVd-infected chrysanthemum cv. Okayama-heiwa plants. CChMVd invaded SAM and leaf primordia in the shoot meristem at a high frequency. In flower buds, CChMVd was present in the bract, receptacle, vascular petals, ovules, and immature stamens and pistils. Further, seed transmission of CChMVd was at a high frequency when CChMVd-infected plants were used as female parents (Ebata et al. 2019). Viroid multiplication and cellular movement are dependent on host plant components. Exposure to low temperature reduced titers of *Chrysanthemum stunt viroid* (CSVd) in chrysanthemum plants. Distribution patterns of CSVd in developing shoot tips of CSVd-infected chrysanthemum plant were subjected to 1-, 3, and 6-month low-temperature treatment. After treatment for 3 months, weak CSVd signals were detected in the apical dome (AD), whereas 6-month treatment of plants resulted in the absence of CSVd signals in AD, but only in tissues within the vascular bundle directly beneath viroid-free AD tissues (Matsushita and Shima 2015).

Potato spindle tuber viroid (PSTVd) induced severe symptoms in the early stages and severity of symptoms decreased later, and almost healthy-looking leaves appeared on the upper portion of tomato plants. PSTVd titers attained peak levels in leaves with severe symptoms and declined in the upper leaves recovering from severe phase of disease development. PSTVd-specific short interfering RNAs (siRNAs), characteristic of RNA silencing, accumulated in all leaves in which PSTVd reached a high level. The results suggested that recovery from severe phase of disease might be associated with RNA silencing via sequence-specific degradation of PSTVd RNA genome (Sano and Matsuura 2004). siRNAs resulting from the host RNA silencing, the defensive response triggered by double-quasi-siRNAs, are considered to induce viroid disease symptoms. Viroid pathogenicity derived from RNA sequences (21–24 nt) during infection might guide RNA silencing of host genes and subsequent induction of disease (Wang et al. 2004). Levels of small RNAs and symptom severity in plants infected by PSTVd and *Avocado sun blotch viroid* (ASBVd) showed positive relationships. Further, development of symptom was correlated with production of small RNAs in tomato transgenic lines expressing nonreplicating hairpin PSTVd RNAs (Wang et al. 2004; Matoušek et al. 2007). Viroids belonging to two families induce posttranscriptional gene silencing (PTGS), based on the presence of siRNA, representing different regions of the viroid genome. Incubation of *Peach latent mosaic viroid* (PLMVd) transcripts of either plus or minus polarity with Dicer enzyme(s) led to detection of siRNA. Hairpin known to be implicated in the replication of PLMVd had the ability to trigger the Dicer enzymes. Viroids may interact with cell factors which redirect them to replicate and traffic throughout the plant, while avoiding the RNA silencing machinery they trigger in their hosts (Landry and Perreault 2005). Alterations in the tomato gene expression following infection by PSTVd and *Tobacco mosaic virus* (TMV) were compared. PSTVd strains altered expression of both common and unique tomato genes. The genes encode products involved in defense/stress response, cell wall structure, chlorophyll function, proven metabolism and other diverse functions. Expression of some genes altered by PSTVd was also altered by TMV infection. Viroids are capable of triggering complex host responses in spite of their simple structure (Itaya et al. 2007).

Coinfection of *Citrus* spp. by *Citrus tristeza virus* (CTV) and viroids was observed often. The effects of interactions between *Citrus dwarfing viroid* (CDVd) and CTV were investigated. CTV encodes three RNA silencing suppressors (RSSs), p23, p20, and p25. Transgenic Mexican lime plants expressing each of the RSSs were included in the investigation. Single and double infections of CDVd and CTV in Etrog citron (*C. medica*) were evaluated. Interactions between the pathogens were host-dependent. CTV enhanced accumulation of CDVd only in Mexican lime, while concomitantly CDVd limited accumulation of CTV. The reduced CTV titer in the presence of CDVd suggested that these pathogens compete for common biological factor. The role of RSS proteins in the accumulation of viroid RNAs (generated from ds-RNAs) in plants coinfected with CTV and CDVd was studied. In Etrog citron CDVd-sRNAs were detected at similar concentration regardless of whether plants were infected with CDVd alone or coinfected with CTV and CDVd. The results indicated that as with accumulation of genomic CDVd, CTV did not have any effect on the accumulation of CDVd-sRNAs. Accumulation of circular and linear CDVd forms was increased in Mexican lime in the presence of CTV. Titers of viroid genomic RNA were positively correlated with sRNAs consistently. Transgenic Mexican lime expressing p23 displayed enhanced CDVd titer, similar to that of nontransformed control plants coinfected with CTV. Transgenic lines expressing p20 contained viroid titer to the same level as nontransformed controls infected singly with CDVd. Transgenic lines expressing p25 contained CDVd titer slightly greater than that of singly infected nontransformed controls. Interaction between CDVd and CTV affected accumulation of viroid, but not symptom expression. On the other hand, CDVd infection significantly reduced the CTV titer (Serra et al. 2014).

REFERENCES

Alfaro-Fernández A, Serrano A, Tornos T, Cebrian MC, Córdoba-Selles MC et al. (2016) *Turnip yellow mosaic virus* in Chinese cabbage in Spain: commercial seed transmission and molecular characterization. Eur J Plant Pathol 146: 433–442.

Ali A, Mohammed O, Khattab A (2012) Distribution of viruses infecting cucurbit crops and isolation of potential new virus-like sequences from weeds in Oklahoma. Plant Dis 96: 243–248.

Allarangaye MD, Traore O, Traore EVS, Mallogo RI, Konate G (2006) Evidence of nontransmission of *Rice yellow mottle virus* through seeds of wild host species. J Plant Pathol 88: 309–315.

Amari K, Burgos L, Pallas V, Sanches-Pena MA (2007) *Prunus necrotic ringspot virus* early invasion and its effects on apricot pollen grain performance. Phytopathology 97: 892–899.

Amari K, Burgos L, Pallas V, Sanchez-Pina MA (2009) Vertical transmission of *Prunus necrotic ringspot virus* hitch-hiking from gametes to seedlings. J Gen Virol 90: 1767–1774.

Anonymous (2011) *Tomato apical stunt viroid* on tomatoes in the Netherlands. Pest Report, Plant Protection Service, The Netherlands, pp. 1–3.

Aramburu J, Galipienso L, Soler S, Rubio L, Lopez C (2015) A severe symptom phenotype in pepper cultivars carrying the *Tsw* resistance gene is caused by a mixed infection between resistance-breaking and non-resistance-breaking isolates of *Tomato spotted wilt virus*. Phytoparasitica 43: 597–605.

Backus EA, Shugart HJ, Rogers, Morgan JK, Shatters R (2015) Direct evidence of egestion and salivation of *Xylella fastidiosa* suggests sharpshooters can be "flying syringes". Phytopathology 105: 608–620.

Badillo-Vargas IE, Baker CA, Turechek WW, Frantz G, Mellinger HC et al. (2016) Genomic and biological characterization of *Tomato necrotic streak virus*, a novel subgroup 2 *Ilarvirus* infecting tomato in Florida. Plant Dis 100: 1046–1052.

Barba M, Ragozzino E, Faggioli F (2007) Pollen transmission of *Peach latent mosaic viroid*. J Plant Pathol 89: 287–289.

Bendahmane A, Kanyuka A, Baulcombe DC (1999) The *Rx* gene from potato controls separate virus resistance and cell death responses. Plant Cell 11: 781–792.

Benedict CA, Mc Morran DW, Inglis DA, Karasev AV (2015) Tuber symptoms associated with recombinant strains of *Potato virus Y* in speciality potatoes under western Washington growing conditions. Amer J Potato Res 92: 593–602.

Benson AP, Singh RP (1964) Seed transmission of *Potato spindle tuber viroid* in seeds and pollen of infected potato plant. Amer Potato J 47: 75–80.

Berkin S, Luigi M, Parrella G, Giorgini M, Davino S et al. (2018) Survey of the distribution of *Bemisia tabaci* (Hemiptera: Aleyrodidae) in Lazio region (Central Italy): a threat for the northward expansion of *Tomato leafcurl New Delhi virus (Begomovirus, Geminviridae)* infection. Phytoparasitica 46: 171–182.

Bhuvitarkorn S, Reanwarakorn K (2019) Pollen and seed transmission of *Columnea latent virus* in eggplants. Eur J Plant Pathol 154: 1067–1075.

Birithia R, Subramanian S, Pappu HR, Muthomi J, Narla RD (2013) Analysis of *Iris yellow spot virus* replication in vector and nonvector thrips species. Plant Pathol 62: 1407–1414.

Blaisdell GK, Cooper ML, Kuhn EJ, Taylor KA, Daane KM et al. (2016) Disease progression of vector mediated *Grapevine leafroll-associated virus 3* infection of mature plants under commercial orchards. Eur J Plant Pathol 146: 105–116.

Boevink P, Oparka KJ (2005) Virus-host interactions during movement process. Plant Physiol 138: 4–6.

Boukari W, Kaye C, Wei C, Hincapie M, La Borde C et al. (2019) Field infection of virus-free *Sugarcane yellow leaf virus* and effect on yellow leaf on sugarcane grown on organic and on mineral soils in Florida. Plant Dis 103: 2367–2373.

Brault V, Uzest M, Monsion B, Jacquot E, Blanc S (2010) Aphids as transport devices for plant viruses. CR Biol 333: 524–538.

Bubici G, Carluccio AV, Cillo F, Stavolone L (2015) Virus-induced gene silencing of pectin methylesterase protects *Nicotiana benthamiana* from lethal symptoms caused by *Tobacco mosaic virus*. Eur J Plant Pathol 141: 339–347.

Bulajić AR, Stankovic IM, Vucurovic AB, Ristic DT, Milojevic KN et al. (2014) *Tomato spotted wilt virus*-potato cultivar susceptibility and tuber transmission. Amer J Potato Res 91: 186–194.

Burgyán J, Hornyik C, Szitty G, Silhavy D, Bisztray G (2000) The ORF 1 products of tombusviruses play a crucial role in lethal necrosis of virus-infected plants. J Virol 74: 10873–10881.

Burrows ME, Caillaud MC, Smith DM, Gray SM (2007) Biometrical genetic analysis of luteovirus transmission in the aphid *Schizaphis graminum*. Heredity 98: 106–113.

Byamukama E, Robertson AE, Nutter FE Jr (2015) *Bean pod mottle virus* time of infection influences soybean yield, yield components and quality. Plant Dis 99: 1026–1032.

Byamukama E, Seifers DL, Hein L, De Wolf E, Tisserat NA et al. (2013) Occurrence and distribution of *Triticum mosaic virus* in the central Great Plains. Plant Dis 97: 21–29.

Byamukama E, Tatineni S, Hein GL, Baenziger PS, French R et al. (2012) Effects of single and double infections of winter wheat by *Triticum mosaic virus* and *Wheat streak mosaic virus* on yield determinants. Plant Dis 96: 859–864.

Byamukama E, Tatineni S, Hein GL, McMechan J, Wegulo SN (2016) Incidence of *Wheat streak mosaic virus, Triticum mosaic virus* and *Wheat mosaic virus* in wheat curl mites recovered from maturing wheat spikes. Plant Dis 100: 318–323.

Carmo-Sousa M, Moreno A, Plaza M, Garzo E, Fereres A (2016) *Cucurbit aphidborne yellows virus* (CABYV) modifies the alighting behavior, settling and probing behavior of its vector *Aphis gossypii* favoring its own spread. Ann Appl Biol 169: 284–297.

Carroll TW (1980) *Barley stripe mosaic virus*: its economic importance and control in Montana. Plant Dis 64: 136–140.

Chapman EL, Prokhnevsky AI, Gopinath K, Dojia VV, Carrington JC (2004) Viral RNA silencing suppressors inhibit the microRNAs pathway at n intermediate step. Genes Develop 18: 1179–1186.

Chen AYS, Walker GP, Carter D, Ng JCK (2011) A virus capsid component mediates virion retention and transmission by its insect vector. Proc Internatl Acad Sci USA 108: 16777–16782.

Chen L-J, Ren H, Deng XG, Li Y-N, Cha W-Q, Lin H-H, Xi D-H (2015) Effects of light intensity on the susceptibility of *Nicotiana tabacum* to *Cucumber mosaic virus*. J Gen Plant Pathol 81: 399–408.

Chen MH, Tian GW, Gafni Y, Citovsky V (2005) Effects of calreticulin on viral cell-to-cell movement. Plant Physiol 138: 1866–1876.

Chen Y, Lu C, Li M, Zhao G, Wei T (2016) Adverse effects of *Rice gall dwarf virus* upon its insect vector *Recilia dorsalis* (Hemiptera: Cicadellidae). Plant Dis 100: 784–790.

Chen Y-J, Zhu F, Liu J, Deng XG, Pu XJ et al. (2014) A *Turnip crinkle virus* isolate lacking the CP counter-defence protein gene providing protection against the wild-type strain is associated with highly localized RNA silencing. J Phytopathol 162: 758–769.

Choi SK, Yoon JY, Ryu KH, Choi JK, Palukaitis P et al. (2002) Systemic movement of a movement-deficient strain of *Cucumber mosaic virus* in zucchini squash is facilitated by a cucurbit-infecting potyvirus. J Gen Virol 83: 3173–3178.

Chou C-N, Wu M-L, Su H-J, Yeh H-H (2009) Biological and molecular characterization of Taiwanese isolates of *Cucumber mosaic virus* associated with banana mosaic disease. J Phytopathol 157: 85–93.

Chowda Reddy RV, Sun H, Chen H, Poysa V, Ling H et al. (2011) Mutations in the P3 protein of *Soybean mosaic virus* G2 isolates determine virulence in *Rsv-4* genotype of soybean. Molec Plant-Micro Interact 24: 37–43.

Coutts BA, Jones RAC (2015) Potato virus Y: contact transmission, stability, inactivation and infection sources. Plant Dis 99: 387–394.

Coutts BA, Kehoe MA, Jones RAC (2013) *Zucchini yellow mosaic virus*: contact transmission, stability on surfaces and inactivation with disinfectants. Plant Dis 97: 765–771.

Cui Z-H, Bi W-L, Hao X-Y, Li P-M, Duan Y et al. (2017) Drought stress enhances upregulation of anthocyanin biosynthesis in *Grapevine leafroll-associated virus 3*-infected in vitro grapevine (*Vitis vinifera*) leaves. Plant Dis 101: 1606–1615.

da Silva RM, de Souto ER, Pedroso JC, Arakava R, Almeida AMR et al. (2008) Detection and identification of TMV infecting tomato under protected cultivations in Parana State. Braz Arch Biol Technol 51: 903–909.

Daimei G, Ravina HS, Devi PP, Saurav GK, Renuka Devi P et al. (2017) Influence of *Groundnut bud necrosis virus* on the life history traits and feeding preference of its vector, *Thrips palmi*. Phytopathology 107: 1440–1445.

Darós JA, Flores R (2002) A chloroplast protein binds a viroid RNA in vivo and facilitates its hammerhead-mediated self-cleavage. EMBO J 21: 749–759.

de Lillo E, Pozzebon A, Valenzano D, Duro C (2018) An intimate relationship between eriophyid mites and their host plants – a review. Front Plant Sci 9: 1786.

Debreczeni DE, Rubio L, Aramburu J, Lopez C, Galipienso L et al. (2013) Transmission of *Tomato spotted wilt virus* isolate able and unable to overcome tomato or pepper resistance by its vector *Frankliniella occidentalis*. Ann Appl Biol 163: 266–276.

Delgado S, Martinez de Alba E, Hernandez C, Flores R (2005) A short double-stranded RNA motif of *Peach latent mosaic viroid* contains initiation and the self-cleavage sites of both polarity strands. J Virol 79: 12934–12943.

Desbiez C, Gal-On A, Girard M, Wipf-Scheibel C, Lecoq HC (2003) Increase in *Zucchini yellow mosaic virus* symptom severity in tolerant zucchini cultivars is related to a point mutation of relative fitness on susceptible plants. Phytopathology 93: 1478–1484.

Diener TO (2001) The Viroid: biological oddity or evolutionary fossil? Adv Virus Res 57: 137–184.

Dieng H, Satho T, Hassan AA, Azis AT, Morales RE et al. (2011) Peroxidase activity after viral infection and whitefly infestation in juvenile and mature leaves of *Solanum lycopersicum*. J Phytopathol 159: 707–712.

Ding B (2009) The biology of viroid-host interactions. Annu Rev Phytopathol 47: 105–131.

Dombrovsky A, Reingold V, Antignus Y (2014) *Ipomovirus*- an atypical genus in the family *Potyviridae* transmitted by whiteflies. Pest Manag Sci. https://doi.org/10.1002/ps.3735

Domingo-Calap ML, Moreno AB, Diaz-Pendon JA, Moreno A, Fereres A et al. (2020) Assessing the impact on transmission and insect vector behavior of a viral mixed infection in melon. Phytopathology 110: 174–186.

Drucker M, Froissart E, Hebrard M, Uzest M, Raallec M et al. (2009) Intracellular distribution of viral gene products regulates a complex mechanism of *Cauliflower mosaic virus* acquisition by its aphid vector. Proc Natl Acad Sci USA 99: 2422–2427.

Du Z, Fu Y, Liu Y, Wang X (2020) Transmission characteristics of *Wheat yellow striate virus* by its leafhopper vector *Psammotettix alierrus*. Plant Dis 104: 222–226.

Dunoyer P, Voinnet O (2005) The complex interplay between plant viruses and host RNA silencing. Curr Opin Plant Biol 8: 415–423.

Ebata M, Matsushita Y, Morimoto M, Mochizuki T (2019) Distribution of *Chrysanthemum chlorotic mottle viroid* in shoot meristem and flower buds of chrysanthemum. Eur J Plant Pathol 154: 555–563.

Eggenberger AL, Hajimorad MR, Hill JH (2008) Gain of virulence on *Rsv1*-genotype soybean by an avirulent *Soybean mosaic virus* requires concurrent mutation in both P3 and HC-Pro. Molec Plant-Micro Interact 21: 931–936.

Elad Y, Rav-David D, Leibman D, Vintal H, Vunsh R et al. (2012) Tomato plants transformed with the inhibitor of virus replication gene are partially resistant to several pathogenic fungi. Ann Appl Biol 161: 16–23.

Ellis MD, Hoak JM, Ellis BW, Brown JA, Sit TL et al. (2020) Quantitative real-time PCR analysis of individual flue-cured tobacco seeds and seedlings reveals seed transmission of *Tobacco mosaic virus*. Phytopathology 110: 194–205.

ElSayed AI, Komor E (2013) Quantitative analysis of transcripts of the open reading frames of *Sugarcane yellow leaf virus* genome by one-multiplex RT-PCR: evidence for a high transcript level of suppressor gene in sink leaves. J Phytopathol 161: 774–783.

Fageria M, Nie X, Gallagher A, Singh M (2015) Mechanical transmission of *Potato virus Y* (PVY) through seed cutting and plant wounding. Amer J Potato Res 92: 143–147.

Fanelli V, De Giovanni C, Saponari M, Leonetti P, Ricciardt L, Montemurro C (2018) A possible role of CTV-20 gene methylation in response to *Citrus tristeza virus* infection. Eur J Plant Pathol 150: 527–532.

Fariña AE, Rezende JAM, Wintermantel WM (2019) Expanding knowledge of the host range of *Tomato chlorosis virus* and host plant preference of *Bemisis tabaci* MEAM 1. Plant Dis 103: 1132–1137.

Favara G, Bampi D, Molina JPE, Rezende JAM (2019) Kinetics of systemic invasion and latent incubation periods of *Tomato severe rugose virus* and *Tomato chlorosis virus* in single and coinfections in tomato plants. Phytopathology 109: 480–487.

Feng J, Wang Y, Lin R, Chen J (2013) Altered expression of microRNAs and target mRNAs in tomato root and stem tissues upon different viral infection. J Phytopathol 161: 107–119.

Fernow KH, Peterson LC, Plasited RL (1970) *Spindle tuber virus* in seeds and pollen of infected potato plant. Amer Potato J 47: 75–80.

Ferriol I, Rubio L, Perez-Panades J, Carbonell EA, Davino S et al. (2013) Transmissibility of *Broad bean wilt virus* 1 by aphids: influence of virus accumulation in plants, virus genotype and aphid species. Ann Appl Biol 162: 71–79.

Flores R, Duran-Vila N, Pallás V, Semancik JS (1985) Detection of viroid and viroid-like RNAs from grapevine. J Gen Virol 66: 2095–2107.

Forster TM, Lough TJ, Emerson SI, Lee RH, Bowman JL et al. (2002) A surveillance system regulates selective entry of RNA into the shoot apex. Plant Cell 14: 1497–1508.

Fox A, Collins LE, Macarthur R, Blackburn LF, Northing P (2017) New aphid vectors and efficiency of transmission of *Potato virus A* and strains of *Potato virus Y* in the UK. Plant Pathol 66: 325–335.

Franz W, van der Wilk F, Verbeek M, Dullemans AM, van den Heuvel JF (1999) *Faba bean necrotic yellows virus* (genus: *Nanovirus*) requires a helper factor for its aphid transmission. Virology 262: 210–219.

Fridborg I, Grainger J, Page A, Coleman M, Findlay K et al. (2003) TIP, a novel host factor linking callose degradation with cell-to-cell movement of *Potato virus X*. Molec Plant-Micro Interact 16: 132–140.

Fuji S, Zen S, Satp O, Kishi H, Furuya H et al. (2015) Population dynamics of Iris yellow spot virus in onion fields and lisianthus greenhouses in Japan. Plant Pathol 64: 808–816.

Gamarra HA, Fuentes S, Morales FJ, Glover R, Malumphy C et al. (2010) *Bemisia afer* sensu lato, a vector of *Sweet potato chlorotic stunt virus*. Plant Dis 94: 510–514.

Gambino G, Vallania R, Gribaudo I (2010) In situ localization of *Grapevine fan leaf virus* and phloem-restricted viruses in embryonic callus of *Vitis vinifera*. Eur J Plant Pathol 127: 557–570.

García-Cámara D, Perez-Brito O, Moreno-Valenzuela A, Magna-Alvarez A, Fernandes PMB et al. (2018) Molecular experimental evidence of watermelon (*Citrullus lanatus*) as host of the Mexican variant of *Papaya meleira virus*. Eur J Plant Path 151: 117–123.

Garcia-Canoe E, Resende RO, Fernandez-Minoz R, Morines E (2006) Synergistic interaction between *Tomato chlorosis virus* and *Tomato spotted wilt virus* results in breakdown of resistance in tomato. Phytopathology 96: 1253–1263.

Garita LC, Tassi AD, Calegairo RF, Kitajima EW, Carbonell SAM et al. (2013) Common bean: experimental indicator plant for *Citrus leprosis virus C* and some other cytoplasmic type *Brevipalpus*-transmitted viruses. Plant Dis 97: 1346–1351.

Garret A, Kerlan C, Thomas D (1996) Ultrastructural study of acquisition and retention of potato leafroll luteovirus in the alimentary canal of its aphid vector *Myzus persicae* Sulz. Arch Virol 141: 1279–1292.

Genda Y, Sato K, Nunomura O, Hirabayashi T, Ohmshi J et al. (2005) Immunolocalization of *Pepper mild mottle virus* in *Capsicum annuum* seeds. J Gen Plant Pathol 71: 238–242.

Genda Y, Sato K, Nunomura O, Hirabayashi T, Tsuda S (2011) Immunolocalization of *Pepper mild mottle virus* in developing seeds and seedlings of *Capsicum annuum*. J Gen Plant Pathol 77: 201–208.

Ghanim M, Brumin M, Popovski S (2009) A simple rapid and inexpensive method for localization of *Tomato yellow leafcurl virus* and *Potato leaf roll virus* in plant and insect vectors. J Virol Meth 256: 75–84.

Ghanim M, Czosnek H (2000) Tomato yellow leafcurl geminivirus (TYLCV-Is) is transmitted among whiteflies (*Bemisia tabaci*) in a sex-related manner. J Virol 74: 4738–4745.

Ghanim M, Morin S, Zeidan M, Czosnek H (1998) Evidence for transovarial transmission of *Tomato yellow leafcurl virus* by its vector the whitefly, *Bemisia tabaci*. Virology 240: 295–303.

Gill-Salas FM, Peters J, Boonham N, Cuadrado IM, Janssen D (2011) Yellowing disease in zucchini squash produced by mixed infections of *Cucurbit yellow stunting disorder virus* and *Cucumber vein yellowing virus*. Phytopathology 101: 1365–1372.

Gottlieb Y, Zchori-Fein E, Mozes-Daube N, Kontsedalov S, Skaljac M et al. (2010) The transmission efficiency of *Tomato yellow leafcurl virus* by the whitefly *Bemisia tabaci* is correlated with the presence of a specific symbiotic bacterium. J Virol 84: 9310–9317.

Gotz M, Popovski S, Kollenberg M, Gorovits R, Brown JK, Cicero JM, Czosnek H, Winter S, Ghanim M (2012) Implication of *Bemisia tabaci* heat shock protein 70 in begomovirus-whitefly interactions. J Virol 86: 13241–13252.

Gouveia P, Dandlen S, Costa A, Marques N, Nolasco G (2012) Identification of an RNA silencing suppressor encoded by *Grapevines leafroll- associated virus 3*. Eur J Plant Pathol 133: 237–245.

Gramazio P, Lerma MD, Villanueva G, Vilanova S, Garcia-Fortea E et al. (2019) Detection of molecular characterization and aspects involving the transmission of *Tomato chlorotic dwarf viroid* in eggplant. Ann Appl Biol 175: 172–183.

Gray S, Gildow FE (2003) *Luteovirus*-aphid interactions. Annu Rev Phytopathol 41: 539–566.

Hafez YM, Baesó R, Kiraláy Z, Kunstler A, Kiraláy L (2012) Upregulation of antioxidants in tobacco by low concentrations of H_2O_2 suppress necrotic disease symptoms. Phytopathology 102: 848–856.

Hajimorad MR, Eggenberger AL, Hill JH (2005) Loss and gain of elicitor function of *Soybean mosaic virus* G7 providing Rsv1-mediated lethal systemic hypersensitive response to maps to P3. J Virol 79: 1215–1222.

Hajimorad MR, Eggenberger AL, Hill JH (2006) Strain-specific P3 of *Soybean mosaic virus* elicits *Rsv-1* mediated extreme resistance, but absence of P3 elicitor function alone is insufficient for virulence on *Rsv-1* genotype soybean. Virology 345: 156–166.

Hajimorad MR, Hill JH (2001) Rsv1- mediated resistance against Soybean mosaic virus-N is hypersensitive response-independent at inoculation site, but has the potential to initiate hypersensitive response-like mechanism. Molec Plant-Micro Interact 14: 587–598.

Hajimorad MR, Wen RH, Eggenberger AL, Hill JH, Maroof MA (2011) Experimental adaptation of an RNA virus mimics natural evolution. J Virol 85: 2557–2564.

Harris KF (1977) In: Harris KF, Maramorosch K (eds.), *Aphids as Vectors*. Academic Press, New York, pp. 166–208.

Harth JE, Simmons HE, Stephenson AG (2017) Vertical infection *of Zucchini yellow mosaic virus* via pollen transmission occurs at a lower frequency than ovule transmission. Eur J Plant Pathol 147: 717–720.

Hogenhout SA, Ammar ED, Whitefield AE, Reddinbaugh MG (2008) Insect vector interaction with persistently transmitted viruses. Annu Rev Phytopathol 46: 327–359.

Honda K, Wei T, Hagiwara K, Higashi T, Kimura L et al. (2007) Retention of *Rice dwarf virus* by descendents of pairs of viruliferous vector insects after rearing for 6 years. Phytopathology 97: 712–716.

Houle JL, Kennedy GG (2017) *Tomato spotted wilt virus* can infect resistant tomato when western flower thrips inoculate blossoms. Plant Dis 101: 1666–1670.

Hu X, Meacham T, Ewing L, Gray SM, Karasev AV (2009) A novel recombinant strain of *Potato virus Y* suggests a new viral genetic determinant of vein necrosis in tobacco. Virus Res 143: 68–76.

Huang AT, Zhang H, Yang J, Chen JP, Hebrard E et al. (2011) Identification, characterization and distribution of *Southern black-streaked dwarf virus* in Vietnam. Plant Dis 95: 1063–1069.

Huo Y, Liu W, Zhang F, Chen X, Li L et al. (2014) Transovarial transmission of a plant virus is mediated by vitellogenin of its insect vector. PLoS Pathog 10(3): e1003949. https://doi.org/10.137/journal.ppat.1003949

Iglesias VA, Meins F Jr (2000) Movement of plant virus is delayed in a ß-1,3-glucanase-deficient mutant showing reduced plasmodesmatal size exclusion limit and enhanced callose deposition. Plant J 21: 157–166.

International Whitefly Genome Initiative (2015) The genome of whitefly *Bemisia tabaci*, an important plant pest and virus vector. International Whitefly Genome Initiative Published online. www.whiteflygenome.org

Itaya A, Zhong X, Bundshchuh R, Qi Y, Wang Y et al. (2007) A structural viroid RNA serves as a substrate for dicer-like cleavage to produce biologically active small RNAs but is resistant to RNA-induced silencing complex-mediated degradation. J Virol 81: 2980–2994.

Jaspars MV, Falloon PG, Pearson MN (2015) Seed and pollen transmission of *Asparagus virus 2*. Eur J Plant Pathol 142: 173–183.

Johansen LK, Carrington JC (2001) Silencing on the spot: induction and suppression of RNA silencing in *Agrobacterium*-mediated transient expression system. Plant Physiol 126: 930–938.

Jones JDG, Dangl IL (2006) Plant immune system. Nature 444: 323–329.

Jossey S, Hobbs HA, Domier LL (2013) Role of *Soybean mosaic virus* endosperm in seed and aphid transmission in soybean. Phytopathology 103: 941–948.

Just K, Arif U, Lui KA, Kvarnheden A (2017) Monitoring infection of tomato fruit by *Tomato yellow leaf curl virus*. Plant Pathol 66: 522–528.

Kalantidis K, Denti MA, Tzortzakaki S, Marinou E, Tabler M et al. (2007) VirP1 is a host protein with a major role in *Potato spindle tuber viroid* infection in *Nicotiana* plants. J Virol 81: 12872–12880.

Kassachau KD, Carrington JC (2001) Long-distance movement and replication maintenance functions correlate with silencing suppression activity of potyviral HC-Pro. Virology 285: 71–81.

Kaur AYS, Chen W, Zheng Y, Hasegawa DK, Ling KS et al. (2016) Transcriptomic studies of *Bemisia tabaci*, the whitefly vector of *Tomato chlorosis virus*. Proceedings of the 2nd International Whitefly Symposium, Arusha, Tanzania, p. 37.

Kawakami S, Watanabe Y, Beachy RN (2004) *Tobacco mosaic virus* infection spreads cell-to-cell as intact replication complexes. Proc Natl Acad Sci USA 101: 6291–6296.

Kawamura R, Shimura H, Mochizuki J, Okhi SI, Masuta C (2014) Pollen transmission *Asparagus virus -2* (Av-2) many facilitate mixed infection by two Av-2 (Av-2) isolates in asparagus plants. Phytopathology 104: 1001–1006.

Kil EJ, Lee YJ, Chor S, Auh CK, Kim D et al. (2015) Identification of natural weed hosts of *Tomato chlorosis virus* in Korea by RT-PCR with root tissues. Eur J Plant Pathol 142: 419–426.

Kil EJ, Park I, Choi EY, Byun HS, Lee KY et al. (2018) Seed transmission of *Tomato yellow leafcurl virus* in sweet pepper (*Capsicum annuum*). Eur J Plant Pathol 150: 759–764.

Kim J, Kil EJ, Kim S, Seo H, Byun HS et al. (2015) Seed transmission of *Sweet potato leafcurl virus* in sweet potato (*Ipomoea batatus*). Plant Pathol 64: 1284–1291.

Koganezawa H, Yang X, Zhu SF, Hashimoto J, Hadidi A (2003) *Apple scar skin viroid* in apple. In: Hadidi A, Flores R, Randles J, Semancik JS (eds.), *Viroids*. CSIRO Publishing, Collingwood, Australia.

Kogovšek P, Kladnik A, Mlakar J, Žnidarič CMT, Dermastia M et al. (2011) Distribution of *Potato virus Y* in potato plant organs, tissues, and cells. Phytopathology 101: 1292–1300.

Kogovšek P, Pompe-Novek M, Baebler Š, Rotter A, Gow L et al. (2010) Aggressive and mild *Potato virus Y* isolates trigger different specific responses in susceptible potato plants. Plant Pathol 59: 1121–1132.

Koziel E, Otulak K, Lockhart BEL, Garbaczewska G (2017) Subcellular localization of proteins associated with *Prune dwarf virus* replication. Eur J Plant Pathol 149: 653–658.

Landry P, Perreault JC (2005) Identification of a *Peach latent mosaic viroid* hairpin able to act as a dicer-like substrate. J Virology 79: 6540–6545.

Lapidot M, Guenoune-Gelbart D, Leibman D, Holdengreber V, Davidovitz M et al. (2010) *Pelargonium zonate spot virus* is transmitted vertically via seed and pollen in tomato. Phytopathology 100: 798–804.

Le Maguet J, Beuve M, Herrbach E, Lemaire O (2012) Transmission of six ampeloviruses and two vitiviruses to grapevine by *Rhenococcus aceris*. Phytopathology 102: 717–723.

Leh V, Jacquot E, Geldreich A, Haas M, Blanc S, Keller M, Yot P (2001) Interaction between the open reading frame III product and the coat protein is required for transmission of *Cauliflower mosaic virus* by aphids. J Virology 75: 100–106.

Lenarčic R, Morisset D, Mehle N, Ravinkar M (2012) Fast real-time detection of *Potato spindle tuber viroid* by RT-LAMP. Plant Pathol 62: 1147–1156.

Liang Z, Dickison V, Singh M, Xiong X, Nie X (2016) Studies of tomato plants in response to infections with PVX and different PVY isolates reveal a remarkable PVX-PVY^{NTN} synergism and diverse expression profiles of genes involved in different pathways. Eur J Plant Path 144: 55–71.

Lin Y, Rundell PA, Xie L, Powell CA (2000) In situ immunoassay for detection of *Citrus tristeza virus*. Plant Dis 84: 937–940.

Liu HW, Luo LX, Li JQ, Liu PF, Chen XY et al. (2014) Pollen and seed transmission of *Cucumber green mottle virus* in cucumber. Plant Pathol 63: 72–79.

Liu J, Li L, Zhao H, Zhou Y, Wang H et al. (2019) Titer variation of *Citrus tristeza virus* in aphids at different acquisition access periods and its association with transmission efficiency. Plant Dis 103: 874–879.

Llave C (2010) Virus-derived small-interfering RNAs of the use of plant virus interactions. Trend Plant Sci 15: 701–707.

Love AJ, Martin T, Graham IA, Milner JJ (2005) Carbohydrate partitioning and sugar signaling in *Cauliflower mosaic virus*-infected turnip and *Arabiopsis*. Physiol Molec Plant Pathol 67: 63–91.

Lozano-Duran R, Garcia I, Huguet S, Palzergue S, Romero LC et al. (2012) *Geminivirus* C2 protein represses genes involved in sulphur assimilation and this effect can be counteracted by jasmonate treatment. Eur J Plant Pathol 134: 49–59.

Luan JB, Li JM, Varela N, Wang YL, Li FF et al. (2011) Global analysis of the transcriptional response of whitefly to *Tomato yellow leafcurl China virus* reveals the relationship of coevolved adaptation. J Virol 85: 3330–3340.

Lucas WJ (2006) Plant viral movement proteins: agents for cell-to-cell trafiking of viral genome. Virology 344: 169–184.

Macharia I, Backhouse D, Wu S-B, Ateka EM (2016) Weed species in tomato production and their role as alternate hosts of *Tomato spotted wilt virus* and its vector *Frankliniella occidentalis*. Ann Appl Biol 169: 224–235.

MacKenzie TDB, Lavole J, Nie X, Singh M (2018) Differential spread of *Potato virus Y* (PVY) strains O, N:O and NTN and NTN in the field: implications for rise of recombinant PVY strains in New Burnswick, Canada. Amer J Potato Res 95: 301–310.

Mackie AE, Coutts BA, Barbetti MJ, Rodoni BC, McKirdy SJ et al. (2015) *Potato spindle tuber viroid* stability on common surfaces and inactivation with disinfectants. Plant Dis 99: 770–775.

Mackie AE, Rodoni BC, Barbetti MJ, McKirdy SJ, Jones RAC (2016) *Potato spindle tuber viroid*: alternative host reservoirs and strain found in a remote subtropical irrigation area. Eur J Plant Pathol 145: 433–446.

Malfitano M, Barone M, Duran-Vila N, Alioto D (2005) Indexing of viroids in citrus orchards of Campania, Southern Italy. J Plant Pathol 87: 115–121.

Malfitano M, Di Serio F, Covelli L, Ragozzino A, Hernandez C et al. (2003) *Peach latent mosaic viroid* variants including peach calico (extreme chlorosis) contain a characteristic insertion that is responsible for this symptomatology. Virology 313: 492–501.

Maliogka VI, Dovas CI, Lesemann DE, Winter S, Katis NI (2006) Molecular identification, reverse transcription-polymerase chain reaction detection, host reactions and specific cytopathology of *Artichoke yellow ringspot virus* infecting onion crops. Phytopathology 96: 622–629.

Mallory AC, Ely L, Smith TH, Marathe R, Anandalakshmi R et al. (2001) HC-Pro suppression of transgene silencing eliminates the small RNAs but not transgene methylation or the mobile signal. Plant Cell 13: 561–83.

Maniataki E, Tubler M, Tsagris M (2003) Viroid RNA systemic spread may depend on the interaction of a 71-nucleotide bulged hairpin with the host protein VirP1. RNA 9: 346–34.

Martinez de Alba AE (2000) *Isolation and Characterization of Viroid-Binding Proteins*. Universidad del Pais Vasco, Bilbao, Spain.

Maruthi MN, Jeremiah SC, Mohammed IU, Legg JP (2017) The role of the whitefly *Bemisia tabaci* (Gennadius) and farmer practices in the spread of cassava brown streak ipomoviruses. J Phytopathol 165: 707–718.

Masiri J, Velasquez NV, Murphy JF (2011) *Cucumber mosaic virus* 2b-deficient mutant causes limited asymptomatic infection of bell pepper. Plant Dis 95: 331–336.

Matoušek J, Kozlova P, Orctova L, Schmitz A, Pesina K et al. (2007) Accumulation of viroid specific small RNAs and increase in nucleolytic activities linked to viroid-caused pathogenesis. Biol Chem 388: 1–13.

Matoušek J, Orctová LC, Škopek J, Pešina K, Steger G (2008) Elimination of *Hop latent viroid* upon developmental activation of pollen nucleases. Biochemistry 389: 905–918.

Matsushita Y, Shima Y (2015) Effect of low temperature on the distribution of *Chrysanthemum stunt viroid* in *Chrysanthemum morifolium*. Phytoparasitica 43: 609–614.

Matsushita Y, Tsuda S (2014) Distribution of *Potato spindle tuber viroid* in reproductive organs of petunia during its developmental stages. Phytopathology 104: 964–969.

Matsushita Y, Tsuda S (2015) Host ranges of *Potato spindle tuber viroid, Tomato chlorotic dwarf viroid, Tomato apical stunt virioid* and *Columnea latent viroid* in horticultural plants. Eur J Plant Pathol 141: 193–197.

Mautino GC, Sacco D, Ciuffo M, Turina M, Tavella L (2012) Preliminary evidence of recovery from *Tomato spotted wilt virus* infection in *Frankliniella occidentalis* individuals. Ann Appl Biol 161: 266–276.

McMechan AJ, Tatineni S, French R, Hein GL (2014) Differential transmission of *Triticum mosaic virus* by wheat curl mite populations collected in the Great Plains. Plant Dis 98: 806–810.

Mederos DC, Giolittin F, Torres C, Portal O (2019) Distribution and phylodynamics of *Papaya ringspot virus* on *Carica papaya* in Cuba. Plant Pathol 68: 234–250.

Medina AC, Grogan RG (1961) Seed transmission of *Bean common mosaic virus*. Phytopathology 51: 452–456.

Mello AFS, Olarte RA, Gray SM, Perry KL (2011) Transmission efficiency of *Potato virus Y* strains PVY^{O} and $PVY^{N\text{-}wi}$ by five aphid species. Plant Dis 95: 1279–1283.

Mituti T, Molina JPE, Rezende JAM (2018) Bioassays on the role of tomato-potato and sweet pepper as sources of *Tomato chlorosis virus* transmitted by *Bemisia tabaci* MEAM1. Eur J Plant Pathol 152: 613–619.

Molnar A, Csorba T, Lakatos L, Varallyay E, Lacomme CL et al. (2005) Plant virus-derived small interfering RNAs originate predominantly from the highly structured single-stranded viral RNAs. J Virol 79: 7812–7818.

Mondal S, Lin YH, Carroll JE, Wenninger EJ, Bosque-Perez NA et al. (2017) *Potato virus Y* transmission efficiency from potato infected with single or multiple virus strains. Phytopathology 107: 491–498.

Mondal S, Wenninger EJ, Hutchinson PJS, Whitworth JL, Shreshtha D et al. (2017) Responses of aphid vector of *Potato leafroll virus* to potato varieties. Plant Dis 101: 1812–1818.

Montero R, Elaououad H, Pacifico D, Marzachi C, Castillo N et al. (2017) Effects of *Grapevine leafroll associated virus 3* on the physiology in asymptomatic plants of *Vitis vinifera*. Ann Appl Biol 171: 155–171.

Morales K, Zambrano JL, Stewart LR (2014) Coinfection and disease severity of Ohio *Maize dwarf mosaic virus* and *Maize chlorotic dwarf virus* strains. Plant Dis 98: 1661–1665.

Moreno-Delafuente A, Garzo E, Moreno A, Fereres A (2013) A plant virus manipulates the behavior of its whitefly vector to enhance its transmission efficiency and spread. PLoS ONE 8: e61543.

Moutinho-Pereira J, Correia CM, Gonsalves B, Bacelar EA, Coutinho JF et al. (2012) Impacts of leafroll-associated viruses (GLRaV-1 and -3) on the physiology of the Portuguese grapevine cultivar 'Touriga Nacional' growing under field conditions. Ann Appl Biol 160: 237–249.

Nagendran K, Venkataravanappa V, Chauhan NS, Kodandaram MH, Rai AB (2018) Viral diseases, a threat for tomato cultivation in Indo-Gangetic eastern plains of India. J Plant Pathol 101: 15–22.

Nakamura K, Yamagishi N, Isogai M, Komori S, Ito T et al. (2011) Seed and pollen transmission of *Apple latent spherical virus* in apple. J Gen Plant Pathol 77: 48–53.

Narayanasamy P (2001) *Plant Pathogen Detection and Disease Diagnosis*, second edition. Marcel Dekker, New York, NY.

Narayanasamy P (2002) *Microbial Plant Pathogens and Crop Disease Management*. Science Publishers, Enfield, USA.

Narayanasamy P (2008) *Molecular Biology in Plant Pathogenesis and Disease Management, Volume 1, Microbial Plant Pathogens*. Springer Science, Heidelberg, Germany.

Narayanasamy P (2011) *Microbial Plant Pathogens: Detection and Disease Diagnosis, Volume 3, Viral and Viroid Pathogens*. Springer Science, Heidelberg, Germany.

Narayanasamy P (2017) *Microbial Plant Pathogens – Detection and Management in Seeds and Propagules*. Wiley-Blackwell, Chichester, UK.

Narayanasamy P (2020) *Soilborne Microbial Plant Pathogens and Their Management*. Taylor & Francis, Hobekken, Florida, USA.

Nasr-Eldin MA, El Dougdoug KA, Othman BA, Ahmed SA, Abdel-Aziz SH (2012) Three viroids frequency naturally infecting grapevine in Egypt. Internatl J Virol 8: 1–13.

Nekoduka S, Sano T (2018) Symptom development in planta distribution and transmission of *Impatiens necrotic spot virus* in gentian evidence for survival in roots and winter buds. J Gen Plant Pathol 84: 279–283.

Nelson MR, Down EE (1933) Influence of pollen and ovule infections in seed transmissions of bean mosaic. Phytopathology 23: 25.

Nelson RS, Citovsky V (2005) Plant viruses: invaders of cells and pirates of cellular pathways. Plant Physiol 138: 1809–1814.

Nikolaeva OV, Karasev AV, Garnsey SM, Lee RF (1998) Serological differentiation of *Citrus tristeza virus* isolates causing stem pitting in sweet orange. Plant Dis 68: 971–975.

Nyalugwe EP, Barbetti MJ, Clode PL, Jones RAC (2016) Programmed cell death pathways induced by early plant-virus infection are determined by isolate virulence and stage of infection. Plant Pathol 65: 1518–1528.

Nyalugwe EP, Wilson CR, Coults BA, Jones RAC (2012) Biological properties of *Potato virus X* in potato: effects of mixed infection with *Potato virus S* and resistance phenotypes in cultivars from three continents. Plant Dis 96: 43–54.

Ogwok I, Alicai T, Ray MEC, Beyene G, Taylor NJ (2015) Distribution and accumulation of cassava brown streak viruses within infected cassava (*Manihot esculenta*) plants. Plant Pathol 64: 1235–1246.

Ohki T, Sasaya T, Maoka T (2019) Cylindrical inclusion protein of *Wheat yellow mosaic virus* is involved in differential infection of wheat cultivars. Phytopathology 109: 1475–1480.

Olesinski AA, Almon E, Navot N, Perl A, Galun E et al. (1996) Tissue-specific expression of *Tobacco mosaic virus* movement protein in transgenic potato plants alters plasmodesmatal function and carbohydrate partitioning. Plant Physiol 111: 541–550.

Oliveira-Hofman C, Wegulo SN, Tatineni S, Hein GL (2015) Impact of *Wheat streak mosaic virus* and *Triticum mosaic virus* coinfection of wheat on transmission rates by wheat curl mites. Plant Dis 99: 1170–1174.

Pacifico D, Cavalieri V, Marzachi C, Bosco D (2016) Transmission of *Grapevine virus A* and *Grapevine leafroll-associated viruses 1* and *3* by *Planococcus ficus* and *Planococcus citri* on mixed-infected plants. Ann Appl Biol 169: 53–63.

Paliwal YC (1980) Relationship of *Wheat streak mosaic virus* and *Barley stripe mosaic virus* to vector and nonvector eriophyid mites. Arch Virol 69: 38–46.

Pan LL, Cui XY, Chen QF, Wang XW, Liu SS (2018) Cotton leaf curl disease: which whitefly is the vector? Phytopathology 108: 1172–1183.

Papayiannis LC, Katis NI, Idris AM, Brown JK (2011) Identification of weed hosts of *Tomato leafcurl virus* in Cyprus. Plant Dis 95: 120–125.

Pérez-Clemente RM, Montoliu A, Vives V, Lopez-Climent MF, Gomez-Cadenas A (2015) Photosynthetic and antioxidant responses of Mexican lime (*Citrus aurantifolia*) plants to *Citrus tristeza virus* infection. Plant Pathol 64: 16–24.

Pérez-Padilla V, Fortes IM, Romero-Rodriguez B, Mateos MA, Castillo AG et al. (2020) Revisiting seed transmission of the type strain of *Tomato yellow leafcurl virus* in tomato plants. Phytopathology 110: 121–129.

Petersen SM, Keith C, Austin K, Howard S, Su L et al. (2019) A natural reservoir and transmission vector of *Grapevine vein clearing virus*. Plant Dis 103: 571–577.

Pilsson O, Drucker M, Blanc S, German-Retana S, Le Gall O et al. (2003) Structural characterization of HC-Pro, plant virus multifunctional protein. J Biol Chem 278: 23753–23761.

Pinheiro PV, Wilson JR, Xu Y, Zhen Y, Rebelo AL et al. (2019) Plant viruses transmitted in two different modes produce differing effects in small RNA-mediated processes in their aphid vector. Phytobiomes 3: 71–81.

Pirone TP, Blanc S (1996) Helper-dependent vector transmission of plant viruses. Annu Rev Phytopathol 36: 227–247.

Plisson C, Uzest M, Drucker M, Foissart R, Dumas C, Conway J, Thomas D, Blanc S, Bron P (2005) Structure of the mature P3 virus particle complex of *Cauliflower mosaic virus* revealed by crypto-electron microscopy. J Molec Biol 346: 267–277.

Pradhan GP, Xue Q, Jessup KE, Hao B, Price JA et al. (2015) Physiological responses of hard red winter wheat to infection by *Wheat streak mosaic virus*. Phytopathology 105: 621–627.

Qi Y, Pelissier T, Itaya A, Hunt E, Wassenegger M et al. (2004) Direct role of a viroid RNA motif in mediating directional RNA trafficking across specific cellular boundary. Plant Cell 16: 1741–1752.

Rakshandehroo F, Behboodi BS, Mohammadi M (2012) Changes in peroxidase activity and transcript level of the *MYB1* gene in transgenic tobacco plants silenced for the *RDR-1* gene after systemic infection with *Potato virus Y^O^*. J Phytopathol 160: 187–194.

Rana VS, Popli S, Saurav GK, Raina HS, Jamwal R et al. (2019) Implication of the whitefly *Bemisia tabaci*, collagen protein in begomoviruses acquisition and transmission. Phytopathology 109: 1481–1493.

Redondo E, Krause-Sakate R, Yang SJ, Lot H, Le Gall O et al. (2001) *Lettuce mosaic virus* (LMV) pathogenicity determinants in susceptible and tolerant varieties map to different regions to viral genome. Molec Plant-Micro Interact 14: 804–810.

Reinbold C, Herrbach E, Brault V (2003) Posterior midgut and hindgut are both sites of acquisition of cucurbit aphidborne yellows virus in *Myzus persicae* and *Aphis gossypii*. J Gen Virol 84: 3473–3484.

Requena A, Simon-Buela L Salcedo G, Garcia-Arenal F (2006) Potential involvement of a cucumber homolog of phloem protein 1 in the long-distance movement of *Cucumber mosaic virus* particles. Molec Plant-Micro Interact 19: 734–746.

Rinne PL, van den Boogaard R, Mensink MG, Kopperud C, Kormelink R et al. (2005) Tobacco plants respond to the constitutive expression of the tospovirus movement protein NS4, with a heat reversible sealing of plasmodesmata that impairs development. Plant J 43: 688–707.

Robbertse N, Omongi BA, Millar IM, Kruger K, Jooste AEC (2019) Nondestructive DNA extraction from aphids: the application in virus-vector studies of *Banana bunchy top virus* (BBTV). Eur J Plant Pathol 153: 571–582.

Roberts AG, Oparka GW, Gafni Y, Citovsky V (2005) Plasmodesmata and control of symplastic transport. Plant, Cell and Environ 26: 103–124.

Rodio ME, Delgado S, Flores R, Di Serio F (2006) Variants of *Peach latent mosaic viroid* inducing peach calico: uneven distribution in infected plants and requirements of the insertion containing the pathogenicity determinants. J Gen Virol 87: 231–240.

Roistacher CN (1991) *Graft-Transmissible Diseases of Citrus: Handbook for Detection and Diagnosis*. FAO, Rome, p. 286.

Rojas MR, Zerbini FM, Allison RF, Gilbertson RL, Lucas W (1997) Capsid protein and helper component proteinase function as potyvirus cell-to-cell movement proteins. Virology 237: 283–295.

Rondon SI, Roster MS, Hamlin LL, Green KJ, Karasev AV et al. (2016) Characterization of *Beet curly top virus* strains circulating in beet leafhoppers (Hemiptera: Cicadellidae) in Northern Oregon. Plant Dis 100: 1586–1590.

Rosario K, Capobianco H, Ng TFF, Breithart M, Polston JE (2014) RNA viral metagenome of whiteflies leads to the discovery and characterization of a whitefly-transmitted *Carlavirus* in North America. PLoS ONE 9: e86748.

Rowhani A, Daubert S, Arnold M, Al Rwahnih M, Klassen V et al. (2018) Synergy between grapevine vitiviruses and grapevine leafroll viruses. Eur J Plant Pathol 151: 919–925.

Ruiz-Ruiz S, Moreno P, Guerri J, Ambros S (2009) Discrimination between and severe *Citrus tristeza virus* isolates with a rapid and highly specific real-time reverse transcription polymerase chain reaction method using TaqMan LNA probes. Phytopathology 99: 307–315.

Ryang BS, Kobori T, Matsumoto T, Kosaka Y, Ohki ST (2004) *Cucumber mosaic virus* 2b protein compensates for restricted systemic spread of *Potato virus Y* in doubly infected tobacco. J Gen Virol 85: 3405–3414.

Sáenz P, Salvador B, Simon-Mateo C, Kasschau KD, Carrington JC et al. (2002) Host-specific involvement of the HC protein in long-distance movement of potyviruses. J Virol 76: 1922–1931.

Salvador B, Saénz P, Yangüez E, Quiot JB, Quiot L et al. (2008) Host-specific effect of P1 exchange between two potyviruses. Molec Plant Pathol 9: 147–155.

Sanchez-Campos S, Navas-Castillo J, Camero R, Soria C, Diaz J (1999) Displacement of *Tomato yellow leafcurl virus* (TYLCV)-Sr by TYLCV-Is in tomato epidemics in Spain. Phytopathology 89: 1038–1043.

Sanchez-Navarro JA, Carmen-Herranz M, Pallas V (2006) Cell-to-cell movement of *Alfalfa mosaic virus* can be mediated by the movement proteins of Ilar, Bromo-, Cucumo-, Tobamo-, and Como-viruses and does not require virion formation. Virology 346: 66–73.

Sano T, Matsuura Y (2004) Accumulation of short interfering RNAs characteristic of RNA silencing precedes recovery of tomato plants from severe symptoms of *Potato spindle tuber viroid* infection. J Gen Plant Pathol 70: 50–53.

Santillan FW, Fribourg CE, Adama IP, Gibbs AJ, Boonham N et al. (2018) The biology and phylogenetics of *Potato virus S* isolates from the Andean Region of South America. Plant Dis 102: 869–885.

Santos RC, Gomes MF, Penaflor V, Sanches PA, Nardi C et al. (2016) The effects of *Gibberella zeae, Barley yellow dwarf virus*, and coinfection of *Rhopalosiphum padi* olfactory preference and performance. Phytoparasitica 44: 47–54.

Sasaki M, Fukamizu T, Yamamoto K, Ozawa T, Kagmi Y, Shikata E (1981) Detection of HSVd in latent-infected hops by mass diagnosis and its application. Ann Phytopathol Soc Jpn 47: 416.

Sasaki N, Shishikura R, Nyunoya HC (2014) Formation and intracellular movement of cytoplasmic bodies of *Tomato mosaic virus* 126-kDa replication protein in association with viral movement protein. J Gen Plant Pathol 80: 272–281.

Saucke H, Uteau D, Brinkman K, Ziebell H (2019) Symptomatology and yield impact of *Pea necrotic dwarf virus* (PNYDV) in faba bean (*Vicia faba* L. *minor*). Eur J Plant Pathol 153: 1299–1315.

Schaad MC, Carrington JC (1996) Suppression of long-distance movement of *Tobacco etch virus* in a nonsusceptible host. J Virol 70: 2556–2561.

Schindler IM, Muhlbach H-P (1992) Involvement of nuclear DNA-dependent RNA polymerase in *Potato spindle tuber viroid* replication: a reevaluation. Plant Sci 84: 221–229.

Sedlar A, Stare BG, Plesko JM, Dolmičar P, Maras M et al. (2018) Expression and regulation of programmed cell death-associated genes in systemic necrosis of PVY^{NTN} susceptible potato tubers. Plant Pathol 67: 1238–1252.

Sekine KT, Ishihara T, Hase S, Kusano T, Shah J et al. (2006) Single amino acid alterations in *Arabidopsis thaliana* RCY1 compromise resistance to *Cucumber mosaic virus*, but differently suppress hypersensitive response-like cell death. Plant Molec Biol 62: 669–682.

Selth LA, Dogra SC, Rasheed MS, Heally H, Randles JW et al. (2005) A NAC domain protein interacts with *Tomato leaf-curl virus* replication accessory protein and enhances viral replication. Plant Cell 17: 311–325.

Sempere RN, Gómez-Aix C, Ruiz-Ramón F, Gómez P, Hasiów-Jaroszewska B et al. (2016) *Pepino mosaic virus* RNA-dependent RNA polymerase POL domain is a hypersensitive response-like elicitor shared by necrotic and mild isolates. Phytopathology 106: 395–406.

Serra P, Hashemian BSM, Fagiaga C, Ruiz-Ruiz S, Gorris MT et al. (2014) Virus-viroid interactions: *citrus tristeza virus* enhances accumulation of *Citrus dwarfing viroid* in Mexican lime virus-encoded silencing suppressors. J Virol 88: 1394–1397.

Shargil D, Smith E, Lachman O, Reingold V, Darzi E et al. (2017) New weed hosts for *Cucumber green mottle mosaic* virus in wild Mediterranean vegetation. Eur J Plant Pathol 148: 473–480.

Simmons HE, Holmes EC, Gildow FE, Bothe-Goralozyk MA, Stephenson AG (2011) Experimental verification of seed transmission of *Zucchini yellow mosaic virus*. Plant Dis 95: 751–759.

Singh RP (1971) A local lesion host for potato spindle tuber virus. Phytopathology 61: 1034–1035.

Stenger DC, Hein GL, Gildow FE, Horken KM, French R (2005) Plant virus HC-Pro is a determinant of eriophyid mite transmission. J Virol 79: 9054–9061.

Stenger DC, Young BA, French R (2006) Random mutagenesis of *Wheat streak mosaic virus* HC-Pro: noninfectious interfering mutations in a gene dispensable for systemic infection of plants. J Gen Virol 87: 2741–2747.

Su Y-L, Li J-M, Li M, Luan JB, Ye X-D et al. (2012) Transcriptome analysis of the salivary glands of an infective whitefly. PLoS ONE 7: e39303.

Sufrin-Ringwald T, Lapidot M (2011) Characterization of a synergism interaction between two cucurbit-infecting begomoviruses: *squash leaf curl virus* and *Watermelon chlorotic stunt virus*. Phytopathology 101: 281–289.

Sui X, Li R, Shamimuzzaman M, Wu Z, Ling KS (2019) Understanding the transmissibility of *Cucumber green mottle mosaic virus* in watermelon seeds and seed health assays. Plant Dis 103: 1126–1131.

Syller J (2003) Inhibition of long-distance movement of *Potato leafroll virus* to tubers in potato genotypes expressing combined resistance to infection, virus multiplication and accumulation. J Phyotpathol 151: 492–499.

Syller J, Grupa A (2014) The effects of coinfections by different *Potato virus Y* (PVY) isolates on virus concentration in solanaceous hosts. Plant Pathol 63: 466–475.

Szajko K, Soltyo-Kalina D, Szarzynsk B, Strzelczyk-Zyta D, Szweykowska-Kulinska Z et al. (2019) A comparative proteomic analysis of the PVY-induced hypersensitive response in leaves of potato (*Solanum tuberosum* L.) plants that differ in NY-1 gene dosage. Eur J Plant Pathol 153: 385–396.

Tatineni S, Hein GL (2018) Genetics and mechanisms underlying transmission of *Wheat streak mosaic virus* by wheat curl mite. Curr Opin Virol 33: 47–54.

Tatineni S, McMechan AJ, Bartels M, Hein GL, Graybosch RA (2015) In vitro transcripts of wild-type and fluorescent protein-tagged *Triticum mosaic virus* (Family *Potyviridae*) are biologically active in wheat. Phytopathology 105: 1495–1505.

Timmerman-Vaughan G, Larsen R, Murray S, McPhee K, Coyne C (2009) Analysis of the accumulation of *Pea enation mosaic virus* genomes in seed tissues and lack of evidence for seed transmission in pea (*Pisum sativum*). Phytopathology 99: 1281–1288.

Todd JC, Ammar E, Redinbaugh MC, Hoy C, Hogenhout SA (2010) Plant host range and leafhopper transmission of *Maize fine streak virus*. Phytopathology 100: 1138–1145.

Tsai CW, Rowhani A, Golino DA, Daane KM, Almeida RPP (2010) Mealybug transmission of grapevine leafroll viruses: an analysis of virus-vector specificity. Phytopathology 100: 830–834.

Tsuda S, Kobota K, Kanda A, Ohki T, Meshi T (2007) Pathogenicity of *Pepper mild mottle virus* is controlled by the RNA silencing suppression activity of its replication protein but not the viral accumulation. Phytopathology 97: 412–420.

Tsushima D, Nishimura M, Toda T, Furuya H, Fuji S (2019) Molecular characterization and eradication of new *Potato spindle tuber viroid* isolates from dahlia plants in Japan. Eur J Plant Pathol 154: 1091–1102.

Tsushima D, Purushothama CRA, Taneda A, Sano TC (2015) Changes in relative expression levels of viroid-specific small RNAs and microRNAs in tomato plants infected with severe and mild symptom-inducing isolates of *Potato spindle tuber viroid*. J Gen Plant Pathol 81: 49–62.

Tufail M, Takeda M (2009) Insect vitellogenin/lipophorin receptors: molecular structures, role in oogenesis and regulatory mechanism. J Insect Physiol 55: 87–103.

Turechek WW, Kousik CS, Adkins S (2010) Distribution of four viruses in single and mixed infections within infected watermelon plants in Florida. Phytopathology 100: 1194–1203.

Vance CB (1999) Synergism: plant viruses. In: Granoff A, Webster RG (eds.), *Encyclopedia of Virology*, second edition. Academic Press, San Diego, CA, pp. 1694–1699.

Vazquez-Iglasias I, Ochoa-Corona FM, Tang J, Robinson R, Clover GRG et al. (2020) Facing *Rose rosette virus*: a risk to European rose cultivation. Plant Pathol 69: 1603–1617.

Verbeek M, Piron PGM, Dullemans AM, Cuperus C, van der Vlugt RAA (2010) Determination of aphid transmission efficiencies for N, NTN, and wild strains of *Potato virus Y*. Ann Appl Biol 156: 39–49.

Verhoeven JTJ, Huner L, Marn MV, Plesko IM, Roenhorst JW (2010a) Mechanical transmission of Potato spindle tuber viroid between plants of *Brugmansia suaveolens, Solanum jasminoides* and potatoes and tomatoes. Eur J Plant Pathol 128: 417–421.

Verhoeven JTJ, Jansen CCC, Botermans M, Roenhorst JW (2010b) Epidemiological evidence that vegetatively propagated solanaceous plant species act as sources of *Potato spindle tuber viroid* inoculum for tomato. Plant Pathol 59: 3–12.

Voinnet O (2005) Induction and suppression of RNA silencing: insight from viral infections. Nature Rev Genet 6: 206–220.

Voinnet O, Lederer C, Baulcombe D (2000) A viral movement protein prevents spread of the gene silencing signal in *Nicotina benthamiana*. Cell 103: 157–167.

Wah YFWC, Symons RH (1999) Transmission of viroid via grapevine seeds. J Phytopathol 147: 285–291.

Waliullah S, Kosaka N, Yaeno T, Ali ME, Sekine K-T et al. (2015) *Cauliflower mosaic virus* Tav protein induces leaf chlorosis in transgenic tobacco through a hot response to virulence function of Tav. J Gen Plant Pathol 81: 261–270.

Wang MB, Bian XY, Wu LM, Liu L, Smith NA et al. (2004) On the role of RNA silencing in the pathogenicity and evolution of viroids and viral satellites. Proc Natl Acad Sci USA 101: 3275–3280.

Webster CG, Frantz G, Reitz SR, Funderburk JE, Mellinger HC et al. (2015) Emergence of *Groundnut ringspot virus* and *Tomato chlorotic spot virus* in vegetables in Florida and the Southern United States. Phytopathology 105: 388–398.

Wei J, He YZ, Guo Q, Guo T, Liu YQ et al. (2017) Vector development and vitellogenin determine the transovarial transmission of begomoviruses. Proc Natl Acad Sci USA 114: 6746–6751.

Weiland JJ, Edwards MC (2011) Linear-motion tattoo machine and prefabricated needle sets for the delivery of plant tissues by vascular puncture inoculation. Eur J Plant Pathol 131: 552d.

Whitefield AE, Campbell LR, Sherwood JL, Ullman DE (2003) Tissue blot immunoassay for detection of *Tomato spotted wilt virus* in *Ranunculus asiaticus* and other ornamentals. Plant Dis 87: 618–622.

Wintermantel WM, Gilbertson RL, McCreight JD, Natwick ET (2016) Host-specific relationship between virus titer and whitefly transmission of cucurbit yellowing disorder. Plant Dis 100: 92–98.

Wosula EN, Davis JA (2012) Effect of host plant, aphid species and virus infection status on transmission of *Sweet potato feathery mottle virus*. Plant Dis 96: 1331–1336.

Wosula EN, Davis JA, Clark CA, Smith TP, Arancibia RA et al. (2013) The role of aphid abundance, species diversity, and virus titer in the spread of sweet potato potyviruses in Louisiana and Mississippi. Plant Dis 97: 53–61.

Wosula EN, McMechan AJ, Oliveira-Hofman C, Wegulo SN, Hein GL (2016) Differential transmission of two isolates of *Wheat streak mosaic virus* by five wheat curl mite populations. Plant Dis 100: 154–158.

Xiaohua H, Tingmu C, Runqing Y, Shulei G, Mengmeng X et al. (2019) A *Rice black-streaked dwarf virus* replication curve model to evaluate maize rough dwarf disease resistance. Plant Dis 103: 868–873.

Xu A, Zhao Z, Chen N, Zhang H, Liao Q et al. (2013) Self-interaction of the *Cucumber mosaic virus* 2b protein plays a vital role in the suppression of RNA silencing and the induction of viral symptoms. Molec Plant Pathol 14: 803–812.

Yanagisawa H, Matsushita Y (2017) Host ranges and seed transmission of *Tomato planta macho viroid* and *Pepper chat fruit viroid*. Eur J Plant Pathol 149: 211–217.

Yoshida N, Tamada T (2019) Host range and molecular analysis of *Beet leaf yellowing virus*, *Beet western yellows virus*-JP and *Brassica yellows virus* in Japan. Plant Pathol 68: 1045–1058.

Yu D, Fan B, MacFarlane SA, Chen Z (2003) Analysis of the involvement of an inducible *Arabidopsis* RNA-dependent RNA polymerase in antiviral defense. Molec Plant-Microbe Interact 16: 206–216.

Zanini AA, Cuellar WJ, Celli MG, Luque AV, Medina RD et al. (2018) Distinct strains of the reemergent *Cassava common mosaic virus* (genus: *Potexvirus*) infecting cassava in Argentina. Plant Pathol 67: 1814–1820.

Zhan B, Lang F, Zhou T, Fan Z (2016) Nuclear import of *Maize chlorotic mottle virus* capsid protein is mediated by importin-α. Eur J Plant Pathol 146: 881–892.

Zhang Y, Liu Y, Wang Y, Wang Q, He S et al. (2019) Transmissibility of *Citrus yellow vein clearing virus* by contaminated tools. J Plant Pathol 101: 169–171.

Zhang Z, Zhang P, Li W, Zhang J, Huang F, Yang J, Bei Y, Liu Y (2013) De novo transcriptome sequencing in *Frankliniella occidentalis* to identify genes involved in plant virus transmission and insecticide resistance. Genomics 101: 296–305.

Zhu Y, Qi Y, Xun Y, Owens R, Ding B (2002) Movement of *Potato spindle tuber viroid* reveals regulatory points of phloem-mediated RNA traffic. Plant Physiol 130: 138–146.

Zhu Y, Woo Y-M, Owens R, Ding B (2001) Cellular basis of *Potato spindle tuber viroid* systemic movement. Virology 279: 69–77.

Ziebell H, Payne T, Berry JO, Walsh JA, Carr JP (2004) A *Cucumber mosaic virus* mutant lacking the 2b counter-defence protein gene provides protection against wild-type strains. J Gen Virol 88: 2862–2871.

Additional References

Blanc S, Drucker M, Uzest M (2014) Localizing viruses in their insect vectors. Annu Rev Phytopathol 52: 403–425.

Boquet S, Delayen C, Couty A, Giordanengo P, Ameline A (2012) Modulation of aphid vector activity by *Potato virus Y* on in vitro potato plants. Plant Dis 96: 82–86.

Caglayan K, Serce CV, Barutcu E, Kaya K, Medina V et al. (2010) Comparison by sequence-based and electrom microscopic analyses of *Figure mosaic virus* isolates obtained from field and experimentally inoculated figure plants. Plant Dis 94: 1448–1452.

Castellano MA, De Stradis A, Minafra A, Boscia D, Martelli GP (2009) Seed transmission of *Figure latent virus 1*. J Plant Pathol 91: 697–700.

Chen X, Liu J, Xu L, Jiang F, Xie X et al. (2010) Inhibiting virus infection by RNA interference of the eight functional genes of the *Potato virus Y* genome. J Phytopathol 158: 776–784.

Damsteegt VD, Stone AL, Kujlmann M, Gildow FE, Domier LL et al. (2011) Acquisition and transmissibility of U. S. *Soybean dwarf virus* isolates by the soybean aphid *Aphis glycines*. Plant Dis 95: 945–950.

Darzi E, Smith E, Shargil D, Lachman O, Ganot L et al. (2018) The honeybees *Apis mellifera* contributes to *Cucumber green mottle mosaic virus* spread via pollination. Plant Pathol 67: 244–251.

Di Bello PL, Thekke-Veetil T, Druciarek T, Tzanetakis IE (2018) Transmission attributes and resistance to *Rose rosette virus*. Plant Pathol 67: 499–504.

Diener TO (1971a) *Potato spindle tuber virus*: a plant virus with properties of a free nucleic acid. III. Subcellular location of PSTV-RNA and the question whether virions exist in extracts or in stiu. Virology 43: 75–89.

Diener TO (1971b) Potato spindle tuber 'virus'. IV. A replicating low molecular RNA. Virology 45: 411–428.

Ding S-W, Li W-X, Symons RH (1995) A novel naturally occurring hybrid gene encoded by a plant RNA virus facilitates long distance virus movement. EMBO J 14: 5762–5772.

Domier LL, Hobbs HA, McCoppin NK, Bowen CR, Steinlage TA et al. (2011) Multiple loci condition seed transmission of *Soybean mosaic virus* (SMV) and SMV-induced seed coat mottling in soybean. Phytopathology 101: 750–756.

Edwards MC, Weiland JJ, Todd J, Stewart LR (2015) Infectious *Maize rayado fino virus* from cloned cDNA. Phytopathology 105: 833–839.

Flores R, Delgado S, Rodio M, Ambros S, Hernandez C, Di Serio F (2006) *Peach latent mosaic viroid*: not so latent. Molec Plant Pathol 7: 202–221.

Flores R, Hernandez C, Martinez de AE, Darós JA, Serio FD (2005) Viroids and viroid host interactions. Annu Rev Phytopathol 43: 117–139.

Ghorbel R, Lopez C, Fagoaga C, Moreno P, Navarro L et al. (2001) Transgenic citrus plants expressing the citrus tristeza p23 protein exhibit virus-like symptom. Molec Plant Pathol 2: 27–36.

Goldberg KB, Kiernan J, Shepherd RJ (1999) A disease syndrome associated with expression of gene VI of caulimoviruses may be a nonhost reaction. Molec Plant-Micro Interact 4: 182–189.

Gonsalvez-Bernal B, Garcia-Castilo S, Pallas V, Sanchez-Pina MA (2006) Distribution of carnation viruses in the shoot-tip exclusion from shoot apical meristem. Physiol Molec Plant Pathol 69: 43–51.

Gotz M, Popovski S, Kollenberg M, Gorovitz R, Brown JK et al. (2010) Implication of *Bemisia tabaci* shock protein 70 in begomovirus-whitefly interactions. J Virol 86: 13241–13252.

Govier DA, Kassanis B (1974) A virus-induced component of plant sap needed when aphids acquire *Potato virus Y* from purified preparation. Virology 61: 420–426.

Gray S, Gildow FE (2003) Glycosylation of *Beet western yellows virus* protein is implicated in the aphid transmission of the virus. Arch Virol 151: 967–984.

Guardo M, Marletta T, Boninelli G, Leonardi A, Sorrentino G et al. (2005) Detection of viroids in forced citrus cuttings. J Plant Pathol 87: 141–144.

Hanćević K, Radic T, Paskovic I, Urlic B (2018) Biochemical and physiological responses to long term *Citrus tristeza virus* infection in Mexican lime plants. Plant Path 67: 987–994.

Hogenhout SA, Redinbaugh MG, Ammar ED (2003) Plant animal rhabdovirus host range – a bug's view. Trends Microbiol 11: 264–271.

Hogenhout SA, van der Wilk F, Verbeek M, Goldbach RW, van den Heuvel JFJM (1998) *Potato leafroll virus* binds to the equatorial domain of the endosymbiotic GroEL homology. J Virol 72: 355–365.

International Committed on Taxonomy of Viruses (2015) Master species List, 2015 v1. ICTV, Published online. https://talk.ictvonline.org/files/master-species.lists/

Jacquot E, Geldreich A, Hermann T, Leclerc D, Cerutti M et al. (1999) Aphid transmission of *Cauliflower mosaic virus* requires the viral PIII protein. EMBO J 18: 7077–7085.

Jaspars EM, Bos L (1980) *Alfalfa Mosaic Virus. No. 229, Description of Plant Viruses*. CMI/AAB, Kew, England.

Jenner CE, Sanches F, Nettleship SB, Foster GD, Ponz F et al. (2000) The cylindrical inclusion gene of *Turnip mosaic virus* encodes a pathogenic determinant to the *Brassica* resistance gene *TURBO1*. Molec Plant-Micro Interact 13: 1102–1108.

Kaplan JM, Zhang L, Palukaitis P (1998) Characterization of *Cucumber mosaic virus*. V. Cell-to-cell movement requires capsid protein but not virions. Virology 246: 221–231.

Kassachau KD, Carrington JC (1998) A counter-defensive strategy of plant viruses: suppression of posttranscriptional gene silencing. Cell 95: 461–470.

Kil EJ, Byun HS, Kim S, Cho S, Cho S et al. (2015) *Tomato yellow leafcurl virus* can overwinter in *Stellaria aquatica*, a winter-hardy TYLCV reservoir weed. Plant Dis 99: 588–592.

Kimura K, Usugi T, Kagiwada S, Nishio T, Tsuda S (2016) Surveys of viruliferous alate aphid of *Plum pox virus* in *Prunus mume* orchards in Japan. Plant Dis 100: 40–48.

Kliot A, Cilia M, Czosnek H, Ghanim M (2014) Implication of the bacterial endosymbiont *Rickettsia* spp. in interactions of the whitefly *Bemisia tabaci* with *Tomato yellow leafcurl virus*. J Virol 88: 5652–5660.

Kubota K, Ng JCK (2016) *Lettuce chlorosis virus* P23 suppressing RNA-silencing and induces local necrosis with increased severity at raised temperatures. Phytopathology 106: 653–662.

Lawrence DM, Jackson AO (2001) Requirements for cell-to-cell movement of *Barley stripe mosaic virus* in monocot and dicot hosts. Molec Plant Pathol 2: 65–75.

Le Maguet J, Fuchs JJ, Chadoeuf J, Beauve M, Herrbach E et al. (2013) The role of the mealybug *Phenococcus aceris* in the spread of *Grapevine leafroll-associated virus 1* (GLRav-1) in two French vineyards. Eur J Plant Pathol 135: 415–427.

Li S, Wang H, Zhou G (2014) Synergism between *Southern rice black-streaked dwarf virus* and *Rice ragged stunt virus* enhances their insect vector acquisition. Phytopathology 104: 794–799.

Liu S, Liu L, Aranda MA, Peng B, Gu Q (2019) Expression and localization patterns of a small heat shock protein that interacts with the helicase domain of *Cucumber green mottle mosaic virus*. Phytopathology 109: 1648–1657.

Maule AJ, Wang D (1996) Seed transmission of plant viruses: a lesson in biological complexity. Trends Microbiol 4: 153–158.

Milkona C, Jelkmann W (2010) Replication of *Grapevine leafroll-associated virus-7* (GLRaV-7) by *Cuscuta* species and its transmission to herbaceous plants. Plant Dis 94: 471–476.

Murakami T, Tayama R, Nakahara KS (2016) Microperforated leaf blotting on polyvinylidene difluoride and nylon membrane to analyze spatial distribution of endogenous viral gene expression in plant leaves. J Gen Plant Pathol 82: 254–260.

Nault LR (1997) Arthropod transmission of plant viruses: a new synthesis. Ann Entomol Soc Amer 90: 521–541.

Nishiguchi M, Motoyoshi F, Oshima N (1978) Behavior of a temperature sensitive strain of *Tobacco mosaic virus* in tomato leaves and protoplasts. J Gen Virol 39: 53–61.

Noël P, Hance T, Bragard C (2014) Transmission of the *Pepino mosaic virus* by whitefly. Eur J Plant Pathol 138: 23–27.

Orfanidou CG, Pappi PG, Efthimiou KE, Katis NI, Maliogka VI (2016) Transmission of *Tomato chlorosis virus* (ToCV) by *Bemisia tabaci* biotype Q and evaluation of four weed species as viral sources. Plant Dis 100: 2043–2049.

Pallas V, Garcia JA (2011) How do plant viruses induce disease? Interactions and interference with host components. J Gen Virol 92: 2691–2705.

Qainoo AK, Wetten AC, Allainguillaume J (2008) Transmission of *Cocao swollen shoot virus* by seeds. J Virol Meth 150: 45–49.

Scheets K (1998) Maize chlorotic mild machlomovirus and wheat streak mosaic rymovirus concentrations increase in the synergistic disease corn lethal necrosis. Virology 242: 23–38.

Seifers DL, Harvey TL, Louie R, Gordon DT, Martin TJ (2002) Differential transmission of isolates of the *High Plains Virus* by different sources of wheat curl mites. Plant Dis 86: 138–142.

Seifers DL, Martin TJ, Fellers JP (2010) An experimental host range for *Triticum mosaic virus*. Plant Dis 94: 1125–1131.

Stewart LR, Medina V, Tian T, Turina M, Falk BW et al. (2010) A mutation in the *Lettuce infectious yellows virus* minor coat protein disrupts whitefly transmission but not in planta systemic movement. J Virol 84: 12165–12173.

Sudarshana MR, Perry KL, Fuchs MF (2015) *Grapevine red blotch-associated virus*, an emerging threat to the grapevine industry. Phytopathology 105: 1020–1035.

Thompson JR, Down S, Perry KL (2006) Compensatory capsid protein mutations in *Cucumber mosaic virus* confer systemic infectivity in squash (*Cucurbita pepo*). J. Virol 80: 7740–7743.

Tian Z, Liu W, Luo C, Li LT (2012) Transmission comparisons of *Cucumber mosaic virus*, subgroup I and II isolates by different aphid species. J Phytopathol 160: 299–303.

Toloy RS, Mituti T, Freitas DMS, Maluta NKP, Silva TNZ et al. (2018) Features of the relationship between tomato severe ruguose begomovirus and *Bemisia tabaci* MEAM 1 reveal that the virus is acquired during a probe lasting only one minute. Eur J Plant Pathol 151: 541–547.

Velasco L, Botal J, Montero R, Certazzo E (2014) Differences of three Ampeloviruses multiplication in plant may explain their incidences in vineyards. Plant Dis 98: 395–400.

Venkataravanappa V, Swarnalatha P, Reddy CNL, Chauhan N, Reddy MK (2016) Association of recombinant *Chilli leafcurl virus* with enation leafcurl disease of tomato, a new host for chilli begomovirus in India. Phytoparasitica 44: 213–223.

Wang L, Wei X, Ye X, Xu H, Zhou X et al. (2014a) Expression and functional characterization of a soluble form of *Tomato yellow leafcurl virus* coat protein. Pest Manag Sci 70: 1624–1631.

Wang Y, Mao Q, Liu W, Mar TT, Wei T et al. (2014b) Localization and distribution of *Wheat dwarf virus* in its vector leafhopper *Psammotettix alienus*. Phytopathology 104: 897–904.

Wefels E, Morin JP, Randles JW (2015) Molecular evidence for a persistent-circulative association between *Coconut foliar decay virus* and its vector *Myndus taffini*. Austr Plant Pathol 44: 283–288.

Workneh F, O'Shaughnessy S, Evelt S, Rush CM (2017) Relationship between early wheat streak mosaic severity levels and grain yield: implications for management decisions. Plant Dis 101: 1621–1626.

Wosula EN, McMechan AJ, Knoell E, Tatineni S, Wegulo SN et al. (2018) Impact of timing and method of inoculation on severity of wheat streak mosaic disease. Plant Dis 102: 645–650.

Zhang YB, Wang J, Xe ZK, Wang RY, Yang G et al. (2019) A study of viral coat protein accumulation in lily chloroplasts form mixed infection of *Lily mosaic virus* and *Cucumber mosaic virus*. Plant Pathol 68: 26–268.

Zhang YH, Liu CH, Wang Q, Wang YL, Zhou CY, Zhou Y (2019) Identification of *Dialeurodes citri* as a vector of *Citrus yellow vein clearing virus* in China. Plant Dis 103: 65–68.

Zhao M, Ho, Wu Y, He Y, Li M (2014) Western flower thrips (*Frankliniella occidentalis)* transmits *Maize chlorotic mottle virus*. J Phytopathol 162: 532–536.

Index

J

K

L

M

Q

R

U

V

W

Z

For Product Safety Concerns and Information please contact our EU representative GPSR@taylorandfrancis.com Taylor & Francis Verlag GmbH, Kaufingerstraße 24, 80331 München, Germany